T0320649

FROM INTERACTING BINARIES TO EXOPLANETS: ESSENTIAL MODELING TOOLS

IAU SYMPOSIUM No. 282

COVER ILLUSTRATION: Môj rodný kraj: My native land

The cover photograph of the High Tatra Mountains in Slovakia is entitled "Môj rodný kraj" ("My native land") and was taken by Stanislav Dubik ©2011 stano dubik [1970] (http://sdphoto.acompp.sk/).

INTERNATIONAL ASTRONOMICAL UNION

UNION ASTRONOMIQUE INTERNATIONALE

FROM INTERACTING BINARIES TO EXOPLANETS: ESSENTIAL MODELING TOOLS

PROCEEDINGS OF THE 282nd SYMPOSIUM OF THE INTERNATIONAL ASTRONOMICAL UNION HELD IN TATRANSKÁ LOMNICA, SLOVAKIA JULY 18–22, 2011

Edited by

MERCEDES T. RICHARDS

Pennsylvania State University, University Park, PA, USA

and

IVAN HUBENY

University of Arizona, Steward Observatory, Tucson, AZ, USA

Shaftesbury Road, Cambridge CB2 8EA, United Kingdom

One Liberty Plaza, 20th Floor, New York, NY 10006, USA

477 Williamstown Road, Port Melbourne, VIC 3207, Australia

314–321, 3rd Floor, Plot 3, Splendor Forum, Jasola District Centre, New Delhi – 110025, India

103 Penang Road, #05–06/07, Visioncrest Commercial, Singapore 238467

Cambridge University Press is part of Cambridge University Press & Assessment, a department of the University of Cambridge.

We share the University's mission to contribute to society through the pursuit of education, learning and research at the highest international levels of excellence.

www.cambridge.org
Information on this title: www.cambridge.org/9781107019829

First published 2012

A catalogue record for this publication is available from the British Library

ISBN   978-1-107-01982-9   Hardback

Table of Contents

Opening Address

1. Multiwavelength Photometry and Spectroscopy of Interacting Binaries

2. Observations and Analysis of Exoplanets and Brown Dwarfs

Panel Discussion I

3. Imaging Techniques

4. Model Atmospheres of Stars, Interacting Binaries, Disks, Exoplanets, and Brown Dwarfs

5. Synthetic Light Curves and Velocity Curves, Synthetic Spectra of Binary Stars and their Accretion Structures

Panel Discussion II

6. Analysis of Spectra and Light Curves

7. Formation and Evolution of Binary Stars, Brown Dwarfs, and Planets

Panel Discussion III

8. Hydrodynamic Simulations of Exoplanets and Mass Transfer in Interacting Binaries

Panel Discussion IV

Summary

Preface

The conference entitled "From Interacting Binaries to Exoplanets: Essential Modeling Tools" is unique because it represents the first joint meeting between exoplanet and interacting binary star astronomers. The goals of the conference were primarily to bring these groups together to discuss the techniques that they have in common; to demonstrate the extent to which current computer programs are effective in modeling observations of interacting binary stars, brown dwarfs, and exoplanets; to identify ways to improve these codes by incorporating more detailed and realistic physics, while maximizing computer capacity; and to examine how to utilize active and proposed survey projects like Kepler, LSST, and Gaia to obtain data of the highest quality that can be modelled to extract optimal physical parameters, specifically to improve our understanding of the physics. The acceleration of discoveries of brown dwarfs and exoplanets and the rapid influx of very precise light curves from programs like CoRoT and Kepler provide additional stimuli for improving our modeling techniques. In summary, this conference focuses on the tools (imaging techniques, modeling codes, computational power) as they are applied to interacting binaries, brown dwarfs, and exoplanets. The study of binary stars is important because well over half of the stars in the sky belong to binaries, and they provide the only means of calculating stellar masses, which provide a direct link to their evolutionary histories. Moreover, some of the most interesting objects in the universe are found in interacting binaries, and some of these objects are used as standard candles to study the scale of the universe. Interacting binaries, including eclipsing and spectroscopic binaries, have also been used to test theories of stellar structure and evolution as well as general relativity. Besides the mass, they provide us with information about sizes, and even the shapes of stars. The star formation process will create binary or multiple star systems instead of planets based on the initial conditions of the nebula; and a recent study has shown that rocky planets tend to form more readily from nebulae containing a lot of dust. Moreover, stars with a planet or brown dwarf companion behave dynamically like binary star systems, so they share a common formation mechanism, differing only in the masses of the components. Hence, the connection between planetary systems and binary star systems is a natural one. Historically, the analytical tools and simulations used to model single stars were advanced for application to stars in binaries, then later to model interacting binaries and accretion disks. These techniques include astrometry, the radial velocity method, transits or eclipses, and timing studies, and they utilize stellar atmospheres models, as well as atomic and molecular data. Many of these procedures have been expanded to model the more complex low mass systems like brown dwarfs and exoplanets, especially those with Jupiter- and Earth-masses, and also to study the physics of the gas and dust involved in the formation of these systems. Recent observations show that planets can exist in binary star systems, and dynamical studies have shown that terrestrial planets around some close and wide binaries (with separations of several AU) can look similar to planets around a single star. So, planets could orbit one member of the pair, or each separately. This meeting will enhance our knowledge of the most effective tools for the study of brown dwarfs and exoplanets. Our understanding of binary and multiple systems has been driven simultaneously by observations and theory. Computational codes were developed to bridge the gap between these approaches and to permit the extraction of physical parameters that could perhaps uniquely describe the data. Over time, the modeling codes have not kept pace with our knowledge of the theory, partly because of computational constraints, and also because of the time and effort required to incorporate more realistic physics into the codes. As the accuracy of the observations increases, carefully selected physical and astronomical constants need to be incorporated in the codes. For example, the solar mass and radius are used as constants in many calculations, so we need to adopt the most recent determinations of these constants. Simultaneously, various imaging tools have provided an important advance in the study of interacting binaries since nearly all of these systems are unresolved. Adaptive optics (e.g., AEOS), interferometry (e.g., CHARA, VLTI, NPOI), polarimetry, and Doppler tomography are already delivering an increasing number of resolved images of the active environments in these systems. Subsequently, these images will play an important role by providing viable constraints on the models. The ability to model single and binary stars has advanced substantially since the advent of the first model atmosphere codes (e.g., Kurucz 1970, SAO Special Report No. 309; Gustaffson 1971, A&A, 10, 187; Mihalas 1972, ApJ, 176, 139) and binary star synthesis codes (e.g., Wilson & Devinney 1971, ApJ, 66, 605). These computer programs have included physical processes that were as realistic as our computer technology could handle at that time. Forty years later, we have better models and computers and we need to take advantage of these advances (e.g., Hubeny, Mihalas & Werner 2003, ASP Conf. Ser. Vol. 288). Model atmosphere codes now include on the order of 108 atomic lines and 109 molecular lines for cool stars; and all 108 atomic lines are treated in non-LTE for hot stars. While these codes have evolved enormously, further improvements should include better microphysics to progress beyond the classical approximation of 1D plane-parallel, horizontally-homogeneous, static atmospheres. In the case of light curve, velocity curve, and spectral synthesis codes for binary star systems, the effects of limb darkening, gravity darkening, and the reflection effect have been incorporated for stars with spherical and Roche geometries. As our understanding of interacting binaries expanded, new codes were developed to model circumstellar structures like accretion disks, gas streams, winds, jets, and spots, in addition to stellar atmospheres and stellar pulsation. However, the distortional effects of rotation on the properties of high mass and rapidly rotating stars still need to be included in the models. Starting with the semi-analytical work of Lubow & Shu (1975, ApJ, 198, 383), numerical modeling of interacting binaries has advanced to full 3D hydrodynamic simulations (e.g., Bisikalo & Matsuda 2007, IAU Symp. No. 240, 356) that describe the accretion process in substantial detail based primarily on the assumption that gravitational forces dominate the gas flows. These simulations have been used effectively to derive the physical properties of the circumstellar gas (e.g., densities, temperatures, and velocities), and to study the processes of mass transfer and mass accretion assuming that the binary evolves under conservative conditions even though there may be substantial mass loss from the system. These processes will be the focus of our discussions, as well as active stages, instability in accretion disks, and oscillations. The effects of magnetic fields on the stars and circumstellar material should also be included in the simulations since there is now sufficient observational evidence that these fields will influence the evolution of the binary. Similar concerns apply to brown dwarfs and exoplanets. Substantial progress in our understanding of the observations can be achieved if we include more physical processes in the computer codes to achieve an enhancement in modeling that would be as monumental as the observational advances achieved in the last few decades. These advances are now feasible because computing power is growing at a rapid rate. Moreover, it will be possible to extract the maximum amount of information from the data once the theoretical models have been enhanced. Simultaneously, data need to be collected in a systematic way, at high resolution (in wavelength, orbital phase, spatial dimensions), within our galaxy and in external galaxies, and at multiple wavelengths to take advantage of observing facilities on the ground and in space (e.g., Gaia and LSST). Coordinated analyses encompassing several independent procedures simultaneously are now being used to demonstrate the consistency of the models; e.g., synchronized photometric and spectroscopic analyses, comparison of data with synthetic models and hydrodynamic simulations. The topics of this conference are so central that they span at least four IAU Divisions, especially Divisions V (Variable Stars), IV (Stars), IX (Optical & Infrared Techniques), and III (Planetary Systems Sciences), as well as eight Commissions: C25 (Stellar Photometry & Polarimetry), C26 (Double & Multiple Stars), C27 (Variable Stars), C29 (Stellar Spectra), C36 (Theory of Stellar Atmospheres), C42 (Close Binary Stars), C53 (Extrasolar Planets), and C54 (Optical & Infrared Interferometry). The Working Groups on Active B Stars, Ap and Related Stars, and Extrasolar Planets are also pertinent to this conference. We received the official support of all four Divisions (V, III, IV, IX), all three WGs, and seven Commissions, including the essential Commissions C53 and C42. This support was influential in gaining the status of an IAU Symposium.

The conference was special because it officially linked the exoplanet and interacting binary star communities for the first time and the program truly reflected a blending of the two disciplines. In addition, the lectures included deeper examinations of the modeling codes than usual. Another innovation was the introduction of daily Panel Discussions to discuss the main lectures and to provide recommendations for future research directions.

The conference location in Slovakia was noteworthy because it is one of the places where codes have been created, and in a region where a substantial amount of the stellar astrophysics is still taught today. This meeting commemorated the 40th anniversary of the first model atmosphere and binary star synthesis codes, as well as the 110th anniversary of the birth of Dr. Antonín Bečvář, founder of the Skalnate Pleso Observatory and author of several famous atlases and catalogues: Atlas Coeli, Atlas Borealis, Atlas Eclipticalis and Atlas Australis which were used nightly by astronomers around the world for almost half a century. The conference was advertised to the public on television, radio, and in the newspapers. The public was specifically invited to attend a public event including a lecture by Ivan Hubeny on "Hledani a studium planet mimo Slunecni soustavu" (Detecting and Studying Exoplanets). The presentation was given in the Town Hall in the nearby city of Poprad, and it was well received by the public.

We are delighted that the goals of the conference were fully achieved and we are grateful to all participants for their contributions to the success of the meeting. One hundred and seventy-seven participants from thirty-one countries attended the meeting and most participants of the conference characterized the meeting as extremely valuable and highly educational. The program contained forty-six invited and contributed lectures, in addition to the opening lecture and two summary lectures, plus fifty-seven mini-talks and one hundred and twenty-one posters. The invited speakers included those who have developed specialized modeling codes or who are active observers of exoplanets, brown dwarfs, and interacting binaries containing normal, chemically peculiar, and active stars. They were carefully selected to reflect gender and age balance and geographic distribution. Over 25% of the invited and contributed speakers were female, which reflects international representation.

We gratefully acknowledge financial support from the IAU, travel support provided to participants by many international organizations, logistical and administrative support from the Astronomical Institute of the Slovak Academy of Sciences, and financial support through discounts from many local businesses in the High Tatras region of Slovakia. We also thank Dr. Richard Komžík for serving as chief editor of the Abstract Book, developing and maintaining the conference web pages, managing the computer and audio-visual equipment for the presentations, and providing Internet access for participants. We are also grateful to Suzanne Richards for transcribing the discussion sheets and to L'ubomír Hambálek and Alexander Cocking for transcribing the audio recordings.

Mercedes T. Richards and Ivan Hubeny, SOC co-chairs,
Theodor Pribulla and Ladislav Hric, LOC co-chairs
Tatranská Lomnica, Slovakia, 19 August, 2011

Dr. Antonín Bečvář (1901 – 1965)
(Courtesy: Academy of Sciences of the Czech Republic)

THE ORGANIZING COMMITTEES

Scientific Organizing Committee

Dmitry Bisikalo (Russia)
Ján Budaj (Slovakia)
Osman Demircan (Turkey)
Gojko Djurašević (Serbia)
Edward F. Guinan (USA)
Petr Hadrava (Czech Republic)
Petr Harmanec (Czech Republic)
Ladislav Hric (Slovakia)
Ivan Hubeny, (co-chair, USA)
Pavel Koubský (Czech Republic)

Panagiotis Niarchos (Greece)
Geraldine Peters (USA)
Theodor Pribulla (Slovakia)
Didier Queloz (Switzerland)
Mercedes T. Richards, co-chair (USA)
Philippe Stee (France)
Paula Szkody (USA)
Juraj Zverko (Slovakia)
Simon Portegies Zwart (Netherlands)

Local Organizing Committee

Anna Bobulova
Ján Budaj
Zuzana Cariková
Drahomír Chochol
L'ubomír Hambálek
Ladislav Hric (co-chair)
Richard Komžík

Emil Kundra
Theodor Pribulla (co-chair)
Matej Sekeráš
Augustín Skopal
Martin Vaňko
Juraj Zverko

Acknowledgements

The symposium was sponsored and supported by IAU Divisions III (Planetary Systems Sciences), IV (Stars), V (Variable Stars), and IX (Optical & Infrared Techniques); by IAU Commissions C25 (Stellar Photometry & Polarimetry), C27 (Variable Stars), C29 (Stellar Spectra), C36 (Theory of Stellar Atmospheres), C42 (Close Binary Stars), C53 (Extrasolar Planets), and C54 (Optical & Infrared Interferometry); and by the IAU Working Groups on Ap & Related Stars and Active B Stars.

Funding by the
International Astronomical Union,
Astronomical Institute, Slovak Academy of Sciences,
American Astronomical Society International Travel Grant Fund,
SÚH - Slovak Central Observatory, Hurbanovo,
TESCO Poprad, Sintra Poprad, Papyrus Poprad,
The Town of Vysoké Tatry, The Town of Poprad,
The High Tatras Tourism Association,
Nord Svit, and Buntavar Svit
is gratefully acknowledged.

Conference photographs

Carlson Chambliss celebrates his 70th Birthday during the Welcome Party.

Participants in the conference hall.

Participants in the conference hall.

Participants in the conference hall.

Panelists (left to right): Robert Wilson, Alan Batten, Virginia Trimble, France Allard, Peter Eggleton, Albert Linnell, Wilhelm Kley, Artie Hatzes, Edwin Budding. Not shown: Edward Devinney, Helmut Lammer, and Ivan Hubeny.

Panelists: Robert Wilson, Alan Batten, and Virginia Trimble.

Rafting on the Dunajec River.

Rafting on the Dunajec River.

Exploring Spiš Castle.

Inside Hotel Stela in Levoča.

Observing performance of Slovak music and dancers at the Conference Reception.

Watching the performance during the Conference Reception.

Showing appreciation for the entertainment at the Conference Reception.

Performers leaving the Conference Reception.

Participants

Albrecht, Simon	USA	**Guinan**, Edward	USA
Alfonso-Garzon, Julia	Spain	**Hadrava**, Petr	Czech Republic
Allard, France	France	**Hambleton**, Kelly	United Kingdom
Allers, Katelyn	USA	**Harmanec**, Petr	Czech Republic
Baluev, Roman	Russian Federation	**Hatzes**, Artie	Germany
Barria, Daniela	Chile	**Hegedus**, Tibor	Hungary
Batten, Alan	Canada	**Hillier**, D. John	USA
Bergfors, Carolina	Germany	**Hinkley**, Sasha	USA
Bisikalo, Dmitry	Russian Federation	**Hric**, Ladislav	Slovakia
Bjorkman, Jon	USA	**Hubeny**, Ivan	USA
Bjorkman, Karen	USA	**Hypki**, Arkadiusz	Poland
Bochkarev, Nikolai	Russian Federation	**Iliev**, Ilian	Bulgaria
Bonanos, Alceste	Greece	**Iliev**, Lubomir	Bulgaria
Bonavita, Mariangela	Canada	**Inlek**, Gulay	Turkey
Bonifacio, Piercarlo	France	**Ionov**, Dmitry	Russian Federation
Budaj, Jan	Slovakia	**Janik**, Jan	Czech Republic
Budding, Edwin	New Zealand	**Jurkic**, Tomislav	Croatia
Burrows, Adam	USA	**Jurkovic**, Monika	Serbia and Montenegro
Carikova, Zuzana	Slovakia	**Kovari**, Zsolt	Hungary
Celik, Lale	Turkey	**Kafka**, Stella	USA
Chambliss, Carlson	USA	**Kalomeni**, Belinda	Turkey
Chochol, Drahomir	Slovakia	**Kang**, Young-Woon	Korea
Chrastina, Marek	Czech Republic	**Karitskaya**, Eugenia	Russian Federation
Christopoulou, Eleftheria	Greece	**Kim**, Eun-Jeong	Korea
Clarke, Cathie	United Kingdom	**Kiss**, Laszlo	Hungary
Cseki, Attila	Serbia and Montenegro	**Kley**, Wilhelm	Germany
Csizmadia, Szilard	Germany	**Kolbas**, Vladimir	Croatia
Daemgen, Sebastian	Germany	**Konacki**, Maciej	Poland
Danehkar, Ashkbiz	Australia, Iran	**Konorski**, Piotr	Poland
Day-Jones, Avril	Chile	**Korcakova**, Daniela	Czech Republic
De Marco, Orsola	Australia	**Kotnik-Karuza**, Dubravka	Croatia
Demidova, Tatiana	Russian Federation	**Koubsky**, Pavel	Czech Republic
Demircan, Osman	Turkey	**Koumpia**, Evgenia	Greece
Devinney, Jr., Edward	USA	**Kreiner**, Jerzy	Poland
Djurasevic, Gojko	Serbia and Montenegro	**Krejcova**, Tereza	Czech Republic
Dobbs-Dixon, Ian	USA	**Kundra**, Emil	Slovakia
Drozdz, Marek	Poland	**Kurfurst**, Petr	Czech Republic
Eggleton, Peter	USA	**Kyurkchieva**, Diana	Bulgaria
Eyer, Laurent	Switzerland	**Lammer**, Helmut	Austria
Freimanis, Juris	Latvia	**Latkovic**, Olivera	Serbia and Montenegro
Funk, Barbara	Austria	**Lee**, Jeong Eun	Korea
Galis, Rudolf	Slovakia	**Lee**, Jae Woo	Korea
Gazeas, Kosmas	Netherlands	**Lee**, Chung-Uk	Korea
Gomes, Joana	United Kingdom	**Lehmann**, Holger	Germany
Gonzalez Hernandez, Jonay	Spain	**Liska**, Jiri	Czech Republic
Groh, Jose	Germany	**Liakos**, Alexios	Greece
Grygar, Jiri	Czech Republic	**Linnell**, Albert	USA
Macenka, Steven	USA	**Sarta Dekovic**, Mariza	Croatia
Maceroni, Carla	Italy	**Schmidt**, Tobias	Germany
Marchev, Dragomir	Bulgaria	**Schwarz**, Richard	Austria
Markakis, Konstantinos	Greece	**Sejnova**, Klara	Czech Republic
Mayama, Satoshi	Japan	**Sekeras**, Matej	Slovakia
Mennickent, Ronald	Chile	**Senavci**, Hakan	Turkey
Miklos, Peter	Slovakia	**Serabyn**, Eugene	USA
Mikulasek, Zdenek	Czech Republic	**Sipocz**, Brigitta	United Kingdom

Milic, Ivan	Serbia and Montenegro	Skoda, Petr	Czech Republic
Mkrtichian, David	Ukraine	Skopal, Augustin	Slovakia
Mochnacki, Stefan	Canada	Stachowski, Greg	Poland
Montgomery, Michele	USA	Stateva, Ivanka	Bulgaria
Morais, Helena	Portugal	Stee, Philippe	France
Nedoroscik, Jozef	Slovakia	Stringfellow, Guy	USA
Neilson, Hilding	Germany	Southworth, John	United Kingdom
Nemravova, Jana	Czech Republic	Sudar, Davor	Croatia
Netopil, Martin	Austria	Szabo, Gyula	Hungary
Neustroev, Vitaly	Finland	Szalai, Tamas	Hungary
Niarchos, Panagiotis	Greece	Tout, Christopher	United Kingdom
Nikolov, Nikolay	Germany	Triaud, Amaury	Switzerland
Ogloza, Waldemar	Poland	Trimble, Virginia	USA
Olah, Katalin	Hungary	Tsvetkova, Tatiana	Russian Federation
Otulakowska, Magdalena	Poland	Vanko, Martin	Slovakia
Pasternacki, Thomas	Germany	Vidotto, Aline	United Kingdom
Paunzen, Ernst	Austria	Vince, Istvan	Serbia and Montenegro
Pavlovski, Kresimir	Croatia	von Essen, Carolina	Germany
Peters, Geraldine	USA	Votruba, Viktor	Czech Republic
Pilat-Lohinger, Elke	Austria	Whittaker, Gemma	United Kingdom
Pilecki, Bogumil	Poland	Wilson, Robert	USA
Pilello, Antonio	Germany	Wolf, Marek	Czech Republic
Plavalova, Eva	Slovakia	Yakobchuk, Taras	Ukraine
Popova, Elena	Russian Federation	Yakut, Kadri	Turkey
Pribulla, Theodor	Slovakia	Zakhozhay, Olga	Ukraine
Prsa, Andrej	USA	Zakrzewski, Bartlomiej	Poland
Queloz, Didier	Switzerland	Zasche, Petr	Czech Republic
Ratajczak, Milena	Poland	Zejda, Miloslav	Czech Republic
Reed, Phillip	USA	Zielinski, Pawel	Poland
Reckova, Valeria	Slovakia	Ziznovsky, Jozef	Slovakia
Richards, Mercedes	USA	Zola, Staszek	Poland
Rode-Paunzen, Monika	Austria	Zucker, Shay	Israel
Rucinski, Slavek	Canada	Zverko, Juraj	Slovakia
Ruzdjak, Domagoj	Croatia		

Address by the Scientific Organizing Committee

Vitajte! This means "Welcome" in Slovak. Thank you for your participation in this conference.

On behalf of the Scientific Organizing Committee for IAU Symposium 282, we thank the Prime Minister of Slovakia, Professor Dr. Iveta Radičová, for her support of this conference, and also Dr. Aleš Kucera, Director of the Astronomical Institute of the Slovak Academy of Sciences. We are grateful also for the support of the mayors of Vysoké Tatry and Poprad.

It is a delightful honor to be standing here today before such a distinguished group of scientists. We have come here to make a bridge between the astronomers who study multiple star systems and those who study systems containing multiple planets. Over 560 exoplanets have already been discovered and now we are on the cusp of discovering true Earth-like exoplanets. The tools we will discuss at this conference will help to make these discoveries a reality, hopefully within the next five years.

The scientific organizing committee has prepared an interesting program for us and we hope that many new collaborations will result from this exchange of ideas.

It is our pleasure to introduce our distinguished panelists who will lead us in some invigorating discussions about unsolved problems: France Allard, Alan Batten, Edwin Budding, Edward Devinney, Peter Eggleton, Artie Hatzes, Ivan Hubeny, Wilhelm Kley, Helmut Lammer, Albert Linnell, Virginia Trimble, and Robert E. Wilson.

The Local Organizing Committee and their families led by Dr. Theodor Pribulla and Dr. Ladislav Hric, have done a lot of work to get us to this moment. We are indebted to them for their contributions and the wonderful events they have planned for us.

Finally, this conference is dedicated to the modeling and analysis tools developed over the past 40 years and also to Dr. Antonín Bečvář, who provided us with his stellar atlases of the sky.

Dakujeme!

Mercedes T. Richards and Ivan Hubeny, SOC co-chairs
Tatranská Lomnica, 18 July 2011

From Interacting Binaries to Exoplanets: Essential Modeling Tools
Proceedings IAU Symposium No. 282, 2011   © International Astronomical Union 2012
Mercedes T. Richards & Ivan Hubeny, eds.   doi:10.1017/S1743921311026780

Shaking the Pot of Modelling Tools: Some Open Problems in the Field

Petr Harmanec

Astronomical Institute of the Charles University, Faculty of Mathematics and Physics,
V Holešovičkách 2, CZ-180 00 Praha 1, Czech Republic
email: hec@sirrah.troja.mff.cuni.cz

Abstract. To inspire and provoke lively discussions, I argue that the accuracy of the basic physical properties of stars, based on analyses of well-observed detached binaries, might be worse than usually believed. I offer some ways how to deal with the situation. I end with a few comments on the studies of extra-solar planets.

Keywords. stars: fundamental parameters (classification, colors, luminosities, masses, radii, temperatures, etc.), (stars:) binaries: general, (stars:) binaries (including multiple): close, (stars:) binaries: eclipsing, stars: planetary systems, stars: evolution, stars: atmospheres

1. Introduction

Let me begin this talk in a somewhat personal tone. When many of us met at the IAU Symp. 240 *Binary Stars as Critical Tools and Tests in Contemporary Astrophysics* in 2006, we had the chance to welcome there Mirek J. Plavec, a Czech stellar astromomer who spent the last few decades of his life in the USA. Mirek was a teacher and friend of several of us, who are present in this audience. In one of his excellent review talks (Plavec 1983) he made a statement, which I believe will be very appreciated by the colleagues studying the wealth of data coming from the space observatories like MOST, CoRoT or Kepler:

"I think it is fair to say that a theory or a model is always the closer to being worshipped the fewer are the observational data."

My intention, as an astronomer who tries to analyze observations in an effort to learn something new about stars, is to inspire lively discussions during this meeting. I shall touch on some problems worth considering and ask various questions, to which I hope to hear answers and/or comments from the experts, who met here.

2. How accurately do we know the masses, radii and other basic physical properties of stars?

Andersen (1991) claimed the errors in stellar masses and radii smaller than 2% and Torres *et al.* (2010) relaxed this accuracy to better than 3% in their excellent review of the properties of 95 well detached binary systems. I am afraid, however, that this estimate is still too optimistic as detailed below. Besides, we should keep in mind the current strong selection effect, namely that we have most observations for the systems seen roughly equator-on. It will be very interesting to study also spectroscopic binaries seen under lower inclination when the interferometry will become a widely used technique to see whether our theoretical models of the gravity and limb darkening are sufficiently sophisticated or not.

2.1. *Radial-velocity curves and stellar masses*

In spite of great progress in the spectral resolution and S/N ratios, there still must be a difference in accuracy between the radial velocities (RVs hereafter) from the photographic and electronic spectra. Torres *et al.* (2010) discuss carefully various techniques of RV determination and orbital solutions and the potential sources of systematic errors. They recalculated the orbital and light-curve solutions for all of their systems, but they did not mention which programs they had used. Furthermore, there was no warning that the accuracy of RV and mass determination inevitably decreases with the increasing projected rotational velocity of the stars and can be affected by different spectral resolutions of different spectrographs. Take the example of hot stars: Different authors use different effective laboratory wavelengths for He I triplet lines and there is always the problem of line blending, especially for stars with non-negligible rotational velocities. Most authors publish only mean RV, which is usually based on different numbers of spectral lines for different spectra, depending on the quality of each spectrogram. If the RV of each spectral line has its own 'γ velocity,' then averaging RVs not always for the same set of lines will inevitably decrease the RV amplitude of the mean RV curve. One should investigate the RV solutions line by line to see if there are such systematic differences or not. Such an approach is possible with the technique of spectral disentangling (Simon & Sturm 1994, Hadrava 1995, 1997, 2004), where one can derive separate solutions in the neighbourhood of all stronger spectral lines. It is not possible, however, for the RV determinations based on the principles of the cross-correlation technique, where the orbital velocity shifts of all available spectra are mutually compared (see, e.g., Simkin 1974, Hill 1993, Zucker & Mazeh 1994). There is, therefore, some danger that the effects such as inaccuracies in the wavelength scale or different sensitivities of different spectral lines to unrecognized slight physical variability of the line profiles can affect the resulting amplitude of the RV curves, and therefore also the estimates of stellar masses. The fact that one can only derive $M_j \sin^3 i$ $(j = 1, 2)$ from the orbital solutions implies that the accuracy of stellar masses depends critically also on the accuracy of the determination of the orbital inclination i from the light-curve or accurate interferometric solution. I admit that there are some very-high accuracy mass determinations for sharp-lined late-type stars (for instance Konacki *et al.* 2010) but the chances are hopelessly low that similar studies could be carried out for a representative sample of detached binaries with the needed spectral resolution.

2.2. *Two different connotations of the effective temperature*

The situation regarding the accuracy of the determination is even worse for the radiative properties of the stars: luminosity and effective temperature. The principal definition of the effective temperature, which is also adopted in stellar evolutionary models is that it is a parameter characterizing the total bolometric luminosity of a star via

$$L = S\sigma T_{\text{eff}}^4 \, , \, (S = 4\pi R^2 \text{ for spherical stars}),$$

where S, R, and σ are the stellar surface area, stellar radius and the Stefan-Boltzmann constant. The problem is that the observer on the Earth can only record the flux $\mathcal{F}_E$ coming from the star in the direction of the observer. At present, the stellar bolometric luminosity and T_{eff} are the quantities that *cannot be directly measured* and are only *deduced*. Denoting $\mathcal{F}_S$ the bolometric flux from the unit area of the stellar surface, d the distance to the star and θ the stellar angular *diameter* and assuming the simplest case of a spherical star with a uniform brightess distribution, one has $L = 4\pi R^2 \mathcal{F}_S$ and $F_S = \frac{d^2}{R^2} \mathcal{F}_E$ or $\sigma T_{\text{eff}}^4 = \frac{4\mathcal{F}_E}{\theta^2}$. However, already for rotating stars, $\mathcal{F}_S$ becomes a function of the stellar latitude. A proper estimate of T_{eff} from the observed $\mathcal{F}_E$ requires the knowledge of the

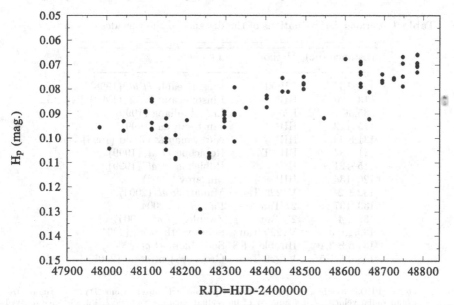

Figure 1. A plot of the Hipparcos H_p photometry (Perryman & ESA 1997) of α Lyr vs. reduced heliocentric Julian days. Only the observations with error flags 0 and 1 were used and the largest rms error of individual data points is 0.005 mag. Clear secular light variations are observed.

inclination of the rotational axis of the star and also proper modelling (which includes the limb and gravity darkening and also reflection in the case of binaries). Regrettably, a parameter, which is also called T_{eff}. is used as one of the basic parameters characterizing a specific model of a stellar atmosphere, the most often a plane-parallel one. A frequent 'logical short-circuit' then is that the T_{eff}. characterizing the stellar atmosphere model providing the best fit of the observed stellar line spectrum, is interpreted as the true stellar T_{eff}. Alternatively, in the cases when one has some idea about the geometry of a non-spherical star, one can find statements like that the T_{eff}. varies from the equator to the pole of the star in question. I maintain that this is a misuse of the term 'effective temperature,' which may complicate the comparison of the observations with the stellar evolutionary models.

There are various calibrations between indices of standard photometric systems and the stellar effective temperature. One, which has been often used in recent years is Flower's (1996) calibration between $B-V$, T_{eff}., and the bolometric correction. Flower undoubtedly did a great job but it is necessary to realize that he only critically collected the published data for 335 stars with published T_{eff}. between 2900 and 52500 K. There is no guarantee the these determinations are mutually consistent. What is in the background are observations and model atmospheres of various degrees of accuracy and sophistication as they developed over several decades. In many cases, the spectral energy distribution of particular stars near its maximum was only approximated by a model atmosphere. Flower tabulates the bolometric corrections to three valid digits but I am afraid that their accuracy might be less than two digits in fact, especially at the hot end of the T_{eff}. range.

Many published studies rely on several published absolute flux calibrations of Vega (α Lyr). Figure 1 shows the good Hipparcos broadband H_p observations of Vega,

Table 1. Various determinations of the distance of the Pleiades cluster

Distance (pc)	Method	Source
129–134	PHOT	Pinsonneault *et al.* (1998)
113–120	HIP	Pinsonneault *et al.* (1998)
135.56± 0.72	DYN	Li & Junliang (1999)
115–122	HIP	van Leeuwen (1999)
131± 11	HIP	Narayanan & Gould (1999)
121–134	PHOT	Robichon *et al.* (1999)
115–121	HIP	Robichon *et al.* (1999)
126–134	HIP	Makarov (2002)
132± 2	V1229 Tau	Munari *et al.* (2004)
133–137	27 Tau	Pan *et al.* (2004)
132± 4	27 Tau	Zwahlen *et al.* (2004)
139.1± 3.5	V1229 Tau	Southworth *et al.* (2005)
134.6± 3.1	Hubble FES	Soderblom *et al.* (2005)
138.0± 1.5	V1229 Tau	Groenewegen *et al.* (2007)

Column "Method": PHOT... estimates from calibrated photometric observations; DYN... estimate from the proper motion and radial velocities of a number of individual cluster members; HIP... distance derived from analyses of Hipparcos satellite astrometric observations; V1229 Tau... distance estimated from combined orbital and light-curve solutions for this eclipsing-binary member of the cluster; 27 Tau... distance estimated from the comparison of the angular size of the semi-major axis of this visual and double-lined spectroscopic binary with the semimajor axis in km derived from the orbital solution; Hubble FES... distance derived from the astrometric observations with the Hubble Space Telescope's Fine Guidance Sensor used as a white-light interferometer.

indicating that Vega is a variable star! This finding must be verified by systematic observations and may have serious consequences for existing calibrations.

2.3. *Stellar radii, luminosities and distances*

It is true that there are several advanced programs for the light-curve solutions or even combined solutions for several different observables, such as the WD program (Wilson & Devinney 1971). Although the flux and temperature distribution are properly modelled in such programs, one still has to assume T_{eff} of the primary, again depending on the existing calibrations. It is true that Wilson (2008) pushed forward eclipsing-binary solutions in physical units, which should in principle lead to direct distance estimates. It seems, however, that the calibrations of photometric systems and theoretical spectral energy distributions vs. T_{eff} will need further improvement before it will be possible to derive individual T_{eff} for both components in binaries from the combined light-curve and RV-curve solutions. Note also the alarming case of the W UMa binary AW UMa, which seems to be a detached system with equatorial disks mimicking the contact configuration (Pribulla & Rucinski 2008), a situation which could not be properly analysed for the stellar radii with the WD program.

The situation is even worse for the stellar luminosities and the distances deduced from them. An incomplete knowledge of the true interstellar reddening forces us to use statistical mean relations, for instance the well-known expression for the reddening of the Johnson V magnitude as a function of the $B-V$ index $A_{\text{V}} = \alpha \, \text{E}(B-V)$. A search in the literature shows that various authors are using α values from 2.9 to 3.2, with certainly not quite negligible effect on the deduced distance moduli. Our current uncertainties can be well illustrated on several determinations of the distance to the Pleiades cluster, especially those based on analyses of V1229 Tau, a double-lined spectroscopic and eclipsing binary, and of 27 Tau, which is a double-lined spectroscopic and interferometrically resolved binary – see Table 1.

3. What could be done to improve the situation?

3.1. *Improving calibrations*

I believe that if we really want to improve our knowledge of basic properties of stars, the whole community must begin to oppose the current attitude of time-allocating committees to allocate observing time at the best ground-based and space observatories only to such science projects, the results of which look great in the prime-time TV news. In particular, we need to select a representative subset of well detached double-lined spectroscopic and eclipsing binaries with low $v \sin i$ values and to launch their systematic observations over virtually the entire electromagnetic spectrum and with a good distribution over orbital phases. Their spectral energy distribution combined with a detailed comparison of disentangled component spectra with the best present-day model atmospheres should allow us to define an accurate and universally adopted scale of T_{eff}. and bolometric corrections. In another part of such a project, these would be calibrated vs. all frequently used standard photometric systems. Then, at least for basically spherical stars, we could really critically compare the results of observations with the prediction of stellar evolutionary models.

The much more complicated problem of the proper modelling of rapid rotators might become solvable later, when optical interferometry becomes a standard tool like spectroscopy and photometry are today.

3.2. *Obligatory constants and units*

As suggested by Harmanec & Prša (2011) and by Prša & Harmanec (this proc.), one thing can be done immediately. We should start using obligatory *nominal* values of the solar radius and luminosity, which would serve as exact, error free conversion factors to express the stellar radii and luminosities from solar to SI units. It would not be wise to do the same at present with the solar mass since the gravitational constant is known with an accuracy, which is full five orders of magnitude worse than that of the product $GM_{\odot}$. Fortunately, Kepler's 3rd Law and the formulas to derive stellar masses, expressed in the units of the solar mass from the RV-curve solutions, both depend on the $GM_{\odot}$ product. One can therefore express the stellar masses in the units of solar mass very accurately while an accurate conversion to SI can be carried out later when a truly exact value of G will be known. See the above-quoted papers for more details but our prescription is

(*a*) Fix the nominal solar radius and luminosity and use the accurately known product $GM_{\odot}^{2010} = 1.32712442099(10) \times 10^{20}\,\text{m}^3\text{s}^{-2}$:

$$1\ M_{\odot}^{2010} = 1.988416 \times 10^{30}\ \text{kg}, \tag{3.1}$$

$$1\ \mathcal{R}_{\odot}^{\text{N}} = 6.95508 \times 10^{8}\ \text{m, and} \tag{3.2}$$

$$1\ \mathcal{L}_{\odot}^{\text{N}} = 3.846 \times 10^{26}\ \text{W}. \tag{3.3}$$

(*b*) Use IAU (CODATA) constants like G or σ and always quote their source (or values) explicitly in your publications.

Note that all of the most frequently used formulas become either error free or depend on the accuracy of *only one* physical constant if our suggestion is applied. For example the stellar equatorial rotational velocity in km s^{-1} becomes error free:

$$V[\text{km s}^{-1}] = 50.57877 \frac{R[\mathcal{R}_{\odot}^{\text{N}}]}{P_{\text{rot}}[\text{days}]}. \tag{3.4}$$

We suggest that exactly the same should be also done with the basic properties of Jupiter and Earth, which are used by the students of brown dwarfs and extrasolar planets.

4. A few comments on the technique of spectral disentangling from the user's point of view

The spectral disentangling invented by Simon & Sturm (1994) and Hadrava (1995), Hadrava (1997) is the most advanced method for the determination of orbital elements and reconstruction of individual line spectra of the components of a binary or even multiple system of stars. According to my experience, the technique performs best for binaries with components of widely different spectral types and not so extreme luminosity ratios. The situation gets more complicated in cases when the secondary is some 3 or more magnitudes fainter than the primary. The sum of squares of the residuals, which is usually used as the minimizer, is dominated by the primary and the noise of the observed spectra in such cases. This decreases the sensitivity of the method to the spectrum of the secondary and a relatively large range of mass ratios might lead to a similar sum of squares of residuals. I hope the experts present here might have some suggestions for a better criterion for the optimal solution in such situations. Also disentangling of the spectra of binaries composed from two fast rotating components of similar spectral types is complicated and need not lead to a unique solution simply because the resulting line widths and semi-amplitudes of the orbital motion are strongly correlated.

One of the widely used programs for spectral disentangling, KOREL by Dr. P. Hadrava (see, e.g., Hadrava 2004) provides an excellent possibility to remove also telluric lines from the spectra via modelling them as a distant component moving with a period of the Earth's orbit around the Sun and allowing for their varying strength from one spectrum to another. The users should be aware, however, that the relative line strength is treated as a single value for each spectrum. Since the strength of the water vapour lines varies much more than that of the atmospheric oxygen, it is not advisable to disentangle the spectral regions containing both types of telluric lines in one program run.

As demonstrated in detail in the contribution by Chadima et al. (in these proceedings), it is always advisable to verify the result of disentangling in some independent way. While preparing a recent detailed study of a large number of electronic spectra of ε Aur by Chadima et al. (2011), we made an attempt to disentangle the spectra of both binary components, keeping the orbital period of 9890.62 d and other orbital elements fixed from the solution derived by Chadima et al. (2010). Solutions with two independent computer programs, KOREL (Hadrava 1995, 2004) and FDBINARY (Ilijić et al. 2001) were obtained and led to very similar disentangled spectra. To our surprise, both programs returned also a clear spectrum of the secondary for all spectral lines also seen in the spectrum of the primary. However, the secondary is hidden in a cool and opaque disk and a very detailed attempt by Bennett et al. (this proc.) to detect any trace of the secondary spectrum failed. Chadima et al. (this proc.) also tried to disentangle the same set of observed spectra of ε Aur for the dominant period of observed physical variations, 66.21 d and for an arbitrarily chosen, non-existent period of several hundreds of days. The secondary spectrum was disentangled in both of these attempts, too. The probable explanation lies in the fact that the semi-amplitude of real RV changes (about 15 km s^{-1}) is so low that the lines of both putative spectra remain completely blended in all orbital phases. The disentangling programs then try to interpret the varying line asymmetries, caused by real physical changes, in terms of the spectrum of the secondary, which is inevitably similar to that of the primary.

5. A common language

In passing, I am going to ask a few questions related to the field of exoplanet studies, in which I am not working actively. So I only hope to learn the answers while listening to the talks during this meeting.

As the so far most fruitful techniques of the exoplanet detection, accurate RV measurements and photometric eclipses of host stars by passing planetary bodies, are methodologically very similar to binary studies, one would expect that the same terms will be used. This has not been quite so. While the binary community talks about *eclipsing binaries*, the students of exoplanets prefer the term *transiting exoplanets*. The binary community distinguishes 'transits' and 'occultations,' depending on the actual geometry of the eclipse and it is true that the eclipses of host stars by planets are indeed 'transits.' Still, I would tend to talk about the time of mid-eclipse, not mid-transit, even in the case of exoplanets.

I also hope to learn how well the very-small amplitudes of the RV curves of cool stars hosting the planets are actually known. Possible small inhomogeneities of the brightness distribution over the stellar surfaces can be reflected differently in individual spectral lines and any cross-correlation technique must invevitably mask such effects, returning some averaged RV curve. Finally, talking about the limits of accuracy of precise RV measurements, I hope to hear how accurately the barycentric RV corrections can be obtained. Should one use the geographic coordinates of the slit of the spectrograph or which particular point inside the dome?

Acknowledgements

I appreciate comments by P. Mayer and S. M. Rucinski. This research was supported by Czech Science Foundation grant P209/10/0715 and a Czech Ministry of Education grant from the Research Program MSM0021620860: *Physical study of objects and processes in the solar system and in astrophysics*. The NASA/ADS database was used in this work.

References

Andersen, J. 1991, *Astron. Astrophys. Rev.*, 3, 91

Chadima, P., Harmanec, P., Yang, S., Bennett, P. D., Božić, H., Ruždjak, D., Sudar, D., Škoda, P., Šlechta, M., Wolf, M., Lehký, M., & Dubovský, P. *IBVS*, No. 5937, 1

Chadima, P., Harmanec, P., Bennett, P. D., Kloppenborg, B., Stencel, R., Yang, S., Božić, H., Šlechta, M., Kotková, L., Wolf, M., Škoda, P., Votruba, V., Hopkins, J. L., Buil, C., & Sudar, D. 2011, *A&A*, 530, A146

Flower, P. J. 1996, *ApJ*, 469, 355

Groenewegen, M. A. T., Decin, L., Salaris, M., & De Cat, P. 1999, *A&A*, 463, 579

Hadrava, P. 1995, *A&AS*, 114, 393

Hadrava, P. 1997, *A&AS*, 122, 581

Hadrava, P. 2004, *Publ. Astron. Inst. Acad. Sci. Czech Rep.*, 92, 15

Harmanec, P. & Prša, A. 2011, *PASP*, 123, 976

Hill, G. 1993, *ASPC*, 38, 127

Ilijić, S., Hensberge, H., & Pavlovski, K. 2001, *Lectures Notes in Physics*, 573, 269

van Leeuwen, F. 1999, *A&A*, 341, L71

Li, C. & Junliang, Z. 1999, *ASPC* 167, 259

Konacki, M., Muterspaugh. W., Kulkarni, S. R., & Hełminiak, K. G. 2010, *ApJ*, 719, 1293

Makarov, V. V. 2002, *AJ*, 124, 3299

Munari, U., Dallaporta, S., Siviero, A., Soubiran, C., Fiorucci, M., & Girard, P. 2004, *A&A*, 418, L31

Narayanan, V. K. & Gould, A. 1999, *ApJ*, 523, 328

Pan, X., Shao, M., & Kulkarni, S. R. 2004, *Nature*, 427, 326

Perryman, M. A. C. & ESA 1997, The HIPPARCOS and TYCHO catalogues, *ESA SP Series*, 1200

Pinsonneault, M. H., Stauffer, J., Soderblom, D. R., King, J. R., & Hanson, R. B. 1998, *AJ*, 504, 170

Plavec, M. 1983, *JRASC*, 77, 283

Pribulla, T. & Rucinski, S. M. 2008, *MNRAS*, 386, 377

Robichon, N., Arenou, F., Mermilliod, J.-C., & Turon, C. 1999, *A&A*, 345, 471

Simon, K. P. & Sturm, E. 1994, *A&A*, 281, 286

Soderblom, D. R., Nelan, E., Benedict, G. F., McArthur, B., Ramirez, I., & Spiesman, W. 2005, *AJ*, 129, 1616

Southworth, J., Maxted, P. F. L., & Smalley, B. 2005, *A&A*, 429, 645

Torres, G., Andersen, J., & Gimenéz, A. 2010, *Astron. Astrophys. Rev.*, 18, 67

Wilson, R. E. 2008, *ApJ*, 672, 575

Wilson, R. E. & Devinney, E. J. 1971, *ApJ*, 166, 605

Zucker, S. & Mazeh, T. 1994, *ApJ*, 420, 806

Zwahlen, N., North, P., Debernardi, Y., Eyer, L., Galland, F., Groenewegen, M. A. T., & Hummel, C. A. 2004, *A&A*, 425, L45

Discussion

I. HUBENY: I do not quite agree with the suggestion that there should be a single effective temperature defined for the whole star. Such a quantity would only be a proxy for the total stellar luminosity and its surface extent. The term effective temperature has a well-defined, and widely used, meaning in the stellar atmosphere theory, namely that it represents total energy flux coming from the stellar interior on the lower boundary of the atmosphere. Since the stellar atmosphere is a passive region where no additional energy is generated or destroyed, this effective temperature also represents a total, frequency integrated outgoing radiation at the upper boundary of the atmosphere. It is perfectly legitimate to assign different values of so defined effective temperature to different positions on the stellar surface for a non-spherical star, or for regions where the total incoming energy is modified by intervening external forces, as for instance for sunspots driven by magnetic field.

R. WILSON: My point is similar to Ivan's. Effective temperature is a fundamental quantity in stellar atmosphere theory and in reality. Teff is a single number that serves as a basic stellar atmospheric parameter and a natural parameter for eclipsing binary analyses, given proper modeling. There is no need to use other temperature-related parameters such as color temperature, etc.

P. HARMANEC: I understand that Teff is one of the parameters characterizing model atmospheres. But you compare the model atmosphere with *the flux* coming in the direction towards the observer; and for non-spherical or any more complicated object the model-atmosphere Teff giving the best fit between the observed and model spectrum will not describe the bolometric luminosity properly. For instance the Be stars have time-variable pseudophotospheres, which can mimic, say, a B6 star at one epoch, and B8 star at another. So I do believe the term Teff should be reserved for a parameter characterizing the bolometric luminosity of a star (to be compared with evolutionary models).

R. WILSON: To define and utilize a consistent effective temperature for EB solutions is straightforward, as has always been the situation in the WD model. The solution parameter is a flux-weighted mean effective temperature over the surface. Starting from that parameter, one can recover the full Teff distribution over the surface, including all modeled physical effects, if the distribution is needed.

Part 1
Multiwavelength Photometry and
Spectroscopy of Interacting Binaries

From Interacting Binaries to Exoplanets: Essential Modeling Tools
Proceedings IAU Symposium No. 282, 2011
Mercedes T. Richards & Ivan Hubeny, eds.
© International Astronomical Union 2012
doi:10.1017/S1743921311026792

Advances in Telescope and Detector Technologies – Impacts on the Study and Understanding of Binary Star and Exoplanet Systems

Edward F. Guinan, Scott Engle, and Edward J. Devinney

Department of Astronomy & Astrophysics, Villanova University, Villanova, PA 19085, USA
email: edward.guinan@villanova.edu

Abstract. Current and planned telescope systems (both on the ground and in space) as well as new technologies will be discussed with emphasis on their impact on the studies of binary star and exoplanet systems. Although no telescopes or space missions are primarily designed to study binary stars (what a pity!), several are available (or will be shortly) to study exoplanet systems. Nonetheless those telescopes and instruments can also be powerful tools for studying binary and variable stars. For example, early microlensing missions (mid-1990s) such as EROS, MACHO and OGLE were initially designed for probing dark matter in the halos of galaxies but, serendipitously, these programs turned out to be a bonanza for the studies of eclipsing binaries and variable stars in the Magellanic Clouds and in the Galactic Bulge. A more recent example of this kind of serendipity is the Kepler Mission. Although Kepler was designed to discover exoplanet transits (and so far has been very successful, returning many planetary candidates), Kepler is turning out to be a "stealth" stellar astrophysics mission returning fundamentally important and new information on eclipsing binaries, variable stars and, in particular, providing a treasure trove of data of all types of pulsating stars suitable for detailed Asteroseismology studies. With this in mind, current and planned telescopes and networks, new instruments and techniques (including interferometers) are discussed that can play important roles in our understanding of both binary star and exoplanet systems. Recent advances in detectors (e.g. laser frequency comb spectrographs), telescope networks (both small and large – e.g. Super-WASP, HAT-net, RoboNet, Las Combres Observatory Global Telescope (LCOGT) Network), wide field (panoramic) telescope systems (e.g. Large Synoptic Survey Telescope (LSST) and Pan-Starrs), huge telescopes (e.g. the Thirty Meter Telescope (TMT), the Overwhelming Large Telescope (OWL) and the Extremely Large Telescope (ELT)), and space missions, such as the James Webb Space Telescope (JWST), the possible NASA Explorer Transiting Exoplanet Survey Satellite (TESS – recently approved for further study) and Gaia (due for launch during 2013) will all be discussed. Also highlighted are advances in interferometers (both on the ground and from space) and imaging now possible at sub-millimeter wavelengths from the Extremely Long Array (ELVA) and Atacama Large Millimeter Array (ALMA). High precision Doppler spectroscopy, for example with HARPS, HIRES and more recently the Carnegie Planet Finder Spectrograph, are currently returning RVs typically better than $\sim$2-m/s for some brighter exoplanet systems. But soon it should be possible to measure Doppler shifts as small as $\sim$10-cm/s – sufficiently sensitive for detecting Earth-size planets. Also briefly discussed is the impact these instruments will have on the study of eclipsing binaries, along with future possibilities of utilizing methods from the emerging field of Astroinformatics, including: the Virtual Observatory (VO) and the possibilities of analyzing these huge datasets using Neural Network (NN) and Artificial Intelligence (AI) technologies.

Keywords. instrumentation: detectors, spectrographs; (stars:) binaries: eclipsing, planetary systems

1. Introduction

The advances in telescopes and detectors, and their impacts on the study of exoplanets and (eclipsing) binary star systems are briefly discussed. Because of page and time limits, we will give a broad overview, with more detailed discussions of several topics closely related to the theme of this symposium. We are lucky to live during wonderful time of discovery – at least for Astronomy and Astrophysics. Nearly 20 years ago at the "New Frontiers in Binary Star Research" conference held in Korea in 1990, I was asked to write a review paper on "New Directions in Eclipsing Binary Research" (Guinan 1993). And now, reviewing the interesting and exciting papers presented at this conference (and appearing in this volume) many of which are based on greater than expected advances in technology, discovery and theory since that time, we wonder what new and exciting (and unanticipated) discoveries await us during the next decade. And what awaits us in binary star and exoplanet research during the next decade (as anticipated from the pace and breath of discovery indicated by papers at this conference) could truly be breathtaking!

2. Evolution of Telescopes and Detectors

In 1609, Galileo's first telescope (Optic Tube) was basically a tube containing two lenses. His first attempt was a 3-power instrument, followed by one that magnified objects $\sim 9\times$, but the quality of the lenses was poor, with many bubbles in the glass. Later refracting telescopes evolved and improved to larger achromatic refractors in the 18th and 19th centuries. This development culminated with the completion, in 1897, of the largest refracting telescope in the world – the Yerkes 40-inch telescope, which remained the largest in the world until 1907. However, since that time (and even before), reflectors (beginning around the time of Isaac Newton) also evolved in size and quality. Classical reflector designs culminated with the Mt. Wilson 100-inch telescope (1917) and the Hale 200-inch (5-m) telescope, operational since 1948. These were followed by light-weight, multi-segmented mirrors, and later adaptive optics innovations, culminating with the very large, powerful telescopes of today (e.g. the 10.3-m Gran Telescopio Canarias (GTC), the twin 10-m Keck telescopes, the twin 10-m Hobby-Eberle/SALT, the twin 8-m Gemini Telescopes, the four 8-m telescopes of the VLTI, and most recently the Large Binocular Telescope (LBT) with its 11.2-m effective aperture, and several other telescopes apertures of 4-m or greater). Also it is important to mention telescopes in space such the UV–Optical Hubble Space Telescope (HST) and the infrared Spitzer Space Telescope (SST). There are now over 100 telescopes with apertures of 1-m or larger; and there are over a dozen with apertures of >4-m. In addition to these ground-based telescopes that cover the visible and near-infrared, there are observatories in space and on the ground that cover nearly the entire electromagnetic spectrum, from gamma-rays/hard X-rays (COMPTON & RXTE), soft X-rays (e.g. XMM-Newton and Chandra), Far-UV–Ultraviolet (e.g. Hubble Space Telescope), Infrared (e.g. the Spitzer Space Telescope, as well as several ground-based telescopes), microwave–mm (e.g. ALMA, GMT), radio–cm (VLA, VLBI, ELVA etc.) and pushing out to the very long wavelength frontier with gravity wave instruments like LIGO (Laser Interferometer Gravitational Wave Observatory) and, in the future, LISA (Laser Interferometer Space Antenna). These and other instruments are having profound impacts on almost all aspects of Astronomy, including the studies of binary stars and exoplanet systems.

2.1. *Detector Development*

Although "developed" for other immediate survival purposes, the human eye (in a sense) is the first astronomical detector. The maximum quantum efficiency of the dark-adapted human eye (in the green-yellow spectral region) is typically ~5-6% (e.g. see Hallett 1987). However, our eyes are not well suited for accurate astronomical photometry due to a small, variable aperture: the pupil. Moreover, the eye has a very limited exposure time (time-constant < 0.05 sec) making long exposures (as carried out in photography and CCD photometry) impossible. The eye also has a non-linear response to light and, when processed by the brain, shows a well-known logarithmic response to stimuli (the Weber-Fechner Law). Nonetheless, visual photometry carried out before (and even after) photography (by F.W. Argelander and others), and currently being done by hundreds of amateur astronomers (e.g. the AASVO members) – has been very important for our understanding of variable and eclipsing binary stars. Typical photometric precisions of ~0.1 mag are returned by experienced amateurs. The historic visual photometry is of great value in many cases, providing otherwise unavailable long baselines (over 100 years in some cases) in monitoring long-period variable stars, searching for nova outbursts, and determining eclipse timings of eclipsing binary stars.

Photographic photometry (with quantum efficiencies usually < 2%) has the advantage of panoramic fields of view for imaging and photometry. Moreover, long exposure times permit faint magnitude limits to be reached using large telescopes. Quantum efficiencies – starting with photo-diodes in the 1920-30s and photoelectric photo-multipliers in the 1940s and onward – improved up to ~10-30%. But the major revolution in astronomical detector technology took place during the 1980s with the introduction of charge-coupled devices (CCDs). CCD photometers have now evolved into high quantum efficiency (up to 95%), and low noise (and fast read-out) panoramic detectors that have changed the whole face of modern astronomy. Today, for example, a < 1.0-m telescope equipped with a CCD can achieve 1% photometric precision for stars as faint as 19th magnitude. This is superior to the earlier photographic photometry accomplished using the largest telescope of its time – the 5-m Hale Telescope at Palomar – during the 1950s and 60s. Now, with arrays of CCDs, wide fields can be covered. Also possible are "giant arrays," such as the 3.2-gigapixel CCD camera for the LSST. Such arrays are becoming much more common.

2.2. *Wide-Field Telescopes*

While increases in telescope size and detector efficiency allowed us to peer much deeper into the Universe (and observe fainter variables and eclipsing binary stars), astronomers began using innovative combinations of telescope sizes and detectors to achieve wide-field surveys where many (1000s) stars can be measured in a single exposure. Some of the pioneers in this field were EROS, MACHO and OGLE. Although primarily microlensing missions to search for dark matter, they resulted in a cumulative wealth of information on all classes of variable stars, both inside and external to the Galaxy. In fact, even though these missions failed to solve the dark matter problem, they proved so useful to the field of observational astrophysics that the OGLE program continues to this day (OGLE-IV). The modern wide-field surveys of today (and tomorrow) seem to come in two flavors: those using small telescopes or camera lenses, and those using medium- and large-aperture telescopes. Notable small scope / lens surveys include ASAS, APASS and KELT. Notable medium-/large-aperture surveys include Pan-STARRS and LSST. Although Pan-STARRS and LSST are not designed for high-cadence observations of variable stars, the scientific community is already preparing itself for the rich, long-term datasets these surveys will be providing for variable stars (including extra-galactic eclipsing binaries) down to 24th magnitude.

2.3. Telescope Networks

Over the last twenty years, networks (some global) of telescopes have developed, many of which utilize commercially available, small telescopes (typically 20–50-cm), while others are wide angle camera lenses mounted on telescope drives (e.g. WASP, HAT-Net, RoboNet I & II). Several are in operation today and more are planned for the future. These networks carry out wide field (almost all sky) photometry for stars as faint as 11–12th mag. Even though the science focus is mainly on planet hunting via transit eclipses, an important side-product is the synoptic photometry of all types of variable and eclipsing binary stars. These networks include HatNet, SuperWasp, TrES and others. The photometric precision is not very high (on the order of a several milli-mag), but is still quite sufficient to discover exoplanets with diameters equal to or greater than the size of Neptune, and in fact these networks have discovered several hot Jupiter-size exoplanets. For smaller red dwarf stars, it is even feasible to detect super-Earth size planets with such telescopes, as is the main goal of the MEarth program (https://www.cfa.harvard.edu/~zberta/mearth/).

Also, a global network of small- to medium-aperture telescopes (10–15 × 0.4-m, 12 × 1-m and 2 × 2-m) is planned by the Los Cumbres Observatory Global Telescope (LCOGT) Network. The major focus of this network is global, uninterrupted time-series photometry of planetary transits, along with variable and eclipsing binary stars.

2.4. Ultra-High Precision Photometry from Space

Of course, some of the most exciting research being carried out recently has been coming from space-borne missions. HST and Spitzer have been involved in high-precision follow-up photometry of known planetary transits, returning excellent transit curves for several systems. The CoRoT satellite (http://smsc.cnes.fr/COROT/), launched in December 2006, was the first satellite dedicated to the detection of transiting exoplanets. As of June 2011, CoRoT has officially discovered 24 exoplanets, but over 100 more candidates are awaiting confirmation. The Kepler Mission was launched in March 2009, to provide near-constant monitoring of ~150,000 stars within a ~100-deg^2 field of view. Since that time, Kepler has become by far the primary contributor to the current database of planetary candidates. In fact, as of December 2011, 2326 planetary candidates have been identified. Of these, 207 are approximately Earth-size, 680 are super Earth-size, 1181 are Neptune-size, 203 are Jupiter-size and 55 are larger than Jupiter. Included are 48 planet candidates orbiting within their host star's habitable zone. Most recently, just before this paper went to press, the Kepler Mission confirmed that one of these candidates, Kepler 22b, is the smallest yet found to orbit within the habitable zone of a star similar to our Sun (Borucki et al. 2011). The planet is about 2.4× the radius of Earth. From the available data, it is not yet known if Kepler 22b has a predominantly rocky, gaseous or liquid composition. But more discoveries will surely follow.

Also, ultra-high precision photometry promises valuable binary star scientific returns, such as:

• Tidally Excited Pulsating Binary Stars ("Heartbeat Stars"): Tidally induced pulsations occur in a binary system when, during a period of time around periastron passage, the proximity of the stars and the gravitational forces at play excite pulsations in either (or both) of the stars. The precise characterization of these events can serve almost as a form of asteroseismology, returning a wealth of new information about the stars involved.

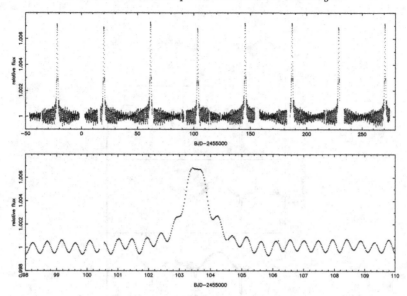

Figure 1. Kepler lightcurve of KOI-54 (from Welsh *et al.*(2011)) showing tidally induced pulsations and brightening events.

Recently, tidally induced activity has been observed in Kepler photometry of the system KOI-54 (Welsh *et al.* (2011) and papers cited within – see Fig. 1). Burkart *et al.* (2011) develop a general framework for interpreting and analyzing high precision light curves from eccentric stellar binaries, such as KOI-54, that display this effect. They refer to the studies of these tidally induced pulsations as "Tidal Asteroseismology." Also in this volume, Kelly Hambleton discusses a case study of an interesting tidally-excited star in the paper "KIC 4544587: An Eccentric, Short Period Binary with δ Scuti Pulsations and Tidally Excited Modes." The topic of tidally excited stars is also discussed by Carla Maceroni in her review paper on the impact of CoRoT and Kepler on eclipsing binary science (in this volume).

• Doppler and Relativistic Beaming (also called light amplification) – motion of stars produces small (<0.2% light variations) from the Doppler Effect, Aberration and Special Relativistic beaming (time dilation effects). Beaming in binary stars is a new and interesting effect recently observed with ultra-high precision photometry. The light variations arising from beaming effects in binary systems are very small (a few millimag) and can only really be studied with CoRoT and Kepler. Doppler beaming in stellar binary systems and star-planet systems has been theoretically discussed by Loeb & Gaudi (2003) and Zucker *et al.* (2007). Additional discussions about beaming in binary stars are provided in this volume in the paper by Carla Maceroni and in Andrej Prsa's paper on "Advances in Modeling Eclipsing Binary Stars in the Era of Large All-Sky Surveys with EBAI and PHOEBE."

• Light variations from Orbital Doppler Beaming behave in a similar way to spectroscopic radial velocities, and their analysis can give the mass ratio, eccentricity, omega and *asini* values for a system. For example see the recent study of the beaming binary KPD 1946+4340 given by Bloemen *et al.* (2011) – see Fig. 2.

• Rotation Beaming: This is analogous to the Rossiter-McLaughlin effect and occurs as a transiting planet covers the advancing/receding hemispheres of the star. The potential

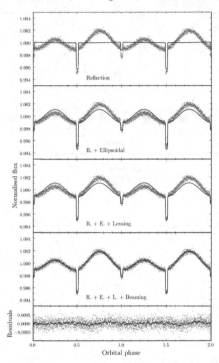

Figure 2. Phase-folded light curve of KPD 1946+4340, with the best-fitting models overplotted. One can clearly see the benefits of including the effects of beaming (bottom lightcurve, just above the residuals) in the model. Figure taken from Bloemen *et al.* (2011).

of studying rotational Doppler Beaming in eclipsing binaries has been recently discussed by Groot (2011).

• Eclipse Mapping: The analysis of high precision photometry during the eclipses of a spotted star in well-suited eclipsing binaries can permit the unambiguous determination of starspot sizes, distributions and motions. See, as an example, Huber *et al.* (2010) for planetary eclipse mapping of CoRoT-2a.

2.5. *Transit Timing Variations (TTVs) – What Can Be Learned*

The study of exoplanet Transit Timing Variations (TTVs) is now a very active field of research since CoRoT and Kepler. For eclipsing binary stars, the study of variations in eclipse timings is over a hundred years old. The detection of third bodies in eclipsing binary systems, from small periodic variations in the system's distance from us that arise from the gravitational pull of the tertiary companion, is made possible by the Light Travel Time Effect (LTTE). For transiting exoplanet systems, TTV studies can reveal the following:

• the presence of addition unseen (low mass) planets (TTVs of several seconds to minutes).

• the presence of hosted moons (Exomoons) – this is a very small effect, best detected in the presence of larger moons (>0.2 $M_{\oplus}$) in short periods (see Kipping 2009).

• changes in the exoplanet's orbital period arising from tidal (and/or magnetic) coupling. The bright transiting exoplanet system HD 189733 could be a good candidate for studying this effect since it's of hot Jupiter-size and short period (P=2.2 days), and appears to be losing its orbital angular momentum to its K-type host star (see Santapaga *et al.* 2011).

Figure 3. An artist's representation of a laser frequency comb. This very promising technology will be used in many of the ultra-high precision, next generation spectrographs. Since 2005, over 100 papers have been published dealing with laser frequency combs.

3. State of the Art Spectrographs – Impacts on Exoplanets and Eclipsing Binary Stars

Modern spectrographs have throughputs of >50% (the amount of incoming photons that reach the detector), where older ones had only ~10%, but spectrographs have benefitted greatly from more advances than throughput alone. When spectrographs are fiber-coupled to the telescopes, and thermally stabilized, amazing things can be accomplished. Most modern spectrographs used to measure the small radial velocities (RVs) of stars arising from hosted planets are very carefully constructed, thermally stabilized, bench mounted and fiber-fed. Some examples include HIRES, HARPS, HERMES and the Carnegie Planet Finder Spectrograph (PFS) which can currently achieve RV measures of 1–2-m/s precision (perhaps even 50-cm/s – Paul Butler, private comm.). The ultimate goal is to develop spectrographs with sufficiently high precision to allow the detection of terrestrial-mass planets.

3.1. *Toward Ultra-high precision (<10-cm/s) radial velocity determinations*

Laser Frequency Combs (see Fig. 3) can produce a spectrum of evenly spaced narrow lines spanning the optical into the IR. Combining a fiber-fed frequency comb with a stable spectrograph can result in RV measures of ~6-cm/s precision, allowing the possibility of discovering Earth-mass exoplanets – e.g. the Earth causes the Sun's RV to shift by ± 10-cm/s (for comparison Jupiter produces a reflex motion of ± 13-m/s). Steinmetz *et al.* (2008) and Murphy *et al.* (2007) give excellent reviews, and the first stellar RV measures using a laser comb are presented in Osterman *et al.* (2011).

ESPRESSO (Echelle SPectrograph for Rocky Exoplanet Stable Spectroscopic Observations) is a new generation, very stable, fiber-fed Echelle spectrograph (R=140,000) designed to operate at visible wavelengths. ESPRESSO is being developed by ESO for use with the 8-m Very Large Telescope. It is designed to secure RV measures of ~6–10-cm/s (you can walk faster than that speed). Installation and commissioning of ESPRESSO at the VLT is expected during 2016. The instrument is designed to operate with a single telescope or using the combined light gathering power of all four 8-m telescopes configured to form a 16-m equivalent telescope. This arrangement will permit ESPRESSO to reach faint stars or achieve very high signal-to-noise measures for brighter targets.

4. Some Impacts of Large Telescopes

It is obvious that large, next generation telescopes will be excellent tools for studying eclipsing binaries and exoplanets. For example, very faint (distant) stars can be effectively observed. Or brighter stars can be observed with very high signal-to-noise ($100,000\times$) as well as with very high dispersion. Most notable among the many large telescopes now available for this work are the Large Binocular Telescope, with the largest light-gathering power of a single instrument (11.3-m effective aperture when the light of the two mirrors are combined), exceeded however by the four 8-m telescopes of the VLTI, with the combined light gathering power of a 16-m telescope. The enhanced light gathering power of these great telescopes will allow us to study in detail the atmospheres of transiting exoplanets, and enable us to determine the physical properties of faint eclipsing binary stars in other members of the Local Group, and further, to secure direct distances to them, as reported in this volume and described briefly below.

4.1. *Exoplanet Systems:*

Today's great telescopes (and even larger ones in the future) will be invaluable in spectral studies of the atmospheres of exoplanets using both spectrum subtraction (spectrum of [planet + star] minus the spectrum of star alone) techniques as well as in planet atmosphere transmission techniques discussed at this Symposium in Adam Burrow's comprehensive paper "Towards a Theory for the Atmospheres, Structure, and Evolution of Giant Exoplanets." In addition, there is a recent book on this topic, "Exoplanet Atmospheres: Physical Processes" by Seager (2010). We can expect such investigations as follows:

• Reflection and Transmission Spectra: providing the chemical compositions of Exoplanet atmospheres, cloud cover etc. (See Burrows in this volume, and Seager (2010) for reviews).

• Exoplanet Dynamics: Rossiter-McLaughlin (R-M) Effect (Simon Albrecht/Amaury Triaud in this volume). As discussed in these papers and references therein, the application of the R-M effect to transiting exoplanets is truly amazing. These studies indicate that many exoplanet star systems (unlike our solar system) have highly inclined orbits (even retrograde orbits in some cases) relative to the rotational planes of their host stars. This result was totally unexpected.

• More precise RV orbits for exoplanet systems leading, ultimately, to Earth-mass exoplanets.

• Direct imaging of planetary accretion disks and protoplanets.

• Imaging and spectra of exoplanets (with sufficiently large ground- and space-based telescopes).

4.2. *Eclipsing Binaries*

Large telescopes are needed to study faint eclipsing binary stars in other galaxies such as M33 and M31. This is discussed by a paper in this volume by Alceste Bonanos. With large telescopes, faint eclipsing binaries in distant galaxies can be studied photometrically and spectroscopically. Moreover, excellent distances to M31 and M33 are being returned using eclipsing binaries (Vilardell *et al.* (2010) and references therein).

• Radial velocity curves of selected extragalactic eclipsing binaries in Local Group galaxies to determine accurate masses, and unbiased distances.

• Observations of faint, astrophysically selected eclipsing systems: e.g. SN Ia progenitors, rare binaries in fast stages of evolution or with black hole and neutron star components, pre- and post-common envelop binaries, and many other interesting or extreme binary systems would be great targets.

Just as the large observatories have negotiated the learning curve on adaptive-optics (AO) technology, interferometry now stands to follow. Interferometry is becoming a powerful tool to "image" stars and proto-planetary disks. There are some interesting papers on imaging techniques in this volume. For example, with the VLTI and/or with Adaptive Optics (AO), protoplanetary disks are being imaged, showing what appear to be planet-forming regions contained within the disks. Recently, Kloppenborg *et al.* (2010) used the CHARA Interferometer to "image" the large dark disk transiting the F-supergiant component during the recent ∼2-yr long eclipse of the ∼27.2-yr eclipsing binary ϵ Aur. Many more additional binary stars (including β Lyrae), proto-planetary disks and even proto-planets are likely interferometry targets, with improved AO methods using larger telescopes.

5. Great Expectations: Looking into the Future, From Very Large to Tiny

The James Webb Space Telescope (JWST) will be a large infrared telescope with a 6.5-m primary mirror. In spite of its financial and technical problems, JWST is working to a 2018 launch date, after which it will be the premier observatory of the next decade, serving thousands of astronomers worldwide. JWST will study every phase in the history of our Universe, ranging from the first luminous glows after the Big Bang, to the formation of solar systems capable of supporting life on planets like Earth. It will also be useful for faint binary stars in other galaxies. Several innovative technologies have been developed for the JWST, including: a folding, segmented primary mirror, adjusted to shape after launch; ultra-lightweight beryllium optics; detectors able to record extremely weak signals; micro-shutters that enable programmable object selection for the spectrograph; and a cryro-cooler for cooling the mid-IR detectors to ∼7K. The long-lead items, such as the beryllium mirror segments and science instruments, are presently made or under construction.

During 2012, the BRITE Constellation satellites, which are 20-cm^3 nano-satellite systems (Schwarzenberg-Czerny *et al.*2010), are expected to be launched. The individual BRITE satellites are essentially orbiting filtered CCD cameras, each having a wide ∼20° field of view. They are designed to carry out continuous millimag photometry of bright stars (brighter than 4th mag) in selected regions for several months at a time. Each of these nano-satellites is equipped with a different filters (the first two will have blue and red wide-band filters). And within the next few years, the ambitious ESA Gaia mission is expected to launch. Gaia will be amazing! Gaia will be carrying out parallax measures, photometry of millions of stars (as well as low dispersion spectroscopy of a subset of these stars). In the process, hundreds of thousands of new eclipsing binaries will be found each with measured parallaxes, colors and spectra. While on the ground, huge telescopes such as the OverWhelmingly Large (OWL) Telescope, the Thirty Meter Telescope (TMT), and maybe others, will be built within the next decade or two. As for exoplanet research, DARWIN and/or the Terrestrial Planet Finder (TPF) may eventually merge and get approved to image and secure spectra of nearby exoplanets (using nulling interferometers or other methods) to search for life-supporting atmospheres or even spectroscopic bio-signatures. Also, to save money, an innovative way of doing this could be to deploy a star occulting shroud several km in front of JWST to use as a coronagraph to block out the star hosting planet. But the logistics of such an arrangement would be very challenging.

Finally we desperately need to address the expected overwhelming amount of data (petabytes) to be pouring in from astronomical missions in the near future. The emerging new field of Astroinformatics (see Borne *et al.* 2009) is being developed to cope with and take advantage of these splendid datasets. Astroinformatics focuses on data organization, taxonomies, data mining, machine learning / artificial intelligence (AI), visualization and cyber infrastructure (e.g. Virtual Observatory). These methods are necessary to scientifically exploit the ever expanding "deluge" of astronomical data. In this volume, the example of the Virtual Observatory (VO) is discussed by Gerrie Peters. In closing, the telescopes, equipment, innovative technologies and supercomputers (available today and in the near-future) offer thrilling opportunities for research in our fields of study. Moreover, with the internet and VO, these important datasets are accessible for use to many more people not directly involved with building the missions; for use not only by professional astronomers, but as exciting resources for "citizen" astronomy projects and in the education and training of high school students with internet access.

Acknowledgements

This research is supported by NSF/RUI Grant AST 05-07542 and grants from NASA which we gratefully acknowledge. We wish to thank Mercedes Richards and the SOC and LOC of the symposium for doing an outstanding job in organizing and successfully conducting such a superb and scientifically memorable meeting.

References

Bloemen, S., Marsh, T. R., Östensen, R. H., *et al.* 2011, *MNRAS*, 410, 1787
Borne, K., Accomazzi, A., Bloom, J., *et al.* 2009, astro2010 : The Astronomy and Astrophysics Decadal Survey, 2010, 6P
Borucki, W. J., *et al.* 2011, arXiv:1112.1640
Burkart, J., Quataert, E., Arras, P., & Weinberg, N. N. 2011, arXiv:1108.3822
Groot, P. J. 2011, arXiv:1104.3428
Guinan, E. F. 1993, *New Frontiers in Binary Star Research*, 38, 1
Hallett, P. E. 1987, *J. Optical Soc. America A*, 4, 2330
Huber, K. F., Czesla, S., Wolter, U., & Schmitt, J. H. M. M. 2010, *A&A*, 514, A39
Kipping, D. M. 2009, *MNRAS*, 392, 181
Kloppenborg, B., Stencel, R., Monnier, J. D., *et al.* 2010, *Nature*, 464, 870
Loeb, A. & Gaudi, B. S. 2003, *ApJL*, 588, L117
Murphy, M. T., Udem, T., Holzwarth, R., *et al.* 2007, *MNRAS*, 380, 839
Osterman, S., Diddams, S., Quinlan, F., *et al.* 2011, *BAAS*, 43, #401.02
Santapaga, T., Guinan, E. F., Ballouz, R., Engle, S. G., & Dewarf, L. 2011, *BAAS*, 43, #343.12
Schwarzenberg-Czerny, A., Weiss, W., *et al.* 2010, 38th COSPAR Scientific Assembly, 38, 2904
Seager, S. 2010, Exoplanet Atmospheres: Physical Processes. By Sara Seager. Princeton University Press, 2010. ISBN: 978-1-4008-3530-0,
Steinmetz, T., Wilken, T., Araujo-Hauck, C., *et al.* 2008, *Science*, 321, 1335
Vilardell, F., Ribas, I., Jordi, C., Fitzpatrick, E. L., & Guinan, E. F. 2010, *A&A*, 509, A70
Welsh, W. F., Orosz, J. A., Aerts, C., *et al.* 2011, *ApJS*, 197, 4
Zucker, S., Mazeh, T., & Alexander, T. 2007, *ApJ*, 670, 1326

From Interacting Binaries to Exoplanets: Essential Modeling Tools
Proceedings IAU Symposium No. 282, 2011
Mercedes T. Richards & Ivan Hubeny, eds.
© International Astronomical Union 2012
doi:10.1017/S1743921311026809

Ground-Based and Space Observations of Interacting Binaries

Panagiotis G. Niarchos

Department of Astrophysics, Astronomy and Mechanics, National and Kapodistrian University
of Athens, Athens, Greece
email: pniarcho@phys.uoa.gr

Abstract. Multi-wavelength observational data, obtained from ground-based and space observations are used to compute the physical parameters of the observed Interacting Binaries (IBs) and study the interactions and physical processes in these systems. In addition, the database of IBs from ground-based surveys and space missions will provide light curves for many thousands of new binary systems for which extensive follow up ground-based observations can be carried out. In certain cases, light curves of superior quality will allow studies of fine effects of stellar activity and very accurate determination of stellar parameters. Moreover, many new discoveries of interesting systems are expected from ground-based all-sky surveys and space missions, including low mass binaries and star-planet binary systems. The most important current and future programs of observations of IBs from ground and space are presented.

Keywords. Surveys, stars: fundamental parameters, (stars:) binaries (including multiple): close-binaries: (stars:) binaries: eclipsing

1. Introduction

Interacting Binaries (IBs) or Close Binaries (CBs) are two stars that do not pass through all stages of their evolution independently of each other, but in fact each has its evolutionary path significantly altered by the presence of its companion. Processes of interaction include: gravitational effects, mutual irradiation, mass exchange and mass loss from the system. The study of CBs provides insights into nearly all areas of astrophysics, including stellar interiors and atmospheres, stellar evolution, nucleosynthesis, plasma physics, magnetic dynamos (in cool stars), and relativistic physics to name a partial list. Recently, the study of Eclipsing Binaries (EBs) in other galaxies and clusters makes it possible to explore stellar evolution and establish mass-luminosity laws for galaxies with vastly different evolutionary and chemical histories from our Galaxy (such as LMC and SMC). Moreover, EBs are beginning to play an important role in cosmology as distant indicators to nearby galaxies.

2. Observational approaches of Interacting Binaries

The ultimate goal for observational astronomers who study the properties of binary stars is to make direct determination of the astrophysical parameters: *masses, radii, shapes, temperatures and luminosities*. These parameters, also called *absolute dimensions*, can be derived from the analysis of light and radial velocity curves, regardless of the distances of the binaries from us. During the last three decades two distinct developments had a great impact in deriving the basic astrophysical quantities describing the close binary systems. The first was the development of the Roche model for light curve analysis, and the second one was the invention of new modern methods in deriving radial velocities for close binary systems.

2.1. *Photometry*

The photometric observations made by modern detectors (CCDs) are expected to have 1-2% precision if they are carefully reduced and well transformed. The photometric light curves are analysed with modern synthetic light curve codes (Prša & Zwitter 2005), based on Roche model, and enable us to derive much more realistic and accurate physical parameters of close binary systems. These codes allow also a simultaneous solution of photometric and radial velocity curves. Among those parameters the one of main interest is the mass ratio (q) of the system, which is necessary for the calculation of the absolute dimensions for single line spectroscopic binaries.

2.2. *Spectroscopy*

Spectroscopic studies of CBs lead to spectral classification, line profile analysis and radial velocity determinations. Low resolution, usually R>1000, is used for spectral classification, higher resolutions are desirable for precision radial velocity work, and highest resolution $R \geqslant 10^5$ is needed for the analysis of spectral line profiles and extra-solar planet research. In any case high level of precision and accuracy in radial velocities is needed for close binary star modeling. The present day precision is better than 1m/s (e.g. HARPS spectrograph at ESO). During the last three decades it has been possible to overcome difficulties in studying spectra of close binaries by reducing the spectra in a digitized form using modern techniques, like *Cross Correlation Technique* (Mazeh & Zucker 1994), *Broadening Function Approach* (Rucinski 2002) and *Spectral Disentangling* (Hadrava 2006). A combination of photometric and spectroscopic observations yields the fundamental source of information about *sizes, masses, luminosities and distances or parallaxes* of stars.

2.3. *Polarimetry*

Almost every class of binary stars can produce observable polarimetry. In these systems, polarimetry can help to: determine the geometry of the circumstellar or circumbinary matter distribution, yield information on asymmetries and anisotropies, identify obscured sources, map star-spots, detect magnetic fields, and establish orbital parameters, particularly the orbital inclination i which is an important parameter for deriving the stellar masses. (See the paper by K. Bjorkman, this symposium).

2.4. *Interferometry*

Interacting binaries typically have separations in the milliarcsecond regime. Recent advances in optical interferometry have improved our ability to discern the components in these systems and have now enabled the direct determination of physical parameters. Application of interferometric observations in the study of binary stars yield individual stellar masses, distances to the systems and provide reliable data for the empirical mass-luminosity relation in a region which is intermediate between visual and purely spectroscopic data. In addition, speckle interferometry is applicable to the determination of the angular diameters of objects, and the development of long baseline techniques allows the achievement of angular resolutions sufficiently high for the determination of the diameters of the individual components of close binaries (e.g. Zavala *et al.* 2010). The mas, or sub-mas separation can be reached only with VLB radio interferometry, or the recent, most powerful optical, near-IR long-baseline interferometric equipment, e.g. CHARA array, Palomar Testbed Interferometer and others. Only the brightest close binaries can be reached by such methods (Coughlin *et al.* 2010; see also the paper by Ph. Stee, this symposium).

3. Why multi-wavelength observations?

Modern astrophysics requires studying an object across the whole EM spectrum, since different physical processes can be studied at different wavelengths. In the optical range, information from the massive companion can be collected. For systems with degenerate components, the compact object is responsible for the emission of high energy photons (X-rays and γ-rays), while IR and UV observations give us information about the interstellar environment and the mass transfer from one component to the other. In some systems, e.g. those containing a BH, radio emission can be expected.

3.1. *UV observations of Interacting Binaries*

The UV is of outmost importance in the study of IBs, as a large part of their luminosity is radiated away in this wavelength range, and the UV hosts a multitude of low and high excitation lines of a large variety of chemical species. UV spectroscopy of IBs obtained with IUE, HST and FUSE have dramatically improved our understanding of IBs and of the physical processes that characterize their emission. UV imaging has made it possible to isolate binaries and the products of binary evolution in old stellar populations and thereby test directly models of binary evolution in dense stellar systems (Gansicke *et al.*, 2008). With the future World Space Observatory-Ultraviolet (WSO-UV) (http://wso.inasan.ru/), powered stellar flares, developed from the complex processes of interaction between the accretion disk and the central star, and other energetic processes accompanied by strong UV radiation will be studied. Monitoring with WSO-UV of extrasolar planet transits will provide important information on the planetary atmosphere and its interaction with the parent star.

3.2. *X-Ray Variability*

X-ray Binaries (XRBs) are variable on many timescales in different ways: (a) X-ray pulsations: are periodic with spin period, due to magnetically funnelled accretion onto the poles, (b) flickering: quasi-periodic oscillations caused by instabilities in the disc (noise), (c) transient accretion events: alternation between phases of high and low accretion rates due to thermal transitions in the accretion disc (in particular for BHs accreting at low rates; also CVs), and (d) thermonuclear explosions, once enough H/He fuel has been accreted. The active X-ray observatory satellites are ESA XMM-Newton (0.1-15 keV), INTEGRAL (15-60 keV), and the NASA Rossi X-ray Timing Explorer (RXTE), Swift and Chandra.

3.3. *Infrared observations of IBs*

IR observations may yield substantial information on the location, size, density and temperature of dust and gas components. Close companions in systems with high mass loss rates may modify the rate of mass loss, flow velocity or grain formation and these effects will change the infrared emission characteristics, a study of which allows a deeper understanding of the underlying mechanisms of the mass loss itself. IR observations from space provide the means to study the mid-IR properties of systems with compact objects. The goals are to establish the mid-IR SED, search for signatures of jets, circumbinary disks, low mass or planetary companions and debris disks, and study the local environment of these sources (Adame *et al.* 2011).

3.4. *Radio observations of IBs*

Radio observations with Interferometric Arrays provide a key tool for a unified understanding of XRBs in the context of accretion powered sources in the Universe. We can learn about the anatomy of X-ray and possibly γ-ray sources. The very sensitive arrays under construction will allow us to address new astrophysical issues about the interaction of relativistic jets with the galactic ISM. They will also resolve and study the ejecta of XRBs. Using these observations we can develop the first semi-quantitative models to interpret how jet production, or suppression, is related to the X-ray spectral states of the accretion disk and corona in the system (Miller-Jones 2008).

4. Ground-based observations

New catalogues of various categories of IBs have been compiled from ground-based and space observations. A catalogue of CVs, LMXBs, and Related Objects (ROs), containing 98 LMXBs, 114 HMXBs, 880 CVs, 312 ROs (7th Edition, rev. 7.15, March 2011) is given by Ritter & Kolb (2003), and a catalogue of symbiotic stars with 188 + 28 (suspected) was presented by Belczyński et al. (2000). An updated catalogue, based on the GCVS, of 6330 EBs was presented by Malkov et al. (2006). That number will increase tremendously over the next couple of decades, as large and smaller-scale surveys are undertaken. Thousands (over 10^4 of 13-14 mag) of new candidates EBs have been discovered through surveys looking for micro-lensing events, like EROS, OGLE, MOA and MACHO, the DIRECT project and others in very crowded fields.

4.1. Microlensing Surveys

The main objective was the search and study of dark stellar bodies, so-called "brown dwarfs" or "MACHOs" in our Galaxy. This is made possible by their gravitational microlensing effects on stars in the Magellanic Clouds. The surveys are:

• *EROS*: EROS-1 (1990-1995) and EROS-2 (1996-2003). EROS has localized about (75+176) eclipsing binaries in LMC.

• *MACHO*: (1993-2001). A catalogue with 4500 binaries in LMC and 1500 in SMC is underway. About 3000 are genuine EBs.

• *SuperMACHO project* (CTIO Blanco 4 m tel.) has surveyed the MCs down to VR 23 mag (some W UMa).

• *OGLE* (since 1992). Two catalogues exist with 1459 EBs detected in SMC and 2580 EBs detected in LMC. A catalogue of 10862 EBs (detected in the galactic bulge fields) was presented in OGLE II (Devor, 2005).

• *MOA* (1996-2004): A catalogue of 167 EBs in SMC was presented by Bayne at al. (2002).

• Two other microlensing surveys toward M31: *Wendelstein Calar Alto Pixellensing Project, WeCAPP*, (2000-2003) (Fliri et al. 2006) discovered 31 EBs in M31; *POINT-AGAPE Survey* (1999-2001, INT+WFC) (An et al. 2004) has released a catalogue with 35000 variables (systematic search for EBs, 20 CNe).

4.2. Other large-scale surveys

• *The Robotic Optical Transient Search Experiment (ROTSE-I, -III)* (1998) was designed to look for the optical counterparts to gamma ray bursts. In the process it has discovered over 1000 EBs in a survey covering about 5% of the sky area that it monitors (Gettel et al. 2006).

• *ASAS (All Sky Automated Survey)*. The ASAS-3 Catalog of Variable Stars contains over 11,099 EBs binaries found among 17,000,000 stars on the sky south of dec. +28 (Paczynski et al. 2006).

• *SuperWASP photometric survey* (Pollacco et al. 2006; Norton, A.J. et al. 2011): 48 EBs (40 W UMa with P<0.23 d and 1 with P<0.20 d)

- *The Sloan Digital Sky Survey (SDSS)*: The result is a catalogue of more than 1200 spectroscopically selected close binary systems observed (Silvestri *et al.*, 2007).
- *The Panoramic Survey Telescope & Rapid Response System (Pan-STARRS)* (2006). Goal: discover and characterize Earth-approaching objects (asteroids & comets).
- The Large Synoptic Survey Telescope (LSST) (See the paper by L. Eyer, this symposium).

4.3. *Specialized Projects*

- The *DIRECT project* (1996-1999). The aim was to determine the distances to nearby galaxies M31, M33 by monitoring for Cepheids and Detached EBs (Stanek *et al.* 1998). Results: 89 EBs were found in 6 fields surveyed in M31, 237 EBs in M33, and 437 EBs in M31 (by INT 2.5 m tel.) (Vilardell *et al.* 2006).
- The *W UMa project*. The aim is the determination of very accurate physical parameters of stars in contact binaries of W UMa-type by using high quality homogeneous photometric and spectroscopic observations. The program is a novel approach and the systems to be studied are > 100 (Kreiner *et al.* 2006).

5. Observations from Space

5.1. *Past missions*

Hipparcos (1989-1993, ESA). In the catalogue released in June 1997, there are 120,000 stars with 1 milliarcsec level astrometry. From these 1034 are EBs with 117 unsolved cases. 35% of these EBs were not previously known.

5.2. *Current space missions*

- The *COROT* Mission (CNES, ESA, Brazil; launched December 2006). The objectives are: (i) a search for extrasolar planets of large terrestrial size, and (ii) perform asteroseismology in solar-like stars. Selected binary systems will be observed in the Additional Program frame as targets of long and continuous pointed observations (see the paper by C. Maceroni, this symposium).
- *Kepler* Mission. Launched in March 2009. The aim is to detect one-Earth radius planets in the habitable zone of solar-like stars. The total number of identified EBs systems in the Kepler FOV has increased to > 2200, 1.4% of the 156,000 Kepler target stars (see the paper by C. Maceroni, this symposium).
- MOST (Microvariability and Oscillations of STars telescope). A Canadian Space Agency mission, in operation since 2003. 16 new EBs have been detected (Pribulla *et al.* 2010).
- STEREO (Solar TErrestrial RElations Observatory) mission (since 2006). Researchers have discovered 122 (!) new EBs and observed hundreds more variable stars in an innovative survey (NASA, Press Release: 19 April, 2011).

5.3. *Future space missions*

The *Gaia* mission. (See the paper by L. Eyer, this symposium). Launch date 2013. Objectives: to build a catalogue of $\sim 10^9$ stars with accurate positions, parallaxes, proper motions, magnitudes and radial velocities. The catalogue will be complete up to V = 20 mag with no input catalogue and therefore no associated bias. The strength of the Gaia mission is in the numbers. Gaia will observe $\sim 4 \times 10^5$ EBs brighter than $V \leqslant 15$ and $\sim 10^5$ of these will be double-lined (SB2) systems. For $V \leqslant 13$ the number of SB2 will be about 16 000 for which Gaia should provide orbital solutions formally accurate to $\sim 2\%$ (Niarchos *et al.* 2006). This is a fantastic number compared to <100 systems studied at similar accuracy by ground-based observations so far (Andersen 1991).

6. Concluding Remarks

There are many theoretical and observational areas in the field of Binary Stars that remain practically unexplored, and the mysteries are challenging and important. Great advances on the observational front are expected with large-scale photometric and spectroscopic surveys (from ground and space) and radio-optical interferometers. The new advanced techniques will allow shallow or marginal stellar eclipses to be detected easily, and complex physical processes to be observed. Binaries of all types can now be studied across the entire EM spectrum. New technologies and instruments used in large-scale surveys will lead to a renaissance and a "Brave New World" of binary star studies.

References

Adame, L. *et al.*, 2011, *ApJ*, 726, L3
Andersen, J. 1991, *A&AR*, vol. 3, No 2, 91
An, J. H. *et al.*, 2004, *MNRAS*, 351, 1071
Bayne, G. *et al.*, 2002, *MNRAS*, 331, 609
Belczyński, K. *et al.*, 2000, *A&AS*, 146, 407
Coughlin, J. L., Harrison, T. E., & Gelino, D. M. 2010, *ApJ*, 723, 1351
Devor, J. 2005, *ApJ*, 628, 411
Fliri, J. *et al.*, 2006, *MmSAI*, 77, 332
Gansicke, B. T. *et al.*, 2006, *Ap&SS*, 306, 177
Gettel, S. J., Geske, M. T., & McKay, T. A. 2006, *AJ*, 131, 621
Hadrava, P. 2006, *Ap&SS*, 304, 337
Kreiner, J. M. *et al.*, 2006, *Ap&SS*, 304, 71
Malkov, O., Yu. *et al.*, 2006, *A&A*, 446, 785
Mazeh, T. & Zucker, S. 1994, *Ap&SS*, 212, 349
Miller-Jones, J. C. A. 2008, *JPhCS* 131, 012057
Niarchos, P., Munari, U., & Zwitter, T. 2007, in: Hartkopf, W. I., Guinan, E. F., & Harmanec, P. (eds.), *Binary Stars as Critical Tools & Tests in Contemporary Astrophysics* (Proceedings of IAU Symposium 240, 22-25 August 2006, Prague), p. 244
Paczynski, B. *et al.*, 2006, *MNRAS*, 368, 1311
Pollacco, D. L. *et al.*, 2006, *PASP*, 118, 1407
Pribulla, T. *et al.*, 2010, *AN*, 331, 397
Prša, A. & Zwitter, T. 2005, *ApJ*, 628, 426
Ritter, H. & Kolb, U. 2003, *A&A*, 404, 301
Silvestri, N. M. *et al.*, 2007, *AJ*, 134, 741
Stanek, K. *et al.*, 1998, *AJ*, 115, 1894
Vilardel, F., Ribas, I., & Jordi, C. 2006, *A&A*, 459, 321
Zavala, R. T. *et al.*, 2010, *ApJ*, 715, L44

Discussion

J. SOUTHWORTH: The SuperWASP survey has discovered thousands of eclipsing binaries in the course of its search for transiting extrasolar planets. We reckon that there are maybe 100,000 in our archive, which is now publicly available. If someone has the time then this is an excellent research project.

P. NIARCHOS: I agree, but we must keep in mind that photometric data alone are not enough to derive the physical parameters of the system's components (with the exception of W UMa systems with complete eclipses). Radial Velocities (RVs) of the components are needed to be used with the photometric data. Even with no RVs available, valuable information about EBs can be extracted from such a large amount of data.

From Interacting Binaries to Exoplanets: Essential Modeling Tools
Proceedings IAU Symposium No. 282, 2011
Mercedes T. Richards & Ivan Hubeny, eds.
© International Astronomical Union 2012
doi:10.1017/S1743921311026810

Techniques for Observing Binaries in Other Galaxies

Alceste Z. Bonanos

National Observatory of Athens, Institute of Astronomy & Astrophysics,
I. Metaxa & Vas. Pavlou, Palaia Penteli 15236, Greece
email: bonanos@astro.noa.gr

Abstract. I present an overview of the techniques used for detecting and following up binaries in nearby galaxies and present the current census of extragalactic binaries, with a focus on eclipsing systems. The motivation for looking in other galaxies is the use of eclipsing binaries as distance indicators and as probes of the most massive stars.

Keywords. binaries: eclipsing, binaries: spectroscopic, stars: distances, stars: fundamental parameters, Local Group

1. Motivation

Eclipsing binaries are not only powerful tools for obtaining fundamental parameters of stars (Andersen 1991, Torres et al. 2010), but also the most accurate tools currently available for measuring masses and radii of massive stars and for probing the upper stellar mass limit. Double-lined spectroscopic binary systems exhibiting eclipses in their light curves provide accurate geometric measurements of the fundamental parameters of their component stars. Specifically, the light curve provides the orbital period, inclination, eccentricity, the fractional radii and flux ratio of the two stars. The radial velocity semi-amplitudes determine the mass ratio; the individual masses can be solved using Kepler's third law. Furthermore, by fitting synthetic spectra to the observed ones, one can infer the effective temperatures of the stars, solve for their luminosities and derive the distance (e.g. Bonanos et al. 2006). In the past two decades, many eclipsing binaries have been discovered in other galaxies and several of these have been subject to follow up studies, resulting in the measurement of their fundamental parameters. The main motivations for observing eclipsing binaries in other galaxies are to study massive stars and to obtain independent distances.

Massive stars are intrinsically rare compared to their lower mass counterparts, due to their shorter lifetimes and the steep initial mass function, which results in the formation of a smaller number of massive stars. Studying massive stars in the Galaxy is challenging, because they are located in the Galactic plane, where they reside in young massive clusters and usually near giant molecular clouds, and are therefore often heavily obscured by dust. Fig. 10 of Mauerhan et al. (2011) demonstrates the small fraction of the Milky Way surveyed for massive stars, by showing the locations of known Wolf-Rayet (WR) stars in the Galaxy. Although the total estimated number of WR stars in the Galaxy is 6500, there are only ~ 600 known and most are located within 5 kpc of the Sun, i.e. only $\sim 10\%$ of the Milky Way has been surveyed. This fraction is slowly increasing, with the recent availability of near-infrared and mid-infrared maps of the Galactic plane (obtained with *Spitzer*), which have been used both to identify new massive clusters (e.g. Davies et al. 2007) and massive evolved stars with nebulae (Gvaramadze et al. 2010, Wachter et al. 2010).

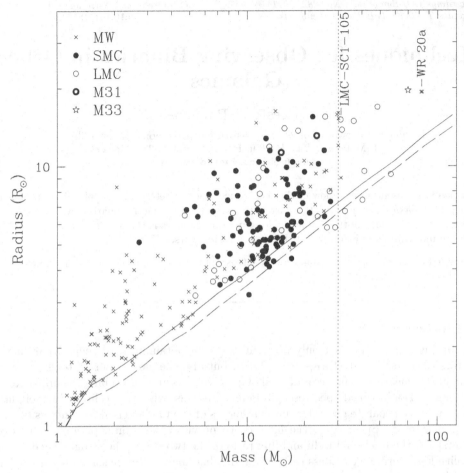

Figure 1. Mass and radius determinations of stars in eclipsing binaries, accurate to ⩽ 10% and complete ⩾ 30 M$_\odot$ (from Bonanos 2009). Since this compilation, measurements of only 2 more massive stars (Stroud *et al.* 2010) satisfy these accuracy requirements, bringing the total to 16.

Another reason why our knowledge of massive stars is incomplete is because their fundamental parameters are not well known, leaving formation and evolution models unconstrained. Masses and radii of massive stars measured from eclipsing binaries remain scarce. Bonanos (2009) compiled a list of the most massive stars accurately measured in eclipsing binary systems and found only 14 stars above 30 M$_\odot$ with mass and radius measurements accurate to 10% or better (see Figure 1). Since this compilation, measurements of only 2 more massive stars (Stroud *et al.* 2010) satisfy these accuracy requirements, bringing the total to 16. Therefore the need for accurate fundamental parameters of very massive stars at a range of metallicities and evolutionary phases remains of primary importance.

There are several advantages to studying massive stars in other galaxies, despite their greater distance from us. The low foreground extinction allows observations in the optical and ultraviolet, where the stars emit the most light, making possible their identification and study with smaller telescopes. The large metallicity range found in Local Group galaxies and beyond allows for a comparative study of the properties of massive stars as a function of metallicity (see e.g. Massey 2003), which is an important factor determining their fate. As variability studies become more widespread, eclipsing binaries are being identified in an increasing number of galaxies. Follow up studies of massive extragalactic systems is crucial to our understanding of massive star evolution.

Studying massive stars in other galaxies also offers the opportunity to obtain a complete census of eclipsing binaries and statistics on the binarity of whole populations of massive stars, a task that is currently impossible in our Galaxy. Specifically, obtaining the complete number of eclipsing systems in a galaxy down to a certain magnitude and within a certain period range will help constrain the binarity fraction of the higher mass population, which is near 50% among massive stars (Sana & Evans 2010).

Finally, another motivation for studying eclipsing binaries in other galaxies is that they are good distance indicators (Paczynski 1997), which can provide independent and accurate distances to Local Group galaxies. Given the radius and effective temperature of the component stars of the system, their luminosities (or absolute magnitude) can be calculated. Armed with both the absolute and apparent magnitude, and after correcting for extinction, one can obtain the distance.

2. Techniques

The most efficient techniques for observing eclipsing binaries in other galaxies, while not vastly different from galactic studies, include photometric variability studies with wide field CCDs and follow-up observations with multi-object spectrographs. Difference imaging or image subtraction (Alard & Lupton 1998, Alard 2000) is a technique widely used in extragalactic variability studies, given the crowded nature of the fields. Bonanos & Stanek (2003) demonstrated it to be a much more efficient method for detecting variables in crowded fields compared with traditional PSF-fitting photometry.

The discovery of extragalactic eclipsing binaries mainly comes from variability surveys of nearby galaxies with 1-2 meter telescopes, such as the DIRECT project (Stanek *et al.* 1998, Bonanos *et al.* 2003) that specifically aimed to discover eclipsing binaries in M31 and M33, or the Araucaria Project (e.g. Pietrzynski *et al.* 2002), which is surveying several nearby galaxies for RR Lyrae, Cepheids and eclipsing binaries to obtain accurate distances. Large numbers of extragalactic binaries have also resulted as side products of microlensing surveys, such as MACHO and OGLE, which have discovered thousands of eclipsing binaries in the Magellanic Clouds (see Derekas *et al.* 2007 and Faccioli *et al.* 2007 for MACHO results, and Wyrzykowski *et al.* 2003, Wyrzykowski *et al.* 2004 for OGLE-II results). Furthermore, the long time baseline of the OGLE project has resulted in the discovery of very long period systems or other rare systems, such as an eclipsing system containing a Cepheid (Pietrzynski *et al.* 2010).

Once the eclipsing binaries have been identified via photometric variability surveys, 6-10 meter class telescopes are needed for follow-up spectroscopic observations. Service mode observing (available e.g. at Gemini, VLT), targeting quadrature phases has been shown to be the most efficient way of obtaining spectroscopy for a small number of targets per galaxy (Gonzalez *et al.* 2005). When the number of targets is large (e.g. Hilditch *et al.* 2005), then multi-object spectrographs, such as FLAMES/VLT or 2dF/AAT, provide the most efficient follow up method. With the currently available telescopes, fundamental parameters of eclipsing binaries can be measured out to a distance limit of about 1 Mpc, as a resolving power $R \geqslant 3000$ and $S/N \geqslant 30$ are necessary for early-type systems and targets typically have $V > 18$ mag.

Last but not least, several multi-epoch spectroscopic surveys have been undertaken to identify spectroscopic binaries (e.g. Foellmi *et al.* 2003), some of which are later found to be eclipsing systems as well (e.g. WR20a, Rauw *et al.* 2004, Bonanos *et al.* 2004).

Table 1. Census of Extragalactic Eclipsing Binaries.

Galaxy	Distance	# of EBs	Source
LMC	50 kpc	4634, 2580	MACHO, OGLE
SMC	60 kpc	1509, 1350	MACHO, OGLE
NGC 6822	460 kpc	3	Araucaria Project
IC 1613	730 kpc	1	Araucaria Project
M31	750 kpc	∼ 500	DIRECT Project & Ribas *et al.* (2004)
M33	960 kpc	148	DIRECT Project
NGC 300	1.9 Mpc	1	Araucaria Project
NGC 2403	2.5 Mpc	1	Tammann & Sandage (1968)

The VLT-FLAMES Tarantula survey (Evans *et al.* 2011) is a recent example of such a multi-epoch spectroscopic survey, with the goal to identify massive binaries via radial velocity variations.

3. Eclipsing Binaries in Other Galaxies

Table 1 presents a census of known extragalactic eclipsing binaries. The first six galaxies are Local Group members, while NGC 300 is in the Sculptor group and NGC 2403 in the M81 group. The eclipsing binary in NGC 2403 was discovered by Tammann & Sandage (1968) and has a B magnitude of 22.

The large number of systems in the Magellanic Clouds is due to the MACHO and OGLE microlensing surveys. Faccioli *et al.* (2007) presented a catalog of MACHO eclipsing binaries, while Wyrzykowski *et al.* (2003) and Wyrzykowski *et al.* (2004) presented the catalogs from the OGLE survey. While some of these are bound to be foreground systems, most are indeed extragalactic. Note, there is some overlap between the catalogs. Moving farther out, the dwarf galaxy eclipsing systems in IC 1613 and NGC 6822 were discovered by the Araucaria project. Finally, the significant number of systems discovered in M31 and M33 is due to the dedicated searches by the DIRECT Project (e.g. Stanek *et al.* 1998) and Ribas *et al.* (2004).

The Local Group eclipsing binaries lend themselves as distance indicators and have been used as such so far to derive distances to the LMC, SMC, M31 and M33. Guinan *et al.* (1998), Ribas *et al.* (2002), Fitzpatrick *et al.* (2002), Fitzpatrick *et al.* (2003) have used early-B type systems to derive eclipsing binary distances to the LMC, while Pietrzynski *et al.* (2009) used a G-giant eclipsing system and Bonanos *et al.* (2011) an O-type eclipsing system. Most systems in the bar of the LMC are found to be at 50 kpc, however the distance to HV 5936 is discrepant, likely due to the 3-dimensional structure of the galaxy. In the SMC, Harries *et al.* (2003) and Hilditch *et al.* (2005) have obtained a distance modulus of 18.91 ± 0.03 mag by measuring 50 OGLE-II binaries with AAT/2dF spectrograph, while North *et al.* (2010) obtained a distance modulus of 19.11 ± 0.03 mag with 33 OGLE-II eclipsing binaries, using VLT/FLAMES. The discrepancy in the distance likely arises from systematic errors associated with lower resolution spectra from 2dF and the estimation of the extinction.

In M31, the eclipsing binary distances of Ribas *et al.* (2005, 772 ± 44 kpc or 24.44 ± 0.12 mag) and Vilardell *et al.* (2010, 724 ± 37 kpc or 24.30 ± 0.11 mag) are in agreement with each other. However, in M33, the long distance derived by Bonanos *et al.* (2006), 960 ± 54 kpc, was not in agreement with most measurements in the literature, and in particular with the HST Key Project measurement (Freedman *et al.* 2001), possibly because of the difficulty in estimating reddening with other methods. Nonetheless, the M33 result has pushed our current capabilities to the limit, measuring fundamental parameters of stars out to 1 Mpc. Overall, eclipsing binary distances are very valuable, because they provide independent distances, which can help evaluate the systematic errors associated with other widely used standard candles (e.g. Cepheids, RR Lyrae, tip of the red giant branch).

4. Future

The potential of eclipsing binaries for obtaining fundamental parameters of massive stars and independent distances to other galaxies is extremely promising. The ongoing OGLE project, now in its phase IV, is surveying even larger areas of the Magellanic Clouds and is bound to discover tens of thousands of eclipsing binaries. Furthermore, transient surveys such as Pan-STARRS and the Palomar Transient Factory, as well as asteroid surveys, such as the Catalina Sky Survey, and in the future, the Large Synoptic Sky Telescope will be including many nearby galaxies in their fields and monitoring them for long periods of time.

In conclusion, wide field surveys and multi object spectrographs are truly revolutionizing extragalactic binary studies. The rate of discovery of such systems is bound to increase and provide ample opportunity for studies of extragalactic massive stars, the determination of their distances, the binarity fraction and finally, statistics on binarity of various populations of stars in nearby galaxies.

Acknowledgements

The author gratefully acknowledges research support from the European Commission Framework Program Seven under the Marie Curie International Reintegration Grant PIRG04-GA-2008-239335, and travel support provided by an IAU Travel Grant.

References

Alard, C. & Lupton, R. H. 1998, *ApJ*, 503, 325
Alard, C. 2000, *A&AS*, 144, 363
Andersen, J. 1991, *A&ARv*, 3, 91
Bonanos, A. Z. & Stanek, K. Z. 2003, *ApJ*, 591, L111
Bonanos, A. Z. & Stanek, K. Z. 2003, *ApJ*, 126, 175
Bonanos, A. Z., Stanek, K. Z., & Udalski, A. 2004, *ApJ*, 611, L33
Bonanos, A. Z., Stanek, K. Z., Kudritzki, R. P. *et al.*, 2006, *ApJ*, 652, 313
Bonanos, A. Z. 2009, *ApJ*, 691, 407
Bonanos, A. Z., Castro, N., Macri, L. M., & Kudritzki, R. P. 2011, *ApJ*, 729, L9
Davies, B., Figer, D. F., Kudritzki, R. P. *et al.*, 2007, *ApJ*, 671, 781
Derekas, A., Kiss, L. L., & Bedding, T. R. 2007, *ApJ*, 663, 249
Evans, C. J., Taylor, W. D., Henault-Brunet, V. *et al.*, 2011, *A&A*, 530, 108
Faccioli, L., Alcock, C., Cook K. *et al.*, 2007, *AJ*, 134, 1963
Fitzpatrick, E. L., Ribas, I., Guinan, E. F., *et al.*, 2002, *ApJ*, 564, 260
Fitzpatrick, E. L., Ribas, I., Guinan, E. F., *et al.*, 2003, *ApJ*, 587, 685
Foellmi, C., Moffat, A. F. J., & Guerrero, M. A. 2003, *MNRAS*, 338, 360
Freedman, W. L., Madore, B. F., Gibson, B. K. *et al.*, 2001, *ApJ*, 553, 47

Gonzalez, J. F., Ostrov, P., Morrell, N., & Minniti, D. 2005, *ApJ*, 624, 946

Guinan, E. F., Fitzpatrick, E. L., Dewarf, L. E. *et al.*, 1998, *ApJ*, 509, L21

Gvaramadze, V. V., Kniazev, A. Y., & Fabrika, S. 2010, *MNRAS*, 405, 1047

Harries, T. J., Hilditch, R. W., & Howarth, I. D. 2003, *MNRAS*, 339, 157

Hilditch, R. W., Howarth, I. D., & Harries, T. J. 2005, *MNRAS*, 357, 304

Massey, P. 2003, *ARA&A*, 41, 15

Mauerhan, J. C., Van Dyk, S. D., & Morris, P. W. 2011, *AJ*, 142, 40

North, P., Gauderon, R., Barblan, F., & Royer, F. 2010, *A&A*, 520, 74

Paczynski, B. 1997, *The Extragalactic Distance Scale*, STScI Series, ed. M. Livio (Cambridge University Press), 273

Pietrzynski, G., Gieren, W., Fouque, P., & Pont, F. 2002, *AJ*, 12, 789

Pietrzynski, G., Thompson, I. B., Graczyk, D. *et al.*, 2009, *ApJ*, 697, 862

Pietrzynski, G., Thompson, I. B., Gieren, W., *et al.*, 2010, *Nature*, 468, 542

Rauw, G., De Becker, M., Naze, Y. *et al.*, 2004, *A&A*, 420, L9

Ribas, I., Fitzpatrick, E. L., Maloney, F. P. *et al.*, 2002, *ApJ*, 574, 771

Ribas, I., Jordi, C., Vilardell, F. *et al.*, 2004, *NewAR*, 48, 755

Ribas, I., Jordi, C., Vilardell, F. *et al.*, 2005, *ApJ*, 635, L37

Sana, H. & Evans, C. J. 2010, *IAU S272 Proceedings*, in press (arXiv:1009.4197)

Stanek, K. Z., Kaluzny, J., Krockenberger, M. *et al.*, 1998, *AJ*, 115, 1894

Stroud, V. E., Clark, J. S., Negueruela, I. *et al.*, 2010, *A&A*, 511, 84

Tammann, G. A. & Sandage, A. 1968, *ApJ*, 151, 825

Torres, G., Andersen, J., & Gimenez, A. 2010, *A&ARv*, 18, 67

Vilardell, F., Ribas, I., Jordi, C. *et al.*, 2010, *A&A*, 509, 70

Wachter, S., Mauerhan, J. C., Van Dyk, S. D. *et al.*, 2010, *AJ*, 139, 233

Wyrzykowski, L, Udalski, A., Kubiak, M. *et al.*, 2003, *AcA*, 53, 1

Wyrzykowski, L, Udalski, A., Kubiak, M. *et al.*, 2004, *AcA*, 54, 1

Discussion

R. WILSON: The way you find distances to eclipsing binaries is very good and logical (using complete optical light curves for most parameters and then the few infrared points for distance, thereby being relatively free of interstellar extinction dependence). However, now one can go a bit further, as the 2010 version of the WD program avoids the spherical star approximation previously used with the infrared points in the distance step. The program also gives options (process the optical and infrared data separately or together, or both ways) and assumes consistency. It is directly absolute, with fluxes in physical units and, since the program does most of the work, it makes the overall process very fast.

From Interacting Binaries to Exoplanets: Essential Modeling Tools
Proceedings IAU Symposium No. 282, 2011
Mercedes T. Richards & Ivan Hubeny, eds.
© International Astronomical Union 2012
doi:10.1017/S1743921311026822

The Impact of Gaia and LSST on Binaries and Exoplanets

L. Eyer[1], P. Dubath[2], N. Mowlavi[2], P. North[3], A. Triaud[1], F. Barblan[1], C. Siopis[4], L. Guy[2], B. Tingley[5], S. Zucker[6], D.W. Evans[7], Ł. Wyrzykowski[7,8], M. Süveges[2] and Z. Ivezic[9]

[1] Geneva Observatory, University of Geneva, Sauverny, Switzerland
[2] ISDC, Geneva observatory, University of Geneva, Versoix, Switzerland
[3] LASTRO, Ecole Polytechnique Fédérale de Lausanne (EPFL), Observatoire de Sauverny, Versoix, Switzerland
[4] IAA, Université Libre de Bruxelles, Bruxelles, Belgium
[5] Instituto de Astrof-sica de Canarias, La Laguna, Spain
[6] Department of Geophysics & Planetary Sciences, Tel Aviv University, Tel Aviv, Israel
[7] Institute of Astronomy, University of Cambridge, UK
[8] Warsaw University Observatory, Warsaw, Poland
[9] Department of Astronomy, University of Washington, Seattle, USA

Abstract. Two upcoming large scale surveys, the ESA Gaia and LSST projects, will bring a new era in astronomy. The number of binary systems that will be observed and detected by these projects is enormous, estimations range from millions for Gaia to several tens of millions for LSST. We review some tools that should be developed and also what can be gained from these missions on the subject of binaries and exoplanets from the astrometry, photometry, radial velocity and their alert systems.

Keywords. (stars:) binaries, surveys, catalogs, astrometry, space vehicles: instruments, methods: data analysis, etc.

1. Introduction

What new knowledge will large surveys such as Gaia and LSST bring to the field of binary stars and exoplanets? Astronomy has evolved into a science of large numbers. With Gaia and LSST, this trend is even accelerating. With these very large numbers of observed objects, it will be possible to (a) describe statistically populations of stars, binary/multiple stars and exoplanets, revealing relations between their physical properties and (b) find very rare objects among the many millions, which may shed light on specific physical processes or capture these objects in a very brief moment of their evolution.

Complementary to the technological challenges that these ambitious projects require to fulfill their stringent requirements, there are software tools which have to be developed to handle the vast amount of data, to compute, search, classify and browse through Terabytes or even Petabytes of data.

To predict the scientific impact of a project on a specific topic can be a very perilous exercise, especially when the project observes domains of astrophysical parameters never explored before. We therefore proceed with caution, starting with a description of the projects and their characteristics, followed by a brief overview of some analysis tools for characterization and classification, and finally we present a general description of binary stars and exoplanets.

Table 1. Technical characteristics of Gaia and LSST telescopes.

	Gaia	LSST
Mirror size	1.45mx0.45m	diameter:8.4m
Field of view (deg^2)	0.7x0.7	9.6
Point Spread Function (arcsec^2)	0.14x0.4	0.7x0.7
Pixel counts (billion)	1	3.2
sky coverage of survey	whole sky	half of the sky
Depth per Observation	$V \simeq 20$	$r \simeq 24$
Bright limit	$V \simeq 6$	$r \simeq 16\text{-}17$
Number of epochs per object	70 in 5 years	1000 in 10 years
Number of photom. obs. per object	70*4=280	1000
Number of RVS obs. per object	40	–

2. The Gaia and LSST projects

Gaia: is a space mission of the European Space Agency, which will be located at the L2 point, 1.5 million km from Earth. It will observe all stars between mag $V \simeq 6$ to 20, amounting to about 1 billion objects. The measurements consist of astrometric, photometric, spectrophotometric and spectroscopic data. The length of the mission is 5 years with a possible one year extension. For 5 years, the average number of measurements will be about 70 per object. The launch is foreseen in 2013. There will be an alert system and intermediate data releases throughout the mission. The final results will be made available by 2020-2021.

LSST: is a ground-based telescope that will observe about half of the sky, with a harvest amounting to 10 billion stars and 10 billion galaxies. The measurements will consist of repeated positions and photometry in the 5 Sloan bands plus the "y" band. The survey ranges from mag $r \simeq 16$ to 24, and can be extended to mag 27 with the stacking of images. The length of the project is planned to be 10 years during which about 1000 visits will be taken for a given region of the sky. First light is planned to be in 2018.

For any astronomy project, the constraints on space-based missions are different from those of ground-based projects. One of the bottlenecks from space is the transmission of data back to Earth. For Gaia, once a star is detected, there is an on-board data processing to compress the information into a line spread function. This ladder information is transmitted to the Earth. Consequently Gaia does not have the full pixel images. With the ground-based LSST project, all the pixel images will be stored. Other constraints from Space are the weight and size, the Gaia mission was downsized in order to fit on a Soyuz rocket, which is less costly than an Ariane launch. The advantages of space is that we get rid of the atmosphere and there is access to full sky from one instrument.

A comparison of the instruments of these two contemporary projects, Gaia from space, and LSST on Earth, are presented in Table 1. For the performance we refer to Figure 1, which presents the astrometric (parallax, proper motion) and photometric performances. In addition to photometric measurements, Gaia has a spectrograph with a resolution 11,500 which will provide radial velocities up to magnitude $V \simeq 17$ at the level of 1-10 kms^{-1} depending on the spectral type of the star.

Descriptions of the science cases can be found for Gaia on the Gaia webpage at ESA (www.rssd.esa.int/Gaia under information sheets) and for LSST in LSST Science Collaborations *et al.* (2009) (or the www.lsst.org).

3. Automated variable source characterization and supervised classification in large-scale surveys

Source classification is an important component of any large-scale survey. The resulting large sample of variable stars is useful both to study stellar population properties and to provide candidates for further detailed investigation of individual cases. This is particularly true for binaries and exoplanets. The numbers involved in recent surveys are however so large that this task requires the development of automated and efficient machine-learning techniques. Different supervised classification schemes are evaluated on controlled samples of well-known stars using cross-validation approaches of two recent studies (Dubath *et al.* 2011 and Richard *et al.* 2011), each carried out in the framework of preparatory work for the Gaia and the LSST missions respectively. Both studies acknowledge the quality and convenience of the random forest method (Breiman 2001), which is shown to perform as well as other competing techniques. As several techniques lead to comparable results, it seems that important improvements are unlikely to come from changes in the classification method itself. The choice of the attributes used in the classification is more critical. They have to describe the source in the most appropriate manner and adding well-designed attributes, introducing a complementary piece of information if possible, is more likely to result in positive changes. The use of principal component analysis with a multicolour photometry (e.g. in Süveges *et al.* 2011) is an attempt to use efficiently the variability in different photometric bands.

The quality and, above all, the relevance of the training sample is another critical factor. Any bias between the training set and the to-be-classified samples is likely to be reflected in the classification results. Additional work is currently being pursued to investigate how to best adapt a given training set to a particular survey.

4. Binaries

For those who work on large-scale surveys, binary stars are a problem due to the complexity and diversity of their signals. This complexity is also what makes them

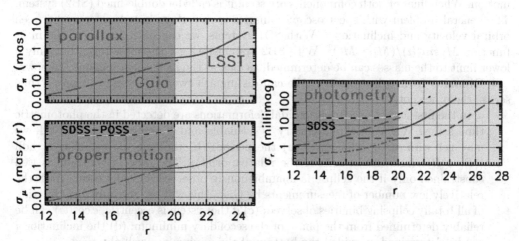

Figure 1. Precision of Gaia and LSST for the astrometry (parallax and proper motion) and photometry as function of the *r* magnitude. In red (light grey) Gaia, in blue (dark grey) LSST. For the panel on photometry, two lines are drawn for each project, one is the precision for "one" measurement (= a transit for Gaia, i.e. one passage on the 9 (or 8) CCDs of the Astrometric Field (AF); = a "visit" for LSST, i.e. a pair of 15-second exposures), the others (lower) lines are the end-of-mission precision. The dashed colored lines are for SDSS and SDSS-POSS data.

interesting. Consequently, significant effort is devoted to signal processing and analysis to extract binaries from large-scale surveys and to derive their orbital parameters as well as the astrophysical parameters of each stellar component.

Let us go back to some basic principles: binary systems can be detected and characterized through astrometry, photometry or spectroscopy. We review in a few words these three "types" of binaries and what can be derived from the observations:

• **Astrometric binaries:** A binary system is called astrometric when the positions on the sky of its components (or at least of the primary) change due to their orbital motion. When the astrometric orbits are determined for both components, we can get the mass ratio and the inclination, in short, nearly all parameters of the binary system, except for the actual size of the orbit and the astrophysical parameters. Furthermore, if the distance is known, we can then obtain the real size of the orbit and hence the total mass from Kepler's third law. If the distance is unknown, it is possible to obtain a good estimate of it by reversing the reasoning: we first estimate the two masses of the components using the mass-luminosity relation (which of course restricts this method to the main sequence), then we use the known orbital period and angular semi-major axis of the relative orbit to get the distance d. The precision of the latter will be fairly good even for a rough estimate of the total mass $M_{\rm tot}$, because $d \propto M_{\rm tot}^{1/3}$. This method was applied to Atlas (a member of the Pleiades) by Pan *et al.* (2004). It should be noted that Gaia and LSST will provide absolute orbits†. There are complications in some cases; for example, when the stars are not resolved, we may only see the motion of the photocentre of the binary system. A further complication arises in the case of a variable component in an unresolved optical double; the photocentre will move (Variability Induced Movers) and will thereby mimic the behaviour of a true binary system.

Within Gaia a whole group on astrometric binaries is led by D. Pourbaix from Université Libre de Bruxelles, Belgium. This group is dedicated to the art of solving double and multiple systems.

• **Spectroscopic binaries:** A binary system is called spectroscopic when at least one component (SB1) shows a radial velocity varying periodically with time due to orbital motion. When lines of both components are seen, it is called a double-lined (SB2) system. The general problem with spectroscopic binaries is that there is a degeneracy between orbital velocity and inclination i. With SB1 systems, we can only determine the mass function $M_2^3 sin^3(i)/(M_1 + M_2)^2$. With SB2 systems, we get the mass ratio, but only a lower limit to the masses can be determined, as long as the inclination remains unknown.

• **Binaries from photometry**: We can distinguish two cases, one requiring a time series, the other not.

○ In the first case, the eclipses or tidal deformations are detected in the photometric time series. When only photometry is available and if eclipses are observed, we may get with good confidence: (a) the period; (b) the sum of the relative radii $((R_1 + R_2)/($semi-major axis of the relative orbit$))$. The ratio of the radii is well constrained only if the eclipse is unambiguously observed as being total. With the relatively low number of measurements by Gaia, this will be true only for a fraction of all totally eclipsing binaries observed; (c) if the system is eccentric, $e\cos(\omega)$ can be reliably determined from the phase of the secondary minimum; (d) the inclination i can be determined, provided the bottom of the minima are well observed.

○ Binary stars can be detected through their special location in colour-colour or colour-magnitude diagrams. For example, a binary sequence in a cluster can be observed in the colour-magnitude diagram, which runs parallel to the main sequence,

† With the wide field CCD, relative orbits are of lesser interest.

as shown by e.g. Mermilliod *et al.* (1992). Another example is that of stars showing up in the Herzsprung gap, because they are binaries consisting of a turn-off star and of a red giant (see e.g. star H110 in NGC 752, Fig. 1 of Mermilliod *et al.* 1998). Binaries can also be detected in a colour-colour diagram, as exemplified by Smolčić *et al.* (2004): binaries made up of a white dwarf and an M dwarf have been detected in the SDSS database.

Obviously, when a binary system is at the same time "eclipsing and spectroscopic," it can be fully characterized (orbit, masses and radii), see for example Barblan *et al.* (1998), North *et al.* (2010), and when it is "astrometric and spectroscopic," then the orbit is completely determined (Zwahlen *et al.* 2004).

Gaia will contribute directly to these three approaches. LSST on the other hand will contribute to the astrometric and photometric detections of binary systems as well as their characterization. Obviously, Gaia and LSST will detect objects that will prove excellent targets for follow-up studies. The number of astrometric binaries that will be detected by Gaia and LSST is not so well constrained.

Gaia and LSST, being extensive photometric surveys, will contribute most significantly to the detection and characterization of new eclipsing binaries. Much effort is devoted to develop software to deal with this type of behaviour. Within Gaia there is a special group, led by C. Siopis (from Université Libre de Bruxelles, Belgium), developing software for the characterization of the detected eclipsing binaries. This software includes a number of light curve simulation and fitting tools which are adapted to work within mission constraints (duration, number of observations, software guidelines, etc.) and to make optimal use of mission strengths (multicolor photometry, spectroscopy, etc.). Several attempts have been made to pin down the number of eclipsing systems that Gaia and LSST are expected to detect. For Gaia, these estimates disagree by a factor of nearly 10, going from half a million to 6 millions. However, we can assert with confidence that the number of known eclipsing systems in our Galaxy will make a prodigious jump. Predictions for LSST have been recently made by Prša *et al.* (2011): the number will amount to 24,000,000 among which 6,700,000 will be well-characterized systems, and 1,700,000 are expected to show up as eclisping double-lined binaries, that could be confirmed as such by follow-ups. The largest homogeneous sample of eclipsing binaries to date comes from the Large Magellanic Cloud OGLE-III data (Graczyk *et al.* 2011), from which 26,121 systems have been extracted. Thanks to Gaia and LSST, these numbers will increase by several orders of magnitude.

There are many other subjects that can't be covered in this short review: the case of AM CVn binary stars, potential source of gravitational waves, with extremely short periods (the Gaia per CCD photometry may be suitable for a detection of eclipses); the binary stars with a pulsating component; etc.

5. Eruptive and cataclysmic phenomena

Cataclysmic and eruptive phenomena are often due to interacting binaries. Detection, classification and rapid response are crucial steps in order that these transient objects be followed-up and studied during unusual states. Both Gaia and LSST are developing alert systems.

Within Gaia the task to detect and deliver photometric alerts is performed at the Institute of Astronomy of the University of Cambridge. From the satellite data acquisition to the ground-based first calibration, this whole process for Gaia is not immediate. In the worst case the received observations might be 48 hours old. In some cases the promptest observations could be accessed within a couple of hours. The alerts will be based on any

Table 2. The detection techniques used to discover exo-planets and the possible contribution from Gaia and LSST

Detection Techniques	Gaia	LSST
Astrometry	bright stars	–
Transits	bright stars	yes (though faint)
Radial velocities	–	–
Microlensing	foreseen	foreseen

changes in the photometry of objects (or appearance of new ones) and then the classification of transients will be performed exploiting all available photometric measurements for a given object, as well as the spectrophotometric one, which will also be available from Gaia. This should assure relatively low rates of false positives in the alerts stream.

The Gaia alert system group also proposes to activate and maintain a watch list, providing data for objects known to be interesting and important to follow-up (e.g. FU Ori-type stars), allowing these objects to be observed at the moment they move out of dormant phase or undergo any unexpected change.

LSST will issue alerts based on photometric and astrometric changes, and any transient event will be posted in less than 60 seconds (with a goal of 30 seconds) via web portals including the Virtual Observatory. LSST with its dense time sampling and very wide photometric system, including a u band, will be very sensitive to eruptive and cataclysmic phenomena.

6. Exoplanets

Exoplanets produce astrometric signals similar to those produced by binary stars. We used the word "similar" because the signals or their duration usually have amplitudes much smaller than those of typical binary stars, and the detection of these signals has required very specific developments either from the instrumental or from the signal analysis point-of-view. In fact, the scientific importance of studying exoplanets has stirred much developments, which are also benefiting the binary star community.

In Table 2 we summarize the way in which exoplanets can be detected and studied, and the possible contribution of Gaia and LSST.

Gaia will be helpful in many topics related to exoplanets, in particular for exoplanet detection and for the characterization of their host stars. One of the essential points is that Gaia has a bright limit at magnitude 6, therefore allowing easy follow-up activities of potentially interesting sources with relatively small telescopes, without requiring large amounts of telescope time. Here is a non-exhaustive list of contributions that the Gaia survey may provide:

• Radius determination: One problem for planetary transit detection is the possible confusion between a secondary star and an exoplanet when the radius of the star is unknown. The luminosity of the primary star, estimated from the parallax, will enable the separation of giants from dwarfs. The spectrophotometric system may give an estimate of surface gravity and temperature and may also eliminate false detections due to the confusion between giant and dwarf parent stars. As pointed out by Triaud (these proceedings), the parallax will also lead to more accurate estimates of stellar masses and ages.

• Radial velocity variations: The precision of the radial velocity instrument on board Gaia, which is at the level of kms^{-1}, is obviously not suited for the detection of exoplanets, which requires a precision at the level of 1 to 10 ms^{-1}. This level of kms^{-1} can be used, however, for brown dwarf detection. Furthermore, the radial velocity instrument

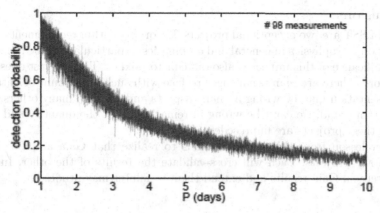

Figure 2. TrES-1 was used as template in order to determine the probability to have 5 points in transit as a function of period (days) given the scarce Gaia sampling, here having 98 Gaia-observations over 5 years.

of Gaia may detect the radial velocity variations from grazing eclipsing binaries and can therefore eliminate another source of false detection.

• Target selection: Gaia, realizing a complete map of the sky for bright stars, will be most useful for target selection for further studies or surveys. For example, a mission such as Plato† will use the information collected by Gaia.

For astrometry, the position precision required to detect planets is very demanding: currently the estimate for the bright stars along the scanning direction is at the level of 20 micro-arcseconds per Gaia-transit (average of 9 CCD positions) cf. de Bruijne (2009). A small degradation of the performance will generate a significant decrease in the number of the exoplanets detected by the astrometry. Sozzetti (2011) presented the following predictions for Gaia: about 1000 planets will be detected and 400 to 500 will have their masses determined at the 10%-20% level. One interesting point about the astrometric method is that it is less sensitive to the orbital inclination than the radial velocity or transit detection methods.

The case of planetary transits is somewhat controversial. Indeed, the estimations range from 0 up to 5,000-30,000 (Robichon 2002). The matter here is to distinguish between the presence of a signal and its unambiguous detection. A knowledge of possible outlying values and technical problems is primordial in such a low signal-to-noise regime. Moreover, the bulk of the exoplanetary targets will be red dwarfs, for which little is known regarding planetary populations.

Within Gaia, Tingley (2011) and Dzigan & Zucker (2011) are preparing two different methods to detect such planetary signals. Dzigan & Zucker, using a Bayesian approach, showed that "5 points in a transit deeper than 0.002" means an almost certain detection. This method also allows the optimization of follow-up observations. What is then the probability that we have 5 points in transit in spite of the scarce sampling of Gaia? Taking as an example the exoplanet TrES-1, Figure 2 shows that this probability can be quite high. Details will be published by Dzigan & Zucker (in preparation).

The detection of planets by microlensing is another challenge especially with Gaia because of its scarce sampling. However this activity is planned. LSST with its denser sampling of a visit every 3 days will be more successful.

† Plato is an ESA project, which aims to detect planetary transits in bright stars.

7. Conclusions

Gaia and LSST are two exceptional projects. If working within requirements, the scientific impact on astrophysics in general and on binaries in particular will be mind-blowing, but the significance of this impact is also difficult to predict. This harvest does not come without effort. There are many challenges to face with such an amount of data.

This presentation may be wrong in many aspects and blind to many others. But with these caveats in mind, we can't be wrong in remarking that the quantity and quality of the data of these projects are unprecedented.

The other lesson from this contribution is to realize that Gaia and LSST are two complementary projects. Each will cross-validate the results of the other. In addition, LSST can rely on Gaia results and extend them to fainter magnitudes.

References

Barblan, F., Bartholdi, P., North, P., Burki, G., & Olson, E. C. 1998, *A&AS*, 132, 367
Breiman L., 2001, *Machine Learning*, 45, 5
de Bruijne, J., 2009, *Gaia Technical Note*, GAIA-CA-TN-ESA-JDB-053
Dubath, P., *et al.*, 2011, *MNRAS*, 414, 2602
Dzigan, Y. & Zucker, S. 2011, *MNRAS*, 415, 2513
Graczyk, D., *et al.*, 2011, *AcA*, 61, 103
LSST Science Collaborations, *et al.*, 2009, arXiv:0912.0201
Mermilliod, J.-C., Rosvick, J. M., Duquennoy, A., & Mayor, M. 1992, *A&A*, 265, 513
Mermilliod, J.-C., Mathieu, R. D., Latham, D. W., & Mayor, M. 1998, *A&A*, 339, 423
North, P., Gauderon, R., Barblan, F., & Royer, F. 2010, *A&A*, 520, A74
Pan, X., Shao, M., & Kulkarni, S. R. 2004, *Nature*, 427, 326
Prša, A., Pepper, J., & Stassun, K. G. 2011, *AJ*, 142, 52
Richards, J. W., *et al.*, 2011, *ApJ*, 733, 10
Robichon, N. 2002, *EAS Publications Series*, 2, 215
Smolčić, V., *et al.*, 2004, *ApJ*, 615, L141
Sozzetti, A. 2011, *EAS Publications Series*, 45, 273
Süveges, M., Bartholdi, P., Becker, A., Ivezic, Z., Beck, M., & Eyer, L. 2011, arXiv:1106.3164
Tingley, B. 2011, *A&A*, 529, A6
Zwahlen, N., *et al.*, 2004, *A&A*, 425, L45

Discussion

P. ZASCHE: Only a short comment. It is a pity that these new surveys have a bright limit. The advantage of Hipparcos was its unlimited range in bright magnitude. Now these surveys cannot observe a 4th magnitude star. Moreover, I cannot imagine how we will observe bright stars in 20 years, when all classical photometers will be replaced with CCD cameras.

L. EYER: When public talks are made, it is indeed somewhat embarrassing to say to the lay public: look at the night sky, and all the stars you see with your eyes, these are exactly the ones that Gaia won't measure... However it concerns only a few thousands of stars, which are often too bright even for moderate-sized telescopes. The magnitude ranges which are covered by Gaia and LSST are exceptional, Gaia through a gating system of the CCDs, which allows us to push the bright limit to $V \simeq 6$ and LSST through stacking images which allows us to go very deep reaching $r \simeq 27$. Furthermore, Gaia and LSST have an extensive overlap. With these two projects we go from mag. 6 to mag. 27, which represents an exceptional dynamical range, never encountered before for such a large number of objects.

From Interacting Binaries to Exoplanets: Essential Modeling Tools
Proceedings IAU Symposium No. 282, 2011 © International Astronomical Union 2012
Mercedes T. Richards & Ivan Hubeny, eds. doi:10.1017/S1743921311026834

The Impact of CoRoT and *Kepler* on Eclipsing Binary Science

Carla Maceroni[1], Davide Gandolfi[2], Josefina Montalbán[3], and Conny Aerts[4]

[1]INAF–Osservatorio Astronomico di Roma,
via Frascati 33, I-00040, Monteporzio C. (RM), Italy
email: `maceroni@oa-roma.inaf.it`

[2]ESA Estec,
Keplerlaan 1, 2201 AZ Noordwijk, Netherlands
email: `dgandolf@rssd.esa.int`

[3]Institut d'Astrophysique et Géophysique Université de Liège,
Allée du 6 Aôut, B-4000 Liège, Belgium
`j.montalban@ulg.ac.be`

[4]Institute of Astronomy, K.U.Leuven,
Celestijnenlaan 200D, B3001 Leuven, Belgium
email: `conny@ster.kuleuven.be`

Abstract. The CoRoT and *Kepler* space missions have opened a new era in eclipsing binary research. While specifically designed for exoplanet search, they offer as by-products the discovery and monitoring of variable stars, in great majority eclipsing binaries (EB). The missions are therefore providing thousands of EB light curves of unprecedented accuracy (typically a few hundred parts per million, ppm), with regular sampling (from 1^s to 29^m), extending over time spans of months, and with a very high duty cycle ($> 90\%$).

Thanks to this excellent photometry, research topics as asteroseismology of EB components are quickly developing, and physical phenomena such as doppler boosting, theoretically predicted but extremely difficult to observe from the ground, have been unambiguously detected. We present the main properties of the Corot and *Kepler* EB samples and briefly review the highlights of the missions in this field.

Keywords. binaries: eclipsing, stars: oscillations, surveys

1. CoRoT and *Kepler* space missions

CoRoT† (COnvection, ROtation and planetary Transits) is a French-led international "small" space mission launched in December 2006. The mission is devoted to the achievement of two parallel "core programs," asteroseismology and extra-solar planet search, which require the same type of observations, i.e. high accuracy photometry and long continuous monitoring. These programs are carried out in two contiguous "seismo" and "exo" fields, with a 27 cm telescope and four CCDs of $1.3° \times 1.3°$ on the sky. After launch each field was covered by two CCDs, the observations were performed in Long and Short runs lasting, respectively, ~ 150 and ~ 30 days, and for each run up to ten bright ($5.7 > V > 9.5$) seismo-targets and up to 12000 exoplanet targets ($11.5 > V > 16.5$) were observed. Unfortunately, the number of observed targets dropped to half after the loss in March 2009 of one of the two data processing units. That event induced as well a

---

† The CoRoT space mission was developed and is operated by the French space agency CNES, with participation of ESA's RSSD and Science Programmes, Austria, Belgium, Brazil, Germany, and Spain; complete information is available at `http://corot.oamp.fr`

change in observing strategy (shorter "Long" runs of ~ 80 days have also been scheduled to increase the total number of targets).

CoRoT has provided high-accuracy $(10^{-3}$–10^{-4} mag) photometry of about 140,000 stars in a broad bandpass spanning 370 – 1000 nm. Chromatic information was also obtained for exo-planet targets brighter than $V = 15$.

Only a handful of EBs were observed in the seismo-field, as binaries are in general rejected in the target selection process. Sometimes, however, binarity is discovered by CoRoT itself (e.g., Maceroni *et al.* 2009). The standard sampling of the seismo field is 32^s, the fast one 1^s, yielding light curves with hundred thousands or millions of points and a point-to-point deviation of the order of 10^{-4} mag.

Most CoRoT EBs are exo-field targets, which are sampled at a standard rate of 8^m or a fast one of 32^s. The light curves contain from 8000 to 300,000 points. An estimate of the white noise level as function of target R-magnitude (Aigrain *et al.* 2009) yields 0.5 mmag for $R = 12$ and 2 mmag for $R = 15$.

The classification of variable stars in the exo-fields is performed by the CoRoT Variability Classifier (CVC, Sarro *et al.* 2009, Debosscher *et al.* 2009), which provides a probabilistic classification in 29 different variability classes. Independent, and somehow different, lists of binaries have also been published for the first runs by the exoplanet search teams (Carpano *et al.* 2010, Cabrera *et al.* 2009, Carone *et al.* 2011) containing the EBs rejected by the planet search algorithms or subsequent follow-up observations.

So far nineteen different fields have been observed by CoRoT (and the data of the first ten are public†), the results we present in this paper, however, refer mainly to ~ 400 EBs from the first CoRoT runs (IRa1, LRc1, LRa1) as for these fields EB samples from both CVC and exoplanet search are available.

CoRoT's trail is being widened by the *Kepler* space mission, thanks to its higher performance instrumentation and longer monitoring of targets. *Kepler*, NASA Discovery mission #10, was specifically designed to discover Earth-size planets and is in operation since March 2009. The details of the mission can be found elsewhere (e.g., Borucki *et al.* 2010, Koch *et al.* 2010). In short, *Kepler* monitors $\sim 156,000$ stars of interest (preferentially late type dwarfs) in a field of view extending over 105 deg^2 in the Cygnus - Lyra region. The standard photometry sampling is 29.4^m ("long cadence") but up to 512 stars can be observed in "short cadence" mode (59^s). The effective dynamic range is 7–17 *Kepler* magnitudes (K_p, in a broad bandpass from 425 to 900 nm). The target continuous monitoring can last up to the mission lifetime (the programmed 3.5 years or longer in case of extension). The estimate of the instrument performance after launch (Koch *et al.* 2010) indicates that the design goal (a photometric precision of 20 ppm for a 6.5 hr exposure of a G2-type $V = 12$ target) is close to being achieved.

Kepler data are delivered in "Quarters" (Q0, Q1, ..., Qn), typically three months long (a quarter ends when the spacecraft rolls to re-align its solar panels). Quarter Q10 is currently in progress, the first quarters (Q0–Q3) are publicly available‡.

In addition to the exoplanet core program, scientific programs devoted to the asteroseismology of *Kepler* targets have been devised and outsourced to the European *Kepler* Asteroseismology Consortium (KASC), which maintains its own archive (KASOC). Subscription to this Consortium (open to collaboration) allows access to all the KASC data.

A comprehensive catalog of 1832 *Kepler* EBs, detected in the first two quarters Q0-Q1, has been published by Prša *et al.* (2011) and updated with the addition of Q2 data by Slawson *et al.* (2011). The current version contains 2165 eclipsing or ellipsoidal binaries.

---

† the Archive is mantained by the CoRoT Data Center at IAS, http://idoc-corot.ias.u-psud.fr/
‡ available from the Multimission Archive at STScI (MAST) http://archive.stsci.edu/Kepler/

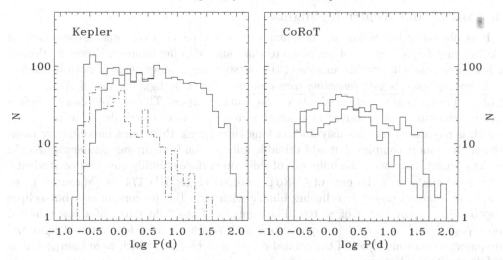

Figure 1. The period distribution of *Kepler* and CoRoT binaries. Left panel, thin line: all systems of Q0-Q2, thick line: detached and semidetached binaries only. For comparison the distribution of OGLE-I binaries with I< 16.5 is shown (dotted-dashed line), its quick decline is due to period dependent selection effects, affecting shorter periods for a ground based survey. Right panel, thin line: CVC sample (IRa1, LRa1, LRc1 fields), thick line: exoplanet sample for the same fields. The difference is related to the adopted filtering methods (see text).

2. CoRoT and *Kepler* eclipsing binaries

CoRoT and *Kepler* have provided light curves of unprecedented precision, sampling and extension. All these assets contribute to the increase of EBs frequency, which is about twice the value typical of ground-based surveys, as EBs with eclipses of smaller amplitude and shorter fractional eclipse duration are detected; see the comparison of CoRoT EB sample with that of OGLE-I in Maceroni (2010). The EB frequency of the first CoRoT fields is 1.2% of all targets. A similar, slightly larger, value of 1.4% is found by Slawson *et al.* (2011) for the first *Kepler* Quarters.

The higher efficiency of discovery of the space missions is evident as well in the orbital period distributions in Fig.1, showing different subsamples of EBs and a comparison with OGLE-I EBs of magnitude I > 16.5.

The difference between the CoRoT sample from CVC and that from exoplanet search is due to the different filter applied to extract binaries. The Fourier-based CVC algorithm fails more frequently for longer period narrow V-shaped eclipses, especially if additional (quasi-) sinusoidal variability is present (the target is classified in another variability class). On the other hand the exo-planet algorithms, designed to detect U-shaped variations, misses short period (W UMa type) binaries. The CVC algorithm has been recently improved, to better handle binary extraction, so one can expect better agreement between the two sources for the following fields.

In both figures, and more markedly in the *Kepler* sample one, the log P distribution shows an almost flat behavior for periods larger than 1^{d}. Besides, the histograms suggest that the higher the precision, the more complete the sample for low amplitude detection, and the wider the "plateau" of the distribution. A flat log P distribution, implying no preferred length scale for the formation of short-period binaries (Heacox 1998) is in agreement with the results of Mazeh *et al.* (2006) for a quite different sample (LMC B-type EBs).

3. CoRoT and *Kepler* highlights

It is obviously impossible to concentrate in a few pages a comprehensive review of CoRoT and *Kepler* results of relevance to close and eclipsing binaries. However, already a few highlights will provide an idea of the outstanding achievements of both missions.

Eclipsing binaries with pulsating components. Most EBs light curves of *Kepler* and CoRoT present additional variability on the top of eclipses. This is often due to surface inhomogeneities (stellar activity) but another frequent cause is intrinsic stellar pulsation, which is monitored continuously and on long time spans. Pulsations undoubtedly make the analysis more complex, but add valuable information from an independent source. In a close binary, moreover, the influence of tides on surface stability can also be studied.

A good example is the case of CoRoT (seismo) target HD 174884 (Maceroni *et al.* 2009), an unusual eccentric eclipsing binary with twin B-type components but eclipse depths differing by factor of ~ 100. The analysis of the light curve of a few hundred ppm precision allowed to detect pulsations with amplitude of a few hundred ppm and frequencies exact multiples of the orbital one (8 and 13 f_{orb}), which were interpreted as tidally excited pulsations.

Another interesting system is CoRoT 102918586, whose light curve is shown in Fig. 2 together with the Fourier spectrum after subtraction of the EB model, see Maceroni *et al.* (2010) for a description of the method. It is as well an example of the difficulties in the analysis when pulsations and eclipses are of comparable amplitude. The first analysis – presented in the above-mentioned paper and based on the CoRoT photometry alone – assumed a configuration with two very similar F0 dwarfs in a circular orbit and a period $P \simeq 8.78^{\rm d}$. The short fractional eclipse duration implied, however, very small fractional radii of the components (and no tidal deformation). On the other hand the light curve residuals, after subtraction of the binary model, contained harmonics of the orbital period, difficult to explain in terms of tidally excited pulsations, the stars being spherical and in circular orbit. The subsequent acquisition of high-resolution time-resolved spectroscopy has solved the issue: the true orbital period is half the value from photometry; the system is still formed by similar components (SB2) but the orbit is eccentric, and because of orientation in space only one eclipse is observed. The harmonics of the orbital period derived from the out of eclipse shape of the binary light curve, while

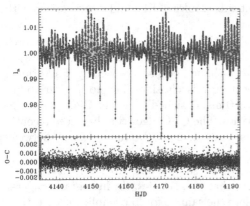 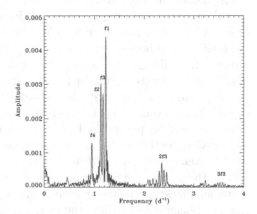

Figure 2. Left: The white-light lightcurve of CoRoT 102918586, time is in HJD-2450000, the continuous line is the fit (eclipsing binary model + pulsations), the lower box shows the corresponding residuals. Right: the power spectrum of the lightcurve after subtraction of the EB model. The main pattern (a multiplet of five frequencies around $f3 = 1.1713$ c/d), repeats at $2 \times f3$ and $3 \times f3$. Frequency $f4$ is $f3 - f_{orb}$.

the analysis of the pulsations suggests the presence of a γ Dor primary component and a rotational splitting of a $\ell = 2$ gravity mode. A complete analysis of this interesting system will be presented elsewhere (Maceroni *et al.* 2011, in preparation).

In the case of solar-like pulsations (present as well in red giants) an estimate of "asteroseismic" mass and radius is possible based on the properties of the power spectrum (frequency of maximum power and large frequency separation) combined to effective temperature from spectroscopy and to scaling laws (Kjeldsen & Bedding, 1995). This is especially valuable in the red giant case, where the estimate from classical methods is problematic. CoRoT has first detected solar-type oscillations in red giants (De Ridder *et al.* 2009), and among the first *Kepler* discoveries is that of a long period ($\sim 400^d$) eclipsing binary with a pulsating red giant component (Hekker *et al.* 2010), which is a promising milestone of such studies.

The *Kepler* satellite also implied the discovery of several compact binaries among its fast pulsators. Kawaler *et al.* (2010) found two compact binaries with gravity-mode pulsations superposed to an irradiation effect typical of sdB stars with a close M-dwarf companion. The orbital periods are less than half a day such that seismic sounding should become possible in the future, once the noise level of the data can be brought down to such a level that tidal and/or rotational splitting can be disentangled from period spacings of the modes. An even more interesting case is 2M1938+4603, an extremely rich pulsating sdB star in a 0.126^d period binary with an eclipsing dM companion (Østensen *et al.* 2010). This star is a hybrid pulsator in that it reveals numerous pressure and gravity modes, offering the potential to probe both the outer layers and the inner core if the modes can be identified.

New discoveries: beaming binaries, tidal brightening. An exciting *Kepler* result is the clear detection of the relativistic beaming effect in the light curve of two targets: KPD 1946+4340 and KOI-74 (Bloemen *et al.* 2011, van Kerkwijk *et al.* 2010).

Beaming (also known as Doppler boosting) is a signature of the component radial velocity in the light curve. This takes the form of a modulation of measured flux according to the radial velocity difference, weighted by the component contribution to the total flux. It is, therefore, best observed in systems with components of very different spectral characteristics. When one star dominates the total flux the beaming effect measurement along the orbit provides information equivalent to that obtained from the radial velocity of a single lined spectroscopic binary.

The effect was predicted by Loeb & Gaudi (2003) and Zucker *et al.* (2007) but its small amplitude (of the order of a few hundreds ppm for systems with periods of a few days) prevented detection in EBs before *Kepler*. For both detections the radial velocity amplitude from beaming is in excellent agreement with that derived from spectroscopy. KPD 1946+4340 and KOI-74 are the first members of the new class of "beaming binaries". The beaming effect has as well been detected in the light curve of CoRoT-3, a 22 Jupiter-mass object, orbiting an F3-star (Mazeh & Faigler 2010).

Another remarkable, ever-observed, effect is the "tidal brightening" characterizing the light curve of KOI-54, a strongly eccentric ($e = 0.83$) *non-eclipsing* system with two similar A-type components and orbital period of 41.8^d (Welsh *et al.* 2011). The light curve shows regular brightenings of a few mmag amplitude and tidally excited pulsations at high harmonics of the orbital frequency (90 and 91 f_{orb}). The orbital phenomena are interpreted as due to tidal distortion and irradiation at periastron. Several systems with brightenings and eclipses are found in the *Kepler* data and will provide new insights in tidal phenomena.

Tertiary eclipses. Finally, it is worth mentioning the *Kepler* discovery of a few triple (or multiple) systems showing tertiary eclipse events. The intersting case of HD 181068, a compact hierarchical triple system with a red giant component, has been fully analyzed by Derekas *et al.* (2011) and appears elsewhere in this volume. Another remarkable object, KOI-126 (Carter *et al.* 2011), shows eclipses from a closer pair formed by two M-dwarfs ($P=1.7^d$) and from this close pair and a wider 1.35 $M_\odot$ third component ($P=33.9^d$).

4. Conclusions

The exploitation of CoRoT and *Kepler* data will require many years, and, for sure, many exciting discoveries are still to come. The quality of the data is a formidable challenge for theoretical models and for the analysis tools, which have to be adapted to comply with the unprecedented accuracy of the data. Besides the excellent photometry has to be complemented by other observations (e.g., spectroscopy, interferometry, multicolor photometry) to take full advantage of its potentials, tasks requiring both time and manpower. The CoRoT community and KASC-WG9 (the working group on EBs) welcome collaboration to fully exploit these gold mines.

References

Aigrain, S., *et al.*, 2009, *A&A*, 506, 425
Bloemen, S., *et al.*, 2011, *MNRAS*, 410, 1787
Borucki, W. J., *et al.*, 2010, *Science*, 327, 977
Cabrera, J., *et al.*, 2009, *A&A*, 506, 501
Carpano, S., *et al.*, 2009, *A&A*, 506, 491
Carone, L., *et al.*, 2011, in preparation
Carter, J. A., *et al.*, 2011, *Science*, 331, 562
Debosscher, J., *et al.*, 2009, *A&A*, 506, 519
Derekas, A., *et al.*, 2011, *Science*, 332, 216
De Ridder, J., *et al.*, 2009, *Nature*, 459, 398
Heacox, W. D. 1998, *AJ*, 115, 325
Hekker, S., *et al.*, 2010, *ApJ*, 713, L187
Kawaler, S. D., *et al.*, 2010, *MNRAS*, 409, 1509
van Kerkwijk, M. H., Rappaport, S. A., Breton, R. P., Justham, S., Podsiadlowski, P., & Han, Z. 2010, *ApJ*, 715, 51
Kjeldsen, H. & Bedding, T. R. 1995, *A&A*, 293, 87
Koch, D. G., *et al.*, 2010, *ApJ* (Letters), 713, L79
Loeb, A. & Gaudi, B. S. 2003, *ApJ*, 588, L117
Maceroni, C., *et al.*, 2009, *A&A*, 508, 1375
Maceroni, C. 2010, *ASP-CS*, 435, 5
Maceroni, C., Cardini, D., Damiani, C., Gandolfi, D., Debosscher, J., Hatzes, A., Guenther, E. W., & Aerts, C. 2010, *arXiv* 1004.1525
Mazeh, T., Tamuz, O., & North, P. 2006, *Ap&SS*, 304, 343
Mazeh, T. & Faigler, S. 2010, *A&A*, 521, L59
Østensen, R. H., *et al.*, 2010, *MNRAS*, 408, L51
Prša, A., *et al.*, 2011, *ApJ*, 141, 83
Slawson, R. W., *et al.*, 2011, *arXiv* 1103.1659
Sarro, L. M., Debosscher, J., López, M., & Aerts, C. 2009, *A&A*, 494, 739
Welsh, W. F., *et al.*, 2011, *arXiv* 1102.1730
Zucker, S., Mazeh, T., & Alexander, T. 2007, *ApJ*, 670, 1326

From Interacting Binaries to Exoplanets: Essential Modeling Tools
Proceedings IAU Symposium No. 282, 2011
Mercedes T. Richards & Ivan Hubeny, eds.
© International Astronomical Union 2012
doi:10.1017/S1743921311026846

The Use of Virtual Observatory Databases in Binary Star Research

Geraldine J. Peters

Space Sciences Center/Department of Physics & Astronomy, University of Southern
California, Los Angeles, CA 90089-1341, USA
email: gjpeters@mucen.usc.edu

Abstract. The rapidly-accumulating archives of ground-based and spacecraft data worldwide that are being linked together through the International Virtual Observatory Alliance (IVOA) provide the binary star community with unparalleled opportunities for research. The main databases that are available to the astronomical community through the IVOA are discussed. Data from long-lasting spacecraft missions such as *IUE* are especially valuable for studying long-term variability. Some examples of current research on close binary stars that is being carried through with UV spectra from the *IUE* archive are presented. Included are the search for O-subdwarf companions to bright Be stars and some results from an ongoing investigation of the Double Periodic Variable phenomenon in Algol binaries.

Keywords. astronomical data bases: miscellaneous, (stars:) binaries: close, stars: emission-line, Be, ultraviolet: stars

1. Introduction

A vast amount of astronomical data of high quality from both spacecraft and ground-based facilities resides in public archives that are accessible to researchers worldwide. Spectroscopic and photometric data that span the electromagnetic spectrum from the gamma ray region to the infrared are available. UV and optical polarimetry also exists. These archived observations are now being linked together through the International Virtual Observatory Alliance (IVOA), an umbrella organization currently with members from 19 international scientific centers. As a result multiwavelength studies of astronomical objects are now commonplace. Especially valuable are data from long-lasting spacecraft missions such as *IUE* that generated a uniform set of FUV & NUV spectra over a period of nearly two decades. The discovery of long-term variability cycles and studies of active binary stars with long periods are now possible. The availability of numerous advanced codes for the interpretation of the data, such as the ones discussed at this meeting, has enabled the astronomical community to discover and study phenomena that were unknown when the spacecraft mission was in operation. The new codes also allow researchers to determine fundamental parameters for stellar atmospheres, winds, disks, and other circumstellar material that represent an immense improvement over the earlier values.

In this paper the developing International Virtual Observatory Alliance and some major databases that are available through it are reviewed. Particularly useful for studying long-term activity in close binary stars are the archives of UV spectra from long-lived spacecraft missions such as *IUE*. Examples of a discovery that could never have been possible without the latter data and current research on the recently-identified Double Periodic variable (DPV)phenomenon in Algol binaries are presented.

Figure 1. The logos for the IVOA and its member organizations.

2. The International Virtual Observatory Alliance (IVOA)

The International Virtual Observatory Alliance (IVOA, http://www.ivoa.net/) formed in 2002 is an umbrella organization currently with 19 members whose purpose is to combine archival data from spacecraft missions and ground-based observatories into one major site than can be accessed worldwide for astronomical research. Logos for the member institutions, which include programs from Argentina, Armenia, Australia, Brazil, Canada, China, Europe, France, Germany, Hungary, India, Italy, Japan, Russia, Spain, the United Kingdom, and the United States and inter-governmental organizations (ESA and ESO), are shown in Fig. 1. Working groups are establishing standards for the data formats and common software that will ultimately be approved by Commission 5 (Astronomical Data) of the International Astronomical Union.

The National Virtual Observatory of the United States (NVO, http://www.us-vo.org/) hosts an evolving website that allows researchers to find and retrieve astronomical data from international archives and data centers. Included are data from spacecraft, optical facilities, and catalogs such as those accessible through the VizieR†. The Virtual Astronomical Observatory (VAO, http://www.usvao.org/) of the United States is developing software tools that will make efficient use of the information gathered through the NVO. Of note are the NVO links to spacecraft data currently in the Multimission Archive at STScI (MAST, http://archive.stsci.edu/), the High Energy Astrophysics Science Archive Research Center (HEASARC, http://heasarc.gsfc.nasa.gov/), and the Infrared Science Archive (IRSA, http://irsa.ipac.caltech.edu/). From the MAST website a researcher can link to the databases from the spacecraft missions listed in Table 1. High resolution UV spectra from *Copernicus* to the *Far Ultraviolet Spectroscopic Explorer* can be accessed through MAST as well as UV, optical, and infrared images from the *Hubble Space Telescope*, the Digitized Sky Survey, and the high precision optical photometry currently being delivered from the *Kepler* spacecraft. NASA's HEASARC is the source of spacecraft data on high energy phenomena. Links can be found to missions that produced gamma-ray to X-ray spectra and fluxes (cf. Table 2). HEASARC's databases also include FUV/NUV images and grism spectra from the *Galaxy Evolution Explorer* (*GALEX*) mission and microwave data from the *Cosmic Background Explorer* (*COBE*) and the *Wilkinson Microwave Anisotropy Probe* (*WMAP*). Databases in the IR and microwave regions are accessed through the NASA/IPAC Infrared Science Archive. Currently sixteen missions/projects are represented including data from the *Infrared Astronomical Satellite* (*IRAS*), *Infrared Space Observatory* (*ISO*), *Spitzer Space Telescope*,

† A joint effort of CDS (Centre de Données Astronomiques de Strasbourg and ESA-ESRIN (Information Systems Division).

Table 1. Datasets Available Through the MAST

Mission	Name	Type of Data
ASTRO/HUT ASTRO/UIT ASTRO/WUPPE	Hopkins Ultraviolet Telescope Ultraviolet Imaging Telescope Wisconsin Ultraviolet Photo-Polarimeter Experiment	FUV spectra FUV images FUV polarimetry
DSS	Digitized Sky Survey	Optical/NIR images
EUVE	Extreme Ultraviolet Explorer	EUV spectra
FUSE	Far Ultraviolet Spectroscopic Explorer	FUV spectra
GALEX	Galaxy Evolution Explorer	FUV/NUV images & LORES spectra
GSC	Guide Star Catalog	Optical positions of stars
HPOL	University of Wisconsin Polarimeter	Optical polarimetry, photometry, and spectra
HST	Hubble Space Telescope	FUV/NUV, Optical, IR spectra, photometry, & images
IUE	International Ultraviolet Explorer	HIRES/LORES FUV & NUV spectra
Kepler	Kepler spacecraft	High precision optical photometry
OAO-3	Copernicus satellite	HIRES FUV/NUV spectra
ORFEUS-SPAS ORFEUS/BEFS ORFEUS/IMAPS ORFEUS/TUES	Orbiting and Retrievable Far and Extreme UV Spectrograph Berkeley Extreme and Far-UV Spectrometer Interstellar Medium Absorption Profile Spectrograph Tübingen Echelle Spectrograph	See below HIRES FUV spectra Very HIRES FUV spectra HIRES FUV spectra
VLA-FIRST	Very Large Array	Faint images of sky at 21 cm
XMM-OM	X-ray Multi-Mirror Telescope Optical Monitor	LORES X-ray spectra

Wide-field Infrared Survey Explorer (*WISE*), and the *Planck* mission to study the cosmic microwave background.

The American Association of Variable Star Observers (AAVSO, http://www.aavso.org/) has been collecting and analyzing ground-based photometry from astronomers worldwide now for over 100 years. Their archive contains more than 20.5 million observations and they have established strong collaborations between professional and amateur astronomers. The All Sky Automated Survey (ASAS, http://www.astrouw.edu.pl/asas/) is a current photometric project in which more than 10^7 stars are being monitored in the Johnson V and I bands from two small telescopes in Hawaii and Chile. Notable are the OGLE (Optical Gravitational Lensing Experiment, http://ogle.astrouw.edu.pl/) and MACHO (search for MAssive Compact Halo Objects, http://wwwmacho.anu.edu.au/) microlensing projects that have produced $> 10^9$ optical observations of stars in the Magellanic Clouds and Galactic Bulge in the past two decades. Other databases include the RAVE (RAdial Velocity Experiment, http://www.rave-survey.aip.de/rave/) which is giving positions, distances and proper motions of more than 10^6 stars, the ELODIE archive (http://atlas.obs-hp.fr/elodie/) of high resolution stellar spectra, and the Be Star Spectra (BeSS) database http://basebe.obspm.fr/basebe/) which contains 54,000 spectra of more than 600 different Be stars.

Perhaps the most valuable databases for investigating short and long-term activity in the close binaries are those from the *International Ultraviolet Explorer* (*IUE*) and *Far*

Table 2. Datasets Available Through NASA's HEASARC

Mission	Name/Description	Type of Data
AGILE	Italian Space Agency (ASI) spacecraft	Gamma-ray/Hard X-ray detectors
ASCA	ASTRO-D, Japanese X-ray satellite	X-ray photometry & LORES spectroscopy
BeppoSAX	ASI/NIVR (Netherlands)/ESA X-ray satellite	Soft/Hard X-ray detectors
COBE	Cosmic Background Explorer	Microwave/ FIR photometry & LORES spectroscopy
CGRO	Compton Gamma Ray Observatory	Gamma-ray photometry
Chandra	Chandra X-ray Observatory	HIRES X-ray imaging/photometry & spectroscopy
EUVE	Extreme Ultraviolet Explorer	EUV spectra
Fermi	Fermi Gamma-ray Space Telescope	Gamma-ray photometry
GALEX	Galaxy Evolution Explorer	FUV/NUV images & LORES spectra
HETE-2	The High Energy Transient Explorer Mission	Gamma-ray/X-ray burst photometry
INTEGRAL	The INTErnational Gamma-Ray Astrophysics Laboratory	Gamma-ray imaging/spectroscopy
ROSAT	The Róntgen Satellite	X-ray imaging/photometry
RXTE	The Rossi X-ray Timing Explorer Mission	Rapid X-ray photometry
Suzaku	The Suzaku Mission (ASTRO-E)	X-ray imaging spectrometer & hard X-ray detector
Swift	The Swift Gamma-Ray Burst Mission	Gamma-ray burst detector, X-ray & optical region telescopes
WMAP	Wilkinson Microwave Anisotropy Probe	Microwave imaging
XMM-Newton	ESA X-ray mission	HIRES X-ray images & spectra

Ultraviolet Spectroscopic Explorer (FUSE). Both produced high resolution, well-calibrated UV spectra over a long duration. *IUE* was launched on January 26, 1978 and operated for 18.75 years. This small Ritchey-Chretien telescope with an aperture of 45 cm was placed in an elliptical geosynchronous orbit with a perigee/apogee of 26000/46000 km that allowed it to be operated in real time by NASA and ESA with the observer present at either operations center. The *IUE* spacecraft yielded FUV (1150–1950 Å, SWP) and NUV (1900–3200 Å, LWR/LWP) spectra at resolutions of 0.1–0.3 Å (HIRES) and 6-8 Å (LORES). Over 120,000 images were produced of which 2374 are of binary stars, including 556 SWP HIRES spectra. The *FUSE* spacecraft, launched on June 24, 1999, was operated for NASA and the French and Canadian Space Agencies for 8.3 years. It produced FUV spectra (950–1188 Å) with a resolution of 0.05 Å. Over 6000 observations were made of which 195 were of binary stars (mostly from two large survey programs).

3. Some Uses of Virtual Databases in Binary Star Research

3.1. *Detection of O-Type subdwarfs*

The *IUE* archive has proven to be especially valuable for detecting the presence of close sdO companions to bright Be stars. More than twenty years ago researchers were beginning to suspect that some Be stars may have been spun up to their very rapid rotation rates through the process of mass transfer. It seemed impossible to confirm the presence

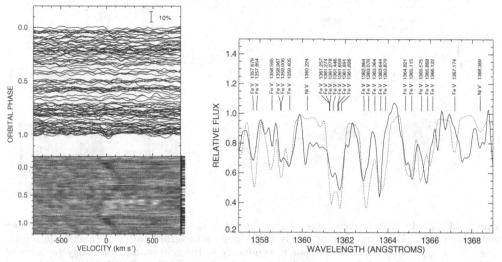

Figure 2. The nature of the subdwarf in the FY CMa system. *Left panel:* The grayscale at the bottom shows the orbital phase variations in the CCFs in the 1310–1385 Å region produced with a hot star template. The orbital motion of the subdwarf (P = $37\overset{d}{.}25$) is apparent. *Right panel:* The reconstructed spectrum of the 45 kK subdwarf. From Peters *et al.* (2008)

of a close hot subdwarf, that would eventually become a massive white dwarf, about a Be star because of the immense difference in their fluxes. But through the analysis of 16 *IUE* HIRES images of the bright Be star ϕ Per (B2 Vpe) with a Doppler tomography code (Bagnuolo *et al.* 1994), Thaller *et al.* (1995) were able to recover the spectrum of a hot sdO secondary with a T_{eff} of ~50 kK. The period of the system is $126\overset{d}{.}696$ and the subdwarf contributes only ~15% of the light in the FUV. We recently employed the same procedures discussed in Thaller *et al.* (1995) and Gies *et al.* (1998) to recover the spectrum of a 45,000 K sdO around the B0.5 IVe star FY CMa (Peters *et al.* 2008). The analysis made use of 97 *IUE* SWP HIRES images that were obtained from 1979–95. We cross-correlated each *IUE* spectrum with a theoretical spectrum computed from TLUSTY/SYNSPEC (Hubeny 1988, Hubeny & Lanz 1995) to produce the run of cross-correlation functions (CCFs) shown in Fig. 2. The recovered spectrum of the sdO object is also displayed. In the case of this system the subdwarf contributes only 4% of the light in the FUV. We are currently in pursuit of other hot subdwarfs that may be hidden in the light of bright Be stars by making use of the extensive *IUE* archive.

3.2. *Study of Cyclic Long-Term Variability in Algol Systems*

The *IUE* database is especially valuable for investigating long-term activity in Algol systems. A problem of current interest is finding the cause for the DPV phenomenon that was first identified by Mennickent *et al.* (2003) from the analysis of OGLE-II observations of Be stars in the LMC. DPV binaries show two photometric periods, one associated with orbital motion and the other a cyclic long-term light modulation that is ~35 times P_{orb}. The DPV phenomenon was first observed in the galactic Algols AU Mon (Lorenzi 1980) and RX Cas (Kalv 1979). The 164 SWP/LWP HIRES & LORES images of AU Mon (B3 Ve + G IV, $P_{orb} = 11\overset{d}{.}11$) that exist in the *IUE* archive are currently being used to shed some light on the DPV phenomenon. Some early results (Peters & Hageman, in prep.) suggest that the long period light cycle is caused from a waxing/waning of an obscuring accretion disk. In the left panel of Fig. 3 we show a series of *IUE* LORES spectra taken throughout a primary eclipse when the star was relatively bright in its

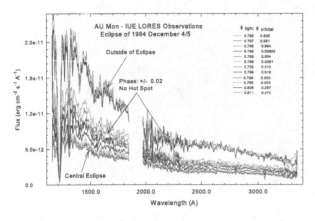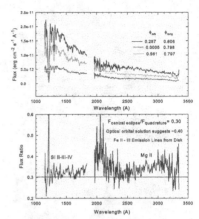

Figure 3. *IUE* LORES observations of AU Mon taken throughout the 1994 December eclipse (*left panel*) when the system was relatively bright in its long-term light cycle (maximum brightness is at $\phi_{\mathrm{long}} = 0.0$) reveal CS emission, symmetry in the eclipse, and the likely presence of significant obscuration of the equatorial region of the B star by the CS disk (*right panel*).

long-term light cycle. It is immediately apparent that the eclipse was symmetrical (no evidence for a hot accretion spot on the *trailing* hemisphere) and that the photospheric absorption lines are less visible at primary minimum. The latter suggests that the accretion disk around the B star not only emits strongly in Hα but also in the lines of ionized species in the UV. Such circumstellar (CS) emission features are clearly seen when the spectrum taken at minimum light is divided by a spectrum taken well outside of eclipse (right panel of Fig. 3). The depth of the primary eclipse in the UV implies that 30% of the B star is uncovered at minimum (the G star contributes essentially no flux in the UV) in contrast to the 40% that is suggested by the orbital solution from optical data published by Desmet *et al.* (2010). This infers the presence of a substantial optically-thick disk, even near light maximum, that obscures the equatorial region. Some other conclusions from our current study are that the optical thickness of the CS disk is greatest at minimum light in the long-term cycle, mass infall is greatest at about $\phi_{\mathrm{long}} \sim 0.25$, mass loss from $\phi_{\mathrm{orb}} \sim 0.5$ is largest when the star is bright, and the photosphere of the B star is substantially deficient in carbon as was implied in the study of the star's high temperature CS plasma (Peters & Polidan 1984).

Support for this project was provided by NASA Grants NNX10AD60G, NNX07AH56G, and NNX04GC48G, the USC WiSE program, and an international travel grant from the American Astronomical Society.

References

Bagnuolo, W. G., Jr., *et al.*, 1994, *ApJ*, 423, 446
Desmet, M., *et al.*, 2010, *MNRAS*, 401, 418
Gies, D. R., *et al.*, 1998, *ApJ*, 493, 440
Hubeny, I. 1988, *Comput. Phys. Commun.* 52, 103
Hubeny, I. & Lanz, T. 1995, *ApJ*, 439, 875
Kalv, P. 1979, *Tartu Astrof. Obs. Teated*, 58, 3
Lorenzi, L. 1980, *A&A*, 85, 342
Mennickent, R. E., Pietrzyński, G., Diaz, M., & Gieren, W. 2003, *A&A*, 399, L47
Peters, G. J., Gies, D. R., Grundstrom, E. D., & McSwain, M. V. 2008, *ApJ*, 686, 1280
Peters, G. J. & Polidan, R. S. 1984, *ApJ*, 283, 745
Thaller, M. L., Bagnuolo, W. G., Jr., Gies, D.. R., & Penny, L. R. 1995, *ApJ*, 448, 878

From Interacting Binaries to Exoplanets: Essential Modeling Tools
Proceedings IAU Symposium No. 282, 2011
Mercedes T. Richards & Ivan Hubeny, eds.
© International Astronomical Union 2012
doi:10.1017/S1743921311026858

Multi-Technique Study of the X-Ray Binary Cygnus X-1

E. A. Karitskaya

Astronomical Institute of RAS, Pyatnitskaya str. 48, 119017 Moscow, Russia
email: karitsk@sai.msu.ru

Abstract. Short review of our 36-year Cyg X-1 study using multi-technique methods and based on our optical photometric, high-resolution spectral and spectropolarimetrical observations.

Keywords. Stars: magnetic fields, stars: atmospheres, stars: abundances, accretion disks, techniques: polarimetric, techniques: spectroscopic, techniques: image processing, stars: early-type, stars: individual (Cyg X-1 = HDE 226868=V1357 Cyg), X-rays: binaries

---

Cyg X-1/HDE226868/V1357 Cyg ($m_V = 9^m$) is an X-ray binary system (the orbital period $P = 5.6^d$) whose relativistic component is the first black hole (BH) candidate. The optical component, an O9.7 Iab supergiant, is responsible for about 95% of the system's optical luminosity. The remaining 5% is due to the accretion structure (disc and surrounding gas) near the BH. The intensive investigations of Cyg X-1 are being carried on for 40 years, but a lot of phenomena in the system remain unclear. Here there is a short review of our 36-year Cyg X-1 study using multi-technique methods and based on our optical photometric, high-resolution spectral and spectropolarimetrical observations.

The main optical photometric variation (the ellipsoidality effect) was studied in detail with the Roche model. Thirty-six years ago, Bochkarev, Karitskaya & Shakura (1975) used the amplitude $A = 0.035 - 0.050^m$, the difference in depth minima $\Delta A < 0.005^m$ to derive the admissible values of parameters for Cyg X-1: inclination $25 < i < 67$; mass ratios $0.2 < q < 0.55$; filling factor $0.9 < \mu < 1$; $M_o > 17 M_{sun}$; $7 M_{sun} < M_x < 27 M_{sun}$. It is interesting that very new research (Orosz *et al.*, 2011) has obtained values within our ranges for the parameters.

In the frame of the 1994-1998 international campaign "Optical Monitoring of Unique Astrophysical Objects" 2258 UBVR observations of Cyg X-1 were made during 407 nights. The main results are reported by Karitskaya *et al.* (2000) and Karitskaya *et al.* (2001). Evidence of irregularities in matter flowing between the components was found. By comparing photoelectric (UBVR) and X-ray ASM/RXTE (3-12 keV) flux variations we found different kinds of variability (orbital variations, various flares, dips, so-called precession period 147/294 days) and a correspondence between optical and X-ray variations. Cross-correlation analysis revealed lags of X-ray (2-10 keV) long-term variations in respect to the optical ones (7^d in 1996 and 12^d in 1997-1998). It allowed determining accretion time which is much shorter than with the standard accretion model.

The results of Cyg X-1 spectral monitoring are presented in Karitskaya *et al.* (2003, 2005, 2006, 2007a), Karitskaya *et al.* (2008). The observations were carried out with the Echelle-spectrographs of the Peak Terskol observatory (altitude 3100 m, North Caucasus) 2 m telescope (spectral resolution R = 13000 or 45000) and BOAO (Korea) 1.8 m telescope (R = 30000) covering most of the optical range. Optical spectral line profile variations were found during the X-ray flare.

Non-LTE modeling of the observed spectra allowed us to put limits on the parameters of Cyg X-1 O-supergiant component: $T_{eff} = 30400 \pm 500$ K, $\log g = 3.31 \pm 0.07$, and the element overabundances: from 0.4 dex to 1.0 dex for He, N, Ne, Mg, Si, that is, the elements affected by CNO- and α-processes (Karitskaya, et al. (2005, 2007b), Karitskaya et al. (2011a)). Tidal distortion of the Cyg X-1 optical component and its illumination by X-ray emission of the secondary are taken into account.

The photometric and spectral variations point to the supergiant parameters' changes on the time scale of tens of years. Line profile non-LTE simulations lead to the conclusion that the star radius has grown $\sim$ 1-4% from 1997 to 2003-2004 while the temperature decreased by $1300 - 2400$ K (Karitskaya, et al. 2006, Karitskaya et al. 2007a). This agrees with the X-ray activity growth in these years. An increase in the Roche lobe filling degree causes greater instability in the accretion process.

The spectral line profile sets permitted us to construct the binary 2D and 3D tomographic maps using HeIIλ4686Å profiles (Karitskaya et al. 2005, 2007a, Sharova et al. 2011). The comparison of the 2D tomographic map with the theoretical calculations allowed us to construct a more precise system model and receive better information on the gas flowing. The hard limits on Cyg X-1 component mass ratio were obtained by such manner: $1/4 < M_X/M_O < 1/3$.

Our VLT 8-m telescope spectropolarimetric observations permitted us to reveal the magnetic field of $\sim$ 100 G on the supergiant and to suspect the magnetic field of $\sim$ 600 G on the outer parts of the accretion structure (Karitskaya et al. 2010, Karitskaya et al. 2009, Karitskaya et al. 2011b, Bochkarev & Karitskaya 2011a, Bochkarev & Karitskaya 2011b). For the first time the existence of magnetic accretion on the black hole has been confirmed. Such magnetic field will be increased during disc accretion and can be responsible for the observed X-ray flickering.

Acknowledgements

The work is supported by RFBR grants 04-02-16924, 06-02-16234, 09-02-01136 and 09-02-00993.

References

Bochkarev N. G. Karitskaya E. A., & Shakura N. I. 1975, *Soviet Astronomy Letters*, 1, 118

Bochkarev, N. G. & Karitskaya, E. A. 2011a, in D.O. Kudryavtsev, & I.I.Romanyuk (eds.), *Magnetic Stars. Proc. Int. Conf.* (N.Arkhyz: SAO RAS publ.), p. 199

Bochkarev, N. G. & Karitskaya, E. A. 2011b, in this volume

Karitskaya, E. A. 2003, *Kinematika i Fizika Nebesnykh Tel, Suppl.*, No.4 230

Karitskaya, E. A., Goranskij V. P., Grankin E. N. *et al.*, 2000, *Astron. Lett.*, 26, 22

Karitskaya, E. A., Voloshina, I. B., Goranskij, V. P. *et al.*, 2001, *Astron. Rep.*, 45, 350

Karitskaya, E. A., Agafonov, M. I., Bochkarev, N. G. *et al.*, 2005, *AApTr*, 24, 383

Karitskaya, E. A., Lyuty, V. M., Bochkarev, N. G. *et al.*, 2006, *IBVS*, 5678, 1

Karitskaya, E. A., Agafonov, M. I., Bochkarev, N. G. *et al.*, 2007a, *AApTr*, 26, 159

Karitskaya, E. A., Shimanskii, V. V., Sakhibullin, N. A. *et al.*, 2007b, *ASP-CS* 378, 123

Karitskaya, E. A., Bochkarev, N. G., Bondar', A. V. *et al.*, 2008, *Astron. Rep.*, 52, 362

Karitskaya, E. A., Bochkarev, N. G., Hubrig, S. *et al.*, 2009, *astro-ph* 0908.2719

Karitskaya, E. A., Bochkarev, N. G., Hubrig, S. *et al.*, 2010, *IBVS*, 5950, 1

Karitskaya, E. A., Bochkarev, N. G., Shimanskii, V. V., & Galazutdinov, G. 2011a, *ASP-CS*, 445, in press

Karitskaya, E. A., Bochkarev, N. G., Hubrig, S. *et al.*, 2011b, in D.O. Kudryavtsev, & I.I. Romanyuk (eds.), *Magnetic Stars. Proc. Int. Conf.*, (N.Arkhyz: SAO RAS publ.), p. 188

Orosz, J. A., McClintock, J. E., Aufdenberg, J. P. *et al.*, 2011, *Astro-ph*, 1106.3689

Sharova, O. I., Agafonov, M. I., Karitskaya, E. A. *et al.*, 2011, in this volume

From Interacting Binaries to Exoplanets: Essential Modeling Tools
Proceedings IAU Symposium No. 282, 2011 © International Astronomical Union 2012
Mercedes T. Richards & Ivan Hubeny, eds. doi:10.1017/S174392131102686X

Light Curve and Orbital Period Analysis of the Eclipsing Binary AT Peg

Alexios Liakos[1], Panagiotis Niarchos[1], and Edwin Budding[2,3]

[1]Department of Astrophysics, Astronomy and Mechanics, University of Athens, Athens, Hellas
email: alliakos@phys.uoa.gr, pniarcho@phys.uoa.gr

[2]Carter Observatory & School of Chemical and Physical Sciences, Victoria University,
Wellington, New Zealand

[3]Department of Physics and Astronomy, University of Canterbury, Christchurch, New Zealand
email: budding@xtra.co.nz

Abstract. CCD photometric observations of the Algol-type eclipsing binary AT Peg have been obtained. The light curves are analyzed with modern techniques and new geometric and photometric elements are derived. A new orbital period analysis of the system, based on the most reliable timings of minima found in the literature, is presented and apparent period modulations are discussed with respect to the Light-Time effect (LITE) and secular changes in the system. The results of these analyses are compared and interpreted in order to obtain a coherent view of the system's behaviour.

Keywords. methods: data analysis, (stars:) binaries (including multiple): close, (stars:) binaries: eclipsing, stars: evolution, stars: individual (AT Peg).

1. Observations and analyses

The system was observed during 6 nights in summer of 2010 at the Athens University Observatory, using a 20-cm Newtonian telescope equipped with the CCD camera ST-8XMEI and B and R Bessell photometric filters.

The light curves (hereafter LCs) have been analysed with the PHOEBE v.0.29d software (Prša & Zwitter 2005). The temperature of the primary component was used as a fixed parameter (T_1=8400 K), based on the classification of Maxted *et al.* (1994). The spectroscopic mass ratio q of Maxted *et al.* (1994) was used as initial value, and then it was adjusted, but always kept inside the spectroscopic error. The rest parameters were either given theoretical values or they were adjusted (for details see Liakos *et al.* 2011). The contribution of a third light was also considered as there are indications of a third star orbiting the eclipsing pair. The absolute parameters of the components were calculated and used for further study of their present evolutionary status.

The least squares method with statistical weights in a MATLAB code (for details see Zasche *et al.* 2009) has been used for the analysis of the O–C diagram. The current O–C diagram of AT Peg includes 276 times of minima taken from the literature. A LITE fitting function, corresponding to a cyclic variation of the O–C points, and a parabola, assuming a mass-transferring configuration, were chosen to fit the times of minima.

The synthetic and observed LCs and the O–C fitting curve and its residuals are shown in Fig. 1, while the derived parameters from both analyses are listed in Table 1.

2. Discussion and conclusions

New photometric and O-C diagram analyses of the eclipsing binary AT Peg were performed. The results were combined with those of previous spectroscopic studies and new geometric elements of the system and absolute parameters of its components were

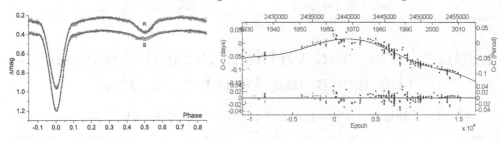

Figure 1. Left panel: Observed (symbols) and synthetic (solid lines) light curves of AT Peg. Right panel: The O–C diagram of AT Peg fitted by the combined function (solid line).

Table 1. Results of light curve and O–C analyses and absolute parameters of the components.

	Light curve parameters					Absolute parameters		
Component:	Primary	Secondary	Filters:	B	R	Component:	Primary	Secondary
T [K]	8400*	5189 (7)	x_1	0.556	0.413	M [$M_\odot$]	2.2 (1)	1.0 (1)
g	1*	0.32*	x_2	0.834	0.596	R [$R_\odot$]	1.70 (3)	2.14 (3)
A	1*	0.5*	L_1/L_T	0.809 (2)	0.727 (2)	L [$L_\odot$]	13.0 (4)	3.0 (1)
Ω	4.49 (1)	2.83	L_2/L_T	0.115 (1)	0.212 (1)	M_{bol} [mag]	2.0 (2)	3.6 (2)
i [deg]	77.54 (5)		L_3/L_T	0.076 (1)	0.062 (2)	a [$R_\odot$]	2.18 (3)	4.61 (9)
q	0.478 (3)					$\log g$ [cm/s^2]	4.31 (3)	3.79 (3)

O–C diagram parameters					

Min.I [HJD]	243803.447 (1)	$\dot{P}$ [days/yr]	$-3.563(1)\times10^{-7}$	ω [deg]	204 (37)	f (M_3)[$M_\odot$]	0.0129 (3)
P [days]	1.1460905 (2)	P_3 [yrs]	49.7 (9)	A [days]	0.018 (1)	$M_{3,\,min}$ [$M_\odot$]	0.57 (1)
C_2 [days/cycle]	$-5.590(1)\times10^{-10}$	T_0 [HJD]	2446308 (1796)	e	0.1 (1)		

*assumed, $L_T = L_1 + L_2 + L_3$

derived. The primary component was found to be a dwarf star and close to the ZAMS limit. On the other hand the secondary is located beyond the TAMS line indicating that it is at the subgiant stage of evolution. Moreover, according to the LC analysis, this component was found to fill its Roche lobe, therefore it may be expected to be a mass loser. Hence, according to these results AT Peg can be considered as a classical Algol type system.

The period changes analysis suggests two main conclusions: **(a):** The LC solution revealed a third light of ∼7%, while the O–C analysis suggested a third body with a minimal mass of 0.57 $M_\odot$. Assuming a MS nature with coplanar orbit and taking into account the Mass-Luminosity relation for dwarfs (L∼$M^{3.5}$) the expected light contribution was found ∼1%, which is much less than the observed one. However, if the third body orbits the eclipsing binary with an inclination of ∼12° (as a MS star), or it is more evolved, then the observed additional luminosity can be justified. **(b):** The orbital period secular change found, contrary to what it was expected (the mass should flow from the less massive to the more massive component), resulted in a decreasing period rate. This discrepancy could be explained with a mass loss rate from the system of 6.6×10^{-7} $M_\odot$/yr, perhaps due to strong stellar winds, or with a magnetic breaking mechanism or even with systemic angular momentum loss, which superimposes the expected mass transfer.

References

Liakos, A., Zasche, P., & Niarchos, P. 2011, *New Astron.*, 16, 530
Maxted, P. F. L., Hill, G., & Hilditch, R. W. 1994, *A&A*, 285, 535
Prša, A. & Zwitter, T. 2005, *ApJ*, 628, 426
Zasche, P., Liakos, A., Niarchos, P., *et al.*, 2009, *New Astron.*, 14, 121

From Interacting Binaries to Exoplanets: Essential Modeling Tools
Proceedings IAU Symposium No. 282, 2011   © International Astronomical Union 2012
Mercedes T. Richards & Ivan Hubeny, eds.   doi:10.1017/S1743921311026871

The Third Body in the Eclipsing Binary AV CMi: Hot Jupiter or Brown Dwarf?

Alexios Liakos[1], Dimitris Mislis[2] and Panagiotis Niarchos[1]

[1] Department of Astrophysics, Astronomy and Mechanics, University of Athens, Athens, Hellas
email: alliakos@phys.uoa.gr, pniarcho@phys.uoa.gr

[2] Institute of Astronomy, Madingley Road, Cambridge CB3 0HA, UK
email: misldim@ast.cam.ac.uk

Abstract. New transit light curves of the third body in the system AV CMi have been obtained. The eclipsing pair's light curves were re-analysed with the W-D code and new absolute elements were derived for the two components. Moreover the new light curves (together with those given by Liakos & Niarchos 2010) of the third body transiting one of the components were analysed with the Photometric Software for Transits (PhoS-T). The results from both analyses are combined with the aim to study the nature of the third component.

Keywords. methods: data analysis, (stars:) binaries (including multiple): close, (stars:) binaries: eclipsing, stars: evolution, stars: individual (AV CMi)

---

1. Observations and analyses

The new transit observations were obtained with a 40-cm Cassegrain telescope equipped with the CCD camera ST-10XME and by using only the I-filter, in order to achieve a better time resolution and higher signal-to-noise ratio. 11 new transit light curves were obtained in the years 2009-2011 increasing the total number to 18.

The light curves (hereafter LCs) of AV CMi were re-analysed with the PHOEBE v.0.29d software (Prša & Zwitter 2005) (for method details see Liakos & Niarchos 2010). The solution presented herein is based on a different assumed spectral type of the system (according to its B-V index (0.14-0.2)) as given in many catalogues (e.g. NOMAD, NPM2, ASCC-2.5 V3). Moreover, the absolute parameters of the eclipsing components were derived in order to check their evolutionary status and they were used for the calculation of some of the third body's characteristics as described below.

Five complete transit LCs were analysed with the PhoS-T (Mislis *et al.* 2011). For a first approach of the third body's parameters, the following hypotheses were adopted: **Case A:** The third body orbits the primary and **Case B:** it orbits the secondary component. For each case the light contribution of the binary component around which the third body is not orbiting (secondary in case A and primary in case B) was subtracted from the total light (of the triple system) by taking into account its fractional luminosity (see Table 1) and the residual light curves were re-normalized. The masses of the components and the period of the third body were used to find the semi-major axis of the tertiary component's orbit (assumed a circular one) for each case. The period and the semi-major axis of the third body, the radius and the limb darkening coefficients (Claret 2004) of the 'host' component were kept fixed in the programme, while the radius R_3 and the inclination i_3 of the orbit of the third body were adjusted. A sample of synthetic and observed transit LCs are shown in Fig. 1, while the derived parameters from both analyses are listed in Table 1.

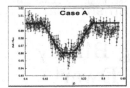 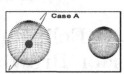 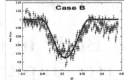 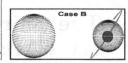

Figure 1. Observed (symbols) and synthetic (solid lines) transit light curve of the third body of AV CMi for each case for HJD 2455538 along with the corresponding 3D plots where the third body and its orbit are shown.

Table 1. Light curve and transit analyses results.

Light curve parameters							Absolute parameters		
Component:	Primary	Secondary	Filters:	V	R	I	Component:	Primary	Secondary
Ω	5.73 (1)	6.76 (1)	x_1	0.527	0.443	0.353	M [M$_\odot$]	1.90*	1.60 (1)
g	1*	1*	x_2	0.522	0.444	0.357	R [R$_\odot$]	2.38 (5)	1.72 (4)
A	1*	1*	L_1/L_T	0.654 (2)	0.646 (2)	0.634 (2)	L [L$_\odot$]	19.8 (8)	10.3 (4)
e		0.11 (1)	L_2/L_T	0.341 (1)	0.337 (1)	0.331 (1)	T [K]	7900*	7897 (8)
i [deg]		83.6 (1)	L_3/L_T	0.004 (2)	0.016 (2)	0.036 (3)	a [R$_\odot$]	5.2 (2)	6.2 (1)
q		0.843 (3)					log$g$ [cm/s^2]	3.96 (2)	4.17 (2)

Transit parameters										
HJD:	2454521		2454783		2455538		2455588		2455601	
Case:	A	B	A	B	A	B	A	B	A	B
R_3 [R$_{Jup}$]	4.1	5.4	4.6	6.1	4.7	6.9	4.7	6.9	4.1	6.6
i_3 [deg]	55.5	61.9	56.1	62.3	56.8	60.0	53.7	58.1	55.6	57.7
χ^2	2.14	9.13	2.70	7.36	2.14	9.96	1.29	5.52	1.32	6.01

*assumed, $L_T = L_1 + L_2 + L_3$

2. Discussion and conclusions

New LC modelling of AV CMi and analysis of five transits of the third body in front of one of the components were obtained. The absolute parameters of the eclipsing components were calculated and showed that they are both MS stars with almost the same temperature and eccentric orbits. Using the 18 transit observations an updated ephemeris of the transits was calculated: $\mathbf{T}_{transit} = \mathbf{HJD\ 2454899.354\ (1) + 0.519215\ (1)}^d \times \mathbf{E}$. The results of transit analyses showed that both R_3 and i_3 for each case are varying, probably due to the non-spherical shape of the components. A mean value for the third body's radius yielded as 4.4(3) R$_{Jup}$ and 6.4(6) R$_{Jup}$ for cases A and B, respectively. However, the χ^2 value of case B was found to be systematically greater than that of case A, indicating that the solution of case A is more realistic. The present results, although they offer a first step for the investigation of the third component in AV CMi, cannot provide a final conclusion about its nature. The 'Hot Jupiter' scenario seems to fail due to the big value of the radius, while, the 'Brown dwarf' hypothesis is probably the solution for the nature of the tertiary companion. Moreover, the LC analysis of AV CMi showed that a third light contribution of ∼2% maybe exists. This value is impossible for planets but not for low-luminosity stars. High accuracy spectral observations are certainly needed for: (a) spectral classification of the eclipsing components, and (b) radial velocity measurements which will help to derive the mass ratio of the system and probably will reveal the third body's motion, and maybe its nature.

References

Claret, A. 2004, *A&A*, 428, 1001
Liakos, A. & Niarchos, P. 2010, *ASP-CS*, 424, 208
Mislis, D., *et al.* 2011, *AIP-CP*, in press
Prša, A. & Zwitter, T. 2005, *ApJ*, 628, 426

From Interacting Binaries to Exoplanets: Essential Modeling Tools
Proceedings IAU Symposium No. 282, 2011   © International Astronomical Union 2012
Mercedes T. Richards & Ivan Hubeny, eds.   doi:10.1017/S1743921311026883

Multiwavelength Photometry of the Young Intermediate Mass Eclipsing Binary TY CrA

M. Ammler-von Eiff[1], M. Vaňko[2], T. Pribulla[2], E. Covino[3], R. Neuhäuser[4], and V. Joergens[5,6]

[1] Thüringer Landessternwarte Tautenburg, Sternwarte 5, 07778 Tautenburg, Germany
email: ammler@tls-tautenburg.de

[2] Astronomical Institute of the Slovak Academy of Sciences, 059 60 Tatranská Lomnica, Slovakia
email: vanko@ta3.sk, pribulla@ta3.sk

[3] INAF - Osservatorio Astronomico di Capodimonte, Salita Moiariello, 16 80131 - Napoli, Italy

[4] Astrophysikalisches Institut und Universitäts-Sternwarte, Schillergäßchen 2-3, 07745 Jena, Germany

[5] Max-Planck-Institut für Astronomie, Königstuhl 17, 69117 Heidelberg, Germany

[6] Zentrum für Astronomie Heidelberg, Institut für Theoretische Astrophysik Albert-Ueberle-Str. 2, 69120 Heidelberg, Germany

Abstract. One of the handful of known PMS eclipsing binaries is a component of the spectroscopic triple TY CrA. Its secondary component is particularly interesting since it is a star of relatively high mass ($1.64\,M_\odot$) which is still on the pre-main sequence. The eclipsing binary was analyzed in the optical wavelength range $\sim$10 years ago, however, the crucial secondary eclipse minimum is very shallow. Therefore, we are obtaining new photometry in both optical and near-IR bands. We present first observations in *(BVRI)* which show that the secondary eclipse depth increases to about 0.1 mag in the *I* band. The increased eclipse depth with respect to other bands will help to better determine the colours and dimensions of the system. Furthermore, we show and discuss first near-IR observations of the primary eclipse. In addition to the light curves we are obtaining radial velocities in order to pin down the orbital parameters of the triple. Our first observations agree with the orbital parameters derived $\sim$10 years ago.

Keywords. binaries: eclipsing, stars: fundamental parameters, stars: individual (TY CrA) etc.

1. Introduction

TY CrA is a hierarchical triple, maybe even quadruple system, embedded in a reflection nebula in the CrA star forming region. Two components form a massive eclipsing double-lined spectroscopic binary with an orbital period of almost 3 days. A third spectroscopic component is in a wide orbit around the eclipsing pair (Casey *et al.* 1995; Corporon *et al.* 1996). A visual fourth component was detected by Chauvin *et al.* (2003).

2. Overview

The eclipsing binary has been analysed in detail previously but uncertainties remain. Casey *et al.* (1998) obtained optical light curves of the eclipsing binary in *uvby*. They derived component masses of $3.16\pm0.02\,M_\odot$ and $1.64\pm0.01\,M_\odot$. The latter is particularly interesting since it has not yet reached the main sequence. Its properties are uncertain since the light curve analysis is challenging.

The secondary minimum is very shallow and the out-of-eclipse variability is larger than the measurement errors (Vaz *et al.* 1998). A few years ago, the benefits of observations

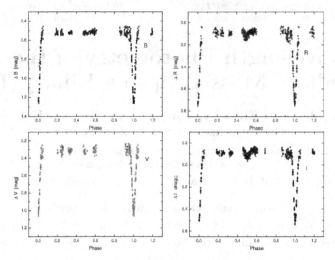

Figure 1. *BVRI* light curves obtained with VYSOS6 at Cerro Armazones, Chile.

in the near-infrared (NIR) have been demonstrated. Covino *et al.* (2004) analysed an eclipsing binary in the optical and NIR and noticed that the depth of the secondary minimum is more pronounced in the NIR. The error bars on the stellar parameters could be reduced by ≈ 80 % compared to previous analysis without NIR data (Covino *et al.* 2000). The goal of the present work is to analyse TY CrA in a way similar to Covino *et al.* (2004) and to get more precise secondary parameters by observations in the NIR. New photometric observations of TY CrA have been obtained by the authors in recent years in both optical and NIR bands. The primary minimum was observed in 2006 with SOFI at La Silla, Chile. New optical data and additional NIR photometry have been obtained from 2009-2011 with VYSOS6 at Cerro Armazones, with REM at La Silla, and ANDICAM at Cerro Tololo, Chile. First results have been obtained. The minimum time derived from the SOFI light curves differs by 7 minutes from the ephemeris presented by Casey *et al.* (1998) which is indicative of additional components. We reproduced the stellar and orbital parameters derived by Casey *et al.* (1998) based on the 2009 VYSOS6 data (Fig. 1) using the methods of Pribulla *et al.* (2008). It can be noticed that the secondary minimum at a phase of ≈ 0.5 already becomes more pronounced in the *R* and *I* bands, compared to the *B* and *V* bands.

References

Casey, B. W., Mathieu, R. D., Vaz, L. P. R., Andersen, J., & Suntzeff, N. B. 1998, *AJ*, 115, 1617

Casey, B. W., Mathieu, R. D., Suntzeff, N. B., & Walter, F. M. 1995, *AJ*, 109, 2156

Chauvin, G., Lagrange, A., Beust, H., Fusco, T., Mouillet, D., Lacombe, F.,Pujet, P., Rousset, G., Gendron, E., Conan, J. M., Bauduin, D., Rouan, D., Brandner, W., Lenzen, R., Hubin, N., & Hartung, M. 2003, *A&A*, 406, L51

Corporon, P., Lagrange, A. M., & Beust, H. 1996, *A&A*, 310, 228

Covino, E., Catalano, S., Frasca, A., Marilli, E., Fernández, M., Alcalá, J. M., Melo, C., Paladino, R., Sterzik, M. F., & Stelzer, B. 2000, *A&A*, 361, L49

Covino, E., Frasca, A., Alcalá, J. M., Paladino, R., & Sterzik, M. F. 2004, *A&A*, 427, 637

Pribulla, T., Baludanský, D., Dubovský, P., Kudzej, I., Parimucha, Š., Siwak, M., & Vaňko, M. 2008, *MNRAS*, 390, 798

Vaz, L. P. R., Andersen, J., Casey, B. W., Clausen, J. V., Mathieu, R. D., & Heyer, I. 1998, *A&AS*, 130, 245

From Interacting Binaries to Exoplanets: Essential Modeling Tools
Proceedings IAU Symposium No. 282, 2011
Mercedes T. Richards & Ivan Hubeny, eds.
© International Astronomical Union 2012
doi:10.1017/S1743921311026895

Eccentricity of Selected Eclipsing Systems

W. Ogłoza[1], J. M. Kreiner[1], G. Stachowski[1], B. Zakrzewski[1], Z. Mikulášek[2], and M. Zejda[2]

[1]Mt. Suhora Observatory, Cracow Pedagogical University,
30-084 Cracow, ul Podchorążych 2, Poland
email: ogloza@up.krakow.pl

[2]Masaryk University, Brno, Czech Republic

Abstract. This paper presents the results of verification of known stars showing evidence of orbital eccentricity and apsidal motion.

Keywords. stars: binaries: eclipsing, ephemerides

1. Introduction

The study of orbital eccentricity and apsidal motion are very important for verification of theories of evolution of binary systems and of the internal structure of stars. One of the methods used to study long time effects is (O-C) analysis of times of minima.

This paper presents the preliminary results of the selection of GCVS eclipsing systems with phase-shifted secondary minima. At present, the Cracow database contains 194993 individual minima of 5931 stars. Examination of the (O-C) diagrams shows about 256 stars with a significant phase shift of the secondary minimum, interpreted as the effect of an eccentric orbit. The poster shows 98 selected systems (with 5800 times of minima) for which the primary and secondary minima were observed at least 4 times over well separated epochs (e.g. the diagram for V799 Cas). About 56% of those systems show the effect of rotation of the line of apsides (see the diagram for the well known system Y Cyg).

2. Results

The orbital period of the eccentric orbit binaries is in the range of ~ 1.2 to ~ 20 days (except OW Gem $P \approx 1200$d V883 Mon $P \approx 50$d, VW Peg $P \approx 21$d, and LV Her $P \approx 18$d), but for long-period systems the number of minima is small.

There are a few short-period variables which show an apparent phase shift of the secondary minimum (LV Vir $P = 0.4094$, TY Men $P = 0.4617$, V407 Peg $P = 0.6369$, LR Car $P = 0.8529$, GI Boo $P = 1.0335$, DU Boo $P = 1.0559$), however this could be the effect of an asymmetrically-shaped minimum.

For several systems (V947 Cyg, RW Lac, RU Mon, ζ Phe, V1094 Tau, AO Vel, DR Vul) the effect of an eccentric orbit on the (O-C) diagram is present together with other phenomena, such as constant period change or the influence of a third body.

3. Database

The full list of eccentric systems, pictures of O-C diagrams and up-to-date linear elements for primary and secondary minima will be continuously published at:
www.as.up.krakow.pl/ephem

W. Ogłoza *et al.*

Reference

Kreiner, J. M. 2004, *AcA*, 50, 247

From Interacting Binaries to Exoplanets: Essential Modeling Tools
Proceedings IAU Symposium No. 282, 2011   © International Astronomical Union 2012
Mercedes T. Richards & Ivan Hubeny, eds.   doi:10.1017/S1743921311026901

Reconstruction of an Accretion Disk Image in AU Mon from CoRoT Photometry

P. Mimica[1] and K. Pavlovski[2,3]

[1] Department of Astronomy and Astrophsycs, University of Valencia, 46 100 Burjassot, Spain

[2] Department of Physics, Faculty of Science, University of Zagreb, 10 000 Zagreb, Croatia

[3] Astrophysics Group, EPSAM, Keele University, Staffordshire, BG5 5JR, UK

Abstract. The long-period binary system AU Mon was photometrically observed on-board the CoRoT satellite in a continuous run of almost 60 days long which has covered almost 5 complete cycles. Unprecedented sub milimag precision of CoRoT photometry reveals all complexity of its light variations in this, still active mass-transfer binary system. We present images of an accretion disk reconstructed by eclipse mapping, and an optimization of intensity distribution along disk surface. Time resolution and accurate CoRoT photometric measurements allow precise location of spatial distribution of 'hot' spots on the disk, and tracing temporal changes in their activity. Clumpy disk structure is similar to those we detected early for another W Serpentis binary W Cru (Pavlovski, Burki & Mimica, 2006, A&A, 454, 855).

Keywords. accretion disks, (stars:) binaries: eclipsing, stars: individual (AU Mon)

---

1. Modelling light curves of the interacting binaries

One of the breakthrough outcome of UV spectroscopy on-board the IUE satellite was the discovery of emission lines of highly-ionized species in the spectra of the rather sparse group of long-period binary stars (Plavec & Koch 1978). The binaries in the sample were known for long for their active nature with large period changes, almost permanent Balmer emission features, peculiar light curves, etc. (Plavec 1980). It was Plavec who named this group after its prominent member *W Serpentis Binaries*.

It is now well-known that light curves of these 'Active Algols' cannot be solved with standard models. The first light curve synthesis model included an optically thick accretion disk surrounding a mass-gaining component; it has been successfuly applied in solving light changes in SX Cas by Pavlovski & Kříž (1985). The same simple disk model has been used in solving complex light variations in RX Cas (Andersen, Pavlovski & Piirola 1989). These authors also account for long-term cycles in light variations superimposed on the orbital light curve, and modeled it by changing the geometry of an accretion disk. Our further development of this disk model has been toward reconstruction of the disk image using eclipse mapping and optimization by genetic algorithm (Mimica & Pavlovski 2003, Pavlovski, Burki & Mimica 2006).

2. Disk image in AU Monocerotis

AU Mon (HD 50846, HIP 33237) is a long-period eclipsing and double-lined spectroscopic binary system with signatures common for W Serpentis interacting binaries. A comprehensive study was undertaken by Desmet *et al.* (2010) initiated by CoRoT space photometry, and complemented by ground-based high-resolution spectroscopy. Their analysis led to the determination of improved and consistent fundamental stellar and orbital properties for AU Mon. The light curve of AU Mon in Desmet *et al.* (2010) is

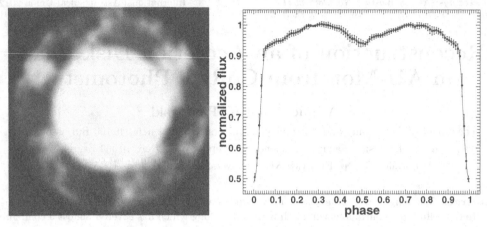

Figure 1. An accretion disk image reconstructed from an average binned light curve (left). The quality of fit of the phased light curve is shown on the right panel by solid line.

solved using a standard model, and does not account for superimposed fine changes in the light from the system. Djurašević *et al.* (2011) attempt to improve the fit of light curve by introducing a disk into the model. Since their disk model is homogenous, the fit was not satisfactory and only after the inclusion of hot spots they were able to improve it. Moreover, they modeled the light curve changes in a long-term cycle of about 416 days, by changes in disk geometry, just in the same way as was done for RX Cas by Andersen *et al.* (1989).

CoRoT photometry of AU Mon was secured in a continuous run of almost 60 days long and almost 5 complete orbital cycles were covered. Unprecedented sub milimag precision of CoRoT photometry reveals all complexity of its light variations in this, still active mass-transfer binary system. About 17 000 measurements were secured. Such quality and quantity of space-born photometry is ideal for an accretion disk image reconstruction along our model developed in Mimica & Pavlovski (2003), and applied for W Crucis in Pavlovski *et al.* (2006). Analysis is performed on binned light curves, and an improved release of our code which now is running on a computer cluster. A rather asymmetrical and clumpy (almost elliptical) disk is revealed. We are planning to publish a detailed paper elsewhere.

Acknowledgements

KP acknowledges receipt of a Leverhulme Visiting Professorship which enabled him to stay at Keele University where the work reported here was started. PM acknowledges the support from the European Research Council (grant CAMAP-259276). The calculations have been performed on the *Lluís Vives* cluster at the University of Valencia.

References

Andersen, J., Pavlovski, K., & Piirola, V. 1989, *A&A*, 215, 272
Desmet, M., *et al.*, 2010, *MNRAS*, 401, 418
Djurašević, G., Latković, O., Vince, I., & Cséki, A. 2011, *MNRAS*, 401, 418
Pavlovski, K. & Mimica, P. 2003, *ASP-CS*, 292, 405
Pavlovski, K. & Kříž, S. 1985, *Bulletin of the Astronomical Institutes of Czechoslovakia*, 35, 135
Pavlovski, K., Burki, G., & Mimica, P. 2006, *A&A*, 454, 855
Plavec, M. J. 1980, *in IAU Symp.*, 88, 251
Plavec, M. J. & Koch, R. H. 1978, *IBVS*, No. 1482

From Interacting Binaries to Exoplanets: Essential Modeling Tools
Proceedings IAU Symposium No. 282, 2011
Mercedes T. Richards & Ivan Hubeny, eds.
© International Astronomical Union 2012
doi:10.1017/S1743921311026913

Multiwavelength Modeling the SED of Strongly Interacting Binaries

Augustin Skopal

Astronomical Institute of the Slovak Academy of Sciences,
059 60 Tatranská Lomnica, Slovakia
email: skopal@ta3.sk

Abstract. The spectrum of strongly interacting binaries, as for example, high and low mass X-ray binaries, symbiotic (X-ray) binaries and/or classical and recurrent novae, consists of more components of radiation contributing from hard X-rays to radio wavelengths. To understand the basic physical processes responsible for the observed spectrum we have to disentangle the composite spectrum into its individual components, i.e. to determine their physical parameters. In this short contribution I demonstrate the method of modeling the multiwavelength SED on the example of the extragalactic super-soft X-ray source RX J0059.1-7505 (LIN 358).

Keywords. stars: fundamental parameters, X-rays: individual (RX J0059.1-7505)

1. Introduction

Strongly interacting binaries consist of an accreting compact object and a low-mass ($\leqslant 1 M_\odot$) main-sequence or slightly evolved late-type star. In special cases, the donor star can be an M-type giant. The mass transfer from the cool component to the compact one, represents principal interaction in these binaries. It can happen via the Roche lobe overflow in short-period binaries, e.g. cataclysmic variables, or via the wind in extended systems as, e.g., symbiotic stars. These objects are subject to occasional outbursts, during which a large amount of radiative and kinetic energy is liberated. A part of the kinetic energy of particles can be converted to radiation due to shocked gas that creates an extremely hot plasma. Its radiation can be detected at hard to soft X-rays. The hot star radiation from, e.g., a burning white dwarf, can be measured directly at the supersoft X-rays and the far-UV. A fraction of its radiation is re-processed via the ionization/recombination acts into the nebular radiation, which signatures are best indicated at the near-UV and the radio. The contribution from the giant dominates usually the optical/near-IR. In some cases, the post-outburst circumstellar material or that produced by a mira-type giant can condense in dust particles, which can re-process the radiation from the binary components into a dust emission, seen in the infrared domain. As a result, the observed spectrum of strongly interacting binaries is composed of different components of radiation. Figure 1 shows an example of the above described components. Their extraction from the composite spectrum can aid us in understanding physical processes giving rise the observed result. Here, I present the observed and model SED for the extragalactic super-soft X-ray sources, RX J0059.1-7505. The method of disentangling the composite spectrum was described by Skopal *et al.* (2009).

2. Multiwavelength model SED of RX J0059.1-7505

This star is classified as a symbiotic X-ray binary. Figure 2 shows its observed and model SED from the supersoft X-rays to the near-IR. The X-ray fluxes were reconstructed

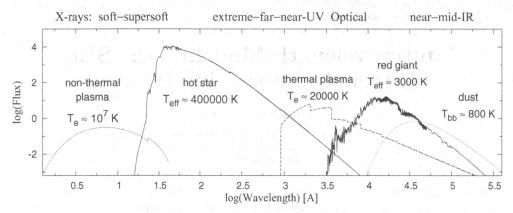

Figure 1. Scheme of main components of radiation in the continuum that can contribute to the X-ray to mid-IR spectrum of strongly interacting binaries.

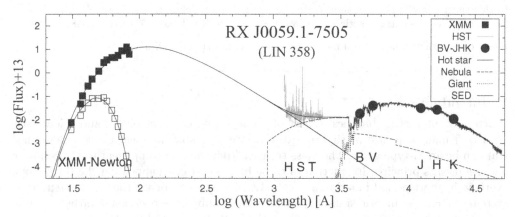

Figure 2. A comparison of the measured (in keys) and modeled (solid heavy line) SED of the symbiotic X-ray binary in the Small Magellanic Cloud, RX J0059.1-7505 (LIN 358). Open/filled squares are the absorbed/unabsorbed X-ray fluxes. Fluxes are in units of $\mathrm{erg\,cm^{-2}\,s^{-1}\,\AA^{-1}}$.

from Fig. 4 of Kahabka & Haberl (2006). The *HST* ultraviolet spectra were obtained from the satellite archive, and the *BVJHK* photometric measurements were published by Mürset *et al.* (1996). The preliminary model SED corresponds to the hot star parameters, $T_{\mathrm{h}} = 250\,000$ K, $R_{\mathrm{h}} = 0.09\,\mathrm{R}_\odot$ and $L_{\mathrm{h}} = 1.1 \times 10^{38}$ erg s^{-1}, which radiation is attenuated with $N_{\mathrm{H}} = 6.1 \times 10^{20}$ cm^{-2}. The nebula has $T_{\mathrm{e}} = 18\,000$ K and its amount corresponds to the emission measure, $EM = 2.4 \times 10^{60}$ cm^{-3}. The giant in LIN 358 can be classified as a K5 Ib supergiant, because its $T_{\mathrm{eff}} = 4\,000 \pm 200$ K, $R_{\mathrm{g}} = 178\,(d/60\,\mathrm{kpc})\,\mathrm{R}_\odot$ and $L_{\mathrm{g}} \sim 7\,300\,(d/60\,\mathrm{kpc})^2\,\mathrm{L}_\odot$, as given by the model SED.

Acknowledgement

This research was supported by a grant of the SAS, VEGA No. 2/0038/10.

References

Kahabka, P. & Haberl, F. 2006, *A&A*, 452, 431
Mürset, U., Schild, H., & Vogel, M. 1996, *A&A*, 307, 516
Skopal, A., Sekeráš, M., Gonzĺez-Riestra, R., & Viotti, R. F. 2009, *A&A*, 507, 1531

From Interacting Binaries to Exoplanets: Essential Modeling Tools
Proceedings IAU Symposium No. 282, 2011
Mercedes T. Richards & Ivan Hubeny, eds.
© International Astronomical Union 2012
doi:10.1017/S1743921311026925

Spectroscopic Study of the Early-Type Binary HX Vel A

Burcu Özkardeş[1,2], Derya Sürgit[1,2], Ahmet Erdem[1,2], Edwin Budding[3], Faruk Soydugan[1,2], and Osman Demircan[1,2]

[1] Astrophysics Research Center and Observatory, Çanakkale Onsekiz Mart University,
Terzioğlu Kampüsü, TR-17020, Çanakkale, Turkey
[2] Department of Physics, Faculty of Arts and Sciences, Çanakkale Onsekiz Mart University,
Terzioğlu Kampüsü, TR-17020, Çanakkale, Turkey
email: burcu@comu.edu.tr, dsurgit@comu.edu.tr, aerdem@comu.edu.tr,
fsoydugan@comu.edu.tr, demircan@comu.edu.tr

[3] Carter National Observatory, PO Box 2909, Wellington, New Zealand
email: budding@xtra.co.nz

Abstract. This paper presents high resolution spectroscopy of the HX Vel (IDS 08390-4744 AB) multiple system. New spectroscopic observations of the system were made at Mt. John University Observatory in 2007 and 2008. Radial velocities of both components of HX Vel A were measured using gaussian fitting. The spectroscopic mass ratio of the close binary was determined as 0.599±0.052, according to a Keplerian orbital solution. The resulting orbital elements are $a_1\sin i=0.0098\pm0.0003$ AU, $a_2\sin i=0.0164\pm0.0003$ AU, $M_1\sin^3 i=1.19\pm0.07$ $M_\odot$ and $M_2\sin^3 i=0.71\pm0.04$ $M_\odot$.

Keywords. methods: data analysis, techniques: spectroscopic, stars: early-type, (stars:) binaries (including multiple): close, stars: individual (HX Vel A)

---

1. Spectroscopic observations and reductions

High-resolution spectra of HX Vel were taken at the Mt John University Observatory (MJUO, New Zealand) in September and October 2007 and in December 2008, using the HERCULES (High Efficiency and Resolution Canterbury University Large Echelle Spectrograph) and a 4k×4k Spectral Instruments 600 series (SI600s) CCD camera attached to the 1-m McLellan telescope. Our observations were made choosing a slitless 100 μm fiber cable with the second option of R=40000, considered to be better adjusted to the mean seeing value (3.5 arcsec) at MJUO as given by Hearnshaw *et al.* (2002). A total of 20 spectra have been collected during six observing nights in 2008, while 7 high-resolution spectra of the binary were obtained during four observing nights between September and October 2007.

For all observations, a thorium-argon lamp spectrum for wavelength calibration was taken before and after each stellar exposure. White lamp spectra for flat-fielding were also taken every night. All spectra obtained were reduced using the HRSP (Hercules Reduction Software Package, HRSP: Skuljan & Wright 2007) software.

2. Radial velocities and orbital solution

Radial velocities (RVs) of both components of HX Vel A binary were measured using gaussian fittings to the selected spectral lines, with the aid of the 'splot' task of IRAF. According to the B1.5V spectral type of the binary (SIMBAD), there should be strong neutral He and hydrogen lines of the Balmer series in its observed spectra. We found four

Table 1. Spectral orders and stellar lines used in RVs measurements of HX Vel A.

Order No	Wavelength Interval (Å)	Dominant Spectral Lines
85	6647-6715	He I (6678.151 Å)
97	5840-5907	He I (5875.989 Å)
116	4884-4939	He I (4921.931 Å)
127	4453-4508	He I (4471.480 Å)

Table 2. Spectroscopic orbital parameters of HX Vel A.

Parameter	Value	Parameter	Value
P (days)	1.12447 (fixed)	V_γ (km/s)	26±3
T_0 (HJD+2448501)	0.0833±0.0046	$M_1 \sin^3 i$ (M$_\odot$)	1.19±0.07
K_1 (km/s)	95±3	$M_2 \sin^3 i$ (M$_\odot$)	0.71±0.04
K_2 (km/s)	158±3	$a_1 \sin i$ (AU)	0.0098±0.0003
q	0.599±0.052	$a_2 \sin i$ (AU)	0.0164±0.0003

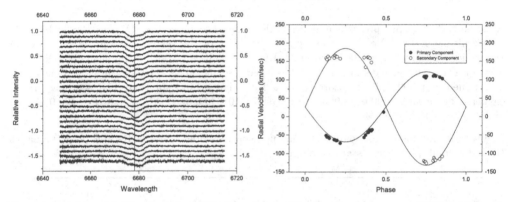

Figure 1. Left panel: The He I lines (échelle order no 85) of both components of HX Vel A were used for RVs measurements. The vertical line represents the laboratory wavelength of He I line (6678.151 Å). Right panel: Best theoretical fit to the radial velocity curves of HX Vel A. Solid line fitted to the radial velocities represents the theoretical fit for the pure Keplerian orbit.

spectral regions (containing HeI lines) where the secondary star's spectral lines could be detected. The information about the spectral regions used is given in Table 1.

The ELEMDR77 program, developed by T. Pribulla, was used to obtain orbital parameters from the radial velocity data presented here. The best fitting orbital elements are given in Table 2; the best fits to the composite spectra (left panel; showing échelle order 85, including He I line, as example) and RVs (right panel) are presented in Fig. 1.

In our forthcoming work, the physical parameters of HX Vel A will be examined, together with available light curves of the system.

References

Hearnshaw, J. B., Barnes, S. I., Kershaw, G. M., Frost, N., Graham, G., Ritchie, R., & Nankivell, G. R. 2002, *ExA*, 13, 59

Skuljan, J. & Wright, D. 2007, *HRSP Hercules Reduction Software Package*, vers. 3, Univ. Canterbury, New Zealand

From Interacting Binaries to Exoplanets: Essential Modeling Tools
Proceedings IAU Symposium No. 282, 2011   © International Astronomical Union 2012
Mercedes T. Richards & Ivan Hubeny, eds.   doi:10.1017/S1743921311026937

Planets in Binary Systems

Antonio Pilello

Institut für Astrophysik, Georg-August-Universität Göttingen,
Friedrich-Hund-Platz 1, D-37077 Göttingen, Germany
E-Mail: pilello@astro.physik.uni-goettingen.de

Abstract. In close eclipsing binary systems, measurements of the eclipse timing variations (ETV), obtained by means of accurate light curves, may be used to find circumbinary additional objects. The presence of these objects causes the motion of the eclipsing binary with respect to the centre of mass of the entire system and it results in advances or delays in the times of eclipses due to the light time effect. The most important issue of this project is to inspect the potential of detecting low mass substellar companions to close eclipsing binaries through the timing method. For this purpose, we use the public data from Kepler and CoRoT spacecrafts, collecting the light curves for a selected sample and analyzing the observed minus calculated (O-C) times of the eclipses in the search for ETVs and characterizing them. A large amplitude of the O-C ETVs can be explained in some cases by the presence of a third body in the system.

Keywords. binaries: eclipsing, methods: data analysis

---

1. Introduction

Kepler and CoRoT light curves are mainly analyzed in the search for planetary transits, but many of eclipsing binary stars (EBs) are also discovered. Sometimes EBs show timing variations (TVs), that are changes in the period of their eclipses. Timing variations of eclipses can provide a comprehension of the physical structure of the system.

2. Data Analysis

We use the public data from Kepler and CoRoT space missions, collecting the light curves for a selected sample and analyzing the observed minus calculated (O-C) times of the eclipses in the search for ETVs and characterizing them. The total number of identified eclipsing binaries in the Kepler field of view is 2165, 1.4% of the Kepler target stars (Prša *et al.* 2011, and Slawson *et al.* 2011, after the Q0, Q1 and Q2 data releases). The EBs with a probability of variability more than 80% in the N2 CoRoT public data are instead 1707. Several IDL routines are applied to remove the thermal peaks and to flatten the Kepler light curves (four examples are shown in Fig. 1).

The large amplitudes of the O-C eclipse time variations seen in some close binary systems can be explained by the presence of a third body in the system. We present here an interesting case of EB from Kepler public data with O−C variations evident in the Q0, Q1 and Q2 data. In Fig. 2 the normalized light curve and the O−C diagram of KID6543674 with a period of 2.390105 days (Slawson *et al.* 2011) is shown. This object is a short period EB with deep eclipses from two nearly equal components seen close to edge-on. This system has eclipse timing variations, that may arise from the light time effect as the third-body orbits that binary (Slawson *et al.* 2011).

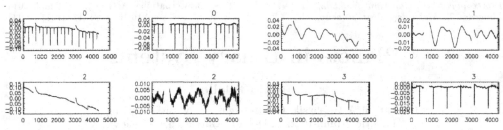

Figure 1. Flatted light curves without thermal effects. The plots with the same index represent the normalized (on the left) and the corrected (on the right) light curve for the corresponding Kepler object.

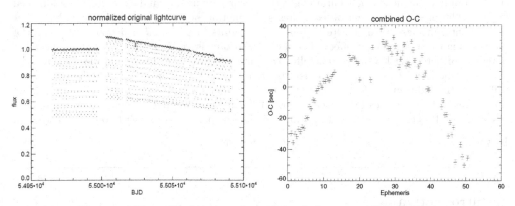

Figure 2. Normalized light curve and O−C diagram of KID6543674.

3. Conclusions

The presence of a third body in the system is one of the possible explanations for TVs in eclipsing systems (light time effect). On the other hand, they can also result from pulsations of the components, magnetic activity, precession of the orbits and mass transfer between the components. The corrected light curves of the Kepler EBs and the new O−C diagrams will be presented soon, in order to confirm that the high photometric precision and duty cycle of space missions allow us to obtain the best results in the detection of low mass substellar objects via ETVs (Schwarz *et al.* 2011).

References
Prša, A., Batalha, N., Slawson, R. W. *et al.*, 2011, *AJ*, 141, 83
Schwarz, R., Haghighipour, N., Eggl, S., Pilat-Lohinger, E., & Funk, B. 2011, *MNRAS*, 414, 2763
Slawson, R. W., Prša, A., Welsh, W. F., *et al.*, 2011, *AJ*, 142, 160

From Interacting Binaries to Exoplanets: Essential Modeling Tools
Proceedings IAU Symposium No. 282, 2011   © International Astronomical Union 2012
Mercedes T. Richards & Ivan Hubeny, eds.   doi:10.1017/S1743921311026949

Northern Binaries in the Evrena Project

V. Bakış[1], H. Hensberge[2], M. Zejda[3], P. de Cat[2], F. Yılmaz[4], S. Bloemen[5], P. Svoboda[6] and O. Demircan[4]

[1] Akdeniz University, Space Sciences & Technologies Dept., Campus, 07058, Antalya, Turkey
email: volkanbakis@akdeniz.edu.tr

[2] Royal Observatory of Belgium, Brussels, Belgium

[3] Masaryk University, Dept. of Theoretical Physics and Astrophysics, Brno, Czech Republic

[4] Çanakkale Onsekiz Mart University, Physics Department and Ulupınar Observatory,
Terzioğlu Campus, TR-17020, Çanakkale, Turkey

[5] Instituut voor Sterrenkunde, K.U.Leuven, Celestijnenlaan 200D, B-3001, Leuwen, Belgium

[6] Vypustky 5, Brno, Czech Republic

Abstract. In the framework of the EVRENA project, high-resolution spectra of northern eclipsing close binaries in stellar groups are obtained with the HERMES Echelle spectrograph at the Mercator telescope (Roque de los Muchachos Observatory). This contribution gives the first results on DV Camelopardalis.

Keywords. (stars:) binaries: eclipsing, (stars:) binaries (including multiple): close

---

1. DV Camelopardalis

DV Camelopardalis (HR 1719, HD 34233, HIP 24836, sp.type B5V) is a Cas-Tau OB association member with a membership probability of 99 per cent (de Zeeuw *et al.* 1999), and is known as a spectroscopic binary since early in the 20th century (Plaskett *et al.* 1921). The radial-velocity variability was confirmed by Petrie (1958), who mentioned an amplitude of 71 km s^{-1} from measurements at 12 epochs, and later by Fehrenbach *et al.* (1997). However, radial velocities were measured as if it was a single-lined binary, while our high-resolution spectroscopy presented here reveals three components (Fig. 1).

The Hipparcos satellite identified DV Cam as an eclipsing binary with an eclipse depth of roughly 0.2 magnitudes. The catalogue gives an orbital period of 1.5295 days. The light curve displays one eclipse. The scatter of the data within eclipse appears larger than expected from the estimated uncertainties in the photometry.

The Hipparcos photometric data set reveals 112 observations made at 21 epochs, of which 4 contain eclipse data. By 'epochs' we mean a series of observations with interruptions not exceeding significantly 0.1 day. The longest 'continuous' observation covers 1.95 days (24 data points) and contains 4 consecutive observations within eclipse, two at each side of mid-eclipse. This set alone already defines the duration of one of the eclipses as 0.15 to 0.2 days and suggests a mid-eclipse time at HJD~2448751.926±0.002. The other 3 observations in eclipse are single observations (the second one, 20 minutes apart, was obviously rejected in the three cases), at the start or the end of a short series of observations. With only 4 epochs in eclipse, the question of whether the orbital period can be determined uniquely is relevant. Therefore, we made a search for possible periodicities in the interval 1.95 to 40 days starting with an analytical computation revealing all periods that (1) put the in-eclipse observations out of the phase intervals covered by series of observations at one epoch; and (2) group the in-eclipse observations in at most two phase intervals without any out-of-eclipse observations between them.

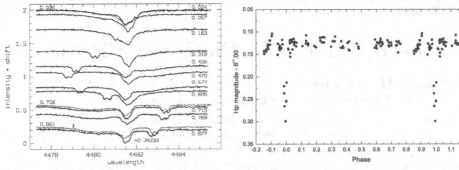

Figure 1. HERMES spectra of DV Cam. Thick lines: spectra taken 2011 Feb.17-27, thin lines: 2011 Jan. 9-11.

Figure 2. DV Cam light curve phased with ephemeris Pri HJD=2448751.926 + 6.67840 × E.

The lower search limit is set by the longest series of observations, that contains only one eclipse. It must be shorter than the orbital period, as a nearly circular shorter period would imply two eclipses. Applying these criteria, 13 possible periodicities between 2.86 and 20.05 days were selected. Inspection of the phase diagrams identifies 6 candidate periods with adequately low scatter of the observations in eclipse(s): 3.0584 days (twice the Hipparcos period), 3.3392 days, and its multiples of 6.6784, 10.0235, 13.356 and 20.047 days. We conclude that the Hipparcos data alone do not determine the orbital period in a unique way.

High-resolution spectroscopy of DV Cam was obtained in 2011 with the HERMES Echelle spectrograph (Raskin *et al.* 2011) attached at the 1.2-m Mercator telescope situated at the Roque de Los Muchachos Observatory. The spectra cover the wavelength range from close to the Balmer jump up to 9000 Å with a resolution of 85000. The high-resolution spectra show clearly a contribution of three B-type stars (Fig. 1): a (quasi)-stationary component, and a close-binary consisting of a sharp-lined and a broad-lined component with opposite Doppler shifts. The velocity range of the less massive sharp-lined stars sufficient to isolate its contribution at favourable orbital phases from those of the other components. The strength of the prominent lines of the (quasi)-stationary component suggests it is intermediate between the more massive component of the close binary and the cooler one.

The spectroscopy uniquely defines the true orbital period as $P_{orb} = 6.6784 \pm 0.0010$ days, with the uncertainty determined from the longer Hipparcos photometric time baseline (Fig. 2). The orbit is significantly eccentric, e ~ 0.48.

Acknowledgements

This work was supported by the Scientific Research Council of Turkey under project No. 109T449. Hermes is funded by the Fund for Scientific Research of Flanders (FWO) grant G.0472.04, the Research Council of K. U. Leuven grant GST-B4443, the Fonds National de la Recherche Scientifique contracts IISN4.4506.05 and FRFC 2.4533.09, and the Royal Observatory of Belgium Lotto(2004) fund.

References

Fehrenbach, C., Duflot, M., Mannone, C., Burnage, R., & Genty, V. 1997, *A&AS*, 124, 255
Petrie, R. M. 1958, *MNRAS*, 118, 80
Plaskett, J. S., Harper, W. E., Young, R. K., & Plaskett, H. H. 1921, *Publ. DAO*, 1, 287
Raskin, G., Van Winckel, H., Hensberge, H., *et al.*, 2011, *A&A*, 526, 69
de Zeeuw, P. T., Hoogerwerf, R., de Bruijne, J. H. J. *et al.*, 1921, *AJ*, 117, 354

From Interacting Binaries to Exoplanets: Essential Modeling Tools
Proceedings IAU Symposium No. 282, 2011   © International Astronomical Union 2012
Mercedes T. Richards & Ivan Hubeny, eds.   doi:10.1017/S1743921311026950

Statistical Investigation of Physical and Geometrical Parameters in Close Binaries using the ASAS Database

J. Nedoroščik[1], M. Vaňko[2] and Š. Parimucha[1]

[1] Institute of Physics, University of P.J. Šafárik, Košice, Slovakia
email: prokyon87@gmail.com

[2] Astronomical Institute of the Slovak Academy of Sciences, Tatranská Lomnica, Slovakia
email: vanko@ta3.sk

Abstract. The main goal of this work was to find dependencies between Fourier coefficients, which were developed by light curve fitting with Fourier polynomials. The light curves were acquired from the ASAS database (All Sky Automated Survey). In this statistical research it was necessary to sort and modify these data, because light curves of eclipsing binaries are just part of a bigger database, which contains the light curves of pulsating variable stars, novas etc. It was required to phase and normalize all of our light curves, that it could be possible to use a program to fit light curves with Fourier coefficients. Thereafter, we were looking for relations between Fourier coefficients.

Keywords. photometry, CCD

1. Introduction

The ASAS† database contains about 11 240 eclipsing binaries or candidates for eclipsing binaries. Light curves of these systems have been phased and normalized, that we can use program, which expands light curves to polynomial according to Fourier coefficients. In order to find Fourier coefficients we have used the following expression:

$$S(x) = a_0/2 + \sum_{n=1}^{N}[a_n \cos(nx) + b_n \sin(nx)] \tag{1.1}$$

For all light curves we obtained file with eleven Fourier coefficients.

2. Results

Some of these Fourier coefficients represent directly physical or geometrical parameters Hambalek (2006). We were interested in the relation between a_2 and a_4. Thereafter, we were looking for boundary which determines detached systems from semi-detached and contact systems for synthetic light curves (see Fig. 1, right) and light curves for real systems (see Fig. 1, left). Selam (2004) published his equation for this boundary as follows:

$$a_4 = a_2(0.125 - a_2) \tag{2.1}$$

† http://www.astrouw.edu.pl/

73

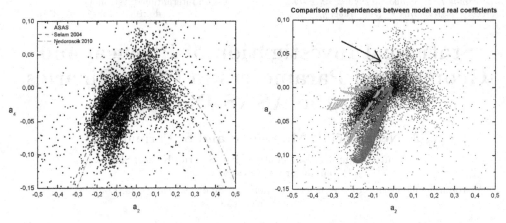

Figure 1. Left Boundary determination between EA types and EB, EW types. The plot represents real systems from ASAS database. **Right** Comparison of fourier coefficients relation for real (points) and model (lines) light curves.

This equation was not correct for our model light curves we needed to derive own boundary:

$$a_4 = -1.06011a_2^2 + 0.21635a_2 + 0.01229 \qquad (2.2)$$

By these boundaries we can determine types of eclipsing binaries. When we compare dependences of these two Fourier coefficients for model and real light curves (see Fig. 1), we can estimate some physical (e.g. mass ratio) or geometrical (e.g. inclination) parameters. It is necessary to model light curves with high precision to obtain correct estimation of physical and geometrical parameters of binary systems. Interesting is the area marked by arrow (Fig. 1, right). We do not know, what kind of binaries represents this area. These stars could be close binaries with a lot of spots, pulsating stars or it could be systems with exoplanets.

Acknowledgements

This work has been supported by project VEGA 2/0094/11

References

Hambalek, L. 2006, *Diploma thesis*
Selam, S. O. 2004, *A&A*, 416, 1097

From Interacting Binaries to Exoplanets: Essential Modeling Tools
Proceedings IAU Symposium No. 282, 2011   © International Astronomical Union 2012
Mercedes T. Richards & Ivan Hubeny, eds.   doi:10.1017/S1743921311026962

Some Developments of the Weak Stellar Magnetic Field Determination Method for the Example of Cygnus X-1

N. G. Bochkarev[1] and E. A. Karitskaya[2]

[1] Sternberg Astron. Inst., Universitetskij prosp. 13, 119991 Moscow, Russia
email: boch@sai.msu.ru

[2] Astronomical Institute of RAS, Pyatnitskaya str. 48, 119017 Moscow, Russia
email: karitsk@sai.msu.ru

Abstract. Some developments of measurements of the weak stellar magnetic fields by the least square technique applied to spectropolarimetric data are proposed and used for the X-ray binary Cyg X-1 = HDE 226868 (the optical counterpart is an O 9.7 supergiant).

Keywords. Stars: magnetic fields, techniques: polarimetric, techniques: spectroscopic, methods: statistical, stars: early-type, stars: individual (Cyg X-1 = HDE 226868 = V1357 Cyg), (stars:) supergiants, (ISM:) extinction, X-rays: binaries, black hole physics

---

In contrast to the stars with strong magnetic fields (mainly A and late B types), luminous O-stars have usually weaker field and significant interstellar / circumstellar linear polarization (up to $\sim 10\%$).

Any spectropolarimeter has cross-talk between linear and circular polarization within the instrument. It creates a spurious circularly polarized wavelength-dependent continual component of radiation for stars with linear polarization.

As a result, more and more often targets for magnetic field measurements have spectra of Stokes parameter V (measuring circular polarization) and spectra of ratios V/I (I is Stokes parameter for total intensity) containing wavelength-dependent continual components $C_V(\lambda)$ and $C_{V/I}(\lambda)$, where λ is the wavelength.

Bochkarev & Karitskaya (2011) proposed some developments of measurements of the weak stellar magnetic fields by the least square technique applied to spectropolarimetric data and used them for the X-ray binary Cyg X-1 = HDE 226868 (the optical counterpart is an O 9.7 supergiant).

The parameters of the object are: magnitude $m_V = 9^m$; > 95% of optical radiation is produced by the O9.7 Iab star; interstellar extinction $A_V = 3.36^m$; interstellar/circumstellar linear polarization $\sim 5\%$; stellar wind $M_{dot} \sim (2-3)*10^{-6}\ M_{sun}$/yr; chemical peculiarities (mainly He, N, Si excess); and rotation velocity $V \sin i = 95$ km/s.

Our observations were made at the Very Large Telescope (VLT) 8.2 m (Mount Paranal, Chile) in the spectropolarimetric mode of the FORS1 spectrograph with resolution $R = 4000$ in the range $3680 - 5129$Å, with signal-to-noise ratio $S/N = 1500 - 3500$ (for I) from June 18 to July 9, 2007 and from July 14 to July 30, 2008 (Cyg X-1 in hard X-ray state). A total of 13 spectra of intensity I and circular polarization V were obtained (Karitskaya *et al.* 2009, Karitskaya *et al.* 2010).

The mean longitudinal magnetic field $\langle B_z \rangle$ was determined by statistical processing of the $V(\lambda)$ and $I(\lambda)$ spectra using the equation: $\frac{V}{I} = -C_Z\, g_{\text{eff}}\, \lambda^2\, \frac{1}{I}\, \frac{dI}{d\lambda}\, \langle B_z \rangle + \frac{V_0}{I_0}$ (e.g., Landstreet 1982), where g_{eff} is the effective Lande factor, $C_Z = \frac{e}{4\pi m_e c^2} = 4.67*10^{-13}$Å^{-1}G^{-1},

$\frac{V_0}{I_0} = const$. The least squares method is used for the $\langle B_z \rangle$ calculation (e.g., Hubrig *et al.* 2004).

The sources of noise in the $\langle B_z \rangle$ measurements, which should be removed from the I- and V/I-spectra, are: 1) interstellar lines and narrow diffuse interstellar bands (DIBs); 2) defects (including residual cosmic ray tracks not removed by the standard observation processing); 3) He II 4686 Å line with profile including the accretion-structure emission component; and 4) the emission components of the lines showing the P Cyg effect. In addition, we removed some λ intervals containing noise only. We found no pollution by telluric lines in our spectra.

The slope value $S = \mathrm{d}C_{V/I}/\mathrm{d}\lambda \sim 10^{-6} \text{Å}^{-1}$ varied irregularly from night to night. The most probable reason for the V/I-spectra sloping is the cross-talk between linear and circular polarization within the FORS1 analyzing equipment.

The application of the least squares method without correcting for the V/I trend results in distorted or even spurious $\langle B_z \rangle$ values and in a distorted accuracy of these values. That happens for at least 2 reasons: 1) strong violation of Gauss statistics by the residuals; 2) appearance of $\langle B_z \rangle$ spurious component $\propto (\Delta\lambda_D/\lambda)^2 * \mathrm{d}C_{V/I}/\mathrm{d}\lambda$. Here $\Delta\lambda_D$ is the spectral line width.

For our Cyg X-1 FORS1 observations, spurious $\langle B_z \rangle$ from single spectral line and sloped V/I continuum without any Zeeman feature is several Gauss. To avoid any influence of the V/I-continuum slope on $\langle B_z \rangle$ measurements, we subtracted the linear trends from V/I spectra. In the case of our Cyg X-1 observations, the uncorrected slopes of the V/I-spectra create $\langle B_z \rangle$ shifts from -20 to -84 G (Bochkarev & Karitskaya 2011).

We normalized the I-spectra using a pseudo-continuum. The wavelength dependence of the I-continuum $C_I(\lambda)$ is produced by: the source energy distribution, interstellar reddening, broad DIBs, atmospheric extinction, and the sensitivity of the detector. The I-spectrum slopes reach $|\mathrm{d}(\log(I(\lambda))/\mathrm{d}(\log(\lambda))| \sim 20$. The removal of the slope gives a $\langle B_z \rangle$ correction up to ~ 20 G. It is usually less than the statistical errors $\sigma(\langle B_z \rangle)$, which are $\geqslant 20 - 30$ G.

The value of the mean longitudinal magnetic field in the optical component (O 9.7 Iab supergiant) changes regularly with the orbital phase, and reaches a maximum of 130 G ($\sigma \approx 20$ G) (Karitskaya *et al.* 2009, 2010).

The measurements based on the Zeeman effect were carried out over all the observed supergiant photosphere absorption spectral lines. Similar measurements over the emission line He II λ4686 Å yielded a value of several hundred Gauss with a smaller significance level. The emission component of this line originates in the outer parts of the accretion structure. So we measure $\langle B_z \rangle$ in this region (Karitskaya *et al.* 2009, 2010).

We got $\langle B_z \rangle \sim 100$ G for the star's photosphere. The gas stream carries the field on to the accretion structure; the gas is compressed by interaction with the structure's outer rim. Gas density is increased by a factor of 6-10 (we obtained several hundred Gauss). According to the Shakura – Sunyaev magnetized accretion disc model at $3R_g$: $B \sim 10^9$ G. Taking into account radiative pressure predominance inside $\sim 10 - 20R_g$, we get: $B(3R_g) \sim (2-3)10^8$ G. Such magnetic fields can be responsible for the observed X-ray flickering.

This study was supported by RFBR grant 09-02-01136.

References

Bochkarev, N. G. & Karitskaya, E. A. 2011, in D. O. Kudryavtsev, & I. I. Romanyuk (eds.), *Magnetic Stars* (N.Arkhyz: SAO RAS publ.), p. 199

Hubrig, S., Szeifert, T., Schöller, M. *et al.*, 2004, *A&A*, 415, 685

Karitskaya, E. A., Bochkarev, N. G., Hubrig, S. *et al.*, 2009, *astro-ph* 0908.2719

Karitskaya, E. A., Bochkarev, N. G., Hubrig, S. *et al.*, 2010, *Inf. Bull. Variable Stars*, 5950, 1

Landstreet J. D. 1982, *ApJ*, 258, 639

From Interacting Binaries to Exoplanets: Essential Modeling Tools
Proceedings IAU Symposium No. 282, 2011
Mercedes T. Richards & Ivan Hubeny, eds.

© International Astronomical Union 2012
doi:10.1017/S1743921311026974

KIC 4544587: An Eccentric, Short Period Binary with δ Scuti Pulsations and Tidally Excited Modes

Kelly Hambleton[1], Don Kurtz[1], Andrej Prša[2], Steven Bloemen[3], and John Southworth[4]

[1] Jeremiah Horrocks Institute, University of Central Lancashire, Preston, PR1 2HE
email: kmh@uclan.ac.uk

[2] Villanova University, Department of Astronomy and Astrophysics, 800 East Lancaster Avenue, Villanova, PA 19085 [3] Instituut voor Sterrenkunde, Katholieke Universiteit Leuven, Celestijnenlaan 200D, B-3001 Leuven, Belgium [4] Astophysics Group, Keele University, Newcastle-under-Lyme, ST5 5BG, UK

Abstract. KIC 4544587 is an eclipsing binary star with clear signs of apsidal motion and indications of tidal resonance. The primary component is early A-type δ Scuti variable, with a temperature of 8270 ± 250 K, whilst the secondary component is an early G-type main sequence star with a temperature of 6500 ± 310 K. The orbital period of this system is $2.18911(1)$ d, with the light curve demonstrating a hump after secondary minimum due to distortion and reflection. The frequency spectrum of the residual data (the original data with the binary characteristics removed) contains both pressure (p) and gravity (g) modes. Eight of the g modes are precise multiples of the orbital frequency, to an accuracy greater than 3σ. This is a signature of resonant excitation.

1. Introduction

In a binary system, where the components are in relatively close proximity to one another, the gravitational forces between the two components can induce tidal interactions. Moreover, if the binary is in an eccentric orbit, the tidal distortions become misaligned with respect to their instantaneous equipotential shapes (Hut 1980). This generates a torque between the two components which causes an exchange of orbital and rotational angular momentum, and thus causes the stellar rotational energy to dissipate.

In an eccentric binary system, the gravitational interactions between the two components vary as a function of phase. In some cases such interactions can cause one or both stellar components to oscillate as a result of the natural free eigenfrequencies of the star resonating with the dynamic tides (Aerts & Hamanec 2004). The signature of these interactions are frequencies that are multiples (harmonics) of the orbital frequency.

2. Overview

KIC 4544587 is an eccentric, $e = 0.28375(5)$, short-period binary system. The primary component is an early A star that is within the δ Scuti instability strip and the secondary component is an early G star. This object was selected as a likely candidate for tidally enhanced pulsations due to its δ Scuti component and the close proximity of the stars ($\sim 8\,R_\odot$) at periastron. The system also has interesting orbital characteristics which include periastron brightening in the *Kepler* photometric light curve after secondary minimum; a feature that is indicative of an eccentric binary with its components in close proximity to each other (Maceroni *et al.* 2009).

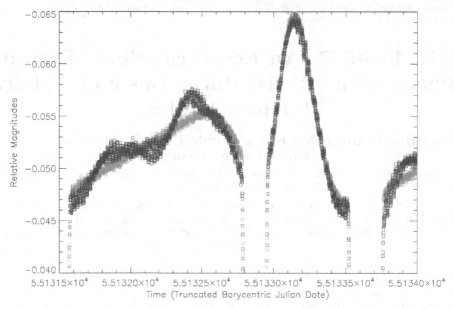

Figure 1. A comparison of the light curve with all the pulsation frequencies removed (green) and with all the frequencies except the harmonics of the orbital period removed (blue).

The *Kepler* photometric light curve that we studied consists of 46 115 data points of both long (29.4 min) and short (58.8 s) cadence data. The orbital period of the binary system was found using the Phase Dispersion Minimization technique (Stellingwerf 1978) on the Quarter 3.2 data and was determined to be 2.1891142(5) d.

Subsequently, the modelling was carried out on the short cadence, phase binned data of Quarter 3.2 (which spans 30.04 d) using the Wilson-Devinney code (Wilson & Devinney 1971). Frequency analysis was performed on the residuals (observed minus computed) by applying a Fourier transform combined with a least squares fit to the data. The identified frequencies were then pre-whitened from the original light curve, which was then remodelled. This enabled the determination of the binary characteristics without interference from the pulsations. This process was then repeated for three iterations.

Forty nine frequencies were identified in the residual data excluding the data taken during eclipse phases. These frequencies consist of 13 modes within the g mode regime ($\nu < 5\,\mathrm{d}^{-1}$) and 37 within the p mode regime ($30\,\mathrm{d}^{-1} < \nu < 50\,\mathrm{d}^{-1}$). Eight of the g mode frequencies were found to be multiples of the orbital frequency to an accuracy of 3 sigma or greater. In Fig. 1 the blue (darker) light curve depicts the original light curve with all the frequencies except the orbital harmonics removed and the green (lighter) light curve depicts the original light curve with all the identified frequencies removed. A plausible explanation for the remnant periodicity in the light curve is the resonant excitation of g modes due to tidal interactions.

References

Abt, H. A. 2009, *ApJ*, 138 28
Aerts, C. & Harmanec, P. 2004, arXiv:astro-ph/0510344, 318 325
Hut, P. 1980, *A&A*, 92 167
Maceroni, C., Montalbán, J., Michel, E., & Harmanec, P. 2009, & Prša, A. *A&A*, 508 1375
Wilson, R. E. & Devinney, E. J. 1971, *ApJ*, 166 605

From Interacting Binaries to Exoplanets: Essential Modeling Tools
Proceedings IAU Symposium No. 282, 2011   © International Astronomical Union 2012
Mercedes T. Richards & Ivan Hubeny, eds.   doi:10.1017/S1743921311026986

Long-Term Variability and Outburst Activity of FS Aurigae: Further Evidence for a Third Body in the System

V. Neustroev[1], G. Sjoberg[2], G. Tovmassian[3], S. Zharikov[3], T. Arranz Heras[4], P. B. Lake[2], D. Lane[5,6], G. Lubcke[2], and A. A. Henden[2]

[1] Astronomy Division, Department of Physics, P.O. Box 3000, 90014 University of Oulu, Finland. email: `vitaly@neustroev.net`
[2] American Assoc. of Variable Star Observers, 49 Bay State Road, Cambridge, MA 02138, USA
[3] Instituto de Astronomia, UNAM, Apdo. Postal 877, Ensenada, Baja California, 22800 Mexico
[4] Observatorio "Las Pegueras", Navas de Oro (Segovia), Spain
[5] Saint Marys University, Halifax, Nova Scotia, Canada
[6] The Abbey Ridge Observatory, Stillwater Lake, Nova Scotia, Canada

Abstract. FS Aurigae is famous for a variety of uncommon and puzzling periodic photometric and spectroscopic variabilities which do not fit well into any of the established sub-classes of cataclysmic variables. Here we present preliminary results of long-term monitoring of the system, conducted during the 2010-2011 observational season. We show that the long-term variability of FS Aur and the character of its outburst activity may be caused by variations in the mass transfer rate from the secondary star as the result of eccentricity modulation of a close binary orbit induced by the presence of a third body on a circumbinary orbit.

Keywords. binaries: close – novae, cataclysmic variables – stars: individual (FS Aurigae)

1. Introduction

FS Aurigae represents one of the most unusual cataclysmic variables (CV) to have ever been observed. The system is famous for a variety of uncommon and puzzling periodic photometric and spectroscopic variabilities which do not fit well into any of the established sub-classes of CVs (Neustroev 2002; Tovmassian *et al.* 2003, 2007).

Based on the short orbital period, FS Aur has been classified as a SU UMa star. Nevertheless, long-term monitoring of the system failed to detect any superoutburst in its light curve. Instead, this monitoring reveals a very long photometric period of ∼900 days. Tovmassian *et al.* (2010) showed that such a long period may be explained by the presence of a sub-stellar third body on a circular orbit around the close binary.

In order to better understand the long-term variability and outburst activity of FS Aur, during the 2010-2011 observational season we have conducted an observing campaign, lasting more than 140 consecutive nights. Here we report the preliminary results of these observations.

2. The 2010-2011 observing campaign

The observations were conducted every clear night from November 26, 2010 until May 3, 2011. The data were taken using telescopes with apertures of 0.28 to 0.5-meters, equipped with CCD cameras and standard Johnson V filters. Depending on the weather conditions, we monitored the star for 6–8 hours per night in the beginning of the campaign and for 3–4 hours at the end. Thus, more than 150 nights of photometry were taken

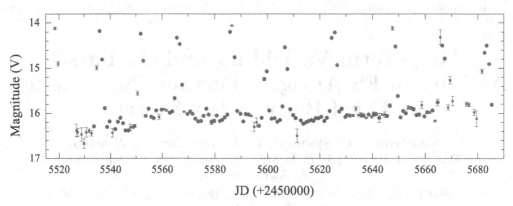

Figure 1. Light curve of FS Aur, from the 2010-2011 observing campaign. Each point is the 1-day average of observations. Blue points represent the observations obtained by the authors, while the red squares represent the AAVSO observations. Note the strong variability during the quiescent state.

(80 nights of time-resolved photometry), and almost 14 000 V-band data points were obtained. The median value of the photometric errors was 0.03 mag in the V filter. In order to reduce the scatter from both random errors and stochastic and short-term variability, we formed 1-day averages of these observations (Fig. 1). Here we note the most prominent features seen in the light curve of FS Aur: (a) during these observations, the average quiescent level increased 0.3−0.4 mag; (b) the system exhibits a strong variability even during the quiescent state; (c) in the interval between the two normal outbursts, two abnormal short low-amplitude outbursts were observed (around JD 2455600).

3. Discussion and Conclusion

The optical flux of a dwarf nova is dominated by the emission from an accretion disk particularly in the short period systems, while the emission from the disk is proportional to the mass-transfer rate. Schreiber *et al.* (2000) also found that the outburst behavior of a dwarf nova is strongly influenced by the variations of the mass-transfer rate. The latter is very sensitive to the Roche lobe size, which is proportional to the binary separation. In hierarchical triple systems, a third body can induce an eccentricity variation in an inner binary (Mazeh & Shaham 1979; Georgakarakos 2009). The long-term modulation is produced by the time-varying tidal force of the perturber upon the binary.

Tovmassian *et al.* (2010) showed that a long ∼900-d period observed in FS Aur may be explained by the presence of a sub-stellar third body on a circular orbit around the close binary. The long-term variability of FS Aur and the character of its outburst activity may also be triggered by variations in $\dot{M}$ from the secondary as the result of eccentricity modulation of a close binary orbit induced by the presence of a third body.

References

Georgakarakos N. 2009, *MNRAS*, 392, 1253
Mazeh T. & Shaham J. 1979, *A&A*, 77, 145
Neustroev V. V. 2002, *A&A*, 382, 974
Schreiber M. R., Gänsicke B. T., & Hessman F. V. 2000, *A&A*, 358, 221
Tovmassian G. H., *et al.*, 2003, *PASP*, 115, 725
Tovmassian G. H., Zharikov S. V., & Neustroev V. V. 2007, *ApJ*, 655, 466
Tovmassian G. H., *et al.*, 2010, *arXiv:1009.5813*

From Interacting Binaries to Exoplanets: Essential Modeling Tools
Proceedings IAU Symposium No. 282, 2011
Mercedes T. Richards & Ivan Hubeny, eds.
© International Astronomical Union 2012
doi:10.1017/S1743921311026998

SPHOTOM – Package for an Automatic Multicolour Photometry

Š. Parimucha[1], M. Vaňko[2], and P. Mikloš[1]

[1] Institute of Physics, University of P.J. Šafárik, Košice, Slovakia
email: stefan.parimucha@upjs.sk

[2] Astronomical Institute of the Slovak Academy of Sciences, Tatranská Lomnica, Slovakia
email: vanko@ta3.sk

Abstract. We present basic information about package SPHOTOM for an automatic multi-colour photometry. This package is in development for the creation of a photometric pipe-line, which we plan to use in the near future with our new instruments. It could operate in two independent modes, (i) GUI mode, in which the user can select images and control functions of package through interface and (ii) command line mode, in which all processes are controlled using a main parameter file. SPHOTOM is developed as a universal package for Linux based systems with easy implementation for different observatories. The photometric part of the package is based on the Sextractor code, which allows us to detect all objects on the images and perform their photometry with different apertures. We can also perform astrometric solutions for all images for a correct cross-identification of the stars on the images. The result is a catalogue of all objects with their instrumental photometric measurements which are consequently used for a differential magnitudes calculations with one or more comparison stars, transformations to an international system, and determinations of colour indices.

Keywords. photometry, CCD

The function of the SPHOTOM can be described in the following steps:

Sorting

It is the first step in command line mode, which creates different lists of images based on information in the FITS header of images as well as names of files. It uses a robust sorting scheme and it is written in Python using the PyFits module. The user can define types of lists, which will be used in next steps. Sorting could be executed also from GUI mode. For a correct functionality it is necessary to have a consistent FITS headers and/or image names, which are observatory dependent.

Master images

Create master images using an average or a median of input files. No other corrections are performed in this step. These images are stored in archive for a future use. The user can define a number of images entering into this procedure.

Photometric reduction

During photometric reduction of raw images we use created or archived master images. We can use, bias, dark frame, flat-field and dark for flat corrections. The procedure automatically controls image dimensions, temperatures and colours. The results are the lists of images based on their colours.

Photometry

The photometric properties are derived from the reduced images using the Sextractor code (Bertin & Arnouts 1996). We can control all photometry options using the parameters files of Sextrator (for more details see Sextrator manual). This package is very effective on uncrowded or semi-crowded fields and allows us to detect all objects above the defined background level on the images and perform their photometry with different user-defined apertures. It can also remove bad or corrupted detections (stars on the edges of images, saturated stars, cosmic ray hits). User can define different types of information in the output file.

Identification

If the FITS headers contain WCS (World Coordinate Systems) information (Calabretta & Greisen, 2002), we perform cross-identification with external catalogs (USNO-A2.0 or Tycho) using up to 20 brightest stars on the image and the astrometric solution of all detected objects is calculated. If we have no WCS information we calculate an astrometric solution with known approximate coordinates of the image center and cross-identification with external catalogs. For each image, a file is created for the detected objects with their coordinates (celestial and image) and their instrumental magnitudes in apertures from photometry.

Output catalogue

After identification of the objects we have 2 possibilities:
(I) In GUI, the user can select up to 9 stars and generate their multicolour light curves with differential magnitudes. No other corrections are performed. The user can manually use procedures in the final corrections step.
(II) In command line mode, we generate differential magnitudes between all pairs of stars, create light curves and store them in a temporary database for an easier manipulation.

Final corrections

In the final step, we can perform several corrections:
1) Heliocentric correction of time.
2) Determination of extinction coefficients and reduction of systematic effects using SARS algorithm (Ofir *et al.*, 2010).
3) Transformation to international system with known transformation coefficients.
4) Calculation of average comparison star from user selected objects.
Finally, we create the differential light curves.

Acknowledgements: This contribution was supported by VEGA project 2/0094/11.

References

Bertin, E. & Arnouts, S., 1996, *A&AS*, 117, 394
Calabretta, M. R. & Greisen, E. W., 2002, *A&A*, 395, 1077
Ofir *et al.*, 2010, *MNRAS* (Letters), 404, L99

From Interacting Binaries to Exoplanets: Essential Modeling Tools
Proceedings IAU Symposium No. 282, 2011
Mercedes T. Richards & Ivan Hubeny, eds.

© International Astronomical Union 2012
doi:10.1017/S1743921311027001

The Photometric Study of Neglected Short-Period Eclipsing Binary BS Vulpeculae

Miloslav Zejda[1], Zdeněk Mikulášek[1,2], Liying Zhu[3,4], Shengbang Qian[3,4], and Jiří Liška[1]

[1]Department of Theoretical Physics and Astrophysics, Masaryk University, CZ-611 37 Brno, Czech Republic, e-mail: zejda@physics.muni.cz, mikulas@physics.muni.cz

[2]VŠB – Technical Univ., Observatory and Planetarium of J. Palisa, Ostrava, Czech Rep.

[3]National Astronomical Observatories/Yunnan Observatory, Chinese Academy of Sciences, P.O. Box 110, 650011 Kunming, China, e-mail: zhuly@ynao.ac.cn, qsb@ynao.ac.cn

[4]Key Laboratory for the Structure and Evolution of Celestial Objects, Chinese Academy of Sciences P.O. Box 110, 650011 Kunming, China

Abstract. The preliminary results of a study of a neglected, relatively bright, short-periodic (P=0.48 d), near contact eclipsing binary BS Vulpeculae is given. We present our new complete *(BVRI)* light curves, and physical parameters of the system based on them, derived by the 2003 version of the Wilson–Van Hamme code.

Keywords. stars: binaries: eclipsing, stars: fundamental parameters, stars: individual BS Vul

1. Introduction

The light variations of BS Vul (α=19$^\mathrm{h}$ 37$^\mathrm{m}$ 27$^\mathrm{s}$, δ=+21° 55' 50", 2000.0) were revealed in 1928. Shaw (1994) included this star in his list of near contact binaries and gives the spectral type F2, apparent magnitude m_V=11 mag with depths of minima 0.7 and 0.2 mag respectively, and the distance of the system as 460 pc. We present the first detailed study of this quite bright, but still neglected eclipsing binary.

2. Observational data and photometric solution

Usually, only data in the vicinity of brightness minima are used for the period analysis. However, to utilize all available information in the entire light curve, we collected as many individual measurements as possible.

Our own CCD measurements were obtained in 4 nights in the autumn of 2009 and another 4 nights in the autumn 2010. We used an 85cm telescope at the Xinglong Station of National Astronomical Observatory, China. We supplemented the observations with V measurements from de Bernardi & Scaltriti (1979), ASAS-3, OMC INTEGRAL and the AAVSO archive, thus we collected and used 7959 measurements for the preliminary period analysis and to determine possible changes in the period. We found the following new light ephemeris:

$$\mathrm{MinJDH} = 2\,452\,500.16969(7) + 0.475970417(10) \times E \qquad (2.1)$$

Our $BV(IR)_c$ CCD light curves were analyzed using the 2003 version of the Wilson–Van Hamme code (Wilson & Van Hamme 2003). Given the F2 spectral type of BS Vul, we assumed an effective temperature of T_1=7 000 K for the primary component (the star eclipsed at primary minimum). We took the gravity-darkening coefficients as g_1=g_2=0.32 and the bolometric albedo as A_1=A_2=0.5, corresponding to the convective envelope of this binary system. Bolometric and bandpass square-root limb-darkening parameters were taken from Van Hamme's paper (1993).

M. Zejda *et al.*

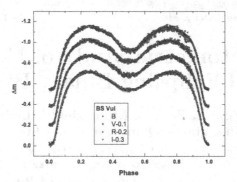

Figure 1. Observed and theoretical $BV(IR)_c$ light curves of BS Vul.

Table 1. Photometric solutions of BS Vul

i (deg)	88.4(9)	r_1 (pole)	0.4459(7)	$\frac{L_{1B}}{L_{1B}+L_{2B}}$	0.9761(3)
T_1 (K)	7 000	r_1 (side)	0.4777(8)	$\frac{L_{1I}}{L_{1I}+L_{2I}}$	0.9202(3)
T_2 (K)	4 644(18)	r_1 (back)	0.5036(8)	$\frac{L_{1V}}{L_{1V}+L_{2V}}$	0.9560(3)
q	0.340(3)	r_2 (pole)	0.2700(32)	$\frac{L_{1R}}{L_{1R}+L_{2R}}$	0.9452(3)
Ω_1	2.5530	r_2 (side)	0.2812(38)		
Ω_2	2.5549(13)	r_2 (back)	0.3136(62)		

After solving for the mass ratio q, the solutions for several assumed values of mass ratio were obtained. For each q, the calculation started at mode 2 (the detached mode). The minimum Σ was achieved at $q{=}0.34$ for BS Vul with mode 4 (the semidetached case with the primary component filling the critical Roche Lobe). Therefore we performed a differential correction so that it converged by making q an adjustable parameter and by choosing $q{=}0.34$ as the initial value. Finally the converged photometric solutions listed in Table 1 were obtained. The theoretical light curves computed with these values are plotted in Fig. 1 with solid lines. Our solution indicates that BS Vul is a near contact binary system. The primary component fills its critical Roche Lobe and the secondary one nearly fills its Roche lobe too.

3. Conclusions

BS Vul belongs to the interesting group of near contact binaries. We determined the parameters of the system. However, for further study spectroscopic observations are highly desired. Continued photometric monitoring and in-depth inspection of all published data are necessary for the description of possible period changes.

Acknowledgements

This study was partly supported by grants GAAV IAA 301630901, GAČR 205/08/ 0003, MUNI/A/0968/2009, and MŠMT ME10099 and by the West Light Foundation of The Chinese Academy of Sciences, Chinese Natural Science Foundation (Nos. 10973037 and 10903026).

References

de Bernardi, C. & Scaltriti, F. 1979, *A&AS*, 35, 63
Shaw, J. S. 1994, *MemSAI*, 65, 95
Van Hamme W., 1993, *AJ*, 106, 2096
Wilson R. E. & Van Hamme W., 2003, Computing Binary Stars Observables, the 4th edition of the W-D program., Ftp: astro. ufl. edu, directory pub /wilson/lcdc2003

From Interacting Binaries to Exoplanets: Essential Modeling Tools
Proceedings IAU Symposium No. 282, 2011   © International Astronomical Union 2012
Mercedes T. Richards & Ivan Hubeny, eds.   doi:10.1017/S1743921311027013

O-C Analysis of Selected 3-Body Systems

W. Ogłoza[1], J. M. Kreiner[1], G. Stachowski[1], M. Winiarski[1], B. Zakrzewski[1], S. Doğru[2] F. Aliçavuş[2], O. Demircan[2], and A. Erdem[2]

[1]Mt. Suhora Observatory, Cracow Pedagogical University,
30-084 Cracow, ul Podchorążych 2, Poland
email: ogloza@up.krakow.pl

[2]Ulupinar Observatory, Çanakkale Onsekiz Mart University, Turkey

Abstract. This paper presents the results of the analysis of (O-C) diagrams of four eclipsing variables. The diagrams are based on times of minima collected in the Cracow database, which contains times of minima found in the literature, from observations at Mt. Suhora and Ulupinar Observatories, or determined using publicly-available photometric surveys (NSVS, ASAS etc).

Keywords. stars: binaries: eclipsing, ephemerides

1. Introduction

The O-C diagrams were prepared using mainly CCD and photoelectric data. Photographic and visual observations were used only for XZ Per, because of the long timescale of the (O-C) effects. A few scattered minima were rejected because of apparent, unidentified errors. All diagrams clearly show light-time effects due to the influence of a third body.

The commonly-used software written by T. Pribulla was used to determine the orbital parameters of the third body. For each star, (O-C) diagrams were calculated using linear elements, and the data were improved for the average period correction and the residuals of the LTE fit.

2. Results

For V1061 Cyg the 3-body effect described by Torres *et al.* (2006) is confirmed by recent observations.

Table 1. Orbital parameters of 3-body systems

	BE And	V1061 Cyg	XZ Per	V482 Per
P(3) (days)	3354 ± 58	5671 ± 105	24630 ± 177	6062 ± 64
e	0.62 ± 0.20	0.43 ± 0.08	0.61 ± 0.03	0.84 ± 0.03
w(rad)	5.20 ± 0.18	4.97 ± 0.19	3.19 ± 0.05	3.57 ± 0.04
T(periast)(JD)	2442694 ± 150	2443058 ± 178	2400094 ± 394	2400747 ± 569
asin(i) (AU)	0.82 ± 0.14	2.46 ± 0.11	4.23 ± 0.11	4.83 ± 0.35
To(HJD)	2434962.4035 ± 0.0018	2426355.2335 ± 0.0043	2425150.4482 ± 0.0009	2428327.7615 ± 0.0025
P (days)	0.462902516 ± 4.58E-08	2.346653857 ± 4.03E-07	1.15163193 ± 4.18E-08	2.4467531 ± 2.49E-07
Sum of Sqr	0.00002	0.00091	0.0117	0.0033
f(m3)	0.0064 ± 0.0032	0.0615 ± 0.0082	0.01665 ± 0.00137	0.4095 0.0893

The period changes of XZ Per were described by Qian (2001) as the result of mass transfer, but recent observations and more complete historic data confirm a 3-body model of the system.

The observed period changes of BD And and V482 Per are currently being analysed.

3. Database

The pictures of O-C diagrams and ephemerides based on the last linear trend on the O-C diagrams are presented at: **www.as.up.krakow.pl/ephem**.

Acknowledgements

This study was supported by TUBITAK (Scientific and Technological Research Council of Turkey) under Grant No. 108T714. A portion of this study is part of the PhD thesis of Serkan Doğru.

References

Kreiner, J. M. 2004, *AcA*, 50, 247
Qian, S. 2001, *AJ*, 121, 1216
Torres, G., Lacy, C. H., Marschall, L. A., Sheets, H. A., & Mader, J. A. 2006, *ApJ*, 640, 1018

From Interacting Binaries to Exoplanets: Essential Modeling Tools
Proceedings IAU Symposium No. 282, 2011
Mercedes T. Richards & Ivan Hubeny, eds.
© International Astronomical Union 2012
doi:10.1017/S1743921311027025

Hard X-ray and Optical Activity of Intermediate Polars

Rudolf Gális[1], Ladislav Hric[2] and Emil Kundra[2]†

[1]Department of Theoretical Physics and Astrophysics, Institute of Physics,
Faculty of Sciences, P. J. Šafárik University, Park Angelinum 9, 040 01 Košice, Slovakia
email: rudolf.galis@upjs.sk

[2]Astronomical Institute of the Slovak Academy of Sciences, 059 60 Tatranská Lomnica,
Slovakia

Abstract. Intermediate polars represent a major fraction of all cataclysmic variables detected by *INTEGRAL* in hard X-rays. Nevertheless, only 25% of all known intermediate polars have been was detected in hard X-rays. This fact can be related to the activity state of these close interacting binaries. Multi-frequency (optical to X-ray) investigation of intermediate polars is essential to understand the physical mechanisms responsible for the observed activity of these objects.

Keywords. stars: novae, cataclysmic variables, accretion, accretion disks

1. Introduction

Cataclysmic variables (CVs) manifest strong activity in the whole spectrum from radio up to γ-rays. CVs are close binary systems consisting of a hot white dwarf (WD) and red main-sequence star, which fills the volume of its inner Roche lobe and transfers matter to the vicinity of the WD (Warner 1995). The mass transfer between components is a main reason for the observed CV activity.

According to the strength of the WD magnetic field, the transferred matter is creating an accretion disk or follows magnetic lines and falls to the surface of the WD. In intermediate polars (IPs), the WD magnetic field ($10^4 - 10^6$ G) is not strong enough to disrupt the disc entirely (as in the case of polars) and simply truncates the inner part of the disc. An accretion flow is channelled down towards the magnetic poles and onto the WD surface.

When the transferred material impacts the WD atmosphere, a shock will form and hard X-ray emission will result from thermal bremsstrahlung cooling by free electrons in the hot post-shock region (King & Lasota 1979). The broad-band spectra (3 - 100 keV) of the studied IPs can be well fitted by a thermal bremsstrahlung model with the post-shock temperature of $kT \approx (20 - 25)$ keV (Gális *et al.* 2009).

No significant modulation has been found so far in the 20 - 30 keV light curves (Barlow *et al.* 2006). Nevertheless, the sample detected by *INTEGRAL* represented only 25% of all known IPs (Gális *et al.* 2009). Some IPs were not detectable even though we had significant exposure time (more than 4Msec) for these sources. This fact can be related to the activity state of these interacting binaries.

We analysed all available observational data from *INTEGRAL*/IBIS for IP V1223 Sgr (Gális *et al.* 2009). Our analysis showed that the fluxes of this object are long-term variable, mainly in (15 - 25) keV and (25 - 40) keV bands. Moreover, this hard X-ray variability is correlated with changes in the optical spectral band. Our analysis revealed a deep flux drop around MJD ≈ 53650 observed in both X-ray and optical bands for this IP.

† This study was supported by the project of the Slovak Academy of Sciences *VEGA* Grant No. 2/0078/10.

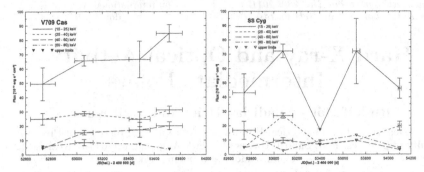

Figure 1. *INTEGRAL*/IBIS flux curves of V709 Cas (left panel) nad SS Cyg (right panel) in the corresponding energy bands. The arrows represent 3σ upper limits.

2. Observations, analysis and results

We used all publicly available observational data from *INTEGRAL*/IBIS to study possible variability of V709 Cas and SS Cyg in hard X-rays. Observational data used in our analysis were processed by *INTEGRAL*'s Offline Standard Analysis Package OSA7.

V709 Cas was recognized as an IP following its detection in the ROSAT All Sky Survey. The broad-band (3 - 100) keV spectrum from *INTEGRAL* was well fitted by a thermal bremsstrahlung model with the temperature of $kT = 24.4^{+1.5}_{-1.4}$ keV (Gális *et al.* 2009).

SS Cyg is an optically bright IP observed to undergo outbursts every ≈ 40 days, characterized by an increase in optical magnitude from V ~ 12 to V ~ 8. Comparison of the *INTEGRAL*/IBIS flux curve with the optical light-curve (validated AAVSO data) over part of the IBIS survey period suggested a strengthening of the 20 - 30 keV flux during optical quiescence (Barlow *et al.* 2006).

Our analysis of all available observational data of V709 Cas and SS Cyg showed that these sources are detectable up to 100 keV. The hard X-ray fluxes are not persistent. The flux curves indicate that the brightness of V709 Cas increased by a factor ≈ 2 from MJD 52700 to MJD 53700 in the (15 - 25) keV energy band (Fig. 1, left panel) and the brightness of SS Cyg increased by a factor ≈ 2 in the (15 - 25) keV and (25 40) keV energy bands during the optical quiescence phases (Fig. 1, right panel).

3. Conclusions

The significant part of the optical emission from IPs is produced by a hot spot, where the matter from the donor star interacts with the outer rim of the accretion disk. X-ray emission is produced by the interaction of the accreting matter with the WD surface. So, the emission in both optical and X-ray bands is related to the mass transfer and therefore observed variations are probably caused by changes in the mass accretion rate.

We are preparing the photometric campaign to obtain long-term homogenous observations (to cover whole activity cycles) as well as sets of observations with high time resolution (to detail coverage of orbital cycles) of selected CVs, mainly as a follow up of the *INTEGRAL* observations. Simultaneous analysis of multi-frequency observation (from optical to X-ray) allows complex study of the physical mechanism related to the mass transfer in these interacting binaries.

References

Barlow, E. J., Knigge, C., Bird, A. J. *et al.*, 2006, *MNRAS*, 372, 224
Gális, R., Eckert, D., Paltani, S., Münz, F., & Kocka M. 2009, *Baltic Astronomy*, 18, 321
King, A. R. & Lasota, J. P. 1979, *MNRAS*, 188, 653
Warner, B. 1995, *Cataclysmic variable stars*, Cambridge University Press, Cambridge

From Interacting Binaries to Exoplanets: Essential Modeling Tools
Proceedings IAU Symposium No. 282, 2011
Mercedes T. Richards & Ivan Hubeny, eds.
© International Astronomical Union 2012
doi:10.1017/S1743921311027037

Is EQ Boo a Quadruple System?

Igor M. Volkov[1], Drahomír Chochol[2], Natalia S. Volkova[1] and Igor V. Nikolenko[3]

[1] Sternberg Astronomical Institute, Moscow University,
Universitetsky Ave., 13, Moscow 119992, Russia
email: imv@sai.msu.ru

[2] Astronomical Institute of the Slovak Academy of Sciences,
059 60 Tatranská Lomnica, Slovakia
email: chochol@ta3.sk

[3] Scientific Research Institute, "Crimean Astrophysical Observatory,"
Community Blue Bay, Shajn str., 1, Ukraine

Abstract. We present the precise multicolour photometry of the eclipsing variable EQ Boo ($P = 5^d.43$, $V = 8^m.8$), which is component "A" of the visual double star ADS 9422 (F7 V+G0 V, $\rho = 1.3$", $\Delta mag = 0.7$). From the analysis of these data, we can propose the existence of the fourth component with a late spectral type.

Keywords. eclipsing binary, multiple system, stellar evolution

1. Introduction

Otero *et al.* (2006) found the position of the secondary minimum of the eclipsing variable EQ Boo at the phase $\phi(II) = 0.399$, so its orbit is eccentric. We observed the star in 2007-2010 at the Zvenigorod, Crimean and Stará Lesná observatories. We obtained observations in two primary and three secondary minima. We always measured the brightness of both components of the visual double star together. Our techniques of the atmospheric extinction correction and the solution of the light curves were the same as those used in our previous papers: Volkov & Volkova (2009), Volkov *et al.* (2010).

2. Absolute parameters of the system

After the solution of our $UBVRI$ light curves by the differential corrections method, we obtained absolute parameters of the system using Kepler's third law and the mass-luminosity relation for main sequence stars (see Table 1). If we accept that the third light in the system is only due to the component "B", which is unresolved in our observations, then the parameters of the three stars do not satisfy the value of the common age $2.0 \cdot 10^9$ years. This contradiction can be resolved by assuming the existence of a fourth star with a later spectral type of K2 V in the system (Fig. 1d). The details of our work are published in Volkov *et al.* (2011).

Acknowledgements

This research was partly supported by VEGA grants 2/0038/10 and 2/0094/11 (D. Chochol), and by Russian Foundation for Basic Research grant 11-02-01213-a (I. M. Volkov and N. S. Volkova). We used in our work the SIMBAD database of the Strasbourg center of astronomical data (France) and the ADS service of NASA (USA).

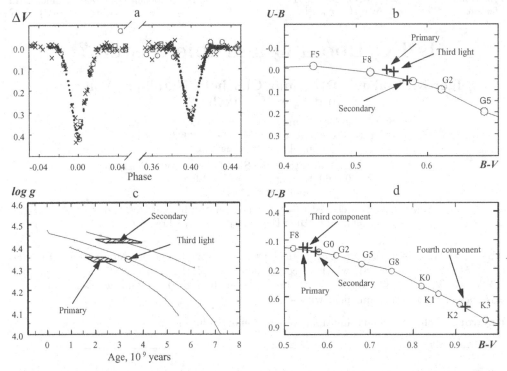

Figure 1. a) The light curve of EQ Boo (points - our *V* photometry, crosses - ASAS (Pojmanski 2002), circles - Hipparcos (Perryman *et al.* 1997), squares - ROTSE (Woźniak *et al.* 2004)). b) The position of the components in the standard two-colour diagram. c) The evolutionary diagram for the stars with the masses from Table 1. d) The position of the components in the standard two-colour diagram taking into account the complicated structure of the third light.

Table 1. Absolute parameters of EQ Boo.

Parameter	Primary	Secondary	The third light
Mass, M$_\odot$	1.15 ± 0.05	1.05 ± 0.04	1.11 ± 0.07
Radius, R$_\odot$	1.19 ± 0.03	1.03 ± 0.03	1.16 ± 0.05
Luminosity, L$_\odot$	1.45 ± 0.08	1.04 ± 0.07	1.26 ± 0.08
T$_{eff}$	6120 ± 100 K	5980 ± 80 K	5940 ± 80 K
Spectral type (from our *UBV* data)	F8 V	G0 V	G0 V

References

Otero, S. A., Wils, P., Hoogeveen, G., & Dubovsky, P. A., 2006, *IBVS*, 5681, 1

Perryman, M. A. C., Lindegren, L., Kovalevsky, J., Hoeg, E., Bastian, U., Bernacca, P. L., Crézé, M., Donati, F., Grenon, M., van Leeuwen, F., van der Marel, H., Mignard, F., Murray, C. A., Le Poole, R. S., Schrijver, H., Turon, C., Arenou, F., Froeschlé, M., & Petersen, C. S., 1997, *A&A*, 323, L49

Pojmanski, G., 2002, *AcA*, 52, 397

Woźniak, P. R., Vestrand, W. T., Akerlof, C. W., Balsano, R., Bloch, J., Casperson, D., Fletcher, S., Gisler, G., Kehoe, R., Kinemuchi, K., Lee, B. C., Marshall, S., McGowan, K. E., McKay, T. A., Rykoff, E. S., Smith, D. A., Szymanski, J., & Wren, J., 2004, *AJ*, 127, 2436

Volkov, I. M. & Volkova, N. S., 2009, *Astron. Rep.*, 53, 136

Volkov, I.M., Volkova, N.S., & Chochol, D., 2010 *Astron. Rep.*, 54, 418

Volkov, I.M., Volkova, N.S., Chochol, D. & Nikolenko, I.V., 2011 *Astron. Rep.*, 55, 824

From Interacting Binaries to Exoplanets: Essential Modeling Tools
Proceedings IAU Symposium No. 282, 2011   © International Astronomical Union 2012
Mercedes T. Richards & Ivan Hubeny, eds.   doi:10.1017/S1743921311027049

Long-Term Monitoring of Polars

B. Kalomeni[1,2]

[1] Department of Astronomy and Space Sciences, University of Ege, 35100, İzmir, Turkey

[2] Department of Physics, İzmir Institute of Technology, 35430, İzmir, Turkey

Abstract. We present long-term observations of magnetic cataclysmic variables AM Her, AN UMa, AR UMa, DP Leo, and V1309 Ori obtained with the ROTSEIIId telescope. All data have been analysed and preliminary results indicate periods of 170 days, 217 days, and 180 days for AM Her, AN UMa, and AR UMa, respectively.

Keywords. cataclysmic variables, stars: individual (AM Her, AN UMa, AR UMa, DP Leo, V1309 Ori

1. Introduction

Magnetic cataclysmic variables are important in stellar evolution studies because of their strong magnetic field intensities and their unpredictable mysterious behaviour. Light variations of polars are characterized by long and short term variations. Unlike the systems with discs, any modulation in mass accretion rate is generally attributed to the donor star (Warner 1988, Richman, Applegate & Patterson 1994). The characteristic time-scale of small amplitude, short period light variations are different for each polar (e.g. AM Her 4.2-4.7 minutes (Bonnet-Bidaud *et al.* 1991, Kalomeni *et al.* 2005) and V1309 Ori 10 minutes (Katajainen *et al.* 2003)). In the literature especially during the last decade long period observations of polars have been intensively studied (Hessman *et al.* 2000, Wu & Kiss 2008, Kalomeni & Yakut 2008, Kafka & Hoard 2009, Sanad 2010). Studies of the long term variations are especially important to reveal any characteristics based on stellar activity and/or mass transfer.

AM Her

Tapia (1976) discovered AM Her as a system showing large and variable polarization. The non-predictable behaviour of the system is apparent after more than 5 years, with 3349 data obtained in this study by using ROTSEIIId at the TÜBİTAK National Observatory (see for details Kalomeni 2011). The long term light curve of the system shows both single and double maxima during the high state.

AN UMa

Hearn & Marshall (1979) by using SAS3 detected AM Her-like properties of AN UMa. The light variation of AN UMa obtained in this study shows almost a sinusoidal variation. However, the system needs more observations to see if this pattern continues.

AR UMa

Wenzel (1993) by studying AR UMa's light variations extending 32 years concluded it to be a cataclysmic variable. Remillard *et al.* (1994) identified AR UMa as a soft X-ray source in the Einstein survey. From ellipsoidal variations they identified the orbital period as 1.932 hr. By using the spectral lines of the secondary star the authors classified its spectral type as ∼M6 and estimated the distance ∼88 pc. AR UMa was observed by

Szkody *et al.* (1999) in the low state with ASCA and in the high state simultaneously with EUVE, RXTE, and ground based optical telescopes. In this study, Fourier analysis of the system yields a period of 180 days.

DP Leo

DP Leo was the first eclipsing polar to be discovered (Biermann *et al.* 1985). DP Leo is an interesting and important polar since it is a member of the post-common envelope binaries group with known/suspected planet companions (Beuermann *et al.* 2011). Recently, Beuermann *et al.* (2011) found the physical parameters of the third body using mid-eclipse times of the WD obtained during the last 21 years. Since the system is faint and requires a long exposure time, we could not deduce an accurate long term light variation (see Kalomeni 2011).

V1309 Ori

V1309 Ori was discovered as a soft X ray source with ROSAT (Beuermann & Thomas 1993) and classified as a magnetic CV by Garnavich *et al.* (1994). V1309 Ori, with ~ 8 hours orbital period, is the longest period polar detected so far. Its light variation was obtained with ROTSEIIId and with RTT150 at TUG during two nights in 2006.

2. Conclusions

We present a preliminary study of the long term light variations of the first detected polar AM Her, the next discovered polar AN UMa, the highest magnetic field polar AR UMa, the first discovered eclipsing polar DP Leo, and the longest period polar V1309 Ori. Our analysis yields periods of 170 days, 217 days, and 180 days for AM Her, AN UMa, and AR UMa. Those variations are probably due to the active secondary star. A more detailed study will be presented in Kalomeni (2011).

Acknowledgements

This study was supported by the Turkish Scientific and Research Council (Project No.109T047), TÜBİTAK National Observatory, and the Ege University Research Fund.

References

Beuermann, K. & Thomas, H.-C. 1993, *AdSpR*, 13, 115
Beuermann, K., Buhlmann, J., Diese, J., Dreizler, S., Hessman, F. V. *et al.*, 2011, *A&A*, 526, 53
Biermann, P., Schmidt, G. D., Liebert, J., Tapia, S., *et al.*, 1985, *ApJ*, 293, 303
Bonnet-Bidaud, J. M., Somova, T. A., & Somov, N. N. 1991, *A&A*, 251L, 27
Garnavich, P. M., Szkody, P., Robb, R. M., Zurek, D. R., & Hoard, D. W., 1994, *ApJ*, 435L, 141
Hearn D. R. & Marshall F. J. 1979, *ApJ*, 232L, 21
Hessman, F. V., Gansicke, B. T., & Mattei, J. A. 2000, *A&A*, 361, 952
Kafka S. & Hoard D. W. 2009, *PASP*, 121, 1352
Kalomeni, B., Pekünlü, E. R., & Yakut, K. 2005, *Ap&SS*, 296, 477
Kalomeni B. & Yakut, K. 2008, *AJ*, 136, 2367
Kalomeni B. 2011, *in preparation*
Katajainen, S., Piirola, V., Ramsay, G., Scaltriti, F., Lehto, H. J., *et al.*, 2003, *MNRAS*, 340, 1
Szkody, P., Vennes, S., Schmidt, G. D., Wagner, R. M., Fried, R., *et al.*, 1999, *ApJ*, 520, 841
Sanad, M. R. 2010, *ApSS*, 330, 337
Remillard, R. A., Schachter, J. F., Silber, A. D., & Slane, P. 1994, *ApJ*, 426, 288
Richman, H. R., Applegate, J. H., & Patterson, J. 1994, *PASP*, 106, 1075
Tapia, S. 1976, *BAAS*, 8, 511
Wenzel, W. 1993, *IBVS*, 3890, 1
Warner, B. 1988, *Nature*, 336, 129
Wu K. & Kiss L. L. 2008, *A&A*, 481, 433

From Interacting Binaries to Exoplanets: Essential Modeling Tools
Proceedings IAU Symposium No. 282, 2011
Mercedes T. Richards & Ivan Hubeny, eds.
© International Astronomical Union 2012
doi:10.1017/S1743921311027050

The Apsidal Motion of the Eclipsing Binary Systems GSC 4487 0347 and GSC 4513 2537

V. S. Kozyreva[1], A. V. Kusakin[2], T. Krajci[3], J. Menke[4] and T. M. Tsvetkova[5]

[1] Sternberg State Astronomical Institute,
13, University avenue, 119992, Moscow, Russia, email: valq@sai.msu.ru

[2] National space agency republic of Kazakhstan,
JSCNCSRT Department "V.G. Fesenkov Astrophysical Institute",
050068, Almaty, Kazakhstan

[3] Astrokolkhoz Observatory,
PO Box 1351, Cloudcroft, 88317, New Mexico, USA

[4] Starlight Farm Observatory,
22500 Old Hudred Rd, Barnesville, MD 20838, USA

[5] Institute of Astronomy Russian Academy of Sciences,
48 Pyatnitskaya st., 119017, Moscow, Russian

Abstract. The eclipsing variable stars GSC 4487 0347 and GSC 4513 2537 are recently discovered binary systems (Otero *et al.*, 2006) with orbital periods $1^d.99$ and $6^d.33$ days. We carried out the photometric observations of these eclipsing binaries from 2009-2010 using a CCD-array at the Tien-Shan Observatory in Kazakstan, at the Crimea Station of the Sternberg Astronomical Institute, at the Astrokolkhoz Observatory in New Mexico (AAVSO), while the spectrophotometric observations were obtained at the Starlight Farm Observatory in Barnesville, USA.

Keywords. eclipsing binary, multicolour photometry, apsidal motion

---

1. The investigations of GSC 4487 0347

The recently discovered binary system GSC 4487 0347 has a neighbor star at a distance of 3.5″ and is fainter than the variable star in the V-band by approximately 2.5^m. We have observations in the V band during Min I and B, V, R observations during Min II, and we obtained the B, V, R magnitudes of the total system (the eclipsing binary with the third star) outside minima. The V-observations obtained in 2009 at Tian-Shan Observatory are shown in Fig. 1.

The photometric elements of the system have been derived. The spectra of GSC 4487 347 including the optical component have been obtained by J. Menke. The most likely hypothesis is also that the spectral classes of all three stars are close to late B or early A.

The moments of minima and periods of this system are:

$$\text{Min I} = \text{JD}_\odot\ 2455122^d.1581(2) + 1^d.988726(1)$$

$$\text{Min II} = \text{JD}_\odot\ 2455121^d.3154(3) + 1^d.988719(5).$$

The apsidal motion of this binary was derived by us: $\dot{\omega}_{obs} = 2.8 \pm 0.6°/year$. The average internal structure constants of stellar components $\bar{k}_2^{obs} = 0.0041 \pm 0.008$ is in agreement with the theory.

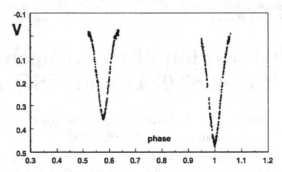

Figure 1. The V- light curves of GSC 4487 0347.

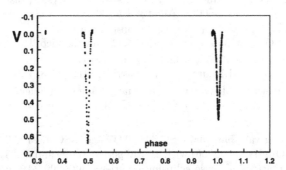

Figure 2. The V- light curves of GSC 4513 2537.

2. The investigation of GSC 4513 2537

GSC 4513 2537 is a rather faint eclipsing binary system with secondary minimum expected at phase $0^p.494$ (Otero *et al.* 2006). However, the latest observation shows that this minimum occurs at phase $0^p.5053$ (see Fig.2). The V-observations obtained in 2009 at Tian-Shan Observatory (Min I) and at the Astrokolkhoz Observatory in New Mexico (Min II) are shown in Fig. 2. Observations relative to a standard comparison star were obtained by G. V. Komissarova at the Crimean Observatory with a Zeiss-600 telescope and a UBV-photometer.

The spectra of stars outside the minimum and within the primary minimum have been obtained by J. Menke. Given the photometric findings, it can be concluded that the spectra of the component stars of this system are G2V + G4V. The estimation of apsidal motion with consideration of data by Otero *et al.* (2006) is: $\dot\omega_{obs} = 1.0 \pm 0.3\ °/year$.

The moments of minima and periods of this system are:

$$\text{Min I} = \text{JD}_\odot\ 2455192^d.84405(20) + 6^d.334436(2)$$

$$\text{Min II} = \text{JD}_\odot\ 2455050^d.3527(2) + 6^d.334427(2).$$

Acknowledgements

We wish to thank G. V. Komissarova for the observations, and A. I. Zaharov and S. E. Leont'ev for their help with the development of computer programs.

References

Otero, S. A., Wils, R., Hoogeveen, G., & Dubovsky, P. A. 2006, *IBVS*, 5681

From Interacting Binaries to Exoplanets: Essential Modeling Tools
Proceedings IAU Symposium No. 282, 2011
Mercedes T. Richards & Ivan Hubeny, eds.
© International Astronomical Union 2012
doi:10.1017/S1743921311027062

Recent Spectral Observations of Epsilon Aurigae in the Near-IR

Lubomir Iliev

Institute of Astronomy, National Astronomical Observatory Rozhen
Bulgarian Academy of Sciences,
72 Tsarigradsko Shossee Blvd., BG-1784, Sofia, Bulgaria
email: liliev@astro.bas.bg

Abstract. High resolution spectral observations of ϵ Aur were carried out in the near-IR spectral range. Observations were obtained with the Coudé-spectrograph of the 2m RCC telescope at National Astronomical Observatory Rozhen and cover all main phases of the current eclipse. Results revealed for the first time absorption components in O I and Ca II triplets and variations of N I lines. Estimation of the electron density was done using lines from the Paschen series of hydrogen.

Keywords. Stars: variable, Stars: binaries: eclipsing, Stars: individual: ϵ Aur

1. Introduction

ϵ Aur exhibits distinguishing features of a classical binary system of Algol-type, but also attracts attention with a long list of yet unsolved problems. Kloppenborg *et al.* (2010) reconstructed 2-D images of ϵ Aur that demonstrate an eclipsing body moving in front of the central star. They estimated the mass of the primary F type star as 3.6 $M_\odot$ and the mass of the eclipsing disk. Analysis of the spectral energy distribution (SED) done by Hoard, Howel & Stencel (2010) reveal that the system of ϵ Aur consists of a post-asymptotic giant branch F type star and a B type MS star with mass 5.9 $M_\odot$ which is surrounded by a disk of gas and dust.

Spectral observations in the near-infrared proved to be a reliable and sensitive instrument for studying evidence of circumstellar matter in B, A and F type stars (see e.g. Slettebak 1986, Munari & Tomasella 1999) and could be used to answer questions about the origin and interaction of different components of the system of ϵ Aur.

2. Observations, results, discussion

All observations were carried out at the Coudé-spectrograph of the Rozhen 2m RCC telescope with a spectral resolution of R=21000 at wavelength 850 nm. They were obtained in 5 characteristic moments of the 2009-2011 eclipse of the ϵ Aur system: ingress and regress parts of the eclipse, first half, central part and second half of the totality. The O I triplet at 777 nm was chosen as the main target of the observing runs as it is a sensitive indicator of luminosity effects. Strong lines of the infrared Ca triplet, Paschen series of the hydrogen and N I were also studied. The investigation of these lines shows no existence of sharp absorption cores or emission features that are usually connected with the presence of cool or hot gas formations in the binary system.

Observations from March 2010 and March 2011 included the hydrogen Paschen series. Lines from this series were resolvable up to quantum number n=25 (see Fig. 1a). Applying

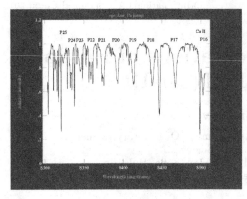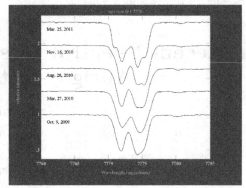

Figure 1. (a) Paschen series jump of eps Aur on frame from March 27, 2010. Continuum rectification was done on the red slop of the series. (b) O I 777nm triplet profiles obtained at different stages of the eclipse. Component structure developed after mid-eclipse.

the Inglis-Teller formula with coefficients given by Allen (1973), we estimate the electron density in the atmosphere of the visible component of ϵ Aur:

$$logN_e = 12.215449$$

This value is higher than expected for the F type giant or supergiant star and could be regarded as evidence of the existence of circumstellar gas formations around the primary, or as is supposed by Chadima *et al.* (2011), around the complex structure of the secondary.

Frames of the O I triplet feature (777.195, 777.418, 777.54 nm) were taken during all sets of observations. From August 2010 additional absorption components started to develop (Fig. 1b) and on March 25, 2011 there are already both redshifted and blueshifted ones. The displacements of the lines are 40.514 km s^{-1} and 44.396 km s^{-1}, respectively. Although not so prominent as those seen in the O I triplet, lines of the Ca II infrared triplet (849.8, 854.2, 866.2 nm) also show the presence of components starting from November 2010. Absorption component structures in these lines are observed in the spectrum of ϵ Aur for the first time and are also evidence for gas structures in the system.

We search our spectra for variations in the lines of Paschen series of H I and some strong lines of N I during the eclipse. In neither of them were there any absorption components.

It should specially be noted that N I lines at 862.9 and 856.7 nm followed the same trend already observed for K I lines. They increased their equivalent length by 15% and 21% respectively during the ingress phase of the eclipse.

References

Allen, C. W. 1973, in: *Astrophysical Quantities*, 3rd ed., The Athlone press

Chadima, P., Harmanec, P., Bennett, P. D., Kloppenborg, B., Stencel, R.,Yang, S., Bozic, H., Slechta, M., Kotkova, L., Wolf, M., Skoda, P., Vortruba, V., Hopkins, L. J., Buil, C., & Sudar, D. 2010, *A&A*, 530, 146

Hoard, D. W., Howel, S. B., & Stencel, R. E. 2010, *ApJ*, 714, 549

Kloppenborg, B., Stencel, R., Monnier, J., Schaefer, G., Zhao, M., Baron, F., McAlister, H., Ten Brummelaar, T., Che, X., Farrington, C., Pedretti, E., Sallave-Goldfinger, P. J., Sturmann, J., Sturmann, L., Thureau, N., Turner, N., & Carroll, S. M. 2010, *Nature*, 464, 870

Munari, U. & Tomasella, L. 1999, *A&ASS*, 137, 521

Slettebak, A. 1986, *PASP*, 98, 867

Part 2
Observations and Analysis of Exoplanets and Brown Dwarfs

From Interacting Binaries to Exoplanets: Essential Modeling Tools
Proceedings IAU Symposium No. 282, 2011   © International Astronomical Union 2012
Mercedes T. Richards & Ivan Hubeny, eds.   doi:10.1017/S1743921311027074

Challenges to Observations of Low Mass Binaries

Stella Kafka

NASA Astrobiology Institute and Department of Terrestrial Magnetism,
Carnegie Institution of Washington,
5241 Broad Branch Road NW,
Washington, DC 20015, USA
email: skafka@dtm.ciw.edu

Abstract. Low mass stars in binaries are frequently used as unique tools to determine and establish fundamental stellar parameters. The need for their study and understanding has led to developing new instruments, new observational techniques and improved theoretical models. The relatively recent discovery of exoplanets and their study as the low-mass constituents around other stars is now opening new horizons in binary research. Here, I examine the most common observational challenges in studying low mass binaries across the electromagnetic spectrum.

Keywords. (stars:) binaries: general, stars: low-mass, brown dwarfs, methods: data analysis

1. Introduction

The launch and operation of space-based observatories revolutionized the field of binary star research and enriched our knowledge of our favorite systems, revealing a wealth of behaviors. With great discoveries, though, came great challenges in our study and understanding of stellar components - especially when it comes to low-mass stars. Also, during the last 10 years with the discovery of various planetary systems, low mass binaries with planet components are at the spotlight of research activity. The knowledge collected over decades from low-mass binary stars, are now applied and used in the discovery and study of exoplanets and their host stars. The unprecedented accuracy in the observed light curves, provided by space based observatories, revealed a suite of phenomena that could not be studied from the ground and require new tools for their study. In this review, I discuss some challenges arising for the study of low mass binaries. This is by no means a comprehensive inclusion of all aspects of low-mass binary star research; it is perhaps a reminder that binaries can be used as tools for understanding significant physical processes which stem from stellar interactions, and that they can provide a means to study nature under conditions hardly encountered on Earth-based laboratories. The star components discussed here have masses of $M \leqslant 0.7\ M_{sun}$ and luminosities less than 1/10 of that of our Sun. They represent the low-mass aspect of all phases of the HR diagram, including stellar remnants such as white dwarfs. The challenges presented here, are relevant to all low-mass stars in binaries - from interacting binaries to exoplanets.

2. Challenges in studying low-mass binaries

Low-mass stars are some of the favorites in the field, since they comprise more than 70% of the galactic stellar population and they spend a substantial time at all stages of their life allowing for a generous insight on stellar evolution. Their presence in eclipsing binary systems, provides an opportunity for the accurate determination of component masses, radii and temperatures. These are fundamental parameters enhancing our understanding of stellar structure in stars with mostly or fully convective interiors.

Binary stars are classified as detached, semi-detached or contact depending on the degree each of the two stellar components fills its respective Roche lobe: The main equation of the Roche geometry (Eggleton 1983) is $R_2/a = [0.49q^{2/3}] / [0.6q^{2/3} + \ln(1+q^{1/3})]$ where the subscripts 1 and 2 refer to the "primary" and "secondary" mass components; the mass ratio of the two stars is $q=M_2/M_1$. Most commonly encountered, the *detached* binaries are those where none of the two stellar components fill their respective Roche lobes. When one of the binary components fills its Roche lobe, mass transfer is enabled through the inner Lagrangian point and the system is a *semi-detached* binary. In the rare case where both stellar components fill their respective Roche lobes, the system is known as a *contact* binary.

The main assumption in this present discussion is that the two stars are in Keplerian orbits, with orbital periods less than 3 days. A natural consequence of this is that one or both components are tidally locked to the orbit, forcing them to a very fast rotation. From an observational point of view, these systems require short exposure times (less than 10% of the orbital period) and larger telescope apertures for their study, which becomes observationally expensive. Nevertheless, if we surpass this obstacle, we are left with a very rich suite of phenomena to account for when it comes to studying those binaries and understanding their properties. For organizational purposes, I will discuss low-mass stars in detached and semi-detached binaries separately. I will treat the latter as a more complex version of the former, therefore all the relevant phenomena present in detached binaries are also present in semi-detached ones.

2.1. *Detached binaries*

Spectral deconvolution

One of the main properties of stars in low-mass binaries is their low temperatures, allowing for the presence of molecules in their atmosphere - such as TiO and VO in M stars, and metal hybrid bands (FeH, but also CaH and CrH) and H_2O in cooler objects. Therefore, their spectral energy distribution peaks at (near-)infrared wavelengths. When it comes to binary stars, however, there is still a great lack of tools for disentangling individual stellar spectra. Spectral templates based on observations of individual stars are often used but they do not always reproduce the observations (e.g., Burgasser *et al.* 2010). A library of theoretical spectra is needed; this requires accurate knowledge of spectral atmospheres and molecular opacities in the near infrared in order to reproduce each spectrum and finally apply it to the data. The lack of such a library hinders understanding of low-mass stellar atmospheres and of variable phenomena (such as the presence of clouds in the lowest mass stars) and stellar evolution properties for the lowest mass systems.

Tidal distortions and rotation

The most obvious effect of gravity in low-mass binaries is the tidal distortion of stellar components because of their gravitational interaction. This results in a teardrop-like shape and tidal locking. The effect is primarily in action when the two stellar components are very close to each other and/or their mass ratio is large (for example, in binaries with a white dwarf): the gravitational potential of one of the two stars is distorting the shape of its companion. Tidal locking results in fast axial rotation of one (or both) of the components, also distorting the shape of stars to being more oblate towards the equator. The net observable effects on the light curves are ellipsoidal variations (e.g., Figure 1),

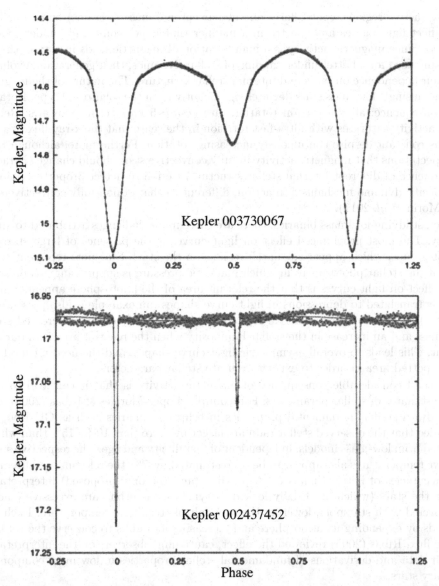

Figure 1. Kepler light curves from Coughlin *et al.* (2011), demonstrating the effect of out-of eclipse light curve distortion because of stellar activity (left) and tidal distortions (right) on the light curve of two different stars.

resulting from observing a larger surface stellar area of the star at quadratures than during conjunctions.

Magnetic activity

Perhaps the largest effect of fast rotation is the enhancement of magnetic activity in stars: there is a well known relationship of magnetic activity with rotation in the sense that all fast rotating stars are active (e.g., Browning *et al.* 2010). Magnetic activity is one of the most well-known but still puzzling challenges in studying low-mass binaries, mostly because of its variable (in time and nature) effects. The usual manifestations of magnetic activity (active regions, spots, flares and stellar coronal mass ejections) on one or both stars affect light curves and spectra in a manner that is not completely understood or modeled. Since magnetic activity is a manifestation of magnetic fields in action, there is a pressing need for a better understanding of stellar dynamos, their generation, evolution and their dependence on mass and internal stellar structure. The nature of the magnetic dynamo in single low-mass stars deviates significantly from the solar $\alpha - \Omega$ type. Rotation is of consequence: after a certain rotation rate ($vsini{\sim}5$ km/sec in M dwarfs; Reiners 2007), activity saturates with ultra-fast rotation in the sense that the x-ray flux of a star remains constant despite continuously increasing rotation. Furthermore, although there are expectations that magnetic activity in fully convective stars should change character, (consequence of different internal stellar structure) we fail to detect properties pointing to a specific dynamo mechanism in action, different to that of non-fully convective stars (e.g., Morin et al. 2011).

When studying low mass binaries, there are three main challenges attributed to stellar activity. The most pronounced effect on light curves is the presence of large starspots or spot groups, which appear as "dark" regions on the star†. Starspots represent cooler parts of the stellar photosphere in which magnetic pressure is suppressing gas pressure. Their effect on light curves is that the relevant area of the photosphere appears fainter, which is translated to depressions in light curve shapes (an example is also presented in Figure 1). At the same time, active regions ("bright" areas) result in increased stellar brightness and an increase in the stellar luminosity when the relevant area is in our line-of-sight. This leads to overall asymmetric light curve shapes, and the need for modeling of the spotted area in order to extract accurate stellar parameters.

A second considerable consequence of magnetic activity is that it can lead to erroneous estimates of stellar parameters. For example, Lopez-Morales & Ribas (2005) studied M dwarf stellar fundamental properties in eclipsing binaries such as GU Boo, and concluded that the observed stellar radii are larger by more than 10% - 15% than what is expected from low-mass models, independent of metallicity and age. The respective stellar effective temperatures also appear to be overestimated by 5% (the relevant measurements have accuracies of better than 3%). A possible (and the only proposed) interpretation is that the stars (which are tidally locked in synchronous orbit) are excessively active and covered with starspots, lowering their overall photospheric temperature. Each star responds by expanding its atmosphere and increasing its radius to conserve the total radiative flux. Ribas (2005) reviewed the effect, cautioning observers on the interpretation of their data and derivations of fundamental stellar properties in low mass components of binary stars.

The third effect is intra-binary emission due to magnetic field interactions between the two stars. This was first observed in the x-ray properties of the K0IVe+G5IV eclipsing binary AR Lac in which x-ray excess in the intra-stellar region attested to an excess of magnetic activity, likely due to magnetic interactions between the two stars (Siarkowski et.al. 1993). These effects are likely induced by the stellar component that has the stronger magnetic field, and they result from magnetospheric interactions between the components of the low-mass binary. Similar effects are now observed in the optical spectra of closed detached binaries and in exoplanets (e.g., Shkolnik et al. 2003), and are prominent on spectral lines such as the Hα emission and CaII H&K lines (e.g., Kafka 2011), which are common diagnostics of magnetic activity.

---

† One could also consider the effect of active regions, which appear as "bright" features on the star.

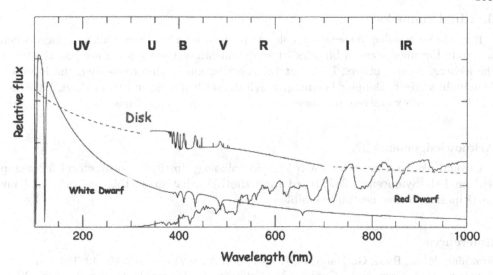

Figure 2. Spectral energy distribution of the various components of a cataclysmic variable from the UV to the near IR demonstrating that accretion-related emission is the main challenge to studying the two stellar components (modified from Hellier 2001).

2.2. *Semi-detached binaries*

Low-mass semi-detached binaries are generally known as cataclysmic variables (CVs), in which a white dwarf (WD) is accreting material from its Roche-lobe filling, low-mass companion (a WD or a low mass, lower main sequence star)†. CVs are widely used for the study and understanding of phenomena connected to a disk or magnetically controlled accretion. At the same time, the presence of accretion is also the main challenge for the study of the two stars in the binary. A pictorial summary of the effect of accretion-generated light, is presented in Figure 2 (Hellier 2001): the spectral energy distribution of the three main components of a cataclysmic variable system (white dwarf and M-dwarf semi-detached binary), are overplotted on a relative intensity scale, from the UV to near-infrared wavelengths. The spectrum of the white dwarf peaks in the UV (as WDs are hot sources) and the K/M dwarf donor star emerges in the near-infrared. In the absence of accretion-generated light, it would be easy to study each of the two stellar components at the wavelength region of their peak emission (UV for the white dwarf and near-infrared for its companion). However, the superimposed spectrum of the accretion disk (or that of the accreting magnetic column in the case of magnetic CVs) onto the composite spectrum of the two stars is dominant at all wavelengths. Therefore, accretion-generated light is overwhelming the light from the binary at all wavelengths and is the main challenge we face while studying semi-detached systems. Related phenomena are reflected on their light curves and spectra, and include outbursts, high and low states, mass transfer variations, magnetic activity, irradiation, accretion disk phenomena (spiral arms, disk rim flaring and accretion hotspot), magnetically-controlled accretion (threading region, atmospheric shocks), outflows and nova explosions.

---

† Some Algols behave like CVs, therefore they present similar phenomena and challenges in their study.

3. Final remarks

It would be possible to spend a whole symposium on only of the challenges faced when we study low mass stars in binaries - the aforementioned phenomena are just the tip of the iceberg. Among others, I left out the contribution of limb darkening, the Rossiter-McLaughlin effect, Doppler beaming, gravitational beaming and irradiation, but these will be covered extensively in other parts of this symposium volume.

Acknowledgements

I would like to acknowledge a NASA Astrobiology Institute postdoctoral fellowship and an IAU Symposium 282 award from the IAU that made this research (and my participation in this meeting) possible.

References

Browning, M. K., Basri, G., Marcy, G. W., West, A. A., & Zhang, J. 2010, *AJ*, 139, 504
Burgasser, A. J., Cruz, K. L., Cushing, M., Gelino, C. R., Looper, D. L., Faherty, J. K., Kirkpatrick, J. D., & Reid, I. N. 2010, *ApJ*, 710, 1142
Coughlin, J. L., López-Morales, M., Harrison, T. E., Ule, N., & Hoffman, D. I. 2011, *AJ*, 141, 78
Eggleton, P. P. 1983, *ApJ*, 268, 368
Hellier, C. 2001, *Cataclysmic Variable Stars: How and Why They Vary* (Springer-Praxis, UK)
López-Morales, M. & Ribas, I. 2005, *ApJ*, 631, 1120
Morin, J., Donati, J.-F., Petit, P., Delfosse, X., Forveille, T., & Jardine, M. M. 2010, *MNRAS*, 407, 2269
Ribas, I. 2005, The Light-Time Effect in Astrophysics: Causes and cures of the O-C diagram, 335, 55
Siarkowski, M. 1995, IAU Symposium, 176, 190P
Shkolnik, E., Walker, G. A. H., & Bohlender, D. A. 2003, *ApJ*, 597, 1092
Kafka, S., Tappert, C., Ribeiro, T., Honeycutt, R. K., Hoard, D. W., & Saar, S. 2010, *ApJ*, 721, 1714
Reiners, A. 2007, *A&A*, 467, 259

Discussion

A. BURROWS: I will remind people that there is not one theoretical radius-mass relation for late M-dwarfs and brown dwarfs, but a range of such models for a given mass and age that depends upon metallicity, and for brown dwarfs upon the (unknown) character of atmospheric cloud opacities. This range in radii for a given mass and age could be $\sim$ 10-15%.

From Interacting Binaries to Exoplanets: Essential Modeling Tools
Proceedings IAU Symposium No. 282, 2011
Mercedes T. Richards & Ivan Hubeny, eds.
© International Astronomical Union 2012
doi:10.1017/S1743921311027086

Brown Dwarf Binaries

Katelyn N. Allers

Department of Physics & Astronomy, Bucknell University
Lewisburg, PA 17837, U.S.A
email: k.allers@bucknell.edu

Abstract. Nearly 500 brown dwarfs have been discovered in recent years. The majority of these brown dwarfs exist in the solar neighborhood, yet determining their fundamental properties (mass, age, temperature & metallicity) has proved to be quite difficult, with current estimates relying heavily on theoretical models. Binary brown dwarfs provide a unique opportunity to empirically determine fundamental properties, which can then be used to test model predictions. In addition, the observed binary fractions, separations, mass ratios, & orbital eccentricities can provide insight into the formation mechanism of these low-mass objects. I will review the results of various brown dwarf multiplicity studies, and will discuss what we have learned about the formation and evolution of brown dwarfs by examining their binary properties as a function of age and mass.

1. Introduction

In this review, I will focus on brown dwarf binaries, where the mass of the primary is less than 0.075 $M_\odot$. Many of the studies presented at this meeting are dealing with samples of thousands, or even tens of thousands of stellar binaries. In contrast, the number of known brown dwarf binaries is ~55. The low number of brown dwarf binaries can be attributed to 3 main properties of brown dwarfs: First, relative to stars, brown dwarfs are rare. The ratio of brown dwarfs to stars is ~20% (Luhman *et al.* 2007). Second, brown dwarfs are intrinsically faint. A 0.06 $M_\odot$ brown dwarf is ~100 times fainter than a 0.10 $M_\odot$ low-mass star. This means multiplicity studies of brown dwarfs are usually limited to

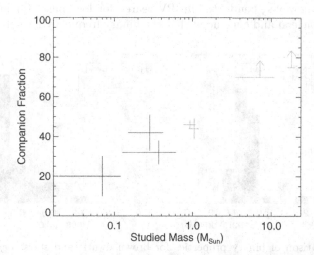

Figure 1. Companion fraction (including binaries and higher order multiple systems) as a function of primary mass. Data points from Mason *et al.* (2009), Kouwenhoven *et al.* (2007), Raghavan *et al.* (2010), Duquennoy & Mayor (1991), Fischer & Marcy (1992), Bergfors *et al.* (2010) and Allen (2007).

the solar neighborhood. Third, the binary fraction of brown dwarfs is significantly lower than that of stars. Figure 1 illustrates the trend of decreasing multiplicity fraction with lower primary mass.

2. Properties of Brown Dwarf Binaries

Given the small number of known brown dwarf binaries, it is very important that these binaries be consistently characterized. For the following analysis, I use the list of brown dwarf binaries from Liu *et al.* (2010), supplemented with new binary discoveries from Artigau *et al.* (2011), Burgasser *et al.* (2011a), Burgasser *et al.* (2011b), Gelino *et al.* (2011) and Liu *et al.* (2011). Where possible, I use values for total masses and semi-major axes determined from orbital monitoring (Dupuy *et al.* 2010, Konopacky *et al.* 2010). Other missing orbital data were taken from www.vlmbinaries.org. Note that this sample does not include known brown dwarf binaries in star forming regions or known young moving groups (which are discussed separately). Figure 2 shows distributions of observational parameters for brown dwarfs, compared to low-mass stars and solar-mass stars. It is important to note that the multiplicity properties of brown dwarfs are an extension of mass-dependent trends.

Compared to stars, brown dwarf binaries exhibit a clear tendency toward equal mass systems, with $\sim$75% of brown dwarf binaries having $M_2/M_1 \geqslant 0.8$. The lack of many low mass ratio systems could, in part, be due to the difficulty in detecting the secondary components. High resolution imaging surveys, however, are complete for $M_2/M_1 \gtrsim 0.6$, so the peak in the mass ratio distribution would be unaffected by incompleteness effects at low mass ratios. Recent advances in image reduction and analysis, however, have led to the recent discovery of a few low mass ratio systems (Burgasser *et al.* 2011b, Liu *et al.* 2011, Liu *et al.* 2010).

The peak of the separation distribution of brown dwarfs is $\sim$3 AU. High resolution imaging surveys are typically not sensitive to binaries with separations smaller than a few AU. As noted by Burgasser *et al.* (2007), the peak of the separation distribution is very close to the incompleteness limit, and thus the actual peak could lie at closer separations. Joergens (2008), however, conducted an RV search for low-mass star and brown dwarf binaries in Chamaeleon and concluded that the binary frequency for sub-AU separations

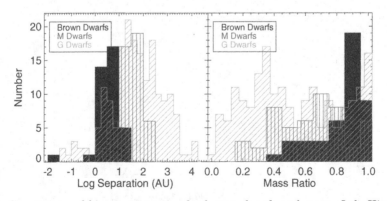

Figure 2. Comparison of binary properties for brown dwarfs and stars. *Left:* Histograms of projected separations for brown dwarf binaries compared to solar-type stars and low-mass stars. *Right:* Histograms of mass ratios (q) for binary brown dwarfs, solar-type stars and low-mass stars. Data for brown dwarf binaries is described in the text. Data for stars comes from Raghavan *et al.* (2010) for solar type stars and Bergfors *et al.* (2010) for M dwarfs.

is small ($\lesssim$10%). Thus, the distribution of brown dwarf binary separations is likely a few AU. Perhaps a more interesting feature of the separation distribution is narrow range of separations for brown dwarf binaries. $\sim$90% of brown dwarf binaries are more closely separated than 10 AU, in contrast to solar-type binaries which have a much broader range of separation.

3. Young Brown Dwarf Binaries

Because the nearest star forming regions are $\sim$125 pc away, high resolution imaging surveys for young brown dwarf binaries, are typically only complete for separations greater than 10 AU. On the other hand, RV searches (e.g. Joergens *et al.* 2008) are sensitive to only small separations (<3 AU). This means that currently, we do not have the observational capability to find 3-10 AU separation brown dwarf binaries (i.e. the peak of the field dwarf distribution) in young clusters. One solution to this problem is to search for nearby, young brown dwarf binaries in the field (e.g. Allers *et al.* 2010), but the low number of known young field brown dwarfs make such studies difficult.

Figure 3 compares the binary properties for "normal" field brown dwarfs and young brown dwarfs. Young brown dwarf binaries appear to have lower mass ratios and larger separations than field brown dwarfs. Both of these effects could result from observational bias and small number statistics. Still, there appears to be a handful of unusually wide separation ($\gtrsim$500 AU) young brown dwarf binaries. It is important to note that these wide systems are very rare (f$\sim$1–2%, Kraus & Hillenbrand 2009), and are unlikely to survive the dispersal of their native cluster (Close *et al.* 2007). Biller *et al.* (2011) determined that the frequency of brown dwarf binaries at separations >10 AU is statistically the same as the wide binary frequency for field brown dwarfs. Clearly, the most important advance in the study of young brown dwarf binaries will be the ability to probe the crucial 3–10 AU separations in star forming regions with JWST or the next generation of large telescopes (e.g. GMT, TMT, ELT).

4. Testing Formation Models

Brown dwarfs present a particular challenge to models of star formation, as their mass is significantly below the typical Jeans mass in star forming clouds. Theories for the

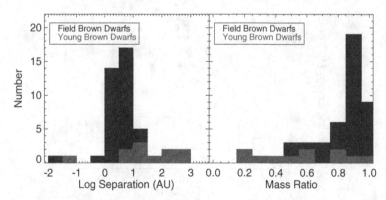

Figure 3. Comparison of binary properties for brown dwarfs in the field (age $\sim$1 Gyr) to young brown dwarfs in star-forming regions and moving groups (age $\lesssim$10 Myr). *Left:* Histograms of projected separations. *Right:* Histograms of mass ratios (q). Data for field brown dwarf binaries is described in the text. Data for young brown dwarf binaries (primary mass < 0.075 M$_\odot$) is taken from Table 1 of Biller *et al.* (2011).

formation of brown dwarfs fall into two categories: 1) lower the Jeans mass via hierarchical or turbulent fragmentation (e.g. Bate 2009) or 2) circumvent the Jeans mass by either ejecting the brown dwarf from its embryo (e.g. Reipurth and Clarke 2001) or forming the brown dwarf from a gravitational instability in a circumstellar disk (Stamatellos & Whitworth 2009). These different formation mechanisms would result in very different binary frequencies, binding energies, and orbital eccentricities.

The close separations and low binary fraction of field brown dwarf binaries initially seemed to support the ejection scenario for forming brown dwarfs. Burgasser *et al.* (2007) present binding energies for several field brown dwarfs and reported that the widest brown dwarf binaries had binding energies 10-20 times higher than field stars. Recent discoveries of less bound field brown dwarfs indicate that the binding energy lower limit for brown dwarf binaries is similar to stars (Figure 4) which argues for a common formation mechanism. Wide, young binaries having extremely low binding energies (Figure 4) could not have formed via ejection.

An additional test of formation models can be made by comparing model predictions for orbital eccentricities. Dupuy & Liu (2011) compare the orbital eccentricities and periods of very low-mass stellar and brown dwarf binaries to the predictions of a numerical simulation of turbulent fragmentation of a cloud and gravitational instability in a massive circumstellar disk. The observed eccentricity distribution agrees quite well with the Bate (2009) model, whereas the Stamatellos & Whitworth (2009) simulation predicts higher eccentricities than observed. As more brown binaries are discovered and well characterized, it will become increasingly important for theoretical models to provide predictions of brown dwarf binary properties.

5. Testing Atmospheric and Evolutionary Models

The atmospheres of brown dwarfs are very complicated, with molecules providing the dominant opacity source. Cloud formation, dust subtended in the photosphere, and non-local chemical equilibrium make brown dwarfs a particularly difficult challenge for modellers (see contributions by F. Allard and A. Burrows). The complexity of the atmospheric models manifests itself in evolutionary models as well (e.g. Burrows *et al.* 2011).

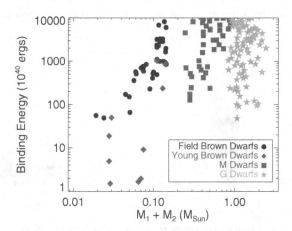

Figure 4. Binding Energies for brown dwarfs in the field (age ∼1 Gyr), young brown dwarfs in star-forming regions and moving groups (age ≲10 Myr), low-mass stars (Bergfors *et al.* 2010) and solar-mass stars (Raghavan *et al.* 2010).

Models atmospheres and isochrones are widely used to estimate the properties of brown dwarfs and extrasolar planets, but the reliability of these results depends on the fidelity of the models. Brown dwarf binaries provide a unique opportunity to empirically determine the fundamental properties of brown dwarfs, which can in turn be used to test the predictions of atmospheric and evolutionary models.

Orbital monitoring has allowed determination of high precision dynamical masses for a handful of brown dwarf binaries (e.g. Dupuy *et al.* 2011, Konopacky *et al.* 2010, Liu *et al.* 2008). One can then compare the log(g) and effective temperatures derived by fitting model atmospheres to the spectra of the binary with those inferred by comparing the luminosity and mass of the binary to evolutionary models. Thus far, the temperatures derived from these two methods differ by ~100-300 K (Dupuy *et al.* 2009, Liu *et al.* 2008). Whether this discrepancy arises from the atmospheric or evolutionary models is unclear. Metallicity could also factor into this difference (Burrows *et al.* 2011), but a large divergence ($\gtrsim 1$ dex) from solar [Fe/H] would be required to account for a 200 K difference in effective temperature.

If the age and/or metallicity of the brown dwarf is independently determined, one can compare the measured luminosity to the luminosity inferred by evolutionary models. To date, the only brown dwarf binary able to provide this test of the models is HD130948BC, which has a known age and metallicity from its G2V primary star (Dupuy *et al.* 2009). In this particular case, the luminosity predicted by the evolutionary models is ~2 times lower than the measured value. Current orbital monitoring surveys should produce dynamical masses for additional brown dwarf binaries in coming years, with the hope of having a large enough sample to start looking at the systematics of these model tests. An additional improvement to our ability to test atmospheric and evolutionary models will come with the publication of parallax data for more brown dwarfs (which will allow precise luminosity measurements).

References

Allen, P. R. 2007, *ApJ*, 668, 492

Allers, K. N., Liu, M. C., Dupuy, T. J., & Cushing, M. C. 2010, *ApJ*, 715, 561

Artigau, É., *et al.*, 2011, *ApJ*, accepted. (arXiv:1107.0768)

Bate, M. R. 2009, *MNRAS*, 392, 590

Bergfors, C., *et al.*, 2010, *A&A*, 520, A54

Biller, B., Allers, K., Liu, M., Close, L. M., & Dupuy, T. 2011, *ApJ*, 730, 39

Burgasser, A. J., Bardalez-Gagliuffi, D. C., & Gizis, J. E. 2011, *AJ*, 141, 70

Burgasser, A. J., Reid, I. N., Siegler, N., Close, L., Allen, P., Lowrance, P., & Gizis, J. 2007, Protostars and Planets V, 427

Burgasser, A. J., Sitarski, B. N., Gelino, C. R., Logsdon, S. E., & Perrin, M. D. 2011, *ApJ*, accepted. (arXiv:1107.1484)

Burrows, A., Heng, K., & Nampaisarn, T. 2011, *ApJ*, 736, 47

Close, L. M., *et al.*, 2007, *ApJ*, 660, 1492

Dupuy, T. J. & Liu, M. C. 2011, *ApJ*, 733, 122

Dupuy, T. J., Liu, M. C., & Ireland, M. J. 2009, *ApJ*, 692, 729

Duquennoy, A. & Mayor, M. 1991, *A&A*, 248, 485

Fischer, D. A. & Marcy, G. W. 1992, *ApJ*, 396, 178

Gelino, C. R., *et al.*, 2011, *AJ*, 142, 57

Joergens, V. 2008, *A&A*, 492, 545

Konopacky, Q. M., Ghez, A. M., Barman, T. S., Rice, E. L., Bailey, J. I., III, White, R. J., McLean, I. S., & Duchêne, G. 2010, *ApJ*, 711, 1087

Kouwenhoven, M. B. N., Brown, A. G. A., Portegies Zwart, S. F., & Kaper, L. 2007, *A&A*, 474, 77

Kraus, A. L. & Hillenbrand, L. A. 2009, *ApJ*, 703, 1511
Liu, M. C., *et al.*, 2011, *ApJ*, in press. (arXiv:1103.0014)
Liu, M. C., Dupuy, T. J., & Ireland, M. J. 2008, *ApJ*, 689, 436
Liu, M. C., Dupuy, T. J., & Leggett, S. K. 2010, *ApJ*, 722, 311
Luhman, K. L., Joergens, V., Lada, C., Muzerolle, J., Pascucci, I., & White, R. 2007, Protostars and Planets V, 443
Mason, B. D., Hartkopf, W. I., Gies, D. R., Henry, T. J., & Helsel, J. W. 2009, *AJ*, 137, 3358
Raghavan, D., *et al.*, 2010, *ApJS*, 190, 1
Reipurth, B. & Clarke, C. 2001, *AJ*, 122, 432
Stamatellos, D. & Whitworth, A. P. 2009, *MNRAS*, 392, 413

Discussion

A. BURROWS: There is not one set of predictions for the luminosity or effective temperature or radius of a brown dwarf at a given age. For a given age, there is a band of possible values of any three due to ambiguities in metallicities and, importantly, this fact needs to temper the zeal of observers when reaching conclusions concerning conflicts with theory. Moreover, ages for stars are not well-determined.

K. ALLERS: It is certainly true that observers need to use caution when comparing models. The example I presented of HD130948BC (Dupuy *et al.* 2009), is a brown dwarf binary with a dynamical mass measurement and with known age and metallicity from its G2V primary.

A. BURROWS: How well can you determine the bolometric luminosity of your component brown dwarfs?

K. ALLERS: Distance determinations are usually the dominant source of error for brown dwarf binaries, as determining bolometric magnitudes by integrating optical to mid-IR photometry and spectroscopy is very straight-forward. For objects without parallax measurements, the uncertainty in L_{bol} can be as large as 50%. The binaries I presented as tests for evolutionary models (HD130948BC and 2MASS J1534-2952) have some of the best distance determinations, yielding small uncertainties in L_{bol} (11% and 5%, respectively).

P. ŠKODA: How many brown dwarfs are known in total? How are they discovered? Is there some systematic survey? And how are the BD binaries discovered? Systematic surveys?

K. ALLERS: To date, we know of ∼500 brown dwarfs (spectroscopically confirmed). Most have been identified from large photometric surveys (2MASS, SDSS, DENIS). Of the ∼55 known brown dwarf binaries, ∼95% have been discovered from high-resolution imaging surveys (HST or laser guide star) targeting nearby brown dwarfs.

R. WILSON: The 2010 version of the WD program gives distances that agree well with HIP parallax distances, so it can stand in for parallax in the few cases where bright and RV curves exist for eclipsing brown dwarfs. Older schemes for EB distance estimation could also do it, of course.

K. ALLERS: Unfortunately, we only know of one eclipsing brown dwarf binary system.

From Interacting Binaries to Exoplanets: Essential Modeling Tools
Proceedings IAU Symposium No. 282, 2011
Mercedes T. Richards & Ivan Hubeny, eds.

© International Astronomical Union 2012
doi:10.1017/S1743921311027098

Circumbinary Planets and the SOLARIS Project

Maciej Konacki[1,2], Piotr Sybilski[1], Stanisław K. Kozłowski[1], Milena Ratajczak[1] and Krzysztof G. Hełminiak[1]

[1] Nicolaus Copernicus Astronomical Center, Department of Astrophysics, ul. Rabianska 8, 87-100 Torun, Poland
e-mail: maciej@ncac.torun.pl

[2] Astronomical Observatory, Adam Mickiewicz University, ul. Sloneczna 36, 60-286 Poznan, Poland

Abstract. Extrasolar planets in binary and multiple star systems have become a noticeable niche with about 50 planets among over 500 known. Here we however focus on a particular subset of exoplanets in binary star systems — circumbinary planets. They have the unique advantage that a search for circumbinary planets does also significantly contribute to the understanding of their parent stars. We review what is currently known about circumbinary planets and then introduce our two projects aimed at detecting circumbinary planets: The TATOOINE project to find circumbinary planets around non-eclipsing double-lined spectroscopic binary stars with precision radial velocities, and the SOLARIS project to detect circumbinary planets with the timing of eclipses of eclipsing binary stars. For the SOLARIS project, we were granted 2.6 million USD to establish a network of at least four robotic 0.5-m telescopes on three continents (Australia, Africa and South America) to carry out precision photometry of a sample of eclipsing binary stars. We expect that both projects will have a large impact also on the observational stellar astronomy.

Keywords. techniques: spectroscopic, photometric, binaries: eclipsing, planetary systems

---

1. Introduction

Searches for planets in close binary systems explore the degree to which stellar multiplicity inhibits or promotes planet formation (Muterspaugh 2005, Muterspaugh *et al.* 2006, Muterspaugh *et al.* 2010). Detection of giant planets orbiting both components of short period ($P < 60$ days) binaries ("circumbinary planets") will have significant consequences for theoretical understandings of how giant planets are formed. The binarity of the central body creates an environment in which the evolution of a protoplanetary disk is substantially different than around single stars (Artymowicz & Lubow, 1994). This must have an effect on the migration of giant planets in a disk as well as on the "parking" mechanism and their final orbit. Likely, also the dynamical interaction between protoplanets and then planets in a multi-planet system should be affected by the central body binarity and presumably result in a different distribution of the orbital elements of planets. Finally, if one assumes that planetary orbits are coplanar with the orbit of an eclipsing binary, then there is an enhanced probability of detecting a circumbinary transiting planet (Konacki 2009a, Ofir 2008, Ofir 2009, Schneider & Chevreton 1990, Schneider 1994).

Recently, there have been a few claims of detecting circumbinary planets around active eclipsing binary stars using the eclipse timing technique. This is however not the first time when substellar companions or planets have been detected with a timing technique. In addition to the confirmed case of the three rocky planets around a millisecond pulsar

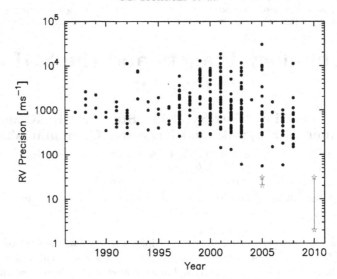

Figure 1. Radial velocity precision (an rms from the orbital fit) for the primaries of double-lined spectroscopic binaries as a function of the publication date based on the ninth catalogue of spectroscopic binary orbits (Pourbaix *et al.* 2004). It is worth noting that already in the early XX century an RV precision of several km/s for double-lined binaries was possible (Plummer *et al.* 1908). Current and previous precision range from our method is delimited with the red stars (figure from Konacki *et al.* 2010).

B1257+12 (Konacki & Wolszczan 2003), planets have been claimed to orbit a pulsar B0329+54 based on the timing of its radio pulses (Demianski & Proszynski 1979, Shabanova 1995). Later, it was demonstrated that the timing variation is quasi-periodic and is not due to planets (Konacki *et al.* 1999). One is left to wonder if these timing variations of active eclipsing binary stars are indeed best explained by a periodic signal due to circumbinary bodies and not an unrecognized quasi-periodic phenomenon (a timing noise) mimicking a periodic planetary signal.

2. The TATOOINE project — high precision radial velocities of double-lined spectroscopic binary stars

Radial velocities (RVs) of double-lined spectroscopic binary stars (SB2s) can be effectively used to derive basic parameters of stars if the stars happen to be eclipsing and one can obtain its light curve or their astrometric relative orbit can be determined. It is quite surprising that the RV precision of double-lined binary stars on the average has not improved much over the last 100 years (see Fig. 1). With the exception of our work (Konacki 2005, Konacki 2009a, Konacki *et al.* 2009b, Konacki *et al.* 2010), the RV precision for such targets typically varies from ~ 0.1 km^{-1} to ~ 1 km^{-1} and clearly is much worse than what has been achieved for stars with planets or single-lined binary stars. The main problem with double-lined binary stars is that one has to deal with two sets of superimposed spectral lines whose corresponding radial velocities change considerably with typical amplitudes of ~ 50-100 kms^{-1}. Consequently, a spectrum is highly variable and obviously one cannot measure RVs by noting a simple shift.

We have developed a novel iodine cell based approach that employs a tomographic disentangling of the component spectra of SB2s and allows one to measure RVs of the components of SB2s with the unique precision on the order of 1-10 m s^{-1} (Konacki 2009a, Konacki *et al.* 2009b, Konacki *et al.* 2010) that is 10 to 100 times more accurate

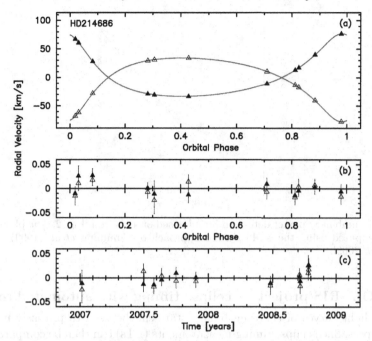

Figure 2. The radial velocities of the spectroscopic binary HD214686 taken with the Shane/CAT (0.9-m !) Hamspec. It is an SB2 with the orbital period of 21.7 days. The 22 measurements (11 for each component) span 2 years (a). The orbital fit residuals are shown as a function of the orbital phase (b) and time (c). The orbital fit is performed simultaneously to the RVs of both components. The rms for the primary is 14.6 ms^{-1} and 14.7 ms^{-1} for the secondary (figure from Konacki *et al.* 2009b).

than archival RVs. Our RV method is described in Konacki (2009a), Konacki *et al.* (2009b), Konacki *et al.* (2010).

The idea to search for circumbinary planets around close *non-eclipsing* double-lined spectroscopic binary stars was born in mid 2003 as a part of our another project (now ended) to search for extrasolar planets in speckle binary star systems with the Keck I/Hires. Over time, it turned out that relatively short period (days to tens of days) double-lined spectroscopic binary stars are actually best suited to measure precision RVs of binary stars as their component's spectra can be disentangled with a tomographic technique.

Our TATOOINE RV project (The Attempt To Observe Outer-planets In Non-single-stellar Enviroments, in collaboration with Matt Muterspaugh) was carried out with the following telescopes/spectrographs: in the years 2003-2007 with the 10-m Keck I/Hires (Keck Obs.), in the years 2006-2007 with the 3.6-m TNG/Sarg (Canary Islands), and in the years 2006-2010 with the 3-m Shane/Hamspec (Lick Obs.). We have also made reconnaissance observations with the 3.9-m AAT/UCLES (Siding Spring Obs.). In a review paper, Udry & Santos (2007) point out that "circumbinary planets offer a completely unexplored field of investigations." However, our data time span for several SB2s is already 8 years long and the first paper from our effort has been published by Konacki (2009a). This publication describes the first ever limits to circumbinary planets thanks to the precision RVs (see Figs. 2-3). We are currently monitoring close to 50 SB2s (now with HARPS). Among them we have at least one curious case of a possible circumbinary planet.

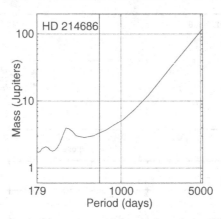

Figure 3. Circumbinary planet detection limits based on RVs from Fig. 2. The planet detection limit was computed using the sophisticated approach of Cumming *et al.* (1999) (figure from Konacki *et al.* 2009b).

3. The SOLARIS project — eclipse timing with automated telescopes

Accurate light curves of eclipsing binary stars can be used to precisely measure the times of eclipses. Such eclipse timing measurements (ETs) can then be compared with the predicted ones and used to infer information on e.g. the presence of an additional body orbiting the eclipsing binary. The presence of an additional body will cause the movement of the eclipsing binary with respect to the center of mass of the entire system and result in advances/delays in the times of eclipses due to the light time effect. This old idea (it dates back to the 17th century and Ole Roemer) has been used to e.g. detect stellar companions to eclipsing binaries. It can also be used to detect circumbinary planets. However, in order to effectively use ET to detect circumbinary planets from the ground, one needs a massive, well designed and coordinated photometric effort employing several telescopes dedicated to just this task.

Why "Solaris"? Solaris is a novel by an outstanding Polish writer Stanislaw Lem (1921-2006). The novel is about a circumbinary planet covered with a supposedly conscious ocean. The ocean is studied by humans (with little success) from a station hovering above its surface. Since the novel was for the first time published in 1961, it precedes Star Wars' Tatooine planet (1977) in terms of the first case of a circumbinary planet in pop culture. Solaris was turned into a movie twice: by a great Russian filmmaker Andrei Tarkovsky (1932-1986) and more recently by a Hollywood filmmaker Steven Soderbergh (starring George Clooney).

The SOLARIS project is part of an effort entitled "Eclipsing binary stars as cutting edge laboratories for astrophysics of stellar structure, stellar evolution and planet formation" funded by the European Research Council through a Starting Independent Researcher Grant 2010 awarded to M. Konacki (2010-2015). This grant enables us to establish a global network of four robotic 0.5-m telescopes (Australia, Argentina and South Africa). These telescopes will be used to detect circumbinary planets via eclipse timing and to provide high precision and high cadence light curves for the spectroscopic part of the effort. Such a combination (RV+photometry) will enable us to derive basic parameters of the components of eclipsing binaries with an unprecedented precision (see Fig. 4). The goal of the Starting Grant project is to survey 200 eclipsing binaries using RVs and another 100 (separate sample) eclipsing binaries using eclipse timing to search for circumbinary planets. The spectroscopic part of the project will be carried out with

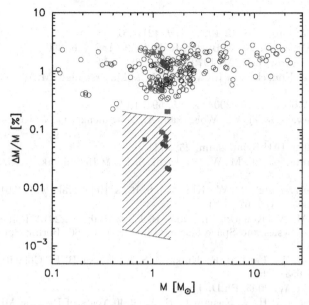

Figure 4. Relative precision in masses for the binary stars with the most accurate mass determinations in the literature. Open circles denote the binaries from a recent review by Torres *et al.* (2010). Filled (blue) circles denote the double neutron star systems B1913+16, B2127+11C, B1534+12, J0737-3039, J1756-2251 and J1906+0746 characterized with radio pulsar timing (Nice *et al.* 2008). Filled (red) rectangles denote the masses for HD78418, HD123999, HD200077 and HD210027 from our recent paper by Konacki *et al.* (2010) and a mass determination for AI Phe from yet another our paper (Helminiak et al 2009). The hatched area is the precision in masses expected to be achieved through this project for double-lined eclipsing binary stars assuming a mass range of the components of 0.5-1.5 $M_\odot$, orbital period range of 3-23 days, orbital inclination range of 85-90°, inclination's error range of 0.05-0.3° and radial velocity amplitude's error range of 1-31 m/s (figure from Konacki *et al.* 2010).

the HRS spectrograph on the 10-m SALT; when HRS becomes available, perhaps from the second half of 2012.

4. The impact of the project on the stellar astronomy.

Modern tests of stellar structure and evolution models require stellar masses accurate to 1% or better (Torres *et al.* 2010). Stellar masses are most often derived from the observations of eclipsing binary stars by combining the photometric (light curves) and spectroscopic data (RVs) for double-lined binaries. The relative error in the masses ($\Delta M/M$) is dominated by the error in the orbital inclination, i, $3\Delta i \cos(i)/\sin(i)$, but the errors in the velocity amplitudes obviously also contribute. In particular, when one seeks high precision (below 1% in the masses). The best results are achieved by targeting eclipsing binaries ($sin(i) \approx 1$, note the dependence of the error in masses on i). Our RV technique allows us to derive the velocity amplitudes, $K_{1,2}$, of the binary components with a precision of several ms^{-1} which corresponds to the relative precision ($\Delta K/K$) of the velocity amplitudes of 0.01% (the RV amplitudes of SB2s are typically $\sim 50 - 100$ kms^{-1}). It easily allows us to derive the mass with a precision up to 0.001%! (see Konacki *et al.* 2010 and Fig. 4). The fact that our stars are eclipsing will hence enable us to combine our precision RVs and our own light curves and determine precise masses, radii and orbital parameters of our targets.

References

[1] Artymowicz, P. & Lubow, S. H. 1994, *ApJ*, 421, 651
[2] Cumming, A., Marcy, G. W., & Butler, R. P. 1999, *ApJ*, 526, 890
[3] Demianski, M. & Proszynski, M. 1979, *Nature*, 282, 383
[4] Hełminiak, K. G., Konacki, M., Ratajczak, M., & Muterspaugh, M. W. 2009, *MNRAS*, 400, 969
[5] Konacki, M. & Wolszczan, A. 2003, *ApJL*, 591, L147
[6] Konacki, M., Lewandowski, W., Wolszczan, A., Doroshenko, O., & Kramer, M. 1999, *ApJ*, 519, L81
[7] Konacki, M. 2009a, IAU Symposium, 253, 141
[8] Konacki, M., Muterspaugh, M. W., Kulkarni, S. R., & Hełlminiak, K. G. 2009b, *ApJ*, 704, 513
[9] Konacki, M., Muterspaugh, M. W., Kulkarni, S. R., & Hełminiak, K. G. 2010, *ApJ*, 719, 1293
[10] Konacki, M. 2005a, *ApJ*, 626, 431
[11] Muterspaugh, M. W., Konacki, M., Lane, B. F., & Pfahl, E. 2010, Planets in Binary Star Systems, Astrophysics and Space Science Library, Vol. 366. Berlin: Springer, ISBN: 978-90-481-8686-0, 77
[12] Muterspaugh, M. W., Lane, B. F., Kulkarni, S. R., Burke, B. F., Colavita, M. M., & Shao, M. 2006, *ApJ*, 653, 1469
[13] Muterspaugh, M. W. 2005, Ph.D. Thesis
[14] Nice, D. J., Stairs, I. H., & Kasian, L. E. 2008, 40 Years of Pulsars: Millisecond Pulsars, Magnetars and More, 983, 453
[15] Ofir, A. 2009, IAU Symposium, 253, 378
[16] Ofir, A. 2008, *MNRAS*, 387, 1597
[17] Plummer, H. C. K., Wright, W. H., & Turner, A. B. 1908, Lick Observatory Bulletin, 5, 21
[18] Pourbaix, D., *et al.* 2004, *A&A*, 424, 727
[19] Shabanova, T. V. 1995, *ApJ*, 453, 779
[20] Schneider, J. 1994, *Planetary and Space Science*, 42, 539
[21] Schneider, J. & Chevreton, M. 1990, *A&A*, 232, 251
[22] Torres, G., Andersen, J., & Gimenez, A. 2010, *A&AR*, 18, 67
[23] Udry, S. & Santos, N. C. 2007, *A&AR*, 45, 397

Discussion

P. STEE: We know for single stars that there is a strong migration of the planets, especially for giants planets. Do you think that the situation can be similar for planets around double or multiple stars?

M. KONACKI: This is certainly one of the things that we would like to find out thanks to our two surveys.

D. QUELOZ: I think one second on the timing of the eclipse is very difficult to get on the ground due to the red noise effect.

M. KONACKI: We are aware of this issue. Our simulations suggest that red noise does have an impact on the timing precision but that it is not very serious. We have also made preliminary observations and compared the eclipse timing precision with our simulations and they are consistent.

S. ALBRECHT: It will be difficult to observe a complete eclipse. Therefore it will be very challenging to get 1 second precision.

M. KONACKI: We will select these detached eclipsing binaries from the quite extensive ASAS catalogue which have possibly short eclipses.

From Interacting Binaries to Exoplanets: Essential Modeling Tools
Proceedings IAU Symposium No. 282, 2011   © International Astronomical Union 2012
Mercedes T. Richards & Ivan Hubeny, eds.   doi:10.1017/S1743921311027104

Probing Bow Shocks Around Exoplanets During Transits

A. A. Vidotto, M. Jardine, and C. Helling

SUPA, School of Physics and Astronomy, University of St Andrews, North Haugh,
St Andrews, KY16 9SS, UK
email: Aline.Vidotto@st-andrews.ac.uk

Abstract. Here, we summarise the conditions that might lead to the formation of a bow shock surrounding a planet's magnetosphere. Such shocks are formed as a result of the interaction of a planet with its host star wind. In the case of close-in planets, the shock develops ahead of the planetary orbit. If this shocked material is able to absorb stellar radiation, the shock signature can be revealed in (asymmetric) transit light curves. We propose that this is the case of the gas giant planet WASP-12b, whose near-UV transit observations have detected the presence of an extended material ahead of the planetary orbit. We show that shock detection through transit observations can be a useful tool to constrain planetary magnetic fields.

Keywords. planets and satellites: general – stars: winds, outflows – stars: coronae

The Earth hosts a bow shock that is formed around its magnetosphere as a result of its interaction with the solar wind. Analogously, from the interaction of an exoplanet with the coronal material of its host star, similar shock structures are expected to develop. A bow shock around a planet is formed when the relative motion between the planet and the stellar corona/wind is supersonic. The shock configuration depends on the direction of the flux of particles that arrives at the planet. If the dominant flux of particles impacting on the planet arises from the (radial) wind of its host star, the normal to the shock always points towards the star ("dayside-shock"). This is the case for a planet orbiting sufficiently distant from its host star, e.g., the Earth. If, on the other hand, the planet orbits very close to the star, the stellar wind velocity is much smaller than the Keplerian velocity u_K of the planet. In this case, the dominant flux of particles impacting on the planet arises from the relative azimuthal velocity between the planetary orbital motion and the ambient plasma. The shock, therefore, forms ahead of the planetary orbit. The velocity of the particles that the planet 'sees' is supersonic if $\Delta u = |u_K - u_\varphi| > c_s$, where u_φ is the azimuthal velocity of the stellar corona.

1. WASP-12b's Magnetic Field

WASP-12b orbits its host star (mass $M_* = 1.35\ M_\odot$, radius $R_* = 1.57\ R_\odot$) at an orbital radius of $a = 3.15\ R_*$ (Hebb *et al.* 2009) and the planet moves at a Keplerian orbital velocity of $u_K = (GM_*/a)^{1/2} \sim 230$ km s^{-1} around the star. At such a close distance, the stellar wind should still present a low velocity. For simplicity, here we assume that this velocity is ≈ 0 (more details can be found in Vidotto *et al.* 2010, 2011a, b). Therefore, stellar coronal material is compressed ahead of the planetary orbital motion, possibly forming a bow shock ahead of the planet. We believe this material is able to absorb enough stellar radiation, causing the early-ingress observed in the near-UV light curve (Fossati *et al.* 2010).

By measuring the phase difference between the beginnings of the near-UV and optical transits, Lai *et al.* (2010) derived the stand-off distance from the absorbing material (shock) to the centre of the planet: $r_M \simeq 4.2\ R_p$. We take this distance to be the extent of the planetary magnetosphere. Pressure balance between the coronal total pressure and the planet total pressure requires that, at r_M,

$$\rho_c \Delta u^2 + \frac{[B_c(a)]^2}{8\pi} + p_c = \frac{[B_p(r_M)]^2}{8\pi} + p_p, \tag{1.1}$$

where ρ_c, p_c and $B_c(a)$ are the local coronal mass density, thermal pressure, and magnetic field intensity, and p_p and $B_p(r_M)$ are the planet thermal pressure and magnetic field intensity at r_M. Neglecting the kinetic term and the thermal pressures in Eq. (1.1), we have that $B_c(a) \simeq B_p(r_M)$. For dipolar stellar and planetary magnetic fields, we have

$$B_p = B_* \left(\frac{R_*/a}{R_p/r_M} \right)^3, \tag{1.2}$$

where B_* and B_p are the magnetic field intensities at the stellar and planetary surfaces, respectively. Using the upper limit of $B_* < 10$ G provided by Fossati et al. (2010b), our model predicts $B_p < 24$ G for WASP-12b.

2. Bow Shocks in Other Exoplanets?

To extend the previous model to other transiting systems, near-UV data must be acquired. In Vidotto et al. (2011a), we presented a classification of the known transiting systems according to their potential for producing shocks that could cause observable light curve asymmetries. Once the conditions for shock formation are met, for it to be detected, it must compress the local plasma to a high enough density to cause an observable level of optical depth. Essentially, this classification is governed by the density n of the medium surrounding the planet. Assuming a hydrostatic, isothermal corona

$$\frac{n}{n_0} = \exp\left\{ \frac{u_K^2}{c_s^2}\left[1 - \frac{a}{R_*}\right] + \frac{v_{\rm rot}^2}{c_s^2}\left[\frac{a}{R_*} - 1\right] \right\}, \tag{2.1}$$

where n_0 is the density at the base of the corona and $v_{\rm rot}$ is the stellar rotation velocity. We illustrate here the case where the coronal material is considered to be corotating with the star, such that $u_\varphi = v_{\rm rot}a/R_*$. Therefore, the maximum temperature that still allow shock formation is such that $c_s = |u_K - v_{\rm rot}a/R_*|$. With these assumptions, we estimate a minimum density required for shock formation through Eq. (2.1). We found that the most promising candidates to present shocks are: WASP-19b, WASP-4b, WASP-18b, CoRoT-7b, HAT-P-7b, CoRoT-1b, TrES-3, and WASP-5b.

We highlight that variations in the medium surrounding the plane, such as due to coronal mass ejections, the star's magnetic cycle, or even due to an eccentric planetary orbit, can cause temporal variations in the shock characteristics. Ultimately, this induces temporal variations in transit light curves (Vidotto et al. 2011b).

References

Fossati, L., Haswell, C. A., Froning, C. S., et al., 2010a, ApJ (Letters), 714, L222
Fossati, L., Bagnulo, S., Elmasli, A., et al., 2010b, ApJ, 720, 872
Hebb, L., Collier-Cameron, A., Loeillet, B., et al., 2009, ApJ, 693, 1920
Lai, D., Helling, Ch., & van den Heuvel, E. P. J. 2010, ApJ, 721, 923
Vidotto, A. A., Jardine, M., & Helling, Ch. 2010, ApJ (Letters), 722, L168
Vidotto, A. A., Jardine, M., & Helling, Ch. 2011a, MNRAS (Letters), 411, L46
Vidotto, A. A., Jardine, M., & Helling, Ch. 2011b, MNRAS, 414, 1573

From Interacting Binaries to Exoplanets: Essential Modeling Tools
Proceedings IAU Symposium No. 282, 2011   © International Astronomical Union 2012
Mercedes T. Richards & Ivan Hubeny, eds.   doi:10.1017/S1743921311027116

The Impact of Red Noise in Radial Velocity Planet Searches

Roman V. Baluev[1,2]

[1] Pulkovo Astronomical Observatory,
Pulkovskoje sh. 65/1, Saint Petersburg 196140, Russia

[2] Sobolev Astronomical Institute, Saint Petersburg State University,
Universitetskij pr. 28, Petrodvorets, Saint Petersburg 198504, Russia
email: roman@astro.spbu.ru

Abstract. We demonstrate that moderate exoplanet radial velocity searches are often subject to the effect of the correlated (red) radial velocity noise. When disregarded, this effect may induce strong distortions in the results of the time series analysis and, ultimately, can even lead to false planet detections. We construct a maximum-likelihood algorithm, which is able to manage this issue rather efficiently.

Keywords. methods: data analysis, methods: statistical, techniques: radial velocities

Traditionally, when analysing radial velocity (RV) time series of exoplanetary surveys, we assume that these measurements possess mutually independent (uncorrelated) errors. Such uncorrelated noise is also called white, since its frequency spectrum is uniform. This is only an assumption, however: various activity effects can easily generate autocorrelated RV noise.

We find that the residual RV noise of many known planet-hosting stars demonstrate non-uniform frequency spectra (see Fig. 1 as example for GJ876). According to the Wiener-Khinchin theorem, such non-white spectra indicate autocorrelated noise. Obviously, this correlated structure affects the results of any data analysis, and should be taken into account.

We find that the shape of the residual power spectrum is well consistent with the so-called red-noise model: we can see a low-frequency band with excessive power (the red noise itself), its diurnal alias band around the frequency of 1 day^{-1}, and a depression in the middle.

The particular examples are the stars GJ876 (Baluev 2011) and GJ581, both are M dwarfs hosting multi-planet extrasolar systems. For these stars, the red noise is seen simultaneously in the time series issued by different observatories, confirming that its "redness" is caused by the star itself, rather than by the instruments.

Typically, the parametric uncertainties are underestimated, if we do not take the noise correlations into account. Also, the estimated values for some of the parameters may shift systematically. Ultimately, the non-white noise may result in false planet detections, which is probably the case of the putative planet candidates GJ581 f and g announced by Vogt et al. (2010). For this exo-system, the periodogram peaks previously interpreted as planets f and g disappear when we take the red noise into account.

We analysed RV data for the 23 HARPS planet-hosting targets currently available in the Vizier catalogue database. Four stars (GJ876, GJ581, HD145377, HD40307) demonstrated clearly non-uniform residual frequency spectra, implying that $\sim 20\%$ of the stars generate this type of RV noise. However, if we select only the stars for which such

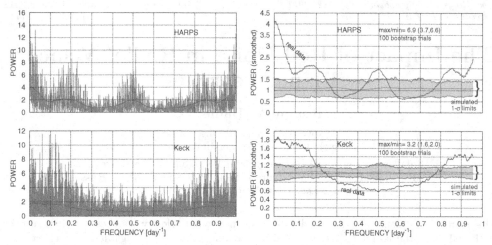

Figure 1. Raw (left) and smoothed (right) residual periodograms for RV data for the four-planet GJ876 extrasolar system. The contribution of the known four planets was subtracted off when constructing these periodograms. The periodograms show systematic deviation from the uniform (white) frequency spectrum: a prominence at low frequencies (long periods), accompanied by its alias at unit frequency (diurnal alias period), with a depression in the middle. These deviations are well above expected statistical limits for the white noise, and are statistically significant. See details in (Baluev 2011).

correlatedness is practically detectable, say only 15 targets for which we have at least 50 RV measurements, then the fraction of "red-noising" stars grows even more, to ~ 25%.

We provide a working maximum-likelihood data analysis algorithm to deal with this issue in the first approximation (Baluev 2011). However, this correlated noise requires a detailed investigation in the future, since we need to have a reliable error model for ongoing and future RV exoplanetary surveys.

This work was supported by the Russian Academy of Sciences (programme "Origin and Evolution of Stars and Galaxies") and by the Russian President programme of support of leading scientific schools (grant NSh.3290.2010.2).

References

Baluev, R. V. 2011, *Celest. Mech. Dyn. Astron.*, 111, 235
Vogt, R. V. 2010, *ApJ*, 723, 954

From Interacting Binaries to Exoplanets: Essential Modeling Tools
Proceedings IAU Symposium No. 282, 2011   © International Astronomical Union 2012
Mercedes T. Richards & Ivan Hubeny, eds.   doi:10.1017/S1743921311027128

Pre-Cataclysmic System V471 Tau with Confirmed Brown Dwarf and Suspected Extrasolar Planet

Ladislav Hric and Emil Kundra†

Astronomical Institute, Slovak Academy of Sciences, 059 60 Tatranská Lomnica, Slovakia
email: hric@ta3.sk

Abstract. We can show that analysis of the (O-C) diagram is a powerful method of detecting new bodies in binary systems. For this purpose we need very symmetric minima with precisely determined shapes. In the case of good covering by observations with high time resolution, it is possible to determine the times of such minima with sufficient accuracy. In the case of V471 Tau, the (O-C) diagram gave us residua which can be explained by the presence of a fourth body with substellar mass in the system.

Keywords. Eclipsing binaries, cataclysmic variables, V471 Tau, exoplanets

1. Introduction and solution

V471 Tau is a binary pre-cataclysmic system of post-common envelope stage consisting of a white dwarf and main sequence star with an orbital period of 0.521183439 days. Existing eclipses in the system are characterized by a fast drop to the minimum as well as by a fast increase in the brightness from the minimum. The change of brightness caused by the eclipse of the white dwarf takes only 55 seconds and the particular eclipse takes only 49 minutes.

Recently we analyzed the (O-C) diagram of this system and solved the long-term problem of (O-C) diagram behaviour and model of the system (Kundra & Hric 2010, Hric, Kundra & Dubovský 2011). We showed that this behaviour can be explained by the presence of a third body in the system. In our case, it is a brown dwarf on an eccentric orbit with a period of 33.2 years.

Moreover, we subtracted the influence of this body from the (O-C) diagram and constructed residua (bottom panel of Fig. 1). We can see that there is still some variability in these residua and after detailed periodic analysis the resulting period is 9.3 years. In our opinion, such behaviour can be explained by another body in the system. It is worth to note that there is an alternative explanation using Applegate's mechanism. If the secondary star has a strong magnetic field with a quadruple shape, the changes of the surface of the secondary component could produce variations in the (O-C) diagram.

The output from period analysis was used as an input parameter to model the fourth body using the 3T code (Pribulla *et al.* 2005). The results are physical and orbital parameters of the fourth body in the V471 Tau system listed in Table 1. The suspected fourth body has planetary parameters and a mass in the interval from 6 to 15 $M_{Jupiter}$, and thus we can announce the presence of a giant extrasolar planet in this triple system.

† This study was supported by the project of the Slovak Academy of Sciences *VEGA* Grant No. 2/0078/10.

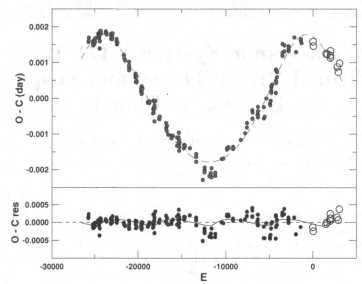

Figure 1. (O-C) diagram of V471 Tau calculated according to ephemeris of close binary. Our new observations are depicted by open circles and points represent all older minima. The bottom plot shows the residua with respect to our model of the third body with variations caused by the suspected fourth body in the system. The curve is the fit of the fourth body.

Table 1. Final physical and geometrical parameters of the fourth component of V471 Tau system. P_4 - orbital period of the fourth body, e_4 - eccentricity of the fourth component, ω_4 - argument of periastron, $T_{periastron}$ - time of periastron passage, $a_{12} \sin(i)$ - main axis orbital projection of the close binary round the center of gravity to the orbital plane; T_0 - time of minimum of the close binary selected for zero epoch, P_0 - period of close binary, $f(M_4)$ - mass function of the fourth component.

Parameter	Value	Unit
P_4	9.3 ± 0.18	year
e_4	0.513 ± 0.027	
ω_4	330 ± 10	$^\circ$
$T_{periastron}$	$2\,454\,923 \pm 202$	HJED
$a_{12} \sin(i)$	0.017 ± 0.004	AU
T_0	$2\,454\,028.452551$	HJED
	$\pm 4.1 \times 10^{-5}$	
P_0	0.521183439	day
	$\pm 2.01 \times 10^{-9}$	
$f(M_4)$	5.99928×10^{-8}	M_{Sun}
	$\pm 1.9 \times 10^{-9}$	

References

Evren, S., Ibanoglu, C., Tunca, Z., & Tunner, O. 1985, *Ap&SS*, 120, 97

Guinan, E. F. & Ribas, I. 2001, *ApJ*, 546, 43

Guinan, E. F. & Sion, E. M. 1984, *AJ*, 89, 1252

Hric, L., Kundra, E., & Dubovský, P. 2011, *Contrib. Astron. Obs. Skalnaté Pleso*, 41, 39

Hric, L., Petrík, K., Niarchos, P., & Gális, R. 2003 *in Stellar Astrophysics - a tribute to Helmut A. Abt, ed.: K.S. Cheng, K.C. Leung, & T.P. Li (Dordrecht: Kluwer Cademic Publishers), Astrophysics and Space Science Library*, 298, 207

Ibanoglu, C., Evren, S., Tas, G., & Cakirlh, O. 2005 *MNRAS*, 360, 1077

Kamiński, K.Z., Ruciński, S.M., Matthews, J.M., Kuschnig, R., Rowe, J.F., Guenther, D.B., Mofat, A.F.J., Sasselov, D., Walker, G.A.H., & Weiss, W.W. 2007 *AJ*, 134, 1206

Kundra, E. & Hric, L. 2010, *Ap&SS*, 331, 121

Pribulla, T., Chochol, D., Tremko, J., & Kreiner, J.M. 2005 *in The Light-Time Effect in Astrophysics, ed.: C. Sterken, PASP*, 103

From Interacting Binaries to Exoplanets: Essential Modeling Tools
Proceedings IAU Symposium No. 282, 2011
Mercedes T. Richards & Ivan Hubeny, eds.
© International Astronomical Union 2012
doi:10.1017/S174392131102713X

The Orbital Period Distribution of Cataclysmic Variables Found by the SDSS

John Southworth[1], Boris T. Gänsicke[2] & Elmé Breedt[2]

[1] Astrophysics Group, Keele University, Staffordshire, ST5 5BG, UK
[2] Department of Physics, University of Warwick, Coventry, CV4 7AL, UK

Abstract. The orbital period is one of the most accessible observables of a cataclysmic variable. It has been a concern for many years that the orbital period distribution of the known systems does not match that predicted by evolutionary theory. The sample of objects discovered by the Sloan Digital Sky Survey has changed this: it shows the long-expected predominance of short-period objects termed the 'period spike'. The minimum period remains in conflict with theory, suggesting that the angular momentum loss mechanisms are stronger than predicted.

Keywords. stars: dwarf novae — stars: novae, cataclysmic variables – stars: white dwarfs

1. The period spike: unveiled

The evolution of all binary systems containing a compact object is driven by the loss of angular momentum from the orbit. Unfortunately, two of the most important mechanisms, common envelope evolution and magnetic braking, are poorly understood. A major advance in our understanding of binary evolution requires the characterisation of a large and homogeneously selected sample of close binaries, such as the population of cataclysmic variables (CVs) spectroscopically discovered by the SDSS (Szkody *et al.*, 2002, and later papers). We are therefore undertaking a project to measure the orbital periods of all SDSS CVs. Of the 291 known systems, 153 now have reliable orbital period measurements and 46 have approximate measurements (Gänsicke *et al.*, 2009; Dillon *et al.*, 2008; Southworth *et al.*, 2006, 2007ab, 2008ab, 2009, 2010ab). Fig. 1 compares the orbital period distributions of SDSS and non-SDSS CVs.

CVs evolve towards shorter orbital periods, reach a minimum value, and then bounce back to longer periods due to structural changes in the low-mass secondary star. A long-standing prediction of CV evolution theory is an accumulation of systems at a minimum period somewhere between 60 and 70 min (Rappaport, Joss & Webbink 1982) where the evolutionary timescale becomes long and the period derivative passes through zero. Unfortunately, the observed population has persistently shown both a longer minimum period of around 80 min and no significant increase in the number of CVs at this period. The population of SDSS CVs, however, shows for the first time a significant excess of objects at a minimum period interval of 80–86 min (Gänsicke *et al.*, 2009). This period spike is composed primarily of CVs which are faint and therefore have been missed by previous surveys. The position of the spike remains at a longer period than predicted, implying that the angular momentum loss is faster than expected.

On recent observing runs with the ESO New Technology Telescope, we have discovered eclipses in four faint CVs within our project. SDSS J075653.11+085831.8 shows 2 mag deep eclipses on a period of 197 min. The system SDSS J093537.46+161950.8 has 1 mag deep eclipses on a period of 92 min, SDSS J105754.25+275947.5 has short and deep eclipses and an orbital period of 90 min, and SDSS J143209.78+191403.5 shows 1.5 mag deep eclipses spaced by 169 min (paper in preparation). Eclipsing CVs are hugely valuable

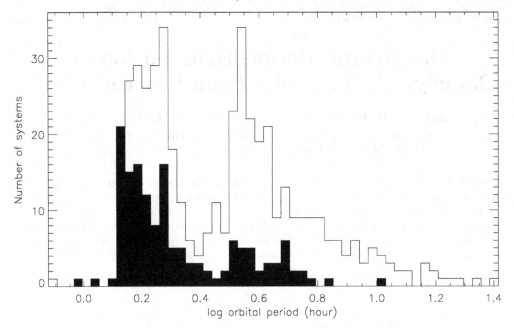

Figure 1. The orbital period distribution of SDSS CVs (black histogram) compared to that of the known non-SDSS CVs (white histogram) as catalogued by Ritter & Kolb (2003).

because they are the only examples whose physical properties can be measured to high precision (e.g. Littlefair *et al.*, 2006, 2008; Southworth & Copperwheat 2011). Follow-up observations of these objects are planned.

References

Dillon, M., *et al.*, 2008, *MNRAS*, 386, 1568

Gänsicke, B. T., *et al.*, 2009, *MNRAS*, 397, 2170

Littlefair, S. P., Dhillon, V. S., Marsh, T. R., Gänsicke, B. T., Southworth, J., & Watson, C. A. 2006, *Science*, 314, 1578

Littlefair, S. P., Dhillon, V. S., Marsh, T. R., Gänsicke, B. T., Southworth, J., Baraffe, I., Watson, C. A., & Copperwheat, C. 2008, *MNRAS*, 388, 1582

Rappaport, S., Joss, P. C., & Webbink, R. F. 1982, *ApJ*, 254, 616

Ritter, H. & Kolb, U. 2003, *A&A*, 404, 301

Southworth, J. & Copperwheat, C. M. 2011, *The Observatory*, 131, 66

Southworth, J., Gänsicke, B. T., Marsh, T. R., de Martino, D., Hakala, P., Littlefair, S., Rodriguez-Gil, P., & Szkody, P. 2006, *MNRAS*, 373, 687

Southworth, J., Gänsicke, B. T., Marsh, T. R., de Martino, D., & Aungwerojwit, A. 2007a, *MNRAS*, 378, 635

Southworth, J., Marsh, T. R., Gänsicke, B. T., Aungwerojwit, A., de Martino, D., & Hakala, P. 2007b, *MNRAS*, 382, 1145

Southworth, J., *et al.*, 2008a, *MNRAS*, 391, 591

Southworth, J., Townsley, D. M., & Gänsicke, B. T. 2008b, *MNRAS*, 388, 709

Southworth, J., Hickman, R. D. G., Marsh, T. R., Rebassa-Mansergas, A., Gänsicke, B. T., Copperwheat, C. M., & Rodriguez-Gil, P. 2009, *A&A*, 507, 929

Southworth, J., Copperwheat, C. M., Gänsicke, B. T., & Pyrzas, S. 2010a, *A&A*, 510, A100

Southworth, J., Marsh, T. R., Gänsicke, B. T., Steeghs, D., & Copperwheat, C. M. 2010b, *A&A*, 524, A86

From Interacting Binaries to Exoplanets: Essential Modeling Tools
Proceedings IAU Symposium No. 282, 2011   © International Astronomical Union 2012
Mercedes T. Richards & Ivan Hubeny, eds.   doi:10.1017/S1743921311027141

Search for the Star-Planet Interaction

Tereza Krejčová[1], Ján Budaj[2] and Július Koza[2]

[1] Dept. of Theoretical Physics and Astrophysics, Masaryk University,
Brno, Czech Republic, email: terak@physics.muni.cz

[2] Astronomical Institute, Tatranská Lomnica, Slovak Republic, email: budaj@ta3.sk

Abstract. We analyse the chromospherical activity of stars with extrasolar planets and search for a possible correlation between the equivalent width of the core of the Ca II K line and orbital parameters of the planet. We found statistically significant evidence that the equivalent width of the Ca II K line reversal, which originates in the stellar chromosphere, depends on the orbital period $P_{\rm orb}$ of the exoplanet. Planets orbiting stars with $T_{\rm eff} < 5\,500\,{\rm K}$ and with $P_{\rm orb} < 20\,{\rm days}$ generally have much stronger emission than planets at similar temperatures but at longer orbital periods. $P_{\rm orb} = 20\,{\rm days}$ marks a sudden change in behaviour, which might be associated with a qualitative change in the star-planet interaction.

Keywords. Ca II K line, exoplanet, star-planet interaction.

1. Introduction

The question of the possible existence of star-planet interactions is currently studied in many ways. Based on the observations in the optical region Shkolnik *et al.* (2005, 2008) discovered the planetary induced variability in the cores of Ca II H & K, Hα and Ca II IR triplet in a few planet hosting stars. Knutson *et al.* (2010) found a correlation between the chromospheric activity of the star and presence of the stratosphere on the planet. Consequently, Hartman (2010) found a correlation between the surface gravity of Hot Jupiters and the stellar activity. Recently Canto Martins *et al.* (2011) searched for correlation between planetary parameters and the $\log R'_{\rm HK}$ parameter but didn't reveal any convincing proof for such a phenomenon.

2. Observation & Statistical Analysis

We used the FEROS instrument on the 2.2m ESO/MPG telescope to obtain spectra of several stars (HD 179949, HD 212301, HD 149143 and Wasp-18) with close-in exoplanets. We also used the publicly available spectra from the HIRES spectrograph archive. Subsequently we measured the equivalent width of the central reversal in the core of Ca II K.

In the first case we divided our data sample into two groups according to the semi-major axis ($a \leqslant 0.15$ and $a > 0.15\,{\rm AU}$). Figure 1 (left-top) shows the dependence of equivalent width on the effective temperature of the star. Subsequently, we performed two statistical tests – Student's t-test and the Kolmogorov-Smirnov test to determine whether the two groups originate from the same population. The resulting probability is a function of temperature and is plotted in the lower part of Figure 1. The tests show that the difference between the two samples is significant for $T_{\rm eff} \leqslant 5\,500\,{\rm K}$. It means that stars with lower temperatures and with planets on closer orbits show more activity as measured in the core of Ca II K line.

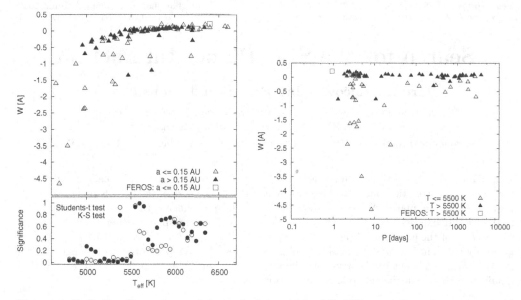

Figure 1. Left *Top*: Dependence of the equivalent width of Ca II K reversal on the temperature of the parent star. Empty triangles are exoplanetary systems with $a \leqslant 0.15$ AU, full triangles are systems with $a > 0.15$ AU. *Bottom*: Statistical Student's t-test (empty circles) and Kolmogorov-Smirnov test (full circles). Red squares are data from FEROS. **Right** Dependence of the equivalent width of Ca II K on the orbital period. Empty triangles are exoplanetary systems with $T \leqslant 5\,500$ K, full triangles are systems with $T > 5\,500$ K and red squares are data from FEROS.

In the second case, we group the data according to the effective temperature of the parent star ($T_{\mathrm{eff}} \leqslant 5\,500$ K and $T_{\mathrm{eff}} > 5\,500$ K) and plot the equivalent width of the Ca II K line reversal as a function of the orbital period (Figure 1–right).

Acknowledgements

This work has been supported by grant GA ČR GD205/08/H005, VEGA 2/0078/10, VEGA 2/0074/09, VEGA 2/0094/11 and the National scholarship programme of Slovak Republic. This research has made use of the Keck Observatory Archive (KOA), which is operated by the W. M. Keck Observatory and the NASA Exoplanet Science Institute (NExScI), under contract with the National Aeronautics and Space Administration. We want to thank Tomáš Henych for fruitful discussions.

References

Canto Martins, B. L., Das Chagas, M. L., Alves, S., *et al.*, 2011, *A&A*, 530, A73
Hartman, J. D. 2010, *ApJ*, 717, L138
Knutson, H. A., Howard, A. W., & Isaacson, H. 2010, *ApJ*, 720, 1569
Shkolnik, E., Bohlender, D. A., Walker, G. A. H., & Collier Cameron, A. 2008, *ApJ*, 676, 628
Shkolnik, E., Walker, G. A. H., Bohlender, D. A., Gu, P., & Kurster, M. 2005, *ApJ*, 622, 1075

From Interacting Binaries to Exoplanets: Essential Modeling Tools
Proceedings IAU Symposium No. 282, 2011
Mercedes T. Richards & Ivan Hubeny, eds.
© International Astronomical Union 2012
doi:10.1017/S1743921311027153

An Extra-Solar Planet in a Double Stellar System: The Modelling of Insufficient Orbital Elements

Eva Plávalová[1], Nina A. Solovaya[2,3] and Eduard M. Pittich[2]

[1] Dept. of Astronomy, Earth's Physics, and Meteorology, Comenius University, Bratislava, Slovak Republic, email: plavala@slovanet.sk

[2] Astronomical Institute, Slovak Academy of Sciences, Bratislava, Slovak Republic email: astrosol@savba.sk, pittich@savba.sk

[3] Sternberg Astronomical Institute, Moscow State University, Moscow, Russia email: solov@sai.msu.ru

Abstract. The modelling of the insufficient orbital elements of extra-solar planets (EPs) revolving around one component in a binary star system is investigated in the present paper. This problem is considered in the frame of the three-body problem using the analytical theory of Orlov & Solovaya (1998). In the general case, the motion is defined by the masses of the components and by the six pairs of the initial values of the Keplerian elements. For EPs, it is not possible to obtain the complete set of elements for the orbit, in particular, the ascending node and the angle of the inclination. So, it is possible the two different variants of orbital evolution of EPs depend on the initial conditions. In one case, the orbit is unstable. Using the stability conditions of Solovaya & Pittich (2004), which are presented by the angle of the mutual inclination of the orbits between the EP and distant star, we varied unknown angular elements and defined the regions with possible values of the elements for which the motion of EP stays stable. We applied these calculations to the particular specific EPs: HD19994b, HD196885Ab and HD222404b.

Keywords. three body problem, orbital stability of extra-solar planets, HD19994b, HD196885Ab and HD222404b

---

1. Application of the theory

We have investigated the motions of EPs for which the Keplerian orbital elements are incomplete. The possible conditions for dynamical stability are presented by the orbital parameters: the mutual inclination and argument of perigee of the EP. For dynamical stability, the eccentricity of the EP orbit remains less than 1. The mutual inclination of the orbits changes within small limits and there is no close approach of the EP to the parent star.

2. Calculations and results

We do not know either the inclination of the orbit or the ascending node for EPs. We changed these unknown angular values in 1° steps, calculated and determined regions where the motion is stable. They are situated in two stable regions with an identical shape. For the exclusion of the trigonometric drift region, we calculated stable values for the inclination of the EP which is orbiting on a circular orbit. The calculated values for the inclination of the EP for a circular orbit have the same values as for one of

E. Plávalová, N.A. Solovaya & E.M. Pittich

Table 1. Orbital elements for EPs and distant stars.

Name	C	Mass (M_{Jup})	Mass (M_{Sun})	a (AU)	e	ω (°)	Ω (°)	i (°)
HD19994	A		1.34^a					
	B		0.022	75.76^b	0.26^b	247.7^b	84.13^b	114.1^b
	b	1.68^a		1.42^a	0.3 ± 0.04^a	41 ± 8^a	67.5 ± 41.5^e	66.5 ± 37.5^e
HD196885	A		1.33^a					
	B		0.45 ± 0.01^c	21.00 ± 0.86^c	0.409 ± 0.038^c	227.6 ± 23.4^c	79.8 ± 0.1^c	116.8 ± 0.7^c
	b	2.98 ± 0.05^a		2.6 ± 0.1^a	0.48 ± 0.02^a	93.2 ± 3^a	47.5 ± 43.5^e	64 ± 31^e
HD222404	A		1.4 ± 0.12^a					
	B		0.362 ± 0.022^d	19.02 ± 0.64^d	0.4085 ± 0.0065^d	160.96 ± 0.40^d	13.0 ± 2.4^d	118.1 ± 2.4^d
	b	1.85 ± 0.06^a		2.05 ± 0.06^a	0.049 ± 0.034^a	94.6 ± 34.6^a	340.5 ± 45.5^e	$5.7^{+15.1}_{-1.9}{}^a$ or 62 ± 38^e

References. ^{a} – Schneider (2011), ^{b} – Hartkopf & Mason (2001), ^{c} – Chauvin, G. *et al.* (2011), ^{d} –Torres (2008), ^{e} – this paper.

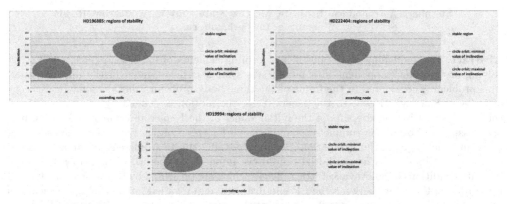

Figure 1. Relation diagram between inclination and ascending node for: HD196885Ab, HD222404b and HD19994b.

the high eccentricity orbits for stable regions (see Fig. 1). This stable region we could consider as a region with possible values for the insufficient Keplerian elements. As an example to illustrate the theory, we took three binary stellar systems with hosting EPs; HD196885Ab, HD222404b and HD19994b. Their known orbital elements for EPs and distant stars are in Table 1. The two possible stable regions are shown on Fig. 1.

We derived the following possible values for the insufficient Keplerian elements: the value of the inclination is $47.5° \pm 43.5°$ and for the node $64° \pm 31°$ for EP HD196885Ab, $66.5° \pm 37.5°$ and $67,5° \pm 41.5°$ for the EP HD19994Ab, and $62° \pm 38°$ and $340.5° \pm 45.5°$ for the EP HD222404b. The published value of the inclination for HD222404b by Schneider (2011) is quite different, $5.7°^{+15.1°}_{-1.9°}$.

References

Hartkopf, W. I. & Mason, B. D. 2001, *Sixth Catalog of Orbits of Visual Binary Stars*

Chauvin,G., Beust, H., Lagrange, A.-M., & Eggenberger, A. 2011, *A&A*, 528, A8

Orlov, A.A. & Solovaya, N.A. 1998, *in The Few Body Problem*, ed. M.J. Valtonen, (Kluwer Acad. Publish., Dordrecht), 243

Schneider, J. 2011, *Extra-solar Planets Catalogue*, http://exoplanet.eu

Solovaya, N.A. & Pittich, E.M. 2004, *Contrib. Astron. Obs. Skalnaté Pleso*, 34, 105

Torres, G. 2008, *arXiv:astro-ph/0609638v1*

From Interacting Binaries to Exoplanets: Essential Modeling Tools
Proceedings IAU Symposium No. 282, 2011   © International Astronomical Union 2012
Mercedes T. Richards & Ivan Hubeny, eds.   doi:10.1017/S1743921311027165

Space-Based Photometry of Eclipsing Binaries

John Southworth

Astrophysics Group, Keele University, Staffordshire, ST5 5BG, UK
email: jkt@astro.keele.ac.uk

Abstract. I briefly review the history and prospects for the study of eclipsing binary star systems from space-based observatories. The benefits of shifting to space satellites lie in the high precision and cadence achievable, as well as the ability to access wavelength regions which are unattainable from the ground. Whilst small amounts of data on eclipsing binaries were obtained by the *Voyager*, IUE, OAO-II, *Hipparcos* and MOST, the more recent CoRoT and *Kepler* missions were the first to provide extensive data on large numbers of systems. The future holds the prospect of the PLATO satellite, which will go bigger, better and brighter.

Keywords. stars: binaries: eclipsing — stars: fundamental parameters

1. Eclipsing binary stars from space

The study of eclipsing binary stars rests on the analysis of extensive high-precision photometry. They are therefore natural candidates to be observed from space, where telescopes are free of the Earth's atmospheric scintillation, extinction, diurnal variations and weather. The resulting light curves can be of extremely high precision and cadence, and are able to spy into wavelength ranges which are inaccessible from the ground.

Early entrants into this field were the two *Voyager* missions, which both possessed ultraviolet imaging instruments among other capabilities, and the IUE and OAO-II satellites (e.g. Cherepashchuk *et al.*, 1984; Kondo *et al.*, 1994). The *Hipparcos* astrometry mission obtained photometry at several hundred epochs for many bright stars, discovering 343 new eclipsing binary systems (Perryman *et al.*, 1997).

The WIRE satellite showed what was possible from space for bright stars. WIRE was launched into a polar orbit in March 1999 in order to obtain infrared photometry of galaxies, but an early failure of the coolant system rendered it unable to attempt its primary mission. Attention then shifted to its 5cm-aperture star tracker, which was in a position to obtain CCD photometry of bright stars at a cadence of 2 Hz. Seven eclipsing binaries were observed which this instrument before communications were lost with the satellite (Bruntt & Southworth, 2007, 2008). These included ψ Centauri, a 4th-magnitude star which was not previously known to eclipse (Bruntt *et al.*, 2006) and β Aurigae, the first known and brightest eclipsing and spectroscopically double-lined binary system (Stebbins 1911, Southworth *et al.*, 2007).

This work has subsequently been continued by MOST (Pribulla *et al.*, 2010), CoRoT (e.g. Maceroni *et al.*, 2009) and now *Kepler*. The *Kepler* satellite was launched in March 2009 and is dedicated to finding new extrasolar planets by the transit method (Borucki *et al.*, 2010). The high-quality photometry it obtains is also excellent for performing asteroseismology and for studying eclipsing binaries. A small number of objects fit into both of these categories, and are of great prospect for furthering our understanding of the physics of stars. The *Kepler* light curve of KIC 10661783 (Fig. 1) shows it to be an oEA system: a semi-detached binary in which the primary star exhibits δ Scuti oscillations

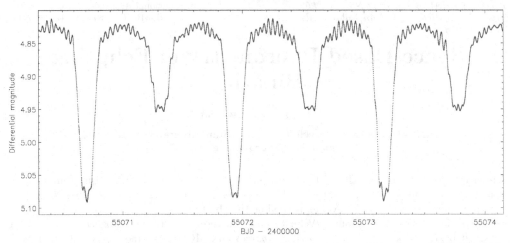

Figure 1. A small section of the *Kepler* satellite light curve of the oEA system KIC 10661783.

(Southworth *et al.*, 2011). These data provided 58 oscillation frequencies, at which point we ran out of frequency resolution. Further observations have been obtained by *Kepler* and will certainly reveal a multitude of new pulsation modes in this system. Other examples include: KIC 8410638, an eclipsing binary containing a pulsating giant star and an F dwarf with an orbital period of at least 1 yr (Hekker *et al.*, 2011); KIC 5952403, a hierarchical doubly-eclipsing triple system comprising a G giant in a 45 d orbit around a 0.9 d binary of two late-type dwarfs (Derekas *et al.*, 2011); and the KIC 5897826 system of an F subgiant and two late-M dwarfs (Carter *et al.*, 2011)

Whilst CoRoT and *Kepler* have furnished the cupboard with excellent light curves of thousands of eclipsing binaries, the European Space Agency PLATO mission will (if selected to fly) provide a much greater banquet. Compared to its immediate predecessors, PLATO is envisaged to have a wider sky coverage (500 deg^2), a larger number of target stars (400 000), and a much improved sampling rate (25 s). It will also aim for brighter stars on average, allowing follow-up observations to be obtained from many ground-based telescopes. PLATO will have a working group dedicated to studying eclipsing binaries: contact me if you are interested!

References

Borucki, W. J., *et al.*, 2010, *Science*, 327, 977
Bruntt, H. & Southworth, J., 2007, *IAU Symp.* 240, 624
Bruntt, H. & Southworth, J., 2008, *JPhys. Conf. Ser.*, 118, 012012
Bruntt, H., *et al.*, 2006, *A&A*, 456, 651
Carter, J. A., *et al.*, 2011, *Science*, 331, 562
Cherepashchuk, A., Eaton, J. A., & Khaliullin, Kh. F. 1984, *ApJ*, 281, 774
Derekas, A., *et al.*, 2011, *Science*, 332, 216
Hekker, S., *et al.*, 2010, *ApJ*, 713, L187
Kondo, Y., McCluskey, G. E., Silvis, J. M. S., Polidan, R. S., McCluskey, C. P. S., & Eaton,
     J. A. 1994, *ApJ*, 421, 787
Maceroni, C., *et al.*, 2009, *A&A*, 508, 1375
Perryman, M. A. C., *et al.*, 1997, *A&A*, 323, L49
Pribulla, T., *et al.*, 2010, *AN*, 331, 397
Southworth, J. Bruntt, H., & Buzasi, D. L. 2007, *A&A*, 467, 1215
Southworth, J., *et al.*, 2011, *MNRAS*, 414, 2413
Southworth, J., *et al.*, 2011, *MNRAS*, 414, 3740
Stebbins, J., 1911, *ApJ*, 34, 112

From Interacting Binaries to Exoplanets: Essential Modeling Tools
Proceedings IAU Symposium No. 282, 2011   © International Astronomical Union 2012
Mercedes T. Richards & Ivan Hubeny, eds.   doi:10.1017/S1743921311027177

Homogeneous Studies of Transiting Planets

John Southworth

Astrophysics Group, Keele University, Staffordshire, ST5 5BG, UK
email: jkt@astro.keele.ac.uk

Abstract. The derived physical properties of the known transiting extrasolar planetary systems come from a variety of sources, and are calculated using a range of different methods so are not always directly comparable. I present a catalogue of the physical properties of 58 transiting extrasolar planet and brown dwarf systems which have been measured using homogeneous methods, resulting in quantities which are internally consistent and well-suited to detailed statistical study. The main results for each object, plus a critical compilation of literature values for all known systems, have been placed in an online catalogue. TEPCat can be found at http://www.astro.keele.ac.uk/~jkt/tepcat/

Keywords. stars: planetary systems — stars: fundamental parameters

1. The *Homogeneous Studies* project

At this point roughly 130 transiting extrasolar planets (TEPs) are known, discovered by over 20 different groups and consortia. The characterisation of these objects is complicated by the fact that the number of measured quantities needed to calculate their physical properties is one greater than the number available directly through photometric and spectroscopic observations. This leads to the requirement to include additional constraints, typically from theoretical stellar models, in order to arrive at a determinate solution. The intricacy of this process leads expectedly to an inhomogeneity in the resulting solutions, and hinders statistical studies of these objects. In order to nullify this problem I am undertaking a project to measure the physical properties of the known TEPs by applying homogeneous methods to published data.

The transit light curves are modelled using the JKTEBOP code (Southworth *et al.*, 2004ab), which represents the star and planet as biaxial spheroids. Careful attention is paid to the calculation of robust statistical errors using Monte Carlo simulations (Southworth *et al.*, 2004c, 2005b), the assessment of systematic errors by a residual-permutation algorithm (Jenkins *et al.*, 2002), issues related to the treatment of limb darkening, contaminating 'third' light, and numerical integration over long exposure times. These aspects are discussed in detail in Papers I and III (Southworth 2008, 2010). New transit light curves for five TEPs have also been obtained in the course of a defocussed-photometry project (Southworth *et al.*, 2009abc, 2010, 2011).

Determination of the physical properties of the TEPs is achieved by using each of five different sets of tabulated predictions from theoretical stellar models as constraints. The input quantities are the orbital velocity amplitude, spectroscopic temperature and metallicity of the parent star, plus the results calculated from the light curve solutions. The output parameters comprise the physical properties of the system (Paper II: Southworth 2009). Statistical errors are supplied via a propagation analysis (Southworth *et al.*, 2005a) and systematic errors from consideration of the spread of results obtained using the five different stellar model predictions. Three quantities are free of systematic errors: the mean density of the star (Seager & Mallén-Ornela 2003), and the surface gravity and equilibrium temperature of the planet (Southworth *et al.*, 2007, Paper III).

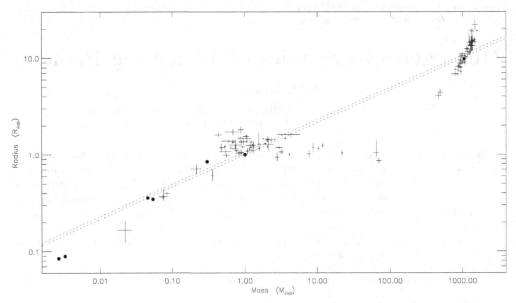

Figure 1. Mass–radius plot of the 58 systems studied within this project. Several Solar system bodies are shown by filled circles. Dashed lines show the densities of Jupiter and the Sun.

In Paper IV (Southworth 2011) I have extended the number of objects to 58, which includes 15 CoRoT systems, 10 observed by *Kepler*, five HAT discoveries, eight WASP objects, and all OGLE, TrES and XO planets. A plot of the masses and radii of the planets and their parent stars is shown in Fig. 1. I have also introduced an online catalogue of transiting planets, at `http://www.astro.keele.ac.uk/~jkt/tepcat/`. TEP-Cat contains the combined results from my project, a critical compilation of the physical properties of all known transiting planetary and brown-dwarf systems, and a summary of the basic observable quantities which are useful for planning follow-up observations.

References

Jenkins, J. M., Caldwell, D. A., & Borucki, W. J. 2002, *ApJ*, 564, 495
Seager, S. & Mallén-Ornelas, G. 2003, *ApJ*, 585, 1038
Southworth, J. 2008, *MNRAS*, 386, 1644 (Paper I)
Southworth, J. 2009, *MNRAS*, 394, 272 (Paper II)
Southworth, J. 2010, *MNRAS*, 408, 1689 (Paper III)
Southworth, J. 2011, *MNRAS*, in press, arXiv:1107.1235 (Paper IV)
Southworth, J. Bruntt, H., & Buzasi, D. L. 2007, *A&A*, 467, 1215
Southworth, J. Wheatley, P. J., & Sams, G. 2007, *MNRAS*, 379, L11
Southworth, J. Maxted, P. F. L., & Smalley, B. 2004a, *MNRAS*, 349, 547
Southworth, J. Maxted, P. F. L., & Smalley, B. 2004b, *MNRAS*, 351, 1277
Southworth, J. Zucker, S., Maxted, P. F. L., & Smalley, B. 2004c, *MNRAS*, 355, 986
Southworth, J. Maxted, P. F. L., & Smalley, B. 2005a, *A&A*, 429, 645
Southworth, J. Smalley, B., Maxted, P. F. L., Claret, A., & Etzel, P. B. 2005b, *MNRAS*, 363, 529
Southworth, J. *et al.*, 2009a, *MNRAS*, 396, 1023
Southworth, J. *et al.*, 2009b, *MNRAS*, 399, 287
Southworth, J. *et al.*, 2009c, *ApJ*, 707, 167
Southworth, J. *et al.*, 2010, *MNRAS*, 408, 1680
Southworth, J. *et al.*, 2011, *A&A*, 527, A8
Torres, G., Winn, J. N., & Holman, M. J., 2008, *ApJ*, 677, 1324

From Interacting Binaries to Exoplanets: Essential Modeling Tools
Proceedings IAU Symposium No. 282, 2011   © International Astronomical Union 2012
Mercedes T. Richards & Ivan Hubeny, eds.   doi:10.1017/S1743921311027189

Light Curves of Planetary Transits: How About Ellipticity?

Carolina von Essen, Klaus F. Huber and Jürgen H. M. M. Schmitt

Hamburger Sternwarte, University of Hamburg,
(21029) Hamburg, Germany
email: `cessen@hs.uni-hamburg.de`

Abstract. The observation of transit light curves has become a key technique in the study of exoplanets, since modeling the resulting transit photometry yields a wealth of information on the planetary systems. Considering that the limited accuracy of ground-based photometry does directly translate into uncertainties in the derived model parameters, simplified spherical planet models were appropriate in the past. With the advent of space-based instrumentation capable of providing photometry of unprecedented accuracy, however, a need for more realistic models has arisen.

Keywords. exoplanets, light curves, modelling, oblateness

1. Motivation

The gas giants in our Solar System are not spherical but oblate. Oblateness, f, is defined as $f = (R_{eq} - R_{pol})/R_{eq}$, where R_{eq} corresponds to the equatorial radius and R_{pol} to the polar radius. Oblateness values for our Solar System gas giants are shown in Table 1.

A relation between oblateness and observables quantities can be obtained by taking into account the Darwin-Radau relation. Defining the parameter $\zeta = \frac{\text{mom. of inertia}}{M \, R_{eq}^2} = \frac{2}{3}\left[1 - \frac{2}{5}\left(\frac{5}{2}\frac{q}{f} - 1\right)^{\frac{1}{2}}\right]$, where the parameter $q = \frac{\Omega^2 \, R_{eq}^3}{G \, M_{Pl}}$ is the ratio between the centripetal and gravitational acceleration, defined by Barnes and Fortney (2003), we obtain the following expression for the oblateness: $f = \frac{4\pi^2 R^3}{G \, M \, P^2}\left[\frac{5}{2}\left(1 - \frac{3}{2}\,\zeta\right)^2 + \frac{2}{5}\right]^{-1}$. Assuming $\zeta \sim 0.25$ (in analogy to gas giants in the Solar System) and tidal locking, we calculated f for different exoplanets (see Table 2).

2. Analytical model: comparing oblate and spherical planets

We assume that the planet shape can be modelled as a rotational ellipsoid and the star as a sphere, and the projections are an ellipse and a circle, respectively. In the case $i = 90°$, the symmetry of the problem allows an analytical model for the flux drop during primary transit. The difference between the models (Fig. 1) can be resolved for instance by the Kepler Telescope, given a good model for limb darkening.

Table 1. Oblateness values for the giant planets in our Solar System.

Planet Name	R_{eq}	R_{pol}	f	Planet Name	R_{eq}	R_{pol}	f
Jupiter	71492 km	66854 km	0.0648	Uranus	25559 km	24973 km	0.0229
Saturn	60268 km	54364 km	0.0979	Neptune	24764 km	24341 km	0.0171

Table 2. Oblateness values for short period exoplanets.

System	P [d]	$< \rho >$ (cgs)	f	System	P [d]	$< \rho >$ (cgs)	f
HD 149026	~ 2.875	~ 1.17	0.00152	WASP - 15	~ 3.752	~ 0.247	0.00422
HD 189733	~ 2.218	~ 0.91	0.00328	CoRoT - 1	~ 1.508	~ 0.38	0.017
HD 209458	~ 3.524	~ 0.38	0.00365	WASP - 19	~ 0.788	~ 0.486	0.048

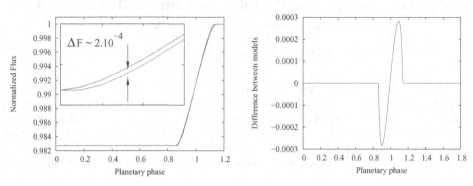

Figure 1. Left: Difference between the two analytical models, considering oblate and spherical planets. Right: Oblateness effects are important only during ingress and egress.

3. Numerical model: Comparison to Kepler data

A realistic model must include an arbitrary inclination and limb darkening, and it has to be compared to real data. We find that models with vanishing and large oblateness fit the data equally well, albeit with different fit parameters. However, one has to take into account that oblateness and inclination are degenerate; an ideal method would be the determination of oblateness and inclination independently. A possible way to minimize the degeneration effects is to take into account the mass - radius relationship for the host-star, such as $R = 1.24 \, M^{0.67}$ (for M > 1.3 $M_\odot$). Fig. 2 shows that the inferred mass of the host star depends on the adopted planetary oblateness. Since KIC 9941662 is of spectral type A3, one clearly favors a low oblateness.

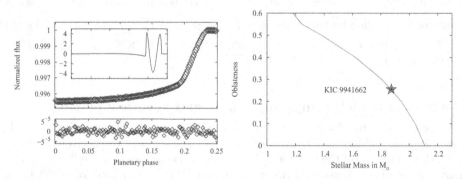

Figure 2. Left: Kepler light curve of KIC 9941662 and fit considering f = 0.5 for χ^2 minimization. Small box: Difference between two models (f = 0 and f = 0.5), multiplied by 10^6. Right: Host-star mass-oblateness relationship for the case of KIC 9941662. ST $\sim$ A3, $T_{eff} \sim$ 8800 K.

References

Mandel, K. & Agol, E. 2002, *ApJ*, 580, 171
Barnes, J. W. & Fortney, J. J. 2003, *ApJ*, 588, 545
Carter, J. A. & Winn, J. N. 2010, *ApJ*, 716, 850

From Interacting Binaries to Exoplanets: Essential Modeling Tools
Proceedings IAU Symposium No. 282, 2011
Mercedes T. Richards & Ivan Hubeny, eds.
© International Astronomical Union 2012
doi:10.1017/S1743921311027190

New Photometric Observations of the Transiting Extrasolar Planet TrES-3b

M. Vaňko[1], M. Jakubík[1], T. Krejčová[2], G. Maciejewski[3], J. Budaj[1], T. Pribulla[1], J. Ohlert[4,5], S. Raetz[6], V. Krushevska[7], P. Dubovsky[8]

[1] Astronomical Institute of the Slovak Academy of Sciences, Slovakia, email: vanko@ta3.sk

[2] Masaryk University, Department of Theoretical Physics and Astrophysics, 602 00 Brno, Czech Republic

[3] Toruń Centre for Astronomy, N. Copernicus University Gagarina 11, 87100, Toruń, Poland

[4] University of Applied Sciences, Wilhelm-Leuschner-Strasse 13, 61169 Friedberg, Germany

[5] Michael Adrian Observatory, Astronomie Stiftung Trebur, Fichtenstrasse 7, 65468 Trebur, Germany

[6] Astrophysikalisches Institut und Universitäts-Sternwarte, Schillergässchen 2-3, 07745 Jena, Germany

[7] Main Astronomical Observatory of National Academy of Sciences of Ukraine, 27 Akademika Zabolotnoho St. 03680 Kyiv, Ukraine

[8] Vihorlat Observatory, Mierová 4, Humenné, Slovakia

Abstract. We present new transit observations of the transiting exoplanet TrES-3b obtained in the range 2009 – 2011 at several observatories. The orbital parameters of the system were redetermined and the new linear ephemeris was calculated. We performed numerical simulations for studying the long-term stability of orbits.

Keywords. exoplanets, fundamental parameters, individual (TrES-3b)

1. Introduction

TrES-3b is one of the more massive transiting extrasolar planets. The planetary system consists of a nearby G-type dwarf and a massive hot Jupiter with an orbital period of 1.3 days. It was discovered by O'Donovan *et al.* (2007) and a discovery-quality light curve has also been obtained by the SuperWASP survey (Collier Cameron *et al.* 2007). Follow-up transit photometry has been presented by Sozzetti *et al.* (2009) and Gibson *et al.* (2009).

2. Observations & Results

All observations used in this study were carried out at several observatories: Stará Lesná (Slovakia), Toruń Center for Astronomy (Poland), Michael Adrian Observatory (Germany), University Observatory Jena (Germany) and Vihorlat Observatory (Slovakia). We used telescopes with diameters of the primary mirrors in the range of 30–120 cm and optical CCD-cameras (RI-bands). In order to have a homogeneous dataset we have used the composition of 560 data points from two nights obtained at one observatory with the same filter (see Fig. 1). To obtain an analytical transit LC, we assumed the quadratic limb darkening law. The limb darkening coefficients c_1 and c_2 were linearly interpolated from Claret (2000) for the following stellar parameters: T_{eff} = 5650 K, log (g) = 4.4 and [Fe/H] = -0.19 (based on the results of Sozzetti *et al.* 2009). Finally, the orbital parameters of the system were redetermined: $R_p/R_* = 0.1819(20)$,

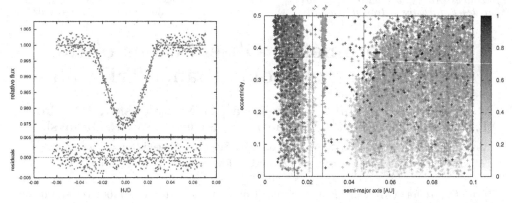

Figure 1. Left Composition of 560 data points from the two nights: 13/07/10 and 20/08/10 (dd/mm/yy) obtained at Michael Adrian Observatory in Trebur (Germany). **Right** Stability plot in the $a - e$ plane showing the maximum eccentricity.

$R_*/a = 0.1785(70)$, $i = 80.9(7)$, $R_* = 0.876^{+0.008}_{-0.016}\ R_\odot$ and $R_p = 1.551^{+0.014}_{-0.028}\ R_J$. To estimate the uncertainties of parameters, we have used the Monte Carlo simulation method. Based on the transits obtained at other observatories, we calculated the following new linear ephemeris for TrES-3b: $T_c(E) = 2454538.5807(1) + E \times 1.30618595(15)$.

In this part of our work, we investigated the gravitational influence of TrES-3b on a potential second planet in the system. Gibson *et al.* (2009) claimed, that their data are sensitive enough to probe small mass planets on circular orbits near the 2:1 mean-motion resonance (MMR) with TrES-3b. We performed numerical simulations for studying the stability of orbits and checking their chaotic behavior using the method of maximum eccentricity (e.g. Dvorak *et al.* 2003). We have generated 10^5 massless particles to represent small planets (Earth-like) in this system. The long-term stability plot in the $a - e$ plane with the maximum eccentricity after 500 years (about 140 000 revolutions of TrES-3b around the parent star) is shown in Fig. 1 (right plot). We found that the region from 0.02 AU to 0.04 AU is almost completely depleted, except the regions near 1:1 and 3:4 MMRs. The region near 2:1 MMR is richly populated, but the particles have relatively high eccentricities and inclinations, thus the stability and also the detection using the TTV method is questionable. More stable regions are beyond the 1:3 MMR, where the gravitational influence of the TrES-3b is weak. A more detailed study of the long-term stability in this system (especially near 2:1 MMR) will be presented in future work.

Acknowledgements This work has been supported by grants VEGA No. 2/0078/10, 2/0074/09, and 2/0094/11. PD received support from APVV grant LPP-0049-06 and LPP-0024-09. TK thanks the grant GA ČR GD205/08/H005. GM acknowledges Iuventus Plus grant IP2010 023070. SR thanks the German National Science Foundation (DFG) for support with project NE 515/33-1.

References

Collier Cameron, A., Wilson, D. M., West, R. G., Hebb, L., *et al.*, 2007, *MNRAS*, 380, 1230
Claret, A. 2000, *A&A*, 363, 1081
Dvorak, R., Pilat-Lohinger, E., Funk, B., & Freistetter, F. 2003, *A&A*, 398, L1
Gibson, N. P., Pollacco, D., Simpson, E. K., Barros, S., *et al.*, 2009, *ApJ*, 700, 1078
O'Donovan, T. F., Charbonneau, D., Bakos, G., Mandushev, G., *et al.*, 2007, *ApJ*, 663, L37
Sozzetti, A., Torres, G., Charbonneau, D., Winn, J. N., *et al.*, 2009, *ApJ*, 691, 1145

From Interacting Binaries to Exoplanets: Essential Modeling Tools
Proceedings IAU Symposium No. 282, 2011
Mercedes T. Richards & Ivan Hubeny, eds.
© International Astronomical Union 2012
doi:10.1017/S1743921311027207

Stellar Wobble Due to a Nearby Binary System

M. H. M. Morais[1] and A. C. M. Correia[2]

[1]Department of Physics, I3N, University of Aveiro,
Campus Universitario de Santiago, 3810-193 Aveiro, Portugal
email: helena.morais@ua.pt

[2] Astronomie et Systemes Dynamiques, IMCCE-CNRS UMR 8028,
77 Avenue Denfert-Rochereau, 75014 Paris, France
email: correia@ua.pt

Abstract. To date, there are several reported exoplanet detections within binary star systems. These findings are based on radial velocity data for the target star. However, the companion star could in turn have a companion of which we are not aware. We describe how this hidden binary system affects the radial velocity of the target star, mimicking a planet in some circumstances We also explain what can be done in practice in order to distinguish between these two effects.

Keywords. (stars:) binaries: general,(stars:) planetary systems, techniques: radial velocities

1. Overview of theory and results

We study a triple system composed of a star with mass $M_\star$, at distance $\vec{r}$ from the centre of mass of a binary with masses $M_1 + M_2$, and inter-binary distance $\vec{r_b}$. We assume that $|\vec{r_b}|/|\vec{r}| \ll 1$ (hierarchical system) thus the motion is, approximately, a composition of two Keplerian orbits described by: $\vec{r_b}$, with semi-major axis a_b, eccentricity e_b, and period $T_b = 2\pi/n_b$; and $\vec{r}$, with semi-major axis a, eccentricity e, and period $T = 2\pi/n$.

The star's radial velocity is $V_R = V_{RK} + V_{RP}$, where V_{RK} is a Keplerian term that describes the motion around a "star", of mass $M_1 + M_2$, located at the binary system's centre of mass, and V_{RP} is a small perturbation[1,2]. Therefore, the radial velocity data of a star with a nearby unresolved binary system is first fitted with a Keplerian curve, V_{RK}. After subtracting V_{RK}, we are left with the perturbation term, V_{RP}, from which we could, in principle, infer the presence of the hidden binary component. However, as we will show next, this is not always possible in practice.

Coplanar circular orbits:

In the case of coplanar circular orbits we have [1]

$$V_{RP} = K_0 \cos(n\,t + \lambda_0) + K_1 \cos((2\,n_b - 3\,n)\,t + \lambda_1) + K_2 \cos((2\,n_b - n)\,t + \lambda_2) \,. \quad (1.1)$$

The term with frequency n is incorporated in the main Keplerian curve, V_{RK}. Since $n_b \gg n$, the 2nd and 3rd terms have very close frequencies that can only be resolved if the observation timespan $t_{obs} \geqslant T/2$. If there is enough resolution and precision, we identify both signals and we conclude that they should not be caused by planets (as such close orbits would be unstable). However, since $|K_1| = 5\,|K_2|$, in practice, due to limited precision, we may only be able to observe the signal with frequency $2\,n_b - 3\,n$. In this case, we may think there is a planet companion to the observed star.

We simulated a coplanar triple system composed of $M_\star = M_\odot$ on a circular orbit of period $T = 22$ y, around binary $M_1 = 0.7\,M_\odot$ and $M_2 = 0.35\,M_\odot$ with circular orbit of period $T_b = 411$ d. We computed radial velocity data points over $t_{obs} = 11$ y at precision

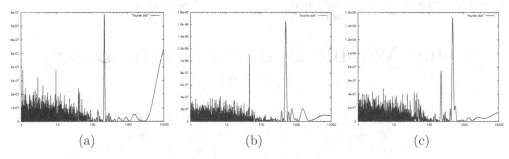

(a)   (b)   (c)

Figure 1: Fourier spectrum of residuals leftover after removing V_{RK}.

0.8 m/s. In Fig. 1(a) we see a signal with frequency $2\,n_b - 3\,n$ (amplitude 0.9 m/s at 223 d) that mimics a planet of $18\,M_E$.

Non-coplanar circular orbits:

In the case of non-coplanar circular orbits, V_{RP} is a combination of 6 periodic terms[2] with frequencies: n, $3\,n$, $2\,n_b \pm n$ and $2\,n_b \pm 3\,n$. Depending on the observation precision and resolution we may observe one or more of these terms. If all these have well separated frequencies we may mistake them for planet(s).

We simulated a triple system composed of $M_\star = M_\odot$ on a circular orbit of period $T = 4.2$ y, around binary $M_1 = M_2 = 0.25\,M_\odot$ with circular orbit of period $T_b = 85$ d. The relative inclination is $i = 30°$. We computed radial velocity data points over $t_{obs} = 11$ y at precision 0.7 m/s. In Fig. 1(b) we see signals with frequency $2\,n_b - 3\,n$ (amplitude 0.8 m/s at 46 d), and frequency $3\,n$ (amplitude 1.4 m/s at 516 d) that mimic planets of $7\,M_E$ and $20\,M_E$.

Eccentric coplanar orbits:

It can be shown that, generally, V_{RP} is a composition of periodic terms [2]. In the eccentric 2D case, up to 1st order in the eccentricities, there are 12 frequencies: n, $2\,n_b - n$, $2\,n_b - 3\,n$, $n_b - n$, $n_b - 3\,n$, $3\,n_b - n$, $3\,n_b - 3\,n$, $n_b + n$, $2\,n$, $2\,n_b$, $2\,n_b - 4\,n$ and $2\,n_b - 2\,n$.

We simulated a coplanar triple system composed of $M_\star = M_\odot$ with $e = 0.1$ and period $T = 22$ y, around binary $M_1 = 0.7\,M_\odot$ and $M_2 = 0.35\,M_\odot$ with $e_b = 0.2$ and period $T_b = 411$ d. We computed radial velocity data points over $t_{obs} = 11$ y at precision 0.8 m/s. In Fig. 1(c) we see signals with frequencies $2\,n_b - 3\,n$ (amplitude 0.8 m/s at 223 d) and $n_b - 3\,n$ (amplitude 1.4 m/s at 487 d). Since $n \ll n_b$ these mimic planets of $15\,M_E$ and $34\,M_E$ at the 2:1 mean motion resonance.

If $t_{obs} \gg T$, signals at or nearby harmonics of n appear [3]. These may be mistaken by planet(s) in mean motion resonance with a companion "star" of mass $M_1 + M_2$. As t_{obs} increases, the short period terms described here become negligible with respect to the orbits' secular evolution [3].

Distinguishing planet from binary:

We showed that, in order to avoid erroneous announcements of new planets, we need to have precise observations over a reasonably long timespan, which is often not possible in practice. However, a signal with frequency, n_{pl}, and amplitude, K_{pl}, mimics a planet with parameters [1,2] $a_{pl} = (G\,M_\star)^{1/3}/n_{pl}^{2/3}$ and $M_{pl}\sin I_{pl} = K_{pl}\,(M_\star + M_1 + M_2)/(n_{pl}\,a_{pl})$. Therefore, we can invert these expressions to predict the binary system parameters that can mimic a given planet and check if they are realistic [1,2].

References

[1] Morais, M. H. M. & Correia, A. C. M. 2008, A&A, 391, 899
[2] Morais, M. H. M. & Correia, A. C. M. 2011, A&A, 525, A152
[3] Morais, M. H. M. & Correia, A. C. M. 2011, MNRAS, submitted

From Interacting Binaries to Exoplanets: Essential Modeling Tools
Proceedings IAU Symposium No. 282, 2011
Mercedes T. Richards & Ivan Hubeny, eds.
© International Astronomical Union 2012
doi:10.1017/S1743921311027219

Asymmetric Transit Curves as Indication of Orbital Obliquity: Clues from the Brown Dwarf Companion in KOI-13

G. M. Szabó[1], R. Szabó[1], J. M. Benkő[1], H. Lehmann[2], G. Mező[1],
A. E. Simon[1], Z. Kővári[1], G. Hodosán[1], Z. Regály[1] and L. L. Kiss[1,3]

[1]Konkoly Observatory of the Hungarian Academy of Sciences, P.O. Box 67, H-1525 Budapest,
Hungary, email: szgy@konkoly.hu
[2]Thüringer Landessternwarte, 07778 Tautenburg, Germany
[3]Sydney Institute for Astronomy, School of Physics A28, University of Sydney, NSW 2006,
Australia

Abstract. Exoplanets orbiting rapidly rotating stars may have unusual light curve shapes. These objects transit across an oblate disk with non-isotropic surface brightness, caused by the gravitational darkening. If such asymmetries are measured, one can infer the orbital obliquity of the exoplanet and the gravity darkened star, even without the analysis of the Rossiter-McLaughlin effect or interferometry. Here we introduce KOI-13 as the first example of a transiting system with a gravity darkened star.

1. Summary

- KOI-13 is a common proper motion binary, with two rapidly rotating components ($v \sin i \approx 65$–70 km/s). The transit curves show significant distortion that was stable over the $\sim$130 days time-span of the data.

- These distortions are due to gravity darkening of a rapidly rotating star. KOI-13 is it *the first example for detecting orbital obliquity for a substellar companion without measuring the Rossiter-McLaughlin effect from spectroscopy.*

- After correcting the *Kepler* light curve to the second light of the optical companion star, we derive a radius of 2.2 R_J for the transiting object, implying that the object is a late-type dwarf. KOI-13 is *also the first example for a late-type dwarf on close-in orbit around an A-type primary.*

See further details in Szabó *et al.* (2011).

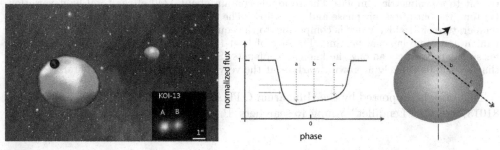

Figure 1. Fig. 1. Left: An artist's concept of KOI-13 (main panel) and its false color (V,I) image (insert) with lucky imaging from Konkoly Observatory, 2011 April 20. Right: Predicted distortion in transit light curves with stellar temperature gradient due to gravity darkening caused by the rapid rotation of the host star. This schematic figure is based on the Barnes (2009) models.

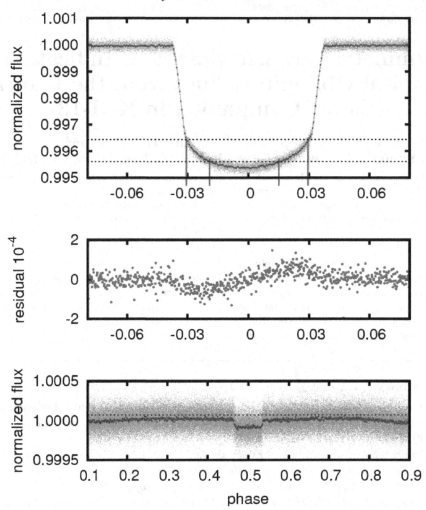

Figure 2. Fig. 2. Fig. 3. Top: folded *Kepler* light curve of KOI-13 (A and B components together; epoch=JD 2454953.56498, period=1.7635892 (Borucki *et al.* 2011) with a grid highlighting the most important asymmetries. The lower grid line intersects the light curve at a relative flux of 0.9956, and the orbital phase of the bottom of the transit is marked. Evidently, the floor of the light curve is shifted towards the ingress phase. Middle: the residuals of the transit to a symmetric template.The residuals contain about 1/40 of the total light variation. Bottom: the out-of-transit phase and the eclipse. The depth of the eclipse results in an effective temperature of 3150 K, which is compatible to the equilibrium temperature of the close-in companion and some internal heating. The size of KOI-13.01 is 2.2±0.1 R_J. There is no example for such large planets among the known exoplanets (Szabó and Kiss, 2011). Thus, we conclude that KOI-13.01 is likely a brown dwarf, or at the very low-end of the red dwarf stars.

This work is supported by the Hungarian OTKA Grants K76816, K83790 and MB08C 81013, and the "Lendület" Young Researchers' Program.

References

Barnes, J. W. 2009, *ApJ*, 705, 683
Szabó, Gy. M. *et al.*, 2011, *ApJ* (Letters), 736, 4
Szabó, Gy. M. & Kiss, L. L. 2011, *ApJ* (Letters), 727, 44

From Interacting Binaries to Exoplanets: Essential Modeling Tools
Proceedings IAU Symposium No. 282, 2011
Mercedes T. Richards & Ivan Hubeny, eds.

© International Astronomical Union 2012
doi:10.1017/S1743921311027220

Multiband Transit Light Curve Modeling of WASP-4

N. Nikolov[1], J. Koppenhoefer[2,3], M. Lendl[4], T. Henning[1] and J. Greiner[3]

[1] Max-Planck-Institut für Astronomie, Königstuhl 17, 69117 Heidelberg, Germany
email: nikolov@mpia.de

[2] Universitäts-Sternwarte München, Scheinerstr. 1, 81679 Munich, Germany
[3] Max Planck Institute for Extraterrestrial Physics, Geissenbachstr., 85748 Garching, Germany
[4] Observatoire de Genève, Universitè de Genève, 51 chemin des Maillettes, 1290 Sauverny, Switzerland

Abstract. We report on the simultaneous g', r', i', z' multiband, high time sampling (18-24s) ground-based photometric observations, which we use to measure the planetary radius and orbital inclination of the extrasolar transiting hot Jupiter WASP-4b. We recorded 987 images during three complete transits with the GROND instrument, mounted on the MPG/ESO-2.2m telescope at La Silla Observatory. Assuming a quadratic law for the stellar limb darkening we derive system parameters by fitting a composite transit light curve over all bandpasses simultaneously. To compute uncertainties of the fitted parameters we employ the Bootstrap Monte Carlo Method. The three central transit times are measured with precision down to 6 s. We find a planetary radius $R_p = 1.413 \pm 0.020 R_{\rm Jup}$, an orbital inclination $i = 88.°57 \pm 0.45°$ and calculate new ephemeris, a period $P = 1.33823144 \pm 0.00000032$ days and reference transit epoch $T_0 = 2454697.798311 \pm 0.000046$ (BJD). The analysis of the new transit mid-times in combination with previous measurements imply a constant orbital period and no compelling evidence for TTVs due to additional bodies in the system.

Keywords. planetary systems, stars: individual (WASP-4b)

1. Introduction

Exoplanetary transits provide a direct access to some of the most interesting physical parameters of these objects. Currently, the transit method is the only tool to measure accurate planetary radii, orbital inclinations, and mean planetary densities, if the radius and the mass of the parent star are known. In this work we aimed at obtaining high precision multiband photometry during three transits of WASP-4b with the ultimate goals to (i) measure accurate system parameters and (ii) to obtain times of minimum light, which we use to refine the transit ephemeris and to search for transit timing variations (TTVs).

2. Observations and analysis

We observed three transits of WASP-4b (Wilson *et al.* 2008) on UT August 26 and 30 and October 8, 2009 using the Gamma Ray Burst Optical and Near-Infrared Detector (GROND), mounted on the MPG/ESO-2.2m telescope at ESO La Silla Observatory (see Fig. 1). We perform circular aperture photometry of 987 calibrated science images. The best apertures are selected by minimizing the magnitude root-mean-square of the out of transit data. We improve the quality of the WASP-4 transit light curves by modeling

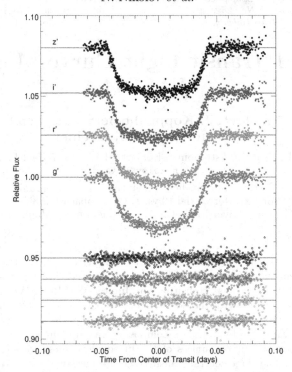

Figure 1. GROND composite transit photometry of WASP-4b, acquired on August 26, August 30 and October 8, 2009. The light curves of each band are shifted with 0.026 in relative flux for clarity. The best-fit transit models are displayed with continuous lines and the residuals from the best-fit modeling are shown at bottom.

the data for two effects: (i) differential air-mass extinction and (ii) correlation of the magnitude with the x,y position of the psf on the detector.

We derive the best-fit transit parameters (mid-transit time, T_0, orbital period, P, planet to star radius ratio, R_p/R_*, normalized semimajor axis, a/R_* and orbital inclination, i) using the light curve models of Mandel & Agol (2002) and minimizing the χ^2 function over the four passbands simultaneously via the down-hill simplex algorithm (Press *et al.* 1992). To compute uncertainties for the fitted parameters we employ the Bootstrap Monte Carlo method (Press *et al.* 1992), taking into account time-correlated noise by (i) rescaling the photometric weights, so that the best-fitting model for each band and run results in a reduced $\chi^2 = 1$ and (ii) by following the "time-averaging" method described in Pont *et al.* (2006).

We also investigated the observed minus calculated $(O - C)$ residuals of our data and the reported transit mid-times at the time of writing for any departures from the predicted values estimated using our ephemeris. We find no compelling evidence for transit timing variations due to additional bodies in the system.

References

Mandel, K. & Agol, E. 2002, *ApJ*, 580L, 171
Pont, F., Zucker, S., & Queloz, D. 2006, *MNRAS*, 373, 231
Press, W. H., Teukolsky, S. A., Vetterling, W. T., & Flannery, B. P. 1992, *Numerical recipes in C. The art of scientific computing*, ed. T.S.A.V.W.T.F.B.P. Press, W. H.
Wilson, D. M., Gillon, M., Hellier, C., *et al.*, 2008, *ApJ*, 675, 113

From Interacting Binaries to Exoplanets: Essential Modeling Tools
Proceedings IAU Symposium No. 282, 2011
Mercedes T. Richards & Ivan Hubeny, eds.

© International Astronomical Union 2012
doi:10.1017/S1743921311027232

New Binary and Exoplanet Candidates from STEREO Light Curves

Gemma Whittaker, Vino Sangaralingam and Ian Stevens

Astrophysics and Space Research, The University of Birmingham, Edgbaston,
Birmingham, B15 2TT, UK
email: gemma@star.sr.bham.ac.uk

Abstract. The Heliospheric Imagers (HI) onboard the STEREO satellites are observing an abundance of background stars as they follow their respective Sun-centered orbits. These are wide-angled CCD cameras with a $20° \times 20°$ field of view, directed $\sim 14°$ from the solar disk. These imagers monitor 20% of the sky over one year, providing light curves for over 500,000 stars down to 12th magnitude and brighter than 7th. We are currently analysing the photometric data from the HI-1 cameras, obtained since March 2007. Following a standard data reduction of the raw photometric images, the resultant light curves underwent a sequence of detrending procedures to minimize systematics in the data, which can contribute to red noise. A transit search was performed using the BLS algorithm, which is sensitive to the box-like shape associated with planetary transits. The resulting candidates were subjected to a number of false-alarm tests to determine the most promising candidates and these were investigated further, visually and using available catalogue data. Possible new exoplanet and binary candidates will now be submitted for follow-up photometric and spectroscopic observations to confirm their nature.

Keywords. techniques: image processing, techniques: photometric, binaries: eclipsing, planetary systems, catalogs, methods: data analysis

1. STEREO Observations

STEREO (Solar TErrestrial RElations Observatory) was launched in March 2006 to study Coronal Mass Ejections (CMEs) in unprecedented detail by providing a stereoscopic view of the Sun. The twin observatories follow a heliocentric orbit, with one moving ahead of the Earth (STEREO-A) and the second lagging behind the Earth (STEREO-B). Among the four, identical suites of instruments onboard the two spacecraft is the SECCHI package (Sun Earth Connection Coronal and Heliospheric Investigation). SECCHI is comprised of an extreme UV imager, two optical coronagraphs and two heliospheric imagers (HI-1 and HI-2), designed (primarily) to study the 3-D evolution of CMEs along the Sun-Earth line.

The HI-1 instruments are optical CCD cameras with a $20° \times 20°$ field of view. Within this field an estimated 29,000 stars between 7th and 12th magnitude may be visible in the background at any one time. These cameras have a 40 minute observing cadence, which provides 36 consecutive frames per day for 20 days after which a star will pass out of the field of view. STEREO has now been fully operational for over 4 years, hence most stars have been observed at least 4 times by each camera. This has provided a wealth of precision, photometric data for many stars which are not well studied at present.

2. Data Processing and Detrending

A standard data reduction is carried out for the raw photometric data, which are down-loaded directly from the STEREO webpage†. This includes corrections for shutterless readout and saturated columns. Background subtraction is then performed to minimise contamination from the F-corona, using a 1-day running, minimum background. The NOMAD catalogue is used to track individual stars over consecutive frames and finally the light curves for these stars are extracted using aperture photometry. The HI-1 cameras have a moderate PSF of 3.7 arcmin (Cf. CoRoTs PSF which is 10.5 arcmin), meaning that the extracted light curves are susceptible to contamination from nearby stars.

3. Transit Analysis and Results

A transit search was performed, within a period range of 0.5 - 10 days, using the BLS transit detection algorithm (Kovács et al., 2002). From an initial study including 82,863 light curves from HI-1A, 952 potential candidates were detected with a Signal Detection Efficiency (SDE) of greater than 9. A series of false-alarm tests were then used to determine the most promising candidates. These included spectral-type information from SIMBAD and the Exoplanet Diagnostic Tool, η^*, (Tingley & Sackett, 2005), which calculates the likelihood of a transit being from a planet as opposed to it being from a star or as a result of noise. The candidates passing all false-alarm tests were visually analysed to enable them to be classified into potential exoplanet or binary candidates.

Some binary targets have recently been submitted for preliminary spectroscopic follow-up using the WIRO telescope in Wyoming, USA. Present and future work is now focussed establishing a reliable set of targets for high resolution follow-up observations. This work will involve refining our search parameters and detrending procedures and the addition of the HI-1B data, which will increase the detection efficiency of BLS.

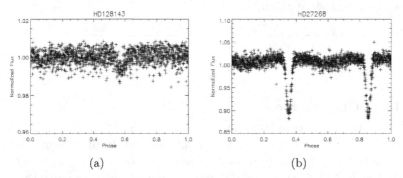

(a)   (b)

Figure 1. Example exoplanet and binary candidates. Figure (a) shows an F5 dwarf, which was detected by BLS with $SDE = 15$, $Period = 0.73$ days and $\eta^* = 0.12$, which lies well below the cutoff value of 1.0 (indicative of a planet). Figure (b) shows an A0 star, which was detected with $SDE = 40$, $Period = 2.1$ days and $\eta^* = 0.96$.

References

Kovács, G., Zucker, S. & Mazek, T. 2002 *A&A*, 391, 369
Tingley, B. & Sackett, P. 2005 *ApJ*, 627, 1011
Sangaralingam, V. & Stevens, I. 2011 *submitted*
Whittaker, G., Sangaralingam, V. & Stevens, I. 2011 *in preparation*

† www.ukssdc.rl.ac.uk/solar/stereo/data.html

From Interacting Binaries to Exoplanets: Essential Modeling Tools
Proceedings IAU Symposium No. 282, 2011   © International Astronomical Union 2012
Mercedes T. Richards & Ivan Hubeny, eds.   doi:10.1017/S1743921311027244

Panel Discussion I

Panel: F. Allard, A. Batten, E. Budding, E. Devinney, P. Eggleton, A. Hatzes, I. Hubeny, W. Kley, H. Lammer, A. Linnell, V. Trimble, and R. E. Wilson

Discussion

M. RICHARDS: Several talks today have expressed fuzzy boundaries to describe the objects called "stars." Is the following classification correct? Are stars restricted to objects that have masses greater than 0.089 solar masses and begin making energy with hydrogen burning? Do we include the stellar remnants: the white dwarfs and neturon stars? Do we include the brown dwarfs because they burn lithium or deuterium. We know that planets are not in this group since they have no energy production.

P. EGGLETON: Well, I'll first admit I don't see this as a tremendously consuming debate and what we call them doesn't matter all that much compared to what they are. I don't think nature is necessarily going to create things that will fit in completely discrete baskets. But I do think with planets we're probably thinking of things that form in disks around stars. However, if we're thinking about brown dwarfs and things that act more like a companion around a binary star, there may not be a clear distinction. The fact that there is nuclear burning is pretty significant for a star. There might be some white dwarfs that are so evolved that they are no longer burning anything nuclear at all, they're just radiating heat, in which case they're not stars according to some people, but I would say they're still stars.

V. TRIMBLE: Planets are chemically differentiated. It makes them very hard to recognize. But that's the line for planets, chemical differentiation.

R. WILSON: Evolved stars are also chemically differentiated.

I. HUBENY: Well I think that it's merely a question of terminology since nature behaves in certain ways. For instance, from the point of view of understanding radiation from those objects it really doesn't matter whether it is a planet or brown dwarf because they are basically the same type of object. Modeling is completely the same for both those topics; the same analysis applies for everything. People like terminology, people like distinctions between objects, but I think that's a secondary problem compared to really understanding the nature of these things.

A. BATTEN: I think I agree with Peter. During the Prague General Assembly when we were all discussing the fate of Pluto, I suggested that astronomers could perhaps sometimes learn from philosophers, and in particular from Emmanuel Kant, who argued that we impose our categories on nature. For some purposes, we do have to define classes of objects, but really there is no essential difference between a star and a brown dwarf at one end of the range and between a brown dwarf and a planet at the other. It's our convenience that imposes these terms, and I don't think it really reflects anything real.

W. KLEY: I would agree that it's usually not necessary to define these categories, but if I say I am working in planet formation sometimes people ask very plainly, "What is a planet?" And then you have to give some answer. It usually helps to have some categories; but I have the opinion that it is not really very useful to distinguish them by their formation mechanism since this is *a priori* not known. If we observe these objects now, we cannot tell whether they were formed by some gravitational instability process, whether they formed in a disk, or whether they formed somewhere else; so this is not really possible. And if we come to the initial definition of a planet where the word came from, like a wanderer in the sky, I think it's something which is not self-luminous and orbits a star that is self luminous, which is burning something. I think if we come back to this original idea of a planet, this would give a quite natural definition of it.

There is some limit, and this is something we argue about - whether we agree with the limit of 75 or 80 Jupiter masses. But there is some limit; and we would agree there is a limit between a fusor and a non-fusor. So, one could set this mass limit essentially, and one could set this as a basic limit between stars and some other objects. I am still a little confused about the limits between brown dwarfs and planets; but we can say we either have a star or we don't have a star. I think all burned out objects, previous stars, are still stars. And Pluto, well, bad luck! But Pluto is still a planet, still a dwarf planet. So it's not a minor object, it's still somewhat of a planet.

V. TRIMBLE: I think the brown dwarfs all became disputatious, because for so long people looked hard for sub-stellar objects and didn't find any. And when finally people began to find something that was perhaps in that regime then it was briefly important to make a clean cut to say: "Yes, we have discovered sub-stellar objects." But now that there is no doubt about their existence, there is no major reason to worry about it. If you do the calculation of course as the gas cloud contracts, you get a bit of fusion and lots of things. And the question is 'Does it stop the contraction?' If it stops the contraction, then you have a star.

E. BUDDING: I've come up with a couple of ideas. One is a kind of built-in argument that applies in science, which you can trace back to Newton on the basis of his analysis of half a dozen planets or planetary bodies. Newton describes his law of gravitation as being universal when speaking of a volume incredibly small compared to the now known volume of the universe. Applying that sort of concept to stars in relation to this meeting, very often we're using our ideas from the analysis of binary systems and saying, "Well, now we know what stars do."

So, let me try and put this a bit more succinctly in relation to say the primary star in the Algols. There's a line of thinking in which an Algol primary has a certain mass and a certain luminosity and therefore it can look like a main-sequence star. So we conclude that it is a main sequence star! Indeed, it looks like a main sequence star, but how can a star which has had this strong history of interactive evolution in a close binary system really be quite the same thing as a single object, which on the face of it may look quite similar. So, we should have some element of caution when we extrapolate ideas which we get from close analysis of binary systems, useful as they are, to stars as a whole. May there not be sometimes special circumstances of binarity which we might overlook?

And now the other point, one of the topics which has cropped up a couple of times today has concerned gravity darkening. The case which pushed itself forward was because of the asymmetry in the transiting minimum, and that happened to be because the axis of rotation was not in the same plane as the orbit, so there's a built-in asymmetry. The thing is that gravity darkening is always there. We also heard a lot about detailed analysis of limb-darkening. So, what about gravity darkening in the same degree of detailed analysis? Here we're on tricky ground, because both theoretically and observationally, it's a difficult parameter to pull out.

Getting back to the point, some years ago at a meeting in Turkey I said, "Well what about Jupiter?" There we have an object which in the far infrared is radiating on its own basis, from its own store of energy. We could check this energy distribution, this flux distribution of the surface in the far infrared and test whether a star-like body is behaving according to our expectations of gravity darkening. But, when I said how that might be done somebody said that this effort would be defeated by meteorological effects, which are more important than the effect I was trying to think about.

A. HATZES: For me, I consider that there are 'stars' and 'the rest.' And in 'the rest' there are the brown dwarfs, the planets, etc. I always liked to show my class the nice evolutionary tracks created by Adam Burrows. You see the stars - they turn up and then they go flat; you know - hydrostatic equilibrium; and 'the rest' go down. I tell my class that 'the rest' are all alike. So, I consider these to be very similar objects.

I also worry about coming up with too much terminology, should we include remnants in the 'stars?' I don't want to get into the issue of calling them 'formerly known as stars.' If they once were stars then I think that entitles them to keep the name of star for the rest of their evolutionary status.

A. LINNELL: I think that inevitably the concept of a star needs to be just a little bit fuzzy. I think that all objects in the Kelvin-Helmholtz phase should be called stars. But hot Jupiters are not stars. So, there has to be a certain degree of fuzziness in the concept and partly it's observationally driven.

E. DEVINNEY: I'd like to just emphasize what Al mentioned. I think that we still need to put things in context when we use each term; based on its origin or maybe the situation in which the object is found. I like the point about the philosophers (I Kant agree more), and I think that the situation reminds me about Principal Component Analysis (PCA). You have a certain coordinate system in which you define things, and then PCAs help to reduce the number of definitions you might need. I think that it's a natural evolution in terminology, you can't escape these problems, and we're not going to come up with a clear answer to that question.

R. WILSON: My idea of a star is an object that believes in the Vogt-Russell theorem at some stage in its youth. That is, it reaches an equilibrium stage early on, and at that time, what you get does not depend on how you put it together. That does not apply to planets and it doesn't apply to brown dwarfs because they do not reach an equilibrium stage.

V. TRIMBLE: A star has an effective temperature? ...

V. TRIMBLE: I have two micro ideas, each will fit in one sentence. One was the remark from the virtual observatory talk about how good the IUE data are. But you will remember (very few of you are as old as me) that it had to be recalibrated many many times before was that good. That is a lesson for future virtual observatories and data archives.

The other thought is: There are now an awful lot of data out there and a finite number of binary stars astronomers and there are two kinds of focuses that one can adopt: one is to look at these neat unique objects that are in fact sextuplets or something and try to understand them. The other is to focus on nice clean spectroscopic eclipsing detached binaries to find out as much as possible about relatively normal stars. Our database of mass-luminosity relationship, age, composition is still very very sparse. And as Dan Popper reached retirement age he worried very much that nobody was going to take over from him. Johannes Andersen did for a while but now he's a director too.

I think there's a real need for people who are willing to work on these rather basic fundamental unexciting things. I cannot resist also mentioning Roger Griffin, whose radial velocity spectrometer has enormously increased our supply of long period SBs with good orbits. It is also an ancestor of the devices that found the first exoplanet from Geneva.

A. BATTEN: I thought the three opening papers by Harmanec, Guinan, and Niarchos were on well-chosen topics by well-chosen speakers. Harmanec asked provocative but fundamental questions. I was sorry that Johannes Andersen was not here as I am sure he would have mounted a spirited defense of his claims for precision in the determination of stellar masses, radii, and luminosities - although, perhaps, the discussion would have gone on all morning!

Petr Harmanec asked some very important questions. Whether or not you agree with him, it is good to ask such questions. Guinan and Niarchos, on the other hand, regaled us with exciting visions of what is to come. Vast quantities of data are expected from new space instruments and interferometry will increase the number of binaries that can be studied completely. Guinan and Niarchos and their generation are doing things my generation could only dream of. In their turn, they are dreaming of things that, very probably, the younger people here will do. We need both their optimism and Harmanec's skepticism. A good motto for this symposium, and for all astronomers, might be: "Reach for the stars, but keep your feet on the ground."

H. LAMMER: I work mainly on 'the rest' but it is still very important to understand the stars. When I listened to all the talks about the missions and telescopes and observations, I found out that it is very important that we need bright stars if we would like to learn something from the rest, or characterizing the rest. There are a lot of very exciting results from missions like CoRoT and Kepler; but the problem in all of these is that the stars are very faint so it is hard to make some follow-up observations.

We know that for instance I am working on the aeronomical part, so I am interested in the interaction between the stellar environment with the upper atmospheres of these planets, and there are a lot of good studies published on this topic. Didier Queloz mentioned the Plato mission where you have not only a simple application of CoRoT and Kepler in finding more planets, but finding planets around bright stars closer to our Sun.

Ed Guinan mentioned that access to UV wavelengths might be gone when the Hubble is not there anymore, and the UV was mentioned again in the second talk about the World Space Observatory. So, if Plato could focus on very bright stars and have a good UV telescope, it would be good for binary star science and also exoplanet characterization. So the 'rest' needs more light.

V. TRIMBLE: European astronomers will have to campaign for these instruments in Europe. The US has declared it is interested in origins, and if you look at citations of papers, the enormous disproportion that I remarked upon a decade ago, I can now say has gotten much worse. If you want your papers cited, you're not at this meeting. The hot stuff is early universe and formation and evolution of large scale structure, and that is where US funding will go.

A. HATZES: Ed Guinan mentioned that we need the great facilities like JWST. However, funding for JWST is in danger, and we should not be complacent about this. This is important even in Europe, because this is one of the great facilities that a lot of people like those who work on 'the rest' are counting on to do some great science, but we should not just assume it will be there.

E. BUDDING : One of the things I found a bit scary about the morning session more than the afternoon was the tremendous amount of information that will come out of the new generation surveys. You mentioned evolution of the universe because I do encounter a bit of this with people at the Australian telescope looking forward to the era of the Square Kilometer Array (SKA), a huge project that will create enormous amounts of information. There are regular monthly meetings going on all the time about planning what will happen with all the data, what will be used, and what science it will address.

So, we've got a lot of data coming, with many possible applications. Maybe it would be useful to think about how to get this into better focus.

A. BATTEN: I'm glad you mentioned that because when I hear predictions of 1,700,000 new spectroscopic binaries, I am relieved that I am no longer compiling catalogues of these objects!

E. DEVINNEY: I'd like to ask the panel this question: I've been thinking about what's been happening since CoRoT and Kepler have been flying and it seems to me that this has given a tremendous boost to our field. Now we can make arguments about the need for more time on bigger telescopes for the spectroscopy that's needed. I wonder what it would have been like if Kepler and CoRoT didn't fly. What would have happened to our field? I was wondering if anyone here has any comments on that.

D. QUELOZ: There is a plan in Europe to build the biggest telescope in the world. It's called the European Extremely Large Telescope (ELT). This plan is about to be discussed today. I just want to mention and maybe get your reaction here that there is no plan to have an Echelle spectrograph, at least in the first generation of instruments. Are you happy with that? If this audience doesn't say anything by the time ELT will be discussed, there will not be an Echelle spectrograph.

This is the situation that is evolving: everything is trapped by infrared, imaging, multi-objects, and low resolution. Is it really useful to continue having a high resolution instrument for ELT? I would certainly love to have some feedback because we're discussing this and we have to make a decision by September or October. And this is not a clear understanding in the committee of the STC whether this needs to be pushed or not. Whatever the wavelength you want to do, is this something the community would require or not?

A. LINNELL: A comment on Petr Harmanec's talk about disentangled spectra and the example of a signal for a secondary star that actually was spurious. I am a strong advocate for numerical experiments. A suggestion: Could you use the derived system parameters to construct a system with and without the contribution of the putative secondary and subject both cases to disentanglement to check the sensitivity of the procedure?

V. TRIMBLE: That's how people correct for poor resolution.

E. BUDDING: I've had also similar concerns about numbers that have been appearing. I saw several times today temperatures of stars like 13,744. A couple of years ago somebody had collected all the different experiments to determine the value of the universal gravitational constant. And most of these values are given to around five or six digits after a decimal place with the corresponding error measure. However, if you looked across the whole table between the different experiment results there is no agreement in the third figure after the decimal place.

Similarly today, we saw the Gaussian gravitational constant, GM in combination, I think it was perhaps to twelve figures. And there was the mass of the Sun given to something like six or seven digits. Does that mean that someone has got the really true value of G? Maybe you can tell me what it is.

V. TRIMBLE: A number of people have determined G to seven significant figures; unfortunately they've got different answers. Which is precisely what you said. There was one discordant one that was significant. 6.6762 (or 3) had been accepted for a long time. I think you can now have four figures, but if you want more than that, you have to go out and orbit the Sun yourself. (I meant as a free flyer.) The temperature of Arcturus went through several thousands of degrees of change, more or less monotonic, over the 1960s and so did the mass. And I don't know where they ended up, except it's a star.

W. KLEY: Coming back to the title of the conference, it seems to be a collection of a variety of fields: Interacting binary stars to exoplanets and essential modelling tools. So, the question arises about planets in binary stars. How do they form, how do they evolve in binary stars? In the last talk we heard from Maciej Konacki that the most close-in planets in interacting binaries don't exist because of some long-term period change and red noise. I thought this was a pity because the closest topic in the combination of the themes of this conference got lost somewhat.

It would have been theoretically very nice to have these planets because they have to survive the common envelope phase of the two stars. Or a second generation of planets comes from the debris of those stars. It was theoretically very nice, very appealing. Not that I have any answers to this or whether I say 'OK, these planets must have been formed by this mechanism or that you can actually form them by this mechanism.' The formation theory is quite critical. I thought this was a very nice connecting topic and I'm a little bit sad that this has gone away, and I am looking for some another connection of the different fields.

H. LAMMER: When I walked around the posters I think even a single star and a hot Jupiter is something like a binary system. You have Roche lobe overflow in one case of the gas of the planet so there are a lot of physical similarities. Getting back to the title, I expected the binary community has a lot of experience and models there, which I think are very useful in this way to be applied to such types of binary systems. That's not a new classification we need because this attaches the first question, what is a star, and what is a planet? From the physics, I think there are a lot of similarities between these unique close-in hot Jupiters and stars.

V. TRIMBLE: If the center of mass of the system is inside one of them, then the other one is a planet.

P. EGGLETON: I am purely a theorist who has never observed anything in my life, except papers written by observers. This might relate to a couple of comments here, when you said let's not ignore the brightest stars, even though it's nice to have data on 10^{10} stars that are faint and 10^4 stars that are really quite bright. The fourth brightest is Capella. I was trying to model it earlier this year, and I relied upon a whole string of papers about Capella. Three of the more recent ones, one published by Roger Griffin who is a well-established figure, and another I think using the TODCOR method with Torres. Each has a velocity semi-amplitude (K) value for one component that differs from the other person's K values by either 9 sigma or 14 sigma, depending on which sigma you are using.

Now, how is that possible? The short answer is that the hotter component of Capella is a rather difficult object with rather broad lines broadened by rotation that makes it rather difficult to determine the parameters. Nevertheless, whatever technique is being used to disentangle the spectra, one would hope that the standard deviation would represent something not much beyond one or two standard deviations. So there is something very fundamentally not right about some methods of getting radial velocity amplitudes.

A. BATTEN: I have not read the papers to which you refer, and I am surprised that the differences between the K values should be so great in modern values. It is, however, as you say, extremely difficult to separate that component out from the composite spectrum for Capella.

V. TRIMBLE: Let me quote the great physicist Fred Reines who said "Half of all three sigma results are wrong." Do you know anybody who ever overestimated his systematic errors?

R. WILSON: One important point is that I appreciate that we have boards sitting around and deciding what we can afford in putting these very expensive machines into orbit and also ground based surveys. But there is one area that I think we should make a point to afford. Recently, I went to look for the response curve of one of these survey missions and all I could find was a graphical result; and in only one publication - no one else had written anything on it. Well, then I went to look for another one and there wasn't even a graphical result; just a comment that "this photometric system is believed to be sort of intermediate between such and such and another Johnson band" - not even a graph. So we wind up with these phenomenally precise observations, and we don't even know what we have an observation of! A reasonably good job has been done with the Johnson bands, Cousins bands, and Stromgren bands. However, response curves should be calculated for the filter, the optics, and what have you the whole throughput thing. It's been done for these various traditional bands, and I think we have to make it our business to have that also done for these new expensive missions.

F. ALLARD: I have a comment on these poor brown dwarfs that are not stars and are not planets. Some new improvement has been made in the modeling of atmospheres for M dwarfs and brown dwarfs. The results are in my paper.

A. BATTEN: We have talked about the problem regarding the accumulation of vast amounts of data. I would like to tell a little story to warn us about relying on vast amounts of digital data. Soon after William, Duke of Normandy, successfully invaded England in 1066 and seized the English Crown, he commissioned a great survey of his new kingdom for taxation purposes. The result, written in the late eleventh century, is known as the "Domesday Book." Some time ago, in celebration of the Book's 900th anniversary, a digital version was created. Two decades later, hardware and software had changed so much that no one could read the digital version. The original Book is still extant and can be consulted by any researcher with a serious interest in its contents!

M. RICHARDS: I want to thank the panel for the interesting discussion this afternoon.

Part 3

Imaging Techniques

From Interacting Binaries to Exoplanets: Essential Modeling Tools
Proceedings IAU Symposium No. 282, 2011
Mercedes T. Richards & Ivan Hubeny, eds.
© International Astronomical Union 2012
doi:10.1017/S1743921311027256

Binaries and Multiple Systems Observed with the CHARA, NPOI, SUSI and VLTI Interferometric Eyes

Philippe Stee

UMR 6525 H. Fizeau, Université de Nice Sophia Antipolis, Centre National de la Recherche
Scientifique, Observatoire de la Côte d'Azur,
Boulevard de l'Observatoire, B.P. 4229 F 06304 Nice Cedex 4 - France.
email: philippe.stee@oca.eu

Abstract. In this review I will present recent results between 2007 and 2011 based on interferometric observations of binaries and multiple systems with the VLTI, NPOI, SUSI and CHARA instruments. I will also explain the kind of constraints an interferometer can provide in order to better understand the physics of multiple systems.

Keywords. stars: binaries (including multiple): close, stars: circumstellar matter, stars: stars: fundamental parameters, stars: imaging, stars: kinematics, stars: winds, outflows, stars: rotation, stars: mass loss, stars: imaging, techniques: interferometric

---

1. Introduction

Interferometry of binaries and multiple systems is a very active and attractive research field and more than 53 papers with the keywords "interferometry + binarity," excluding radio observations, were found between 2007 and 2011 using the Astrophysics Data System (ADS). These papers were arbitrarily "classified" in various themes, namely orbits (10 papers), fundamental parameters (13), model fitting (14), interacting binaries (11) and image reconstruction (5). These five themes are actually certainly biased due to angular resolution of a few milli-arsecond (mas) and magnitude limitations on the order of 6-7 of current ground-based optical interferometers. Note also that interferometric techniques are really fruitful and seem now in a mature age with growing results on direct imaging of circumstellar disks and stellar surfaces.

The first question to answer is: what can you do with an interferometer when observing a multiple system? We have identified 6 main topics, from the simplest case to more elaborate and complicated studies:

- Astrometry with precise orbits determination.
- Measurement of uniform disk, i.e. stellar photospheres and fundamental parameters.
- Measurements of limb darkened stellar surfaces.
- Kinematics of interacting binaries or circumstellar disks.
- Model fitting.
- Image reconstruction.

We must also stress that binary or multiple-system observations are strongly wavelength and time dependent, and also the important role played by amateur astronomers in the spectroscopic and photometric long-term monitoring, or "alert" mode, of some interesting targets, see for instance the BeSS database†. In this review paper, I will present an arbitrary selection of recent results obtained thanks to the CHARA, NPOI, SUSI and VLTI interferometers by following these 6 previously defined topics. Thus, the paper has the following structure: In Sect. 2, we present results of orbit determinations using very accurate interferometric measurements. In Sect. 3, we describe how an interferometer can be used to determine fundamental stellar parameters. Sect. 4 is dedicated to the model fitting of interferometric data. In Sect. 5, we illustrate recent image reconstructions of β Lyrae and ϵ Aurigae binary systems; and in Sect. 6, the main conclusions are drawn.

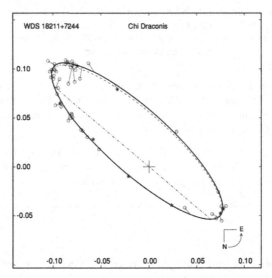

Figure 1. Orbit plot for χ Draconis by Farrington *et al.* (2010) using the CHARA interferometer

2. Orbit determination

On of the more obvious ways of determining a precise orbit for multiple systems is the use of interferometric data. The first measurements were done thanks to speckle-interferometry, see for instance Labeyrie (1970), but this technique is roughly limited to separations $\geqslant$ 30 mas, and more recently long baseline interferometry was successfully used for the same purpose. For instance, the precise orbits of χ Draconis HD 170153 SB2 (F8IV-V + late-G/early-K dwarf) with visual orbit, $m_v = 3.57$, $\pi = 124.11$ mas, P $\sim$ 281 days, was obtained by Farrington *et al.* (2010) using the CHARA interferometer. Using the orbital elements calculated in this paper and the most recent mass ratio calculated from Nordström *et al.* (2004), they obtained masses of $M_P = 0.96 \pm 0.03$ $M_\odot$ and $M_S = 0.75 \pm 0.03$ $M_\odot$, which fall below expected ranges for even the latest F-stars (see Fig. 1). This technique was also used for more complicated systems as σ^2 Coronae Borealis by Raghavan *et al.* (2009) with the CHARA interferometer. This system is composed of two Sun-like stars of roughly equal mass in a circularized orbit with a period of 1.14 days. The long baselines of the CHARA array have allowed them to resolve the visual orbit for this pair, the shortest-period binary yet resolved interferometrically, and enabling them to determine component masses of 1.137 ± 0.037 $M_\odot$ and 1.090 ± 0.036 $M_\odot$. This pair is the central component of a quintuple system, along with another similar mass star, σ^1

† http://basebe.obspm.fr

CrB, in a $\sim$ 730 year visual orbit, and a distant M-dwarf binary, σ CrB C, at a projected separation of $\sim$ 10'. Finally to illustrate this technique, it is also possible to accumulate measurements from various interferometers, which was done by Kraus *et al.* (2009) by combining IOTA, NPOI and VLTI data to obtain a very precise orbit of the nearby high-mass star binary system θ^1 Ori C. Their new astrometric measurements have shown that the companion has nearly completed one orbital revolution since its discovery in 1997. The derived orbital elements imply a short-period (P $\sim$ 11.3 yr) and high-eccentricity orbit (e $\sim$ 0.6) with periastron passage around 2002.6.

3. Fundamental parameters

In the following, I will present a selection of recent results on the determination of fundamental parameters. Nevertheless, we first need to define what are these parameters. From our point, there are three fundamental parameters, namely:
- Radius: it requires sub-mas of spatial resolution, without forgetting that it may be wavelength dependent.
- Effective temperature: it can be determined from the angular size of the stellar source with a sufficient S/N ratio spectral energy distribution curve.
- Mass: it can be determined from dynamical masses measured thanks to precise orbital data with radial velocity measurements.

Other fundamental parameters can also be determined:
- Distance: thanks to parallactic measurements of stellar distances which can calibrate stellar luminosities.
- Temperature structure: from precise limb-darkening measurements. It requires large baselines in order to obtain data points in the 2d or 3d lobes of the visibility function.

A nice example of fundamental parameters determination is the work done by Mérand *et al.* (2011) on the eclipsing binary δ Velorum. From the modeling of the primary and secondary components of the δ Vel A eclipsing pair, they have derived their fundamental parameters with a typical accuracy of 1%. They found that they have similar masses, i.e. 2.43 ± 0.02 M$_\odot$ and 2.27 ± 0.02 M$_\odot$. The physical parameters of the tertiary component (δ Vel B) was also estimated. They obtain a parallax $\pi = 39.8 \pm 0.4$ mas for the system, in good agreement with the Hipparcos value, i.e. $\pi_{Hip} = 40.5 \pm 0.4$ mas. Zavala *et al.* (2010) have observed the Algol triple system with the NPOI interferometer and have produced, for the first time, images resolving all three components. They have separated the tertiary component from the binary and have simultaneously resolved the eclipsing binary pair, which represents the nearest and brightest eclipsing binary in the sky. On the same system, Borkovits *et al.* (2010) carried out interferometric observations both in the optical/near-infrared and in the radio regime. The optical interferometric observations were done with the CHARA array in the K$_s$ band and they were able to test the Kozai resonance/cycles, see Kozai (1962) for more details, by determining the mutual inclination of this triple system, i.e. i=95° $\pm$ 3°. Another interesting work was done by Bruntt *et al.* (2010) who was able, using the CHARA/FLUOR and VLTI/NACO instruments to determine the radius and T$_{eff}$ of the binary (ro)Ap star β CrB. This was also a way to calibrate the T$_{eff}$ scale for Ap stars which is very helpful to constrain asteroseismic models. Another example of asteroseismic constraints thanks to interferometric measurements of multiple systems is the work done by Kervella *et al.* (2008) who were able to infer the radii and the evolutionary status of the 61 Cyg A & B K5V and K7V stars from CHARA/FLUOR data. These new radii were able to constrain efficiently the physical parameters adopted for the modeling of both stars, allowing them to predict asteroseismic frequencies (namely $\delta\,\nu_{n,l}$ & $\Delta\,\nu_{n,l}$) based on their best-fit models. Moreover, their CESAM2k, see Morel (1997), evolutionary models indicate an age around 6 Gyr and are compatible with small values of the mixing length parameter.

Interferometers were recently used to determine very precise angular diameters and T_{eff} of exoplanet host stars following the work done by Baines et al. (2008) and Baines et al. (2009) whereas Duvert et al. (2010) were able to detect a 5-mag fainter companion at a distance of 4 R_* around HD 59717 with VLTI/AMBER using the phase closure nulling technique. They found that errors on the secondary measurements are mainly due to the uncertainty in the flux ratio between both components rather then from the interferometric data. Finally, Absil et al. (2010) have carried out a deep near-infrared interferometric search for low mass companions around β Pictoris. Their results exclude the presence of brown dwarfs with m > 20 M_{Jup} (resp. 47 M_{Jup}) at a 50% (resp. 90%) completeness level within the few nearby AU (2-60 mas). They also exclude the presence of companions with K band contrast <5 × 10^{-3} and finally their best data fit was obtained for a binary model at 14.4 mas from β Pic with a contrast of 1.8 × 10^{-3}.

4. Model fitting

Spectroscopic and photometric techniques are usually used to constrain various models but since they are providing spatially integrated observables, the solution or "best model" obtained is very often not unique since different geometrical models can produce identical spectroscopic and photometric observables. Thanks to interferometry and more especially to spectrally resolved interferometry, see Mourard et al. (2009), it is possible to overcome this degeneracy and put very strong constraints on modern modeling. In the following, I will illustrate how models can be constrained by interferometric measurements for the three well known binaries: ζ Tau, δ Sco and γ^2 Vel.

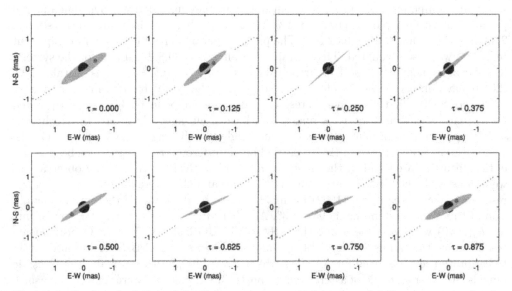

Figure 2. Cartoon depiction of the disk precession variations around ζ Tau as seen in the sky from Schaefer et al. (2010)

4.1. ζ Tau

This binary Be star was observed by Gies *et al.* (2007) with the CHARA interferometer in the K-band. The size of the circumstellar disk, nearly edge-on, was found to be smaller than in the Hα line which is mainly due to a larger Hα opacity and a relatively larger neutral H fraction with increasing disk radius. They were able to estimate the density distribution as a function of the stellar radius and Gies *et al.* (2007) found that the "n" parameter describing the density distribution "slope", i.e. $\rho(r) = \rho_0 [\frac{R_*}{r}]^n$, was smaller in binaries with smaller semimajor axies. This was direct evidence that binary companions do influence the disk properties. An asymmetry was also detected by Schaefer *et al.* (2010) with the CHARA/MIRC interferometer. They have observed a change in the position angle of the disk over time. Moreover a correlation between the position angle and the V/R phase of the Hα emission line variations suggests that the tilt of the disk around ζ Tau is precessing. They have also measured an asymmetry in the light distribution of the disk that roughly corresponds to the expected location of the density enhancement of the spiral oscillation model. Meilland *et al.* (2009) observed this star with the VLTI/MIDI instrument between 8 and 12 μm and found that the size of the disk does not vary strongly with wavelength between 8-12 μm which can be due to a disk truncation by the companion. Finally a global 3D and NETL model of ζ Tau was done by Carciofi *et al.* (2009) using the HDUST code. This model, based on a 2D global disk oscillation that is propagating within the equatorial disk, was in agreement with simultaneous VLTI/AMBER data and V/R variations in the Hα and Brγ spectral lines.

4.2. δ Sco

δ Sco is also a binary Be star already observed by Millan-Gabet *et al.* (2010) in 2007 with CHARA and resolved in the H continuum, Brγ, HeI and Hα lines. This is an interesting system since the companion passage at periastron is so close that tidal effects have strong effects on the formation/dissipation of the circumstellar disk around the primary. Since the rediscovery of its binarity in 1974, δ Sco has been observed several times using speckle-interferometry, mainly published in McAlister & Harkopft (1988). New speckle-interferometric measurements allowed Hartkopf *et al.* (1996) to derive more accurate and significantly different parameters. Using the same dataset complemented by radial velocity measurements close to the 2000 periastron, Miroshnichenko *et al.* (2003) refined the previous analysis. Tango *et al.* (2009) have used the same dataset but with a more consistent model-fitting algorithm from Pourbaix (1998) to estimate the orbit. One of the most recent analyses of δ Sco orbit was done by Tycner *et al.* (2011) using 96 measurements from the Navy Prototype Optical Interferometer (NPOI). Finally, Meilland *et al.* (2011) have obtained new VLTI/AMBER and CHARA/VEGA data and conclude that the next periastron passage should take place around July 5, 2011 ($\pm$ 4 days). They also found that the rotation appears to be Keplerian, with an inner boundary (photosphere/disk interface) rotating at the critical velocity. The expansion velocity within the circumstellar equatorial disk was found to be negligible and considering an outburst scenario, should be on the order of 0.2 kms^{-1}. With the measured vsini of 175 kms^{-1} and the measured inclination angle of 30.2 $\pm$ 0.7°, the star rotates at about 70% of its critical velocity. This could indicate that the stellar rotation is not the main process driving the ejection of matter from the stellar surface. However, taking into account possible underestimation of the vsini due to gravity darkening, see Townsend *et al.* (2004), the star may rotate faster, up to 0.9 of its critical velocity.

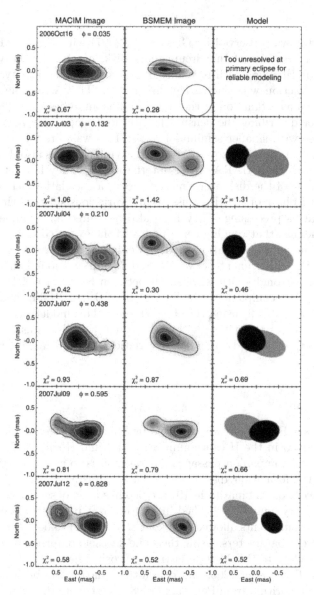

Figure 3. Reconstructed images in the H band and two-component models of β Lyrae with the CHARA/MIRC instrument from Zhao *et al.* (2008) .

4.3. γ^2 *Vel*

γ^2 Velorum is an interacting binary with a primary Wolf-Rayet star and an O type companion. This system was observed by Millour *et al.* (2007) with the VLTI/AMBER and by North *et al.* (2007) with the SUSI interferometer. Millour *et al.* (2007) using a relatively restrained data set have set tight constraints on the geometrical parameters of the γ^2 Velorum orbit. A nice example of the constraints brought by interferometry is the work done by Millour *et al.* (2007) where they have drastically reduced the error box derived from the classical radial velocity method by using an interferometric fit to retrieve the geometrical parameters of the binary star. Moreover, the smaller separation measured by interferometry means that the distance of the system must be reevaluated to 368 pc, in agreement with recent spectrophotometric estimates, but significantly larger than the Hipparcos value of 258 pc.

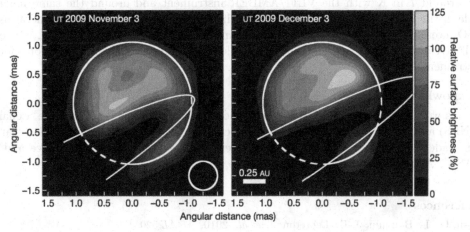

Figure 4. Synthesized images from the 2009 observations of the ϵ Aurigae with the CHARA/MIRC instrument from Kloppenborg *et al.* (2010).

5. Image reconstruction

Nowadays, it is not only possible to constrain "ad-hoc" or more physical models using interferometric data but it is also possible to directly reconstruct images without any a-priori model. This is only possible with interferometers with at least 3 telescopes in order to recover the true phase of the object onto the skyplane which is blurred by the atmosphere with only 2 telescopes. This phase can be determined thanks to the "closure phase" technique, see for instance Roddier (1986), which is the case of the CHARA/MIRC (combining up to 6 telescopes), the VLTI/AMBER (up to 3 telescopes) and the CHARA/VEGA (up to 4 telescopes) instruments. A very spectacular example is the direct image reconstruction in the H band of the interacting binary system β Lyrae done by Zhao *et al.* (2008) with the CHARA/MIRC instruments. They were able to reconstruct the image of the binary system within a few mas but also to follow its time evolution (see Fig. 3). Last but not least, another very spectacular example is the image reconstruction of the transiting disk in the ϵ Aurigae system again with the powerful CHARA/MIRC interferometer by Kloppenborg *et al.* (2010). ϵ Aur is a visually bright, eclipsing binary star system with a period of 27.1 years. Using direct imaging of the system they have demonstrated the validity of the disk model for the previously unseen companion. The elliptical appearance of the disk seems to be more consistent with a model of a tilted thin disk rather than one of a thick disk seen edge on. With these images and a simple model, they were able to estimate the dimensions and masses of the components in the system (See Fig. 4). An animation of the disk passage in front of the stellar disk obtained from the reconstructed images can be seen on YouTube†.

6. Conclusion

It is clear that interferometry in the visible and infrared is now a mature technique. Direct images are possible and spatial and spectral resolution are available with instru-

---

† http://www.youtube.com/watch?v=LfKMmCkV1xE

ments using spectrally resolved observations such as the CHARA/VEGA interferometer. The main drawback is certainly the limited magnitude of the current interferometers, i.e. around 7 in K with the VLTI/AMBER instrument and around the same magnitude in the visible/infrared for the CHARA interferometer. The VLTI is a general user ESO interferometer open to the general community through call for proposals‡ whereas CHARA instruments¶ are generally accessible via collaborations with the PI of a focal instrument.

Acknowledgements

This work has been supported by the French Program National en Physique Stellaire (PNPS) and by the Centre National de la Recherche Scientifique (CNRS). This research has made use of the SIMBAD database, operated at CDS, Strasbourg, France and of NASA's Astrophysics Data System.

References

Absil, O., Le Bouquin, J.-B., Lebreton, J. *et al.*, 2010, *A&AL* 520, L2
Baines, E. K., McAlister, H. A., Brummelaar, T. A. *et al.*, 2008, *ApJ* 682, 577
Baines, E. K., McAlister, H. A., Brummelaar, T. A. *et al.*, 2009, *ApJ* 701, 154
Borkovits, T., Paragi, Z. & Csizmadia, S. 2010, *J. Phys.: Conf. Ser.* 218, 012005
Bruntt, H., Kervella, P. Mérand, A. *et al.*, 2010, *A&A* 512, A55
Carciofi, A. C., Okazaki, A. T., Le Bouquin, J.-B. *et al.*, 2009, *A&A* 504, 915
Duvert, G., Chelli, A., Malbet, F. *et al.*, 2010, *A&A* 509, A66
Farrington, C. D., Brummelaar, T. A., Mason, B. D. *et al.*, 2010, *AJ*, 139, 2308
Gies, D. R., Bagnuolo, W. G., Baines, E. K. *et al.*, 2007, *ApJ*, 654, 527
Hartkopf W. I., Mason B.D. & McAlister H.A. 1996, *AJ*, 111, 370
Kervella, P., Mérand, A., Pichon, B. 2008, *A&A*, 488, 667
Kloppenborg, B., Stence, R., Monnier, J. D. *et al.*, 2010, *Nature*, Vol. 464, p 870
Kraus, S;, Weigelt, G., Balega, Y. Y. *et al.*, 2009, *A&A*, 497, 195
Kozai, Y. 1962, *AJ*, 67, 591
McAlister H. A. & Hartkopf, W. I. 1988, *2nd Catalogue of Interf. Meas. of Binary Stars*
Meilland, A., Stee, Ph., Chesneau, O. *et al.*, 2009, *A&A*, 505, 687
Meilland, A., Delaa, O., Stee, Ph. *et al.*, 2011, *A&A*, 532A, 80M
Merand, A., Kervella, P., Pribulla, T. 2011, *A&A*, 532, A50
Millan-Gabet R., Monnier J. D., Touhami Y. *et al.*, 2010, *ApJ*, 723, 544
Millour, F., Petrov, R. G., Chesneau, O. *et al.*, 2007, *A&A*, 464, 107
Miroshnichenko A. S., Bjorkman K. S., *et al.*, 2003, *A&A*, 408, 305
Mourard, D., Clausse, J.-M., Marcotto, A. *et al.*, 2009, *A&A*, 508, 1073
Morel, P. 1997, *A&AS*, 124, 597
Nordstrom, B *et al.*, 2004, *A&A*, 418, 989
North, J. R., Tuthill, P. G., Tango, W. J. *et al.*, 2007, *MNRAS*, 377, 415
Pourbaix D. 1998, *A&AS*, 131, 377
Raghavan, D., McAlister, H. A., Torres, G. *et al.*, 2009, *ApJ*, 690, 394
Roddier, F. 1986, *Optics Communications*, Vol 60, issue 3, p 145
Schaefer, G. H., Gies, D. R., Monnier, J. D. *et al.*, 2010, *AJ*, 140, 1838
Townsend, R. H., Owocki, S. P. & Howarth, I. D. 2004, *MNRAS*, 350, 189
Zhao, M., Gies, D., Monnier, J. D. *et al.*, 2008, *ApJ*, 684 L95
Zavala, R. T., Hummel, C. A., Boboltz, D. A., *et al.*, 2010, *ApJL*, 715, L44

‡ http://www.eso.org/sci/observing/phase1/proposalsopen.html
¶ http://www.chara.gsu.edu/CHARA/

From Interacting Binaries to Exoplanets: Essential Modeling Tools
Proceedings IAU Symposium No. 282, 2011
Mercedes T. Richards & Ivan Hubeny, eds.
© International Astronomical Union 2012
doi:10.1017/S1743921311027268

Observing Faint Companions
Close to Bright Stars

Eugene Serabyn

Jet Propulsion Laboratory, California Institute of Technology,
4800 Oak Grove Drive, Pasadena, CA 91109, USA
email: `gene.serabyn@jpl.nasa.gov`

Abstract. Progress in a number of technical areas is enabling imaging and interferometric observations at both smaller angular separations from bright stars and at deeper relative contrast levels. Here we discuss recent progress in several ongoing projects at the Jet Propulsion Laboratory. First, extreme adaptive optics wavefront correction has recently enabled the use of very short (i.e., blue) wavelengths to resolve close binaries. Second, phase-based coronagraphy has recently allowed observations of faint companions to within nearly one diffraction beam width of bright stars. Finally, rotating interferometers that can observe inside the diffraction beam of single aperture telescopes are being developed to detect close-in companions and bright exozodiacal dust. This paper presents a very brief summary of the techniques involved, along with some illustrative results.

Keywords. adaptive optics, coronagraphy, interferometry

1. Introduction

Although many exoplanets have now been discovered through both the radial velocity and transit techniques, to date very few exoplanets have been resolved from their host stars and imaged directly, because of their combination of close proximity to their host stars and relative faintness. However, a number of technical developments in both imaging and interferometry are bringing about rapid progress in the direct detection area as well. Here, a few recent developments that provide access to new observational regimes are summarized.

To reach small angular separations from stars using a single telescope, both imaging and interferometry have roles to play. The key in both cases is a high degree of wavefront correction, as provided by either current generation adaptive optics (AO) systems, or even more capable extreme adaptive optics (ExAO) systems that are beginning to come on line.

2. Imaging

The theoretical Strehl ratios predicted to be provided by typical current-generation AO systems and by more capable next-generation ExAO systems are compared in Fig. 1. This plot makes it clear that extreme adaptive optics can be made use of in two ways: shorter wavelengths can be used to achieve higher angular resolution, and high infrared Strehl ratios can enable effective coronagraphy (i.e., diffraction control). Both cases enable observations closer to bright stars. Figs. 2 and 3 give examples of observations of both types obtained with the Palomar "well-corrected subaperture" (WCS), a 1.5 m off-axis subaperture on the Hale telescope that is corrected to ExAO levels (Serabyn *et al.* 2007). Fig. 2 shows a close binary resolved in the B band, and Fig. 3 shows a faint secondary star near ϵ Ceph. In the latter image, the primary star has been largely extinguished with a vector vortex coronagraph (Mawet *et al.* 2010). Even higher contrast performance has allowed the Palomar vortex coronagraph to be used to image the three outer exoplanets in the HR8799 system using the WCS's 1.5 m aperture (Serabyn, Mawet, & Burruss 2010).

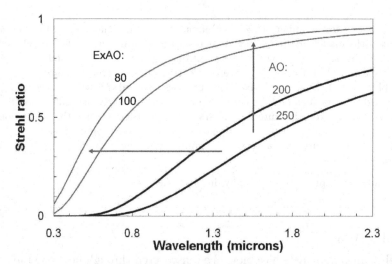

Figure 1. Strehl ratios expected with conventional AO (dark curves; shown for 200 - 250 nm of wavefront error) and ExAO (light curves; shown for 80 - 100 nm of wavefront error). ExAO enables very high Strehl-ratio (and high contrast) observations in the infrared (vertical arrow), and high-resolution observations at shorter (visible) wavelengths (horizontal arrow).

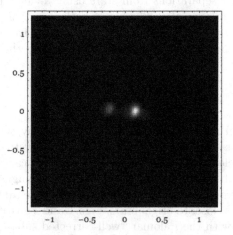

Figure 2. Palomar WCS image of the binary SAO37735 in the B band (Serabyn *et al.* 2007). The separation is 0.34".

3. Interferometry

However, interferometry within a single-aperture telescope allows observations even closer to the center than is possible with coronagraphic imaging using the full telescope aperture. To take advantage of this, rotating nulling interferometers akin to those suggested for space-based interferometry (Bracewell 1978), but operating within single telescope apertues, are under development (Serabyn *et al.* 2010). At the Palomar Observatory, our rotating fiber-based nuller has already put stringent constraints on potential companions and dust near Vega (Mennesson *et al.* 2011), by using a novel data reduction method that can retrieve deep astrophysical nulls from the distribution of the null depth fluctuations (Hanot *et al.* 2011). Moreover, an initial test observation on one of the Keck telescopes has begun to demonstrate the applicability of rotating baseline interferometers to the detection of close companions (Fig. 4).

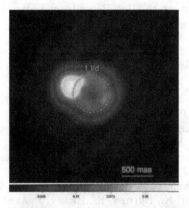

Figure 3. Detection of a potential companion to ϵ Ceph with our vector vortex coronagraph on the Palomar WCS (Mawet *et al.* (2011)). The primary residual appears as a dim ring about the center, and the secondary is located is at approximately 1.1 λ/D.

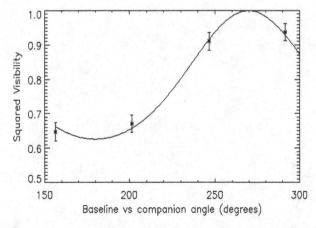

Figure 4. Detection of a companion to HIP87895 with a rotating-baseline interferometer on one of the Keck telescopes (Serabyn *et al.* 2010). The brightness difference is $\Delta K = 2.1$ mag and the separation is 39 mas.

4. Conclusions

New techniques are allowing for significant progress in both high-contrast imaging and interferometry. In the imaging area, these include extreme adaptive optics wavefront correction, including non-common-path speckle reduction with phase retrieval algorithms (Burruss *et al.* 2010), novel coronagraphs with small inner working angles (Guyon *et al.* 2006), and new data reduction methods, such as the locally optimized combination of images (LOCI) algorithm (Lafreniere *et al.* 2007). Regarding interferometry, deeper nulls can be measured by employing fiber optic beam combiners (Haguenauer and Serabyn 2006), and by carrying out a statistical analysis of the measured null depth fluctuations (Hanot *et al.* 2011). Rotating nulling baselines also allow effective exploration of the interferometric visibility plane, providing information on the spatial distribution of emission sources very close to bright stars, i.e., inside the single-aperture diffraction beam (Serabyn *et al.* 2010).

This work was carried out at the Jet Propulsion Laboratory, California Institute of Technology, under contract with NASA.

References

Bracewell, R. N. 1978, *Nature*, 274, 780
Burruss, R. S., Serabyn, E., Mawet, D. P., Roberts, J. E., Hickey, J. P., Rykoski, K., Bikkannavar, S., & Crepp, J. R. 2010, *Proc. SPIE*, 7736, 77365X
Guyon, O., Pluzhnik, E. A., Kuchner, M. J., Collins, B., & Ridgway S. T. 2006, *ApJS*, 167, 81
Haguenauer, P. & Serabyn E. 2006, *Appl. Opt.*, 45, 2749
Hanot, C., Mennesson, B., Martin, S., Liewer, K., Loya, F., Mawet, D., Riaud, P., Absil, O., & Serabyn, E. 2011, *ApJ*, 729, 110
Lafreniere, D., Marois, C., Doyon, R. Nadeau, D., & Artigau, E. 2007, *ApJ*, 660, 770
Mawet, D., Serabyn, E., Liewer, K., Burruss, R., Hickey, J., & Shemo, D. 2010, *ApJ*, 709, 53
Mawet, D., Mennesson, B., Serabyn, E., Stapelfeldt, K., & Absil, O. 2011, *ApJ*, 738, L12
Mennesson, B., Serabyn, E., Hanot, C., Martin, S. R., Liewer, K., & Mawet, D. 2011, *ApJ*, 736, 14
Serabyn, E., Mawet, D., & Burruss, R. 2010, *Nature*, 464, 1018
Serabyn, E., Mennesson, B., Martin, S., Liewer, K. & Mawet, D., Loya, F., & Colavita, M. 2010, *Proc. SPIE*, 7734, 77341E
Serabyn, E., Wallace, K., Troy, M., Mennesson, B., Haguenauer, P. & Gappinger, R. 2007 *ApJ*, 658, 1386

From Interacting Binaries to Exoplanets: Essential Modeling Tools
Proceedings IAU Symposium No. 282, 2011
Mercedes T. Richards & Ivan Hubeny, eds.

© International Astronomical Union 2012
doi:10.1017/S174392131102727X

Tomography of Interacting Binary Systems: Algols to Gamma-Ray Binaries

Mercedes T. Richards

Pennsylvania State University, Department of Astronomy & Astrophysics, 525 Davey Lab,
University Park, PA 16802, USA
email: mrichards@astro.psu.edu

Abstract. Three-dimensional Doppler tomography of interacting binaries has now provided some interesting perspectives of the gas flows beyond the central plane corresponding to the orbital plane. These images suggest that the magnetic field of the mass losing star influences the gas flows in some cases. Earlier 2D tomograms displayed evidence of gas flows associated with the gas stream, accretion disks, accretion annuli, and hot spots as well as evidence of magnetic flows associated with the mass loser. These indirect images have revealed the active environments that exist in the slow-mass-transfer Algols, cataclysmic variables, polars, x-ray binaries, and gamma ray binaries.

Keywords. accretion, accretion disks, (stars:) binaries (including multiple): close, (stars:) binaries: eclipsing, (stars:) binaries: general, (stars:) circumstellar matter, stars: imaging, techniques: image processing, X-rays: binaries, gamma rays: observations, stars: individual (β Per, U CrB, RS Vul, V711 Tau, TT Hya, AU Mon, Cyg X-1, o Ceti)

---

1. Introduction

We can resolve images of distant galaxies and yet most nearby stars remain unresolved. We have high resolution images of the Sun in 2D and 3D, but not for other stars. In the case of interacting binaries, it is challenging to resolve the images of the stars since the angular separations are in the milli-arcsecond regime, with separations comparable to the Mercury-Sun distance. However, spectra provide us with information about the Doppler motions of the gas between and around the stars. Moreover, most interacting binaries contain a magnetically active star like our Sun, and recent images of the Sun from the Solar Dynamics Observatory demonstrate the immense level of activity associated with the Sun; yet we often treat cool stars in binaries as dormant objects when their magnetic activity is more powerful than that of the Sun. The magnetically-controlled structures on the cool sun-like mass-losing star (e.g., prominences) could influence how the gas flows between the stars in a binary.

Well-resolved images of single stars have been achieved in only a few cases, e.g., Betelgeuse, Sirius, Altair, Vega (Haubois *et al.* 2009). However, the CHARA and NOI interferometers have now produced the first resolved optical images of the brightest and nearest eclipsing binary and triple-star system in the sky: Algol, β Per (Czismadia *et al.* 2009; Zavala *et al.* 2010). The Navy Optical and Infrared Interferometer (NOI) is composed of six 1.8m mirrors with a resolution of 0.2 milli-arcsecond resolution while the Center for High Angular Resolution Astronomy Interferometer (CHARA) is composed of six 1m mirrors with a resolution of 0.15 milli-arcsecond resolution. At radio wavelengths, Peterson *et al.* (2010) used a global very long baseline interferometer array to find evidence of asymmetric magnetic structures that are not confined to the orbital plane in a resolved 15 GHz radio image of β Per.

Hydrodynamic simulations have been used to illustrate how the gas should be distributed between stars (e.g., in Algol binaries: Blondin, Richards & Malinkowski 1995; Richards & Ratliff 1998). However, direct images showing the flow of gas between stars in an interacting binary were first obtained by Karovska *et al.* (2005) from soft X-ray Chandra and HST images of a system called Mira (o Ceti). In all other cases, we have to resort to indirect techniques to study the emission sources and gas flows in the binary. Tomography has been used effectively to generate indirect images of these gas flows, and it will be beneficial until interferometers can provide detailed direct images.

2. Tomography

The image reconstruction technique of tomography has been used successfully in many fields, including medicine, geophysics, archaeology, oceanography, and astronomy. The basic procedure can be implemented in two steps: (1) acquire a set of views of the object at directions around the object over the 360° range of angles. These views are called slices or projections that can be represented mathematically by the Radon transform (Radon 1917). (2) The 3D image is recovered through a summation process called back projection; accomplished by taking each image projection and returning it along the path from which it was acquired. The quality of the reconstruction depends on the number of views acquired and the angular distribution of these views. In general, tomography uses (n-1)-dimensional projections to calculate an n-dimensional image of an object (e.g., make a 3D image using 2D projections, like CAT scans in medicine).

The mathematical formulations for 2D and 3D tomography are summarized in Agafonov, Sharova & Richards (2009). In 2D Doppler tomography, the Radon transform of a function, $I = f(v_x, v_y)$, is a set of 1D projections, $p(v_r, \phi)$, which is the *line profile* with Doppler shift, v_r, at each orbital phase, ϕ, after correction for the systemic velocity.

$$ f(v_x, v_y) = \int_0^{2\pi} \int_{-\infty}^{\infty} \int_{-\infty}^{\infty} p(v_r, \phi) \, |\omega| \, e^{2\pi i \omega(-v_x \cos \phi + v_y \sin \phi - v_r)} \, dv_r \, d\omega \, d\phi $$

This filtered back-projection formula can be extended to the next dimension to calculate $f(v_x, v_y, v_z)$ (see Appendix of Marsh & Horne 1988).

The extension to 3D has been achieved with a technique that is familiar to radio astronomers (e.g., Bracewell & Riddle 1967). The Radioastronomical Approach (RA) developed by Agafonov & Sharova (2005) solves the convolution equation $g(x, y, z) = f(x, y, z) * * * h(x, y, z) + n(x, y, z)$, where $g(x, y, z)$ is the summarized image or "dirty map," $f(x, y, z)$ is the brightness distribution of the unknown object, $h(x, y, z)$ is the summarized point spread function, and $n(x, y, z)$ is the noise. This technique can be used for any kind of 3D reconstruction: Cartesian or Doppler. A comparison of the FBP technique with the RA method shows that the RA method produces better results than FBP and with fewer projections (Agafonov, Richards & Sharova 2006).

The basic assumptions and constraints of Doppler tomography were summarized by Richards, Sharova & Agafonov (2010). The main assumption is that the line profiles are broadened only by Doppler motions; which is good to first approximation, although turbulent motions may also contribute. We need numerous spectra distributed around the orbit of the binary, collected at high wavelength resolution and with good coverage in orbital phase. The technique has been applied to spectra that are dominated by emission, but the generality of the equations suggests that the method can be applied with caution to absorption spectra as well. In the case of Algol binaries with weak emission-line spectra, we assume that the circumstellar gas is optically thin; which is good to first order.

The relative gas opacity can be estimated from a comparison between images derived from observed spectra and then from difference spectra. Here the difference spectra are calculated by subtracting the composite spectrum of the stars from the observed spectrum (e.g., Richards 1993). When the RA method is used to derive 3D images, the stretching of the images in the V_z direction relative to V_x and V_y depends on orbital inclination, i. The resolutions are the same if $i = 45°$, however the resolution in V_z degrades if $i > 45°$ up to 90° (Richards, Sharova & Agafonov 2010). The stretching effect in the V_z direction is linked to the effect of orbital inclination on the Point Spread Function (PSF). Finally, the transformation of the emission intensity from velocity space to Cartesian space is difficult because we first need to know the velocity fields of all emission sources to make the conversion. These velocity fields can be estimated using hydrodynamic simulations.

3. Application to Interacting Binary Star Systems

Tomography can be applied only if there exist different angular views of a system. A notable reconstruction of the surface of Venus in Cartesian coordinates was achieved with the use of radar data since the surface is obscured by thick clouds; the reconstruction was achieved with data collected by the *Magellan* satellite which orbited the planet for four years. However, applications to eclipsing binaries and rotating stars (which provide us with changing views of the system) require the use of spectra, and the technique is termed "Doppler tomography" since the gas motions detected through Doppler shifts are used to provide an image of the gas flows in velocity coordinates (e.g., Marsh & Horne 1988).

The structures identified from optical and ultraviolet spectra of interacting binaries include classical and transient accretion disks, the gas stream, emission from the chromosphere and other magnetic structures on the donor star (e.g., prominences), shock regions, an accretion annulus, and an absorption zone (where hotter gas is located in the optical tomograms). Accretion disks are found in binaries that contain white dwarfs and neutron stars (e.g., cataclysmic variables, nova-like systems, soft X-ray binaries, gamma-ray binaries), as well as non-compact main sequence stars (e.g., Algol binaries). The main differences between these systems are: (1) the donor star (compact vs. non-compact) influences the resulting accretion structures; (2) the accretion structures in the compact systems are bright relative to the stars, while the structures in the Algols are faint relative to the luminous main sequence primary star; (3) the large size of the mass gainer in the Algols leads to the direct impact of the gas stream onto the stellar surface in the short-period Algols, while this type of impact does not occur in the long-period Algols or the compact systems. The result is that a complex set of accretion structures is formed in the short-period Algols compared to the classical accretion disks in the other systems.

4. Indirect 2D and 3D Images of Interacting Binaries

Numerous time-resolved spectra of the target systems are required for the tomography calculation. Most of the lines are in the optical regime (e.g., for CVs and X-ray binaries: He I λ5015, He II λ4686, Hβ; and for Algols and gamma-ray binaries: Hα, Hβ, He I λ6678, Si II λ6371). Ultraviolet tomograms were based on the Si IV λ1063,1073 doublet and Si IV λ1394 line. Doppler tomograms of over 33 CVs have been produced (e.g., Marsh 2001; Morales-Rueda 2004; Steeghs 2004; Schwope *et al.* 2004). Other images include X-ray binaries (e.g., Vrtilek *et al.* (2004) for SMC X-1 and Her X-1); the nova-like binary V3885 Sgr (Pinja *et al.* 2012); and the gamma-ray binary LS I+61 303 (McSwain &

Richards 2011). The first 3D image of a black-hole candidate Cyg X-1 was made by
Sharova *et al.* (2012).

The first tomograms of systems with non-compact stars (e.g., the Algols) appeared in
1995. Richards (2004) provided a summary of the 2D images of 20 Algols and V711, a
detached RS CVn system used as a template to interpret the tomogram of the cool donor
star. In 2006, Agafonov, Richards & Sharova (2006) created the first 3D tomograms for
the entire class of interacting binaries using the U CrB alternating Algol system, which
displays disk-like emission in the tomogram at some epochs and stream-like emission
at other epochs. The 3D tomograms of the disk and stream states of this binary were
described in Agafonov, Sharova & Richards (2009), and compared to RS Vul by Richards,
Sharova & Agafonov (2010).

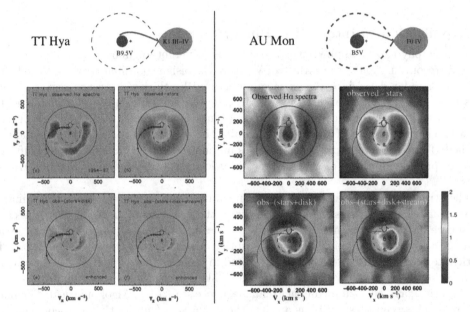

Figure 1. 2D Tomograms of the accretion disk systems TT Hya and AU Mon. The images are
based on observed spectra (top left), and when various accretion structures are subtracted from
the observed spectra: (1) stellar spectra removed (top right), (2) stars and accretion disk removed
(bottom left), and (3) stars, disk and gas stream removed (bottom right). The SHELLSPEC code
is effective in isolating the separate parts of the accretion structure, namely accretion disk and
gas stream (left frame: Miller *et al.* 2007; right frame: Atwood-Stone *et al.* 2011).

5. Using Synthetic Spectra to Isolate the Sources of Emission

An innovative technique was developed to isolate the various accretion sources with
the help of a spectrum synthesis code called SHELLSPEC (Budaj & Richards 2004; Budaj,
Richards & Miller 2005). This code was used to model the spectra of the accretion disk
and gas stream in the case of TTHya, and 2D tomography was applied to demonstrate
how the separate contributors to the tomogram can be isolated (Miller *et al.* 2007). The
results are illustrated in Figure 1. The left-over emission in the TT Hya image (bottom
right) represents the asymmetric portion of the disk that was not included in the synthetic
spectra; hence the disk in TT Hya is asymmetric. The left-over emission in the AU Mon
image represents gas moving at velocities below the Keplerian value.

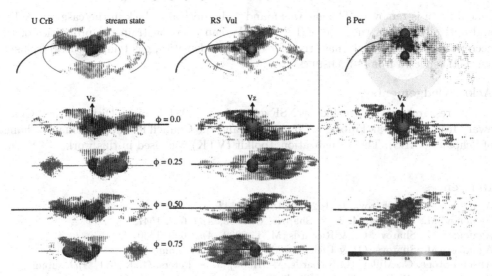

Figure 2. 3D tomograms of U CrB, RS Vul, and β Per. The top frames show a tilted view of the velocity distribution of the gas beyond the central plane of each binary, relative to the V_z axis. The other views are shown in order of increasing orbital phase, ϕ. The brightest emission sources shown are the large red points and the fainter sources are small green points (left frame: Richards, Sharova, & Agafonov 2010; right frame: Agafonov, Sharova, & Richards 2011).

6. Summary of Results from 2D and 3D Tomography

Tomography has been used to make indirect images of interacting binaries at a time when it is not possible to resolve the accretion structures in these systems by any other means. These images provide us with a glimpse of the fascinating structures that form as a result of Roche lobe overflow from a magnetically active star onto a variety of objects from normal stars to compact stellar remnants. We hope to see the direct images of these systems firsthand, perhaps within the next ten years.

The technique has provided new information about interacting binary star systems. These images illustrate the distribution of gas flows in the orbital plane (2D images) and beyond that plane (3D images), and demonstrate the range and complexity of the emission sources in these binaries: from accretion disks and gas streams, to shock regions and magnetic structures. Specific results from the 2D images include evidence of (1) accretion disks around the mass gaining star in the CVs, nova-likes, X-ray binaries, gamma-ray binaries, and long-period Algols; (2) gas streams and gas flowing along magnetic field lines in the magnetic CVs; (3) a combination of Keplerian disks, gas streams flowing along the predicted gravitational path, shock regions where the stream and disk interact, chromospheres, and other magnetic structures in Algol binaries; (4) regions where the gas slows down after circling the mass gainer in direct-impact systems (e.g., Algol, β Per); and (5) asymmetric accretion disks that maintain their asymmetry on long timescales (e.g., TT Hya). Additional results from 3D tomography include evidence that (6) prominent and extensive gas flows exist beyond the central plane of the binary in the z direction (e.g., U CrB, RS Vul, β Per); (7) the accretion disk is tilted or precesses (e.g., U CrB), (7) loop prominences and coronal mass ejections associated with the magnetic field of the donor star also contribute to the gas flows (e.g., RS Vul).

These images have also been used to estimate the gas opacity by comparing the images made from the observed spectra with those made from difference spectra when the stellar contribution is removed. The gas was found to be optically thick in the case of CX Dra and optically thin in the case of TT Hya. We can also constrain the properties of the accretion structures (e.g., mass transfer rates, gas densities, gas temperatures) by using synthetic spectra from SHELLSPEC.

Acknowledgements

This research was supported by NSF grant AST-0908440 and a Distinguished Chairs award from the Slovak Fulbright Commission and the Council for International Exchange of Scholars (CIES). The Visualization Toolkit (VTK) was used in this work.

References

Agafonov, M. I. & Sharova, O. I. 2005, *AN*, 326, 143
Agafonov, M., Richards, M. T., & Sharova, O. 2006, *ApJ*, 652, 1547
Agafonov, M., Sharova, O., & Richards, M. T. 2009, *ApJ*, 690, 1730
Agafonov, M., Sharova, O., & Richards, M. T. 2011, *ApJ*, submitted
Atwood-Stone, C., Miller, B., Richards, M., Budaj, J., & Peters, G. J. 2011, *ApJ*, submitted
Blondin, J. M., Richards, M. T., & Malinkowski, M. 1995, *ApJ*, 445, 939
Bracewell, R. N. & Riddle, A. C. 1967, *ApJ*, 150, 427
Budaj, J. & Richards, M. T., 2004, *Contrib. Astron. Observatory Skalnaté Pleso*, 34, 167
Budaj, J., Richards, M. T., & Miller, B. 2005, *ApJ*, 623, 411
Csizmadia, S., Borkovits, T., Paragi, Z., *et al.*, 2009, *ApJ*, 705, 436
Haubois, X, Perrin, G. Lacour, S., *et al.*, 2009, *A&A*, 508, 923
Karovska, M., Schlegel, E., Hack, W., Raymond, J., & Wood B. 2005, *ApJ* (Letters), 623, 137
Marsh, T. R. 2001, *Lecture Notes in Physics*, 573, 1
Marsh, T. R. & Horne, K. 1988, *MNRAS*, 235, 269
McSwain, M. V. & Richards, M. T. 2011, in preparation
Miller, B., Budaj, J., Richards, M. T., Koubský, P., & Peters, G. J. 2007, *ApJ*, 656, 1075
Morales-Rueda, L. 2004, *AN*, 325, 193
Peterson, W. M., Mutel, R. L., Gudel, M., & Goss, W. M. 2010, *Nature*, 463, 207
Radon, J. 1917, *Berichte Schsische Akademie der Wissenschaften Leipzig Math. Phys. Kl.*, 69, 262 (reprinted in 1983: *Proceedings of Symposia in Applied Math*, 27, 71)
Prinja, R. K., Long, K. S. Long, Richards, M. T., Witherick, D. K., & Peck, L. W. 2012, *MNRAS*, 419, 3537
Richards, M. T. 1993, *ApJ. Suppl.*, 86, 255
Richards, M. T. 2004, *AN*, 325, 229
Richards, M. T., Sharova, O., & Agafonov, M. 2010, *ApJ*, 720, 996
Richards, M. T. & Ratliff, M. A. 1998, *ApJ*, 493, 326
Schwope, A. D., Staude, A., Vogel, J., & Schwarz, R. 2004, *AN*, 325, 197
Sharova, O. I., Agafonov, M. I., Karitskaya, E. A., Bochkarev, N. G. Zharikov, S. V., Butenko, G. Z., & Bondar, A. V. 2012, these proceedings
Steeghs, D. 2004, *AN*, 325, 185
Vrtilek, S. D., Quaintrell, H., Boroson, B., & Shields, M. 2004, *AN*, 325, 209
Zavala, R. T., Hummel, C. A., Boboltz, D. A., Ojha, R. Shaffer, D. B., Tycner, C., Richards, M. T., & Hutter, D. J. 2010, *ApJ* (Letters), 715, L49

Discussion

V. NEUSTROEV: Please explain how the third dimension can be recovered from 2D observations. What kind of info, observations, in addition to velocities and orbital phases, we need to have in order to get the 3rd dimension in Doppler tomography.

M. RICHARDS: We assume a grid of V_z velocities beyond the central plane since we have a fixed view of the binary as set by the orbital inclination.

From Interacting Binaries to Exoplanets: Essential Modeling Tools
Proceedings IAU Symposium No. 282, 2011   © International Astronomical Union 2012
Mercedes T. Richards & Ivan Hubeny, eds.   doi:10.1017/S1743921311027281

Polarimetry of Binary Stars and Exoplanets

Karen S. Bjorkman

Ritter Observatory, Department of Physics & Astronomy, University of Toledo, USA
email: karen.bjorkman@utoledo.edu

Abstract. Polarimetry is a useful diagnostic of asymmetries in both circumstellar environments and binary star systems. Its sensitivity to asymmetries in systems means that it can help to uncover details about system orbital parameters, including providing information about the orbital inclination. Polarimetry can probe the circumstellar and/or circumbinary material as well. A number of significant results on binary systems have been produced by polarimetric studies. One might therefore expect that polarimetry could similarly play a useful role in studies of exoplanets, and a number of possible diagnostics for exoplanets have been proposed. However, the application of polarimetry to exoplanet research is only in preliminary stages, and the difficulties with applying the technique to exoplanets are non-trivial. This review will discuss the successes of polarimetry in analyzing binary systems, and consider the possibilities and challenges for extending similar analysis to exoplanet systems.

Keywords. instrumentation: adaptive optics, instrumentation: interferometers, instrumentation: polarimeters, techniques: image processing, techniques: polarimetric, (stars:) binaries (including multiple): close, (stars:) planetary systems

---

1. Introduction

The most useful purpose for polarimetry is in concert with all of the other kinds of observational image processing techniques. Polarimetry often is a non-unique solution in many cases, but when combined with spectroscopy, photometry, and interferometry it adds an extra dimension to what we can learn about all kinds of astronomical objects. I will review some of the basics about polarimetry, including the polarizing mechanisms in astronomy and different types of polarimetry measurements as applied to close binary stars and exoplanets. To begin, light is an electromagnetic wave and it oscillates. If we can measure the direction of that oscillation then we can measure its polarization. Every individual photon has a particular polarization. First, you need a source to polarize the light in a particular direction. Most of the light we see is not polarized, or at least it's just randomly polarized, so we measure no net polarization. If we have a system where the polarization has been preferentially selected for in one direction, then when we measure even an unresolved system we see a net polarization in the light from that system. An analyzer is needed to collect the information. This is often a half-wave plate that we put into a spectrograph, or an imager, to measure the degree of polarization, i.e., the amount of polarization and the position angle of the polarization. Both of these are critical bits of information.

An extremely useful tool for studying binary stars is called a Q–U diagram, which is simply a plot of the Stokes U vs. Stokes Q parameters in the data that we've measured. The total intensity, I, is the sum of all the light that we see, including all light that is polarized in one direction plus all the light that is polarized in a perpendicular direction to that. If we take the difference between the polarization in one direction and the orthogonal polarization, we get the Stokes Q parameter, $Q = P \cos(2\theta)$, where P is the

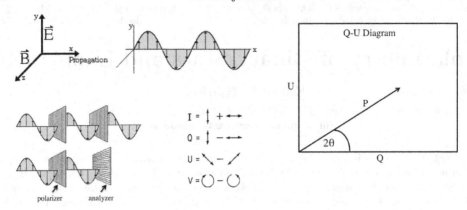

Figure 1. Basics of Polarization (left) and the Q-U Diagram (right)

polarization and θ is the position angle. The Stokes U is essentially 45 degrees away, and again these two are orthogonal. This gives rise to the Stokes U parameter, $U = P \sin(2\theta)$. So the polarization, $P = (Q^2 + U^2)^{1/2}$, and the position angle is $\theta = 0.5 \arctan(U/Q)$. The polarization is really the length of this vector; if we have zero polarization at the origin $(0,0)$, we have no net polarization, and so we have $P = 0$ and no vector. But when we have some net polarization in a preferred direction, then we will measure some level of polarization P at some position angle which is 2θ on the sky. Typically in stellar systems the value of P can be as low as several tenths of a percent. That's because the star which is providing the light which is being scattered or polarized is also emitting light which does not get scattered, which dilutes the polarized light. So, you get very low levels of polarization in stellar polarimetry typically. In obscured sources, for example in brown stars or in AGNs, you can get polarization levels that are quite high (e.g., 20% to 50%) because the central source is obscured and you don't get any direct light that hasn't been scattered. You may wonder about the 2θ factor; that's simply because polarization in one direction is indistinguishable from polarization in the opposite direction; essentially because $0°$ and $180°$ produce the same results.

2. Polarizing Mechanisms

The list of polarizing mechanisms consists of (1) electron scattering, (2) dust scattering, (3) interstellar dichroic extinction, (4) thermal emission by aligned elongated dust grains (important in IR), (5) magnetic fields (circular polarization), and (6) line scattering.

Certainly electron scattering will polarize light, but if the scatterers are all symmetrically distributed, spherically distributed, as we'll see, the net polarization in an unresolved system will be zero because it will all average out.

We can have dust scattering; that is scattering off of dust grains. If those dust grains are spherically symmetric, they will not produce polarization. If they are elongated, as we think many dust grains are typically aligned because of the galactic magnetic field or stellar magnetic fields, they can produce a preferred orientation of the polarization. So, polarization is really a probe of asymmetries in many systems.

Interstellar dichroic extinction is produced when interstellar dust grains polarize light. It is unfortunate that many people think interstellar dust just gets in the way of our intended astronomical sources. In fact, one of the most difficult challenges in using polarimetry is being able to disentangle the contribution to the net polarization that you measure from the interstellar medium compared to the source that you're trying to study.

So you have to be able to do a careful job of removing the contributions by the interstellar medium if you want to know what's actually happening in the your intended source. That can be a little tricky sometimes. Thermal emission by aligned elongated dust grains is also important. This is particularly important in young stars where there are large dust envelopes that may be affected by magnetic fields, and there is extra emission from the dust. Magnetic fields are a major component in polarization. In fact, circular polarization, the Stokes V parameter, is typically used to measure magnetic fields or to attempt to measure magnetic fields in stars where the Zeeman splitting is not strong enough to be measured. Usually this is the case in hot stars where the lines are broadened because of rotation. So, circular polarization is a very nice diagnostic of magnetic fields in stars, particularly in hot stars. The final polarizing mechanism is line scattering. Although many of us like to assume that spectral lines are unpolarized, in fact they may not be. We should be aware that there are processes in the line production that could polarize light. The models so far are relatively simplified and don't include line scattering polarization for the most part; so this mechanism is being actively worked on because we will need it in order to interpret the data accurately.

3. The Effect of Asymmetries

Here are some examples of how the asymmetries of the scatterers (e.g., electrons, dust grains, etc.) may affect what you see. Where polarimetry can really play a role is in systems they are far away, or too faint, so we can't resolve them directly yet. We have heard that interferometry and tomography are ways to begin to resolve these systems. Polarimetry is a way to analyze the asymmetries and the directionality of a system without actually being able to resolve it. I always find it amazing that I can tell you the position angle of a disk around a star when I can't even resolve the disk, even with interferometry or tomography. So, it is a powerful technique that we can exploit when we're trying to look for systems that are too far away to resolve.

Let us assume the very simple case of an unresolved system with a large number of electrons surrounding it in a circumstellar envelope; with the electrons uniformly distributed in density (no clumps) and uniformly distributed about the star (Fig. 2, left image). If light from the star scatters an electron, it becomes polarized and the direction of the polarization is perpendicular to the central plane (i.e., to the line between the source and the scatterer, for electron scattering). We can map put the polarization if the source is resolved. However, if the source is not resolved, we essentially sum over the entire envelope; hence the net polarization that we measure is zero, since there are as many vectors in one direction as in the perpendicular direction, so they cancel each other out. So the net polarization in a perfectly spherical system would be zero.

In reality, the systems are not spherical. If we have blobs, we may have scattering into different blobs, we may have different densities in these blobs, and we may have different sizes and orientations. In the case of an unresolved system, the net polarization will depend on the relative contribution from each of the blobs doing the scattering (Fig. 2, middle image). This gets complicated because we have to include more parameters, like the blob distribution or the distribution of the scatterers. If you have a disk or a jet (Fig. 2, right image), then all of the scatterers are confined to a relatively thin disk (e.g., accretion disks or Be star disks). In this case, the net polarization is perpendicular to the disk or to the jet. So, if we measure the polarization of an unresolved star, we can still measure the position angle of the disk from the polarization.

This theoretical prediction was confirmed in a simple test case in which polarmetric observations were combined with interferometric observations of a resolved disk. We

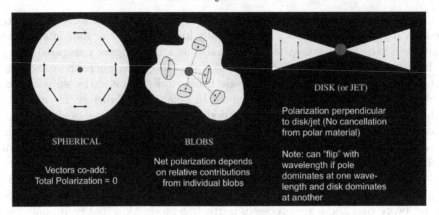

Figure 2. Effect of asymmetries in unresolved systems. The polarization at each point is perpendicular to the radius vector.

found that the position angle derived from the polarization was exactly in agreement with the interferometry image. Hence, the prediction holds true. This is very useful because it means that we can determine position angles of disks that we cannot resolve. We have discovered that the position angles can be wavelength dependent if the scatterers near the pole dominate at one particular wavelength while the disk dominates at another wavelength. We see this effect in stars with thick stellar envelopes, like Herbig Ae/Be stars.

4. Common Polarization Measurements

There are several polarimetric measurements in common use, including (1) broadband photo-polarimetry (UBVRI etc), (2) imaging polarimetry, (3) spectropolarimetry, and (4) circular polarimetry of spectral lines. The simplest measurement is broadband photo-polarimetry, which is essentially photometry using a polarimetric analyzer in conjunction with a photometric filter. For example, the polarization of a star can be measured at the different wavelengths in the UVBRI bands to provide the wavelength dependence for the polarization.

Imaging polarimetry is possible if the analyzer (e.g., a half wave plate) is placed in front of an imaging device. For example, if we are studying many stars in a cluster, we can take a direct image of that cluster or we could place a polarizing analyzer in the light path. By doing so we measure both the ordinary and extraordinary polarization for each star in the field; hence with imaging polarimetry we can measure many stars at once. Another application is in resolved sources, e.g., young stars. If the sources are nearby, we can map out the polarization distribution within the young star envelope. This can be very interesting in terms of understanding the structure within that envelope. Spectropolarimetry is much more complicated because it involves spectroscopy with a polarizing analyzer. This procedure creates a spectrum of the ordinary polarized light and a second spectrum from the extraordinary polarized light. By comparing the two spectra we can measure the polarization as well as the polarization as a function of wavelength, just as we do in spectroscopy. Instead of just measuring the flux, we simultaneously measure both the flux and the polarization at each wavelength. At this conference, we have been talking about the challenges of observing bright stars and faint stars and the need for bigger telescopes; spectropolarimetry is even more delicate because it is clearly a photon-starved, photon-limited technique, and it will be quite challenging to get measurements at the 1% level.

Circular polarimetry of magnetically-sensitive spectral lines is used to get information about the magnetic field by measuring the Stokes V parameter in cases where the Zeeman splitting is weak.

5. Applications of Polarimetry

Imaging polarimetry was used to observe many stars at once in clusters in the Large Magellanic Cloud (e.g., NGC 1948). This was a program to identify unresolved disk systems in the LMC (Wisniewski *et al.* 2005). The procedure creates a double image for every star because the polarizing analyzer is placed in front of the image. One is the extraordinary and the other is the ordinary image. Instead of just doing photometry on one image of a star, we do photometry on both of images and then compare them. Then, we rotate the polarizing analyzer and take another image, and repeat the process for additional rotations of the polarizing analyzer, until we get a measurement of the polarization for each star in the field. The analysis is more complicated when the field is crowded because of confusion with overlapping images; instead higher spatial resolution techniques are needed to separate the cores of crowded clusters. However, the procedure works well in the general field far from the cluster core.

An amazing example of a resolved source is FS Tau, a 0."2-separated classical T Tauri binary system, taken with the Coronagraphic Imager with Adaptive Optics (CIAO) on the Subaru Telescope and superimposed on the HST image (Hioki *et al.* 2011). This is a young stellar object in a binary system. The polarization is represented by little vectors across the image, and the orientation of the vector gives the direction of the polarization at that point in the field. The length of the vector indicates the strength of the polarization; the longer the vector the larger the polarization. In FS Tau, there is a real centro-symmetric pattern, similar to what is shown in Fig. 2 (left frame). Hence, the scatterers are distributed about the star in a very particular way. FS Tau is actually a highly-polarized object because the central source (the star) is obscured in the image. It was possible to use the HST to measure the polarization in this case because of its strength. It is not commonly known that the Hubble Space Telescope originally had polarizing optics in it; but the COSTAR upgrade affected the instrumental polarization and it became very difficult to make measurements at low polarization levels. However, it is still quite useful for high polarization levels.

An nice example of spectropolarimetry is shown in Wood, Bjorkman, & Bjorkman (1997) for the classical Be star ζ Tau, which is also a binary (see Fig. 3). This figure looks like a spectrum, but it is a measurement of polarization as a function of wavelength from the ultraviolet to the infrared (10,000 Å). It shows the polarization Balmer jump, the Paschen Slope, the Paschen Jump, and so on. There are relatively few polarization observations of objects in the ultraviolet. HST could make these observations for a while, and it can still do so. However, the Wisconsin Ultraviolet Photo-Polarimeter Experiment (WUPEE) was the only dedicated instrument even flown, other than balloon flights, to do broadband spectropolarimetry in the ultraviolet. WUPEE flew twice on the Space Shuttle during the ASTRO-1 (1990) and ASTRO-2 (1995) missions, and it is the only current source of spectropolarimetry in the ultraviolet. WUPPE collected UV spectropolarimetry and spectra for 121 objects.

The appearance of the WUPEE spectra were surprising because the models predicted that the polarization would rise towards the UV, but this was not the case. The explanation was that the polarized light, which has been scattered in the disk, still has

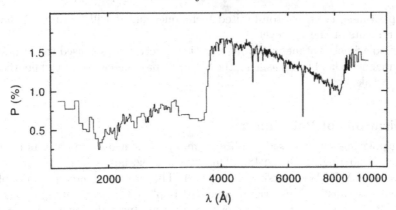

Figure 3. Spectropolarimetry of ζ Tau before correction for interstellar reddening and interstellar polarization taken from Wood, Bjorkman, & Bjorkman (1997) (their Figure 3).

to get out of that disk. So, as it travels through the disk, either before or after it gets scattered, some of that light gets absorbed by the opacity in the disk. So the spectrum of the disk can be obtained directly if you look at the polarized light, even though the disk is unresolved in most cases with interferometry. We also get some physics from the polarimetry because we can measure opacities and temperatures and structures of the disk from the polarized spectrum.

The photon counting error bars and systematic errors associated with the data are very small. Zeta Tau has very little interstellar polarization in its direction, so no correction was needed for those effects. The effect of the lines can clearly be seen in Figure 3 (e.g., Balmer lines, Paschen jump), so the polarization mimics the hydrogen opacity, and line blanketing from the disk itself is revealed in the UV part of the spectrum (e.g., in the Fe III and Fe II lines). This is just an example of the applications if you have bright enough stars and high enough signal to noise.

6. Polarimetry of Binaries

A lot of the early work on the polarimetry of binaries was done in the 1960s and 1970s. The most commonly used models are by Brown, McLean, & Emslie (1978) and Rudy & Kemp (1978). In these models, they assume two stars in a circular orbit surrounded by a co-rotating envelope that was optically thin and undergoing single scattering. This is a simple analytical approach, but it predicted that Q-U diagrams could diagnose both inclination angles and eccentricities for binary orbits. Brown, McLean, & Emslie (1978) showed that the binary traces out a loop or a double loop in the Q-U diagram as the phase advances; and this behavior depends on the inclination and eccentricity of the orbit.

We know that binary stars, especially close binaries, are more complex than described in these models. Interacting binaries contain mass transfer streams, hot spots where the stream falls into the accretion disk, the disk around the gainer star, and an asymmetric donor star. Each of these components represents an asymmetry that contributes to the net polarization. There have been several attempts to disentangle all of these parts using spectropolarimetry (e.g., Hoffman *et al.* 1998, Hoffman *et al.* 2003; for β Lyr). There are changes in the polarized flux as a function of wavelength (the spectropolarimetric flux) at phases before, during, and after primary and secondary eclipse. These changes can be used to diagnose some of the physics. Several models use the UV to look at the jet, and use the optical to look at the disk, and then combine the separate results; so it is quite challenging. Barbara Whitney and collaborators have also performed Monte Carlo modeling of optically thick disks. The latest models now include the hot spot (e.g., Lomax & Hoffman 2011; β Lyr).

Polarimetry of other binary systems include the pre-main-sequence binaries (e.g., Manset & Bastien 2000, Manset & Bastien 2001a, Manset & Bastien 2001b); Wolf-Rayet binaries (e.g., St.-Louis *et al.* 1987, St.-Louis *et al.* 1988); and Algols (e.g., Pfeiffer & Koch 1977).

7. Polarimetry of Exoplanets

Exoplanets are like little binaries. Sara Seager and her collaborators (Seager, Whitney & Sasselov 2000) predicted the level of polarization you might expect from scattering off the atmosphere of an exoplanet; and their levels were on the order of 10^{-6}. This was not at a level that we could measure with current polarimeters, so this really discouraged many people from trying to apply polarimetry to exoplanets. However, Berdyugina *et al.* (2011) have claimed a detection at about a level of 10^{-4}. There is still some discussion about this result, so they collected multiwavelength UBV observations and it seems that the detection is real.

Another method is to search for limb polarization during a transit (e.g., Davidson *et al.* 2010). Carciofi & Magalhaes (2005) predicted various polarization levels that could be observed for these limb transits of exoplanets, and they are at the level of 10^{-6}. However, this limit can be increased to around 10^{-4}, which is in fact observable.

To detect stellar polarization at levels of at least 10^{-4} requires extremely high S/N. The number is very very big. It requires that we use large telescopes or observe bright stars, or preferable both, especially if you want to do spectropolarimetry. Differential polarization, like differential photometry, can help because absolute calibration is not needed but the instrumental polarization has to be very carefully calibrated and removed, and it has to be stable. It is important to remove the interstellar polarization, especially for time variable phenomena, and it is not an easy thing to do.

8. Future Directions

Polarimetry can now be used to detect non-transiting exoplanets at all orbital inclinations. There is hope that we may even be able to use spectropolarimetry to measure the composition of the atmospheres of exoplanets, and there are polarimeters being designed for large telescopes that may achieve this goal. Work is now being done to predict the spectropolarimetry of an exoplanet with light scattering from a star off the atmosphere for a hot Jupiter model. In this case, the polarized spectrum can essentially serve as a surrogate for the spectrum of the atmosphere of the planet. It will take a lot of effort to achieve this kind of detection. Finally, Q-U plots are useful tools because they provide the same kinds of diagnostics for exoplanet systems as they do for binary systems; so we can derive parameters like orbital inclinations and eccentricities for them.

Discussion

R. WILSON: About 15 to 20 years ago I made a computational model for time-wise variability of polarization in close binaries, including limb polarization and disk polarization. That model remains the only one that does the problem quantitatively (both direct and inverse. It solves for the parameters by least squares, and does it without unnecessary intermediaries such as Fourier series fitting). The model would have been developed further, but there were no more observations to process after the J. Kemp data on Algol. Why has there been an emphasis on surveys to the almost complete exclusion of polarization curves for individual binaries? Any limitation due to photon noise should apply equally to surveys and individual variation.

K. BJORKMAN: Thank you for reminding us of these existing models. You are correct that what is needed is more time-monitoring observations of polarimetry for these systems. Unfortunately, as we know, it can be difficult to obtain sufficient observing time for monitoring observations. There are some databases of existing polarimetric data (e.g., the HPOL data available through the NASA MAST web site); however, additional data and polarimetric instruments are still needed.

R. WILSON: Why has there been an emphasis on surveys while time-wise variations of basic simple systems are ignored?

K. BJORKMAN: Long-term monitoring is being done on small telescopes. For example, most of the HPOL data were obtained on a 1m telescope at Wisconsin, where we monitored for 15 years and we had students observing in a Q mode. We had a lot of data for particular sets of stars, but you are right that we need to do a lot more monitoring. The time variability is much more crucial than the surveys, in a sense.

References

Berdyugina, S. V., Berdyugin, A. V., & Piirola, V. 2011, A&A, submitted
Brown, J. C., McLena, I. S., & Emslie, A. G. 1978, A&A, 68, 415
Carciofi, A. C. & Magalhães, A. M. 2005, ApJ, 635, 570
Davidson, J. W., Bjorkman, K. S., Magalhaes, A. M., Carciofi, A. C., Bjorkman, J. E., Seriacopi, D. B., & Wisniewski, J. P. 2010, BAAS
Rudy, R. J. & Kemp, J. C. 1978, ApJ, 221, 200
Hioki, T., Itoh, Y., Oasa, Y., Fukagawa, M., & Hayashi, M. 2011, PASJ, 63, 543
Hoffman, J. L., Nordsieck, K. H., & Fox, G. K. 1998, AJ, 115, 1576
Hoffman, J. L., Whitney, B. A., & Nordsieck, K. H. 2003, ApJ, 598, 572
Lomax, J. R. & Hoffman, J. L. 2011, Bull. Soc. Roy. Sci. de Liege, 80, 689
Manset, N. & Bastien, P. 2000, AJ, 120, 413
Manset, N. & Bastien, P. 2001a, AJ, 122, 2692
Manset, N. & Bastien, P. 2001b, AJ, 122, 3453
Pfeiffer, R. J. & Koch, R. H. 1977, PASP, 89, 147
Seager, S., Whitney, B. A., & Sasselov, D. D. 2000, ApJ, 540, 504
St.-Louis, N., Drissen, L., Moffat, A. F. J., Bastien, P., & Tapia, S. 1987, ApJ, 322, 870
St.-Louis, N., Moffat, A. F. J., Drissen, L., Bastien, P., & Robert, C. 1988, ApJ, 330, 286
Wisniewski, J. P., Bjorkman, K. S., Magalhes, A. M., Bjorkman, J. E., & Carciofi, A. C. 2005, ASP-CS, 337, 333
Wood, K., Bjorkman, K. S., & Bjorkman, J. E. 1997, ApJ, 477, 926

From Interacting Binaries to Exoplanets: Essential Modeling Tools
Proceedings IAU Symposium No. 282, 2011
Mercedes T. Richards & Ivan Hubeny, eds.
© International Astronomical Union 2012
doi:10.1017/S1743921311027293

Adaptive Optics Observations of Exoplanets, Brown Dwarfs, and Binary Stars

Sasha Hinkley

Sagan Fellow, California Institute of Technology, Mail Code 249-17, 1200 E. California blvd.,
Pasadena, CA 91125
email: shinkley@astro.caltech.edu

Abstract. The current direct observations of brown dwarfs and exoplanets have been obtained using instruments not specifically designed for overcoming the large contrast ratio between the host star and any wide-separation faint companions. However, we are about to witness the birth of several new dedicated observing platforms specifically geared towards high contrast imaging of these objects. The Gemini Planet Imager, VLT-SPHERE, Subaru HiCIAO, and Project 1640 at the Palomar 5m telescope will return images of numerous exoplanets and brown dwarfs over hundreds of observing nights in the next five years. Along with diffraction-limited coronagraphs and high-order adaptive optics, these instruments also will return spectral and polarimetric information on any discovered targets, giving clues to their atmospheric compositions and characteristics. Such spectral characterization will be key to forming a detailed theory of comparative exoplanetary science which will be widely applicable to both exoplanets and brown dwarfs. Further, the prevalence of aperture masking interferometry in the field of high contrast imaging is also allowing observers to sense massive, young planets at solar system scales ($\sim$3-30 AU)— separations out of reach to conventional direct imaging techniques. Such observations can provide snapshots at the earliest phases of planet formation—information essential for constraining formation mechanisms as well as evolutionary models of planetary mass companions. As a demonstration of the power of this technique, I briefly review recent aperture masking observations of the HR 8799 system. Moreover, all of the aforementioned techniques are already extremely adept at detecting low-mass stellar companions to their target stars, and I present some recent highlights.

Keywords. instrumentation: adaptive optics, instrumentation: high angular resolution, instrumentation: spectrographs, techniques: high angular resolution, techniques: interferometric, (stars:) binaries: general, (stars:) binaries (including multiple): close

---

1. Introduction

In recent years, astronomers have identified more than 400 planets outside our solar system, launching the new and thriving field of exoplanetary science (Marcy *et al.* 2005). The vast majority of these objects have been discovered indirectly by observing the variations induced in their host star's light. The radial velocity surveys can provide orbital eccentricity, semi-major axes, and lower limits on the masses of companion planets while observations of transiting planets can provide fundamental data on planet radii and limited spectroscopy of the planets themselves. However, studying those objects out of reach to the radial velocity and doppler methods will reveal completely new aspects of exoplanetary science in great detail. *Direct imaging* of exoplanets provides a complementary set of parameters such as photometry (and hence luminosity), as well as detailed spectroscopic information.

Moreover, the direct imaging of planetary mass companions will allow researchers to more fully probe the mass-separation parameter space (Figure 1, right panel) occupied by

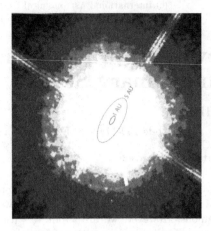 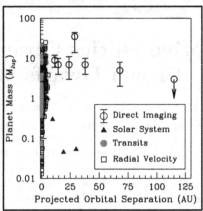

Figure 1. *Left:* A schematic demonstration taken from Oppenheimer & Hinkley (2009) demonstrating the challenges associated with high contrast imaging. Hypothetical orbits of a Jupiter and an Earth have been overlaid on the Point Spread Function of a nearby star. Planetary mass companions typically have angular separations of a fraction of an arcsecond, and contrast ratios spanning several orders of magnitude ($\sim 10^4$-10^7). *Right:* The Mass vs. Orbital separation for planetary mass companions detected through transits (filled circles), radial velocity measurements (squares), solar system planets (triangles), as well as those objects detected through direct imaging (open circles).

these objects (Oppenheimer & Hinkley 2009). At young ages (§3), the orbital placement of planetary mass companions serves as a birth snapshot, lending support to models (Pollack *et al.* 1996) that may be more efficient at building a massive core at only a few AU, or models that allow for the formation of massive objects at tens of AU though fragmentation of the gaseous disk (Boss 1997). Finally, imaging of multi-planet systems at any age will serve as dynamical laboratories for studying the planetary architectures.

Perhaps just as essential, spectroscopy provides clues to the atmospheric chemistry, internal physics, and perhaps may even shed light on non-equilibrium chemistry associated with these objects. More robust classification schemes for planets in general will arise from observing as many planets as possible at different ages, in different environments, and with a broad range of parent stars. Figure 2 provides an illustration of the diversity of the spectra of late-M, L and T dwarfs as well as a spectrum of Jupiter showing broad absorption bands due to e.g. H_2O, CH_4, and NH_3. Spectral characterization of such objects can be accomplished even with the somewhat low spectral resolution ($\lambda/\Delta\lambda \sim 30$-50) of the instruments described in this review. These pieces of information not only reveal the detailed properties of the objects themselves, they serve as key benchmarks for competing evolutionary models describing the thermal and atmospheric properties of these objects.

It is often overlooked, but should be emphasized, that the recent spectacular images (Marois *et al.* 2008, Kalas *et al.* 2008, and Lagrange *et al.* 2010) of Jovian planets were obtained with Adaptive Optics, hereafter "AO", systems and infrared cameras not specifically designed for the task of overcoming the high contrast ratio between the planets and their host stars. These were obtained using existing instrumentation, but with observing and data reduction strategies customized for high contrast imaging. Further, the recent L-band images of HR 8799 and β Pic b (Marois *et al.* 2010, Lagrange *et al.* 2010) were obtained without the use of a coronagraph! These studies have demonstrated that direct imaging of planetary mass companions as well as disks (e.g. Oppenheimer *et al.* 2008, Hinkley *et al.* 2009) is now a mature technique and may become routine using ground-based observatories. More so, the handful of coming instruments dedicated to detailed

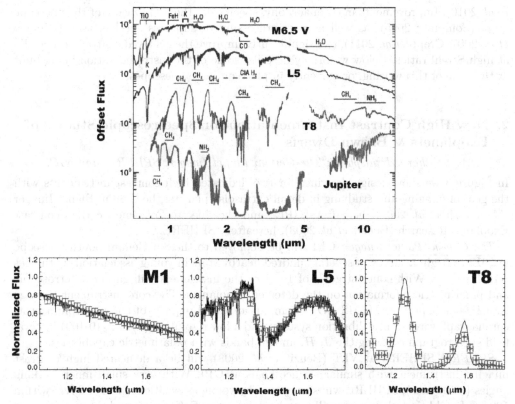

Figure 2. *Top Panel:* Example Spectra for M, L, and T-dwarfs, as well as Jupiter with prominent absorption bands marked, taken from Marley *et al.* (2009). *Lower Panels:* The points with error bars are simulated measurements gathered with a low resolution ($\lambda/\Delta\lambda\sim45$) spectrograph, shown along with higher resolution spectra from NASA IRTF (black curves) for the labelled spectral types. Such low spectral resolution measurements will be obtained by GPI, SPHERE and Project 1640, and can easily sample broad absorption features present in late type stars and exoplanets. Lower panels are taken from Rice *et al.* (2011) in prep. Images used courtesy of Michael Cushing, Mark Marley, and Emily Rice.

spectroscopic characterization of planetary mass companions may make these kinds of discoveries routine, initiating an era of comparative exoplanetary science.

1.1. *The Challenge of High Contrast Imaging*

The major obstacle to the direct detection of planetary companions to nearby stars is the overwhelming brightness of the host star. If our solar system were viewed from 20 pc, Jupiter would appear $10^8 - 10^{10}$ times fainter than our Sun in the near-IR (Barraffe *et al.* 2003) at a separation of 0.25″, completely lost in its glare (See Figure 1, left panel). The key requirement is the suppression of the star's overwhelming brightness through precise starlight control (Oppenheimer & Hinkley 2009).

A promising method for direct imaging of stellar companions involves two techniques working in conjunction. The first, high-order AO, provides control and manipulation of the image by correcting the aberrations in the incoming stellar wave front caused by the Earth's atmosphere. AO has the effect of creating a nearly diffraction-limited point spread function, with the majority of the stellar flux concentrated in this core. Second, a Lyot coronagraph (Sivaramakrishnan *et al.* 2001) suppresses this corrected light. Together, these two techniques can obtain contrast levels of 10^4-10^5 at 1″ (Leconte *et al.* 2010). Improvements in coronagraphy, specifically the apodization of the telescope pupil (Soummer 2005), as well as post-processing to suppress speckle noise (Hinkley *et al.* 2007, Crepp *et al.* 2011), can significantly improve the achieved contrast, especially at high Strehl ratios. Below we briefly describe some of the instrumentation being built at the time of this writing, or currently in place on large telescopes.

2. New High Contrast Instrumentation for Spectroscopic Studies of Exoplanets & Brown Dwarfs

2.1. *Southern Hemisphere: The Gemini Planet Imager, SPHERE, and NICI*

In Figure 3 we show design drawings for two dedicated high contrast instruments with the goal of imaging and studying in detail extrasolar planets: the Gemini Planet Imager (Macintosh *et al.* 2008), hereafter "GPI", and the Spectro-Polarimetric High-contrast Exoplanet Research (Beuzit *et al.* 2008), hereafter "SPHERE."

The Gemini Planet Imager: GPI will be deployed to the 8m Gemini South telescope in 2012 to begin a survey of several hundred nearby stars in young associations with ages ~10-100 Myr. With contrast goals of 10^7-10^8, the instrument will gather spectroscopic and polarimetric information on the detected exoplanets. The core instrument-package is: a 1500-subaperture MEMS AO system; an apodized-pupil Lyot coronagraph; a post-coronagraph wave front calibration system; and a low resolution ($\lambda/\Delta\lambda$~10-100) integral field spectrograph covering the J, H, and K-bands with polarimetric capabilities.

SPHERE: SPHERE at VLT (Beuzit *et al.* 2008) will be a dedicated high contrast imaging instrument with similar science goals as GPI. With very small inner working angles (~100 mas), SPHERE will survey nearby young associations (10-100 Myr) within 100 pc. In addition, the survey will target young active F/K dwarfs and the nearest stars within 20 pc to sense reflected light directly from the planets. The project also intends to monitor stars showing long-term radial velocity trends, astrometric planet candidates, as well as known multi-planet systems.

NICI: The Near Infrared Coronagraphic Imager, "NICI" (Liu *et al.* 2010), at the 8m Gemini-South telescope is using a variety of high contrast techniques to monitor ~300 young and nearby stars. NICI utilitzes an 85-element curvature wave front system, Lyot coronagraphy, Simultaneous Differential Imaging ("SDI") around the methane band, and Angular Differential Imaging ("ADI") for a variety of observing options. NICI will be an effective forerunner for the next generation of high contrast imaging platforms in the Southern Hemisphere (GPI, SPHERE) with spectroscopic capabiliites.

2.2. *Northern Hemisphere: Project 1640 at Palomar & Subaru HiCIAO*

Project 1640 at Palomar: We briefly describe a new instrument which forms the core of a long-term high contrast imaging program at the 200-in Hale Telescope at Palomar Observatory. The detailed descriptions of the instrumentation can be found in Hinkley *et al.* (2008), and Hinkley *et al.* (2011b), and we show an image of the instrument mounted on the 200-in telescope in Figure 3. The primary scientific thrust is to obtain images and low-resolution spectroscopy ($\lambda/\Delta\lambda \sim 45$) of brown dwarfs and young exoplanets of several Jupiter masses in the vicinity of stars within 50 pc of the Sun.

The instrument is comprised of a microlens-based integral field spectrograph (hereafter "IFS"), an apodized-pupil Lyot coronagraph, and a post-coronagraph internal wave front calibration interferometer mounted behind the Palomar adaptive optics system. The spectrograph obtains imaging in 23 channels across the J and H bands (1.06 - 1.78 μm). The Palomar AO system is undergoing an upgrade to a much higher-order AO system

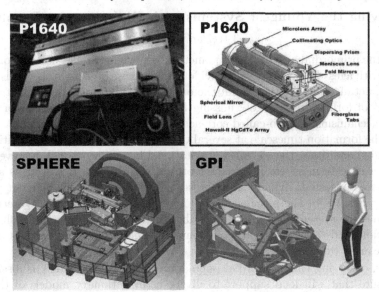

Figure 3. *Top left:* The Project 1640 high contrast imager as described in Hinkley *et al.* (2011c), and installed at the 200-in Hale Telescope. *Top right:* the internal optical layout of the Project 1640 Integral Field Spectrograph. The SPHERE and GPI designs are shown on the lower left and right, respectively. Each of these instruments are designed to image exoplanetary mass companions at contrasts of 10^7 or greater, as well as return spectroscopic information.

("PALM-3000"): a 3388-actuator tweeter deformable mirror working together with the pre-existing 241-actuator mirror. This system will allow correction with subapertures as small as 8.1cm at the telescope pupil using natural guide stars. As part of the project's first phase, the AO system, coronagraph, and IFS achieved contrast of 2×10^{-5} at $1''$ (Crepp *et al.* 2011, Hinkley *et al.* 2011b). We anticipate this instrument will make a lasting contribution to high contrast imaging in the Northern Hemisphere for years.

Among the early science results, we briefly highlight some intriguing findings regarding massive stellar companions to A-stars. The mid-A star ζ Virginis, has been discovered to host a mid-M dwarf companion. As we discuss in Hinkley *et al.* (2010), this newly discovered object may explain the anomalous X-ray emission from this A-star. A comparable scenario involves the star Alcor, a bright mid-A dwarf. A mid-M dwarf companion was found to orbit this star, and was identified through the novel technique of common parallax measurements (Zimmerman *et al.* 2010). Lastly, Hinkley *et al.* (2011a) analyze the α Oph system: a rapidly rotating A-star with a 0.77 $M_\odot$ companion. Monitoring this system over several years allowed us to fit an orbit to the companion astrometry, thereby placing an important dynamical constraint on the mass of the rapidly rotating primary.

The Subaru SEEDS Survey: In addition to Project 1640 at Palomar, the SEEDS survey at Subaru will utilize instrumentation dedicated to observations of planetary mass companions. Sitting at a Nasmyth focus, and integrated with the HiCIAO camera, the AO system consists of a 188-actuator curvature system, plus a MEMS-based wave front control system, with control and correction of the focal plane wave front in the near-infrared. The techniques of Phase-Induced Amplitude Apodization and aperture masking interferometry will allow access to inner working angles of 20-40 mas (Martinache & Guyon 2009). The platform is intended to be highly flexible, and will allow the instrumentation to adapt as techniques and understanding advance in the field.

3. Observing the Youngest Systems with High Angular Resolution

3.1. *Background*

Observing the primordial population of planets in very young stellar associations or moving groups with ages of a few Myr will provide data on these objects in their pristine state. Specifically, detailed information about the orbital distribution of these giant planets at the moment of formation is largely missing from many of the discussions of planetary formation mechanisms. The so-called core-accretion model (e.g. Pollack *et al.* 1996) dictates that the formation timescale at several tens of AUs may become prohibitively long. Under certain conditions, the so-called gravitational instability model, however, may be able to form objects more easily at larger orbital separations. However, as several researchers have suggested, dynamical processes as discussed e.g. by Veras *et al.* (2009), could be responsible for the placement of wide separation companions. Observing these nascent planetary systems as early as possible then will serve as a "snapshot" of the true formation landscape and greatly eliminate any confusion about the system's initial conditions caused by dynamical processes.

Moreover, observing this primordial population, just as the gas disk is dissipating, is a key measure that will lend support to divergent evolutionary models of the planets themselves. Two of the leading evolutionary models, the so-called "hot-start" models of Baraffe *et al.* (2003) and the more recent "cold-start" models as discussed by Fortney *et al.* (2008) are largely unconstrained at very young ages—the evolutionary window where they are most divergent. Observations of young planetary systems will serve as key benchmark measurements against which these models can be compared. However, the youngest stellar systems reside in star forming associations ($\sim$140pc), where a Jupiter-like orbital separation of $\sim$6 AU subtends an angle of 40-50 mas—about the size of the diffraction limit for a 10m telescope. *Hence, probing within the diffraction limit of the telescope is absolutely critical.*

3.2. *Aperture Masking Interferometry*

The technique of aperture masking interferometry (e.g. Tuthill *et al.* 2000, Ireland & Kraus 2008), achieves contrast ratios of $\sim$10^2 - 10^3 at very small inner working angles, usually within $\sim\frac{1}{3}\lambda/D - 4\lambda/D$ ($\sim$ 20-300 mas for Keck L'-band imaging). Other high contrast techniques, such as polarimetry (e.g. as discussed in Graham *et al.* 2007, Oppenheimer *et al.* 2008, and Hinkley *et al.* 2009), can achieve impressively small inner working angles in theory comparable to aperture masking. Applications of aperture masking, (e.g. Tuthill *et al.* 2000, Ireland & Kraus 2008, Hinkley *et al.* 2011c), employ AO with the use of an opaque mask containing several holes, constructed such that the baseline between any two holes samples a unique spatial frequency in the pupil plane. Moreover, a coronagraphic mask is not used in this observing mode, since it would obscure much of the accessible parameter space, as well as providing a great deal of uncertainty in the astrometry of the host star as discussed e.g. in Digby *et al.* (2006).

This technique is particularly well suited for studies of stellar multiplicity and the brown dwarf desert (e.g. Kraus *et al.* 2008, Ireland & Kraus 2008), or detecting giant planets orbiting young stars ($\lesssim$ 50-100 Myr). As an example of the power of this technique, we describe the recent results of Hinkley *et al.* (2011c), targeting the HR 8799 system. We are able to place upper limits on additional massive companions to the HR 8799 system that would otherwise have been veiled by the telescope diffraction or the associated uncorrected quasi-static speckle noise (e.g. Hinkley *et al.* 2007) using more conventional direct imaging techniques.

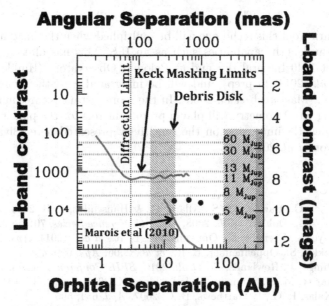

Figure 4. *The inner 10 AU of HR 8799:* The figure shows detection limits for the HR 8799 system taken from Hinkley *et al.* (2011c). The thick curves show the minimum brightness required for a 99% confidence detection in our HR 8799 aperture masking data, labelled "Keck Masking Limits". Also shown are the Marois *et al.* (2010) sensitivities (C. Marois and B. Macintosh, private communication 2011). We also indicate the theoretical diffraction limit at L'-band on a 10m telescope (black dotted line), as well as the L'-band brightnesses of the four known companions to HR 8799 (black points). The hatched regions represent the debris disk structures as defined by Su *et al.* (2009)

3.3. *The inner 10 AU of HR 8799*

HR 8799, a $\sim$30 Myr A5V star hosting several planets, presents a challenge for formation models of massive exoplanets. The recent identification of four $\sim$5-7 M_{Jup} exoplanets (Marois *et al.* 2008, 2010) reinforces that this system is characterized by a complicated and intriguing architecture. Gaining a more complete understanding of additional companions in the HR 8799 system will be essential for understanding the dynamical stability of the system, as well as performing an overall mass census.

Although we detect no other candidate companions interior to 14 AU, the location of HR 8799e, in Figure 4 we show our ability to place constraints on companions more massive than $\sim$11 M_{Jup} within this region. Details on the calculation of the detection limits can be found in Hinkley *et al.* (2011c). We achieve our best contrast of 8 mags at the L'-band diffraction limit of the Keck telescopes, corresponding to a 3 AU projected orbital separation for HR 8799. This orbital separation is comparable to the ice line boundary for a 30 Myr mid-A star, and these limits place a stringent constraint on any massive inner perturbers that might be responsible for the placement of the four planetary mass companions.

In addition, the study of the HR 8799 debris disk carried out by Su *et al.* (2009), indicates the presence of an inner warm ($\sim$150 K) debris belt (see Figure 4). The outer boundary of this disk is most likely sculpted by the "e" component at 14 AU. Su *et al.* (2009) state that few, if any, dust grains exist interior to 6 AU, and postulate that additional planets may be responsible for the clearing. Hence, any planetary mass companions responsible for sculpting this inner edge must be $\lesssim$11 M_{Jup}

3.4. *The future: multi-wavelength studies*

The science return from this technique will be multiplied when the imaging science camera has multi-wavelength capabilities such as an IFS. This has already been achieved with the Project 1640 Integral IFS and at Palomar Observatory (Hinkley *et al.* 2011b, Zimmerman *et al.* 2011, in prep), and will be integrated with the Gemini Planet Imager coronagraph (Macintosh *et al.* 2008). In the more distant future, multi-wavelength aperture masking interferometry will play a prominent role for the planet-finding efforts of JWST, significantly improving on the sensitivity presented here (Sivaramakrishnan *et al.* 2010, Doyon *et al.* 2010).

References

Baraffe, I., Chabrier, G., Barman, T. S., Allard, F., & Hauschildt, P. H. 2003, *A&A*, 402, 701.
Beuzit, J.-L., Feldt, M., Dohlen, *et al.*, 2008, *SPIE Conference Series*, 7014, pp. 701418.
Crepp, J. R., Pueyo, L., Brenner, D., Oppenheimer, B. R., *et al.*, 2011, *ApJ*, 729, 132.
Digby, A. P., Hinkley, S., Oppenheimer, B. R., *et al.*, 2006, *ApJ*, 650, 484.
Doyon, R., Hutchings, J., Rowlands, N., *et al.*, 2010, *SPIE Conference Series*, 7731.
Fortney, J. J., Marley, M. S., Saumon, D., & Lodders, K. 2008, *ApJ*, 683, 1104.
Graham, J. R., Kalas, P. G., & Matthews, B. C. 2007, *ApJ*, 654, 595.
Haniff, C. A., Mackay, C. D., Titterington, D. J., & Sivia, D., *et al.*, 1987, *Nature*, 328, 694.
Hinkley, S., Oppenheimer, B. R., Soummer, R., *et al.*, 2007, *ApJ*, 654, 633.
Hinkley, S., *et al.*, 2008, Vol. 7015, SPIE Conference Series, pp. 701519.
Hinkley, S., Oppenheimer, B. R., Soummer, R., Brenner, D., *et al.*, 2009, *ApJ*, 701, 804.
Hinkley, S., Oppenheimer, B. R., Brenner, D., Zimmerman, N., *et al.*, 2010, *ApJ*, 712, 421.
Hinkley, S., Monnier, J. D., Oppenheimer, B. R., Roberts, L. C., *et al.*, 2011a, *ApJ*, 726, 104.
Hinkley, S., Oppenheimer, B. R., Zimmerman, *et al.*, 2011b, *PASP*, 123, 74.
Hinkley, S., Carpenter, J. M., Ireland, M. J., & Kraus, A. L. 2011c, *ApJ*, 730, L21.
Ireland, M. J. & Kraus, A. L. 2008, *ApJ*, 678, L59.
Kalas, P., Graham, J. R., Chiang, E., Fitzgerald, M. P., *et al.*, 2008, *Science*, 322, 1345.
Kraus, A. L., Ireland, M. J., Martinache, F., & Lloyd, J. P. 2008, *ApJ*, 679, 762.
Lagrange, A., Bonnefoy, M., Chauvin, G., Apai, D., *et al.*, 2010, *Science*, 329, 57.
Leconte, J., Soummer, R., Hinkley, S., Oppenheimer, B. R., *et al.*, 2010, *ApJ*, 716, 1551.
Liu, M. C., Wahhaj, Z., Biller, B. A., *et al.*, 2010, *SPIE Conference Series*, 7736, pp. 77361K.
Macintosh, B. A., Graham, J. R., Palmer, *et al.*, 2008, *SPIE Conference Series*, 7015, pp. 701518.
Marcy, G., Butler, R. P., Fischer, D., Vogt, S., *et al.*, 2005, *Prog. of Theo. Phys. Supp.*, 158, 24.
Marley, M. S., *et al.*, 2009, *The Future of Ultracool Dwarf Science w/ JWST*, pp. 101.
Marois, C., Macintosh, B., Barman, T., Zuckerman, B., *et al.*, 2008, *Science*, 322, 1348.
Marois, C., Zuckerman, B., Konopacky, Q. M., Macintosh, B., *et al.*, 2010, *Nature*, 468, 1080.
Martinache, F. & Guyon, O. 2009, *SPIE Conference Series*, 744000–744000–9.
Oppenheimer, B. R., Brenner, D., Hinkley, S., Zimmerman, N., *et al.*, 2008, *ApJ*, 679, 1574.
Oppenheimer, B. R. & Hinkley, S. 2009, *ARA&A*, 47, 253.
Pollack, J. B., Hubickyj, O., Bodenheimer, P., Lissauer, J. J., *et al.*, 1996, *Icarus*, 124, 62.
Sivaramakrishnan, A., Koresko, C. D., Makidon, R. B., *et al.*, 2001, *ApJ*, 552, 397.
Sivaramakrishnan, A., Lafrenière, *et al.*, 2010. *SPIE Conference Series*, 7731, pp. 77313W.
Soummer, R. 2005, *ApJ*, 618, L161.
Su, K. Y. L., Rieke, G. H., Stapelfeldt, K. R., Malhotra, R., Bryden, *et al.*, 2009, *ApJ*, 705, 314.
Tuthill, P. G., Monnier, J. D., Danchi, W. C., Wishnow, E. H., *et al.*, 2000, *PASP*, 112, 555.
Veras, D., Crepp, J. R., & Ford, E. B. 2009, *ApJ*, 696, 1600.
Zimmerman, N., Oppenheimer, B. R., Hinkley, S., Brenner, *et al.*, 2010, *ApJ*, 709, 733.

From Interacting Binaries to Exoplanets: Essential Modeling Tools
Proceedings IAU Symposium No. 282, 2011
Mercedes T. Richards & Ivan Hubeny, eds.
© International Astronomical Union 2012
doi:10.1017/S174392131102730X

Mass Determination of Sub-stellar Companions Around Young Stars - The Example of HR 7329

Tobias O. B. Schmidt[1], Ralph Neuhäuser[1] and Andreas Seifahrt[2]

[1] Astrophysikalisches Institut und Universitäts-Sternwarte, Friedrich-Schiller-University
Schillergässchen 2-3, 07745 Jena, Germany
email: tobi@astro.uni-jena.de

[2] Physics Department, University of California
Davis, CA 95616, USA

Abstract. Lowrance *et al.* (2000) found a faint companion candidate about 4 arcsec south of the young A0-type star HR 7329. Its spectral type of M7-8 is consistent with a young brown dwarf companion. Here we report spectroscopic J band observations using the integral field spectrograph SINFONI at VLT, enabling a new estimation of effective temperature, extinction and surface gravity of the object and hence its mass. Although the data were reduced carefully, the presence of a spike within the point spread function of the object in each spectral image hampered the precise estimation of the properties of HR 7329. Nevertheless, we will show with the example of this sub-stellar companion how mass estimates independent of evolutionary models of directly imaged sub-stellar companions can be obtained, after removing most of the strong influence of the spike in the present data, and present a new mass estimation of HR 7329 B/b based on the values gained.

Keywords. stars: low-mass, brown dwarfs, imaging, atmospheres, techniques: spectroscopic

---

1. Introduction

The A0-type star HR 7329 [also called η Tel at a distance of 47.7 ± 1.5 pc (Perryman *et al.* 1997), V = 5.0 mag according to SIMBAD] is a member of the β Pic moving group (Zuckerman *et al.* 2001). It has an age of 12 Myr (Zuckerman *et al.* 2001); see Torres *et al.* (2008) for a review about this association.

Lowrance *et al.* (2000) discovered a 6 mag fainter companion candidate $\sim$ 4 arcsec south of HR 7329 with coronagraphic images using NICMOS at the Hubble Space Telescope (HST) and also obtained a spectral type of M7-8. Guenther *et al.* (2001) confirmed the spectral type with an infrared (IR) H band spectrum obtained with the Infrared Spectrograph and Array Camera (ISAAC) at the European Southern Observatory (ESO) 8.2-m Very Large Telescope (VLT) Antu (Unit Telescope 1, UT 1), in April 2000.

Further information on the primary HR 7329 and its sub-stellar companion can be found in Neuhäuser *et al.* 2011 and references therein. While the significance for common proper motion did not exceed $\sim 1\sigma$ in Guenther *et al.* (2001), using an acquisition image and measuring the separation and position angle (PA) between HR 7329 A and B, Neuhäuser *et al.* 2011 presented 10 new astrometric imaging observations, which allowed us to confirm the common proper motion by more than 21σ (see Neuhäuser *et al.* 2011 for additional information on possibly detected orbital motion and further information on e.g. eccentricity or periastron distance from a possible interaction with the debris disc of HR 7329).

2. Observations

We used the adaptive-optics integral-field spectrograph SINFONI, mounted at UT 4 of the ESO VLT, to obtain spectra that do not suffer from wavelength-dependent slit losses that occur on normal spectrographs with narrow entrance slits in combination with AO.

The observations of the HR 7329 companion were carried out in H+K (resolution 1500) and J (resolution 2000) band in the night of 18 May 2009. Two nodding cycles with an integration time of 300s per frame were obtained each. We chose the maximum possible pixel scale (50 x 100 mas) of the instrument leaving HR 7329 A as the AO guide star outside of the FoV (3 arcsec × 3 arcsec). All observations were done at good airmass (~ 1.2) and good seeing conditions of ~ 0.6 arcsec (optical DIMM seeing).

HIP 94378, a B5 V star, was used as the telluric standard for the J band. In order to correct for the features of this standard star, the Pa-β absorption at $\sim 1.282~\mu$m of HIP 94378 was fitted by a Lorentzian profile and removed by division.

3. Data reduction

We used the SINFONI data reduction pipeline version 2.0.5 offered by ESO (Jung *et al.* 2006) with reduction routines developed by the SINFONI consortium (Abuter *et al.* 2006). After standard reduction, all nodding cycles were combined to a final data cube. We used the *Starfinder* package of IDL (Diolaiti *et al.* 2000) and an iterative algorithm to remove the halo of HR 7329 A, both described in detail in Seifahrt *et al.* (2007). See Fig. 1 for an image of this algorithm, including the removal of most of the strong influence of a spike of the primary star, superimposed onto the point spread fuction of the companion and having to much influence in H+K for its usage here.

4. Results

We have used the combination of the non-equilibrium, stationary cloud model DRIFT (Helling *et al.* 2008) with the general-purpose model atmosphere code PHOENIX

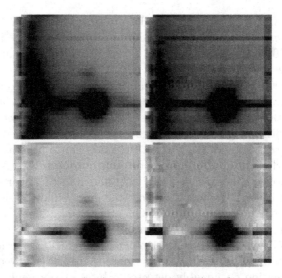

Figure 1. Cleaning of the J band (left) and H+K band (right) cubes. Always shown are the median along the wavelength of the cubes with and without cleaning at the same cut levels from top to bottom, respectively. See text for more information.

(Hauschildt & Baron 1999), see e.g. Witte *et al.* (2011) and references therein for additional information. We used an almost complete grid of DRIFT-PHOENIX models in the range of $T_{eff} = 1000 \ldots 3000$ K, $\log g = 3.0 \ldots 5.5$, and [M/H]=-0.6 \ldots 0.3 in steps of 100 K, 0.5, and 0.3, respectively. Moreover, we still need to account for reddening of our spectra, see Schmidt *et al.* (2008) for more details on the method.

We find a best fit for the companion candidate of HR 7329 at $T_{eff} = 2800 \pm 400$ K, a visual extinction $A_V = 2.75 \, ^{+7.25}_{-2.75}$ mag, $\log g = 4.25 \pm 0.25$, and [Me/H]=-0.3, shown in Figs. 2 & 3. All values are derived from our χ^2 minimization analysis. Unfortunately not all of the influence of the superimposed spike could be removed, hence not allowing for a more precise determination of the properties of the companion given above.

5. Implications

Using the newly derived value of extinction we can correct the luminosity of the companion of $\log(L_{bol}/L_\odot) = -2.627 \pm 0.087$ (Neuhäuser *et al.* 2011) to $\log(L_{bol}/L_\odot) = -2.504^{+0.220}_{-0.151}$, giving R$= 0.238^{+0.104}_{-0.077}$ R$_\odot$ or R$= 2.33^{+1.015}_{-0.748}$ R$_{Jup}$ using the derived temperature value, and finally, using the derived surface gravity, we find a mass of HR 7329 B/b of M$= 38.6^{+45.1}_{-30.0}$ M$_{Jup}$, very well consistent with the value of 20 – 50 M$_{Jup}$ derived from evolutionary models (Neuhäuser *et al.* 2011). A more precise mass estimate could be

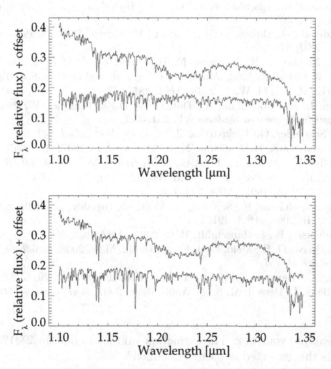

Figure 2. J band spectra. *From top to bottom in each panel:* Our SINFONI spectrum of the HR 7329 companion in spectral resolution 1500 in comparison to the best-fitting Drift-Phoenix synthetic spectra (same spectral resolution) of $T_{eff} = 2800$ K, $\log g = 4.0$, [Me/H]=0.0 and a visual extinction of $A_V = 3.35$ mag (*top panel*) and of $T_{eff} = 2800$ K, $\log g = 4.5$, [Me/H]=-0.6 and a visual extinction of $A_V = 2.14$ mag (*bottom panel*). Note that the surface gravity sensitive features, as e.g. the equivalent width of the alkali lines of K I at ~ 1.25 μm, of the observed spectrum lie inbetween the surface gravity values of the two synthetic model spectra shown here, giving a surface gravity of $\log g = 4.25 \pm 0.25$.

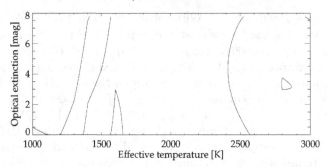

Figure 3. Result of the χ^2 minimization analysis for the HR 7329 companion. Plotted are the best value (*point like on the right hand side*) and the 1, 2, and 3 sigma error contours (*towards lower temperatures*) for effective temperature $T_{\rm eff}$ and optical extinction A_V, determined from comparison of our SINFONI J band spectrum and the Drift-Phoenix model grid, yielding a best fitting temperature of $T_{\rm eff} = 2800 \pm 400\,$K and an extinction of $A_V = 2.75\,^{+7.25}_{-2.75}$ mag.

found if new spectra are observed without the influence of a spike, since the uncertainties of temperature and possibly surface gravity could be reduced.

References

Abuter, R., Schreiber, J., Eisenhauer, F., Ott, T., Horrobin, M., & Gillesen, S. 2006, *New Astron. Revs*, 50, 398

Diolaiti, E., Bendinelli, O., Bonaccini, D., Close, L. M., Currie, D. G., & Parmeggiani, G. 2000, *Proc. SPIE*, 4007, 879

Guenther, E. W., Neuhäuser, R., Huélamo, N., Brandner, W., & Alves, J. 2001, *A&A*, 365, 514

Hauschildt, P. H. & Baron, E. 1999, *Journal of Computational and Applied Math.*, 109, 41

Helling, C., Woitke, P., & Thi, W.-F. 2008, *A&A*, 485, 547

Jung, Y., Lundin, L. K., Modigliani, A., Dobrzycka, D., & Hummel, W. 2006, *Astronomical Data Analysis Software and Systems XV*, 351, 295

Lowrance, P. J., Schneider, G., Kirkpatrick, J. D., *et al.*, 2000, *ApJ*, 541, 390

Mohanty, S., Jayawardhana, R., Huélamo, N., & Mamajek, E. 2007, *ApJ*, 657, 1064

Neuhäuser, R., Ginski, C., Schmidt, T. O. B., & Mugrauer, M. 2011, *MNRAS*, 1135

Patience, J., King, R. R., de Rosa, R. J., & Marois, C. 2010, *A&A*, 517A, 76

Perryman, M. A. C., *et al.*, 1997, *A&A*, 323, L49

Schmidt, T. O. B., Neuhäuser, R., Seifahrt, A., Vogt, N., Bedalov, A., Helling, C., Witte, S., & Hauschildt, P. H. 2008, *A&A*, 491, 311

Seifahrt, A., Neuhäuser, R., & Hauschildt, P. H. 2007, *A&A*, 463, 309

Torres, C. A. O., Quast, G. R., Melo, C. H. F., & Sterzik, M. F. 2008, *Handbook of Star Forming Regions, Volume II*, 757

Witte, S., Helling, C., Barman, T., Heidrich, N., & Hauschildt, P. H. 2011, *A&A*, 529, A44

Zuckerman, B., Song, I., Bessell, M. S., & Webb, R. A. 2001, *ApJ* (Letters), 562, L87

Discussion

K. ALLERS: I enjoyed your talk. The extinction values you find for 2M1207 A and b are very different. Is this expected or easily explained?

T. SCHMIDT: As suggested in Mohanty *et al.* (2007), 2M1207 b might have an edge-on disk, as also found for the primary 2M1207 A. This would also explain the underluminosity of the planetary mass companion candidate 2M1207 b in comparison to evolutionary model predictions for the effective temperature found from spectroscopy (Mohanty *et al.* 2007, Patience *et al.* 2010).

From Interacting Binaries to Exoplanets: Essential Modeling Tools
Proceedings IAU Symposium No. 282, 2011
Mercedes T. Richards & Ivan Hubeny, eds.
© International Astronomical Union 2012
doi:10.1017/S1743921311027311

Lucky Imaging Survey for Binary Exoplanet Hosts

Carolina Bergfors[1], Wolfgang Brandner[1], Sebastian Daemgen[2] and Thomas Henning[1]

[1] Max-Planck-Institut für Astronomie, Königstuhl 17,
69117 Heidelberg, Germany
email: **bergfors@mpia.de**

[2] European Southern Observatory, Karl-Schwarzschild-Strasse 2,
85748 Garching, Germany

Abstract. Binary or multiple stars are common in our neighbourhood, and many of the exoplanets we know of belong to a star in such a system. The influence of a second star on planet formation can be probed by comparing properties of planets in binary/multiple-star systems with those of single-star planets. We present some of the results from our Lucky Imaging survey for binary companions to hosts of transiting exoplanets.

Keywords. Techniques: high angular resolution – Binaries: visual – Planetary systems

1. Introduction

The presence of a close stellar companion is expected to affect the formation of planets. System characteristics such as the occurrence of a stellar companion and the binary orbital separation together with planetary properties provide valuable constraints on models of planet formation and dynamical evolution.

Exoplanets that transit their parent star can provide essential properties such as radius and mass, from which the mean density and surface gravity can be derived. Previous large studies of the physical properties of planets in binary star systems have primarily included exoplanets found with radial velocity spectroscopy, since the number of transiting exoplanets (TEP's) has been small until lately. Almost half of the known TEP's have been found during the last year, mainly thanks to large ground- and space-based programs. The increasingly large number of transiting exoplanets allows for this study of how binarity/multiplicity affects their properties.

2. The AstraLux survey for binary TEP hosts

We searched for faint, close stellar companions to 21 known TEP host stars using the two Lucky Imaging instruments *AstraLux Norte* at the 2.2 m telescope at Calar Alto, Spain, for the northern sky targets, and *AstraLux Sur* mounted at the ESO 3.5 m New Technology Telescope (NTT) at La Silla, Chile, for the southern sky TEP hosts.

Two previously unknown companion candidates were discovered at ~ 1 arcsec separation to the TEP host stars HAT-P-8 and Wasp-12 (Bergfors *et al.* 2011, in prep.). We could also confirm the common proper motion of the candidate binary exoplanet host TrES-4 which was discovered earlier within this survey (Daemgen *et al.* 2009).

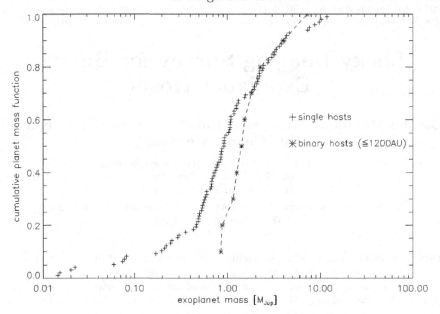

Figure 1. Cumulative planet mass function. Plus-signs represent the masses of transiting exoplanets in single-star systems, and asterisks the masses of TEP's in binary/multiple star systems closer than 1200 AU. Almost half of the planets in single-star systems have masses smaller than the lowest-mass planet in the binary/multiple star sample, indicating that a stellar companion might influence the formation of the planets.

While the sample of close binary TEP hosts is still very small, we found that

• The distributions of planetary radii, orbital periods and surface gravities do not differ significantly between TEP's in binary/multiple systems and those belonging to a single star.

• The division into two classes of hot Jupiters based on Safronov number, $\theta = 0.5 \times (v_{esc}/v_{orb})^2$, and equilibrium temperature, that was suggested by Hansen & Barman (2007) based on the properties of a small number of hot Jupiters, can no longer be discerned for the much larger number of TEP's known today. Additionally, we can no longer see the correlation between binary separation and Safronov number suggested by Daemgen *et al.* (2009) in the larger sample of binaries.

• The transiting exoplanets closer than ~ 1200 AU appear to be on average more massive than single-star TEP's. The cumulative mass function (Fig. 1) shows that almost half of the single-star TEP's have masses smaller than the least massive binary/multiple TEP. A Kolmogorov-Smirnov test shows that the hypothesis of both samples being drawn from the same parent distribution can be rejected with a significance of $\approx 96.5\%$.

References

Daemgen, S., Hormuth, F., Brandner, W. *et al.*, 2009, *A&A*, 498, 567
Hansen, B. M. S. & Barman, T. 2007, *ApJ*, 671, 861
Hippler, S., Bergfors, C., Brandner, W. *et al.*, 2009, *The Messenger*, 137, 14
Hormuth, F., Hippler, S., Brandner, W. *et al.*, 2008, *SPIE*, 7014

From Interacting Binaries to Exoplanets: Essential Modeling Tools
Proceedings IAU Symposium No. 282, 2011
Mercedes T. Richards & Ivan Hubeny, eds.
© International Astronomical Union 2012
doi:10.1017/S1743921311027323

Pipeline for Making Images of Gas Flows in Binary Stars

Mercedes T. Richards[1], Elena Slobounov[1,2], Marshall Conover[1], John Fisher[1], and Alexander Cocking[1]

[1]Pennsylvania State University, Department of Astronomy & Astrophysics, 525 Davey Lab, University Park, PA 16802, USA
email: mtr@astro.psu.edu, mjc5336@psu.edu, jgf5013@psu.edu

[2]Pennsylvania State University, Research Computing and Cyberinfrastructure, 224G Computer Building, University Park, PA 16802, USA
email: ess3@psu.edu

Abstract. The data collection and data analysis pipeline for the study and imaging of interacting binaries is outlined. This process includes the systematic collection of time-resolved spectra of individual systems, data reduction including subtraction of the stellar spectra, application of tomography codes to reveal images of the gas flows in 2D and 3D, comparison of the observed spectrum with synthetic spectra of the accretion disk and gas stream, and application of 3D visualization techniques.

Keywords. (stars:) binaries: eclipsing, circumstellar matter, image processing, stellar dynamics

1. Introduction

A multi-dimensional process has been used to study the evolutionary behavior of interacting binary stars, from Algols to nova-like systems to gamma ray binaries. The focus has been on the visualization of the gas flowing between and around the stars in these binaries since nearly all of these systems are unresolved.

The pipeline for making and interpreting the images includes (1) the systematic collection of time-resolved multiwavelength spectra of individual systems over time-spans of 1-2 weeks, and over multiple epochs. (2) The data reduction stage includes subtraction of the stellar spectra to isolate the spectrum of the non-photospheric gas (e.g., Richards 1993). The stellar spectra have been represented by model atmospheres calculations (e.g., Richards & Albright 1999) or by a representative spectrum based on the observed spectra (e.g., Prinja *et al.* 2011). (3) Tomography codes were applied to create 2D and 3D images of the gas flows (e.g., Richards 2004; Richards, Sharova, & Agafonov 2010). This stage requires the same wavelength dispersion for all spectra so they can be merged into a single file for processing with the tomography codes. (4) The 2D tomograms are displayed using MATLAB. The 3D tomograms are created as discrete sets of 2D (V_x, V_y) images over a grid of V_z velocities. Hence 3D visualization techniques have to be applied to display the image in true 3D format; the Visualization Toolkit (VTK) was the most effective tool in this case.

The SHELLSPEC spectral synthesis code (Budaj & Richards 2004) was created to model the accretion structures, e.g., the disk and gas stream. (5) It was tested on a well-known accretion disk system, TT Hya (e.g., Budaj, Richards & Miller 2005). The observed spectra were compared with synthetic spectra of the accretion disk and gas stream, and tomography was used to illustrate how the separate accretion structures can be isolated. Also, (6) hydrodynamic simulations (e.g., Richards & Ratliff 1998) can be performed to examine the suitability of the models and images derived from the observations.

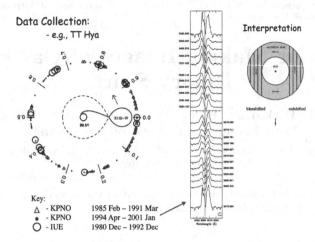

Figure 1. Data collection procedure illustrated in the case of TT Hya (Miller *et al.* 2007).

Data Reduction and Analysis

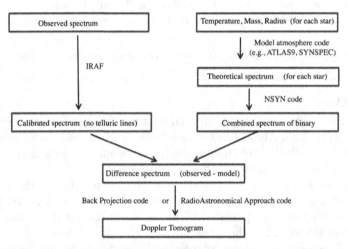

Figure 2. Data processing pipeline.

Acknowledgements

This research was supported by NSF grant AST-0908440 and an award from the Slovak Fulbright Commission and the Council for International Exchange of Scholars (CIES).

References

Budaj, J. & Richards, M. T. 2004, *Contrib. Astron. Obs. Skalnaté Pleso*, 34, 167
Budaj, J., Richards, M. T., & Miller, B. 2005, *ApJ*, 623, 411
Miller, B., Budaj, J., Richards, M. T., Koubský, P., & Peters, G. J. 2007, *ApJ*, 656, 1075
Prinja, R. K., Long, K. S. Long, Richards, M. T., Witherick, D. K., & Peck, L. W. 2011, *MNRAS*, in press
Richards, M. T. 1993, *ApJ. Suppl.*, 86, 255
Richards, M. T. 2004, *AN*, 325, 229
Richards, M. T. & Albright, G. E. 1999, *ApJ. Suppl.*, 123, 537
Richards, M. T. & Ratliff, M. A. 1998, *ApJ*, 493, 326
Richards, M. T., Sharova, O., & Agafonov, M. 2010, *ApJ*, 720, 996

From Interacting Binaries to Exoplanets: Essential Modeling Tools
Proceedings IAU Symposium No. 282, 2011
Mercedes T. Richards & Ivan Hubeny, eds.
© International Astronomical Union 2012
doi:10.1017/S1743921311027335

Differential Rotation in Two RS CVn Systems: σ Gem and ζ And

Zsolt Kővári[1], János Bartus[1,2], Levente Kriskovics[1], Katalin Oláh[1], Krisztián Vida[1], Orsolya Ribárik[1], and Klaus G. Strassmeier[2]

[1]Konkoly Observatory, Konkoly Thege út 15-17., H-1121 Budapest, Hungary
email: kovari, kriskovics, olah, vida, wennot@konkoly.hu

[2]Leibniz Institute for Astrophysics Potsdam,
An der Sternwarte 16, 14482 Potsdam, Germany
email: jbartus, kstrassmeier@aip.de

Abstract. The differentially rotating convective envelope is an indispensable element of the magnetic dynamo working in RS CVn-systems. Tidal coupling is responsible for maintaining fast rotation, and also the observed high level of magnetic activity. In this work, we compare the physical properties of two well known RS CVn-type binaries, that is the long-period system σ Gem and the ellipsoidal variable ζ And. For the comparison, we use the results obtained from processing time-series Doppler images. We also aim at understanding how differential rotation could be affected by tidal forces in such close binaries.

Keywords. stars: activity, stars: imaging, stars: individual (σ Gem, ζ And), stars: spots, stars: late-type

1. Binarity and activity

The differential rotation (DR) is of utmost importance in understanding the magnetic activity, since it is a key element of the dynamo mechanism, which has a controlling influence over the strength of magnetic fields generated, thus that of the activity itself. DR on stars with convective envelopes puts constraints on the large scale topology of the magnetic field, therefore has important information on the working of the dynamo beneath the surface. It is well known that in close binaries tidal effects help maintain the fast rotation and also magnetic activity at higher levels. In this interaction, tidal coupling between the star with the differentially rotating envelope and its companion star is essential, cf. Scharlemann (1981, 1982), Schrijver & Zwaan (1991), Holzwarth & Schüssler (2000, 2002). Although, the theoretical background is still under development and earlier observational techniques gave only a poor chance for justification of this interaction, with Doppler imaging this can be observationally studied.

2. Method and results

Measuring DR by means of time-series Doppler imaging was introduced and demonstrated on different targets by Kővári *et al.* (e.g., in 2004, 2007a,b, etc.). The employed method called ACCORD (acronym from 'Average Cross-CORrelation of consecutive Doppler images') is based on averaging cross-correlation function (ccf) maps of subsequent Doppler images in a way to enhance the DR pattern in the ccf maps, while suppressing the unwanted effect of stochastic spot changes. For this study, the stellar and system parameters of our targets, σ Gem and ζ And, are adopted from Kővári *et al.* (2001, 2007a).

Distortion. From our extended DI code we determine the gravitational distortion (for the method see Kővári *et al.* 2007a). The star is approximated with a rotational ellipsoid, that is elongated towards the secondary, where $\epsilon = (1 - (\frac{a}{b})^2)^{0.5}$ parameterizes the distortion (a and b are the long and short radii, respectively). Scanning through a meaningful part of the $(\epsilon - v\sin i)$ parameter plane, while all other parameters are held constant in the imaging process, yields the likely best estimate when the χ^2 of the line-profile fits reaches a minimum. This way for σ Gem we get $\epsilon \leqslant 0.12$, while in the case of ζ And the formal O-C minimum suggests $\epsilon = 0.27 \pm 0.02$ (filling 82% of the Roche-volume).

Surface differential rotation. Our time-series data were obtained during a 70-night long observing run at NSO in 1996/97. From that we reconstruct 34 and 36 time-series Doppler images for σ Gem and for ζ And, respectively. Applying ACCORD for σ Gem yields equatorial deceleration, i.e. anti-solar-type DR with an average surface shear of $\alpha = -0.07 \pm 0.026$. In the case of ζ And, we get solar-type DR with a surface shear of $\alpha = +0.05 \pm 0.02$ (Kővári *et al.* 2007a). This value is consistent with our new result of $\alpha \approx +0.053$ derived from new ζ And data (see the forthcoming paper by Kővári *et al.*).

3. Discussion

From the scaled graphs of the two binaries (see Fig. 1 in Schrijver & Zwaan 1991), a striking difference is seen: i.e., the center of gravity of σ Gem is outside the limb of the giant component, while in the case of ζ And, the center of mass lies well within the star. This difference may reshape the DR inside the convective bulk, which can explain the observed differences. It is also interesting to compare the corotating latitudes (β_{cor}, i.e. the latitude of the differentially rotating component that rotates synchronously with the system). Scharlemann's (1982) theoretical calculations showed that in a given close binary system the stellar and system parameters determine the developing corotation latitude. Using his assumption, we get $\beta_{cor} \approx 20°$ for σ Gem, which is near the value of 22° derived from ACCORD. However, the more distorted ζ And with synchronized equatorial belt ($\beta_{cor} \approx 10°$) performs a different subsurface scenario: Holzwarth & Schüssler (2000, 2002) showed that the tidal forces and the distortion of the active component in an RS CVn-type binary can explain the emergence of magnetic flux at preferred longitudes. Observations showed the existence of spots concentrating at quadrature positions in ζ And (e.g. Kővári *et al.* 2007a, Korhonen *et al.* 2010). Likewise, the deformation could also account for the disparate DR laws obtained for ζ And and for σ Gem.

Acknowledgements

This work is supported by the Hungarian science research grant OTKA K81421 and by the "Lendület" Young Researchers' Program of the Hungarian Academy of Sciences.

References

Holzwarth, V. & Schüssler, M. 2000, *AN*, 321, 175
Holzwarth, V. & Schüssler, M. 2002, *AN*, 323, 399
Korhonen, H., Wittkowski, M., Kővári, Zs., *et al.*, 2010, *A&A*, 515, A14
Kővári, Zs., Strassmeier, K. G., Bartus, J., *et al.*, 2001, *A&A*, 373, 199
Kővári, Zs., Strassmeier, K. G., Granzer, T., *et al.*, 2004, *A&A*, 417, 1047
Kővári, Zs., Bartus, J., Strassmeier, K. G., *et al.*, 2007a, *A&A*, 463, 1071
Kővári, Zs., Bartus, J., Strassmeier, K. G., *et al.*, 2007b, *A&A*, 474, 165
Scharlemann, E. T. 1981, *ApJ*, 246, 292
Scharlemann, E. T. 1982, *ApJ*, 253, 298
Schrijver, C. J. & Zwaan, C. 1991, *A&A*, 251, 183

From Interacting Binaries to Exoplanets: Essential Modeling Tools
Proceedings IAU Symposium No. 282, 2011
Mercedes T. Richards & Ivan Hubeny, eds.
© International Astronomical Union 2012
doi:10.1017/S1743921311027347

Modelling of an Eclipsing RS CVn Binary: V405 And

Krisztián Vida, Katalin Oláh, and Zsolt Kővári

Konkoly Observatory,
Konkoly Thege út 15-17., H-1121 Budapest, Hungary
email: vida, olah, kovari@konkoly.hu

Abstract. V405 And is an ultrafast-rotating ($P_{\rm rot} \approx 0.46$ days) eclipsing binary. The system consists of a primary star with radiative core and convective envelope, and a fully convective secondary. Theories have shown that stellar structure can depend on magnetic activity, i.e., magnetically active M-dwarfs should have larger radii. Earlier light curve modelling of V405 And indeed showed this behaviour: we found that the radius of the primary is significantly larger than the theoretically predicted value for inactive main sequence stars (the discrepancy is the largest of all known objects), while the secondary fits well to the mass-radius relation. By modelling our recently obtained light curves, which show significant changes of the spotted surface of the primary, we can find further proof for this phenomenon.

Keywords. stars:activity, binaries: eclipsing, stars: fundamental parameters, stars: late-type, stars: spots

1. Introduction

V405 And is an X-ray emitting active binary detected by the ROSAT satellite (Voges *et al.* 1996). The first detailed study of the system was done by Chevalier & Ilovaisky (1997), who detected an orbital period of $P_{\rm orb} = 0.465$ days, and a small, near grazing eclipse. The authors found that the primary and the secondary have spectral types of M0V and M5V, and both of them are active, as both show Hα emission. Vida *et al.* (2009) presented photometric $BV(RI)_C$ data, analysed optical spectroscopic measurements, and found, that the light curve modulation is caused by the combined effect of spottedness and binarity. Using an iterative modelling method to separately describe these two effects, the authors determined the physical properties of this binary system. The primary and the secondary component was found to have masses of $0.49\,M_\odot$ and $0.21\,M_\odot$ respectively, thus the primary is supposed to consist of a radiative core and a convective envelope, while the secondary is probably fully convective. The radii of the two components are $0.78\,R_\odot$ and $0.24\,R_\odot$. Plotting these values on the theoretical mass-radius diagram of Baraffe *et al.* (1998) together with other binaries, we find that the secondary fits well to this relation, but the primary has a significantly larger radius than the theoretically predicted value.

2. Observations and analysis

We have obtained new $BV(RI)_C$ photometry with the 1 m RCC telescope at Piszkéstető between JDs 2455148 and 2455531 (2009 November–2010 November, about 700 days after the light curves modelled in Vida *et al.* 2009). Previously we found the light curve to be stable (Vida *et al.* 2009), but during the time of the new observations the surface seemed to evolve significantly. The main spotted area moved from phase ≈ 0.5

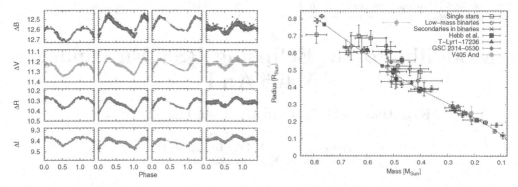

Figure 1. Left: Fits to $BV(RI)_C$ light curves of V405 And. The first column shows the results from Vida *et al.* (2009), the rest show new results. Right: Mass-radius diagram for 5 Gyr stars from Baraffe *et al.* (1998) (with continuous line). Dots show measurements of López-Morales (2007), Blake *et al.* (2008) and López-Morales *et al.* (2006). Filled symbols denote V405 And, GSC 2314-0530 (Dimitrov & Kjurkchieva 2010), T-Lyr-17236 (Devor *et al.* 2008), and 2MASS 04463285+1901432, a binary in NGC 1647 (Hebb *et al.* 2006).

to phase ≈ 1 in the first two new seasons, and in the last season two active nests were observed: around phases 0.1 and 0.5.

Using the same modelling method described in Vida *et al.* (2009), we modelled the new observations using PHOEBE (Prša & Zwitter 2005) and SpotModeL (Ribárik *et al.* 2003). The light curves and the fits are plotted in Fig. 1. The models fitted to the new observations left the system parameters unchanged. This indicates, that the radius of the primary is indeed much larger than expected. The two known similar binaries with similar structure, GSC 2314-0530 (Dimitrov & Kjurkchieva 2010) and the one from Hebb *et al.* (2006) does not show this behaviour, although the primary of 2MASS 04463285+1901432 (Hebb *et al.* 2006) has somewhat larger radius. This indicates that V405 And is a currently unique system, definitely worthy of further studies.

Acknowledgements

The authors acknowledge support from the Hungarian research grant OTKA K81421 and the "Lendület" Program of the Hungarian Academy of Sciences.

References

Baraffe, I., Chabrier, G., Allard, F., & Hauschildt, P. H. 1998, *A&A*, 337, 403
Blake, C. H., Torres, G., Bloom, J. S., & Gaudi, B. S. 2008, *ApJ*, 684, 635
Chevalier, C. & Ilovaisky, S. A. 1997, *A&A*, 326, 228
Devor, J., *et al.*, 2008, *ApJ*, 687, 1253
Dimitrov, D. P. & Kjurkchieva, D. P. 2010, *MNRAS*, 406, 2559
Hebb, L., Wyse, R. F. G., Gilmore, G., & Holtzman, J. 2006, *AJ*, 131, 555
López-Morales, M. 2007, *ApJ*, 660, 732
López-Morales, M., Orosz, J. A., Shaw J. S., Havelka L., Arevalo, M. J., McIntyre T., & Lazaro, C. 2006, *ArXiv Astrophysics e-prints*, 0610225
Mullan, D. J. & MacDonald, J. 2001, *ApJ*, 559, 353
Prša, A. & Zwitter, T. 2005, *ApJ*, 628, 426
Ribárik, G., Oláh, K., & Strassmeier, K. G. 2003, *AN*, 324, 202
Vida, K., Oláh, K., Kővári, Zs., Korhonen, H., Bartus, J., Hurta, Zs., & Posztobányi, K. 2009, *A&A*, 504, 1021
Voges, W., Gruber, R., Haberl, F., Kuerster, M., Pietsch, W., & Zimmermann, U. 1996, *VizieR Online Data Catalog*, 9011, 0

From Interacting Binaries to Exoplanets: Essential Modeling Tools
Proceedings IAU Symposium No. 282, 2011   © International Astronomical Union 2012
Mercedes T. Richards & Ivan Hubeny, eds.   doi:10.1017/S1743921311027359

Doppler Tomography in 2D and 3D of the X-ray Binary Cyg X-1 for June 2007

O. I. Sharova[1], M. I. Agafonov[1], E. A. Karitskaya[2], N. G. Bochkarev[3], S. V. Zharikov[4], G. Z. Butenko[5], and A. V. Bondar[5]

[1]Radiophysical Research Institute, B.Pecherskaya str, 25/12a, Nizhny Novgorod 603950 Russia
[2]Institute of Astronomy RAS, Moscow 119017 Russia
[3]Sternberg Astronomical Institute 13 Universitetskij pr., Moscow 119991 Russia
[4]Mexican National Astronomical Observatory, Institute of Astronomy, UNAM, Mexico
[5]International Center for Astronomical, Medical and Ecological Research, Terscol,
Kabarda-Balkarian Republic, 361605 Russia

Abstract. The 2D and 3D Doppler tomograms of X-ray binary system Cyg X-1 (V1357 Cyg) were reconstructed from spectral data for the line HeII 4686Å obtained with 2-m telescope of the Peak Terskol Observatory (Russia) and 2.1-m telescope of the Mexican National Observatory in June, 2007. Information about gas motions outside the orbital plane, using all of the three velocity components V_x, V_y, V_z, was obtained for the first time. The tomographic reconstruction was carried out for the system inclination angle of 45°. The equal resolution ($50 \times 50 \times 50$ km/s) is realized in this case, in the orbital plane (V_x, V_y) and also in the perpendicular direction V_z. The checkout tomograms were realized also for the inclination angle of 40° because of the angle uncertainty. Two versions of the result showed no qualitative discrepancy. Details of the structures revealed by the 3D Doppler tomogram were analyzed.

Keywords. Accretion, accretion disks – binaries: x-ray – binaries: imaging – stars: individual (Cyg X-1) – techniques: image processing

---

The realization of the 3D Doppler tomography became possible due to the development of the Radioastronomical Approach for reconstruction in the case of few projections (Agafonov & Sharova 2005a; Agafonov & Sharova 2005b). A direct comparison was made between the observed spectra and those computed from the constructed 3D Doppler tomograms. Chi-square statistics show the good quality of the reconstruction. The number of Mexican observation profiles (83) was larger than the 51 spectra from Terskol, so we used mainly the results of the 3D Mexican Doppler tomogram, and the Terskol results were used only as a test. However, there was good similarity in the discovered features.

Two-Dimensional standard Doppler tomograms for 2007 June are similar to the earlier reconstructed tomograms for 1997, 2003 and 2004 (Karitskaya *et al.*, 2005; Karitskaya *et al.*, 2007). They show: A) the emission component of the HeII 4686Å line is generated mainly in the outer parts of the accretion structure closest to the donor star (O-supergiant) and from the optically thick accretion disk; B) the absorption component is the feature of the O-supergiant atmosphere.

Three-Dimensional tomograms are reconstructed in 3D velocity space (V_x, V_y, V_z). Their structure shows that the formation of He II 4686Å line profiles is also connected both with the area of accretion structure (see Fig. 1 and 2) and with the donor-star (supergiant). However, some additional features have been discovered.

A. The first predominant feature is the emission component located around the central slice. This area consists of individual feature components with different V_z. But all these V_z values lie within the limits of −200 to +160 km/s. Here we can see a combination of three main emission feature components: 1) the emission of the outer part of the accretion

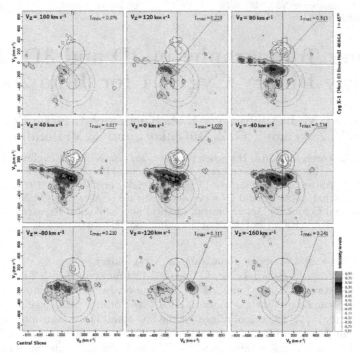

Figure 1. Nine central slices of the Cyg X-1 3D Doppler tomogram in the (V_x, V_y) plane for different V_z. The outlines of the Roche lobe of the donor star and the pattern of the outer parts of the accretion disk are plotted on the slices. We used the mass ratio of $q = M_x/M_O = 1/3$ and two versions of the disk size.

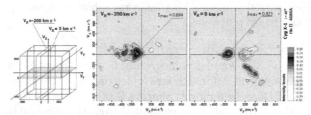

Figure 2. Cross-sections of of the Cyg X-1 3D Doppler tomogram in the (V_y, V_z) plane for two different V_x. The cube shows the geometry of the arrangement of the slices in 3D velocity space.

structure near the donor-star; 2) the elongated feature of the stream emerges from the L1 point; 3) the emission produced by the stream - accretion structure interaction.

B. The other predominant feature is visible in absorption and is associated with the supergiant. It is a compact structure. The maximum of absorption corresponds to the central slice (V_x, V_y) of the tomogram with $V_z = 0$ km/s. There is also an interesting emission feature identified with a structure related with the supergiant (in the co-rotation coordinate system). It has a $V_z \sim -200$ to -400 km/s and an intensity $I \sim 20 - 30\%$ of the maximum 3D tomogram intensity (see Fig. 2). That is probably a stream from the donor-star visible in emission flowing almost perpendicularly to the orbital plane.

This work was partially supported by RFBR (grants 09-02-01136 and 09-02-00993).

References

Agafonov, M. I. & Sharova, O. I. 2005a, *Radiophys. Quant. Electron.*, 48, 5, 329
Agafonov M. I. & Sharova, O. I. 2005b, *AN*, 326, 143
Karitskaya, E. A., Agafonov, M. I., Bochkarev, N. G. *et al.*, 2005, *A&A Trans.*, 24, 383
Karitskaya, E. A., Agafonov, M. I., Bochkarev, N. G. *et al.*, 2007, *A&A Trans.*, 26, 159

From Interacting Binaries to Exoplanets: Essential Modeling Tools
Proceedings IAU Symposium No. 282, 2011   © International Astronomical Union 2012
Mercedes T. Richards & Ivan Hubeny, eds.   doi:10.1017/S1743921311027360

PTPS Candidate Exoplanet Host Star Radii Determination with CHARA Array

Paweł Zieliński[1], Martin Vaňko[2], Ellyn Baines[3], Andrzej Niedzielski[1] and Aleksander Wolszczan[4]

[1] Toruń Centre for Astronomy, Nicolaus Copernicus University, Gagarina 11, 87100 Toruń, Poland, email: pawziel@astri.umk.pl

[2] Astronomical Institute, Slovak Academy of Sciences, 05960 Tatranská Lomnica, Slovakia

[3] Naval Research Laboratory, Remote Sensing Division, 4555 Overlook Ave. S.W., Washington, DC 20375

[4] Department for Astronomy and Astrophysics & Center for Exoplanets and Habitable Worlds, Pennsylvania State University, 525 Davey Laboratory, University Park, PA 16802

Abstract. We propose to measure the radii of the Penn State - Toruń Planet Search (PTPS) exoplanet host star candidates using the CHARA Array. Stellar radii estimated from spectroscopic analysis are usually inaccurate due to indirect nature of the method and strong evolutionary model dependency. Also, the so-called degeneracy of stellar evolutionary tracks due to convergence of many tracks in the giant branch decreases the precision of such estimates. However, the radius of a star is a critical parameter for the calculation of stellar luminosity and mass, which are often not well known especially for giants. With well determined effective temperature (from spectroscopy) and radius, the luminosity may be calculated precisely. In turn also stellar mass may be estimated much more precisely. Therefore, direct radii measurements increase precision in the determination of planetary candidates masses and the surface temperatures of the planets.

Keywords. techniques: interferometric, stars: fundamental parameters, stars: late-type, planetary systems

1. Motivation

Within the Penn State - Toruń Planet Search (PTPS, Niedzielski & Wolszczan 2008), which is a radial velocity project to search for and characterize planets around stars more massive than the Sun, the stellar integral properties (M, R and L) are currently determined from a combination of atmospheric parameters (surface gravity, $T_{\rm eff}$ and [Fe/H]) with the evolutionary tracks (Girardi *et al.* 2000). Unfortunately, this method is uncertain due to difficulties in accurate placement of individual stars on the H-R diagram.

Fig. 1 presents the comparison between the determinations of stellar radii from our spectroscopic analysis (Zieliński *et al.* 2011) and the empirical calibration by Alonso *et al.* (2000). Both results are in general agreement, nevertheless, large discrepancies for individual objects are visible. The uncertainties of our radii determinations based on stellar $T_{\rm eff}$ and L was found to be 0.8 $R_\odot$ on average, but due to missing parallaxes the results are strongly dependent on adopted stellar evolutionary models as well as on photometric data quality. We plan to improve this by using the CHARA Array (ten Brummelaar *et al.* 2005) for direct measurements of stellar radii and, in turn, using the relation $g = \gamma M R^{-2}$, constrain better the stellar mass. On the other hand, the luminosity can be calculated with high accuracy when using the Stefan-Boltzmann law directly.

Two stars from the PTPS survey were already observed with the CHARA Array: HD 17092 (Baines *et al.* 2009) and HD 214868 (Baines *et al.* 2010). The results obtained from direct measurements agree well with the spectroscopic values estimated by us (see

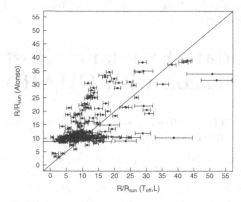

Figure 1. The comparison of the $R/R_\odot$ obtained in the spectroscopic study of PTPS subsample of 332 stars and from the empirical calibration of Alonso *et al.* (2000). Only the uncertainties in PTPS determinations are shown. The solid line denotes the one to one relation.

Niedzielski *et al.* 2007, Zieliński *et al.* 2011). However, the accuracy of interferometry-based measurements is significantly better.

2. Targets selection

Our observing list was derived from the PTPS survey of evolved stars with RV planet candidates and comprises 212 stars suitable for CHARA observations. These stars have already been observed spectroscopically using the Hobby-Eberly Telescope, so high-resolution spectra are available. The targets are sufficiently bright (V < 8.5 mag and K < 6 mag) and nearby ($\pi \geqslant 4.03$ mas) giants with large radii that can be easily directly measured interferometrically.

3. Expected results

Using new CHARA measurements we will be able to obtain high-precision radius measurements and in turn better estimates of the stellar masses. With the angular diameter precision of 2%, the stellar radius can be determined with a precision better than 5%. The largest contributor to the error budget is the parallax, therefore, after taking into account 10% parallax uncertainty, the final mass will then be precise within 0.1 $M_\odot$. This is significantly better compared to our current estimates based on moderate quality photometry and evolutionary models. For PTPS targets studied in detail, we estimated masses between 1 and 3 $M_\odot$ and even after critical assessment, an uncertainty of 30% remains. Hence, the CHARA data will improve the precision in mass sufficiently for the subsequent analysis.

We acknowledge support from the European Community's Seventh Framework Programme Grant 226604; the Polish Ministry of Science and Higher Education grant N N203 510938 (PZ and AN); VEGA 2/0094/11 (MV); NASA grant NNX09AB36G (AW).

References

Alonso, A., Salaris, M., Arribas, S., *et al.*, 2000, *A&A*, 355, 1060
ten Brummelaar, T. A., McAlister, H. A., Ridgway, S. T., *et al.*, 2005, *ApJ*, 628, 453
Baines, E., McAlister, H. A., ten Brummelaar, T. A., *et al.*, 2009, *ApJ*, 701, 154
Baines, E., Döllinger, M. P., Cusano, F., *et al.*, 2010, *ApJ*, 710, 1365
Girardi, L., Bressan, A., Bertelli, G., & Chiosi, C. 2000, *A&AS*, 141, 371
Niedzielski, A., Konacki, M., Wolszczan, A., *et al.*, 2007, *ApJ*, 669, 1354
Niedzielski, A. & Wolszczan, A. 2008, in Proc. IAU Symposium No. 249, 43
Zieliński, P., Niedzielski, A., Wolszczan, A., Adamów, M., & Nowak, G. 2011, *A&A*, submitted

From Interacting Binaries to Exoplanets: Essential Modeling Tools
Proceedings IAU Symposium No. 282, 2011   © International Astronomical Union 2012
Mercedes T. Richards & Ivan Hubeny, eds.   doi:10.1017/S1743921311027372

The MUNI Photometric Archive

Marek Chrastina, Miloslav Zejda, and Zdeněk Mikulášek

Department of Theoretical Physics and Astrophysics, Faculty of Science, Masaryk University,
Kotlářská 2, CZ-611 37 Brno, Czech Republic, homepage: http://astro.physics.muni.cz

Abstract. In the 1990s of the last century, CCD cameras became more reachable. Due to many advantages of CCD cameras, astronomers began using them as the primary detector for photometry of stellar objects. A typical observatory, which operates one telescope at a time, obtained 0.5 TB of raw data during two decades, that means one million about 500 kB-sized files. There are several observatories in the Czech Republic and Slovakia (taking into account all scientific, public as well as private ones). A rough estimate of the total amount of this photometric data is 10 TB, which could be a very interesting source of observational data. Unfortunately, these data are not available online. These data are stored in observatory archives in arbitrary format. Often it is not even possible to find requested data. We have decided to change this state by establishing a common archive of raw photometric data, which would be available online together with tools for searching, listing etc. We already defined the data format, file and directory structure of our archive. We developed sophisticated tools for archive maintenance as well. Our goal is to provide data storage with simple and straightforward access and we are ready to interconnect with the VO right after the IVOA Photometry Data Model will be released.

Keywords. astronomical data bases, techniques: photometric, methods: data analysis

---

1. Introduction

CCD photometry was initiated at Masaryk University Observatory in 1996, when the main 0.62m reflector was equipped with an SBIG ST-8 camera. Most of the observations were focused on variable star measurements. However, observations of comets, minor planets, extrasolar planets, star clusters and GRB afterglows were also obtained. All data were stored in FITS files on hard disks, CDs and DVDs. We obtained approximately 0.5 TB raw photometric data (including calibration images) during 15 years of camera operation. In other words, we have circa one million about 500 kB-sized files in our archive. Unfortunately, the data are not located in any kind of database, therefore they are without any online access.

There are several observatories in Czech Republic and Slovakia (taking into account all scientific, public as well as private ones). Unfortunately, the situation with their raw data is more or less the same, that means no database, no online access. These data are stored in observatory archives in arbitrary format. Often it is not even possible to find requested data.

Our rough estimate of the total amount of photometric data obtained in our region is about 10 TB. Just for comparison, SDSS DR7 contains 15.7 TB of images [e1]. Each observatory has a different scientific programme, but each of them worked over quite a long timeline. Often they provide long-term monitoring of several objects. In addition, the CCD provides measurements of the whole sky field, thus we can take advantage of the field overlapping. The field can also contain objects, that may be interesting for someone else.

2. MUNI Photometric Archive

The MUNI Archive can be a very interesting source of observational data. Data are not available online from a single location. That is the problem which prevents further usage of the data. For that reason, we have decided to establish a single common archive of raw photometric data.

Due to the arbitrary format of a particular data source, we have to convert data to one specific uniform data format. We used restricted alternation of standard FITS format. We proposed an explicitly defined set of keywords. Furthermore, we propose a strict naming convention for FITS-file names and directory structure. There are many FITS files, thus we have developed a package of useful scripts for converting FITS files into the uniform format and structure (Chrastina et al. 2010).

Acknowledgements

This work has been supported by grants MUNI/A/0968/2009, GAAV IAA301630901 and MEB051018. Thanks also to Gabriel Szász and Filip Hroch for their cooperation.

References

e1, SDSS DR7 (The Sloan Digital Sky Survey Data Release 7), http://www.sdss.org/dr7/
Chrastina, M., Zejda, M., & Mikulášek, Z. 2010, in "Binaries Key to Comprehension of the Universe," eds. A. Prsa et al., ASP-CS, 435, 83

From Interacting Binaries to Exoplanets: Essential Modeling Tools
Proceedings IAU Symposium No. 282, 2011
Mercedes T. Richards & Ivan Hubeny, eds.

© International Astronomical Union 2012
doi:10.1017/S1743921311027384

Eclipsing Binaries Within Visual Ones: Prospects of Combined Solution

Petr Zasche

Astronomical Institute, Faculty of Mathematics and Physics, Charles University Prague,
CZ-180 00 Praha 8, V Holešovičkách 2, Czech Republic
email: `zasche@sirrah.troja.mff.cuni.cz`

Abstract. The study of eclipsing binaries as members of multiple systems can provide us important information about their origin, evolution, mutual inclination of the orbits, independent distance and mass determination, as well as the stellar multiplicity in general. We are carrying out a long-term photometric monitoring of several eclipsing binaries within the visual multiples and, besides the complete light curves, we are trying to detect the period changes due to the orbital motion around a common barycenter.

Systems like DN UMa, V819 Her, LO Hya, or VW Cep are typical examples of eclipsing binaries orbiting around the barycenter of the multiple system, while their respective periods are on the order of years or decades. However, the expected period variation is only hardly detectable and there is still uncertainty about which of the components is the eclipsing one. Precise spectroscopy would be of great benefit, but detecting the changes in the gamma velocity is still problematic, and spectral disentangling of such complicated systems like sextuple VV Crv (periods 1.46, 3.14, and 44.51 days) is also rather difficult. However, the detection of the changing depths of the eclipses in the latter system would be interesting.

Keywords. (stars:) binaries: eclipsing, (stars:) binaries: spectroscopic, (stars:) binaries: visual, stars: fundamental parameters

---

There are still known only a few systems where an eclipsing binary is a member of a more complex visual multiple system, and exhibits periodic modulation of its orbital period as a result of orbital motion around a common barycenter. There are only eight such systems known nowadays – i Boo, VW Cep, KR Com, V772 Her, V819 Her, QS Aql, ζ Phe, and V505 Sgr. Such systems are usually rather close to the Sun (often < 100pc), and their angular separations are on the order of a few mas only. There are many possible effects which can be studied and moreover there are also several physical constraints which have to be satisfied in such systems (e.g. stability criteria as a function of ratio of long and short periods).

Three different systems studied recently (namely DN UMa, VV Crv, and V2083 Cyg) indicate that there could be hidden components in these multiples, but each of the additional components was discovered by a different method. Detailed analyses will be published elsewhere in a separate paper. Discovering additional components in a particular system shifts the object to even higher multiplicity and also affects the statistics of such systems. Eggleton & Tokovinin (2008) published a study of multiplicity among bright (well-studied) systems. They presented a correlation between the number of systems of a particular multiplicity versus the number of components in the system. As we have pointed out, such statistics are nowadays still very incomplete and many systems are of higher multiplicity than listed in catalogues.

Radial velocity residuals, spectral disentangling, or long-term photometric monitoring of these systems is very fruitful, but to obtain a reliable and complete picture of the system is still difficult. A study of dynamical effects in these systems should be of interest, for instance a slow precession of the orbits or changing the inclination between them.

In some of the systems, discovering additional components could be a less difficult task due to several reasons. One of them is quite surprising, but it is the fact that most of these stars are relatively bright. Very bright stars (< 6 mag) are photometrically observed only very rarely nowadays due to larger telescopes and CCD detectors used (these targets easily saturate the detectors). New observations can reveal additional components simply by observing the minima of eclipsing binaries and detecting the period changes. Many of the bright stars are very neglected nowadays. Furthermore, discovering additional components has other aspects, for example most of the detected components are of similar spectral types as the eclipsing pair itself. This is understandable, because it is much more difficult to discover a low-mass companion near a pair of B stars.

Another issue is the detection method of such bodies. As an illustrative example we present here three different cases, where three different methods of detection were applied:

• DN UMa - This well-known eclipsing binary is a member of a more complex multiple system with two distant components ($63''$ and $4''$). The light curve (LC) was analyzed by García & Giménez (1986), and the radial velocity curve (RV) by Popper (1986). Moreover, another component is orbiting around the eclipsing pair with period about 118 yr. Many observations of minima times for the EB pair obtained during the last 30 years were used for a period analysis. After subtracting the 118 yr trend, we discovered another variation with period about 640 days only, which can be attributed to another component in the system. Therefore, we have a sextuple star system.

• VV Crv - This rather neglected multiple system was studied by Massarotti *et al.* (2008) by means of radial velocities. Two different spectroscopic binaries were discovered (periods 44.5 and 1.46 day). Furthermore, two other distant visual components are present in the system ($60''$ and $5''$). And finally, there is also a 3.14 day eclipsing pair, which cannot be related to either of the spectroscopic periods. Therefore, the system is a typical hierarchical sextuple. Moreover, a very favorable configuration of the system allows us to hope to detect a change of inclination (i.e. short nodal period).

• V2083 Cyg - A typical triple system consisting of an EB pair on a wide orbit with period of about 372 yr (Seymour *et al.* 2002). However, our new detailed analysis of LC, RV and visual orbit reveals that the system is probably of higher multiplicity. We found that the total mass of the system as derived from the visual orbit is much higher than the one found from the LC+RV analysis. A possible solution is to move the system closer to the Sun, but this leads to a value of parallax well outside of the error bars of the Hipparcos data. Another possible solution is that the distant component is also a double.

We can also study the mutual inclinations of the orbits in these binaries (eclipsing-pair orbit versus visual-pair orbit). As pointed out by Zakirov (2008), the inclination angles are not randomly oriented and such systems tend to be coplanar. This indicates something about a common origin of the system. However, we found a completely different picture. The mutual inclinations are high ($> 30°$) and for V2083 Cyg the orbits are almost perpendicular to each other ($i_{2-3} = 82°$).

Acknowledgements: Supported by the Czech Science Foundation grant P209/10/0715.

References

Eggleton, P. P. & Tokovinin, A. A. 2008, *MNRAS*, 389, 869

García, J. M. & Giménez, A. 1986, *Ap&SS*, 125, 181

Massarotti, A., Latham, D. W., Stefanik, R. P., & Fogel, J. 2008, *AJ*, 135, 209

Popper, D. M. 1986, *PASP*, 98, 1312

Seymour, D. M., Mason, B. D., Hartkopf, W. I., & Wycoff, G. L. 2002, *AJ*, 123, 1023

Zakirov, M. M. 2008, *Kinematics and Physics of Celestial Bodies*, 24, 25

From Interacting Binaries to Exoplanets: Essential Modeling Tools
Proceedings IAU Symposium No. 282, 2011
Mercedes T. Richards & Ivan Hubeny, eds.
© International Astronomical Union 2012
doi:10.1017/S1743921311027396

Polarimetry of Exoplanetary System CoRoT-2

N. M. Kostogryz, T. M. Yakobchuk, and A. P. Vidmachenko

Main Astronomical Observatory of NAS of Ukraine,
27, Zabolotnoho str., Kyiv 03680, Ukraine
email: kosn@mao.kiev.ua yakobchuk@mao.kiev.ua

Abstract. We present the results of modelling the polarization resulting from the planetary transits and stellar spots in the system Corot-2 using the Monte Carlo method. The planetary transit was estimated to produce a polarization maximum at the limb of $\sim 5 \times 10^{-6}$, adopting solar center-to-limb polarization. Assuming different parameters of the spots, we evaluated the flux and polarization changes due to the stellar activity.

Keywords. methods: numerical, techniques: polarimetric, planetary systems, stars: spots

1. Introduction

It is known that radiation from an unresolved, centro-symmetric star is normally unpolarized as the average of the polarization directions of all photons from the star detected by the observer results in zero polarization. However, an intrinsic polarization can occur if the star is not centro-symmetric, e.g. during planetary transit. This effect is also known as the Chandrasekhar effect (Chandrasekhar 1950). In order to model it we used the Monte Carlo method as proposed by Carciofi & Magalhães (2005). For a known set of parameters that describe the configuration of the star-planet system, we adopted a photon-by-photon procedure with weights according to the chosen point on the stellar surface. If the photon packet was emitted in the opposite hemisphere with respect to the observer or its trajectory crossed the planet, the weight was set to zero. Otherwise, the weight was chosen according to the limb-darkening law (see also Kostogryz *et al.* 2011). The spot weights were calculated differently, accounting for their lower temperature.

Results and discussions

We considered a young and active star Corot-2, the second planet-hosting star discovered by the CoRoT satellite. It has a highly active and intrinsically variable host star. Its rotation period is $P_{star} = 4.52d$ (Lanza *et al.* 2009), while the orbital planetary period is only three time less ($P_{pl} = 1.74d$). Being interested in the occultation effects, we particularly chose the system for the high ratio of planetary to star radii $R_{planet}/R_{star} = 0.167$ (Schneider 2011).

The upper panels in Figure 1 illustrate the size of the planet compared to the star and the path of the transit across the stellar disk. For cases b) and c), positions and radii of the spots are shown. The lower panels present the time dependence of the stellar flux, Stokes parameters and polarization degree. The maximum polarization observed on the stellar limb is estimated at $\sim 5 \times 10^{-6}$ (Fig. 1a). While the shapes of the curves for polarization parameters remain essentially the same, the average values grow considerably when spots are added (Fig. 1b). The planet's contribution to the polarization also dominates in one stellar rotation period, as is seen from Fig. 1c.

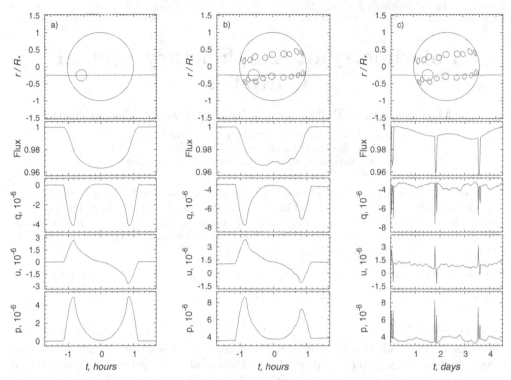

Figure 1. Modelled curves for the fluxes and polarization parameters for CoRoT-2: a) one planet transit; b) one planet transit with stellar spots; c) planet transits and spots for one stellar rotation period.

Conclusions

In this study the two possible cases of symmetry breaking of the host star Corot-2 were investigated. The first one appears as a result of an occultation of the star by a transiting planet; the second one is the total effect of the planetary transit and spots on the stellar surface. The planetary transit was found to produce a polarization maximum at the limb of $\sim 5 \times 10^{-6}$, adopting the solar center-limb polarization. At the same time, it was shown that the polarization value can vary substantially depending on the positions and sizes of the stellar spots.

Acknowledgement

The authors gratefully acknowledge Dr. Sci. Nataliya Shchukina for providing the data on solar polarization. This work has been partially funded by National Academy of Sciences of Ukraine through project 1.4.6/5-261B.

References

Chandrasekhar, S. 1950, *Radiative Transfer. Dover Press, New York*
Carciofi, A. C. & Magalhães, A. M. 2005, *ApJ*, 635, 570
Kostogryz N. M., Yakobchuk T. M., Vid'machenko A. P. & Morozhenko, O. V. 2011, *MNRAS*, 415, 695
Lanza, A. F., Pagano, I., Leto, G., Messina, S., *et al.*, 2005, *A&A*, 493, 193
Schneider, J. 2011, *on-line http://exoplanet.eu*

Part 4
Model Atmospheres of Stars, Interacting Binaries, Disks, Exoplanets, and Brown Dwarfs

From Interacting Binaries to Exoplanets: Essential Modeling Tools
Proceedings IAU Symposium No. 282, 2011
Mercedes T. Richards & Ivan Hubeny, eds.
© International Astronomical Union 2012
doi:10.1017/S1743921311027402

LTE Model Atmospheres: MARCS, ATLAS and CO5BOLD

P. Bonifacio[1], E. Caffau[2,1], H.-G. Ludwig[2,1], and M. Steffen[3,1]

[1] GEPI, Observatoire de Paris, CNRS, Université Paris Diderot
Place Jules Janssen, 92190 Meudon, France

[2] Zentrum für Astronomie der Universität Heidelberg, Landessternwarte
Königstuhl 12, 69117 Heidelberg, Germany

[3] Leibniz-Institut für Astrophysik Potsdam
An der Sternwarte 16, D-14482 Potsdam, Germany

Abstract. In this talk, we review the basic assumptions and physics covered by classical 1D LTE model atmospheres. We will focus on ATLAS and MARCS models of F-G-K stars and describe what resources are available through the web, both in terms of codes and model-atmosphere grids. We describe the advances made in hydrodynamical simulations of convective stellar atmospheres with the CO^{5}BOLD code and what grids and resources are available, with a prospect of what will be available in the near future.

Keywords. stars: atmospheres, stars: abundances, radiative transfer, hydrodynamics, convection

1. Introduction

A model atmosphere is a numerical model that describes the physical state of the plasma in the outer layers of a star, and is used to compute observable quantities, such as the emerging spectrum or colours. Different degrees of complexity lead to different classes of models. The first simplification that is made in the models we shall describe is that of Local Thermodynamic Equilibrium (LTE). Although we know that a stellar atmosphere cannot be in thermodynamic equilibrium, since we see that radiation is escaping in open space, we make the assumption that *locally* we are very close to thermodynamic equilibrium. In practice, this means that we assume that at each point in the atmosphere, which we identify by some suitable coordinates X, Y, Z and in a volume $\Delta X \Delta Y \Delta Z$ around it, we can define a temperature T. This temperature can be used to describe the velocity distribution of the particles, that will be a Maxwellian distribution, to compute the occupation numbers of the atomic levels of the different species, through Boltzmann's law and the ionisation equilibria through Saha's law. The LTE hypothesis greatly simplifies the computation of model atmospheres, by providing us a quick and easy way to compute all the micro-physics of the plasma, just by knowing the temperature (and gas pressure) at any given point.

The next simplification concerns the dimensionality of the problem. Do we really need to treat this as a three-dimensional problem? If we simplify our model by assuming that the atmosphere is homogeneous in two directions and shows variations of the physical quantities only along one direction (vertical, for plane-parallal atmospheres, radial for spherical atmospheres), we have reduced our problem to one dimension.

In principle, all the physical quantities in the plasma may change with time. A simplification is to assume that the atmosphere is stationary and then the problem becomes time-independent.

All these simplifications are adopted in the widely used MARCS and ATLAS model atmospheres. In spite of what can appear at first sight as an oversimplification, these model atmospheres are capable of reproducing a wide range of observable quantities and have a high predictive power.

The CO^{5}BOLD model atmospheres make the hypothesis of LTE, however are fully three dimensional (two dimensional models may also be computed) and time-dependent. In this respect, they are more realistic since they can describe effects that cannot be accounted for by 1D models. This extra realism, however comes at a price and we shall discuss this later.

In the following, we shall give only a very sketchy description of the basic principles that are the basis for the computation of model atmospheres, since these are well described in the relevant publications. We shall instead try to adopt an "end user" approach pointing out how such tools can be used and what resources are available.

2. ATLAS and MARCS

2.1. *Basic principles*

Both the ATLAS (Kurucz 1970,1993,2005) and MARCS (Gustafson *et al.* 1975; Plez *et al.* 1992; Edvardssoon *et al.* 1993; Asplund *et al.* 1997; Gustafsson *et al.* 2003,2008) models are one dimensional and static. MARCS can deal either with plane-parallel or spherical geometry, ATLAS only with plane-parallel, although SATLAS (Lester & Neilson 2008) can compute spherical models. Elsewhere in this volume, Neilson (2011) talks about SATLAS and spherical models. In what follows, we shall assume plane-parallel geometry for both MARCS and ATLAS.

Both codes assume that the atmosphere is in hydrostatic equilibrium, this provides the first basic equation necessary to compute a model atmosphere. The equation of hydrostatic equilibrium states that the gas pressure gradient is balanced by the difference between gravity and the sum of turbulent pressure gradient and radiative acceleration. In MARCS the turbulent pressure gradient is treated defining an "effective gravity" g_{eff}, see Gustafsson *et al.* (2008). In ATLAS, there is an explicit term for the turbulent pressure, proportional to density and the square of turbulent velocity. The equation of hydrostatic equilibrium must be coupled with the equation of radiative transfer, and the condition of energy conservation. The atmosphere simply transports the energy; there is no net absorption or creation of energy within the atmosphere. A convenient variable is the mass column density $dm = -\rho dx$, RHOX (pronounced "rocks" for those who like reading the `FORTRAN`). The goal of the model atmosphere computation is to define the run of the temperature $T(m)$, and the energy conservation provides a mean to modify a trial value of T in order to satisfy the condition.

The global parameters that define the model are the surface gravity g, the energy/ surface = integral of flux over all frequencies = σT_{eff}^4, where the latter can be taken as a definition of "effective temperature" and the "chemical composition," that affects the opacities and the equation of state. At each step in the computation the hypothesis of LTE allows us to compute occupation numbers and ionization fractions that allows us to compute the opacity.

ATLAS has two ways of dealing with line opacities, either through the Opacity Distribution Functions (ODFs, version 9 of ATLAS) or through Opacity Sampling (version 12 of ATLAS). Early versions of MARCS also used ODFs, but the currently used version is OSMARCS, that uses opacity sampling. We stress that tests conducted with ATLAS 9 and ATLAS 12 show that the differences between a model computed with ODFs and one computed with Opacity Sampling are minor and can be ignored for all practical purposes. The choice on whether to use ATLAS 9 or ATLAS 12 is a matter of convenience, if a large number of models needs to be computed with the same chemical composition, then ATLAS 9 is the obvious choice. If several models with slightly different chemical composition are required, ATLAS 12 is more handy.

In the atmospheres of cool stars, a complication is that in the deep layers energy is mainly transported by convection. Both MARCS and ATLAS use a "mixing length" approximation. However MARCS uses the Henyey *et al.* (1965) formulation, while ATLAS uses essentially the Mihalas (1970) formulation, in a way that is detailed by Castelli, Gratton & Kurucz (1997). At large optical depths where convection dominates, a consequence is that a MARCS and an ATLAS model will be slightly different, even if they have been computed with the same mixing length parameter α_{MLT}. The effect is illustrated in Bonifacio *et al.* (2009), appendix A, Figure A.2. We wish to give here a warning: ATLAS has an option for an "approximate treatment of overshooting," well described by Castelli *et al.* (1997), that we recommend users to switch off. This is also the recommendation of Castelli *et al.* (1997), but in our view the most convincing reason is that this option produces a temperature structure that is inconsistent with the mean temperature structure of hydrodynamical simulations.

2.2. *Availability*

The ATLAS code has always been distributed publicly, on an "as is" basis, with a "do not use blindly" clause. The main site is Kurucz's site `kurucz.harvard.edu` where you can find all the source codes of versions 9 and 12 of ATLAS, as well as the spectrum synthesis suite SYNTHE, the abundance analysis code WIDTH and a lot more. We call attention to a code called `binary` that allows us to combine two synthetic spectra to synthesize the spectrum of an SB2 binary. The site also contains a large choice of ODFs, atomic data to compute new ODFs, atomic and molecular data for spectrum synthesis. There is also a large grid of computed ATLAS 9 models, that can be used "off the shelf."

One difficulty faced by ATLAS users is that Kurucz uses DEC computers running under the VMS operating system. While such systems were widely spread in the eighties, and still existent in the nineties, most researchers have to Unix work stations, and a large fraction use Linux systems. Sbordone *et al.* (2004) presented a port of ATLAS and the other codes for Linux, see also Sbordone (2005).

Fiorella Castelli is very active in updating and improving the codes, the latest version of the Linux version of the codes can always be found at wwwuser.oats.inaf.it/castelli. On her web site, you can also find a large grid of computed ATLAS 9 models and fluxes, as well as ODFs to compute further models you might need.

With Luca Sbordone and Fiorella Castelli, we also provided a site were the codes are nicely bound in tar-balls and come with a `Makefile` that allows an easy installation. The site attempts also to collect available documentation and example scripts to run the various programmes, to provide a starting point for beginners. The site was initially hosted by the Trieste Observatory, but has now moved to the Paris Observatory `atmos.obspm.fr`. We try to keep the source codes always aligned with those on the site of Fiorella Castelli.

For users of ATLAS and related codes there are discussion and announcement lists that are mantained at the University of Ljubljana:
`list.fmf.uni-lj.si/mailman/listinfo/kurucz-discuss`.

The MARCS code is not publicly available, so if you need a particular MARCS model you have to ask one of its developers. However there is a web site on which a large grid of computed models and fluxes is publicly available: `marcs.astro.uu.se`. You must register on the site, but registration is free. You find both plane-parallel and spherical models, as well as programs to read the models and interpolate in the grid. If you use a spherical model, make sure that the spectrum synthesis code you use is capable of properly treating the spherical transfer. For example, SYNTHE is not capable of doing it; it will nevertheless run happily interpreting the spherical model as if it were plane-parallel, which is inconsistent. The `turbospectrum` code by B. Plez Alvarez & Plez (1998) is capable of treating correctly both spherical and plane-parallel models.

Both ATLAS and MARCS are "state of the art" 1D model atmospheres codes. In the range of F-G-K stars, the differences between the two kinds of models are immaterial, as shown e.g. in Bonifacio *et al.* (2009). For the very cool models (below 3750 K), MARCS models are probably more reliable because all the relevant molecular opacities are included. ATLAS can compute models also for A-B-O stars, although for stars hotter than 20 000 K the hypothesis of LTE clearly becomes questionable. MARCS only computes models up to 8000 K, therefore if you are dealing with A-F stars, ATLAS is preferable.

3. CO⁵BOLD models

CO⁵BOLD stands for COnservative COde for the COmputation of COmpressible COnvection in a BOx of L Dimensions with L = 2 or 3 and is developed by B. Freytag and M. Steffen with contributions of H.-G. Ludwig, W. Schaffenberger, O. Steiner, and S. Wedemeyer-Böhm (Freytag *et al.* 2002, 2011).

What it does is to solve the time-dependent equations of compressible hydrodynamics in an external gravity field coupled with the non-local frequency-dependent radiation transport. It can operate in two modes: 1) the "box-in-a-star" mode, in which the computational domain covers a small portion of the stellar surface; 2) the "star-in-a-box" mode, in which the computational domain includes the whole convective envelope of a star.

The need to go to the hydrodynamical model is obvious: all the phenomena that are linked to the motions in the stellar atmospheres, like line-shifts, line asymmetries, microvariability etc., cannot be described by 1D static models like MARCS or ATLAS.

There are cases in which the use of 1D models will provide the wrong results, thus making the use of 3D models mandatory. One case is the measurement of ^{6}Li in metalpoor warm dwarf stars. Cayrel *et al.* (2007) have shown that the line asymmetry due to convection can mimic the presence of up to 5% ^{6}Li, thus an analysis of the spectra based on 1D model atmospheres, that provide only symmetric line profiles, will result in a ^{6}Li abundance that is spurious, even if no ^{6}Li is present. In fact, this is a misinterpretation of the convective line asymmetry. Another case is the measurement of the abundance of thorium in the solar photosphere. The only line suitable for this purpose is the 401.9 nm ThII resonance line, that lies on the red wing of a strong blend of Ni and Fe. Caffau *et al.* (2008) have shown that neglect of the line asymmetry of this Ni-Fe blend would result in an over-estimate of the Th abundance by about 0.1 dex. Another effect that is very important in metal-poor stars is the so called "overcooling" (Asplund *et al.* 1999; Collet *et al.* 2007; Caffau & Ludwig 2007; González Hernández *et al.* 2008; Bonifacio *et al.* 2010). The hydrodynamical models predict much cooler temperatures in the outer layers than do the 1D models, resulting in important differences in the computed line strength for all the lines that form in these outer layers. This is the case for all the molecular species usually used for abundance determinations (Behara *et al.* 2009), but also for some atomic species (Bonifacio *et al.* 2010).

The larger amount of information provided by the 3D models comes at a larger computational cost and the need to simplify the treatment of some physical effects. The first necessary simplification is the treatment of opacity. In the computation of a hydrodynamical model, we cannot afford a treatment of opacity as detailed as we can in 1D. For this reason the opacities are grouped into a small number (less or equal to 14) of opacity bins (Nordlund 1982; Ludwig 1992; Ludwig *et al.* 1994; Voegler *et al.* 2004). Further simplifications are currently an approximate treatment of scattering in the continuum, and neglecting effects of line shifts in the evaluation of the line blocking.

3.1. *The CIFIST grid*

A hydrodynamical model is not easy to perform since it can need several months for dwarf stars, and up to more than a year for giants (on PC-like machines). Typically, one ends up with a time series covered by about 100 snapshots representing a couple of convective turn-over times. These occupy several GB of storage space. The computational and human effort to compute a hydrodynamical model is such that we cannot expect to be able to compute a model for any set of input parameters on a working time scale of a few weeks.

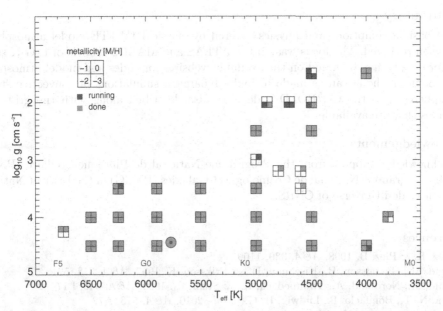

Figure 1. The current status of the CIFIST grid. This is an update of the figure shown by Ludwig *et al.* (2009) Symbols mark the location of a model in the T_{eff}-log g-plane. Green (gray) color indicates completed model runs, red (black) ongoing calculations. Each square is split into four sub-squares indicating solar, 1/10, 1/100, and 1/1000 of solar metallicity. The solar position is indicated by the round symbol. In addition to the models shown in this plot there are a few low gravity models that are being computed in Vilnius by Arunas Kučinskas and his collaborators.

Furthermore, not all snapshots are statistically independent. It would thus be not wise to attempt a "brute force" approach and compute the emerging spectrum from each and every snapshot. A preferable strategy is to select a sub-sample of the snapshots that has the same global statistical properties as the total ensemble (Caffau 2009). However, even selecting only 15 to 20 snapshots, the line formation computations for a large number of lines, as are usually employed in the analysis with 1D models, is computationally demanding.

For these reasons, in the course of the CIFIST project (Cosmological Impact of the FIrst STars, `cifist.obspm.fr`), we decided to attempt the computation of a complete grid of hydrodynamical models. The foreseen use of this grid is that observable quantities are computed on the grid, so that they can then be conveniently interpolated for any value within the grid points. An example of this is provided in Sbordone *et al.* (2010) where a fitting function is applied to the curves of growth of the Li I doublet computed on the grid. One inputs the measured equivalent width, effective temperature, surface gravity and metallicity, and the fitting function provides the lithium abundance. Another example is Gonzàlez Hernàndez *et al.* (2010), where the OH lines have been computed on the grid and the results may be used for abundance analysis.

Such ready-to-use tools are probably more useful to researchers than providing the 3D models "as is." We are thus concentrating on computing a meaningful set of observable quantities so that the CO^{5}BOLD models can be widely used for abundance analysis.

4. Summary

The basic assumptions and physics covered by classical 1D LTE model atmospheres have been reviewed. The focus was on the ATLAS and MARCS models of F-G-K stars. Further details may be found on the available websites for codes and model-atmosphere grids. Some of the advances made in hydrodynamical simulations of convective stellar atmospheres with the CO^{5}BOLD code were also described along with the grids and resources that are available.

Acknowledgements

I acknowledge support from the Programme National de Physique Stellaire (PNPS) and the Programme National de Cosmologie et Galaxies (PNCG) of the Institut National de Sciences de l'Universe of CNRS.

References

Alvarez, R. & Plez, B. 1998, *A&A*, 330, 1109
Asplund, M., Gustafsson, B., Kiselman, D., & Eriksson, K. 1997, *A&A*, 318, 521
Asplund, M., Nordlund, A., Trampedach, R., & Stein, R., 1999, *A&A*, 346, L17
Behara, N. T., Bonifacio, P., Ludwig, H.-G., *et al.*, 2010, *A&A*, 513, A72
Bonifacio, P., *et al.*, 2009, *A&A*, 501, 519
Bonifacio, P., Caffau, E., & Ludwig, H.-G. 2010, *A&A*, 524, A96
Caffau E., 2009 *Abondances des éléments dans les soleil et dans les étoiles de type F-G-K avec des modèles hydrodynamiques d'atmosphères stellaires*, thèse doctorale, Observatoire de Paris
Caffau E. & Ludwig H.-G., 2007, *A&A*, 467, L11
Caffau, E., Sbordone, L., Ludwig, H.-G., *et al.*, 2008, *A&A*, 483, 591
Castelli, F., Gratton, R. G., & Kurucz, R. L. 1997, *A&A*, 318, 841
Cayrel, R., *et al.*, 2007, *A&A*, 473, L37
Collet, R., Asplund, M., & Trampedach, R., 2007, *A&A*, 469, 687
Edvardsson, B., Andersen, J., Gustafsson, *et al.*, 1993, *A&A*, 275, 101
Freytag, B., Steffen, M., & Dorch, B. 2002, *AN*, 323, 213
Freytag, B. *et al.*, 2011, "Realistic simulations of stellar convection", *Journal of Computational Physics: special topical issue on computational plasma physics*, ed. Barry Koren

González Hernández, J., Bonifacio, P., Ludwig, H.-G., *et al.*, 2008, *A&A*, 480, 233
González Hernández, J. I., Bonifacio, P., Ludwig, H.-G., *et al.*, 2010, *A&A*, 519, A46
Gustafsson, B., Bell, R. A., Eriksson, K., & Nordlund, A. 1975, *A&A*, 42, 407
Gustafsson, B., Edvardsson, B., Eriksson, K., *et al.*, 2003, *Stellar Atmosphere Modeling*, 288, 331
Gustafsson, B., Edvardsson, B., Eriksson, K., *et al.*, 2008, *A&A*, 486, 951
Henyey, L., Vardya, M. S., & Bodenheimer, P. 1965, *ApJ*, 142, 841
Kurucz, R. L. 1970, *SAO Special Report*, 309,
Kurucz, R. 1993, *ATLAS9 Stellar Atmosphere Programs and 2 km/s grid. Kurucz CD-ROM No. 13. Cambridge, Mass.: Smithsonian Astrophysical Observatory, 1993.*, 13,
Kurucz, R. L. 2005, *Mem. SAIt Suppl.*, 8, 14
Lester, J. B. & Neilson, H. R. 2008, *A&A*, 491, 633
Ludwig, H.-G. 1992, *PhDT*, University of Kiel
Ludwig, H.-G. & Jordan, S., Steffen M. 1994, *A&A*, 284, 105
Ludwig, H.-G., Caffau, E., Steffen, *et al.*, 2009, *Mem. Soc. Astron. It.*, 80, 711
Mihalas, D. 1970, *Stellar Atmospheres*, San Francisco: Freeman, —c1970
Neilson H. 2011, this volume
Nordlund, Å. 1982, *A&A*, 107, 1
Plez, B., Brett, J. M., & Nordlund, A. 1992, *A&A*, 256, 551
Sbordone, L. 2005, *Mem. SAIt Suppl.*, 8, 61
Sbordone, L., Bonifacio, P., Castelli, F., & Kurucz, R. L. 2004, *Mem. SAIt Suppl.*, 5, 93
Sbordone, L., *et al.*, 2010, *A&A*, 522, A26
Vögler, A., Bruls, J. H. M. J., & Schüssler, M. 2004, *A&A*, 421, 741

Discussion

C. TOUT: Can you not get around the time taken to complete the full 3-D calculation by something like a two-stream model with large slowly rising cells and smaller rapidly falling cells?

P. BONIFACIO: No. This was the approach tried by R.L. Kurucz in ATLAS 11, a version of ATLAS that was never released for public use, and employed by him to assess the effect of granulation on the line formation of the lithium doublet in Pop II stars (Kurucz 1995, ApJ 452, 102). His result was that the Li abundance should be higher than that derived from an analysis with ATLAS 9 by almost one order of magnitude. This result is totally wrong. We now know that when treated in LTE the effect of granulation is to *lower* the Li abundance by 0.2 to 0.3 dex (Asplund *et al.* 1999 A&A 346 L17). The two stream model was certainly worth exploring; at the time it looked to me a brilliant idea. In retrospect, we can see why it failed: to compute each stream in a 1D approach you have to assume energy conservation, *for each stream*. This condition is clearly violated and with this approach you cannot take into account the energy exchanges between the streams.

P. HARMANEC: I enjoyed your excellent talk. Do you think that the 3D model atmosphere models will get to the level that non-experts would be able to use them for a reliable comparison with real observations?

P. BONIFACIO: Indeed we hope 3D models may become of general use, we think that the most promising approach is to have a grid of 3D model atmospheres, like the CIFIST grid, and provide to users observable quantities computed across the grid and means to interpolate in between grid points. A good example is the fitting function for the Li abundance provided by Sbordone *et al.* (2010). I stress again that in the computation of 3D model atmospheres and associated line formation we make some simplifications. Notably line opacity and scattering are treated in more detail in a 1D computation.

A. PRŠA: Your 3D models are computed for spherical stars. If one wanted to compute a spectrum of a distorted star, could one compute the intensities for each tile of the discretized surface and then sum them up?

P. BONIFACIO: The spherical geometry is dealt with in the star-in-a-box models. It looks to me more correct to treat distorsion to sphericity directly in this kind of computation, rather than tiling several box-in-a-star models. This has never been done so far, to my knowledge. I suggest you contact B. Freytag, who has been computing many star-in-a-box CO^5BOLD models and can answer your question more fully than myself. I am not a CO^5BOLD developer, I am just an end user.

From Interacting Binaries to Exoplanets: Essential Modeling Tools
Proceedings IAU Symposium No. 282, 2011
Mercedes T. Richards & Ivan Hubeny, eds.
© International Astronomical Union 2012
doi:10.1017/S1743921311027414

Basic Tools for Modeling Stellar and Planetary Atmospheres

Ivan Hubeny

Steward Observatory and Dept. of Astronomy, University of Arizona, Tucson, Arizona, USA
email: hubeny@as.arizona.edu

Abstract. Most popular computer codes for calculating model stellar and planetary atmospheres are briefly reviewed. A particular emphasis is devoted to our universal computer program TLUSTY (model stellar atmospheres and accretion disks), COOLTLUSTY (a variant of TLUSTY for computing model atmospheres of substellar-mass objects such as giant planets and brown dwarfs), and SYNSPEC (an associated spectrum synthesis code). We show the highlights of actual applications of these codes which include extensive grids of fully line-blanketed non-LTE model atmospheres of O and B stars, and grids of model atmospheres of extrasolar giant planets and L and T dwarfs.

Keywords. stars: atmospheres, planets and satellites: general, radiative transfer

1. Introduction

The term stellar (planetary) atmosphere refers to any medium connected physically to a star (planet) from which the photons escape to the surrounding space. In other words, it is a region where the radiation, observable by a distant observer, originates. Since in the vast majority of cases the radiation is the only information about a distant astronomical object that can be obtained (exceptions being a direct detection of solar wind particles, neutrinos from the Sun and SN 1987a, or gravitational waves), all the information that is gathered about stars is derived from analysis of their radiation.

Stellar atmospheres are an example of an astrophysical medium where radiation is not only a *probe* of the physical state, but is in fact an *important constituent*. In other words, radiation in fact *determines* the structure of the medium, yet the medium is probed *only* by this radiation. This leads to a mathematical complexity of the modeling procedure, because the radiation field has to be determined self-consistently with the atmospheric structure, in order to fully extract the wealth of information encoded in an observed spectrum.

The aim of this paper is to provide a brief overview of the problems and approximations in modeling atmospheres of stars and sub-stellar mass objects – brown dwarfs and extrasolar giant planets.

2. Model Atmospheres

By the term *model atmosphere* we mean a specification of all the atmospheric state parameters as functions of position. These parameters are obtained by solving appropriate structural equations. Which equations are to be solved, and what form they attain, depends sensitively on the adopted assumptions and approximations.

2.1. *Approximations*

In order to make the overall problem tractable, one has to make a number of simplifications by invoking various approximations. The quality of an appropriate model, and consequently its applicability to the individual stellar types, is closely related to the degree of approximation used in the construction of the model. Needless to say, the degree of approximation critically influences the amount of computational effort to compute it. It is fair to say that the very art of computing model stellar atmospheres is to find such physical approximations that allow the model to be computed with a reasonable amount of numerical work, yet the model is sufficiently realistic to allow its use for a reliable interpretation of observed stellar data. The adopted approximations are therefore critical. There are several types of approximations that are typically made in the model construction; we shall describe the most important types in turn.

Approximations of the geometry

By the geometrical simplification we mean that either some prescribed geometrical configuration is assumed, or some special kind of overall symmetry is invoked. The goal of those simplifications is to reduce the dimensionality of the problem from a spatially 3-dimensional problem to a 1-D or 2-D problem. The most popular approximations are, from simplest to more complex:

– Plane-parallel geometry, with an assumption of horizontally homogeneous layers. This decreases the number of dimensions to one: the depth in the atmosphere. This approximation is typically quite reasonable for stellar photospheres, which indeed are by several orders of magnitude thinner than the stellar radius, so the curvature effects are negligible. In the presence of horizontal inhomogeneities (such as stellar spots, accretion belts, etc.), 1-D models still have their value since in many cases one may construct different 1-D models for the individual "patches" on the surface.

– Spherical symmetry. Again, the problem is one-dimensional. The approach is used for extended atmospheres, for which the atmospheric thickness in no longer negligible with respect to the stellar radius. These models are appropriate for giants and supergiants.

– Multi-dimensional geometry. This field is enjoying a period of rapid development. We will not discuss it here (see contributions by Bonifacio, this volume; Allard, this volume).

Approximations of the dynamical state of the atmosphere

This is basically a specification of the realism of the treatment of the macroscopic velocity fields. From the simplest to the most complex the approaches are the following:

– Static models, in which the macroscopic velocity field is set to zero. These models describe a *stellar photosphere.*

– Models with an *a priori* given velocity field. In these models the velocities are taken into account explicitly, and their influence upon other state parameters, in particular the emergent radiation, is studied in detail. The most successful computer programs for computing such models are CMFGEN (Hillier & Miller 1998; Hillier, this volume), PHOENIX (Hauschildt *et al.* 1997; Allard, this volume), Munich codes (FASTWIND – Santolaya-Rey *et al.* 1997; WM-BASIC – Pauldrach *et al.* 2001), and the Kiel-Potsdam code (Hamann 1985; Koesterke *et al.* 2002).

– Models where the velocity field is determined self-consistently by solving the appropriate hydrodynamical equations. This problem is very complicated because the wind driving force is given by the absorption of photons in thousands to millions of metal lines so the hydrodynamical equations should be solved together with at least an approximate treatment of radiative transfer in spectral lines. Such a fully self-consistent model is yet to be constructed.

Approximations of the opacity sources

In real stellar atmospheres, there is an enormous number of possible opacity sources. The light elements (H, He, C, N, O) have a comparatively small number of lines per ion (say 10^2 to 10^4) because of a relatively simple atomic level structure. The number of lines generally increases with increasing atomic number, and for the iron-peak elements (Fe and Ni being the most important ones), we have on the order of 10^6 to 10^7 spectral lines per ion. Models that deal with such a large number of lines are called *metal line-blanketed models*. There are two basic approaches to compute them, using i) *Opacity Distribution Functions* (ODF), routinely used in LTE, and generalized to NLTE by Anderson (1989) and Hubeny & Lanz 1995; or ii) *Opacity Sampling* (OS), which is a simple Monte Carlo-like sampling of frequency points (Anderson 1989; Dreizler & Werner 1993). In fact, an "exact" method is essentially a variant of the OS with a sufficiently high resolution. An explicit comparison between results using the ODF and the OS approaches, and with various frequency resolutions in the latter, is presented e.g. in Lanz & Hubeny (2003).

Approximations concerning the thermodynamic equilibria

Here, the issue is whether the approximation of LTE is adopted or not. In practice, LTE models may be useful only for stellar photospheres, because for extended atmospheres and/or stellar winds this approximation breaks down completely and its application would yield erroneous and misleading results.

The models that take some kind of departure from LTE into account are called non-LTE (or NLTE) models. This term is rather ambiguous because it is not a priori clear what is actually allowed to depart from LTE in a given model. In early models, the populations of only a few low-lying energy levels of the most abundant species, like H and He, were allowed to depart from LTE; the rest were treated in LTE. During the development of the field, progressively more and more levels were allowed to depart from LTE.

A commonly accepted rule of thumb is that for solar-type and cooler atmospheres, LTE provides an acceptable approximation, while for hotter stars, and particularly for, A, B, O stars, NLTE effects are important. Such a division is, however, misleading and even dangerous, because in essentially any kind of atmosphere there are geometric and spectral regions (outer layers; cores of strong lines), where the NLTE effects are important or even crucial.

3. Available modeling codes

As follows from the above, computing atmospheric structure is rather complicated. However, from the point of view of spectroscopic diagnostics of observed objects, the most important quantity is not so much a detailed structure, but rather a predicted emergent radiation. Therefore, a computed structure may be, to some extent, approximated, while the emergent radiation needs to be known as accurately as possible.

This suggests a two-step strategy. First, one computes an atmospheric structure, which may (and, in fact, has to) be computed using various approximations. Second, taking the computed structure, one calculates the emergent radiation in detail. (For instance, for normal main-sequence stars – not chemically peculiar stars – Cr and Mn do not have to be taken into account in detail or can be disregarded when computing a global atmospheric structure, while they can be taken into account when computing the emergent spectrum.) The codes that perform the first step are called the *model atmosphere* codes, and those performing the second step are called *spectrum synthesis* codes. We will briefly describe them in turn.

3.1. *Model atmosphere codes*

There are several publicly available codes for computing model stellar atmospheres. The following list is by no means exhaustive.

For LTE models, the most popular code is ATLAS (Kurucz 1970; 1993), and MARCS (Gustafsson *et al.* 1975). It should be noted that most of the NLTE codes listed below can be used to calculate LTE models as well. Appropriately modified versions of PHOENIX and TLUSTY were actually used for generating grids of LTE models – see Sect. 4.1.

In the context of NLTE static models, the first publicly available NLTE model atmosphere code was the "NCAR code" (Mihalas, Heasley, & Auer, 1975). More recently, popular and widely used codes are TMAP – Tuebingen Model Atmosphere Package (Werner 1986, 1989; Dreizler & Werner 1993; Werner *et al.* 2003), and TLUSTY (Hubeny 1988; Hubeny & Lanz 1992, 1995; Hubeny, Hummer, & Lanz 1994). Static models are also being constructed by codes originally designed for expanding atmospheres (by setting the expansion velocity to a very low value), such as CMFGEN (Hillier & Miller 1998) or PHOENIX (Hauschildt, Baron, & Allard 1997; Hauschildt *et al.* 1999a).

3.2. TLUSTY

The model atmosphere computer program TLUSTY has been described in several papers: Hubeny (1988) – the original version based on Complete Linearization (CL; originally developed by Auer & Mihalas 1969); Hubeny & Lanz (1992) – implementation of Ng and Kantorovich accelerations; Hubeny, Hummer & Lanz (1994) – treatment of level dissolution and occupation probabilities; Hubeny & Lanz (1995) – hybrid complete-linearization/accelerated Lambda Iteration (CL/ALI) method, concept of superlevels and superlines; and Lanz & Hubeny (2003) – opacity sampling method.

The program solves the basic equations (radiative transfer, hydrostatic equilibrium, radiative equilibrium, statistical equilibrium, charge and particle conservation). The previously separate variant called TLUSDISK was combined into one universal TLUSTY, which thus allows one to compute either a model stellar atmosphere, or the vertical structure of a given annulus in an accretion disk. Accretion disk models are described in detail by Hubeny & Hubeny (1998), and Hubeny *et al.* (2001). Recent upgrades contain an improved treatment of convection (with several variants of the mixing-length formalism); external irradiation; Compton scattering (described in Hubeny *et al.* 2001); dielectronic recombination; and X-ray opacities, including the inner-shell (Auger) ionization (described in Hubeny *et al.* 2001). The program is fully data-oriented as far as the choice of atomic species, ions, energy levels, transitions, and opacity sources is concerned, with no default opacities built in. Both options, ODF and Opacity Sampling (OS), are offered for a treatment of metal line blanketing, but with increasing computer power the Opacity Sampling option becomes largely preferable.

Recently, we have developed a variant of TLUSTY called COOLTLUSTY (described briefly in Hubeny, Burrows, & Sudarsky 2003; Burrows, Sudarsky, & Hubeny 2006) designed to compute model atmospheres of sub-stellar mass objects – brown dwarfs and giant planets. The code uses pre-calculated opacity and state-equation tables that include effects of extensive molecular opacities. Other significant upgrades with respect to TLUSTY are including the cloud formation and associated cloud opacity and scattering, and departures from chemical equilibrium (Hubeny & Burrows 2007).

3.3. *Spectrum synthesis codes*

The spectrum synthesis codes take a previously computed or stored atmospheric structure and solve, frequency-by-frequency, the radiative transfer equation, with a sufficiently high resolution in the frequency space to provide a reliable predicted spectrum to be compared with observations. From the physical point of view, such a problem is much simpler than the model atmosphere construction. From the computational point of view, the only problem is connected to a necessity of dealing with extended line lists containing on the order of 10^7 - 10^8 spectral line data; in the case of cool atmospheres where molecules are present, the line lists are even more extended (e.g. a line list for water lines contain over 10^9 lines).

Here, we only briefly mention two codes. The first is Kurucz's SYNTHE, specfically designed to provide detailed emergent spectra for Kurucz model atmospheres.

A spectrum synthesis code for providing detailed synthetic spectra for TLUSTY models is called SYNSPEC. The code is also able to provide synthetic spectra for Kurucz and TMAP models, and with a slight modification of input routines, for any other models. The original reference is Hubeny, Štefl, & Harmanec (1985); these days it is available on the web either at the TLUSTY website, `http://nova.astro.umd.edu`, or, its latest versions, on `http://aegis.as.arizona.edu/~hubeny/pub/synspec49.tar.gz`.

There is also a useful, IDL-based graphical package called SYNPLOT, which is a wrapper around SYNSPEC to enable an easy calculation of synthetic spectra based on setting appropriate keywords without the necessity of preparing the corresponding input files. It also provides a graphical user interface for a synthetic spectrum and its fitting to the observed spectrum. When used in conjunction with the OSTAR2002 and BSTAR2006 grids (see below), SYNPLOT allows for an easy evaluation of synthetic spectra for any values of basic stellar parameters, not just the gridpoint values. The program is available on `http://aegis.as.arizona.edu/~hubeny/pub/synplot.tar.gz`.

4. Existing model atmosphere grids

4.1. *LTE models*

The most extensive grid of LTE plane-parallel line-blanketed models is that of Kurucz (1979; 1993), which is widely used by the astronomical community. The grid covers effective temperatures between 3500 K and 50,000 K, $\log g$ between -1 and 5, and for several metallicities.

Using the MARCS code, Gustafsson *et al.* (1975) generated their original grid of models for cool stars, with T_{eff} between 3750 and 6000 K, $\log g$ between 0.75 and 3.0, and metallicities $-3.0 \leqslant [M/H] \leqslant 0$. Recently, Gustafsson *et al.* (2008) made public a new, very extensive grid of MARCS model atmospheres, with T_{eff} between 2500 and 8000 K, $\log g$ between -1 and 5, and metallicities $-5 \leqslant [M/H] \leqslant 1$. They also include "CN-cycled" models with C/N = 4.07 (solar), 1.5, and 0.5; and C/O from 0.09 to 5, which represents stars of spectral types R, S, and N.

Hauschildt *et al.* (1999a, b) used their code PHOENIX to generate a grid of LTE spherical models for cool stars, called NextGen, with T_{eff} between 3000 and 10000 K, with step 200 K; $\log g$ between 3.5 and 5.5, with step 0.5; and metallicities $-4.0 \leqslant [M/H] \leqslant 0$. Another grid (Allard *et al.* 2000) is for pre-main-sequence cool stars with T_{eff} between 2000 and 6800 K, $\log g$ between 2 and 3.5, with step 0.5; stellar mass $M = 0.1 M_{\odot}$; and metallicities $-4.0 \leqslant [M/H] \leqslant 0$. The models are available on-line at `www.hs.uni-hamburg.de/EN/For/ThA/phoenix/index.html`. A detailed comparison between the ATLAS and NextGen models was performed by Bertone *et al.* (2004).

4.2. *NLTE models*

There are several partial grids of NLTE models for various stellar types, mostly of hot stars. The models constructed by the TMAP code for very hot white dwarfs, sub-dwarfs, and pre-white dwarfs (also known as the PG 1159 stars) are available on-line at `http://astro-uni-tuebingen.de/~rauch/TMAP/TMAP.html`. Rauch & Werner (2009) describe the so-called *Virtual Observatory*, which is a web-based interface that enables a user either to extract already computed models, or generate specific models using TMAP, for very hot objects (hottest white dwarfs; super-soft X-ray sources).

Our effort culminated in the construction of a grid of NLTE fully-blanketed model atmospheres for O stars (OSTAR2002; Lanz & Hubeny 2003) and early B stars (BSTAR2006; Lanz & Hubeny 2007). We believe that these grids, which each took several years of computer time on several top-level workstations, represent a more or less definitive grids of models in the context of 1-D plane-parallel geometry, with hydrostatic and radiative equilibrium, and without any unnecessary numerical approximations.

The basic characteristics are as follows: The OSTAR2002 grid contains 680 individual model atmospheres for 12 values of $T_{\rm eff}$ between 27,500 and 55,000 K, with a step of 2,500 K, and 8 values of $\log g$, and for 10 metallicities: 2, 1, 1/2, 1/5, 1/10, 1/30, 1/50, 1/100, 1/1000, and 0 times the solar metal composition. The following species are treated in NLTE: H, He, C, N, O, Ne, Si, P, S, Fe, Ni, in all important stages of ionization; which means that there are altogether over 1000 (super)levels to be treated in NLTE, about 10^7 lines, and about 250,000 frequency points to describe the spectrum.

The BSTAR2006 grid is similar. It contains 1540 individual models for 16 values of $T_{\rm eff}$ between 15,000 and 30,000 K, with a step of 1000 K, and for 6 metallicities: 2, 1, 1/2, 1/5, 1/10, and 0 times solar. The species treated in NLTE are the same as in OSTAR2002, adding Mg and Al, but removing Ni. There are altogether about 1450

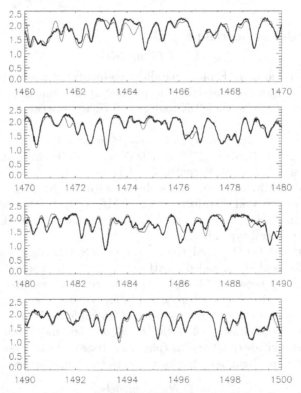

Figure 1. A comparison of the observed HST/GHRS flux for 10 Lac (heavy line) and the predicted flux from the fully blanketed NLTE model atmosphere with $T_{\rm eff} = 33,500$ K, $\log g = 3.85$, and for the solar abundances of all species (thin line). The abscissa is the wavelength in Å, and the ordinate is the flux in 10^{-9} erg cm^{-2} s^{-1} Å^{-1}. Most spectral features are lines of Fe IV, Fe V, Ni IV, and Ni V. A difference between theory and predictions is hardly seen on the plots.

(super)levels treated in NLTE, about 10^7 lines, and about 400,000 frequency points. The models for both grids are available on-line at http://nova.astro.umd.edu.

As an example, we present in Fig. 1 a sample of the predicted flux for a model for $T_{\mathrm{eff}} = 33,500$ K, log $g = 3.85$, and a high-resolution, high signal-to-noise observation of a late-O main-sequence star 10 Lac secured by the *Goddard High Resolution Spectrograph* (GHRS) aboard the *Hubble Space Telescope* (Lanz *et al.*, in prep). The agreement between observations and predictions is excellent, and demonstrates the power of the present-day model atmospheres of early-type stars.

5. Atmospheres of Substellar-Mass Objects

The two most interesting and important classes of the substellar-mass objects (SMO) are L and T dwarfs (sometimes referred to as brown dwarfs), and extrasolar giant planets (EGP). We stress that from the point of view of atmospheric modeling, these two types of objects are identical.

Comparing the effort needed to model the atmospheres of SMOs and classical stars, there are advantages and disadvantages. On the plus side, the atmosphere of the substellar mass objects can be treated in LTE (because these are rather cool and dense), and the radiation pressure is negligible. On the minus side, there are several serious complications: i) Molecular opacities. ii) Condensation of certain compounds and a formation of clouds. This involves several related problems, such as the distribution of cloud particle sizes, the position and extent of the cloud deck, cloud opacity and scattering cross-sections, and a strong anisotropy of cloud (Mie) scattering. iii) Complicated state equation that needs to solve a complex chemical network, even in a relatively simple case of chemical equilibrium. iv) There may be departures from chemical equilibrium, both because of photochemistry, and because of advection of certain species resulting from a long chemical reaction time compared to an atmospheric mixing time (e.g., Hubeny & Burrows 2007); and v) strong irradiation, particularly in the case of close-in EGPs.

There are several computer programs that are being currently used to produce models of SMO atmospheres; the following three are used most actively: i) a code used by Fortney and coworkers (e.g., Fortney *et al.* 2005); ii) the PHOENIX-based code (e.g., Barman *et al.* 2001); and our code COOLTLUSTY, which is a variant of TLUSTY. It is described briefly in Hubeny, Burrows, & Sudarsky 2003; and Sudarsky, Burrow, & Hubeny 2003. It solves equations of radiative+convective equilibrium, state equation, (hydrostatic equilibrium is satisfied automatically), self-consistently with the radiative transfer equation. An upgrade to treat departures from chemical equilibrium is described in Hubeny & Burrows (2007). The code has been used to generate extensive grids of EGP model atmospheres (Sudarsky, Burrows, & Hubeny 2003), L and T dwarfs both with chemical equilibrium (Burrows, Sudarsky & Hubeny 2006), and with departures from equilibrium (Hubeny & Burrows 2007).

References

Allard, F. Hauschildt, P. H., & Schweitzer 2000, *ApJ*, 539, 366

Anderson, L. S. 1989, *ApJ*, 339, 588

Auer, L. H. & Mihalas, D. 1969, *ApJ*, 158, 641

Barman, T. S., Hauschildt, P. H., & Allard, F. 2001, *ApJ*, 556, 885

Bertone, E., Buzzoni, A., Chavez, M., & Rodriguez-Merino, L. H. 2004, *AJ*, 128, 829

Burrows, A., Sudarsky, D., & Hubeny, I. 2006, *ApJ*, 640, 1063

Dreizler, S. & Werner, K. 1993, *A&A*, 278, 199

Gustafsson, R., Bell, R. A., Eriksson, K., & Nordlund, Å. 1975, *A&A*, 42, 407

Gustafsson, R., Edvardsson, B., Eriksson, K., Jørgensen, U. G., Nordlund, Å., & Plez, B. 2008, *A&A*, 486, 951

Fortney, J. J., Marley, M. S., Lodders, K., Saumon, D., & Freedman, R. 2005, *ApJ* (Letters), 627, L69

Hamann, W.-R. 1985, *A&A*, 148, 364

Hauschildt, P. H., Baron, E., & Allard, F. 1997, *ApJ*, 483, 390

Hauschildt, P. H., Allard, F., & Baron, E. 1999a, *ApJ*, 512, 377

Hauschildt, P. H., Allard, F., Ferguson, J., & Baron, E. 1999b, *ApJ*, 525, 871

Hillier, D. J. & Miller, D. L. 1998, *ApJ*, 496, 407

Hubeny, I. 1988, *Comp. Phys. Commun.*, 52, 103

Hubeny, I., Blaes, O., Krolik, J. H., & Agol, E. 2001, *ApJ*, 559, 680

Hubeny, I. & Burrows, A. 2007, *ApJ*, 669, 1248

Hubeny, I. Burrows, A., & Sudarsky, D. 2003 *ApJ*, 594, 1011

Hubeny, I. & Hubeny, V. 1998, *ApJ*, 505, 558

Hubeny, I., Hummer, D. G., & Lanz, T. 1994, *A&A*, 282, 151

Hubeny, I. & Lanz, T. 1992, *A&A*, 262, 501

Hubeny, I. & Lanz, T. 1995, *ApJ*, 439, 875

Hubeny, I., Štefl, S., & Harmanec, P. 1985, *Bull. Astron. Inst. Czechosl.*, 36, 214

Koesterke, L., Hamann, W.-R., & Graefener, G. 2002, *A&A*, 384, 562

Kurucz, R. L. 1970, *Smithsonian Astrophys. Obs. Spec. Rep.* No. 309

Kurucz, R. L. 1979, *ApJS*, 40, 1

Kurucz, R. L. 1993, *ATLAS9 Stellar Atmosphere Programs and 2 km/s Grid*, Kurucz CD-ROM 13 (Cambridge, Mass: SAO)

Lanz, T. & Hubeny, I. 2003, *ApJS*, 146, 417

Lanz, T. & Hubeny, I. 2007, *ApJS*, 169, 83

Mihalas, D. 1978, *Stellar Atmospheres*, (2nd ed., Freeman, San Francisco)

Mihalas D., Heasley J. N., & Auer L. H. 1975, *A Non-LTE model stellar atmospheres computer program*, NCAR-TN/STR-104, (NCAR, Boulder)

Pauldrach, A. W. A., Hoffmann, T. L., & Lennon, M. 2001, *A&A*, 375, 161

Rauch, T. & Werner, K. 2009, in: I. Hubeny *et al.*, (eds.) *Recent Directions in Astrophysical Quantitative Spectroscopy and Radiation Hydrodynamics*, AIP Conf. Proc, 1171, p. 85

Santolaya-Rey, A. E., Puls, J., & Herrero, A. 1997, *A&A*, 323, 488

Sharp, C. S. & Burrows, A. 2007, *ApJS*, 168, 140

Sudarsky, D., Burrows, A., & Hubeny, I. 2003, *ApJ*, 588, 1121

Werner, K. 1986, *A&A*, 161, 177

Werner, K., Deetjen, J. L., Dreizler, S., Nagel, T., Rauch, T., & Schuh, S. L. 2003, in: I. Hubeny, D. Mihalas, & K. Werner (eds.), *Stellar Atmosphere Modeling*, ASP Conf. Ser. vol. 288, (ASP, San Francisco 2003), p. 31

Discussion

K. ALLERS: Are the model grids (particularly the COOLTLUSTY models) available on-line?

I. HUBENY: Yes, they are available on Adam Burrows' website at http://www.astro.princeton.edu/~burrows/.

From Interacting Binaries to Exoplanets: Essential Modeling Tools
Proceedings IAU Symposium No. 282, 2011
Mercedes T. Richards & Ivan Hubeny, eds
© International Astronomical Union 2012
doi:10.1017/S1743921311027426

Hot Stars with Winds: The CMFGEN Code

D. John Hillier

University of Pittsburgh, Department of Physics and Astronomy,
100 Allen Hall, Pittsburgh, PA 15260, USA
email: hillier@pitt.edu

Abstract. CMFGEN is an atmosphere code developed to model the spectra of a variety of objects – O stars, Wolf-Rayet stars, luminous blue variables, A and B stars, central stars of planetary nebula, and supernovae. The principal computational aim of CMFGEN is to determine the temperature and ionization structure of the atmosphere, and the atomic level populations. Toward this end, we have developed several different radiative transfer modules that (a) solve the transfer equation for spherical geometry in the comoving frame, (b) solve the static transfer equation in the plane parallel approximation without, or with, a vertical velocity field, (c) solve the static transfer code for a spherical atmosphere allowing for all relativistic terms, (d) solve the time-dependent spherical transfer equation to first order in v/c for a homologous expansion, and (e) solve the time-dependent spherical transfer equation allowing for all relativistic terms. To achieve consistency between the radiation field, temperature structure, and level populations we use a linearization technique. Line blanketing is accurately treated while complex photoionization cross-sections, containing numerous resonances, can also be handled. Spectra, for comparison with observation, are computed using CMF_FLUX. Several other auxiliary programs have also been developed – these include diagnostic tools as well as programs that allow the effect on spectra of rotation and departures from spherical geometry to be investigated. In this presentation we will briefly describe CMFGEN and auxiliary codes.

Keywords. stars: atmospheres, emission line, line: formation

---

1. Introduction

CMFGEN is a model atmosphere code originally developed to study the spectra of massive stars, although it can also be used to study a variety of other objects. A common feature of massive stars is that they exhibit stellar winds. These stellar winds are driven by radiation pressure (acting through bound-bound transitions in the UV and EUV). Because of the stellar winds, the atmospheres are not plane-parallel – the simplest geometry we can adopt is spherical geometry. Further, because of the velocity field associated with the wind, it is more convenient to solve for the radiation field in a frame moving with the gas – the comoving frame. In this frame, the opacities and emissivities can be assumed to be isotropic. Also, because of the intense radiation fields, and low photosphere and wind densities, we cannot assume Local Thermodynamic Equilibrium (LTE) to evaluate atomic and ion level populations. Rather, the atmospheres are said to be in non-LTE, and we have to solve the statistical equilibrium equations (which describe rates into, and out of, each atomic/ion level). Because of the strong coupling between the radiation field and the level populations, simple iterative techniques cannot be used.

2. CMFGEN

The basic purpose of CMFGEN is to determine the star's atmospheric structure: the temperature and ionization structure, and the atomic level populations for all atoms/ions.

CMFGEN simultaneously solves the radiative transfer equation (RTE) for spherical geometry in the comoving frame in conjunction with the statistical equilibrium equations (SEEs) and radiative equilibrium equation (REE) (Hillier 1990; Hillier & Miller 1998). Since the equations are non-linear and coupled – the radiation field depends on the level populations which in turn depend on the radiation field – an iterative technique must be used. In CMFGEN we linearize the RTE which allows us to solve for the δJ_ν (the linearized mean intensity) as a function of the (unknown) corrections to the opacities and emissivities, and hence as a function of the unknown correction (δn) to the populations (and temperature and electron density). We then eliminate the δJ_ν from the linearized SEEs and REE to obtain a set of simultaneous equations in the δn. In eliminating the δJ_ν, we only retain the influence of the local populations (diagonal operator), or that of populations at the local and neighboring depths (tridiagonal operator). With the diagonal operator, we have a set of N_V simultaneous equations at each depth, where N_V is the total number of unknown populations (for all species) at each depth. For the tridiagonal operator, we have a block-tridiagonal system (with each block $N_V \times N_V$) which is solved efficiently using the Thomas algorithm. For reasons of stability, we rescale the simultaneous equations so that we solve for $\delta n/n$ (Hillier 2003). The technique yields similar convergence properties to approximate lambda operators (see Hillier 1990).

At present, the mass-loss rate and the velocity law above the sonic point must be specified when modeling stellar winds. The structure below the sonic point is obtained by solving the equation of hydrostatic equilibrium.

As CMFGEN has evolved we have developed several different radiative transfer modules:

(a) A module to solve the transfer equation for spherical geometry in the comoving frame: this is the mode used when studying massive stars and their stellar winds.

(b) A module to solve the static transfer equation in the plane parallel approximation.

(c) A module to solve the transfer equation in the plane parallel approximation with a vertical velocity field.

(d) A module to solve the static transfer code for a spherical atmosphere allowing for all relativistic terms.

(e) A module to solve the time-dependent spherical transfer equation to first order in v/c for a homologous expansion.

(f) A module to solve the time-dependent spherical transfer equation for all orders in v/c (undertaken by a graduate student, Chendong Li).

The latter three modules were developed to facilitate the analysis of supernova (SN) spectra.

3. Applications of CMFGEN

CMFGEN has been used to determine fundamental stellar parameters, and investigate the stellar winds of Wolf-Rayet (W-R) stars (e.g., Herald *et al.* 2001), Luminous Blue Variables (LBVs) (e.g., Groh *et al.* 2009), CSPNs (e.g., Marcolino *et al.* 2007), O Stars (e.g., Martins *et al.* 2002; Hillier *et al.* 2003; Bouret *et al.* 2003) and A and B stars. It is also being used to study Type I and Type II SN ejecta (e.g., Dessart & Hillier 2010; Dessart *et al.* 2011; Dessart & Hillier 2011). CMFGEN has many uses:

(a) To provide accurate abundances and stellar parameters for comparison with evolution calculations.

(b) To provide EUV radiation fields for input to nebula photoionization calculations.

(c) To provide "limb darkening laws" for interferometry (see, e.g., Figure 1).

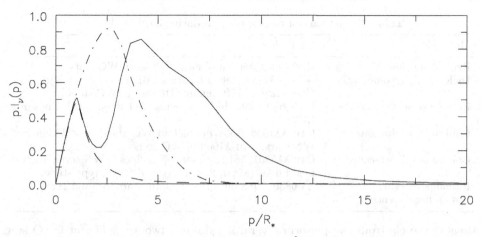

Figure 1. Illustration of $pI_\nu(p)$ at 3 wavelengths for a γ^2 Velorum (WC8)-like model with an effective temperature of $\sim$52,500 K. Shown is $pI_\nu(p)$ at line center for C III λ5696 (solid line), the continuum (λ5750; dashed) and near line center for the C IV $\lambda\lambda$5801, 5812 (dash-dot) doublet. Notice how the line formation regions are more extended than the line emitting region. The dip in $pI_\nu(p)$ for C III λ5696 is a consequence of the ionization structure.

(*d*) To provide data for the study of starbursts, star formation in galaxies, etc.

(*e*) To provide a better understanding of the hydrodynamics of winds.

(*f*) To assist in the development and testing of approximate methods that can be used in more complex geometries and inhomogeneous media.

(*g*) To derive distances to Type II SNe using the expanding-photosphere method (EPM), and variants thereof.

(*h*) To allow SN spectra to be used as an evolutionary and explosion diagnostic tool.

4. Line Formation in Winds

While CMFGEN can be run without understanding all of the complex physics involved, a rudimentary knowledge of the basic mechanisms of continuum and line formation in stellar winds is useful if one is to interpret model results, and to understand the sensitivities of various lines to different model parameters. Many different processes contribute to the formation of emission lines in stellar winds, and some of these, with examples, are summarized in Table 1. In some cases more than one mechanism may contribute, while in other cases (as you go closer to LTE, for example) it may not be possible to determine a specific mechanism. Further information on these processes, with references, can be found in Hillier (2011).

5. CMF_FLUX

For the computation of the observed spectrum, we use the comoving frame/observer's frame code CMF_FLUX (Busche & Hillier 2005). This code was designed to allow the accurate computation of observed spectra with full allowance for different types of broadening mechanisms (e.g., Doppler, Stark, Voigt). To facilitate spectral interpretation, we can compute spectra where we include only selected species, or alternatively, omit selected species (e.g., Fe II). This is particularly useful for examining the direct influence of lines from a given species on the observed spectrum.

The code has two parts. First, as in CMFGEN, we solve the radiative transfer equation in the comoving frame. To treat incoherent electron scattering, arising from the thermal

Table 1. Summary of mechanisms producing emission lines.

Mechanism	Example
Recombination	H, He emission lines; most C lines in WC stars
Dielectronic recombination	N III λ4640 triplet in Of stars (Mihalas *et al.* 1972); C III λ2296, λ6740; many C II lines in WC stars
Resonance scattering	P Cygni profiles in UV spectra of O stars; Na I D lines in LBVs
Collisional excitation	C IV $\lambda\lambda$1548, 1551 doublet in WC stars; N IV λ1486 in WN stars; C III λ1909 in WC stars
Continuum fluorescence	C IV $\lambda\lambda$5801, 5812 and some N IV lines in WN stars; many metal lines (Mg II, Si II, Fe II) in P Cygni type stars
Raman scattering Bowen fluorescence	$\lambda\lambda$ 6830, 7088 lines in symbiotic stars (Schmid 1989)

motions of the electrons, we perform several iterations – two are sufficient for O stars. This calculation provides emissivities, opacities, and mean intensities for the observer's frame calculation.

Second, an observer's frame calculation is performed using the usual (p, z) co-ordinate system. Along each ray, we choose a fine spatial grid such that $\Delta V < 0.25 V_{\mathrm{Dop}}$ to allow adequate sampling of the rapidly varying opacities and emissivities. On this revised spatial grid, we map the comoving-frame opacities and emissivities — this requires interpolation in both space and in frequency. Along each ray, we perform a formal integration to compute $I_\nu(p)$, and from the $I_\nu(p)$ we can compute the observed flux using standard numerical quadrature. Optionally, the $I_\nu(p)$ (effectively giving the limb darkening law), can also be output.

6. Diagnostic Tools/Files

CMFGEN is a complicated code, with many different capabilities and control options. While care has been taken developing the code, it uses many different numerical techniques and approximations. These techniques and approximations may only work in a limited parameter range, and it is important that users perform checks to ensure code accuracy when moving to different parameter ranges. CMFGEN, like all complicated codes, should not be used as a black box.

To facilitate checking and interpretation of results, we have developed two display packages, DISPGEN and PLT_SPEC. The first, DISPGEN, is designed to investigate the structure of the model atmosphere. It allows populations and non-LTE departure coefficients, the ionization structure, temperature structure, line and continuum opacities, and optical depth scales, for example, to be examined. PLT_SPEC is used for comparing model spectra with observation.

In addition to the display packages, many diagnostic files are created. For example, for every ionization stage we create a file which lists photoionization rates and recombination rates for each level, and the recombination coefficient, as a function of depth. Another file contains a check for the electron energy balance. As noted by Hillier (2003), the assumptions of radiative equilibrium and electron energy balance are related by the SEEs. In the presence of super levels they are not equivalent, and an examination of the errors in the electron energy balance provides a check on both program coding, and inaccuracies introduced by the use of super levels.

Another file, OBSFLUX, provides the observed spectrum, and the luminosity as a function of depth. The latter also provides a check on code consistency. In stellar models, the luminosity is specified only at the lower boundary — its constancy with depth follows only when radiative equilibrium is satisfied at each depth.

STEQ_VALS lists the current error in the SEEs and the fractional corrections $(\delta n/n)$ for each population at each depth. The latter is particularly useful for investigating convergence issues.

7. Obtaining the code

CMFGEN is available at **http://kookaburra.phyast.pitt.edu/hillier**. Register to receive emails with updated information about CMFGEN. Also available, at the same website, is documentation, and a small grid of models that provides test cases and starting models for use with CMFGEN.

7.1. *Technical Support*

CMFGEN is a complex code, and it takes time to understand all its features and nuances. New users should first read the supplied manual to at least familiarize themselves with some of the basic philosophies behind the code. When problems occur, check the input files for errors, check OUTGEN for diagnostic messages, and the manual for possible solutions. If you are unable to solve the problem, you can email me for assistance. If needed, you can also visit me in Pittsburgh. Base models for O stars and for some W-R stars are on my website. There are numerous CMFGEN models around, so ask me, or other CMFGEN users for access to these models.

I appreciate feedback. If you find an error, it is *absolutely essential* that you inform me so that other users are not also affected. Suggestions for improvements are also welcome.

Acknowledgements

DJH gratefully acknowledges support from STScI theory grant HST-AR-11756.01.A and NASA theory grant NNX10AC80G.

References

Bouret, J.-C., Lanz, T., Hillier, D. J., Heap, S. R., Hubeny, I., Lennon, D. J., Smith, L. J., & Evans, C. J. 2003, *ApJ*, 595, 1182
Busche, J. R. & Hillier, D. J. 2005, *AJ*, 129, 454
Dessart, L. & Hillier, D. J. 2010, *MNRAS*, 405, 2141
—. 2011, *MNRAS*, 410, 1739
Dessart, L., Hillier, D. J., Livne, E., Yoon, S.-C., Woosley, S., Waldman, R., & Langer, N. 2011, *MNRAS*, 414, 2985
Donati, J.-F., Babel, J., Harries, T. J., Howarth, I. D., Petit, P., & Semel, M. 2002, *MNRAS*, 333, 55
Georgiev, L. N., Hillier, D. J., & Zsargó, J. 2006, *A&A*, 458, 597
Groh, J. H., Hillier, D. J., Damineli, A., Whitelock, P. A., Marang, F., & Rossi, C. 2009, *ApJ*, 698, 1698
Herald, J. E., Hillier, D. J., & Schulte-Ladbeck, R. E. 2001, *ApJ*, 548, 932
Hillier, D. J. 1990, *A&A*, 231, 116
Hillier, D. J. 2003, in Astronomical Society of the Pacific Conference Series, Vol. 288, Stellar Atmosphere Modeling, ed. I. Hubeny, D. Mihalas, & K. Werner, p. 199
—. 2011, *Ap&SS*, (in press)
Hillier, D. J., Lanz, T., Heap, S. R., Hubeny, I., Smith, L. J., Evans, C. J., Lennon, D. J., & Bouret, J.-C. 2003, *ApJ*, 588, 1039
Hillier, D. J. & Miller, D. L. 1998, *ApJ*, 496, 407
Kurosawa, R., Hillier, D. J., & Pittard, J. M. 2002, *A&A*, 388, 957

Luehrs, S. 1997, *PASP*, 109, 504

Marcolino, W. L. F., Hillier, D. J., de Araujo, F. X., & Pereira, C. B. 2007, *ApJ*, 654, 1068

Martins, F., Schaerer, D., & Hillier, D. J. 2002, *A&A*, 382, 999

Mihalas, D., Hummer, D. G., & Conti, P. S. 1972, *ApJL*, 175, L99+

Schmid, H. M. 1989, *A&A*, 211, L31

Wade, G. A., Fullerton, A. W., Donati, J.-F., Landstreet, J. D., Petit, P., & Strasser, S. 2006, *A&A*, 451, 195

Discussion

O. DE MARCO: Stellar winds are disturbed by companion stars. Can wind models model the effects of this disturbance on the line profiles? Have such efforts already been undertaken?

J. HILLIER: An obvious manifestation of binarity in O stars is wind-wind collisions which produce hard X-rays. Phase-locked profile variability is also seen, especially in O star and W-R systems (e.g., V444 Cygni). A semi-quantitative description of the influence of the interaction region on line profiles is provided by Luehrs (1997). Limited modeling of wind-wind collision processes is being attempted by various groups but the multi-dimensional nature (at least 2D) makes it a difficult task. For example, Kurosawa *et al.* (2002) used a Monte-Carlo code, in conjunction with 1D CMFGEN models, to interpret the line-profile and polarization variations in V444 Cygni. With the advent of improved hydrodynamic models, which can predict the structure of the interaction region, the demand for improved modeling of interacting binary systems will increase.

M. MONTGOMERY: Have you considered modeling oblate stars for rapid rotators as you consider spherical geometry?

J. HILLIER: To fully model rapid rotators with stellar winds, we would need to go to 2D. Going to 2D is computationally expensive, requiring $> 10^3$ times the computational effort for a model of similar complexity to a 1D CMFGEN model (due to the extra spatial dimension and the loss of symmetry in the radiation field). For stars without winds, the plane-parallel approximation is valid, and we can create spectra by tiling the surface of the rapidly rotating star with the appropriate model atmospheres. I have a code to do this — it was developed by an undergraduate student, Charles Warren. For stars with winds, it is a little more difficult because we now have the additional variation of the (uncertain) wind properties with latitude. For slow/moderate rotators, we have a 2D synthesis code (Busche & Hillier 2005) where we can apply latitude scaling laws (or potentially string a series of 1D models together) to investigate the influence of rotation. Finally, I note that we have developed a 2D code (Georgiev *et al.* 2006), but it is much less sophisticated than CMFGEN.

J. FREIMANIS: What about the inclusion of magnetic fields, Zeeman effect, Hanle effect, and so on, into your models?

J. HILLIER: Most O stars have only very weak magnetic fields ($< 100\,G$), and these are probably not crucial for understanding their spectra. However a few O stars have very intense magnetic fields [e.g., θ^1 Orionis C, $B > 1\,kG$; Donati *et al.* (2002)]. In these cases the magnetic field will control the flow of wind material from the star's surface producing shocks, etc. In such cases 2D or 3D modeling is required, which, as noted above, is difficult. However, researchers have successfully shown that an oblique rotator model can match the emission-line variability seen in θ^1 Orionis C (Wade *et al.* 2006).

From Interacting Binaries to Exoplanets: Essential Modeling Tools
Proceedings IAU Symposium No. 282, 2011
Mercedes T. Richards & Ivan Hubeny, eds.

© International Astronomical Union 2012
doi:10.1017/S1743921311027438

Stellar to Substellar Model Atmospheres

France Allard, Derek Homeier, and Bernd Freytag

Centre de Recherche Astrophysique de Lyon,
UMR 5574, CNRS, Université de Lyon,
École Normale Supérieure de Lyon,
46 Allée d'Italie, F-69364 Lyon Cedex 07, France,
email: france.allard@ens-lyon.fr

Abstract. The spectral transition from Very Low Mass stars (VLMs) to brown dwarfs (BDs) and planetary mass objects (Planemos) requires model atmospheres that can treat line, molecule, and dust-cloud formation with completeness and accuracy. One of the essential problems is the determination of the surface velocity field throughout the main sequence down to the BD and planemo mass regimes. We present local 2D and 3D radiation hydrodynamic simulations using the **CO5BOLD** code with binned **Phoenix** gas opacities, forsterite dust formation (and opacities) and rotation. The resulting velocity field vs depth and Teff has been used in the general purpose model atmosphere code **Phoenix**, adapted in static 1D spherical symmetry for these cool atmospheres. The result is a better understanding of the spectral transition from the stellar to substellar regimes. However, problems remain in reproducing the colors of the dustiest brown dwarfs. The global properties of rotation can change the averaged spectral properties of these objects. Our project for the period 2011-2015 is therefore to develop scaled down global 3D simulations of convection, cloud formation and rotation thanks to funding by the Agence Nationale de la Recherche in France.

Keywords. stars, brown dwarfs, planets, model atmospheres, hydrodynamic simulations.

1. Introduction

Since infrared observations of M dwarf stars (late 80s), brown dwarfs (mid 90s), and extrasolar planets (mid 2000s) became available, one of the most important challenges in modeling their atmospheres has become the understanding of molecular opacities at high temperatures (where the high energy transitions can only be captured by ab initio calculations) and dust cloud formation. Their compositions, assumed solar in general, take the form of molecular hydrogen, N_2 and CO, while the leading opacities in their spectra is due to titanium oxide and water vapor bands, with smaller contributions from vanadium oxide and various hydrides in the optical spectra range, and CO bands in the infrared. Model atmospheres and synthetic spectra have been improving along with the quality of these molecular opacities. The treatment of water vapor alone — which begins to form in early type M dwarfs (around Teff=3500K) and is present in Very Low Mass stars (VLMs), brown dwarfs, and planets including Hot Jupiters reaching an equilibrium temperature with the star of around 1600K — has seen an important evolution throughout the years: from band model approximations, to straight means based on the hot flames experiments (Ludwig 1971) used in our early model atmosphere grids (Allard 1990, Allard & Hauschildt 1995), to ab initio computations by Partridge & Schwenke (1997) (see also Allard, Hauschildt & Schwenke 2000), and to the current ab initio line lists by Baber *et al.* (2006) used in the most recent models. Nevertheless, the models atmosphere have systematically failed to reproduce the relative strength and shapes of the water bands, as shown for the case of an M dwarf in Fig. 1.

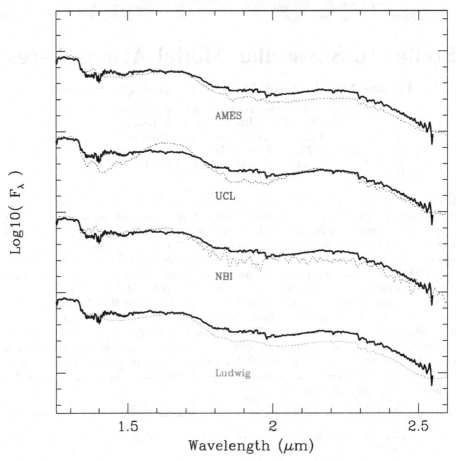

Figure 1. The infrared SED of the M8e dwarf star VB10 (thick full line) is compared to **Phoenix** synthetic spectra using different water vapor opacity profiles (dotted lines) from bottom to top: the hot flames experiments by Ludwig (1971), labeled as Ludwig, used in the early model grids for VLMs by Allard (1990) and Allard & Hauschildt (1995); the ab initio though incomplete line list by Schryber, Miller, & Tennyson (1995), labeled as UCL, used in the revised opacity-sampling NextGen model grid for VLMs by Hauschildt *et al.* (1999a,b), and the very complete ab initio line list by Partridge & Schwenke (1997), labeled as AMES, used in the limiting case for dust formation AMES-Cond/Dusty model grids for brown dwarfs by Allard *et al.* (2001). Results obtained for the line list by Jørgensen *et al.* (2001), labeled NBI, is also shown for comparison.

2. The impact of revised solar abundances on VLMs properties

The modeling of atmospheres has also evolved with the development of computing capacities from an analytical treatment of the transfer equation using moments of the radiation field (e.g. Allard, 1990), to a line-by-line opacity sampling in spherical symmetry (e.g. Hauschildt *et al.* 1999a,b), and to 3D radiation transfer (e.g. Seelmann, Hauschildt & Baron 2010). In parallel to detailed radiative transfer in an assumed static environment, hydrodynamical simulations have been developed to reach a realistic representation of the granulation and the line profiles shifted and shaped by the hydrodynamical flow of the Sun and Sun-like stars (see a review by Freytag *et al.* in a special issue of the Journal of Computational Physics to appear) by using a non-grey (multi-group opacities) radiative

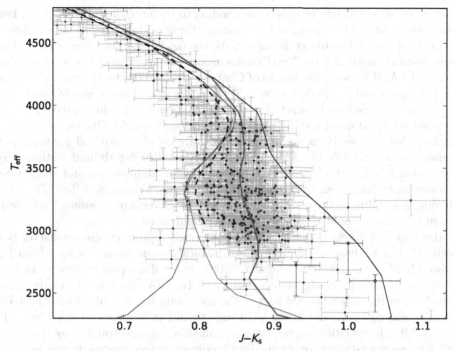

Figure 2. Estimated Teff for M dwarfs by Casagrande, Flynn, & Bessell (2008) (dots with grey error bars) and brown dwarfs by Golimowski *et al.* (2004) (dots with black error bars) are reported as a function of J-Ks . Over-plotted are the NextGen model isochrones for 5 Gyrs (Baraffe *et al.* 1998) using various generations of model atmospheres for the interpolation into the observational plane, starting with the NextGen (full line redwards of the bulk of stars), the limiting case AMES-Cond/Dusty grids by Allard *et al.* (2001) (full lines bluewards of the bulk of M dwarfs and splitting below 3300K bluewards and redwards respectively), and with the BT-Settl models using the Asplund *et al.* (2009) solar abundances (full line running through the bulk of M dwarfs). A version of the BT-Settl models, called BT-NextGen, using the Grevesse, Noels & Sauvals (1993) abundances are shown (dashed line) to illustrate the effects of the revised solar abundances.

transfer with a blackbody source function (scattering is neglected). Using such local 3D simulations and revised signal to noise and spectrally resolved observations, Asplund *et al.* (2009) and Caffau *et al.* (2011) have been able to redefine the solar abundances. These two independent groups agree in favoring a reduced (by a factor 2) solar oxygen abundance compared to previous values (Grevesse, Noels & Sauval 1993).

The revised lower oxygen abundances help models to reproduce the spectral properties of M dwarfs compared to earlier models based on previous estimates of the solar abundances by Grevesse, Noels, & Sauval (1993). This can be seen from Fig.2, where we compare models to the empirical determinations of Teff of individual stars by Golimowski *et al.* (2008). The J bandpass corresponds to the peak of the Spectral Energy Distribution (or SED) of M dwarfs, and is relatively insensitive to changes in atmospheric composition; while the Ks bandpass is very sensitive, with a broad sodium doublet, H_2 Collision-Induced Absorption (or CIA), CO bands in M dwarfs, and methane bands in late-type brown dwarfs. We compare the published NextGen and AMES-Cond/Dusty models, as well as the new calculations: i) the new BT-Settl models using the Asplund *et al.* (2009) solar abundances, ii) the new BT-NextGen models using solar abundances by Grevesse, Noels, & Sauval (1993) to allow isolating the effect of the revised oxygen abundance, and

iii) the new BT-Cond/Dusty models corresponding to the published AMES-Cond/Dusty models in terms of the treatment of the limiting effects of dust formation to isolate the effects of the revised **Phoenix** code version. All the models use the same published interior and evolution models: i) the NextGen isochrones of Baraffe et al. (1998) in the stellar regime, ii) the AMES-Dusty isochrones of Chabrier et al. (2000) for the transition regime between M dwarfs and brown dwarfs, and iii) the AMES-Cond isochrones of Baraffe et al. (2003) in the late type brown dwarf to planetary mass regime. Evolution models are currently being prepared using the BT-Settl model atmosphere grid. One can see from this comparison that the NextGen models systematically overestimate Teff throughout the lower main sequence, while the AMES-Cond/Dusty models slightly underestimate Teff. This situation is relieved when using the revised solar abundances, and the BT-Settl models now agree fairly well with most of the empirical estimations of Teff. The scatter in this diagram is due to the presence of dwarfs from several populations (metallicity), and an improvement would be to identify these populations.

We also find (not shown†) that the BT-Settl models reproduce also satisfactorily the complete SED of M dwarfs, solving a major historical issue as can be seen from Fig.1 (see also Allard, Hauschildt & Schwenke 2000). Some discrepancy remains in the K band which can be due to slightly overestimated H_2 CIA opacities. The agreement is also excellent in the optical to red part of the spectrum, in particular in the FeH Wing Ford bands near 0.99 μm, and in the VO bands thanks to line lists provided by B. Plez (GRAAL). Missing opacities are, however, still affecting the spectral distribution (CaOH bands), and current titanium oxide line lists become now too incomplete and inaccurate compared to the precision of the current water vapor opacities. Revised opacities are expected from the UCL group thanks to an ERC grant to J. Tennyson (UCL) for the computation of ab initio molecular opacities.

3. Dust formation in late type VLMs and brown dwarf atmospheres

An important additional difficulty has been identified in the greenhouse or backwarming effects due to dust cloud formation in late-type M dwarf and brown dwarf atmospheres by Tsuji (1996a,b). Oxygen-rich dust grains (silicates) deplete as much as 20% of the oxygen content of the gas, increasing the apparent C/O ratio. The formation of dust clouds and its associated greenhouse effect cause the infrared colors of late M and early L dwarfs to become extremely red compared to the colors of low-mass stars. The cloud composition, according to equilibrium chemistry, is going from zirconium oxide (ZrO_2), to refractory ceramics (perovskite and corundum; $CaTiO_3$, Al_2O_3), to silicates (e.g. forsterite; Mg_2SiO_4), to salts (CsCl, RbCl, NaCl), and finally to ices (H_2O, NH_3, NH_4SH), as brown dwarfs cool down with time from M through L, and T spectral types and beyond (Allard et al. 2001; Lodders & Fergley 2006). Many cloud models have been constructed to address this problem in brown dwarfs over the past decade (see the review by Helling et al. 2008 for a comparison of the various models). However, none treated the mixing properties of the atmosphere and the resulting diffusion mechanism realistically enough to reproduce the properties of the spectral transition from M through L and T spectral types without changing cloud parameters (e.g. Ackerman & Marley 2001). It is in this context that we have decided to address the issue of mixing and diffusion by 2D radiation hydrodymanic simulations of VLMs and brown dwarfs atmospheres (Teff=2800K to 900K), using the **Phoenix** gas opacities in a multi-group, a two

---

† These results will be demonstrated at the occasion of the detailed publication of the final BT-Settl model grid.

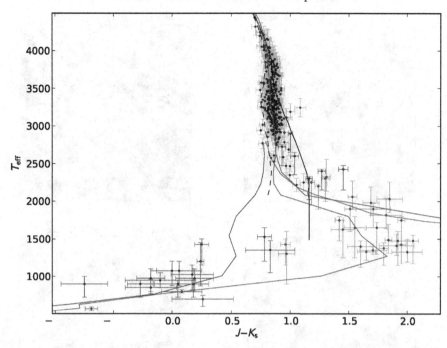

Figure 3. Same plot as Fig.2 but zooming out and extending into the brown dwarf region of the diagram. The J band corresponds to the peak of the SED while the Ks band is sensitive to the composition of the atmosphere (see text). The region between 2500K and 1400K is dominated by dust formation (essentially forsterite and other silicates) and CO absorption in the K band, while cooler brown dwarfs show methane formation and a turn to the blue of the J-Ks color. The limiting case AMES-Cond/Dusty models atmosphere provide a description of the span in colors of the brown dwarfs in this diagram. We use isochrones of 5 Gyrs for the purpose of illustration.

bin (monomers and dust) cloud model and forsterite geometric cross-sections (Freytag *et al.* 2010). We discovered the formation of gravity waves as one driving mechanism for the formation of clouds in these atmospheres. Around Teff $\lesssim 2200$ K, the cloud layers become optically thick enough to initiate cloud convection, which participate in the global mixing. Overshoot is also important in the mixing of the largest dust particles.

We derived a rule for the velocity field versus atmospheric depth and Teff, which is relatively insensitive to gravity. This rule was then used in the construction of the new model atmosphere grid BT-Settl. The cloud model used in the BT-Settl models is based on the condensation and sedimentation timescales from a study of planetary atmospheres by Rossow (1978). However, one improvement compared to previous model versions (see e.g. Helling *et al.* 2008), is that we compute supersaturation from our pre-tabulated equilibrium chemistry in order to obtain the correct amount of dust formation (as opposed to the value suggested by Rossow). We solve the cloud model and equilibrium chemistry layer by layer outward to account for the history of grain formation as a function of the cooling of the gas. One can see from Fig. 3 that the late-type M and early type L dwarfs behave as if dust is formed nearly in equilibrium with the gas phase with extremely red colors, in good agreement with the Dusty models. At the low Teff regime dominated by T-type dwarfs, the AMES-Cond models also appear to provide a good limitation for late-type brown dwarfs. The improved BT-Settl cloud model shows promising results in describing the stellar to substellar transition. It appears, however, that not enough dust

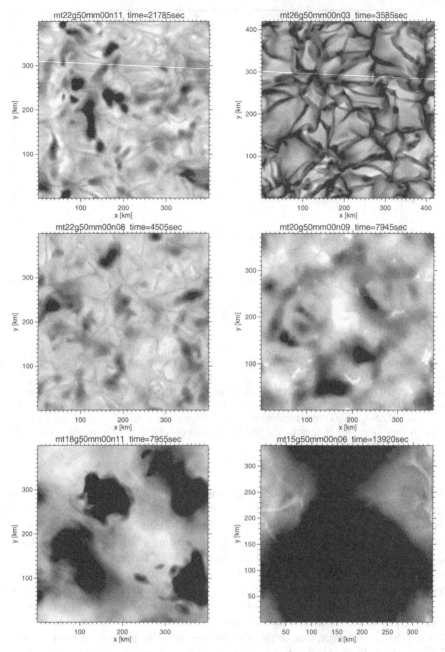

Figure 4. Snapshots of the intensity of local 3D RHD simulations at the surface of, from top to right to bottom right, 2600K, 2200K, 2000K, 1800K and 1500K atmospheres of logg=5.0, solar metallicity and no rotation, except for the 2200K case shown (top left) using a rotational period of 15 minutes to highlight the effects on the convection cells and the cloud formation.

is predicted at the lower gravities predicted by evolution models for 5 Gyrs (log g=4.0 to 3.5) to reproduce the reddest brown dwarfs with these isochrones. Revised interior and evolution models using the new solar abundances and the BT-Settl atmospheres as surface boundary are expected by mid-2012.

While the distribution of dust with depth is important to understand the SED and possible spectral variability of VLMs, brown dwarfs and planets, the surface distribution of dust also matters. For this, 3D global or "star in the box" RHD simulations with rotation are required. This is our current project supported by the French "Agence Nationale de la Recherche" for the period 2010-2015. Rotation is already included in **CO5BOLD** for a scaled down model of the Sun (Steffen & Freytag (2007) and can be applied to brown dwarf and planet simulations. In Fig.4, 3D RHD simulation boxes at the surface of the star) are shown for dwarfs with Teff ranging across the stellar to brown dwarfs transition. The 2600 K case shows no or negligible dust formation, while dust formation progresses to reach optically thick density at around 2200K, before sedimenting out again towards the 1300K regime (not shown). The Teff = 1500 K case illustrates the importance of gravity waves, where the minima of the waves reach condensation levels while the maxima remain in condensed phase. Also shown is the case of a 2200 K simulation using a wind at the equator corresponding to a 15 mins period rotation of the star). This period of rotation is very extreme compared to the averaged period of rotation of brown dwarfs (2 hrs), and illustrates the stability of the dust formation process against rotation. The simulation boxes are, however, too small compared to the stellar radius to show the overall variability. For this purpose, scaled down global 3D simulations of red dwarfs, brown dwarfs, and planets are currently developed using **CO5BOLD**.

4. Summary

We present in this paper, a brief overview of the main preliminary highlights of the BT-Settl model atmosphere grid for stars, brown dwarfs, and young planets computed using the atmosphere code **Phoenix**, which has been updated since Allard *et al.* (2001), with: i) spherical symmetry and non-LTE for most elements, ii) alpha element enrichment, iii) the Barber *et al.* (2006) BT2 water opacity line list, iv) solar abundances revised by Asplund *et al.* (2009), and v) a cloud model accounting for an improved treatment of supersaturation and RHD mixing. These models will be published in greater detail shortly in A&A. The grid spans $100,000\,K < T_{eff} < 400\,K$; $-0.5 < \log g < 5.5$; and $+0.5 < [M/H] < -1.5$. The models are available at the **Phoenix** simulator website "http://phoenix.ens-lyon.fr/simulator/" and are in preparation for publication. They will serve the GAIA, MUSE, and PLATO missions for which a detailed comparison of models from various authors is planned.

We find that the BT-Settl models with reduced oxygen abundance show a significant improvement in reproducing M dwarfs, while the improved cloud model allows an improved reproduction of the turn to the red due to dust formation before predicting a turn over to blue below 1700 K in J-Ks colors. The models, however, appear not to be forming enough high altitude dust to reproduce the most extreme dusty brown dwarfs in this diagram. Improvements to our computed supersaturation and porosity of grains are possible venues of refinement. The RHD simulations of Freytag *et al.* (2010) have shed light on the importance of mixing in the atmospheres of VLMs, brown dwarfs, and possibly planets. These results imply that the gas is out of equilibrium. Beyond molecular opacities, model atmospheres require, therefore, reaction rates for the most abundant molecules and most important absorbers.

To address the new source of spectroscopic information on Hot Jupiters and young imaged planets to come, we have developed the BT-Settl model atmosphere grid to encompass the parameter regime of these objects (low gravity around log g=4.0, Teff below 2000 K). These models are those of single objects, but the development of the binary star function of the Phoenix simulator will shortly allow the study of impinging irradiation effects on planetary atmospheres.

Acknowledgements

We would like to thank the French "Agence Nationale de la Recherche" (ANR) and "Programme National de Physique Stellaire" (PNPS) of CNRS (INSU) for their financial support. The computations of dusty M dwarf and brown dwarf models were performed at the *Pôle Scientifique de Modélisation Numérique* (PSMN) at the *École Normale Supérieure* (ENS) in Lyon.

References

Ackerman, A. S. & Marley, M. S. 2001, *ApJ*, 556, 872
Allard, F. 1990, Ph.D. thesis, PhD thesis. Ruprecht Karls Univ. Heidelberg, (1990)
Allard, F. & Hauschildt, P. H. 1995, *ApJ*, 445, 433
Allard, F., Hauschildt, P. H., Alexander, D. R., Tamanai, A., & Schweitzer, A. 2001, *ApJ*, 556, 357
Allard, F., Hauschildt, P. H., & Schwenke, D. 2000, *ApJ*, 540, 1005
Asplund, M., Grevesse, N., Sauval, A. J., & Scott, P. 2009, *ARAA*, 47, 481
Baraffe, I., Chabrier, G., Allard, F., & Hauschildt, P. H. 1997, *A&A*, 327, 1054
— 1998, *A&A*, 337, 403
Baraffe, I., Chabrier, G., Barman, T. S., Allard, F., & Hauschildt, P. H. 2003, *A&A*, 402, 701
Barber, R. J., Tennyson, J., Harris, G. J., & Tolchenov, R. N. 2006, *MNRAS*, 368, 1087
Caffau, E., Ludwig, H.-G., Steffen, M., Freytag, B., & Bonifacio, P. 2011, *Solar Phys.*, 268, 255
Casagrande, L., Flynn, C., & Bessell, M. 2008, *MNRAS*, 389, 585
Chabrier, G., Baraffe, I., Allard, F., & Hauschildt, P. 2000, *ApJ*, 542, 464
Freytag, B., Allard, F., Ludwig, H., Homeier, D., & Steffen, M. 2010, *A&A*, 513, A19
Golimowski, D. A. & collaborators, 2004, *AJ*, 127, 3516
Grevesse, N., Noels, A., & Sauval, A. J. 1993, *A&A*, 271, 587
Hauschildt, P. H., Allard, F., & Baron, E. 1999a, *ApJ*, 512, 377
Hauschildt, P. H., Allard, F., Ferguson, J., Baron, E., & Alexander, D. R. 1999b, *ApJ*, 525, 871
Helling, C., Ackerman, A., Allard, F., Dehn, M., Hauschildt, P., Homeier, D., Lodders, K., Marley, M., Rietmeijer, F., Tsuji, T., & Woitke, P. 2008, *MNRAS*, 391, 1854
Jørgensen, U. G., Jensen, P., Sørensen, G. O., & Aringer, B. 2001, *A&A*, 372, 249
Lodders, K. & Fegley, Jr. B. 2006, In Astrophysics Update 2, editor: Mason J. W. Spinger Verlag, p1.
Ludwig, C. B. 1971, *Applied Optics*, 10, 1057
Partridge, H. & Schwenke, D. W. 1997, *Journal for Computational Physics*, 106, 4618
Rossow, W. B. 1978, *ICARUS*, 36, 1
Schryber, J. H., Miller, S., & Tennyson, J. 1995, *JQSRT*, 53, 373
Seelmann, A. M., Hauschildt, P. H., & Baron, E. 2010, *A&A*, 522, A102
Steffen, M. & Freytag, B. 2007, *AN*, 328, 1054
Tsuji, T., Ohnaka, K., & Aoki, W. 1996a, *A&A*, 305, L1
Tsuji, T., Ohnaka, K., Aoki, W., & Nakajima, T. 1996b, *A&A*, 308, L29

Discussion

W. KLEY: Why do you have to use a very small radius in global convection models for dwarf stars?

F. ALLARD: Because we need to resolve the convective cells which are so much smaller compared to the radius of the star.

From Interacting Binaries to Exoplanets: Essential Modeling Tools
Proceedings IAU Symposium No. 282, 2011   © International Astronomical Union 2012
Mercedes T. Richards & Ivan Hubeny, eds.   doi:10.1017/S174392131102744X

Comparison of Limb-Darkening Laws from Plane-Parallel and Spherically-Symmetric Model Stellar Atmospheres

Hilding R. Neilson

Argelander-Institut für Astronomie, Bonn Universität,
Auf Dem Hügel 71, Bonn, D-53121, Germany
email: hneilson@astro.uni-bonn.de

Abstract. Limb-darkening is a fundamental constraint for modeling eclipsing binary and planetary transit light curves. As observations, for example from *Kepler*, *CoRot*, and *Most*, become more precise then a greater understanding of limb-darkening is necessary. However, limb-darkening is typically modeled as simple parameterizations fit to plane-parallel model stellar atmospheres that ignores stellar atmospheric extension. In this work, I compute linear, quadratic and four-parameter limb-darkening laws from grids of plane-parallel and spherically-symmetric model stellar atmospheres in a temperature and gravity range representing stars evolving on the Red Giant branch. The limb-darkening relations for each geometry are compared and are found to fit plane-parallel models much better than the spherically-symmetric models. Assuming that limb-darkening from spherically-symmetry model atmospheres are more physically representative of actual stellar limb-darkening than plane-parallel models, then these limb-darkening laws will not fit the limb of a stellar disk leading to errors in a light curve fit. This error will increase with a star's atmospheric extension.

Keywords. stars: atmospheres, stars: fundamental parameters, (stars:) supergiants

1. Introduction

Stellar limb-darkening is the observed change of intensity from the center of the stellar disk to the observable edge, where the intensity decrease is due to the geometric projection of the line-of-sight relative to the radius of the star. This effect is an important challenge for the interpretation of observations of binary stars (e.g. Claret 2008), and planetary transits (e.g. Knutson *et al.* 2007, Croll *et al.* 2011), as well as interferometric (e.g. Wittkowski *et al.* 2004) and microlensing (e.g. Zub *et al.* 2011) observations. Typically, limb-darkening is treated as a parameterization or relation as a function of the cosine of the angle formed by the radius and line-of-sight, called μ to simplify the analysis.

Stellar atmosphere models and binary/transit observations are complementary tools for understanding limb-darkening and stellar astrophysics in general because observed limb-darkening can help constrain models. There are numerous articles describing different limb-darkening relations (Al-Naimiy 1978, Wade & Rucinski 1985, Claret *et al.* 1995, Claret 2000), limb-darkening coefficients from predicted intensity profiles for a number of stellar atmosphere codes such as ATLAS and PHOENIX (e.g. Howarth 2011, Sing 2010, Claret & Hauschildt 2003), and different fitting methods (Wade 1985, Heyrovsky 2003, 2007, Claret 2008). In this work, I focus on a small number of limb-darkening laws and compare predicted fits for intensity profiles from plane-parallel and spherically symmetric model stellar atmospheres. In the next section, I describe the stellar atmosphere code and three limb-darkening laws of interest: a linear, quadratic, and four-parameter (Claret 2000). In Sect. 3, I present results of the fitting of the limb-darkening laws using model atmospheres, and how the errors of the fit depend on assumed geometry. I summarize this work in Sect. 4.

2. Stellar atmosphere code and Limb-darkening Laws

For this analysis, I use a new Fortran 90 version of the Kurucz ATLAS code (Lester & Neilson 2008). The code computes opacities using opacity distribution functions, and atmospheres are assumed to be in local thermodynamic equilibrium and hydrostatic equilibrium. Each atmosphere model outputs intensity profiles as a function of wavelength, for an equal spacing of μ for 1000 points. Typical calculations for the ATLAS code compute intensity profiles for 10 - 17 μ-points. The program computes models for either plane-parallel or spherically-symmetric geometries, where the plane-parallel model is described by two fundamental parameters such as $T_{\rm eff}$ and $\log g$, while spherical models require an additional parameter such as stellar mass. Neilson & Lester (2008) fit model intensity profiles to interferometric observations from Wittkowski *et al.* (2004) and predicted similar fundamental parameters as those authors. Also, Neilson & Lester (2011) predicted limb-darkening coefficients for a specific limb-darkening law from spherical models, compared them to results for microlensing observations from Fields *et al.* (2003), and found better agreement than the authors did using plane-parallel models and spherical models that had intensity profiles clipped to remove the extended limb.

I have computed approximately 2000 model stellar atmospheres in spherical symmetry for the parameter range $T_{\rm eff} = 3000$ - 8000 K in steps of 100 K, $\log g = -1$ - $+3$ in steps of 0.25 in cgs units, and $M = 2.5$ - 10 $M_\odot$ in steps of 2.5 $M_\odot$. Plane-parallel models are computed for the same values of $T_{\rm eff}$ and $\log g$.

For this work, I compute least-squared fits to three laws:

$$\frac{I(\mu)}{I(1)} = 1 - a(1-\mu) \qquad\qquad \text{Linear,} \qquad\qquad (2.1)$$

$$\frac{I(\mu)}{I(1)} = 1 - b(1-\mu) - c(1-\mu)^2 \qquad\qquad \text{Quadratic,} \qquad\qquad (2.2)$$

$$\frac{I(\mu)}{I(1)} = 1 - \sum_{i=1}^{4} d_i(1-\mu^{i/2}) \qquad\qquad \text{Four Parameter,} \qquad\qquad (2.3)$$

where intensities are computed in the *Kepler* white light passband. All fits are computed using least-square fitting of the limb-darkening coefficients. The quality of the fit may be measured in a number of ways; here, I test the quality of the fit of limb-darkening laws by checking how well they conserve stellar flux, $\Delta F/F = (F_{\rm Model} - F_{\rm Law})/F_{\rm Model}$.

3. Results & Summary

In Fig. 1, I show the predicted intensity profiles for a $T_{\rm eff} = 4000$ K and $\log g = 2$ model atmosphere for both geometries. There is a significant difference between the model intensity profiles such that the intensity near the limb of the spherically-symmetric model atmosphere is much smaller than the plane-parallel model atmosphere. The plane-parallel model does not appear to go to zero, though the equation of transfer suggests that as the intensity in the limit $\mu \to 0$ then $I(\mu) \to 0$ (Mihalas 1978). This may be an issue with the resolution of μ. The plane-parallel model is also much better fit by the limb-darkening laws than the spherically-symmetric model atmosphere because the spherically-symmetric intensity profile is more complex.

The relative difference between the model intensity profiles and stellar fluxes predicted by the limb darkening laws are shown in Fig. 2. The fits to plane-parallel model atmo-

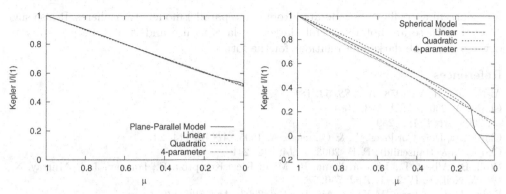

Figure 1. Intensity profiles for a plane-parallel (left) and spherically-symmetric (right) model atmosphere with $T_{\text{eff}} = 4000$ K and $\log g = 2$, along with the best-fit limb-darkening laws.

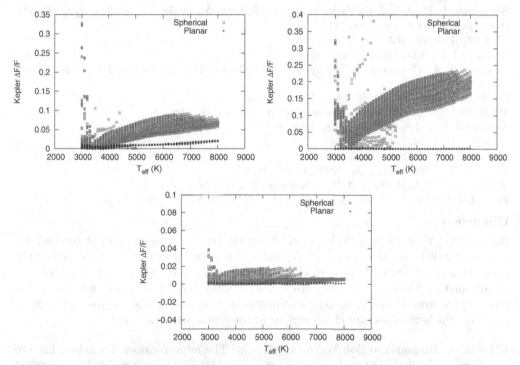

Figure 2. Relative difference of between model and fit stellar fluxes for the three limb-darkening laws in the *Kepler* passband, linear (upper-left), quadratic (upper-right), and Claret (2000) four-parameter (bottom) laws for plane-parallel and spherically-symmetric model stellar atmospheres.

spheres appear to be better; the average error is $< 5\%$ for the linear law, and is much smaller for the other laws. The quality of the fit for plane-parallel model atmospheres is also apparently independent of effective temperature. The results for the spherically-symmetric model atmospheres are strikingly different. The flux errors are much larger, 5-10% for the linear law, 0-20% for the quadratic law, and 0-5% for the four parameter limb-darkening law. The difference in fits due to model atmosphere geometry suggests a significant problem for understanding stellar limb-darkening. It is reasonable to assume that a spherical geometry is a more physical representation of an actual star than a plane-parallel stellar atmosphere, hence people should be hesitant when using limb-darkening coefficients generated from plane-parallel model atmospheres. It also suggests that these are not ideal limb-darkening laws to use and it may be necessary to develop new limb-darkening relations for the future.

References

Al-Naimiy, H. M. 1978, *Ap&SS*, 53, 181

Claret, A. 2000, *A&A*, 363, 1081

—. 2008, *A&A*, 482, 259

Claret, A., Diaz-Cordoves, J., & Gimenez, A. 1995, *A&AS*, 114, 247

Claret, A. & Hauschildt, P. H. 2003, *A&A*, 412, 241

Croll, B., Albert, L., Jayawardhana, R., Miller-Ricci Kempton, E., Fortney, J. J., Murray, N., & Neilson, H. 2011, *ApJ*, 736, 78

Fields, D. L., Albrow, M. D., & An, J., *et al.* 2003, *ApJ*, 596, 1305

Gustafsson, B., Edvardsson, B., Eriksson, K., Jørgensen, U. G., Nordlund, Å., & Plez, B. 2008, *A&A*, 486, 951

Hauschildt, P. H., Allard, F., Ferguson, J., Baron, E., & Alexander, D. R. 1999, *ApJ*, 525, 871

Heyrovský, D. 2003, *ApJ*, 594, 464

—. 2007, *ApJ*, 656, 483

Howarth, I. D. 2011, *MNRAS*, 413, 1515

Knutson, H. A., Charbonneau, D., Noyes, R. W., Brown, T. M., & Gilliland, R. L. 2007, *ApJ*, 655, 564

Lester, J. B. & Neilson, H. R. 2008, *A&A*, 491, 633

Mihalas, D. 1978, Stellar atmospheres /2nd edition/, ed. Hevelius, J.

Neilson, H. R. & Lester, J. B. 2008, *A&A*, 490, 807

—. 2011, *A&A*, 530, A65

Sing, D. K. 2010, *A&A*, 510, A21

Wade, R. A. & Rucinski, S. M. 1985, *A&AS*, 60, 471

Wittkowski, M., Aufdenberg, J. P., & Kervella, P. 2004, *A&A*, 413, 711

Zub, M. & Cassan, A., Heyrovský *et al.*, 2011, *A&A*, 525, A15

Discussion

R. WILSON: Your starting explanation of why the intensity goes to zero at the limb for a plane parallel case is not correct. Actually, the intensity does not go to zero at the limb. This result comes from neglect of emission along the line of sight; it is not just an attenuation problem, but has both a source factor and an attenuation factor in the intensity integral. If one looks into an infinite uniform region, the received intensity is not zero, but is the intensity characteristic of the region's temperature.

I. HUBENY: Response to Bob Wilson's comment: The intensity does not indeed have to go to zero at the limb, but such a case is not covered by a 1-D plane-parallel treatment of the transfer equation anyway because, in this case, the medium is infinite with no natural boundary condition. Comment on the talk: The Eddington factor (the ratio of the K-moment and the mean intensity) is not necessarily equal to 1/3 in the plane-parallel case. Such a quantity is usually called a variable Eddington factor, and depends on depth and frequency; it goes to 1/3 only deep in the atmosphere.

A. PRŠA: I understand why one would use an analytic model for a Mandel-Agol type approach, but perhaps the systematic error from the simple fit may be avoided simply by linearly interpolating (or looking up) $I(\mu)$.

P. STEE: I did not understand why you used the first lobe of the visibility function to fit the LD instead of the second lobe, especially since you may fit the first lobe with a uniform disk?

From Interacting Binaries to Exoplanets: Essential Modeling Tools
Proceedings IAU Symposium No. 282, 2011
Mercedes T. Richards & Ivan Hubeny, eds.
© International Astronomical Union 2012
doi:10.1017/S1743921311027451

Modeling of Circumstellar Dust by the DUSTY Code

Tomislav Jurkić and Dubravka Kotnik-Karuza

Department of Physics, University of Rijeka, Omladinska 14, HR-51000 Rijeka, Croatia
email: tjurkic@phy.uniri.hr

Abstract. We present a circumstellar dust model around the symbiotic Mira RR Tel obtained by modeling the near-infrared JHKL magnitudes and ISO spectra. In order to follow the evolution of infrared colours in time, the published JHKL magnitudes were corrected by removing the Mira pulsations. The RR Tel light curves show three obscuration events in the near-IR. Using the simultaneously available JHKL magnitudes and ISO spectra in three different epochs, we obtained SEDs in the near- and mid-IR spectral region (1-20 μm) in epochs with and without obscuration. The DUSTY numerical code was used to solve the radiative transfer and to determine the circumstellar dust properties of the inner dust regions around the Mira, assuming a spherical dust temperature distribution in its close neighbourhood. The physical properties of the dust, mass loss and optical depth during intervals with and without obscuration have been obtained. Both JHKL and ISO observations during the obscuration period can be reproduced with a spherical dust envelope, while ISO spectra outside obscuration show a different behaviour. The dynamical behaviour of the circumstellar dust was obtained by modeling the JHKL magnitudes observed during the span of more than 30 years.

The DUSTY code was also successfully applied in the modeling of circumstellar dust envelopes of young stellar objects, such as Herbig Ae/Be stars.

Keywords. stars: binaries: symbiotic, stars: circumstellar matter, stars: AGB and post-AGB, infrared: stars, methods: numerical, radiative transfer

1. Introduction

Symbiotic stars are interactive binary systems consisting of a cool red giant or a Mira star, and a hot compact star, such as white dwarf or neutron star. Around 20% of symbiotic stars, called D-type symbiotics, consist of a Mira component with a circumstellar dust shell at the temperature of around 1000 K (Mikolajewska *et al.* 1988, Kenyon 1986). Obscuration events in the near-IR seen in most of the D-type symbiotics are probably caused by a change in optical depth of the dust shell (Kotnik-Karuza *et al.* 2006, Whitelock 2003).

RR Tel is a symbiotic nova that underwent its last outburst in 1944 (Whitelock 2003). A complete model of the circumstellar dust environment around RR Tel capable of explaining all near and mid-IR spectroscopic and photometric features is still to be found. Up to now, most attempts to explain the circumstellar dust environment have included only simple thick dust shell models acting as a blackbody, without determination of any physical dust properties (Kotnik-Karuza *et al.* 2006, Whitelock 1987). The recent colliding wind model of Angeloni *et al.* (2010) has proposed the existence of two blackbody dust shells at 400 K and 1000 K, with maximum emission of the 400 K dust shell at around 10 microns, typical for silicate emission. Generally, only few authors used radiative transfer in other single objects to model the dust properties.

Young stellar objects such as T Tau or Herbig Ae/Be stars show the presence of the dust in the form of strong near and mid-IR emission (Chiang & Goldreich 2001). Some of these stars, especially Herbig Ae/Be stars, have significant near-IR excess that cannot be explained with simple flared disk models, and require introduction of complex disk geometry (Dullemond *et al.* 2001).

2. Observational data and DUSTY code

Multiwavelength and long-term infrared observations are needed to completely determine the dynamical model of symbiotic Miras. In order to fulfill this demand, near-IR JHKL photometric magnitudes were collected and spectral energy distributions (SED) from 1.2 μm to 20 μm reconstructed. We have used all available published JHKL magnitudes from the South African Astronomical Observatory (Gromadzki *et al.* 2009), corrected for Mira pulsations and interstellar reddening using our own developed algorithm. The JHKL light curves show three distinct obscuration intervals (Kotnik-Karuza *et al.* 2006). Spectral energy distributions during and outside obscuration epochs were reconstructed using ISO short wavelength infrared spectra together with the near-IR JHKL observations at three epochs when both JHKL magnitudes and ISO spectra were simultaneously available.

The numerical code DUSTY (Ivezic, Nenkova & Elitzur 1999) was used to model the dust properties of the circumstellar dust shell around the cool Mira component. DUSTY solves the radiative transfer through the circumstellar dust, assuming a spherical temperature distribution and using scaling invariance and self-similarity theory to constrain the number of free input parameters (Ivezic & Elitzur 1997). Scaling invariance implies only the scaled distance y with respect to the inner dust shell radius (sublimation radius), r_{in}, as a relevant parameter:

$$y = \frac{r}{r_{in}} \qquad (2.1)$$

Only the shape of the dust density distribution and the spectral shape of the input heating radiation is needed; absolute stellar luminosity does not enter the equations. Starting from the input parameters, the code models the output radiation reprocessed by dust and gives physical and geometrical properties of the dust shell.

DUSTY can handle various analytical forms for the dust density distribution, including full dynamical calculation for radiatively driven winds. Various dust chemistry and optical properties are supported (silicates, carbon, olivine, forsterite, etc.) or can be entered in the user-supplied files. The dust grain size distribution, the minimum and maximum grain sizes, the dust sublimation temperature, the shell size, and the optical depth can be varied in order to produce an acceptable model. The variation in power index of the dust density distribution gives an insight into the dust geometry (spherical shell, flattened halo, disk-like structure).

The circumstellar model with a Mira in the centre of the dust shell was used, while input stellar radiation was approximated with a Mira blackbody at a temperature between 2200 K and 2800 K, in agreement with its spectral class. As Miras are well known to have strong stellar winds which drive the expansion of their envelopes, an analytical approximation for the dust density distribution, η, enhanced by radiatively driven winds (Ivezic *et al.* 1999) was used:

$$\eta \propto \frac{1}{y^2} \sqrt{\frac{y}{y - 1 + (v_i/v_e)^2}} \qquad (2.2)$$

where v_i and v_e are initial and terminal wind velocities respectively. A typical MRN dust grain size distribution (Mathis, Rumpl, & Nordsieck 1977): $n(a) \propto a^{-q}$ ($a_{min} \leqslant a \leqslant a_{max}$) was applied in the model, with the power index q = 3.5, the minimum grain size a_{min}

Table 1. Dust parameters of RR Tel derived from DUSTY modeling (T_{mira} - mira temperature; T_{dust} - dust sublimation temperature; a_{max} - maximum grain size; τ_V - visual optical depth; A_K - extinction at K band; $\dot{M}$ - mass-loss rate; v_e - terminal wind velocity)

Parameters	Maximum Obscuration	Minimum Obscuration thin shell	thick dust
T_{mira} [K]	2500 ± 300	2200	
T_{dust} [K]	1200 ± 100	1000-1200	1200
a_{max} [μm]	4.0 ± 1.0	0.5-1.0	1.5 ± 0.5
τ_V	5.2 - 5.8	0.4	
A_K	0.58 - 0.64	0.04	
$\dot{M}$ [$10^{-6} M_\odot / yr$]	11 ± 1	0.4	
v_e [km/s]	26 ± 2	28	

fixed at 0.005 μm, and the maximum grain size a_{max} determined by modeling. The dust composition typical for Miras containing 100% warm silicates has been assumed (Ossenkopf, Henning & Mathis 1992). The outer dust shell radius was fixed at 20 r_{in}, while the inner dust shell radius r_{in} was obtained by fitting, together with the dust sublimation temperature T_{dust}.

3. Results

Reconstructed SEDs and two-colour near-IR diagrams during obscuration epochs are very well reproduced by a single circumstellar dust shell model of consistent sublimation temperature at 1200 K and a grain size of 4.0 μm (Table 1). On the contrary, reconstructed SEDs in an epoch without obscuration cannot be explained by a simple single shell model as the Mira blackbody is itself almost unattenuated, while the dust emission is still strongly present. Such behaviour suggests an optically thin dust shell around the Mira component and an optically thick dust of undetermined geometry outside the line of sight, probably in a kind of toroidal configuration or something similar. Modeling of such a two-shell model yielded a very thin dust shell with optical depth around 0.4 and thick dust outside the line of sight, both at sublimation temperature around 1200 K and grain size of about 1.5 μm.

Our model shows no difference in dust sublimation temperatures between epochs with and without obscuration inside the error margin. Obscuration events can be explained by an increase in optical depth, likely to be caused by higher mass loss leading to a larger amount of dust condensed around the Mira and to the formation of a thicker dust shell. The only way to model the reconstructed SEDs and the near-IR colours requires a significant change in the maximum dust grain size from around 1.5 μm to 4.0 μm in transition from epochs with obscuration to epochs without obscuration. This can be explained by grain growth in the dust probably originating from the Mira itself.

In young stellar objects such as Herbig Ae/Be stars, the near-IR excess requires a larger near-IR emitting area, which can be fulfilled by use of the spherical dust shell instead of the complex disk geometry. Simultaneous modeling of SEDs and interferometric observations of different Herbig Ae/Be stars ranging in luminosities from 25 L_{Sun} to 90 000 L_{Sun}, have shown consistently that an optically thin dusty halo around the inner disk regions can very well explain the near-IR excess, with no need for complex disk geometry (Vinkovic *et al.* 2006).

References

Angeloni, R., Contini, M., Ciroi, S., & Rafanelli, P. 2010, *MNRAS*, 402, 2075

Chiang, E. I. & Goldreich, P. 2001, *ApJ*, 490, 368

Dullemond, C. P., Dominik, C., & Natta, A. 2001, *ApJ*, 560, 957

Gromadzki, M., Mikolajewska, J., Whitelock, P., & Marang, F. 2009, *AcA*,

Ivezic, Z. & Elitzur, M. 1997, *MNRAS*, 287, 799

Ivezic, Z., Nenkova, M., & Elitzur, M. 1999, *User Manual for DUSTY*, University of Kentucky Internal Report, accessible at http://www.pa.uky.edu/~moshe/dusty

Kenyon, S. J. 1986 *The symbiotic stars* (Cambridge & New York: Cambridge Univ. Press), p.295

Kotnik-Karuza, D., Friedjung, M., Whitelock, P. A., Marang, F., Exter, K., Keenan, F. P., & Pollacco, D. L. 2006, *A&A*, 452, 503

Mathis, J. S., Rumpl, W., & Nordsieck, K. H. 1977, *ApJ*, 217, 425

Mikolajewska, J., Friedjung, M., Kenyon, S. J., & Viotti, R. 1988, in: J. Mikolajewska, M. Friedjung, S. J. Kenyon & R. Viotti (eds.) *The symbiotic phenomenom*, Proceedings of IAU Colloq. 103 (Dordrecht: Kluwer), vol. 145, p. 381

Monnier, J. D., Berger, J.-P., Millan-Gabet, R., *et al.*, 2006, *ApJ*, 647, 444

Ossenkopf, V., Henning, Th., & Mathis, J. S. 1992, *A&A*, 261, 567

Vinkovic, D., Ivezic, Z., Jurkic, T., & Elitzur, M. 2006, *ApJ*, 636, 348

Vinkovic, D. & Jurkic, T. 2007, *ApJ*, 658, 462

Whitelock, P. A. 1987, *PASP*, 99, 573

Whitelock, P. A. 2003, in: R. L. M. Corradi, J. Mikolajewska & T. J. Mahoney (eds.) *Symbiotic stars probing stellar evolution*, *ASP-CS*, 303, 41

Discussion

K. ALLERS: Your point that the puffed-up disk is not necessary if you have a dust shell does not always apply. T Tauri stars show no obscuration but do show the near-IR bump.

T. JURKIC: Yes, that is true. On the other hand, the dusty halo in the innermost parts of the CS environment is usually optically thin ($\tau_V \sim 0.5$) and can reproduce T Tau spectra very well. The presence of a dusty halo does not exclude the existence of a disk which can be seen at larger wavelengths. Near-IR interferometry of T Tau stars speaks in favor of dusty halos in the regions up to 10 AU (Vinkovic *et al.* 2006).

O. DE MARCO: Sub-mm observations can discriminate between shell and disk/torus dusty CS geometries. Have such observations been carried out for RR Tel or other symbiotics?

T. JURKIC: Such observations have been used to detect disks embedded in dusty halos around YSOs such as Herbig Ae/Be stars. Unfortunately, there was no sub-mm imaging of RR Tel. Sub-mm spectra are affected not only by the dust emission but also by the free-free emission, which further complicates analysis.

P. STEE: We have also tried to fit the SED of HAeBe stars using a disk model and it is not difficult to fit the SED, especially if you have two different dusty disks. Regarding the interferometric data, you are mainly constraining the flux ratio between the central star and the disk rather than the global geometry since you need a better plane coverage.

T. JURKIC: It is true that better plane coverage will constrain dust geometry better. Monnier *et al.* (2006) detected very small deviations from centrally symmetric images using near-IR interferometry, which is not consistent with inclined disk models. Modeling of the near-IR visibilities of 40 YSO (Vinkovic *et al.* 2007) showed that T Tau and low luminosity Herbig Ae/Be stars are best explained by an inner dusty halo.

From Interacting Binaries to Exoplanets: Essential Modeling Tools
Proceedings IAU Symposium No. 282, 2011   © International Astronomical Union 2012
Mercedes T. Richards & Ivan Hubeny, eds.   doi:10.1017/S1743921311027463

Modeling Supernova Spectra

D. John Hillier[1], Luc Dessart[2], and Chendong Li[1]

[1]University of Pittsburgh, Department of Physics and Astronomy,
100 Allen Hall, Pittsburgh, PA 15260, USA
email: `hillier@pitt.edu`; `chl85@pitt.edu`

[2]Laboratoire d'Astrophysique de Marseille, Université de Provence, CNRS, 38 rue Frédéric
Joliot-Curie, F-13388 Marseille Cedex 13, France
email: `Luc.Dessart@oamp.fr`

Abstract. We highlight results from a series of investigations into modeling spectra of core-collapse supernovae (SNe). We have explored the accuracy of the expanding-photosphere method, and found that it can be used to obtain distances to Type IIP SNe with an accuracy of $\lesssim 10\%$. We confirm the result of Utrobin and Chugai (2005) that time-dependent terms must be included in the statistical equilibrium equations in order to model H I line evolution in Type II SNe, and show that time-dependent terms influence other spectral features (e.g., He I lines). We have initiated a study of polarization signatures from aspherical but axially-symmetric Type II SN ejecta. Hillier and Li acknowledge support from STScI theory grant HST-AR-11756.01.A and NASA theory grant NNX10AC80G. Dessart acknowledges financial support from grant PIRG04-GA-2008-239184.

Keywords. stars: supernovae general – stars: supernovae individual (SN 1987A, SN 1999em) – stars: distances – radiative transfer

---

With the advent of blind wide-angle surveys, supernova (SN) research is poised to make huge advances. These surveys will discover thousands of SNe, potentially identifying new subtypes, and affirm their relative occurrence versus class and host galaxy properties. In addition, forthcoming large-aperture telescopes will facilitate the direct identification of SN progenitors and provide spectral monitoring of SNe throughout their temporal evolution. With these new observations comes the need for theoretical advances in modeling SN spectra. Over the last several years, we have improved our radiative transfer code, CMFGEN, so that we can accurately model the formation of SN spectra.

Initially we modeled SNe as we would a stellar atmosphere. That is, we ignored time-dependent terms, imposed a lower boundary condition and adjusted this boundary condition, and photospheric abundances, until our model spectrum matched that observed. We explored the physics of continuum and line formation in Type II SNe (Dessart & Hillier 2005b). Additionally, we explored the accuracy of the expanding-photosphere method (EPM) and related techniques to determine distances to the host galaxies of Type II SNe. We showed that accurate distances to SNe could be obtained using the EPM technique, although there were several sources of inaccuracies which could affect distance determinations (Dessart & Hillier 2005a). Using spectral fitting, rather than photometry, overcomes many of these limitations (see also Baron *et al.* 2004).

A problem with our early investigations was that we could not match the Hα line strength during the recombination epoch – the model Hα was always weaker and narrower than observed. Following a suggestion by Utrobin & Chugai (2005), we explored the effect of time-dependent statistical equilibrium calculations. We showed that time-dependent terms substantially influence the strength of Hα during the recombination epoch, and their inclusion allows much better agreement of spectra with observation for the Type IIP

SN 1999em (Dessart & Hillier 2008). Later, we confirmed the work of Utrobin & Chugai (2005) that time-dependent statistical equilibrium calculations allow us to accurately model the He I and H I lines in SN 1987A (Dessart & Hillier 2010). As the time-dependent term changes the H ionization state, and hence the electron density, lines of other species, such as the Na I D lines, are also influenced.

As SNe are inherently time-dependent phenomena, we improved CMFGEN so that we could undertake fully time-dependent radiative transfer modeling of SN ejecta. We now start all our simulations on full SN ejecta resulting from radiation-hydro explosion models of physically-consistent pre-SN star evolution models. This approach allows a direct confrontation of synthetic observables with ejecta and progenitor properties, and thus provides constraints on the progenitor star. Using this technique, we modeled the spectral evolution of SN 1987A, and found superb agreement with observation (Dessart & Hillier 2010). The modeling was based on the hydrodynamical model lm18a7Ad (Woosley, priv. comm.), and contained *no* free parameters. We then applied the technique to explore spectral formation for two RSG progenitors which produce classic Type IIP SNe. Our models, while showing strong spectral similarities to observations, suggested that the initial radii of the two RSG models were too large (Dessart & Hillier 2011a).

We applied the fully time-dependent technique to a sample of Type Ib/Ic SN models (Dessart *et al.* 2011), and confirmed that most Type Ib/Ic SNe come from low mass progenitors. While trace amounts of hydrogen can be readily detected in early spectra of Ib/Ic SNe, helium may not be seen even if it makes up 50%, by mass, of the outer ejecta. This occurs because of the low effective temperature and the high ionization energy of He I. Eventually ionizations by non-thermal electrons, produced from the decay of high energy electrons, might excite He I lines (Lucy 1991). Thus we have incorporated non-thermal processes into CMFGEN. Such processes are also important for producing nebular-phase Type II SN spectra.

Hydrodynamical simulations of core-collapse SNe suggest that they are inherently 3D. Observational polarization studies support this scenario as most core-collapse SNe show intrinsic polarization. In order to investigate the constraints polarization places on the geometry of SN ejecta we have studied polarization signatures from axially-symmetric Type II SN ejecta (Dessart & Hillier 2011b). At early times, before the end of the plateau stage, we find that strong cancellation effects inhibit a large polarization, even when the intrinsic asymmetry is large. Further, the polarization can show a strong wavelength dependence, and even show 90° changes in the polarization angle. Similar effects can occur across strong lines such as Hα. Interestingly, our results indicate that even with constant asymmetry, a large change in the magnitude of the polarization can occur at the end of the plateau phase. Usually, this has been thought to indicate that the H-rich envelope is roughly spherically symmetric, while the He core has strong asymmetries.

References

Baron, E., Nugent, P. E., Branch, D., & Hauschildt, P. H. 2004, *ApJ*(Letters), 616, L91
Dessart, L. & Hillier, D. J. 2005a, *A&A*, 439, 671
—. 2005b, *A&A*, 437, 667
—. 2008, *MNRAS*, 383, 57
—. 2010, *MNRAS*, 405, 2141
—. 2011a, *MNRAS*, 410, 1739
—. 2011b, *MNRAS*, (in press), ArXiv:1104.5346
Dessart, L., Hillier, D. J., Livne, E., *et al.*, 2011, *MNRAS*, 414, 2985
Lucy, L. B. 1991, *ApJ*, 383, 308
Utrobin, V. P. & Chugai, N. N. 2005, *A&A*, 441, 271

From Interacting Binaries to Exoplanets: Essential Modeling Tools
Proceedings IAU Symposium No. 282, 2011
Mercedes T. Richards & Ivan Hubeny, eds.
© International Astronomical Union 2012
doi:10.1017/S1743921311027475

Polarized Radiative Transfer Equation in Some Geometries of Elliptic Type

Juris Freimanis[1,2]

[1] Ventspils International Radio Astronomy Centre, Ventspils University College,
Inzenieru iela 101a, LV-3600 Ventspils, Latvia
email: `jurisf@venta.lv`

[2] Institute of Mathematical Sciences and Information Technologies, Liepaja University,
Liela iela 14, LV-3401 Liepaja, Latvia

Abstract. A general method, which allows us to derive explicit expressions for the differential operator of stationary quasi-monochromatic polarized radiative transfer equation in Euclidean space, with piecewise homogeneous real part of the effective refractive index, is applied to ellipsoidal, oblate spheroidal, prolate spheroidal and elliptic conical coordinate systems.

Keywords. Polarization, radiative transfer, methods: analytical

---

1. Introduction

While modeling polarized radiative transfer in stellar and planetary atmospheres, circumstellar disks with exoplanets orbiting inside them, close binary systems etc., generally the radiative transfer equation (RTE) should be written and solved in such a coordinate system which reflects the symmetry of the physical problem - at least the symmetry of matter distribution; the symmetry of radiation field is usually somewhat lower.

Let us consider a homogeneous isotropic host medium in Euclidean space, with polydisperse scatterers (e.g. circumstellar dust) of volume concentration, n_0. The conditions of validity of stationary quasi-monochromatic RTE (Mishchenko *et al.* 2006, paragraphs 8.11 and 8.15; Mishchenko 2008a, Mishchenko 2008b) are satisfied, with possible addition of internal primary radiation sources as in Freimanis (2011). The effective refractive index (host medium with scatterers) is only weakly anisotropic so that birefringence can be neglected. The real part of the effective refractive index is piecewise homogeneous; radiation propagates along straight lines in each homogeneous part of the space.

We denote the Stokes 4-vector in the point of observation, $\mathbf{r}$, by $\mathbf{I}(\mathbf{r}, \vartheta, \varphi)$, where the spherical angles (ϑ, φ) describe the direction of propagation with respect to spatial basis vectors, the imaginary part of the wavenumber in the host medium by k'', the statistically averaged particle extinction matrix by $\mathbf{K}(\vartheta, \varphi)$, the statistically averaged particle phase matrix by $\mathbf{Z}(\vartheta, \varphi; \vartheta', \varphi')$, and the primary source function by $\mathbf{I}_0(\mathbf{r}, \vartheta, \varphi)$. The polarized RTE is as follows (see Mishchenko *et al.* 2006, Mishchenko 2008b, Freimanis 2011):

$$\frac{d\mathbf{I}(\mathbf{r}, \vartheta, \varphi)}{ds} - \mathbf{U}_1 \mathbf{I}(\mathbf{r}, \vartheta, \varphi) \frac{d\psi}{ds}$$
$$= -k'' \mathbf{I}(\mathbf{r}, \vartheta, \varphi) - n_0 \mathbf{K}(\vartheta, \varphi) \mathbf{I}(\mathbf{r}, \vartheta, \varphi) \tag{1.1}$$
$$+ n_0 \int_{4\pi} \mathbf{Z}(\vartheta, \varphi; \vartheta', \varphi') \mathbf{I}(\mathbf{r}, \vartheta', \varphi') \sin \vartheta' d\vartheta' d\varphi' + \mathbf{I}_0(\mathbf{r}, \vartheta, \varphi),$$

where $d\mathbf{I}/ds$ is the derivative of the Stokes vector along the path of propagation, $d\psi/ds$ is the speed of rotation of polarization reference basis vectors around the direction of propagation, and $\mathbf{U}_1$ is the transformation matrix of Stokes vector upon infinitesimal rotation of the reference system. It is assumed that the polarization reference plane is that going through the spatially variable polar axis $\vartheta = 0$ and the direction of propagation of radiation; as a result, in curvilinear coordinate system this plane generally rotates.

The left-hand side of Eq. (1.1) is a differential operator with special expression in each particular coordinate system; the right-hand side is one and the same in all coordinate systems. The aim of this study is to find explicit expressions for the differential operator of RTE in some coordinate systems of elliptic type mentioned in Korn & Korn (1968), applying the general method developed in Freimanis (2011).

2. Summary of the results

Standard procedures described in Freimanis (2011) were applied to the following orthogonal coordinate systems:

(*a*) Ellipsoidal system with the longest axis of ellipsoids coinciding with Descartes' *x* axis, and the shortest axis coinciding with Descartes' *z* axis. An alternative parameterization of the ellipsoidal coordinates was introduced;

(*b*) Standard oblate spheroidal system (Korn & Korn 1968), as a particular case of the system mentioned above;

(*c*) Standard prolate spheroidal system (Korn & Korn 1968);

(*d*) Elliptic conical system, with alternative parameterization.

Clear expressions for the differential operator of polarized RTE in all these cases were found. The respective formulae are very long in the case of a triaxial ellipsoidal system but only moderately long in the other cases. Besides, alternative parameterization for an ellipdoidal coordinate system with the longest axis of ellipsoids coinciding with Descartes' *z* axis, and the shortest axis coinciding with Descartes' *y* axis is offered as well. (The only important feature for the offered redefinitions of ellipsoidal coordinate systems is that the *z* axis must be the longest or the shortest; the *x* and *y* axis can be mutually exchanged. The given redefinition is not valid if the *z* axis coincides with the middle axis of ellipsoids.)

Acknowledgements

This study was financed from the basic budget of Ventspils International Radio Astronomy Centre, as well as from co-sponsorship by Ventspils City Council and from Latvian Council of Science grant No. 11.1856. Some financing was provided by the Institute of Mathematical Sciences and Information Technologies of Liepaja University as well. The participation of the author at the Symposium was financed from European Regional Development Fund's project SATTEH (2010/0189/2DP/2.1.1.2.0/10/APIA/VIAA/019) being implemented in Ventspils University College. The author expresses his gratitude to all these entities.

References

Mishchenko, M. I., Travis, L. D., & Lacis, A. A. 2006, *Multiple Scattering of Light by Particles. Radiative Transfer and Coherent Backscattering.* (Cambridge *et al.*: Cambridge University Press)

Mishchenko, M. I. 2008a, *Optics Express*, 16, 2288

Mishchenko, M. I. 2008b, *Journal of Quantitative Spectroscopy and Radiative Transfer*, 109, 2386

Freimanis, J. 2011, *Journal of Quantitative Spectroscopy and Radiative Transfer*, 212, 2134

Korn, G. A. & Korn, T. M. 1968, *Mathematical Handbook for Scientists and Engineers.* (New York *et al.*: McGraw-Hill Book Company)

From Interacting Binaries to Exoplanets: Essential Modeling Tools
Proceedings IAU Symposium No. 282, 2011   © International Astronomical Union 2012
Mercedes T. Richards & Ivan Hubeny, eds.   doi:10.1017/S1743921311027487

Influence of Rotation Velocity Gradient on Line Profiles of Accretion Discs of CVs

D. Korčáková[1], T. Nagel[2], K. Werner[2], V. Suleimanov[2,3], and V. Votruba[4]

[1]Astronomical Institute, Charles University, V Holešovičkách 2, 18000 Praha 8, Czech Republic; email: kor@sirrah.troja.mff.cuni.cz

[2]Institut für Astronomie und Astrophysik, Universität Tübingen, Sand 1, 72076 Tübingen, Germany

[3]Department of Astronomy, Kazan Federal University, Kremlevskaya 18, 420008 Kazan, Russia

[4]Astronomical Institute AV ČR, Fričova 298, 25165 Ondřejov, Czech Republic

Abstract. We show the influence of the Keplerian velocity shear on the line profiles of cataclysmic variable discs. The complete disc structure is taken into account. The radial disc structure follows the alpha disc approximation. Based on this assumption, the vertical structure is computed using the detailed non-LTE code *AcDc*. The obtained opacities and source functions are interpolated in the 2D grid, where the radiative transfer is calculated with the inclusion of the velocity field gradient.

Keywords. cataclysmic variables, accretion disks, line: formation, methods: numerical

Introduction

The influence of the velocity gradient on the line formation in discs of cataclysmic variables (CV) is usually neglected. However, there are works which show, that this effect can be important for high inclination angles. Nevertheless, some fundamental simplifications had to be used in the past considering the computation efficiency. The current technique allows us to describe the problem in more detail. We try to include the 2D nature and non-LTE effects in order to evaluate the influence of the rotation velocity gradient on the line formation. A detailed discussion of our results can be found in Korčáková *et al.* (2011).

Model

Consistent models of accretion discs are still beyond current computation capabilities. Therefore, we use the following approach. The radial disc structure is calculated using the α-disc approximation (Shakura & Sunyaev 1973). The vertical disc structure and the opacity and source function of the material are obtained from the non-LTE *AcDc* code (Nagel *et al.* 2004). The disc is divided into a set of concentric rings, where the radiative transfer equation (RTE), hydrostatic equilibrium and energy balance equations, and non-LTE rate equations are consistently solved using the Accelerated Lambda Iteration (ALI, Werner & Husfeld 1985). The resulting opacities and source functions are interpolated onto a 2D grid, where the RTE is solved with the inclusion of the velocity field (Korčáková & Kubát 2005).

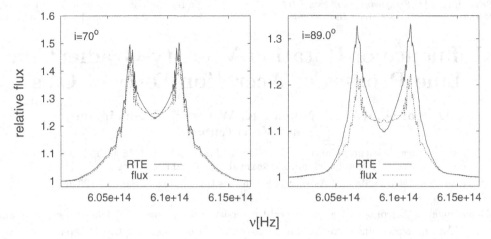

Figure 1. HeI 4923 Å line profiles from accretion discs of an AM CVn system (solid line – the velocity gradient in the RTE, dashed line – the velocity field only in the flux calculation).

Results and conclusion

The influence of the velocity field on the line profile depends on the ratio of the Doppler shift, caused by macroscopic motion, and line width in the line forming region. Therefore, we investigated the optically thinner quiescent phase of CV discs.

AM CVn systems offer a good possibility to test the velocity shear influence, since the discs are relatively cold, transparent in He lines, and rotation velocity reaches high values (3360 km s^{-1}). The results of our calculations are plotted in Fig. 1. The solid line is the solution, where the RTE is solved with the inclusion of the velocity field. The dashed line indicates the result, where the RTE is solved through the static disc and the velocity field is taken into account only in the flux calculation. The line profiles are shown for different inclination angles i. Previous works suggested that the change of the line profiles can be important for $i > 60°$. Our calculations show that even for such an extreme case, the difference is negligible for $i = 70°$ (Fig. 1, left panel). The change in the line profile can be seen just above the disc edge (Fig. 1, right panel). However, the line-profile differences were negligible for SS Cyg, whose properties are closer to a typical CV disc.

The Keplerian velocity shear can be neglected in many cases. The common approximation, where the radiative transfer is calculated through a static disc and only the outgoing flux incorporates the velocity field, is valid to high accuracy. However, the detailed calculation of the radiative transfer is necessary for optically thin lines in discs seen under high inclination angles and for situations where we see the disc rim.

Acknowledgements

This research is partly financed by grants C7 of SFB/Transregio 7 "Gravitational Wave Astronomy," NSh-4224.2008.2, 228398 (the European Commission), 205/09/P476 (GA ČR), and projects MSM0021620860 and AV0Z10030501.

References

Korčáková, D. & Kubát, J. 2005, *A&A*, 440, 715
Korčáková, D., Nagel, T., Werner, K., Suleimanov, V., & Votruba, V. 2011, *A&A*, 529, 119
Nagel, T., Dreizler, S., Rauch, T., & Werner, K. 2004, *A&A*, 428, 109
Shakura, N. I. & Sunyaev, R., A. 1973, *A&A*, 24, 337
Werner, K. & Husfeld, D. 1985, *A&A*, 148, 417

From Interacting Binaries to Exoplanets: Essential Modeling Tools
Proceedings IAU Symposium No. 282, 2011
Mercedes T. Richards & Ivan Hubeny, eds.
© International Astronomical Union 2012
doi:10.1017/S1743921311027499

Hydrodynamics of Decretion Disks of Rapidly Rotating Stars

Petr Kurfürst

Department of Theoretical Physics and Astrophysics, Masaryk University
Kotlářská 2, CZ-611 37 Brno, Czech Republic
email: petrk@physics.muni.cz

Abstract. During the evolution of hot stars, the equatorial rotational velocity can approach its critical value. Further increase in rotation rate is not allowed, consequently mass and angular momentum loss is needed to keep the star near and below its critical rotation. The matter ejected from the equatorial surface forms the outflowing viscous decretion disk. Models of outflowing disks of hot stars have not yet been elaborated in detail, although it is clear that such disks can significantly influence the evolution of rapidly rotating stars. One of the most important features is the disk radial temperature variation because the results will help us to specify the mass and angular momentum loss of rotating stars via decretion disks.

Keywords. Rotation, mass loss, hydrodynamics

1. Basic theoretical considerations

In contrast to the usual stellar wind mass loss we study the role of mass loss via an equatorial outflowing viscous decretion disk evolution in massive stars (Krtička *et al.* 2011). Evolutionary contraction brings massive star to critical rotation: it leads to the formation of the disk. Further increase in rotation rate is not allowed ($\dot{\Omega} = 0$), net loss of angular momentum is given by $\dot{L} = \dot{I}\Omega_{\mathrm{crit}}$, where $\Omega_{\mathrm{crit}} = \sqrt{GM/R_{\mathrm{eq}}^3}$ is the critical rotation frequency. The viscous coupling in a decretion disk can transport angular momentum outward to some outer disk radius, R_{out}. When a Keplerian disk is present, in comparison with the case where mass decouples in a spherical shell just at the surface of the star, the mass loss is then reduced by a factor

$$\frac{3}{2}\sqrt{R_{\mathrm{out}}/R_{\mathrm{eq}}}. \tag{1.1}$$

Key point of the analysis: the angular momentum loss from the decretion disk can greatly exceed the angular momentum loss from the stellar wind outflow.

2. Numerical approach

For the numerical modelling, it is necessary to solve the system of hydrodynamic equations in cylindrical coordinates (Krtička *et al.* 2011). Except for the mass conservation (continuity) equation, we have to include the equations for stationary conservation of R and ϕ components of momentum, supplemented by appropriate boundary conditions.

For further calculations, the Shakura-Sunyaev α viscosity parameter is introduced (Shakura & Sunyaev 1973); it expresses the quantity $\tilde{v}/a$ (somewhat simplified since there are also effects of the magnetic field), where a is the sound speed ($a^2 = kT/\mu m_H$) and we have $v_R(R_{\mathrm{crit}}) = a$.

The temperature distribution in the radial direction is assumed as $T = T_0(R_{\mathrm{eq}}/r)^p$, where p is a free parameter (power law). Some of recent models (e.g. Carcioffi *et al.*

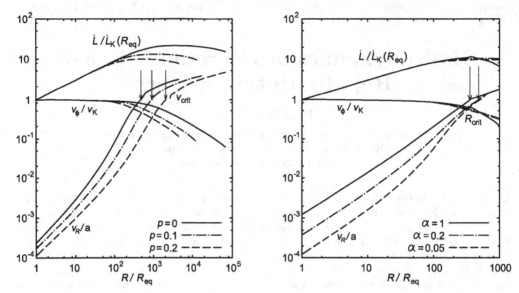

Figure 1. The graphs of dependence of relative radial and azimuthal velocity and relative angular momentum loss on radius.

2008) calculate the temperature distribution in the inner region of the disk as nearly isothermal ($T_0 = \frac{1}{2}T_{\text{eff}}$, $p = 0$). But, for the calculations of the structure of outer parts of the disk, it is necessary to consider also the power law temperature decline ($p > 0$).

The system of three hydrodynamic equations (continuity equation, equations of stationary conservation of R and ϕ components of momentum) is numerically approximated by differentiation at selected radial grid points with the use of the Newton-Raphson method (Krtička 2003).

3. Results of numerical solution

The selected stellar parameters are: spectral type B0, $T_{\text{eff}} = 30\,000\,\text{K}$, $M = 14.5\,M_\odot$, $R = 5.8\,R_\odot$ (Harmanec 1988). The left graph shows the dependencies of relative radial and azimuthal velocities and the relative loss of angular momentum on radius, the isothermal disk with various viscosity parameters α is considered ($T_0 = \frac{1}{2}T_{\text{eff}}$, $p = 0$). At the critical radius, R_{crit}: $v_R = a$ (sonic point).

The right graph shows the same dependence for fixed viscosity parameter $\alpha = 0.1$ and for different temperature profiles.

At large radii, the disk is not rotating as a Keplerian one; in the supersonic region ($v_R > a$) the rotation velocity rapidly decreases as a consequence of the adopted α viscosity parameter. The increase of parameter p (which means added cooling) implies an increase in the critical radius (the sonic point radius) and angular momentum loss. It will, therefore, be useful to calculate the models using different basic expressions for the viscous coupling.

References

Carciofi, A. C. & Bjorkman, J. E. 2008, *ApJ*, 684, 1374
Harmanec, P. 1988, *BAICz*, 39, 329
Krtička, J. 2003, *ASPC*, 288, 259
Krtička, J., Owocki, S. P., & Meynet, G. 2011, *A&A*, 527, 84
Shakura, N. I. & Sunyaev, R. A. 1973, *A&A*, 24, 337

From Interacting Binaries to Exoplanets: Essential Modeling Tools
Proceedings IAU Symposium No. 282, 2011
Mercedes T. Richards & Ivan Hubeny, eds.
© International Astronomical Union 2012
doi:10.1017/S1743921311027505

Multi-dimensional Modeling of Massive Binary Interaction in Eta Carinae

J. H. Groh

Max-Planck-Institut für Radioastronomie, Auf dem Hügel 69, D-53121 Bonn, Germany
email: jgroh@mpifr.de

Abstract. We summarize recent efforts from our group to constrain the nature of both stars in the Eta Carinae binary system and its orbital parameters by studying the influence of the companion star on the spectrum of the primary star. We find that the cavity in the dense wind of the primary star strongly affects multi-wavelength diagnostics such as the ultraviolet spectrum, the optical hydrogen lines, and the shape of the near-infrared continuum region. These diagnostics have been previously interpreted as requiring a latitude-dependent wind generated by a fast-rotating primary star, but the effects of the companion on them provide tenuous evidence that the primary star is a rapid rotator.

Keywords. stars: winds, outflows; stars: binaries; stars: emission line

---

Eta Carinae is one of the most luminous objects in the Galaxy, and its study provides crucial constraints on the evolution and death of the most massive stars. Extensive observational monitoring of Eta Car in the optical and X-rays revealed that Eta Car is a massive binary system consisting of two very massive stars, η_A (primary) and η_B (secondary), with a total system mass amounting to at least 110 $M_\odot$ (Hillier *et al.* 2001). Although most orbital parameters of the system are uncertain, the wealth of multi-wavelength observations are consistent with a high eccentricity ($e \sim 0.9$) and an orbital period of 2022.7 ± 1.3 d (Damineli *et al.* 2008).

To understand and quantify the effects of η_B on the wind of η_A, we developed a two-dimensional radiative transfer code to model massive binaries with extended winds. In the case of Eta Car, our models are based on the spherically symmetric models of η_A (Hillier *et al.* 2001), but use the 2-D code of Busche & Hillier (2005) to study the influence of the low-density cavity, and dense interaction-region walls, on the spectrum. Further details are given in Groh *et al.* (2010) and Groh *et al.* (2011, in prep.).

We approximate the cavity as a conical surface with half-opening angle α and interior density 0.0016 times lower than that of the spherical wind model of η_A. We include cone walls of angular thickness $\delta\alpha$ and, assuming mass conservation, a density contrast in the wall of $f_\alpha = [1 - \cos(\alpha)]/[\sin(\alpha)\delta\alpha]$ times higher than the wind density of the spherical model of η_A at a given radius. The conical shape is justified since the observations we model here were taken at orbital phases sufficiently before periastron, when such a cavity has an approximately 2-D axisymmetric conical form (Okazaki *et al.* 2008). Based on the expected location of the cone apex during these phases (Okazaki *et al.* 2008), we place the cavity at a distance $d_{\rm apex}$ from the primary star. We assume that the material inside the cavity and along the walls has the same ionization structure as the wind of η_A. Thus, at this point we explicitly neglect the ionization changes that might occur in the wind-wind interacting region. We also do not account for the ionizing flux of η_B. Despite these limitations, our implementation should be adequate for understanding how the line profiles of η_A are modified by the carving of its wind by η_B.

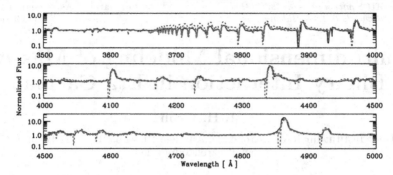

Figure 1. Optical spectrum of η_A observed with *HST/STIS* at $\phi \sim 0.6$ (solid black line) compared to the spherically-symmetric CMFGEN model (blue dashed line) and with our 2-D model (red dashed line). See text for the model parameters.

Here, we illustrate how the optical spectrum is affected by the cavity created by the companion star in the wind of the primary. Assuming a single-star scenario with a spherically-symmetric stellar wind, Hillier *et al.* (2001) obtained a reasonable fit to the observed *HST*/STIS optical spectrum obtained right after periastron, at orbital phase $\phi \sim 0.04$. However, the spherically-symmetric CMFGEN model overestimates the amount of P-Cygni absorption. The comparison is even worse as one moves toward apastron ($\phi \sim 0.5$), when the observations show little or no P-Cygni absorption components in H and Fe II lines (Fig. 1). This has been interpreted as if η_A would be a rapid rotator and its wind, latitude dependent (Hillier *et al.* 2001, Smith *et al.* 2003).

Using our 2-D radiative transfer model, which takes into account the cavity in the wind of η_A caused by η_B, we computed the synthetic optical spectrum of Eta Car at $\phi \sim 0.6$ (Fig. 1). We assumed the same parameters described above for η_A (Hillier *et al.* 2001) and a standard geometry of the cavity at apastron as predicted by 3-D hydrodynamical simulations (Okazaki *et al.* 2008): $d_{\mathrm{apex}} = 25$ AU, $\alpha = 54°$, $\delta\alpha = 3°$ (i.e., $f_\alpha = 9.7$), $b = 0.0016$. For a viewing angle of $i = 41°$ and longitude of periastron of $\omega = 270°$, the 2-D model produces a much better fit to the P-Cygni absorption line profiles of H and Fe II lines than the 1-D CMFGEN model, while still fitting the emission line profiles. The improved fit to the P-Cygni absorption line profiles yielded by the 2-D model is due to the cavity in the wind of η_A, which reduces the H and Fe II optical depths in the line-of-sight to η_A when the observer looks down the cavity.

We find that η_B significantly affects the H line profiles, which is one of the main diagnostics of rapid rotation in η_A. Therefore, an intrinsic latitude-dependent wind generated by fast rotation of η_A may not be the only explanation for existing observations, but this does not mean that η_A is not a rapid rotator. As the presence of the cavity and walls affects also the available interferometric observables (Groh *et al.* 2010), which were the other diagnostics supporting fast rotation, tenuous evidence supports a fast-rotating primary star.

References

Busche, J. R. & Hillier, D. J. 2005, *AJ*, 129, 454

Damineli, A., Hillier, D. J., Corcoran, M. F., *et al.*, 2008, *MNRAS*, 384, 1649

Groh, J. H., Madura, T. I., Owocki, S. P., Hillier, D. J., & Weigelt, G. 2010, *ApJ* (Letters), 716, L223

Hillier, D. J., Davidson, K., Ishibashi, K., & Gull, T. 2001, *ApJ*, 553, 837

Okazaki, A. T., Owocki, S. P., Russell, C. M. P., & Corcoran, M. F. 2008, *MNRAS*, 388, L39

Smith, N., Davidson, K., Gull, T. R., Ishibashi, K., & Hillier, D. J. 2003, *ApJ*, 586, 432

From Interacting Binaries to Exoplanets: Essential Modeling Tools
Proceedings IAU Symposium No. 282, 2011
Mercedes T. Richards & Ivan Hubeny, eds.

© International Astronomical Union 2012
doi:10.1017/S1743921311027517

Modeling of the Be Stars

K. Šejnová[1,2] V. Votruba[1,2] and P. Koubský[1]

[1] Astronomical Institute AV ČR v.v.i, Ondřejov, Czech Republic

[2] Masaryk University, Faculty of Science, Brno, Czech Republic

Abstract. The Be stars are still a big unknown in respect to the origin and geometry of the circumstellar disk around the star. Program `shellspec` is designed to solve the simple radiative transfer along the line of sight in three-dimensional moving media. Our goal was to develop an effective method to search in parameter space, which can allow us to find a good estimate of the physical parameters of the disk. We also present here our results for Be star 60 Cyg using the modified code.

Keywords. Be stars, shellspec, genetic algorithms

1. Introduction

Program `shellspec` is designed by J. Budaj (Budaj & Richards 2004). Optional (non)-transparent objects such as spot, disk, shell, etc. may be defined in 3D and their composite synthetic spectrum calculated. The main input file is called shellspec.in. We can set input parameters of an object here. In the current version of the code, the parameters have to be set manually, which is not an effective method, so we decided to make the program more automatic.

2. Modification of `shellspec`

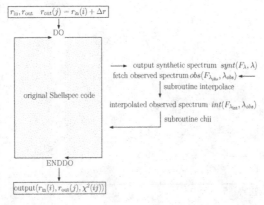

Figure 1. Procedure to modify `shellspec` program

The first part of the modification includes a grid method that systematically searches through the field of control parameters. Then the calculation is repeated, and as a result we get the synthetic spectrum, which we compare with the observed one. On the basis of the known observed spectrum, we find the value of χ^2 from the match. The last part of the calculation consists of choosing the best combination of control parameters that describes the observed spectrum in the best way when the chosen criterium is the χ^2 value.

3. Results for 60 Cyg

We used a modified version of the `shellspec` code to study 60 Cyg. The results are shown in Table 1 and Fig. 2.

Table 1. Final parameters of disk of 60 Cyg star

	i	$R_{in}\,[R_\odot]$	$R_{out}\,[R_\odot]$	$T\,[K]$	$\rho\,[\times10^{-13}\,\mathrm{g\,cm^{-3}}]$	$v_{trb}\,[\mathrm{km\,s^{-1}}]$
P1	30	5,202	12,681	13000	9	90
P2	30	6,979	7,748	13000	1	60
P23	30	5,202	6,742	14000	1	80
P233	30	5,455	5,982	14600	1	70
P3	30	5,202	5,532	14500	2	80

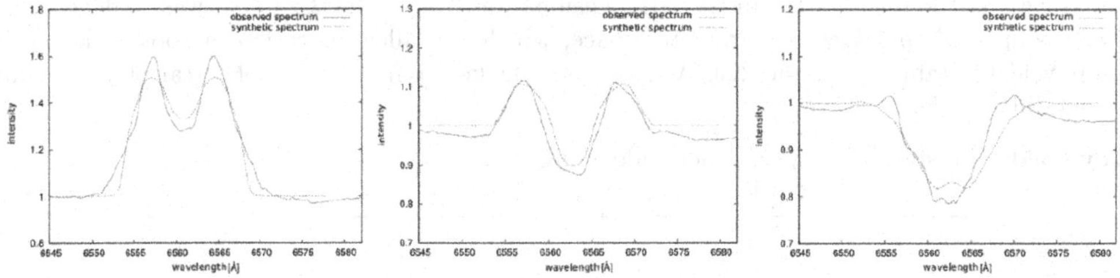

Figure 2. Observed and synthetic spectrum for three different Hα profiles of 60 Cyg

4. PIKAIA subroutine

The grid method is not one of the most effective methods due to its computational heftiness. A genetic algorithm can be used to improve the computational efficiency because it would allow us to increase the number of control parameters. Genetic Algorithms are search and optimization procedures that are motivated by the principles of natural genetics and natural selection. PIKAIA is a general purpose optimization subroutine based on a genetic algorithm (Charbonneau & Knapp 1995).

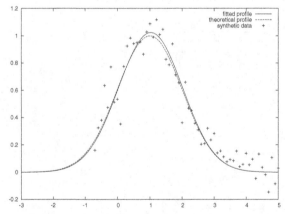

Figure 3. Test example: gaussian curve fit where theoretical curve has shape $f(x) = e^{-\frac{(x-1)^2}{2}}$

References

Šejnová, K. 2010, Shell modeling of the Be stars, Diploma Thesis, MU Brno

Budaj, J. & Richards, M. T. 2004, A description of the Shellspec code, *Contrib. Astron. Obs. Skalnate Pleso* 34, 1–30

Koubský, P., *et al.* 2000, *ASP-CS*, 214, 280

Charbonneau, P. & Knapp, B. 1995, A User's Guide To PIKAIA 1.0, Vol. NCAR/TN-418+IA

From Interacting Binaries to Exoplanets: Essential Modeling Tools
Proceedings IAU Symposium No. 282, 2011   © International Astronomical Union 2012
Mercedes T. Richards & Ivan Hubeny, eds.   doi:10.1017/S1743921311027529

Fitting of Silicon lines in UV and Balmer Hδ Line in Optical Spectra of B supergiant HD 198478

Tomislav Jurkić, Mariza Sarta Deković and Dubravka Kotnik-Karuza

Department of Physics, University of Rijeka, Croatia
email: tjurkic@phy.uniri.hr, msarta@phy.uniri.hr

Abstract. Atmospheric parameters of the Galactic early B-supergiant HD 198478 (55 Cyg) were determined from the UV silicon lines and optical Balmer Hδ 4101 Å line. TLUSTY synthetic spectra were broadened using the ROTIN numerical code in order to determine effective temperature, surface gravity, rotational and macroturbulent velocity.

Keywords. line: profiles, stars: fundamental parameters, stars: supergiants

1. Introduction

Massive B supergiants (BSGs) are mostly investigated in the optical region, while UV spectra have been avoided as this region is dominated by a vast number of spectral lines which make spectral analysis very difficult. HD 198478 is a well known early BSG, subject to numerous studies mainly in the optical region (Crowther *et al.* 2006, Markova & Puls 2008, Searle *et al.* 2008). We have chosen this object as a probe star from a large sample of galactic BSGs. Our aim is to establish a consistent method for determination of atmospheric parameters from International Ultraviolet Explorer (IUE) spectra of BSGs, as IUE is in many cases the only source of UV spectra of these stars.

2. Observations and methods

We have identified and used UV Si III multiplet lines at 1294 Å, 1299 Å, 1301 Å and 1303 Å, Si II doublet lines at 1264 Å and 1309 Å, Si III singlet lines at 1312 Å and 1417 Å, and the optical Balmer Hδ 4101 Å line. High-resolution UV spectra were obtained by IUE between 1200 Å and 1450 Å with a resolution of 0.16 Å at 1400 Å, as well as by the more recent HST STIS in the range 1150-1360 Å with a resolution of 0.012 Å. High-quality optical spectra from 4100 Å to 4900 Å with a resolution of 0.4 Å were obtained by Markova & Puls (2008). Atmospheric parameters of HD 198478 were determined by line profile fitting of the observed spectra with the available BSTAR grid model spectra obtained by TLUSTY and SYNSPEC (Lanz & Hubeny 2003, Lanz & Hubeny 2007) numerical codes. Synthetic spectra were broadened with instrumental profiles, and rotational and macroturbulent velocities by using the ROTIN code. Following Markova & Puls (2008) and Monteverde *et al.* (2000), we have adopted the TLUSTY grid with Si abundance od 7.55. In order to reduce the number of free fitting parameters, we used the TLUSTY grid value with a microturbulence of 10 km/s in accordance with McErlean *et al.* (1999). After determination of $\log g$ from the Hδ line, silicon UV lines were fitted with effective temperature, macroturbulent and projected rotational velocities as free parameters.

Table 1. Atmospheric parameters od HD 198478 determined by different authors.

Author	T_{eff} (K)	Log g	v_{tot} (km/s)
Crowther *et al.* (2006)	16500	2.15	61
Markova & Puls (2008)	17500	2.10	66
Searle *et al.* (2008)	17500	2.25	61
Jurkić *et al.* (2011)	17000	2.25	62

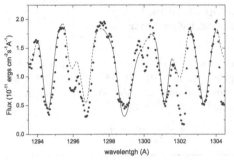

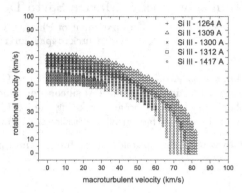

Figure 1. Observed IUE and synthetic spectra for Si III 1294-1303 Å multiplet lines. Only parts of the spectra unambiguously belonging to the spectral lines are fitted (full line).

Figure 2. Plot of the pairs (v_{macro}, v_{rot}) that produces acceptable fits for all measured silicon lines.

3. Discussion

There is a very good agreement of our results with the results of other authors, which were mainly obtained from the optical spectra by use of plane-parallel or wind models (Table 1). Our analysis shows that the plane-parallel TLUSTY atmosphere model, despite the fact that it does not include wind effects, can successfully reproduce Si line profiles and determine atmosphere parameters. Note that Searle *et al.* (2008) failed to reproduce UV spectra of photospheric silicon lines using the CMFGEN wind model.

Our results have shown that the line broadening cannot be explained by the rotational velocity only, but that an additional macroturbulent velocity component should be taken into account, with a significant degeneracy between macroturbulent and projected rotational velocities (Fig. 2). Comparison between the IUE UV spectra and the HST STIS spectra of higher quality have shown that IUE spectra can be used for modeling the stellar atmosphere in the UV (Fig. 1). We can conclude that the Si lines are photospheric and unaffected by the stellar wind, suggesting that line-blanketing, line-blocking, and detailed treatment of ionization levels as found in TLUSTY are essential in treating photospheric UV lines.

References

Crowther, P. A., Lennon, D. J., & Walborn, N. R. 2006, *A&A*, 446, 279
Jurkić, T., Sarta Deković, M., Dominis Prester, D., & Kotnik-Karuza, D. 2011, *Ap&SS*, accepted
Lanz, T. & Hubeny, I. 2003, *ApJS*, 146, 417
Lanz, T. & Hubeny, I. 2007, *ApJS*, 169, 83
Markova, N. & Puls, J. 2008, *A&A*, 478, 823
McErlean, N. D., Lennon, D. J., & Dufton, P. L. 1999, *A&A*, 349, 553
Monteverde, M. I., Herrero, A., & Lennon, D. J. 2000, *ApJ*, 545, 813
Searle, S. C., Prinja, R. P., Massa, D., & Ryans, R. 2008, *A&A*, 481, 777

From Interacting Binaries to Exoplanets: Essential Modeling Tools
Proceedings IAU Symposium No. 282, 2011   © International Astronomical Union 2012
Mercedes T. Richards & Ivan Hubeny, eds.   doi:10.1017/S1743921311027530

Creation of Neutral Disk-like Zone Around the Active Hot Star in Symbiotic Binaries

Zuzana Cariková and Augustin Skopal

Astronomical Institute, Slovak Academy of Sciences,
059 60 Tatranská Lomnica, Slovakia
email: zcarikova@ta3.sk

Abstract. We investigated the ionization structure of symbiotic binaries during their active phases. We found that a neutral disk-like zone around the hot star can be created as a result of its enhanced wind and fast, 200 - 300 km s^{-1}, rotation. Calculated column densities of the neutral hydrogen atoms throughout the neutral zone and emission measures of the ionized part of the wind from the hot star are in a good agreement with those derived from observations.

Keywords. Stars: binaries: symbiotic, stars: winds: outflows

1. Introduction

Symbiotic stars are long-period interacting binary systems, which comprise a late-type giant and a hot compact star, which is in most cases a white dwarf. Accretion from the giant's wind makes the white dwarf to be a strong source of ionizing radiation ($T \sim 10^5$ K). During so called quiescent phases, it ionizes a fraction of the giant's wind giving rise to a dense nebula extended within the binary. However, during active phases, the ionization structure changes significantly. Modeling the spectral energy distribution of symbiotic binaries with high orbital inclination indicated the presence of a two-temperature type UV spectrum (Skopal 2005). The cooler component is produced by a relatively warm stellar source ($T \sim 22\,000$ K), while the hotter one is represented by the highly ionized emission lines and a strong nebular continuum. Skopal (2005) explained this type of the UV spectrum by a model in which a neutral disk-like structure surrounds the accretor in the orbital plane and hot emitting regions are located above/below the disk. The rapid creation of such an ionization structure during the first days/weeks of outbursts is connected with the enhanced wind from the hot star. Due to the fast rotation of the accretor, the wind particles are compressed more to the equatorial plane, where they create a neutral zone in the form of a flared disk. We tested the latter by applying the wind compression model. The presence of such disks is transient, being connected with the active phases of symbiotic binaries.

2. Ionization structure in the wind from the hot star

Modeling of the broad Hα wings showed that the stellar wind from the hot star is significantly enhanced to $\sim 10^{-7}$ - 10^{-6} M$_\odot$yr^{-1} during active phases. However, during quiescent phases, the mass loss rate of the hot star is $\sim 10^{-8}$ M$_\odot$yr^{-1} (Skopal 2006). Rotation of the hot star leads to compression of the outflowing material towards the equatorial plane due to the conservation of angular momentum. The mechanism is described by the so-called wind compression model, which was developed by Bjorkman & Cassinelli (1993). If the streamlines of gas from both hemispheres do not cross the equatorial plane then we are talking about the wind compressed zone model described

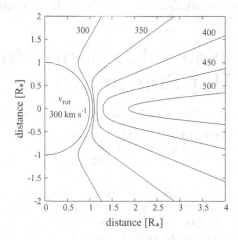

Figure 1. Calculated ionization boundaries in the stellar wind from the rotating hot star with the rotational velocity 300 km s^{-1}. Individual ionization boundaries are denoted by the value of the parameter X. On the left from each ionization boundary there is the ionized zone and on the right there is the neutral zone. Distances are given in units of the radius of the hot star.

by Ignace *et al.* (1996). We calculated the ionization boundary in the compressed wind from the hot star using the equation for photoionization equilibrium. For simplicity, we assumed that the wind contains only hydrogen atoms. We calculated models for different rotational velocities of the hot star from 100 to 350 km s^{-1}. Figure 1 shows some examples of the ionization boundaries which were calculated for the different values of the ionization parameter X (given by the parameters of the hot star and its wind) resulting in the creation of the neutral disk-like zone near the equatorial plane of the hot star.

3. Discussion and conclusion

We showed that during active phases, the compression of the enhanced stellar wind from the rotating hot star towards the equatorial regions can lead to the creation of the neutral disk-like zone around the hot star. In symbiotic binaries with high orbital inclination, we are looking through the neutral disk-like zone up to a high optical depth, at which the material simulates a pseudophotosphere which radiates at significantly lower temperature than the hot central star. Further, we calculated column densities of the neutral hydrogen atoms in the disk-like zone above the pseudophotosphere and emission measures of the ionized part of the wind from the hot star, which are in a good agreement with the observed quantities. During quiescent phases, there is no neutral disk-like zone around the hot star, because of insufficient mass loss rate of the hot star.

Acknowledgement

This research was supported by a grant from the Slovak Academy of Sciences, VEGA No. 2/0038/10. Z. C. acknowledges support from the IAU.

References

Bjorkman, J. E. & Cassinelli, J. P. 1993, *ApJ*, 409, 429
Ignace, R., Cassinelli, J. P., & Bjorkman, J. E. 1996, *ApJ*, 459, 671
Skopal, A. 2005, *A&A*, 440, 995
Skopal, A. 2006, *A&A*, 457, 1003

From Interacting Binaries to Exoplanets: Essential Modeling Tools
Proceedings IAU Symposium No. 282, 2011   © International Astronomical Union 2012
Mercedes T. Richards & Ivan Hubeny, eds.   doi:10.1017/S1743921311027542

New Galactic Candidate Luminous Blue Variables and Wolf-Rayet Stars

Guy S. Stringfellow[1], Vasilii V. Gvaramadze[2], Yuri Beletsky[3], and Alexei Y. Kniazev[4]

[1] Center for Astrophysics and Space Astronomy,
University of Colorado, 389 UCB, Boulder, CO 80309-0389, USA
email: Guy.Stringfellow@colorado.edu

[2] Sternberg Astronomical Institute, Moscow State University, Universitetskij Pr. 13, Moscow
119992, Russia
email: vgvaram@iki.rssi.ru

[3] European Southern Observatory, Alonso de Cordova 3107, Santiago, Chile
email: ybialets@eso.org

[4] South African Astronomical Observatory, PO Box 9, 7935 Observatory, Cape Town, South
Africa
email: akniazev@saao.ac.za

Abstract. We have undertaken a near-infrared spectral survey of stars associated with compact mid-IR shells recently revealed by the MIPSGAL (24 μm) and GLIMPSE (8 μm) *Spitzer* surveys, whose morphologies are typical of circumstellar shells produced by massive evolved stars. Through spectral similarity with known Luminous Blue Variable (LBV) and Wolf-Rayet (WR) stars, a large population of candidate LBVs (cLBVs) and a smaller number of new WR stars are being discovered. This significantly increases the Galactic cLBV population and confirms that nebulae are inherent to most (if not all) objects of this class.

Keywords. stars: emission-line, Be, stars: mass loss, stars: winds, outflows, stars: Wolf-Rayet

1. Introduction

Despite intensive search efforts over the last several decades the Galactic Luminous Blue Variable (LBV) population has remained sparse. This paucity is difficult to reconcile with our understanding of stellar evolution of the most massive stars. Until the last year or two there were only 12 confirmed Galactic LBVs known, and 23 candidate-LBVs (Clark *et al.* 2005). LBVs display rather unique rich infrared emission line spectra, including contributions from H, He, Mg II, Na I, and Fe II. Visual inspection of the *Spitzer* GLIMPSE (Benjamin *et al.* 2003) and MIPSGAL (Carey *et al.* 2009) Galactic plane surveys have produced catalogues of previously unknown 8 μm and 24 μm nebulae with concentric point sources that can be traced back to 2MASS K-band or even optical sources as possible progenitors of the associated nebulae (Gvaramadze *et al.* 2010a, Wachter *et al.* 2010), especially with new imaging (Stringfellow *et al.* 2012). We are conducting a near-IR spectral survey to identify new cLBVs and WRs that produced these shells.

2. Observations and Results

We have obtained spectra of $\sim$50 stars associated with newly discovered mid-IR nebulae using SpeX on the NASA IRTF 3m, Triplespec on APO 3.5m and Palomar Hale 5m, and ISAAC on the ESO-VLT. A few of the K-band spectra are shown in Figure 1. The

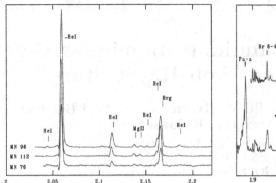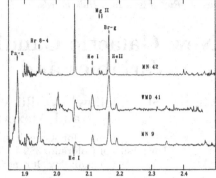

Figure 1. Normalized K-band spectra of newly identified cLBVs and WRs. See Figure 1 of Stringfellow *et al.* 2012 for images identifying MN 96, including optical recovery in the I-band.

left panel shows three newly identified cLBVs that have Fe II emission absent in their IR spectra. Prominent line emission arise from He I, Br γ, and Mg II. A spectrum of MN 96 (WMD 54, Wachter *et al.* 2010), is discussed in Wachter *et al.* (2011), who notes the similarity between both LBV and WR late-type WN spectra for this particular star. Our spectrum clearly indicates the absence of any He II 2.189 μm line emission. Comparison of an optical spectrum of MN 112 with that of P Cyg rendered classification as a cLBV (Gvaramadze *et al.* 2010b); both spectra display numerous optical Fe III lines, but no Fe II line emission. The absence of the 2.089 μm Fe II line in the MN 112 K-band spectrum is consistent with a higher temperature in this line emitting region. MN 76 (WMD 38) was classified as a Be star (Wachter *et al.* 2011), though no spectrum was shown. Clearly the K-band spectra for these three stars - MN 96, MN 112, and MN 76 - are nearly identical (barring small differences in line widths and strengths), and should render the same IR spectral classification. These stars could be transitional between the LBV and late-WN stars, or have spectral types varying between minimum contraction to maximum expansion, corresponding to hot and cool temperature phases, respectively. VLT spectra for two WRs, WMD 41 (WN8-9h) and MN 9 (WN7-9h), are displayed in the right panel of Figure 1 along with the VLT spectrum for the cLBV MN 42 (WMD 15). The WR spectra lack Mg II emission and display broader H and He lines than the cLBVs. MN 42 was classified as B[e]/LBV by Wachter *et al.* (2011) though no spectrum was shown. MN 42 resembles those cLBVs shown in the left panel, lacking Fe II emission, strengthening the case that the FeII-deficient cLBVs may be transitioning to late-WN stars. We designate MN 42, MN 76, MN 96, and MN 112 as currently FeII-deficient cLBVs.

GSS thanks the AAS for receipt of an ITG and SmRG, and the IAU for support.

References

Benjamin, R. A., *et al.*, 2003, *PASP*, 115, 953
Carey, S. J., *et al.*, 2009, *PASP*, 121, 76
Clark, J. S., Larionov, V. M., & Arkharov, A. 2005, *AA*, 435, 239
Gvaramadze, V. V., Kniazev, A. Y., & Fabrika, S. 2010a, *MNRAS*, 405, 1047
Gvaramadze, V. V., *et al.*, 2010b, *MNRAS*, 405, 520
Stringfellow, G. S., *et al.*, 2012, in *Four Decades of Research on Massive Stars: A Scientific Symposium in Honour of Anthony F.J. Moffat*, ASP Vol., eds L. Drissen, N. St-Louis, C. Robert, & A.F.J. Moffat (in press)
Wachter, S., *et al.*, 2010, *AJ*, 139, 2330
Wachter, S., *et al.*, 2011, *BSRSL*, 80, 291

Part 5
Synthetic Light Curves and Velocity Curves, Synthetic Spectra of Binary Stars and their Accretion Structures

From Interacting Binaries to Exoplanets: Essential Modeling Tools
Proceedings IAU Symposium No. 282, 2011
Mercedes T. Richards & Ivan Hubeny, eds.

© International Astronomical Union 2012
doi:10.1017/S1743921311027554

Advances in Modeling Eclipsing Binary Stars in the Era of Large All-Sky Surveys with EBAI and PHOEBE

A. Prša[1], E. F. Guinan[1], E. J. Devinney[1], P. Degroote[2], S. Bloemen[2] and G. Matijevič[3]

[1] Villanova University, Dept. of Astronomy & Astrophysics, 800 Lancaster Ave,
Villanova PA 19085, USA; email: aprsa@villanova.edu
[2] Instituut voor Sterrenkunde, K.U.Leuven, Celestijnenlaan 200D, B-3001 Leuven, Belgium
[3] University of Ljubljana, Dept. of Physics, Jadranska 19, SI-1000 Ljubljana

Abstract. With the launch of NASA's Kepler mission, stellar astrophysics in general, and the eclipsing binary star field in particular, has witnessed a surge in data quality, interpretation possibilities, and the ability to confront theoretical predictions with observations. The unprecedented data accuracy and an essentially uninterrupted observing mode of over 2000 eclipsing binaries is revolutionizing the field. Amidst all this excitement, we came to realize that our best models to describe the physical and geometric properties of binaries are not good enough. Systematic errors are evident in a large range of binary light curves, and the residuals are anything but Gaussian. This is crucial because it limits us in the precision of the attained parameters. Since eclipsing binary stars are prime targets for determining the fundamental properties of stars, including their ages and distances, the penalty for this loss of accuracy affects other areas of astrophysics as well. Here, we propose to substantially revamp our current models by applying the lessons learned while reducing, modeling, and analyzing Kepler data.

Keywords. methods: data analysis, methods: numerical, binaries: eclipsing, stars: fundamental parameters, stars: statistics

---

1. Introduction

A thorough understanding of the fundamental stellar parameters (masses, radii, luminosities, ages, chemical compositions and distances) and processes (energy transport mechanisms, nucleosynthesis, etc.) in stars across the Hertzsprung-Russell diagram is the core of stellar astrophysics. To rigorously study stars, we need to determine their properties as accurately as possible, using the minimum number of underlying assumptions. Eclipsing binary stars (hereafter EBs) are ideal astrophysical laboratories to achieve this goal: the favorable alignment of the line of sight with the orbital plane and the basic principles of classical dynamics that govern the motion of the components in a binary, reduce the determination of principal parameters to a simple geometric problem. This unique property, contrasted with other means of determining stellar radii that either apply to only a handful of objects (such as resolving the disk of a star) or are encumbered with a larger uncertainty (i.e. *P-L-R* relationships), promoted eclipsing binaries as key calibrators of stellar properties and distance gauges. Binarity allows us to determine the masses of individual components, and the alignment of a system's orbit with the line of sight and consequent eclipses allow us to determine their radii to better than a few percent (Andersen 1991). To perform such accurate modeling, both photometric and spectroscopic observations are required. An excellent overview of the state of the field is given by Torres, Andersen & Gimenez (2010).

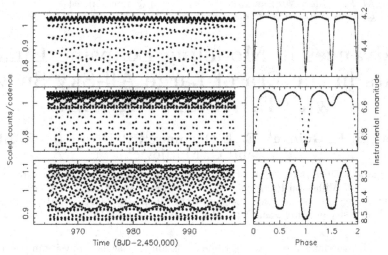

Figure 1. Kepler observations of a detached EB KIC 5513861 (top; P=1.51012-d), a semi-detached EB KIC 8074045 (middle; P=0.53638-d), and an overcontact EB KIC 3127873 (bottom; P=0.67146-d).

NASA's Kepler mission (Borucki *et al.* 2010) revolutionized two crucial aspects of our ability for detailed EB modeling: the unprecedented photometric accuracy, and an essentially uninterrupted observing mode (cf. Fig. 1). Prša *et al.* (2011, paper I) and Slawson *et al.* (2011, paper II) cataloged 2165 EBs in the Kepler field found in the public Quarter 1 and 2 data, with periods ranging from an hour to several months. The EBs in the catalog are classified by morphology as overcontact, semi-detached, detached, and ellipsoidal. A preliminary analysis of these EBs was done by EBAI (Eclipsing Binaries via Artificial Intelligence; Prša *et al.* 2008), an approach based on back-propagating neural networks that yields principal parameters for all binaries. These estimates serve as a basis for detailed modeling.

Having superb quality data at hand has clear repercussions on our modeling capability. For the first time, we are observing astrophysical phenomena uninterruptedly, and to an amazing level of detail. State-of-the-art models such as the renowned Wilson-Devinney code (Wilson & Devinney 1971; Wilson 1979, 1993, 2007), ELC (Orosz & Hauschildt 2000), and PHOEBE (Prša *et al.* 2005) are showing systematics in a whole range of binary light curves. This is partly due to approximations embedded in these models, and partly due to missing and/or inadequate physics that has yet to be accounted for. Here, we identify the most striking deficiencies of our models that hinder the reliability and extensiveness of binary star solutions. Suitably rectified models allow us to determine fundamental stellar parameters to 1% or better. These are subsequently used to calibrate stars across the H-R diagram (Harmanec 1988), determine accurate distances (Guinan *et al.* 1998), and study a range of intrinsic phenomena such as pulsations, spots, accretion disks, etc. (Olah 2007).

2. Eclipsing Binaries via Artificial Intelligence (EBAI)

The EBAI project (Prša *et al.* 2008) employs backpropagating neural networks to rapidly estimate principal parameters from light curves. For many, Artificial Neural Networks (ANN) invoke a veil of suspicion, sometimes because they seem so intangible, at other times because they are a purely mathematical construct deprived of any physical context. In reality, ANNs are very simple algorithms that hardly involve anything beyond summation and multiplication and are trained on a physical content.

In its basic form, an ANN is a system of three layers. Each layer consists of a given number of independent units. Each unit holds a single value. These values are propagated from each unit on the current layer to all units on the subsequent layer by weighted connections. Propagation is a simple linear combination $y_i = \sum_j w_{ij} x_j$, where x_j are the values on the current layer, w_{ij} are weighted connections, and y_i are the values that enter the subsequent layer. Before they are stored in their respective units, y_i are first passed through the activation function, A_f. This function, typically a sigmoid function $A_f(y_i) = 1/[1+exp(-(y_j\mu)/\tau)]$, introduces non-linear mapping properties to the network. Coefficients μ and τ selected so that $A_f(y_i)$ fall in the $(1,1)$ interval. This value is stored in the i-th unit on the subsequent layer. Layers in the three-layer network are usually denoted input, hidden, and output layer. In a nutshell, ANN is a non-linear mapping from the input layer to the output layer. In the domain of EBs, the ANN maps the input light curves to the output set of principal physical parameters. Training the network implies determining the weights, w_{ij}, on weighted connections. The back-propagation algorithm relies on a sample of LCs with known physical parameters; these are called exemplars. All LCs are propagated through the network and their outputs are compared to the known values. The weights are then modified so that the discrepancy between the two sets is minimized. This is an iterative process that needs to be done only once. After training, the network is ready to process any input LC extremely quickly. For example, solving 10,000 LCs on a single 2GHz processor takes around 5 seconds.

3. PHysics Of Eclipsing BinariEs (PHOEBE)

Our current understanding of the properties and processes in binary stars shape our ability to model observed data. PHOEBE (Prša *et al.* 2005) is a physical model based on the Roche geometry; it includes more than 40 physical and geometric parameters that determine light and radial velocity (RV) curve properties. The most important effects implemented in the model are:

- an analytic description of binary star orbits, including apsidal motion;
- iterative solution of the Kepler problem that governs binary star dynamics;
- shape distortion due to eccentricity, tides and (asynchronous) rotation;
- radiative properties of binary star components, including gravity brightening, limb darkening and reflection effect;
- spots, circumbinary attenuation clouds, and third light.

Furthermore, PHOEBE provides minimization algorithms that fit the model curves to the data. This highly non-linear problem suffers from non-unique solutions: the *right* combination of the *wrong* parameters can fit the observed data very well. The algorithms currently in use, namely Differential Corrections (DC), Nelder & Mead (1965)'s Simplex method (NMS), Powell's direction set method, and genetic algorithms, have all met with success, but cannot be run robustly without experienced human intervention.

The level of detail in PHOEBE and its predecessor WD was sufficient for several decades, however this has been superseded by the Kepler mission. We are now seeing phenomena that have been theorized but never observed before, and we are seeing them systematically. The approach of modeling the EB baseline, while assuming that all neglected physical factors are buried deep in noise, is no longer applicable. The residuals are nowhere near Gaussian, and we cannot assume that these effects are only perturbations; rather, we need to account for their signatures in light and radial velocity curves rigorously.

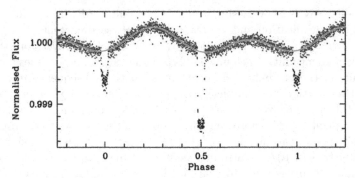

Figure 2. Kepler observations of KOI-74. The model fit consists of two sine waves, at the orbital and half-orbital period, corresponding to Doppler beaming and ellipsoidal variations, respectively. Adapted from Van Kerkwijk *et al.* (2010).

4. Challenges

In this section we present some of the challenges being addressed in the field of eclipsing binary stars.

4.1. *Doppler Beaming*

The required fidelity of models for EBs has been set historically by the photometric precision of ground-based observations, typically a few milli-mag per datapoint. MOST, CoRoT, and now Kepler, have attained a revolutionary photometric precision of several parts per million per datapoint. At this level, the photometric phenomenon of Doppler beaming was predicted to be observable (Loeb & Gaudi 2003). Doppler beaming (also boosting) refers to the shift in bolometric luminosity of a star accruing from its radial velocity. The resulting Doppler-shifted bolometric flux, F, is related to the stationary flux F_0 by $F = F_0(1 + 4v_r/c)$. At a given frequency ν, the observer sees $F_\nu = F_{\nu,0}(1 + (3 - \alpha)v_r/c)$, where α is a parameter that depends on the bandpass and the slope of the spectrum of the observed star. Essentially, beaming makes an object appear brighter on approach and fainter on recession, thus modulating the light curve of a binary system (cf. Fig. 2). Analogous to radial velocity information, disentangling orbital beaming in a light curve can yield the masses of the binary components.

Until recently, beaming was just a footnote for eclipsing binaries. However, the effect has now been detected for two Kepler binaries, KOI 74 and KOI 81, both likely white dwarfs (Van Kerkwijk *et al.* 2010), and for CoRoT-3, a massive planet/low-mass brown dwarf orbiting an F-star (Mazeh & Faigler 2010), as well as in a subdwarf B–white dwarf pair by Bloemen *et al.* (2010). Groot (2011) computed the observability of the rotational Doppler beaming effect for EBs, a photometric phenomenon directly analogous to the classical spectroscopic Rossiter-McLaughlin effect, which allows a photometric determination of the projected radial velocity of the eclipsed star as a function of phase. The effect is shown to be detectable for binaries ranging from double white dwarfs to massive O-stars, as well as for the WASP-33b-like transiting system, for which it could in principle reveal the host star's rotational obliquity.

To date, most studies predicted and confirmed detections of Doppler beaming based on simplified, ad hoc theoretical estimates of the magnitude of the effect for binary stars (or star+planet). While this approach is adequate to confirm the expected physics, it clearly indicates the need to incorporate Doppler beaming, both orbital and rotational, into rigorous eclipsing binary models. Neglecting this effect would cause a substantial

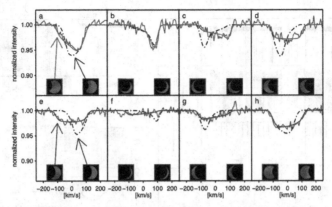

Figure 3. Comparison between observed spectra during the primary eclipse (top) and the secondary eclipse (bottom). Solid black lines are observations, the dash-dotted lines are the model fit of co-aligned stars, and the solid lines are model fits that allow for spin-orbit misalignment. Adapted from Albrecht *et al.* (2011). This is known as the Rossiter-McLaughlin effect, and has been observed in a number of stars and planets.

systematic discrepancy for objects of significantly different luminosities; for such objects, beaming provides constraints on RV semiamplitude without the need for spectroscopy.

4.2. *Spin-Orbit Misalignment*

Most current EB models are based on the Roche model, which assumes perfect alignment between the orbital and the rotational axes and handles each component's tidal distortion as due to the companion's point mass. However, careful ground-based studies have been able to discern that the components of close binaries in a number of systems show misaligned rotational and orbital axes (Albrecht *et al.* 2011). Observationally, this is most easily discerned via the Rossiter-McLaughlin effect, depicted in Fig. 3. The non-alignment of a host star spin-orbit has also been found for a number of hot Jupiters (Hebrard *et al.* 2008). Consequently, EBs and transiting exoplanets bring into question the initial conditions for the formation of both binaries and planetary systems.

The generalized Roche potential for binary systems, in which the stellar rotation is not aligned with the orbital revolution, is fundamentally different from the currently implemented co-aligned potential. Modifications to the potential have been derived by Limber (1963), Kruszewski *et al.* (1966), and Kopal (1978). The properties of the critical equipotential lobe and Lagrangian points for circular orbits have been studied in detail by Avni & Schiller (1982).

In binary star modeling, a Cartesian coordinate system is usually set at the center of the primary star, where x-axis points to the center of the secondary star, and z-axis points along the orbital revolution axis. Misalignment may be fully described by two angles: axial deflection from the z-axis (pitch), and rotation from the x-axis (yaw). The third angle (roll) may be dropped because of symmetry. A simple rotation of the coordinate system about the x-axis reduces this to a single-parametric problem, with a misalignment parameter θ' denoting the angle between the rotated z-axis and the spin axis, which due to the transformation now lies in the $x'z'$ plane. This one parameter determines the location and properties of the Lagrangian points, and hence the shapes of both components and instantaneous force fields acting on them (Avni & Schiller 1982).

A fully numerical search for equipotential surfaces in misaligned binaries needs to be implemented. These surfaces determine the shapes and radiative properties of components in binary stars. Since 3-D minimization is a computationally expensive task,

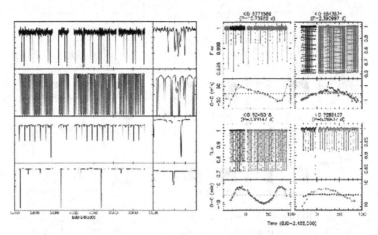

Figure 4. Left: 4 EB targets with tertiary events detected in the public Kepler data. These are indicative of tertiary components transiting the binary star. Right: eclipse timing variations for 4 interesting targets. Colors in the light curves denote data quarters (Q0: magenta, Q1: black, Q2: red), and primary and secondary eclipse timing variations (red and black, respectively).

the search will be initiated only when the misalignment parameter θ' changes its value. Equipotential surfaces will otherwise be stored for subsequent use.

4.3. *Multiple Stellar Systems: Extraneous Bodies in Binary Systems*

In an EB, one would expect that primary eclipses and secondary eclipses are uniformly spaced in time. However, mass transfer from one star to the other, or rotation of the line of apsides (apsidal motion), or the presence of a third star in the system can give rise to changes in the orbital period, which in turn change the spacing in time between consecutive eclipse events. The eclipse times will no longer be described by a simple linear ephemeris, and the deviations from the linear ephemeris (usually shown in the O-C diagram) will contain important clues to the origin of the period change. Systematic measurements of the times of primary and secondary eclipses for the Kepler sample of EBs have been conducted (Orosz *et al.* 2011, Prša *et al.* 2012, in preparation). This is a tedious task, owing to a host of intrinsic variabilities and systematic problems. These include large spot modulations that may or may not be in phase with the eclipses, pulsations and/or noise in the out-of-eclipse regions, thermal events and cosmic ray hits that make the normalization of the light curves hard to automate, and eclipses falling partially or completely in data gaps. Fig. 4 presents some interesting cases of EBs with O-C variations evident in the public Kepler data.

The modeling codes can deal with the simplest case of apsidal motion, where the argument of periastron changes linearly in time ($d\omega/dt = $ const.). We found, and successfully modeled, a number of EBs with strong apsidal motions, and the models successfully predicted the shapes of light curves without a substantial increase in systematics.

Our adopted model follows the basic concept laid out by Carter *et al.* (2011) that was applied to KOI-126, a hierarchical stellar triple system observed by Kepler. A hierarchical (or Jacobi) coordinate system is used when calculating the positions of the three bodies. In this system, r_1 is the position of star A relative to star B (the inner pair) and r_2 is the position of star C relative to the center of mass of (A, B). We may specify r_1 and r_2 in terms of osculating Keplerian orbital elements (period, eccentricity, argument of pericenter, inclination, longitude of the ascending node, and the mean anomaly). Newton's equations of motion, which depend on r_1, r_2, and the masses, may be specified for the accelerations $\ddot{r}_1$ and $\ddot{r}_2$ (Soderhjelm 1984; Mardling & Lin 2002). An additional term may be added to the acceleration of r_1 due to the post-Newtonian potential of the inner binary (Soffel 1989). Further perturbing accelerations may be added to the acceleration of r_1 corresponding to the non-dissipative equilibrium tidal potential between stars A and B and the potential associated with the rotationally-induced oblate distortion of stars A and B (Soderhjelm 1984). In this approximation, the axial spins of stars A and B follow the evolving orbit, staying normal to the orbit and spinning at a rate synchronous with the orbit. Both the accelerations due to tides and rotations depend on r_1, the masses, and the radii of both components. The acceleration due to rotation also depends on the angular axial spin rate of both stars A and B. The spin rates and apsidal constants are assumed to be the same for both stars. This system provides a coupled system of differential equations that are solved using the Bulrisch-Stoer algorithm. The positions of the three objects are then projected to the barycentric plane, which is necessary to account for the finite speed of light and predict eclipse timing variations. Carter *et al.*'s method assumes spherical stars, whereas our implementation will feature tidally deformed stars whose shapes will be derived from the generalized Roche formalism presented in the previous section.

4.4. *Markov Chain Monte Carlo heuristics and Bayesian Error Estimates*

Perhaps one of the worst plagues of EB modeling is the inherent non-linearity of the parameter space and thus a high degree of solution degeneracy. Coupled with that is a classical approach to data fitting by least squares and estimating errors from the co-variance matrix. This has two important and dire consequences. First, essentially any minimizer used is bound to get stuck in local minima. Ways around this have been proposed, most notably by heuristic Monte Carlo scanning and parameter perturbations (Prša *et al.* 2005; Prša & Zwitter 2007), or by utilizing global search algorithms such as the Metropolis-Hastings simulated annealing, which, however, dramatically increases computing time. Second, chi-square fitting is done on a data curve level rather than on the parameter level. Any deviation between the model and the data will penalize all parameters marked for adjustment. To see why this is problematic, consider adjusting the semi-major axis or the mass ratio of a well detached binary. There is absolutely no information contained in a light curve on either of the two parameters; all information lies in radial velocity curves. Hence, if a simultaneous LC+RV least squares fit is employed, the minimizer will fit the correct RV signature for the two parameters, but it will essentially fit numerical noise in the EB light curve. The number of data points in a light curve is typically an order of magnitude larger (several orders in case of Kepler), so the determination of these two parameters will be heavily weighted towards light curves where there is little to no oversight when determining their values.

Markov Chain Monte Carlo (MCMC) proves to be a powerful tool for Bayesian inference because it provides more statistical information and makes better use of data than chi-square fitting. The goal is to determine what configurations of physical and geometric parameters are consistent with given light and radial velocity observations. For each parameter, we provide an initial estimate – a prior – that does not need to be close. After running a given set of chains, MCMC returns a posterior distribution for each parameter: a histogram of attained values that determines the most likely value as well as the uncertainty in the statistical sense. Thus, MCMC provides a best-fit solution and the corresponding error estimates by assessing them on a parametric level rather than on a data-set level. The latter is exactly the culprit that we would like to eliminate, and MCMC is a clear path to doing so. The approach has been implemented and used successfully in BLENDER (Torres *et al.* 2011), a tool devised to discriminate between bona fide extra-solar planets and background EBs in the Kepler data-set. Other examples of a successful MCMC application include (Bloemen *et al.* 2010) for Doppler beaming, and Hou *et al.* (2011) for fitting radial velocities of stars that host extra-solar planets.

5. Conclusion

Ultra-high precision space missions MOST, CoRoT, and Kepler revolutionized observations, and our tools need to keep up with them. The two striking issues are: (1) the amount of data acquired and in need of automated processing, and (2) the inadequacies of our models that cause systematics in the solutions. Both need to be addressed immediately, otherwise our analysis capabilities will be inferior to the data. To address the first issue, we propose to use the artificial intelligence based engine EBAI; for the second issue, significant effort in restating the physics and geometry of the model is underway.

References

Albrecht, S., *et al.*, 2011, *ApJ*, 726, 68
Andersen, J. 1991, *A&ARv*, 3, 91
Avni, Y. & Schiller, N. 1982, *ApJ*, 257, 703
Bloemen, S., *et al.*, 2010, *MNRAS*, 407, 507
Borucki, W. J., *et al.*, 2010, *Science*, 327, 977
Carter, J., *et al.*, 2011, *Science*, 331, 562
Groot, P. J. 2011, *arXiv*, 1104:3428v3
Guinan, E. F., Fitzpatrick, E. L., & Dewarf, L. E., *et al.*, 1998, *ApJ*, 509, L21
Harmanec, P. 1988, *Bulletin of the Astronomical Institutes of Czechoslovakia*, 39, 329
Hebrard, G., *et al.*, 2008, *A&A*, 488, 783
Hou, F., *et al.*, 2011, *arXiv* 1104:2612
Kopal, Z. 1978, *Dynamics of Close Binary Systems*, ISBN 9-789-02770-820-5
Kruszewski, A. 1966, *Adv. Astr. ap.*, 4, 233
Limber, D. N. 1963, *ApJ* 138, 1112
Loeb, A. & Gaudi, S. B. 2003, *ApJ*, 588, 117
Mardling, R. A. & Lin, D. N. C.. 2002, *ApJ*, 573, 829
Mazeh, T. & Faigler, S. 2010, *A&A*, 521, L59
Nelder, J. A. & Mead, R. 1965, *Computer Journal*, 7, 308
Olah, K., *IAUS*, 240, 442
Orosz, J. A. & Hauschildt, P. H. 2000, *A&A*, 364, 265
Prša, A. & Zwitter, T. 2005, *ApJ*, 628, 426
Prša, A. & Zwitter, T. 2005, *ASPC*, 370, 175P
Prša, A., *et al.*, 2008, *ApJ*, 687, 542
Prša, A., *et al.*, 2011, *AJ*, 141, 83
Slawson, R. W., *et al.*, 2011, *AJ*, 142, 160
Soderhjelm, S. 1984, *A&A*, 141, 232
Soffel, M. H. 1989, *Relativity in Astrometry, Celestial Mechanics and Geodesy*, ISBN 3-540-18906-8
Torres, G., Andersen, J., & Gimenez, A. 2010, *A&ARv*, 18, 67
Torres, G., *et al.*, 2011, *ApJ*, 727, 24T
Van Kerkwijk, M. H., *et al.*, 2010, *ApJ*, 715, 51
Wilson, R. E. & Devinney, E. J. 1971, *ApJ*, 166, 605
Wilson, R. E. 1979, *ApJ*, 234, 1054
Wilson, R. E. 1993, *New Frontiers in Binary Star Research*, 38, 91
Wilson, R. E. 2007, *IAUS*, 240, 188

From Interacting Binaries to Exoplanets: Essential Modeling Tools
Proceedings IAU Symposium No. 282, 2011   © International Astronomical Union 2012
Mercedes T. Richards & Ivan Hubeny, eds.   doi:10.1017/S1743921311027566

ROCHE: Analysis of Eclipsing Binary Multi-Dataset Observables

Theodor Pribulla

Astronomical Institute, Slovak Academy of Sciences, 059 60 Tatranská Lomnica, Slovakia
email: pribulla@ta3.sk

Abstract. Code *ROCHE* is devoted to modeling multi-dataset observations of close eclipsing binaries such as radial velocities, multi-wavelength light curves, and broadening functions. The code includes circular surface spots, eccentric orbits, asynchronous or/and differential rotation, and third light. The program makes use of synthetic spectra to compute observed *UBVRIJHK* magnitudes from the surface model and the parallax. The surface grid is derived from a regular icosahedron to secure more-or-less equal (triangular) surface elements with observed intensities computed from synthetic spectra for supplied passband transmission curves. The limb-darkening is automatically interpolated from the tables after each computing step. All proximity effects (tidal deformation, reflection effect, gravity darkening) are taken into account. Integration of synthetic curves is improved by adaptive phase step (important for wide eclipsing systems).

The code is still under development. It is planned to extend its capabilities towards low mass ratios and widely different radii of components to facilitate modeling of extrasolar planet transits. Another planned extension of the code will be modeling of spatially-resolved eclipsing binaries using relative visual orbits and/or interferometric visibilities.

Keywords. stars:binaries:eclipsing, stars:binaries:spectroscopic, techniques:interferometric

1. Introduction

The study of eclipsing binaries is the principal way to determine reliable stellar masses, radii, and luminosities needed to test stellar structure and evolutionary models.

The beginning of the realistic interpretation of the light curves (hereafter LCs) of close eclipsing binaries started with the seminal papers of Lucy (1968ab) leading to a plethora of codes in the 1970s (by e.g., Wilson & Devinney, 1971; Mochnacki & Doughty, 1972; Rucinski, 1973; Binnendijk, 1977).

At present, simultaneous solutions of two or more kinds of observations (LCs, radial-velocity (RV) curves, spectral lines profiles, timing variations, and even polarimetric data) are coming into more frequent usage. The morphology of binaries today includes extensions such as non-synchronous rotation, eccentric orbits, or effects of the radiation pressure (Drechsel *et al.*, 1995). Complicated models computing radiation transfer and including accretion disks have been developed (e.g., SHELLSPEC of Budaj & Richards, 2004).

Parameter adjustment uses advantageous methods such as the Levenberg-Marquardt method or Simplex algorithm. Comprehensive overview of close binaries, including determination of orbits and absolute parameters, as well as imaging of stellar surfaces, can be found in Hilditch (2001).

The program most widely used to analyze eclipsing binary data is the Wilson & Devinney (hereafter W&D) code, which was described in a series of papers (for references see Wilson, 1994). A user-friendly interface to the W&D code (PHOEBE) has been developed by Prša & Zwitter (2005). The EBAI project (see Prša *et al.*, 2008) uses artificial intelligence to cope with the enormous increase in the number of LCs of eclipsing binaries resulting from several sky surveys (e.g., OGLE, ASAS, MACHO, NSVS).

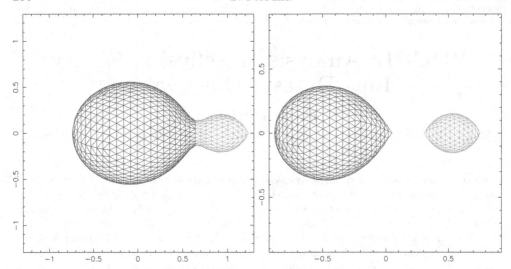

Figure 1. Surface grid based on a regular icosahedron documented in the case of a contact binary (left) and a double-contact binary with the secondary component rotating at break-up velocity with the rotation factor $F = 6$ (right). View from the orbital plane.

2. The ROCHE code

A new code, ROCHE, is based on Roche geometry. In the case of asynchronous rotation and eccentric orbits, the assumptions of Wilson (1979) are used (volume of components is preserved in eccentric orbits). The routines of *ROCHE* are, however, completely independent of the W&D code. It is not planned to add additional structures (such as accretion or circumbinary disks, gas streams) in addition to the binary.

The code uses extensive tables of passband-specific and bolometric limb darkening coefficients of van Hamme (1993). Local passband-specific intensities are interpolated from extensive model-atmosphere tables of Lejeune *et al.* (1997). The model takes into account mutual irradiation of the components and gravity darkening as well. The surface grid is based on the Platonic solid with the largest number of faces - icosahedron. Each of its 20 faces is broken to smaller triangles. In the case of spherical stars, the elements are equal to about 15% (Fig. 1). The relative density of grids for primary and secondary component can be automatically scaled according to the ratio of radii.

The model uses an advanced treatment of the third light: the passband specific third light contributions are taken either independently or it is assumed that the third component is a star. In the latter case, its temperature and radius (with respect to the semi-major axis of the binary) can be optimized.

For the computation of the LCs, the flux from all visible surface points is summed. In the case of the BF synthesis, the summing is done in the RV domain. Intervals of exposures are taken into account when synthetizing BFs. The visibility is tested by the routine based on the principles set by Djurašević (1993). The terminator is modeled by a broken line (the terminator breaks a surface triangular element into a triangle and a quadrangle). The *UBVRIJHK* apparent magnitudes are computed using supplied trigonometric parallax. Modeling of the interstellar reddening/extinction is not included in the current version.

The observables are synthetized in 360 phase points but, in the case of detached eclipsing binaries, the code enables sub-dividing the phase step during the eclipses. Another approach to run the code more efficiently is to use an adaptive phase step, depending on

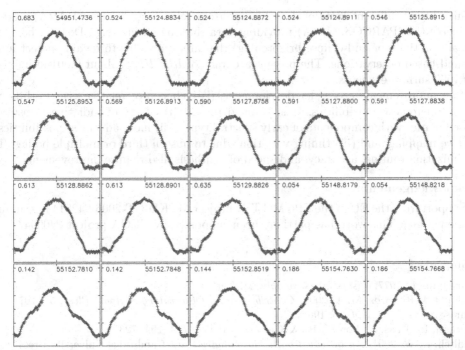

Figure 2. Global fit (black solid line) to BFs (red crosses) of δ Velorum (Pribulla *et al.*, 2011)

the phase derivative of the synthetised LC (yet to be implemented). The observables are then interpolated in the grid of phases.

The code enables simultaneous optimization of up to 7 passband-specific LCs, 2 RV curves, 200 BFs (see Rucinski, 1992), one set of squared interferometric visibilities, and the relative visual orbit (under development).

Optimization is performed using a damped differential correction method. It is possible to perform a grid search for the best χ^2 for a user-defined grid of mass ratios and/or inclination angles.

The code uses the PGPLOT graphic routines, which enables visualization of the input data and fits (or residuals) after each optimization step. Different styles in plotting 3D surfaces including spots can be used. Best fits as well as optimal parameters including standard errors are also automatically saved. A new input parameters file is produced after each computation step.

3. Prospects for further improvements of the code

Although the code has successfully been used to analyze observations of several eclipsing systems (see e.g., case of δ Velorum, Pribulla *et al.* 2011 and Fig. 2; or MOST eclipsing binaries, Pribulla *et al.* 2010) it is still being improved and extended. The code has been extensively tested to synthetize observables for a large range of parameters. The major challenge is its extension to mass ratios below $m_2/m_1 < 0.02$ and ratio of radii below $r_2/r_1 \sim 0.10$. The 3D surfaces are correctly generated by *ROCHE* for the mass ratios down to $q \sim 10^{-5}$, but the major problem poses correct representation of the terminator for systems with components of widely different radii. Modeling LCs and BFs of transiting close-in hot Jupiters assuming Roche model is very important (see Budaj, 2011).

Major challenges in eclipsing-binary data modeling are connected with increasing photometric precision of LCs (CoRoT, KEPLER, MOST, STEREO, SMEI etc.) and availability of different sorts of data like interferometry (e.g., VLTI, CHARA), and astrometry (HIPPARCOS, GAIA) providing data showing effects (e.g., Doppler beaming, dynamical tides, or mid-eclipse brightening) that are impossible to reliably detect in the ground-based observations. The present accuracy of *ROCHE* is about 0.001 in intensity for 1000 surface elements.

Two improvements are under development: (i) inclusion of relative visual orbits and (ii) interferometric visibilities. It is planned to add (i) effects of radiative pressure in close binaries with components of early spectral types, (ii) more advanced possibilities in surface mapping, and (iii) timing variations due to unseen third or multiple bodies. The optimization routine, accuracy, and speed of synthesis also require improvements.

Acknowledgements

Support from the EU in the FP6 MC ToK project MTKD-CT-2006-042514 is gratefully acknowledged. This work has partially been supported by VEGA project 2/0094/11.

References

Binnendijk, L. 1977, *Vistas in Astronomy*, 21, 359
Budaj, J. & Richards, M. T. 2004, *Contrib. Astron. Observatory Skalnaté Pleso*, 34, 167
Djuraševic, G. 1992, *Ap&SS*, 196, 241
Drechsel, H., Haas, S., Lorenz, R., & Gayler, S. 1995, *A&A*, 294, 723
Hilditch, R. W. 2001, *An introduction to close binary stars*, Cambridge University Press
Lejeune, T., Cuisinier, F., & Buser, R. 1997, *A&AS*, 125, 229
Lucy, L. B. 1968a, *ApJ*, 151, 1123
Lucy, L. B. 1968b, *ApJ*, 153, 877
Mochnacki, S. W. & Doughty, N. A. 1972, *MNRAS*, 156, 51
Pribulla, T., Merand, A., Kervella, P. *et al.*, 2011, *A&A*, 528, 21
Pribulla, T., Rucinski, S. M., Latham, D. W. *et al.*, 2010, *AN*, 331, 397
Prša, A., Guinan, E. F., Devinney, E. J. *et al.*, 2008, *ApJ*, 687, 542
Prša, A. & Zwitter, T. 2005, *ApJ*, 628, 426
Rucinski, S. M. 1992, *AJ*, 104, 1968
van Hamme, W. 1993, *AJ*, 106, 2096
Wilson, R.E. 234, 1054
Wilson, R. E. 1994, *PASP*, 106, 921
Wilson, R. E. & Devinney, E. J. 1971, *AJP*, 166, 605

Discussion

P. ZASCHE: Is the code ROCHE somehow publicly available?

T. PRIBULLA: Yes, please write to me to get more information about the code.

C. MACERONI: Concerning future improvements: a trivial, but important one, for analysis of Kepler LCs is to include the long integration time (for long cadence data). That is important in some cases as shown by A. Prša. As far as I know, only PHOEBE (A. Prša) and JKTEBOP (J. Southworth) include that.

T. PRIBULLA: My code does not take long integration times into account in the case of LC synthesis. The code is developed according to the data which are available and systems that are solved. It should be, however, rather simple to implement the "phase smearing" in the case of LCs.

From Interacting Binaries to Exoplanets: Essential Modeling Tools
Proceedings IAU Symposium No. 282, 2011   © International Astronomical Union 2012
Mercedes T. Richards & Ivan Hubeny, eds.   doi:10.1017/S1743921311027578

The BINSYN Program Suite

Albert P. Linnell

Department of Astronomy, University of Washington Seattle, WA, USA
email: linnell@astro.washington.edu

Abstract. The BINSYN program suite has been ported to a Linux-based operating system. The new program structure is a major revision from the original version and a public version is nearing completion. This paper describes research areas where the program suite is particularly applicabile.

1. Introduction

BINSYN is a general purpose program package for analysis of binary stars with or without an optically thick accretion disk. It simulates either photometric or spectroscopic data or both. The original version simulated light curves using a black body approximation; the basic design features are in Linnell (1984). A differential corrections capability was added in Linnell (1989). Linnell & Hubeny (1994) added the capacity to calculate synthetic spectra of binary star systems and Linnell & Hubeny (1996) extended the simulation to include binary stars with optically thick accretion disks.

Until 2009, the suite ran on a Windows operating system compiled under Microsoft Powerstation Fortran. At that time, the decision was made to convert to a VMWare virtual workstation running Linux Ubuntu. The conversion entailed major restructuring of the program intercommunications and a plan was implemented to develop a publicly-accessible version. A fortuitous contact with Mr. Paul DeStefano, a local Linux expert, led to adoption of the Git Version Control System as the appropriate vehicle to accomplish the objective. Mr. DeStefano has been instrumental in promoting this objective and he remains a development participant. As of the date of this report, the software is stable with 71 programs and 89,000 lines of code. The programs have been written in FORTRAN 77 and compiled with GNU gfortran. Several included tutorial programs remain to be completed and a Users Guide is in preparation.

2. Some Program Suite Features

The primary design approach is to divide the simulation task into a series of consecutive program units linked by scripts. To illustrate this, an initial program called CALPT produces photospheric potentials for the two component stars on the Roche model. A following program, PGA, defines the photospheres with grids of colatitude and longitude contours with assignable angular resolution. Subsequent programs produce plane of the sky projections for one or more orbital positions, determine T_{eff} values at grid nodes for both components, calculate light intensity values toward the observer at each grid node, and finally sum the intensity values with due allowance for eclipse effects. The T_{eff} values include allowance for gravity darkening and mutual irradiation and the intensity values are for a list of wavelengths.

Additional program units produce a synthetic spectrum of the binary system including individual spectra of the components and a system spectrum. These units require input spectra, either from the program SYNSPEC (Hubeny, Lanz & Jeffery 1994) or elsewhere, and interpolate to each grid node in T_{eff} and log g. If the input spectra are obtained from SYNSPEC, the user has the advantage of specifying the wavelength range and resolution. In effect, these latter programs attach a limb-darkened, Doppler-shifted synthetic spectrum to each grid node.

Separate program sequences simulate accretion disk systems and include the simulation of the two stars. The program models an accretion disk as an assembly of concentric annuli. The default model for the accretion disk is Keplerian rotation; a different rotation rate can be specified for individual annuli. The simulations include eclipse effects by either star or the accretion disk and optionally include irradiation of the secondary star by the accretion disk, irradiation of the accretion disk rim by the secondary star, and a hot spot on the rim from impact by the mass transfer stream. The simulation uses an expression for the accretion disk temperature profile more general than the Standard Model but includes the Standard Model as the default model. The basic model uses a black body approximation. An interface to the program TLUSTY (Hubeny 1990) produces annulus models on the TLUSTY approximation and, via the program SYNSPEC (Hubeny, Lanz & Jeffery 1994), corresponding synthetic spectra. The interface program produces a system synthetic spectrum and separate synthetic spectra of the two stars, the accretion disk face, and the accretion disk rim at each tabular orbital longitude.

The program suite can calculate and store synthetic spectra for multiple orbital longitudes, permitting the generation of light curves by synthetic photometry. Differentials correction optimization of synthetic photometry light curves has been tested, so far for binary systems without accretion disks.

In the Linux version, each program resides in an individual directory immediately below the binsyn root directory. A script in each program directory copies necessary input files from a separate input file directory, runs the program, and copies output files to a separate output file directory. The input files use a template format that makes them convenient for use. Each main program has an output file, among several, that logs the running time and calculated parameter values in an easy-to-read tabular format.

A differentials correction optimization requires partial derivatives of system light with respect to system parameters as a function of orbital longitude. BINSYN uses a numerical analysis procedure including second differences. A set of programs receives specified central reference values of system parameters and upper and lower values for each parameter with symmetric offsets from the central reference value. The corresponding light curves provide first differences and second differences at the tabular orbital longitudes. This program set uses a pre-established file of orbital longitudes that will be used in the simulation. Inclusion of second difference tabulation permits calculation of derivatives anywhere within the bounding interval for a given derivative without recalculation of the first and second differences.

Linnell (1989) discusses the differentials correction solution. The current version of this program module includes explicit choice of what parameters to include, calculation of the covariance matrix, a test for normal distribution of residuals, production of a histogram of residuals, and a criterion for stopping iterations.

3. Input and output file directories

There is a general purpose input file directory immediately below binsyn and at the same level as the separate program directories. The motivation for this directory is that it acts as an intermediary between individual program directories and directories for particular binary star systems. Thus, a directory for a particular binary system will contain all the input files needed for a solution. A script in the binary system directory copies the complete set of the necessary input files to the general purpose input file directory and starts a run of the solution. The programs in the solution chain successively access the input file directory for the files they need. This procedure permits multiple binary star directories to be set up, each with its appropriate set of parameters. A solution for a given system can be run, suspended, and a run for another system initiated without requiring the user to set the input files of the individual programs. Similarly, there is a general purpose output file directory. On completion of a particular program run, the controlling script in that program directory copies the output files to the general purpose output file directory. At the level of the individual star directory, the master local script controlling the complete solution copies the preset input files to the general purpose input file directory, runs the solution program, and copies the final output files from the general purpose output file directory to the specific binary star directory. The user can remain within the binary directory for the entire solution process and examine the final output files without leaving the directory.

4. Illustrations of BINSYN applications

Three papers illustrate the application of BINSYN under diverse circumstances.

Sion *et al.* (2011) is a study of five AM CVn systems. These are doubly degenerate CV systems with nearly pure He components. We choose EM Ceti for illustration. There is a Hipparcos parallax that sets the scaling factor to superpose a system synthetic spectrum on the available IUE spectrum. We used TLUSTY to calculate an atmospheric model for the WD with a He/H number ratio of 10^4. Our study established a need for a narrow accretion disk with a T_{eff} matching the 40,000K WD. It had been suspected that EM Ceti is a direct impact accretor. We calculated the trajectory of the mass transfer stream and find that it would not intersect the WD; the stream does intersect the postulated accretion disk.

Linnell *et al.* (2010a) is a study of WX LMi, a CV polar whose B light curve is strongly affected by cyclotron radiation. Synthetic B,V,R,I photometry established a distance to the system. A particular problem with this system was synthetic spectra representing a white dwarf, dominating the system spectrum in the UV, combined with a relatively cool (3300K) secondary star that dominates the system in the IR.

Linnell *et al.* (2010b) is a study of the CV RW Sex, a nova-like with an accretion disk which dominates the system spectrum. The study demonstrated the inability of the so-called Standard Model to represent the observed spectrum; the study found a model with a modified radial temperature gradient which fitted the observations to within observational error. It is unknown whether the different model has a more general application to nova-like cataclysmic variables. In this system the fit of the final model to the observed spectra confirmed independent determination of the system distance. BINSYN provides the capacity for distance determination either by calibrated synthetic photometry or by fitting synthetic spectra to calibrated observed spectra.

References

Hubeny, I. 1990, *ApJ*, 351, 632
Hubeny, I., Lanz, T., & Jeffery, C. S. 1994, in Newsletter on Analysis of Astronomical Spectra No. 20, ed. C. S. Jeffery (CCP7; St. Andrews: St. Andrews Univ.), 30
Linnell, A. P. 1984, *ApJS*, 54, 1
Linnell, A. P. 1989, *ApJ*, 342, 449
Linnell, A. P. & Hubeny, I. 1994, *ApJ*, 434, 738

Linnell, A. P. & Hubeny, I. 1996, *ApJ*, 471, 958
Linnell, A. P., *et al.*, 2010a, *ApJ*, 713, 1183
Linnell, A. P., *et al.*, 2010b, *ApJ*, 719, 271
Sion, E. M., *et al.*, 2011, *ApJ*, submitted

Discussion

A. PRSA: What is the computational time cost for the light curve synthesis?

A. LINNELL: Using the black body option with the illustration grid resolution, the time per iteration is about a minute. Using the synthetic photometry option, the time per iteration is about one hour. That will improve considerably with projected modification of the grid.

From Interacting Binaries to Exoplanets: Essential Modeling Tools
Proceedings IAU Symposium No. 282, 2011
Mercedes T. Richards & Ivan Hubeny, eds.
© International Astronomical Union 2012
doi:10.1017/S174392131102758X

Application of the GDDSYN Method in the Era of KEPLER, CoRoT, MOST and BRITE

Stefan W. Mochnacki

Dept. of Astronomy & Astrophysics, University of Toronto,
50 St. George St. Rm. 101, Toronto ON, Canada M5S 3H4
email: stefan@astro.utoronto.ca

Abstract. The precision of observations using observatories in space exceeds by a factor of 100 the accuracy of the light curve and line profile synthesis methods developed decades ago. Furthermore, physical effects too small to detect using ground based observations, such as aberration and Doppler beaming, become important when observing from space.

The GDDSYN method, developed by Hendry and Mochnacki, is both accurate and efficient, and is useful in the new context of space-based observations. Using a geodesic distribution of triangular surface elements varying little in size, it provides an alternative to the Wilson-Devinney code used at the heart of PHOEBE, and is adaptable to the new physical effects which are now observable. Tests and improvements are discussed.

Keywords. stars: binaries: eclipsing, methods: numerical

1. Introduction

It is 43 years since the first synthesis of eclipsing binary light curves based on the Roche model (Lucy 1968), although the first detailed fits to actual observations had to wait for Wilson & Devinney (1971) and Mochnacki & Doughty (1972), Lucy (1973) and soon many others, as cited in the comprehensive textbook by Kallrath & Milone (1998), p.17 ff. Just as the earliest papers were being published, a remarkable conference, IAU Colloquium 16, "Analytical Procedures for Eclipsing Binary Light Curves", was held in Philadelphia, September 8-11, 1971. It was a confrontation between the old analytical methods using simplified geometrical models and the new synthesis computer codes using realistic equipotentials and astrophysics to model close binary stars. I recall that at lunch one day a number of us young synthesizers were sitting with John Merrill, who had devoted much of his life to producing the tables for using the Russell-Merrill method (Merrill 1950 and Russell & Merrill 1952). At some point a visibly energized Dr. Merrill admonished us all:

"You young guys should pay attention to the Tables!"

It had become clear to him that what us "young guys" had done was to make his grand tool for solving eclipsing binary light curves essentially irrelevant, that the elegance of analytical functions and the vast effort of calculation by hand was supplanted by brute force calculations taking a few minutes on a room-sized mainframe computer. This moment, and everything else at the conference, were crystallized for me later when I read Kuhn's "Structure of Scientific Revolutions" (Kuhn 1962). In a small specialized community, we had experienced a "paradigm shift", with many of the sociological phenomena described by Kuhn. In fact, this advance could have taken place almost a decade earlier, since Lucy's computations were done in 1964 (Moyd 1971) and sufficient computing power was available even earlier. Most of the people attempting the new approach were outsiders to the field, or felt that way. The breakthrough had not come from the centres of eclipsing binary research of the time, and it was somewhat painful for those who had invested so much in the analytical methods.

The advent of "second-generation" computers in the early 1960's allowed realistic physics to be accurately modelled, without needing to simplify to allow analytical solutions. This also allowed sophisticated fitting methods to be developed, so that the parameters of the model, or elements of the binary, could be determined with an estimate of their accuracy (e.g. see Chapter 4 of Kallrath & Milone (1998) and Prsa & Zwitter (2005) for reviews). The accuracy of the light curves produced by the synthesis codes needed only to to be somewhat better than the photo-electric observations, which had standard deviations typically between 0.003 and 0.01 magnitudes. Although the advent of charge-coupled devices as detectors in the 1980's improved the quality, consistency and quantity of data, errors of less than a millimagnitude or two have rarely if ever been produced from ground-based observations, and therefore the codes developed around 1970 have served admirably, with appropriate refinement (e.g. Wilson 1994).

Two developments in particular have demonstrated the need for improvement of the binary synthesis codes. The first was the application of methods such as Maximum Entropy to determine spot distributions over close binaries (e.g. Maceroni, van't Veer & Vilhu 1991). This required that a surface element distribution be chosen such that elements did not vary excessively in size, and allowed the determination of the light curve of each element. The second was the advent of photometry obtained using observatories in space, such as MOST (Walker *et al.* 2003), CoRoT (Baglin *et al.* 2007) and KEPLER (Koch *et al.* 2010). These observatories can produce photometry accurate to the order of 10^{-5}, as opposed to the ground-based limit of about 10^{-3}. This means that photometry in space is about 100 times more accurate than the main synthesis codes in use since 1970. Not only must the methods be made more precise, but many kinds of hitherto ignored physical effects must be considered to improve accuracy. A "paradigm shift" is once more taking place, and Merrill's words are poignantly apropos, this time for those of us who developed the first generation of synthesis codes.

2. Accuracy of binary star codes

The accuracy of binary star light curve and line profile synthesis codes is affected by (a) geometrical factors (including gravity), (b) stellar atmosphere and interior physics, and (c) computational precision (algorithms). The "local" physical issues have been well covered by others at this Symposium and here I will concentrate on (a) and (c), for binaries without accretion disks and other complications beyond their photospheres.

2.1. *Geometrical Issues*

(A) *Roche model validity.* The traditional Roche model assumes that each component star's mass acts as a point at its center, that there is fully synchronous rotation and that the orbit is circular. Martin (1970) showed that the central point mass assumption was valid to about 1% in radius, while Limber (1963) and Wilson (1979) showed how to modify the Roche potential for non-synchronous rotation and orbital eccentricity, respectively. However, all of these approaches make simplifying assumptions which should be re-examined in the new accuracy regime.

(B) *Relativity.* Recently, so-called *Doppler boosting* has been successfully used to derive radial velocity curves from KEPLER photometry (van Kerkwijk *et al.* 2010 and references therein). This is caused by the aberration due to redshift, and therefore not only are fluxes and apparent temperatures affected, but also *angular sizes*. In contact binaries, radial velocity variations of 300-400 km/s are observed, which creates effects of order 0.003 magnitudes, far above the spaced-based limit of detection. The whole appearance of the binary is distorted, so that the simple point-source model used by van Kerkwijk *et al.* (2010) is inadequate for contact or near-contact systems. The Roche model needs to be transformed into special relativity, perhaps by applying the special relativistic theory of Penrose (1959) or general relativity as applied by Kopeikin & Ozernoy (1999), to obtain the image of the system as it would be seen by an observer with infinite angular resolution. This includes light time effects, gravitational self-lensing (e.g. Rahvar *et al.* 2011) and gravitational redshift.

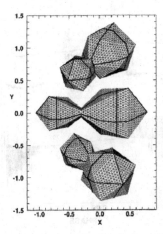

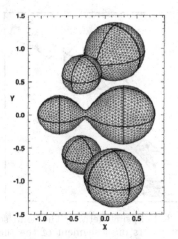

Figure 1. Stretched icosahedron basis, with icosa-faces subtriangulated. Radii are produced from the center of each component though each triangular vertex until they meet the equipotential surface, seen in Fig.2. From Hendry & Mochnacki (1992).

Figure 2. The Roche equipotential surface generated from Fig. 1. This model has mass ratio q = 0.41, fill-out F = 1.1 and inclination i = 67 degrees. Note that detached, semidetached and over-contact configurations can all modelled by GDDSYN. From Hendry & Mochnacki (1992).

2.2. *Astrophysical Issues*

These include limb darkening (see discussions by Prsa, Nielsen and others in this volume), gravity "darkening" and convection effects, computation of intensities and intensity line profiles, irradiation effects, star spots, and that unsolved mystery, the O'Connell effect.

3. The GDDSYN Method and Computational Precision

The GDDSYN method was devised by Hendry & Mochnacki (1992) to compute the "kernel" for Maximum Entropy Method (MEM) fitting of starspots on close binary systems (Hendry *et al.* 1992). Later, a method was developed to simultaneously fit the elements and spots of contact binaries using contemporaneous photometric and spectroscopic observations (Hendry & Mochnacki 2000). Neither the author's GENSYN method (Mochnacki & Doughty 1972) nor the WD-LC method (Wilson & Devinney 1971) was used due to numerical inaccuracy and inappropriate pixel geometry.

The GDDSYN code uses a geodesic distribution of elements, determined by producing a radius vector from the center of mass of each component through the sub-vertices on stretched icosahedrons out to the Roche equipotential (Figure 1).

This distribution of elements has many good features, especially because the areas of the elements range by at most a factor of about two, which is very well suited for techniques such as MEM. Ordering of vertices in triangles is counter-clockwise; the limb of an eclipsing component is defined as a convex polygon with about 50-150 vertices projected counter-clockwise on the sky, which allows for easy and rapid determination of the visibility of all elements, including those partially obscured by the limb (Fig. 3).

The triangularity of all elements also means that all elements are always convex polygons, making visibility computation very stable. Thus the light curve of each element can be accurately computed and stored; the set of all element light curves is the "kernel" for spot fitting. As in the Wilson-Devinney and Mochnacki-Doughty methods, recursive irradiation of each element is easily computed, which is not really possible in the original method of Lucy (1968). GDDSYN is faster, more precise and more self-consistent than the Wilson-Devinney code, at least in its older versions (Hendry & Mochnacki 1992). The inherent smoothness and stability of GDDSYN is

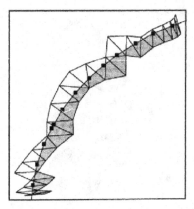

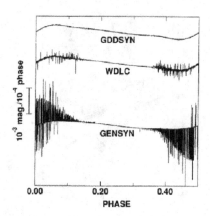

Figure 3. Eclipsing details. The solid squares represent points on a segment of the the limb of the eclipsing component, while the triangles are elements on the eclipsed component. The full limb of the eclipsing component is represented as a convex polygon with order 10^2 points. The intersection of each partially obscured triangular element and the polygonal limb is determined.

Figure 4. The slopes of light curves evaluated in steps of 10^{-4} in phase for a system such as in Fig. 2, comparing GDDSYN with the Mochnacki-Doughty and Wilson-Devinney codes (Hendry & Mochnacki 1992).

demonstrated in Fig. 4, showing that the basic method has the potential of meeting the precision requirements of space-based photometry. The issue now is whether it is also *accurate* enough. What are the systematic errors associated with this and other binary synthesis methods?

4. The Computational Accuracy of Synthesis Techniques

In Section 2, I discussed fundamental physical issues affecting accuracy. There are also systematic computational issues which affect synthesis codes, since all such codes involve discretization of the radiating surfaces. These are surprisingly large (Hendry & Mochnacki 1992), and fall into the following classes:

4.1. *Perpendicular to the line of sight*

In GDDSYN, the limb is approximated as a polygon of straight line segments rather than as a curve. Hendry & Mochnacki (1992) showed that the fractional underestimate of the projected area of the eclipsing component is,

$$\Delta A_{GDD_1} = 1 - \frac{n}{2\pi} \sin \frac{2\pi}{n} \approx \frac{2\pi^2}{3n^2} + \mathcal{O}(n^{-4}) \tag{4.1}$$

where n is the number of line segments defining the outline of the eclipsing component.

The simplest way to reduce this error is to increase the number of points defining the limb where eclipsing takes place, using polynomial interpolation.

4.2. *Along the line of sight*

A problem common with other codes is that along the line of sight, the limb is defined by the first invisible point or the last visible. This means that such a point is not at the true horizon (or limb) as seen by the observer, but $-\theta$ to $+\theta$ from it, where θ is the mean angular distance between points on the photosphere as seen from the star's centre. Thus the *projected* radius at the limb will be less, and the eclipsing star's area is underestimated. In the case of GDDSYN, Hendry & Mochnacki (1992) showed that this error is,

$$\Delta A_{GDD_2} \approx \frac{\pi^2}{3n^2} + \mathcal{O}(n^{-4}) \tag{4.2}$$

Hendry & Mochnacki (1992) for the Wilson-Devinney method estimated an error of,

$$\Delta A_{WD} \approx \frac{4\pi^2}{3n^2} + \mathcal{O}(n^{-4}) \tag{4.3}$$

where $n = \frac{2\pi}{\theta}$ and θ is the mean angular distance between the centres of surface elements.

With n in the range of 50-70, the total error (eqns. 4.1 plus 4.2) in the estimate of the projected area of the eclipsing component is about 0.002-0.004 magnitudes.

Recently, I have corrected the "horizon problem" by employing a routine devised by Mochnacki & Doughty (1972) to compute eclipse contact angles and the projected outline of a binary, but not used in the synthesis code. This routine, SBOUND, is described in Appendix 2 of that paper, and uses the Newton-Raphson technique to solve for two unknowns. In GDDSYN, it is now used to refine the limb, so that the points defining the limb are now much closer to the true horizon and not just the projection of surface elements which happen to be closest to the horizon. The difference between corrected and uncorrected models is shown in Fig. 5 (for the configuration in Fig. 2), and and is below the estimated maximum error above. For a totally eclipsing system, the error is larger, but does not exceed the estimated maximum error.

4.3. *Surface element areas*

The triangular surface elements of GDDSYN can be relatively large and still produce good computational precision (the tests presented here use only 3660 elements). However, if only flat element areas are considered, the surface area of each component is underestimated by about 0.3% compared with the accurate calculations of Mochnacki (1984). GDDSYN corrects the element areas for curvature, and also corrects the centroid position of each element so it lies on the Roche equipotential above (or below) the centroid of its vertices. Appendix B of Hendry & Mochnacki (1992) presents the rather messy element area correction function.

4.4. *Future Prospects*

GDDSYN is currently being incorporated into the PHOEBE code (Prsa & Zwitter 2005), as an alternative to the Wilson-Devinney back-end code PHOEBE currently uses. The aim is to use the atmospheric physics and fitting procedures within Phoebe, but with the faster and more accurate GDDSYN code providing a better tool for fitting models to space-based photometric observations.

5. Conclusions

The GDDSYN code was developed originally for maximum entropy method fitting of starspots on contact binaries (Hendry, Mochnacki & Collier 1992; Hendry & Mochnacki 2000), but it

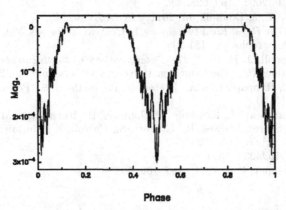

Figure 5. Differences between light curve with horizon refinement and without, for a configuration as in Fig. 2. Refinement increases the projected area of the eclipsing component, hence the minima are deeper.

now is now highly relevant to the fitting of models to space-based photometric observations and ground-based absorption-cell spectroscopy. The geodesic distribution of triangular surface elements allows computation of light curves and spectral line profiles much more accurately and precisely than older methods based on quadrilateral surface elements. It is being incorporated into PHOEBE.

However, the hundred-fold improvement in photometric precision afforded by space-based observatories such as KEPLER, CoRoT, MOST and others means that not only must the modelling computations be more *precise*, they also need to be more *accurate*. This means that special and general relativity must be factored into the model, and that deviations from the classical Roche model must now be considered. This is in addition to the improvement of atmospheric physics, including limb darkening and irradiation, which other authors in this volume are actively working on. There is much important work to be done, and we can truly say that the advent of such precise observations has forced a "paradigm shift" as disturbing as the one around 1971 when synthesis methods replaced the analytical "Royal Road to Eclipses" (Russell 1948).

References

Baglin, A. *et al.*, 2007, *AIP Conf. Proc.*, 895, 201
Hendry, P. D. & Mochnacki, S. W. 1992, *ApJ*, 388, 603
Hendry, P. D. & Mochnacki, S. W. 2000, *ApJ*, 531, 467
Hendry, P. D., Mochnacki, S. W., & Collier, A. C. 1992, *ApJ*, 399,246
Kallrath, J. & Milone, E. F. 1998, *Eclipsing Binary Stars: Modeling and Analysis* (New York: Springer-Verlag)
Koch, D. *et al.*, 2010, *ApJ* (Letters), 713, L79
Kopeikin, S. M. & Ozernoy, L. M. 1999, *ApJ*, 523, 771
Kuhn, T. S. 1962, *The Structure of Scientific Revolutions* (Chicago: University of Chicago Press)
Limber, D. N. 1963, *ApJ*, 138, 1112
Lucy, L. B. 1968, *ApJ*, 153, 877
Lucy, L. B. 1973, *Ap&SS*, 22, 381
Maceroni, C., van't Veer, F., & Vilhu, O. 1991, *ESO Messenger* 1991, No. 66, p. 47
Martin, P. G. M. 1970, *Ap&SS*, 7, 119
Merrill, J. E. 1950, *Tables for solution of light curves of eclipsing binaries*, Contributions from the Princeton University Observatory, no. 23
Mochnacki, S. W. & Doughty, N. A. 1972, *MNRAS*, 156, 51
Mochnacki, S. W. 1984, *ApJS*, 55, 551
Moyd, K. 1971, *private communication*
Penrose, R. 1959, *Mathematical Proceedings of the Cambridge Philosophical Society*, 55, 137
Prsa, A. & Zwitter, T. 2005, *ApJ*, 628, 426
Rahvar, S., Mehrabi, A., & Dominik, M. 2011, *MNRAS*, 410, 912
Russell, H. N. 1948, *The Royal Road to Eclipses*, in Centennial Papers, Vol. 7 of Harvard College Observatory Monographs, pp. 181-209
Russell, H. N. & Merrill, J. E. 1952, *The Determination of the Elements of Eclipsing Binary Stars*, Contributions from the Princeton University Observatory, no. 26, pp. 1-96
van Kerkwijk, M. H., Rappaport, S. A., Breton, R. P., Justham, S., Podsiadlowski, P., & Han, Z. 2010, *ApJ*, 715, 51
Walker, G. A. H., Matthews, J., Kuschnig, R., Johnson, R., Rucinski, S. M., Pazder, J., Burley, G., Walker, A., Skaret, K., Zee, R., Grocott, S., Carroll, K., Sinclair, P., Sturgeon, H., & Harron, J. 2003, *PASP*, 115, 1023
Wilson, R. E. 1979, *ApJ*, 234, 1054
Wilson, R. E. & Devinney, E. J. 1971, *ApJ*, 166, 605

From Interacting Binaries to Exoplanets: Essential Modeling Tools
Proceedings IAU Symposium No. 282, 2011
Mercedes T. Richards & Ivan Hubeny, eds.

© International Astronomical Union 2012
doi:10.1017/S1743921311027591

Synthetic Spectra and Light Curves of Interacting Binaries and Exoplanets with Circumstellar Material: SHELLSPEC

Ján Budaj

Astronomical Institute, Tatranská Lomnica, Slovakia, email: budaj@ta3.sk

Abstract. Program SHELLSPEC is designed to calculate light-curves, spectra and images of interacting binaries and extrasolar planets immersed in a moving circumstellar environment which is optically thin. It solves simple radiative transfer along the line of sight in moving media. The assumptions include LTE and optional known state quantities and velocity fields in 3D. Optional (non)transparent objects such as a spot, disc, stream, jet, shell or stars may be defined (embedded) in 3D and their composite synthetic spectrum calculated. The Roche model can be used as a boundary condition for the radiative transfer. Recently, a new model of the reflection effect, dust and Mie scattering were incorporated into the code.

ϵ Aurigae is one of the most mysterious objects on the sky. Prior modeling of its light-curve assumed a dark, inclined, disk of dust with a central hole to explain the light-curve with a sharp mid-eclipse brightening. Our model consists of two geometrically thick flared disks: an internal optically thick disk and an external optically thin disk which absorbs and scatters radiation. Shallow mid-eclipse brightening may result from eclipses by nearly edge-on flared (dusty or gaseous) disks. Mid-eclipse brightening may also be due to strong forward scattering and optical properties of the dust which can have an important effect on the light-curves.

There are many similarities between interacting binary stars and transiting extrasolar planets. The reflection effect which is briefly reviewed is one of them. The exact Roche shape and temperature distributions over the surface of all currently known transiting extrasolar planets have been determined. In some cases (HAT-P-32b, WASP-12b, WASP-19b), departures from the spherical shape can reach 7-15%.

Keywords. binaries: eclipsing, planets and satellites: general

---

1. Introduction

There are sophisticated computer codes for calculating and inverting light curves or spectra of binary stars with various shapes or geometry including the Roche model (Lucy 1968; Wilson & Devinney 1971; Wood 1971; Mochnacki & Doughty 1972; Rucinski 1973; Hill 1979; Popper & Etzel 1981; Djurasevic 1992; Drechsel *et al.* 1994; Hadrava 1997; Bradstreet & Steelman 2002; Pribulla 2004, Pavlovski *et al.* 2006, Tamuz *et al.* 2006). The Wilson & Devinney code is most often used and is continuously being improved or modified (Kallrath *et al.*1998; Prša & Zwitter 2005). The main focus of these codes is to deal with the stars, to determine their properties and their orbit. However, it is often the case that stellar objects are embedded in some moving optically thin environment and/or are accompanied by discs, streams, jets or shells which give rise to various emission spectra. We address these objects with Shellspec.

The SHELLSPEC code is described in more detail in Budaj & Richards (2004). There has been a lot of progress since that manual was published and an updated version of the manual, with examples of input, output and test runs is available within each new release. A convenient overview may also be found in Budaj & Richards (2010).

Major changes since that time include a new model of the reflection effect applicable to tidally distorted and strongly irradiated cold objects (Budaj 2011a), and dust and angle dependent Mie scattering (Budaj 2011b). It includes extinction due to the absorption and scattering as well as the thermal and scattering emission. The original SHELLSPEC code was written in Fortran77 and does not solve the inverse problem of finding the best fit parameters. There are versions of this code in Fortran90 which solve some particular restricted inverse problems (see Tkachenko *et al.* 2010 and Šejnová *et al.* 2011). Chadima *et al.* (2011a) modeled $H\alpha$ emission V/R variations caused by discontinuous mass transfer in interacting binaries. Ghoreyshi *et al.* (2010, 2011) applied the code to the eclipsing binary AV Del. Miller *et al.* (2007) modeled UV and optical spectra of TT Hya, an Algol-type binary. The sections below describe applications of SHELLSPEC to ϵ Aur and extrasolar planets. Output from the codes SYNSPEC and COOL-TLUSTY (Hubeny 1988; Hubeny, Lanz & Jeffery 1994; Hubeny, Burrows & Sudarsky 2003) is used as a default input of spectra of nontransparent objects, which serves as a boundary condition for the radiative transfer in the interstellar medium.

2. ϵ **Aurigae**

ϵ Aur is an eclipsing binary with the longest known orbital period, 27.1 yr. A very rare eclipse that lasted for two years is over. However, the object, its origin and, in particular, the secondary component of this binary star remain mysterious. Huang (1965) proposed that the secondary is a dark disk seen edge-on. Wilson (1971) and Carroll *et al.* (1991) argued that the observed sharp mid-eclipse brightening (**MEB**)† can only be explained by a tilted disk with a central opening. Ferluga (1990) suggested that the disk is a system of rings.

There has been a wealth of studies during the current eclipse. Orbital solutions were recently revisited by Stefanik *et al.* (2010) and Chadima *et al.* (2010). Kloppenborg *et al.* (2010, 2011) confirmed the dark disk with interferometric observations but they did not confirm the hole in the disk. The spectral energy distribution was studied by Hoard *et al.* (2010). These authors favor the post AGB+B5V model. Wolk *et al.* (2010) analyzed X-ray observations. Sadakane *et al.* (2010) carried out the abundance analysis. Recently, Chadima *et al.* (2010, 2011b) questioned the presence of sharp mid-eclipse brightening and suggested that the photometric variability seen during eclipse is intrinsic to the F-star. Extremely precise observation of the light-curve of ϵ Aur were obtained with the Solar Mass Ejection Imager by Clover *et al.* (2011). These data clearly show a shallow mid-eclipse brightening.

Observations that were used for comparison with our calculations were taken from the AAVSO database (Henden 2011). They were obtained by many observers who contributed to the database during the current eclipse of ϵ Aur. Only the observations in the V-filter were considered here. These observations also indicate the presence of a shallow mid-eclipse brightening but it is not as sharp and pronounced as some might have anticipated. Ingress is not as steep as egress, which indicates that the disk is not perfectly symmetric, but suffers from some small disturbance. Its leading part might be more extended or disk slightly inclined out of the orbital plane (warped?).

To explain these observations we suggested an alternative model of ϵ Aur. Our model of ϵ Aur consists of two geometrically thick flared disks: an internal optically thick disk

---

† A mid-eclipse brightening in ϵ Aur is interpreted by some as only a relatively sharp local maximum near the middle of the eclipse. We proposed a slightly more general definition of the MEB or a new term: mid-eclipse excess (**MEE**). Our MEB or MEE is a convex feature near the middle of an eclipse bed. A common eclipse has a concave eclipse bed.

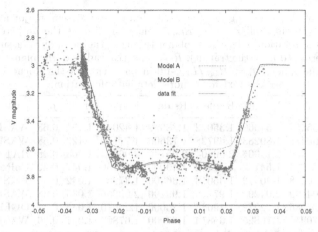

Figure 1. Eclipse of ϵ Aur by a dark, geometrically thick, flared disk of dust. The disk consists of two parts: (1) The flared optically thick part that causes most of the eclipse, and (2) a flared optically thin part that causes additional absorption, scattering and mid-eclipse brightening. Model (A): Disk has only one part (1). Model (B): disk has both parts part (1) and part (2). Mid-eclipse brightening arises mainly because the edges of the flared disk are more effective in the attenuation of the stellar light than the central parts of the disk. Thin dotted line is a best-fit quadratic function to the eclipse bed. Crosses - observations from AAVSO (Henden 2011).

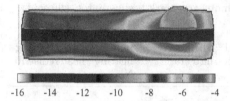

Figure 2. 2D image of ϵ Aur during eclipse. Black region is a dark, geometrically and optically thick, edge-on flared disk of dust. This disk causes the most of the eclipse. Colored regions correspond to the optically thin flared disk which scatters and absorbs the light from the F-star. Notice that this optically thin disk may produce artificial spots on the surface of the F star.

and an external optically thin disk, which absorbs and scatters radiation. Disks are in the orbital plane and are almost edge-on. The model is based on optical properties of dust grains. It takes into account the extinction due to the Mie scattering and absorption as well as thermal and scattering emission. We argue that there is no need for a highly inclined disk with a hole to explain the current eclipse of ϵ Aur, even if there is a possible shallow mid-eclipse brightening (see Fig. 1). For more details, kindly see Budaj (2011b). Fig. 2 displays a 2D image of ϵ Aur as calculated in the V band.

3. Shapes of transiting extrasolar planets

Transiting extrasolar planets are very close to their parent stars. Most of them have circular orbits or very small eccentricity. This indicates that their rotation is synchronous and the classical Roche model can be applied to them. Budaj (2011a) calculated the shape of all transiting exoplanets known at that time. By the shape, we mean the relative proportions of the object. Shape does not change much if the absolute dimensions change within typical measurement errors. Consequently, the new absolute dimensions can be easily linearly rescaled e.g. if the measured cross-section of the planet changes. I use R_{sub}/R_{pole} as a measure of the departure from the spherical shape, where R_{sub} is the

Table 1. Shapes of the transiting exoplanets. Columns are: a -semi-major axis in [AU], R_{sub} -planet radius at the sub-stellar point, R_{back} -planet radius at the anti-stellar point, R_{pole} -planet radius at the rotation pole, R_{side} -planet radius at the side point, (assumed equal to the planet radius determined from the transit), R_{eff} -effective radius of the planet, $rr = R_{sub}/R_{pole}$ -departure from the sphere, $f_i = R_{sub}/L_{1x}$ -fill-in parameter of the Roche lobe, Radii are in units of Jupiter radius. See the text for a more detailed information.

a	Rsub	Rback	Rpole	Rside	Reff	rr	f_i	name
0.01655	1.55474	1.54069	1.35022	1.38600	1.42045	1.151	0.63	WASP-19 b
0.02293	1.90438	1.89283	1.69728	1.73600	1.77110	1.122	0.58	WASP-12 b
0.03440	2.14496	2.14005	2.00804	2.03700	2.06019	1.068	0.48	HAT-P-32 b
0.02540	1.54395	1.54167	1.47462	1.49000	1.50166	1.047	0.42	CoRoT-1 b
0.02312	1.40758	1.40572	1.35013	1.36300	1.37264	1.042	0.41	WASP-4 b
0.05150	2.04439	2.04295	1.97515	1.99100	2.00266	1.035	0.38	WASP-17 b
0.02250	1.23086	1.22972	1.19084	1.20000	1.20670	1.033	0.37	OGLE-TR-56 b
0.03444	1.70957	1.70825	1.65814	1.67000	1.67862	1.031	0.36	WASP-48 b

Table 2. Overview of the reflection effect. Scattering refers to the refletion of light off the surface of one of the objects. A_{bol} is bolometric albedo (Rucinski 1969), A_B is Bond albedo, A_ν is monochromatic albedo. MR stands for muliple reflection between surfaces of the two objects.

Property	Interacting Binaries	Exoplanets	Hybrid model
Shape	Roche (limb+grav.dark.)	sphere, rot. ellipsoid	Roche (limb+grav.dark.)
Scattering	No	Yes	Yes
Albedo	A_{bol}	A_B, A_ν	A_B, A_ν
(A) Meaning	absorbed=>heating	reflected=>no heating	reflected=>no heating
(1-A) Meaning	penetrates into star	heating+heat redistrib.	heating+heat redistrib.
MR	Yes	No	No

radius at the sub-stellar point and R_{pole} is the radius at the rotational pole. I updated these calculations and recalculated the shape of all 138 transiting exoplanets known as of end of June, 2011. Most of the transiting exoplanets have small departures from a sphere on the order of 1%. However, a fraction of them have departures more than 3% and these are listed in Table 1. HAT-P-32 b, WASP-12 b, and WASP-19 b are exceptions with departures of about 7, 12, and 15% respectively. This means that we are observing only a cross section of these planets during the transit. This cross-section is also not spherical, it is characterized by the R_{side}/R_{pole} parameter but this is not as high as the R_{sub}/R_{pole}. Table 1 lists also the effective radius of the planet (radius of the sphere with the same volume) which one could use e.g. for comparison with the theoretical radius calculations. Table 1 illustrates that spherical shape of close-in exoplanets cannot be taken for granted. Moreover, the table demonstrates that although the cross section of the planet might be approximated by an ellipse, the overall Roche shape cannot be approximated by a *rotational* ellipsoid.

4. Reflection effect

The reflection effect operates in very different environments and there are different approaches to model this effect. Table 2 presents an overview of how the models work in the field of interacting binaries and exoplanets. One can see that the models of the reflection effect and definition of the fundamental quantities are very different in these fields. The question is: Is there any transition between the two approaches? What is the amount of the heat redistribution? What fraction of light gets reflected off the surface and does not produce heat? Which approach (model) should one choose for a tidally distorted and strongly irradiated cold object? To address these questions and these objects Budaj (2011a) proposed a sort of hybrid model of the reflection effect whose properties are also listed in the table.

Acknowledgements

JB thanks the VEGA grants No.s 2/0074/9, 2/007810, 2/0094/11.

References

Bradstreet, D. H. & Steelman, D. P. 2002, *BAAS*, 34, 1224

Budaj, J. 2011a, *AJ*, 141, 59

Budaj, J. 2011b, *A&A*, 532, L12

Budaj, J. & Richards, M. T. 2004, *Contrib. Astron. Obs. Skalnaté Pleso*, 34, 167

Budaj, J. & Richards, M. T. 2010, *ASP-CS*, 435, 63

Carroll, S. M., Guinan, E. F., McCook, G. P., & Donahue, R. A. 1991, *ApJ*, 367, 278

Chadima, P., Harmanec, P., Yang, S. *et al.*, 2010, *IBVS*, 5937

Chadima, P., Firt, R., Harmanec, P. *et al.*, 2011a, *AJ*, 142, 7

Chadima, P., Harmanec, P., Bennett, P. D. *et al.*, 2011b, *A&A*, 530, A146

Clover, J., Jackson, B. V., Buffington, A., Hick, P. P., Kloppenborg, B., & Stencel, R. 2011, AAS Meeting 217, 257.02

Djurasevic, G., 1992, *Ap&SS*, 197, 17

Drechsel, H., Haas, S., Lorenz, R., & Mayer, P. 1994, *A&A*, 284, 853

Ferluga, S. 1990, *A&A*, 238, 270

Ghoreyshi, S. M. R.., Ghanbari, J., & Salehi, F. 2010, *Pub. Astron. Soc. of Australia*, 28, 38

Ghoreyshi, S. M. R.., Ghanbari, J., & Salehi, F. 2011, arXiv:1108.3646

Hadrava, P. 1997, *A&AS*, 122, 581

Henden, A. A., 2011, Observations from the AAVSO International Database, private communication.

Hill, G. 1979, *Publ. Dom. Ap. Obs. Victoria*, 15, 297

Hoard, D. W., Howell, S. B., & Stencel, R. E. 2010, *ApJ*, 714, 549

Hubeny, I. 1988, *Computer Physics Comm.*, 52, 103

Hubeny, I., Burrows, A., & Sudarsky, D. 2003, *ApJ*, 594, 1011

Hubeny, I., Lanz, T., & Jeffery, C. S.: 1994, in Newsletter on Analysis of Astronomical spectra No.20, ed. C.S. Jeffery (CCP7; St. Andrews: St. Andrews Univ.), 30

Huang, S. 1965, *ApJ*, 141, 976

Kallrath, J., Milone, E. F., Terrell, D., & Young, A. T. 1998, *ApJ*, 508, 308

Kloppenborg, B., Stencel, R., Monnier, J. D. *et al.*, 2010, *Nature*, 464, 870

Kloppenborg, B. K., Stencel, R., Monnier, J. D. *et al.*, 2011, AAS Meeting 217, 257.03

Lucy, L. B. 1968, *ApJ*, 153, 877

Miller, B., Budaj, J., Richards, M., Koubský, P., & Peters, G. 2007, *ApJ*, 656, 1075

Mochnacki, S. W. & Doughty, N. A., 1972, *MNRAS*, 156, 51

Pavlovski, K., Burki, G., & Mimica, P. 2006, *A&A*, 454, 855

Popper D. M., Etzel P. B. 1981, *AJ*, 86, 102

Pribulla, T. 2004, Spectroscopically and Spatially Resolving the Components of Close Binary Stars, R. W. Hidlitch, H. Hensberge **and** K. Pavlovski, *ASP-CS*, 318, 117

Prša, A. & Zwitter, T. 2005, *ApJ*, 628, 426

Rucinski, S. M. 1969, *AcA*, 19, 245

Rucinski, S. M. 1973, *AcA*, 23, 79

Sadakane, K., Kambe, E., Sato, B., Honda, S., & Hashimoto, O. 2010, *PASJ*, 62, 1381

Stefanik, R. P., Torres, G., Lovegrove, J. *et al.*, 2010, *AJ*, 139, 1254

Šejnová, K., Votruba, V., & Koubský, P. 2011, these proceedings

Tamuz O., Mazeh T., North P. 2006, *MNRAS*, 367, 1521
Tkachenko, A., Lehmann, H., & Mkrtichian, D. 2010, *AJ*, 139, 1327
Wilson, R. E. 1971, *ApJ*, 170, 529
Wilson, R. E. & Devinney, E. J. 1971, *ApJ*, 166, 605
Wolk, S. J., Pillitteri, I., Guinan, E., & Stencel, R. 2010, *AJ*, 140, 595
Wood, D. B. 1971, *AJ*, 76, 701

Discussion

R. WILSON: Did you look into the possible importance of multiple scattering? The scattering geometry affects the angular distribution of scattered light and, to a small extent, even whether a scattered photon escapes the atmosphere. The effect is small but might not be negligible.

J. BUDAJ: There is a scattering (reflection) process in two environments. One is the scattering in the atmosphere of stars or nontransparent objects. I do not solve the radiative transfer in the atmospheres of these objects. I take the flux from the model atmosphere calculation and use it to calculate the boundary condition for the radiative transfer in the moving interstellar matter. Models of the atmosphere take the multiple scattering into account. The other is the scattering process in the interstellar medium. In this case, I assume that the medium is optically thin and that it is irradiated by one or two sources. I take into account only the first scattering event. In the optically thin medium, the probability of the second and higher scattering events rapidly decreases. To take them into account, one would have to solve the 3D radiative transfer in the 3D moving medium.

A. BURROWS: Have you thought of using your code to calculate light curves for WASP-12b at the warm Spitzer bands (3.6 m and 4.5 m), instead of at 8 m? There should be some data soon from Spitzer for WASP-12b at those bands and such models would be relevant.

J. BUDAJ: No. However, this would be indeed very interesting. Thank you for the suggestion/information.

S. RUCINSKI: I congratulate you for reorganizing the definitions of albedo as applied to stars and planets. The old definition was directly related to the so-called "reflection effect" in eclipsing binaries.

J. BUDAJ: Thank you.

J. SOUTHWORTH: You find that the difference in radius between a spherical planet and the substellar point of a Roche-model planet is as large as 15%. When a planet is transiting, we do not see its substellar point but we do measure its equatorial/ polar radius. This will be closer to the radius of a spherical planet so the problem is not as bad. Can you put a figure on this?

J. BUDAJ: Yes. That is true (if one deals with transits and not with other phases). R_{side}/R_{pole} is not as big. I do not remember the exact numbers but they are in my paper for all the transiting planets. This ratio might be about 3% in the worst case. The effective radius of the planet (radius of the sphere with the same volume) is also tabulated there.

From Interacting Binaries to Exoplanets: Essential Modeling Tools
Proceedings IAU Symposium No. 282, 2011   © International Astronomical Union 2012
Mercedes T. Richards & Ivan Hubeny, eds.   doi:10.1017/S1743921311027608

Search for Tidally Driven Anomalies in the Atmospheres of Am Stars

Ivanka Stateva[1], Ilian Iliev[1] and Ján Budaj[2]

[1]Institute of Astronomy with NAO, Bulgarian Academy of Sciences,
Sofia 1784, Bulgaria
email: stateva@astro.bas.bg
[2]Astronomical Institute, Slovak Academy of Sciences,
059 60 Tatranska Lomnica, The Slovak Republic

Abstract. We present here the systematic study of the chemical abundances of Am stars in order to search for possible abundance anomalies driven by tidal interaction in these binary systems. These stars were put into the context of Am binaries with $10 < P_{\mathrm{orb}} < 180d$ and their abundance anomalies discussed in the context of possible tidal effects. There is clear anti-correlation of the Am peculiarities with $v \sin i$. However, there seems to be also a correlation with eccentricity and orbital period.

Keywords. diffusion, stars: abundances, stars: chemically peculiar, stars: binaries: close

1. Introduction

The Am stars is a subgroup of Chemically Peculiar stars on the upper MS. The spectra are characterized by unusually weak spectral lines of light elements like C, Mg, Ca and Sc and contrary, abnormally strong lines of the iron peak and heavier elements. The obtained peculiarities appear to be due to microscopic selective diffusion driven mainly by the radiation pressure and the gravity. Rotation was found to play a key role in this process as it induces a large scale mixing which could disturb the slow diffusion process. The Am peculiarity seems to depend on the orbital elements in a binary system as well. Iliev *et al.* (1998) studied the dependencies between the rotational velocity, orbital period, eccentricity and the abundances in the Am stars. They concluded that there are a number of subtle effects that are difficult to understand if the rotation, stellar age, and mass are the only agents determining the Am peculiarity. The Am phenomenon seems more pronounced in binaries with eccentric orbits and longer orbital periods provided that the binary components are still within the reach of tidal effects.

We started a systematic study (Budaj & Iliev 2003, hereafter Paper I); and Iliev *et al.* 2006, hereafter Paper II) of the Am peculiarity in binary stars in order to search for abundance anomalies driven by the tidal interaction in these systems.

2. Observations, atmospheric parameters and spectral synthesis

Observations were carried out with the 2m RCC telescope of NAO-Rozhen. The Photometric AT200 camera with a SITE SI003AB 1024x1024pxs CCD chip was used to obtain spectra in the spectral region around 6439 Å. The typical S/N ratio was about 300, the resolving power R=30 000. Standard IRAF procedures were used for bias subtraction, flat-fielding, and wavelength calibration.

The atmospheric parameters were derived from both $uvby\beta$ and Geneva photometry. In the case of $uvby\beta$ photometry, we used the TEFFLOGG code of Moon & Dworetsky (1985); and for Geneva photometry, we used the calibration of Künzli *et al.* (1997).

A detailed spectrum synthesis of the spectral regions was accomplished using the code SYNSPEC (Hubeny, Lanz & Jeffery 1994; Krtička 1998). Model atmospheres were interpolated from Kurucz (1993). The VALD atomic line database (Kupka *et al.* 1999) was used to create a line list for the spectrum synthesis.

3. Results

We studied the dependences between the chemical abundances and the orbital elements of the binary systems, $v \sin i$ and $T_{\rm eff}$. The main advantage of our analysis is the homogeneity of the observational material we used - all data were obtained at one telescope with the same spectrograph and detector. The Am peculiarities are mainly manifested by a Ca deficit and Fe overabundances and they could be represented through the values of [Ca/Fe]. This ratio would multiply the effect of Am peculiarities because these two elements have displayed opposite behavior.

Up to now, 15 stars from our sample have been fully processed. Two of the stars were found not to be Am stars.

First, we investigated the dependences of [Ca/Fe] on the eccentricity. Despite some scatter in the data, there seems to be a trend: the correlation coefficient is -0.55±0.05. One star, HD 198391, is distinguished from the common trend. This star is the hottest star amongst the sample and it is not a typical Am star. Namely, the Am peculiarities increase ([Ca/Fe] decreases) with increasing eccentricity.

The dependence of [Ca/Fe] on the orbital period is not as clear, but still there is a tendency for the metalicity to increase towards the longer periods. The eccentricity and the orbital period, are not fully independent because of the synchronization and circularization of the orbits. That was the reason for analysing only stars with $10 < P_{\rm orb} < 180d$.

We studied also the dependence of the Am peculiarities on the projected rotational velocity of the Am stars. In general, there is a clear trend of increasing the peculiarity (decreasing of [Ca/Fe]) towards small values of $v \sin i$; the correlation coefficient is +0.85±0.04. Besides the hottest star already mentioned, two other stars, both marginal Am stars, do not follow the common trend. The departure of these two stars from the clear smooth correlation of metallicity and $v \sin i$ that the rotation is not the only agent responsible for this peculiarity.

The dependence of the Am peculiarity on the effective temperature was also checked. There are might be a trend of decreasing the peculiarity with the temperature.

Acknowledgments

This work is partially supported by grant DO 02-85 from the Bulgarian NSF and by VEGA grants 2/0074/09, 2/0078/10, 2/0094/11 from the Slovak Academy of Sciences.

References

Budaj J. & Iliev I. K.h., 2003, *MNRAS*, 346, 27 (Paper I)
Hubeny I., Lanz T., & Jeffery C. S. 1994, *in Jeffery C.S., ed, Newsletter on Analysis of Astronomical spectra, No.20, CCP7*. St. Andrews Univ., St.Andrews, p. 30
Iliev I. K.h., Budaj J., Zverko J., & Barzova I. S., Žižňovský J. 1998, *A&AS*, 128, 497
Iliev I. K.h., Budaj J., Feňovčik M., Stateva I., & Richards M. T., 2006, *MNRAS*, 370,819 (Paper II)
Krtička J. 1998, *in Dušek J., Zejda M., eds, Proc. 20-th Stellar Conf. Nicholaus Copernicus Observatory and Planetarium*, Brno, p. 73
Kupka F., Piskunov N. E., Ryabchikova T. A., Stempels H. C., & Weiss W. W. 1999, *A&AS*, 138,119
Künzli M., North P., Kurucz R. L., & Nicolet B. 1997, *A&AS*, 122, 51
Kurucz R. 1993, *ATLAS9 Stellar Atmosphere Programs and 2km/s Grid (Kurucz CD-ROM 13)*
Moon T. T. & Dworetsky M. M. 1985, *MNRAS*, 217, 305

From Interacting Binaries to Exoplanets: Essential Modeling Tools
Proceedings IAU Symposium No. 282, 2011
Mercedes T. Richards & Ivan Hubeny, eds.
© International Astronomical Union 2012
doi:10.1017/S174392131102761X

RaveSpan - Radial Velocity and Spectrum Analyzer

Bogumil Pilecki[1,2], Piotr Konorski[1] and Marek Gorski[1]

[1] Warsaw University Observatory,
Al. Ujazdowskie 4, 00-478 Warszawa, Poland
email: (pilecki,piokon,mgorski)@astrouw.edu.pl

[2] Department of Astronomy, Universidad de Concepcion,
Casilla 160-C, Concepcion, Chile
email: bpilecki@astro-udec.cl

Abstract. The RV analysis tool integrates widely used methods of radial velocity determination (CCF, TODCOR, BF) in an easy to use graphical environment. No advanced knowledge of these methods is required to use it. The obtained velocities may be immediately analyzed with the same tool as it comprises flexible fitting of orbital parameters, which includes the third body influence and pulsational velocities of the components. These features together help to establish the most accurate combination of templates, spectrum range, and method. Scripting functionality is to be implemented in the future.

Keywords. methods: data analysis, techniques: radial velocities, spectroscopic

1. Overview

RaveSpan is an easy to use graphical application that brings together three major velocity extraction methods: CCF, TODCOR, and Broadening Function (BF). All extracted velocities are instantly plotted in the RV curve window. Selected orbital parameters may be fitted afterwards.

RaveSpan is composed of several components. There is a spectrum viewer, where you can inspect the collected spectra, compare them with templates, and choose a wavelength range for your analysis. In the orbit viewer, one can see extracted velocities and a model orbit. There is also an orbit fitting tool, with which one can fit selected orbital parameters.

The radial velocity analysis tool allows users to see the output of CCF, TODCOR or BF and interactively fit several profiles of different types.

2. Software features

Spectrum analysis. Currently three methods for velocity determination from spectra are implemented. The simple cross-correlation method (CCF; Simkin 1974, Tonry & Davis 1979), two dimensional cross-correlation (TODCOR; see Zucker & Mazeh94) and the broadening function technique (BF; Rucinski 2002).

Profile fitting. Once one of these functions is calculated, the maxima (or maximum) are automatically detected and velocities evaluated. To improve accuracy, one of a few types of profiles may be fitted. For CCF and BF response functions, we can fit up to 4 (it is just set to 4 now, but may be increased easily) profiles, either Gaussian, rotational or a simple polynomial one. With TODCOR, there is only one two-dimensional polynomial surface fitted.

Radial velocity curve. All extracted velocities instantly appear in the radial velocity curve window, where we can directly see the orbit shape and quality.

Orbital parameters. Once we are ready with velocities, we can fit selected orbital parameters. Third body influence and pulsational velocities of one or both components can be included in the analysis and separated from the basic orbital motion.

Other features. This software allows for selection of different (preset by the user) wavelength ranges for analysis. To allow for radial velocity determination, a normalization is applied to the spectrum, unless a user disables the feature. For convenience, a Calculate All function was implemented to explore different templates and spectrum ranges. This function attempts to extract velocities from all the collected spectra in an automated way.

Technical details. RaveSpan is written in pure Python using the PyQt4 graphical library with Matplotlib as a plotting tool.

3. First results

This software is now being tested in the ARAUCARIA project and it has already proved to be useful both as an easy tool for simple objects and as a powerful tool for more complex spectra, eg. with a third light or pulsations. Several binary systems in the Magellanic Clouds were analyzed, including those with cepheid variables as one of the components.

The code is now in testing phase and is not yet public, but we are going to publish it once the tests are finished.

Acknowledgements

We gratefully acknowledge financial support for this work from the BASAL Centro de Astrofisica y Tecnologias Afines (CATA) PFB-06/2007 and the FOCUS and TEAM subsidies of the Foundation for Polish Science (FNP).

References

Rucinski, S. M. 2002 *AJ*, 124, 1746
Simkin, S. M. 1974 *A&A*, 31, 129
Tonry, J. & Davis, M. 1979, *AJ*, 84, 1511
Zucker, S. & Mazeh, T. 1994, *ApJ*, 420, 806

From Interacting Binaries to Exoplanets: Essential Modeling Tools
Proceedings IAU Symposium No. 282, 2011
Mercedes T. Richards & Ivan Hubeny, eds.
ⓒ International Astronomical Union 2012
doi:10.1017/S1743921311027621

Chemical History of Algol and its Components

V. Kolbas[1], K. Pavlovski[1,2], J. Southworth[2], C.-U. Lee[3], J. W. Lee[3], S.-L. Kim[3] and H.-I. Kim[3]

[1] Department of Physics, Facult of Science, University of Zagreb, Croatia

[2] Astrophysics Group, Keele University, Staffordshire, ST5 5BG, UK

[3] Korea Astronomy and Space Science Institute, Yuseong-yu, Daejeon, Korea

Abstract. We present a new observational project to study the hierarchical triple stellar system Algol, concentrating on the semidetached eclipsing binary at the heart of the system. Over 140 high-resolution and high-S/N spectra have been secured, of which 80 are from FIES at the Nordic Optical Telescope, La Palma, and the remainder were obtained with BOES at the Bohyunsan Optical Astronomy Observatory in Korea. All three components were successfully detected by the method of spectral disentangling, which yields the individual spectra of the three stars and also high-quality spectroscopic elements for both the inner and outer orbits. We present a detailed abundance study for the mass-accreting component in the inner orbit, which holds information on the history of mass transfer in the close inner binary system. We also reveal the atmospheric parameters and chemical composition of the tertiary component in the outer orbit.

Keywords. stars: binaries: eclipsing — stars: fundamental parameters — stars: chemically peculiar — stars: binaries: spectroscopic — stars: early-type

1. New spectroscopy of Algol

Algol (β Persei, HD 19356) is on of the most important stellar systems for studying mass transfer and mass loss in an interacting binary because it contains the brightest eclipsing binary with deep eclipses. It is the prototype of the class of close binaries in which mass reversals occurred due to rapid mass transfer, resulting in ongoing mass transfer from the G-K sub-giant secondaries to the B5-A5 main-sequence primaries. In spite of a rich observational history, Algol is still lacking definitive understanding although progress has been made in the last decade (see Retter, Richards & Wu 2005). Its brightness is favourable for new observational techniques, recently allowing Zavala *et al.* (2011) to fully spatially resolve its components. However, Algol has persistently lacked extensive high-precision photometric and spectroscopic observations.

Therefore, in 2006 we started a new Échelle spectroscopic programme to observe Algol (Pavlovski *et al.* 2010). Over 80 high-resolution and high-S/N Échelle spectra were secured in two observing runs with FIES at the Nordic Optical Telescope on La Palma, Spain. The spectral coverage of these observations is 3640–7380 Å, and the resolving power is $R = 47\,000$. Over 60 additional Échelle spectra were secured since 2009, at the Bohyunsan Optical Astronomy Observatory in Korea. Two different fibres of the Bohyunsan Optical Echell Spectrography (BOES) have been used, giving $R = 45\,000$ and $R = 30\,000$. The spectral range is 3500–10 000 Å and the typical S/N is 400.

2. Chemical composition of the components of the Algol system

Spectral disentangling has been performed on these data as described in Pavlovski *et al.* (2010). Algol is a hierarchical triple system with a third component in a wide orbit of period 640 d. The contribution of the third component to the total light of the system is uncertain (Richards *et al.* 1988), but is significantly larger than that of the sub-giant component in the inner semidetached binary. The best ground-based photometry (Kim 1989) is not of sufficient quality to give the precise light contributions of the three stars, so we must seek these light dilution factors from spectroscopy.

Previous estimates of the spectral type of the third component cover a broad range, from a mid-F to a late A-type star with metallic characteristics. Therefore, we calculated a grid of synthetic spectra covering $T_{\mathrm{eff}} = 7000\text{--}8600\,\mathrm{K}$ in steps of 200 K. We fixed $\log g = 4.3$. By varying the light dilution factor, we calculated the sum of squared residuals between theoretical and disentangled (separated) spectra. The best match is found for $T_{\mathrm{eff}} = 7600\,\mathrm{K}$, and light dilution factor 0.07. The depth of some spectral lines is almost independent of T_{eff}, making them sensitive indicators of the light dilution factor. We corroborate the weakness of the Ca II lines in the spectrum of the third component, and a slight underabundance of Sc, another classical indicator of a metallic-lined stars. The third component may be mildly metallic in nature, and a detailed spectroscopic abundance analysis will be able to confirm or refute this.

Armed with an improved measurement of the contribution of the third component to the total luminosity of the system, we turned back to the B and V light curves from Kim (1989). We now find that the primary is contributing 90.5% of the total system's light outside eclipse. The renormalized disentangled spectrum of the primary will next be used to estimate its T_{eff} (its $\log g$ is known from the dynamics of the inner system), and derive the chemical composition of its photosphere.

From fitting the Balmer lines and then fine-tuning using Fe II lines, we find for the primary star $T_{\mathrm{eff}} = 12\,950\,\mathrm{K}$. Profiles of several prominent He I lines have been calculated in NLTE, and indicate a solar helium abundance for this star. NLTE calculations for the CNO elements indicate a slight deficiency of C relative to N, in line with the results of Cugier & Hardrop (1988) and Tomkin *et al.* (1993). Abundances for other elements have been calculated in LTE, and indicate a slight underabundance of -0.1 dex relative to solar. A more detailed examination is needed before definitive conclusions could be drawn on this issue.

Acknowledgements

KP acknowledges receipt of a Leverhulme Visiting Professorship which enabled him to stay at Keele University where a part of this work was performed.

References

Cugier, H. & Hardrop, I. 1988, *A&A*, 202, 101
Kim, H.-I. 1989, *ApJ*, 342, 1061
Pavlovski, K., Kolbas, V., & Southworth, J. 2010, *ASP-CS*, 435, 247
Retter, A., Richards, M. T., & Wu, K. 2005, *ApJ*, 621, 417
Richards, M. T., Mochnacki, S. W., & Bolton, C. T. 1988, *AJ*, 96, 326
Tomkin, J., Lambert, D. L., & Lemke, M. 1993, *MNRAS*, 265, 581
Zavala, R. T., Hummel, C. A., Boboltz, D. A., Ojha, R., Shaffer, D. B., Tycner, C., Richards, M. T., & Hutter, D. J. 2011, *ApJL*, 715, L44

From Interacting Binaries to Exoplanets: Essential Modeling Tools
Proceedings IAU Symposium No. 282, 2011
Mercedes T. Richards & Ivan Hubeny, eds.
© International Astronomical Union 2012
doi:10.1017/S1743921311027633

Infinity – A New Program for Modeling Binary Systems

Cséki Attila and Olivera Latković

Astronomical Observatory of Belgrade
Volgina 7, 11000 Belgrade, Serbia
email: attila@aob.rs

Abstract. INFINITY is a new program for modeling binary systems. The model is based on Roche geometry with asynchronous rotation, including an assortment of effects like gravity and limb darkening, mutual irradiation, bright and dark spots and so on. However, INFINITY brings innovations in the modeling of accretion disks, and introduces the modeling of radial and non-radial oscillations on one or both components of the system.

At this stage of development, INFINITY can produce light curves, spectra and radial velocity curves; solving the inverse problem is still a work in progress. In terms of programming, INFINITY is being developed in the object-oriented language C#, and great care is taken to produce readable, easily extensible and verifiable code. INFINITY is fully optimized to take advantage of modern multi-core CPUs, and the code is thoroughly covered with unit-tests. We expect to make a public release during 2012.

Keywords. stars: binaries: general, stars: oscillations (including pulsations), accretion disks

1. Modeling the binary system

The physics of INFINITY follows closely the physics of the model developed by Djurašević (1992a). The shape and the basic temperature distribution of the stellar surface are determined by the Roche potential and Von Zeipel's law; this base distribution can subsequently be modified by various effects, like dark and bright spots, stellar pulsations and mutual irradiation. At the moment, we treat the stars as black bodies and calculate the emergent fluxes from the Plank law, but we will make the transition to using models of stellar atmospheres in the near future.

INFINITY uses the geodesic mesh to represent the stellar surface. The mesh is generated from an icosahedron and subdivided into elementary triangles, which are then positioned so as to match the Roche equipotential corresponding to the surface of the star. This approach was inspired by the model of Hendry & Mochnacki (1992). One of the main advantages of using this type of surface division is that the triangles comprising the geodesic mesh are all of fairly similar areas. The smoothness of the mesh is controlled by the level of geodesic subdivision; at level 5, there are about 20 000 triangles per mesh, and at level 6, about 80 000. We are looking into alternative subdivision algorithms that will allow a gentler increase of the number of elementary surfaces per level.

The visibility of an elementary triangle in any given phase is determined by a method similar to the Painter's algorithm in computer graphics (Foley et al. 1990). A triangle is either visible (nothing in front), or invisible (some part covered); in other words, INFINITY doesn't account for partial visibility. Instead, INFINITY implements "adaptive subdivision" that automatically increases the number of elementary surfaces in the region of the eclipse, allowing for arbitrary accuracy.

2. Modeling the accretion disk

Our model of the binary system can include an optional, geometrically and optically thick accretion disk in the orbital plane. Following the model of Djurašević (1992b), the disk may be flat or conical, and the conical disk may further be modeled as concave (outer edge thicker than the inner edge) or convex (outer edge thinner than the inner edge). INFINITY expands the geometrical model of the accretion disk to allow almost any shape, illustrated by the example of a toroidal disk that is already implemented: the shape of the disk in INFINITY is determined by a parameterized cross-section in the meridional plane; the cross-section for the flat disk is a rectangle, for the conical disk, a trapeze, and for a toroidal disk, an ellipse. In addition, INFINITY can model disks that are eccentric and inclined to the orbital plane.

3. Modeling the pulsations

One of the most important aspects of INFINITY is the ability to model radial and non-radial stellar pulsations on one or both components of the binary system. The pulsation model that we are using was originally developed by Bíró & Nuspl (2011) for the purpose of mode identification in eclipsing binaries using the technique of eclipse mapping. The modes are approximated by spherical harmonics, and the periodic changes in the observables are produced by perturbing the emergent flux from the spherical stellar surface. We made slight improvements to this model; first, in the sense that the pulsation modes are modeled as perturbations in both the shape and the temperature of the stellar surface, and second, in the sense that the pulsations can be modeled on both quasi-spherical and gravitationally/rotationally distorted stars.

4. Fitting & mode identification

In addition to fitting the system and component parameters, INFINITY will be able to fit the parameters of stellar pulsations: the amplitudes and quantum mode numbers ℓ and m. This method of mode identification, the direct fitting of spherical harmonics has been tested on artificial data and a model with spherical stars by Bíró & Latković (2009), but we will attempt to apply it on distorted stars in Roche geometry. The fitting algorithm we're working on is a variation of the Nelder-Mead simplex with constraints and parameter mutations.

5. Availability

As of this writing, INFINITY is still a work in progress. We expect to publish the details of the methods and example applications by the end of 2012, after which we will make INFINITY available for public use.

References

Bíró, I. B. & Latković, O. 2009, *Communications in Asteroseismology*, 159, 127
Bíró, I. B. & Nuspl, J. 2011, *arXiv:1101.5162v1, MNRAS*, in press
Djurašević G. 1992, *Ap&SS* 196, 241
Djurašević G. 1992, *Ap&SS* 197, 17
Foley, J., van Dam, A., Feiner, S. K., & Hughes, J. F. 1990, *Computer Graphics: Principles and Practice* p.1174
Hendry, P. D. & Mochnacki, S. W. 1992, *ApJ* 388, 603

From Interacting Binaries to Exoplanets: Essential Modeling Tools
Proceedings IAU Symposium No. 282, 2011
Mercedes T. Richards & Ivan Hubeny, eds.
© International Astronomical Union 2012
doi:10.1017/S1743921311027645

Fundamental Parameters of Four Massive Eclipsing Binaries in Westerlund 1

E. Koumpia and A. Z. Bonanos

National Observatory of Athens, Institute of Astronomy & Astrophysics,
I. Metaxa & Vas. Pavlou St., Palaia Penteli GR-15236 Athens, Greece

koumpia@astro.noa.gr, bonanos@astro.noa.gr

Abstract. We present fundamental parameters of four massive eclipsing binaries in the young massive cluster Westerlund 1. The goal is to measure accurate masses and radii of their component stars, which provide much needed constraints for evolutionary models of massive stars. Accurate parameters can further be used to determine a dynamical lower limit for the magnetar progenitor and to obtain an independent distance to the cluster. Our results confirm and extend the evidence for a high mass for the progenitor of the magnetar.

Keywords. open clusters and associations: individual (Westerlund 1), stars: fundamental parameters, stars: early-type, binaries: eclipsing, stars: Wolf-Rayet

1. Introduction

Westerlund 1 (Wd1) is one of the most massive young clusters known in the Local Group of galaxies, with an age of 3-5 Myr. It was discovered by Westerlund (1961), but remained unstudied until recently due to the high interstellar extinction in its direction. It contains an assortment of rare evolved high-mass stars, such as blue, yellow and red supergiants, Wolf-Rayet stars, a luminous blue variable, many OB supergiants, as well as four massive eclipsing binary systems (Wddeb, Wd13, Wd36, WR77o, see Bonanos 2007). Furthermore, the magnetar CXO J164710.2-455216 was discovered in the cluster by Muno *et al.* (2006) in X-rays. This magnetar is a slow X-ray pulsar that is assumed to have formed from a massive progenitor star.

Eclipsing binaries provide the only accurate way for the measurement of masses and radii of stars. Thus, the study of these systems in the cluster is important for the following reasons: (1) the determination of fundamental parameters of the component stars (mass, radii, etc.), in order to increase the small sample of massive stars with well known physical parameters, (2) the test of stellar models for the formation and the evolution of massive stars, (3) the determination of a dynamical lower limit for the mass of the magnetar progenitor and (4) EBs present a great opportunity for an independent measurement of the distance, based on the expected absolute magnitude.

2. The individual binaries

We have analyzed spectra of all four eclipsing binaries, taken in 2007-2008 with the 6.5 meter Magellan telescope at Las Campanas Observatory, Chile. The spectra were reduced and extracted using IRAF. For the determination of the radial velocities, we adopted a χ^2 minimization technique, which finds the least χ^2 from the observed spectrum and fixed synthetic TLUSTY models (Lanz & Hubeny 2003). For this purpose, we used the narrow Helium lines ($\lambda\lambda6678, 7065$), as they are less sensitive to systematics, rather than the broader hydrogen lines.

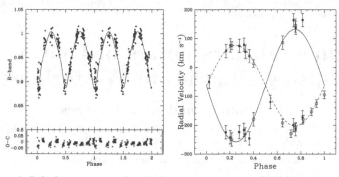

Figure 1: Phased R light curve and radial velocity curve of Wd13, respectively. Radial velocity points from Ritchie *et al.* (2010) are also included (black squares).

The physical and orbital parameters of the members of the four systems (period, masses, radii, surface gravities of the components, eccentricity, inclination etc.), resulted from modelling light and radial velocity curves using PHOEBE software (Prsa & Zwitter 2005). Our final results will be presented in Koumpia & Bonanos (in prep.).

Wd13: is a semi-detached, double-lined spectroscopic binary (B0.5Ia$^+$/WNVL and O9.5-B0.5I types; Ritchie *et al.* 2010) in a circular orbit (see Fig. 1). Its spectra show both absorption and emission lines. It provides the first dynamical constraint on the mass of the magnetar progenitor and has important implications for the threshold mass which gives rise to black holes versus neutron stars.

Wd36: is an overcontact, double-lined spectroscopic binary system (OB-type) in a circular orbit.

WR77o: is probably a double-contact binary system in an almost circular orbit. It is a single-lined spectroscopic binary. The spectroscopic visible star is a Wolf-Rayet star of spectral type WN6-7 (Negueruela & Clark 2005), with line widths of 2000 km s^{-1}.

Wddeb: is a double-lined spectroscopic binary system (OB-type) with an eccentricity of almost 0.2. Being a detached system, it contains the first main-sequence stars detected in the cluster, yielding masses for two of the most massive unevolved stars in Wd1.

3. Conclusions

Having obtained the physical and orbital parameters of each system, we confirm the high mass of Wd13 and therefore the high mass of the magnetar progenitor, also found by the independent study of Ritchie *et al.* (2010). We also compared our results with stellar models of evolution of single stars (Claret 2004).

Acknowledgements: We acknowledge research and travel support from the IAU and the European Commission Framework Program Seven under the Marie Curie International Reintegration Grant PIRG04-GA-2008-239335. IRAF is distributed by the NOAO, which are operated by AURA, under cooperative agreement with the NSF.

References

Bonanos, A. Z. 2007, *AJ*, 133, 2696
Claret, A. 2004, *A&A*, 428, 1001
Lanz, T. & Hubeny, I. 2003, *ApJS*, 146, 417
Muno, M. P., Clark, J. S., & Crowther, P. A. *et al.*, 2006, *ApJ*, 636, 41
Negueruela, I. & Clark, J. S. 2005, *A&A*, 436, 541
Negueruela, I., Clark, J. S., & Ritchie, B. W. 2010, *A&A*, 516, 78
Prsa, A. & Zwitter, T. 2005, *ApJ*, 628, 426
Westerlund, B. 1961, *AJ*, 66, 57

From Interacting Binaries to Exoplanets: Essential Modeling Tools
Proceedings IAU Symposium No. 282, 2011
Mercedes T. Richards & Ivan Hubeny, eds.

© International Astronomical Union 2012
doi:10.1017/S1743921311027657

Is the B[e] Star V2028 Cyg a Binary?

Jan Polster[1,2]**, Daniela Korčáková**[2]**, Viktor Votruba**[1,3]**, Petr Škoda**[3]**, Miroslav Šlechta**[3] **and Blanka Kučerová**[1]

[1]Faculty of Science, Masaryk University, Kotlářská 2, 611 37 Brno, CZ

[2]Astronomical Institute, Charles University, V Holešovičkách 2, 180 00 Praha 8, CZ

[3]Astronomical Institute of the AV ČR, Fričova 298, 251 65 Ondřejov, CZ

email: polster@physics.muni.cz, kor@sirrah.troja.mff.cuni.cz

Abstract. V2028 Cyg shows a B[e] phenomenon. Due to the presence of both cool (K7III) and hot (B4III) components in the spectra, it is supposed to be a binary. Our modelling of the time variability of Hα line bisectors shows, however, that this hybrid spectrum can originate in one star, which is surrounded by a disc.

Keywords. stars: emission-line, B[e], binaries: spectroscopic, stars: individual (V2028 Cyg)

Introduction

V2028 Cyg is a star showing the B[e] phenomenon, i.e. lines from the forbidden transitions are present in its spectra. This is indicative of a very extended circumstellar envelope. Due to this reason, the nature of this object is unknown. The determination of the stellar parameters is very uncertain, since the commonly used synthetic spectra are not applicable. Therefore, we focused on a study of time dependencies of the spectral features. Our aim was also to better specify the role of the binarity in this system. The spectrum of V2028 Cyg is very atypical among stars with the B[e] phenomenon, since it is composed two kinds of spectra: K7III and B4III (Zickgraf 2001; Bergner *et al.* 1995).

We obtained a series of 88 spectra (Polster *et al.* 2010; Polster *et al.* 2011a,b) from the Ondřejov 2 m telescope in the spectral range $6250 - 6770$ Å (Hα line, R ∼ 12500). These data were used for the description of the spectral-line variability. The time dependencies of equivalent width, line intensities, radial velocities and bisectors of the Hα line were modelled.

Modelling

In order to narrow the set of possible geometries, we constructed a simple numerical model. We assumed an optically thin medium, which allows us to add (subtract) radiation along the line-of-sight. Another simplification was a Gaussian profile of the line in every cell. The grid was defined with respect to the velocity field. Such a model is too simple for the B[e] star description, however, it can be used in this case. We do not derive the parameters of the system, but we only compare the relative time variations of certain quantities – equivalent width, line intensities, and line bisectors.

Based on previous work, we investigated the following geometries *i)* a disc with a spot, an arm, an oscillating arm (stellar wind focused by a close compact companion), *ii)* a symbiotic star, *iii)* cool star with a wind and hot compact companion (a possible disc around it), *iv)* pulsations, and *v)* a disc with a dust ring (Fig. 1, right panel).

Our observations defined restrictions on the V2028 Cyg model. The peak of the line must be shifted redward. Radial velocities and bisectors vary differently in the wings

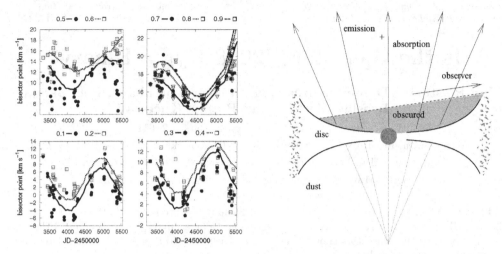

Figure 1. Model corresponding to the observations. The measured (points) and fitted (lines) Hα line bisectors are plotted for the relative highs 0.1 (wings) – 0.9 (peak).

and peak. The absolute value of the equivalent width is correlated with the maximum of relative flux and anti-correlated with the radial velocity of the wings.

Results and conclusion

It is possible to fit the line-profile shape with all the models, however, the observed time dependencies agreed only with a star surrounded by a gaseous disc with a dust ring. The best fit is plotted in Fig. 1 (left panel). The disc with a dust ring must obscure a significant part of the disc itself, otherwise the dependencies will be broken.

The resulting model of the disc with the dust ring implies an alternative explanation of the two-component spectra of V2028 Cyg. The spectrum with features of a hot star originates in the internal part of the disc and wind region. The outer parts of the disc, which we see edge-on, have low temperatures close to the dust condensation value. These conditions allow the formation of an absorption spectrum of the K type.

In order to verify this hypothesis, we compared the radial velocity measurements of the Hα line and the absorption K component. We found a similar trend in these dependencies (Polster *et al.* 2011a). This supports the hypothesis of the geometrically thick disc. However, the accuracy of our measurements is low. There still remains the possibility, that V2028 Cyg is a binary. Considering earlier work (Zickgraf 2001) and our observations, the orbital period of the system in this case must be larger than 25 years.

Acknowledgements

This research is supported by grants 205/09/P476 (GA ČR), 205/08/H005 (GA ČR), MUNI/A/0968/2009, and projects MSM0021620860 (MŠMT ČR) and AV0Z10030501 (AV ČR).

References

Bergner, Yu. K., Miroshnichenko, A. S., Yudin, *et al.* 1995, *A&AS*, 112, 221
Polster, J., D. Korčáková, V. Votruba *et al.*, 2010, *ASP-CS*, 435, 399
Polster, J., D. Korčáková, V. Votruba *et al.*, 2011a, *A&A*, in preparation
Polster, J., D. Korčáková, V. Votruba *et al.*, 2011b, in IAU Symposium 272, in press
Schneider, D. P. & Young, P. 1980, *ApJ*, 238, 946
Zickgraf, F. J. 2001, *A&A*, 375, 122

From Interacting Binaries to Exoplanets: Essential Modeling Tools
Proceedings IAU Symposium No. 282, 2011
Mercedes T. Richards & Ivan Hubeny, eds.
© International Astronomical Union 2012
doi:10.1017/S1743921311027669

Peculiarities in the Spectrum of the Early-type System MY Ser

P. Mayer[1], H. Drechsel[2] and M. Brož[1]

[1]Astronomical Institute, Charles University, V Holešovičkách 2, 180 00 Praha 8,
Czech Republic
[2] Dr. Remeis-Sternwarte, Astronomical Institute of the University Erlangen-Nürnberg,
Sternwartstr. 7, D-96049 Bamberg

Abstract. MY Ser is an eclipsing early-type contact system. Both components are of spectral type O6 III. Using ESO archive spectra, we show that the radial velocity of the third body, which contributes about half of the total luminosity, changed from 2006 to 2009. The line profiles of the eclipsing system have peculiar shapes and strengths; namely around conjunctions, they are affected by circumbinary matter.

Keywords. (stars:) binaries: eclipsing, stars: early-type, stars: individual (MY Ser)

The eclipsing binary MY Ser (HD 167791) belongs to the rare class of early-type contact systems. It has a period of 3.32 days, its magnitude in maximum is V = 7.3, and both minima have depths of 0.3 mag. The integral spectrum was classified as O8 Ib(f)p by Walborn (1972). Leitherer *et al.* (1987) found that the spectrum is dominated by lines of a third component, probably of spectral type O8 Ib, and that the lines of the eclipsing components belong to stars of earlier type. Mayer *et al.* (2010) solved the UBV light curves and derived an overcontact solution with a large filling factor.

The ESO archive contains 150 FEROS spectra of MY Ser, taken in the years 2008 and 2009 (plus one spectrum from 2006). These spectra are available in a pipeline processed form. We chose He I 5876 and He II 5411 Å lines for this study, selected about half of the available spectra and averaged groups of them to represent 22 different phase points. We fitted Gaussian profiles to obtain the positions, widths, and depths of all line components.

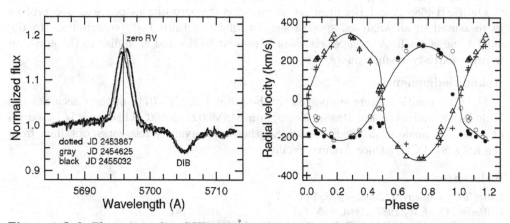

Figure 1. Left: The emission line C III 5696 Å proves the change of the third body radial velocity. Right: The radial velocity curves. Open circles - primary, line 5411 Å; full circles - primary, line 5876 Å; triangles - secondary, line 5411 Å; crosses - secondary, line 5876 Å.

Table 1. Parameters of the binary.

	K (km/s)	Vγ	a sin i (R$_\odot$)	M sin^3 i (M$_\odot$)
Primary	284	28	18.8	36.7
Secondary	305	10	20.1	34.2

The third body lines behave independently of the eclipsing binary lines. The most pronounced spectral features of the third component are He I lines. Their FWHM should not be correlated with the binary orbital phase, but the measured depths of the third component line depend on phase; they are larger around minima. According to our measurements, the depths of the lines 5411 and 5876 Å would be 0.151 and 0.287 if the contribution of the eclipsing binary was negligible. The corresponding FWHMs are 3.1 and 3.3 Å, and the EWs 0.47 and 0.95 Å, respectively. Such values agree with spectral type O8 I (according to OSTAR synthetic spectra by Lanz & Hubeny 2003 and O-star parameters by Martins et al. 2005).

Leitherer et al. (1987) measured the radial velocity of the He II 5411 Å line of the third component as +21 km/s in the year 1986. According to our RVs for the years 2006 to 2009 the velocity is smaller (close to zero) in the years 2008–2009. The decreasing velocity of the third body is confirmed by the C III emission line (Fig. 1). In the year 2006, its velocity was ∼ 20 km/s more positive than in 2008. The correctness of the wavelength scale can be checked by the unchanged position of the 5905.1 Å DIB.

To show the eclipsing component line profiles, lines of the third star were subtracted. The 5411 Å profiles look as expected at phases 0 to 0.5, i.e., the primary line is stronger than the secondary; however, at phases 0.5 to 1, the primary line is weaker. The 5876 Å line has an asymmetric shape with a stronger blue wing around both conjunctions; at other phases, these lines behave similarly as 5411 Å, but at phases 0.5 to 1, the primary is yet weaker. Given the weakness of the primary lines and their enhancement at the blue side one has to assume that circumbinary matter causes these effects.

Radial velocities are presented in Fig. 1. In spite of deviations from the expected RV curve around conjunctions, the velocities at quadratures should represent the orbital motion; different gamma velocities provide a better fit than does a common velocity.

With $i = 83.6°$, the masses are 37.4 and 34.8 M$_\odot$, radii are 16.5 and 15.4 R$_\odot$. According to Martins et al. (2005), such parameters correspond to a spectral type O6 III.

The He II 4686 Å and Hα lines are in emission. Superpositions of these line profiles as measured in all analyzed spectra suggest that the third body contribution to the 4686 Å line is small. A stronger emission is present in Hα and is similar to the emission in other early-type supergiants.

Acknowledgements

The Czech authors were supported by the grant P209/10/0715 of the Czech Science Foundation and from the Research program MSM0021620860. This work is based on observations made with the European Southern Observatory telescopes obtained from the ESO/ST-ECF Science Archive Facility.

References

Lanz, T. & Hubeny, I. 2003, *ApJS*, 146, 417
Leitherer, C. et al., 1987, *A&A*, 185, 121
Martins, F., Schaerer, D., & Hillier, D. J. 2005, *A&A*, 436, 1049
Mayer, P., Božić, H., Lorenz, R., & Drechsel, H. 2010, *AN*, 331, 274
Walborn, N. R. 1973, *AJ*, 78, 1067

From Interacting Binaries to Exoplanets: Essential Modeling Tools
Proceedings IAU Symposium No. 282, 2011
Mercedes T. Richards & Ivan Hubeny, eds.
© International Astronomical Union 2012
doi:10.1017/S1743921311027670

Period Changes of the Algol System SZ Herculis

J. W. Lee, C.-U. Lee, S.-L. Kim, H.-I. Kim, J.-H. Park and T. C. Hinse

Korea Astronomy and Space Science Institute, Daejeon 305-348, Korea
email: jwlee@kasi.re.kr

Abstract. New CCD photometric observations of SZ Her were obtained between February and May 2008. More than 1,100 times of minimum light spanning more than one century were used for the period analysis. We find that the orbital period of SZ Her has varied due to a combination of two periodic variations, with cycle lengths of P_3=85.8 yr and P_4=42.5 yr and semi-amplitudes of K_3=0.013 days and K_4=0.007 days, respectively. The most reasonable explanation for them is a pair of light-time-travel (LTT) effects driven by the existence of two M-type companions with minimum masses of M_3=0.22 $M_\odot$ and M_4=0.19 $M_\odot$, located at nearly 2:1 mean motion resonance. Then, SZ Her is a quadruple system and the 3rd and 4th components would stay in the stable orbital resonance.

Keywords. binaries: close, binaries: eclipsing, stars: individual (SZ Herculis)

1. Introduction

Although the orbital period of SZ Her has been examined several times, its detailed study was made by Szekely (2003) and Soydugan (2008). The former reported that the period change can be fitted to a sine curve with a period of 66 yr, while the latter concluded that it can be represented by a single light-time-travel (LTT) ephemeris, caused mainly by a third body with a cycle length of 72 yr and a minimum mass of 0.25 $M_\odot$. Soydugan (2008) also suggested that the timing residuals from the LTT fit indicate an additional short-term oscillation with a period within about 20 yr. Nonetheless, the period variation still has not been studied as conclusively as can be desired. Here, we present that SZ Her is probably a quadruple system.

2. Observation and result

We performed new CCD photometry of SZ Her on 13 nights from 2008 February 28 through May 17. The observations were taken with a SITe 2K CCD camera and a *BVRI* filter set attached to the 61-cm reflector at Sobaeksan Optical Astronomy Observatory (SOAO) in Korea. The instrument and reduction method are the same as those described by Lee *et al.* (2007). GSC 2610-1116 and GSC 2610-0821, imaged on the chip at the same time as the program target, were selected as comparison and check stars, respectively. In addition to these, we observed two eclipse timing on both 2004 June and 2011 May, using *B* and *BV* filters attached to the 61-cm reflector at SOAO, respectively.

Seventy-nine times of minimum light were determined from our observations and the WASP (Wide Angle Search for Planets) archive (Butters *et al.* 2010). For ephemeris computations, we have collected a total of 1050 timings from the literature to add to our measurements. First of all, we examined whether the period variations of SZ Her could be represented by the single-LTT effect as previous researchers have suggested.

313

Table 1. Parameters for the LTT orbits of SZ Her.

Parameter	τ_3	τ_4	Unit
T_0	2,434,987.39933(79)		HJD
P	0.818095788(46)		days
$a_{12}\sin i_{3,4}$	2.31(18)	1.24(20)	AU
ω	88.8(7.5)	286(11)	deg
e	0.722(90)	0.41(20)	
n	0.01149(13)	0.02320(17)	deg day^{-1}
T	2,422,642(310)	2,406,168(321)	HJD
$P_{3,4}$	85.8(1.0)	42.5(1.1)	yr
K	0.0134(10)	0.0071(11)	days
$f(M_{3,4})$	0.00168(13)	0.00106(17)	M$_\odot$
$M_{3,4}\sin i_{3,4}$	0.221(10)	0.188(15)	M$_\odot$

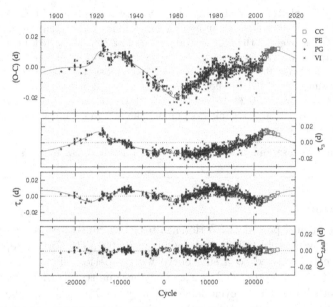

Figure 1. O–C diagram of SZ Her. In the top panel, the continuous curve represents the full contribution of the ephemeris. The second and third panels display the long- and short-term LTT orbits, respectively, and the bottom panel the residuals from the complete ephemeris.

Fitting all timing residuals to that ephemeris form failed to give a satisfactory result. After testing several other forms, we found that the O–C variation is best fitted by the two-LTT ephemeris, $C = T_0 + PE + \tau_3 + \tau_4$. Final results are summarized in Table 1 and plotted in Figure 1.

Because the LTT periods suggest nearly 2:1 resonant capture, the 3rd and 4th components would stay in the stable orbital resonance with a period ratio of 2.02 ± 0.06. If the two objects are on the main sequence and in the orbital plane ($i = 87°.6$) of the eclipsing pair, the masses of the 3rd and 4th bodies become $M_3 = 0.22$ $M_\odot$ and $M_4 = 0.19$ $M_\odot$, respectively. Following the empirical relations, their radii and temperatures are calculated to be $R_3 = 0.23$ $R_\odot$ and $T_3 = 3017$ K, $R_4 = 0.20$ $R_\odot$ and $T_4 = 3008$ K.

References

Butters, O. W., West, R. G., Anderson, D. R., *et al.*, 2010, *A&A*, 520, L10

Lee, J. W., Kim, C.-H., & Koch, R. H. 2007, *MNRAS*, 379, 1665

Soydugan, F. 2008, *AN*, 329, 587

Szekely, P. 2003, *IBVS*, 5467, 1

From Interacting Binaries to Exoplanets: Essential Modeling Tools
Proceedings IAU Symposium No. 282, 2011
Mercedes T. Richards & Ivan Hubeny, eds.

© International Astronomical Union 2012
doi:10.1017/S1743921311027682

The Role of Electron Scattering in Probing the Wind from the Hot Star in Symbiotic Binaries

Matej Sekeráš and Augustin Skopal

Astronomical Institute of the Slovak Academy of Sciences,
Tatranská Lomnica, Slovakia
email: msekeras@ta3.sk, skopal@ta3.sk

Abstract. We modeled the broad wings of the OVI 1032,1038Å resonance lines and HeII 1640Å line in the spectra of some symbiotic stars by the electron-scattering process. We determined an empirical relationship between the emission measure of the symbiotic nebula and the electron optical depth. This allowed us to determine a contribution from the electron-scattering also to emission lines, which originate in a more extended, low density part of the nebula. For example, subtracting the electron-scattering contribution from the $H\alpha$ line profile makes it possible to determine more precisely the mass loss rate via the wind from the hot star in symbiotic binaries.

Keywords. binaries: symbiotic, scattering, line: profiles

1. Introduction

Symbiotic stars are interacting binaries with the longest known orbital period on the order of years to decades, consisting of a cool giant and a hot compact star. As a result, in their spectra we observe contributions from the red giant, the hot star, and the nebula, which represents ionized winds from the binary components. During quiescent phases, the hot star releases its energy at a constant rate and temperature, while during active phases, the energy in the spectrum is redistributed and the brightness of the star increases by about 2 - 3 mag in the optical. Active phases are also connected with an increase of the nebular component of radiation. In this work, we investigated the effect of the electron-scattering on the strongest emission lines in the spectra of symbiotic stars.

2. The contribution of the electron-scattering to the $H\alpha$ wings

We assumed, that the observed broad wings of the OVI 1032, 1038Å resonance line and HeII 1640Å emission line are created only by the electron-scattering process. According to Castor, Smith & van Blerkom (1970), we applied the model in which the photons are transferred throughout the layer of free electrons (the symbiotic nebula) in the direction of the observer, characterized by the electron optical depth, τ_e and electron temperature T_e. An example of the observed and modeled wings is shown in Fig. 1.

By modeling the spectral energy distribution of the symbiotic star AG Dra, Skopal *et al.* (2009) showed that the stellar wind of the hot star is enhanced during the active phases, with an increase in the particle density (and thus free electrons) at the vicinity of the hot star and consequent increases in the electron optical depth of the symbiotic nebula. Concentration of the free electrons and the volume of the nebula determine the so-called emission measure (EM), that changes with the star's activity. According to the definition of τ_e and EM, $EM \propto \tau_e{}^2$. To demonstrate this relationship, we used τ_e from the models of the line profiles observed at different stages of activity for different objects

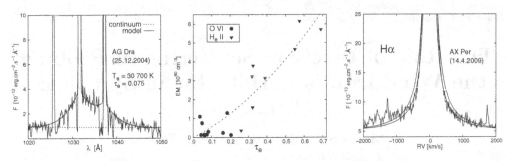

Figure 1. Left: Modeled (heavy line) profile of the OVI 1032, 1038Å lines. In the middle: The relation between τ_e and EM for different levels of symbiotic star's activity. Right: The models of Hα wings before (thin line) and after (heavy line) extracting the electron-scattering contribution.

and corresponding EM, determined from the continuum flux in the near-UV region. In this way, we determined this relationship as $EM = 6.6\,\tau_e{}^2 + 4.9\,\tau_e$ (Fig.1). We didn't take into account the dependence of EM and τ_e on the orbital phase and also neglected the contribution of the enhanced stellar wind of the hot star in the line wings during the active phase, which leads to uncertainties in determination of the $EM(\tau_e)$ relationship.

According to Skopal (2006), the broad wings of the Hα line profile originate in the ionized wind from the hot star. According to the $EM(\tau_e)$ relationship, there is also a contribution due to the electron scattering in the Hα wings. To determine more precisely the mass loss rate, we have to subtract its contribution to the wings. We demonstrate this approach on the spectrum of the symbiotic star AX Per, observed during its last activity increase in 2007-2010 (Skopal et al. 2011). We determined EM from the luminosity of the Hα line using the total volume emission coefficient in Hα line for $T_e = 20\,000$ K. We estimated $\tau_e = 0.059$ from the $EM(\tau_e)$ relationship, which allowed us to subtract the contribution of the electron-scattering from the Hα wings. The corresponding mass loss rate decreased by about 12% from the value of $\dot{M}_h = 3.25 \times 10^{-6} M_\odot \mathrm{yr}^{-1}$ to $\dot{M}_h = 2.88 \times 10^{-6} M_\odot \mathrm{yr}^{-1}$ (Fig.1).

3. Conclusion

Electron-scattering can explain the very broad wings of the strongest emission lines of highly ionized elements, which are created in densest parts of the nebula at a vicinity of the hot star. The increase of τ_e and EM indicate the increase in the number of free electrons in the line of sight, which is in major part due to the enhanced stellar wind from the hot star during active phases. Despite the illustrative character of the $EM(\tau_e)$ relationship, the estimated reduction in the mass loss rate of the hot star suggests that electron-scattering is not the dominant contributor to the broad Hα wings.

Acknowledgements

This research was supported by the Slovak Academy of Sciences under a grant VEGA No. 2/0038/10.

References

Castor, J. I., Smith, L. F., & van Blerkom, D. 1970, *ApJ*, 159, 1119
Skopal, A. 2006, *A&A*, 457, 1003
Skopal, A., Sekeráš, M., González-Riestra, R., & Viotti, R. F. 2009, *A&A*, 507, 1531
Skopal, A., et al., 2011, *A&A*, submitted

From Interacting Binaries to Exoplanets: Essential Modeling Tools
Proceedings IAU Symposium No. 282, 2011   © International Astronomical Union 2012
Mercedes T. Richards & Ivan Hubeny, eds.   doi:10.1017/S1743921311027694

Optical Spectroscopy of V393 Scorpii During its Long Cycle

Ronald E. Mennickent[1], Zbigniew Kołaczkowski[2], Gojko Djurasevic[3], M. Diaz[4] and Ewa Niemczura[2]

[1]Departamento de Astronomía, U. de Concepción, Casilla 160-C, Concepción, Chile
[2]Instytut Astronomiczny Uniwersytetu Wrocławskiego, Wrocław, Poland
[3]Astronomical Observatory, Volgina 7, 11060 Belgrade 38, Serbia
[4]Departamento de Astronomia, IAG, U. de Sao Paulo, Rua do Matao, 1226,
Butanta 05508-900, Sao Paulo, SP, Brasil

Abstract. V393 Scorpii is a bright Galactic Double Periodic Variable showing a long photometric cycle of $\approx$253 days. The ASAS V-band light curve has been disentangled into an orbital and long cycle component. The orbital light curve was modeled with two stellar components and a circumprimary accretion disk. Based on this model, and the careful choice of a template spectrum for the donor, the contribution of the donor to the line+continuum spectrum was removed at every orbital phase. The remaining residual spectra were analyzed. Notable findings are the larger line emissivity observed during the long cycle maximum that is concentrated to low velocities and the presence of discrete absorption components in the wings of the OI 7773 line, whose visibility strongly depends on the orbital phase. In addition, weak emission is observed in donor metallic absorption lines. Finally, we present the first Hα Doppler map for V 393 Scorpii. A modulated wind explains many observational features.

Keywords. stars: binaries, stars: early-type, stars: mass loss, stars: winds, outflows

---

This is a brief report on our recent investigation of the Double Periodic Variable (DPV) V393 Sco. DPVs are Algol-related binaries showing a long photometric periodicity besides the orbital photometric periodicity, they were first mentioned by Mennickent *et al.* (2003). The eclipsing binary V393 Sco resembles β Lyrae in some aspects (Mennickent *et al.* 2010), but the amplitude of their long photometric cycle of 253 days is larger and the orbital light curves are more stable than in β Lyrae. In V393 Sco, we observe in many lines the presence of discrete absorption components (DACs), mainly in the blue wing of the absorption lines, like in OI 7773 (Fig. 1). DACs are mostly visible during the second part of the orbital cycle and in the blue wing of the line. DACs are usually observed in the UV lines of some B-type stars and have been associated with condensations in an expanding stellar wind. Similarly, DACs in V393 Sco could be associated with an anisotropic wind emerging somewhere inside the gainer Roche lobe in the third or fourth quadrant.

We constructed Hβ difference spectra by subtracting a reference spectrum taken at similar orbital phase (first number in Fig. 2 labels) but different long phase (numbers in parentheses in same figure). We find that when the system goes to long cycle maximum, emission increases, especially at low velocities; similarly for Hα and Hγ. We interpret this as evidence for a polar wind whose modulation produces the long photometric cycle.

The Doppler map for the Hβ emission lines can be interpreted in terms of the existence of an optically thin polar wind plus a circumprimary optically thick region (Fig. 2).

In V393 Scorpii, the cooler stellar component is detected in hydrogen lines as well as in a characteristic A-type metallic spectrum. The new finding is the detection of emission

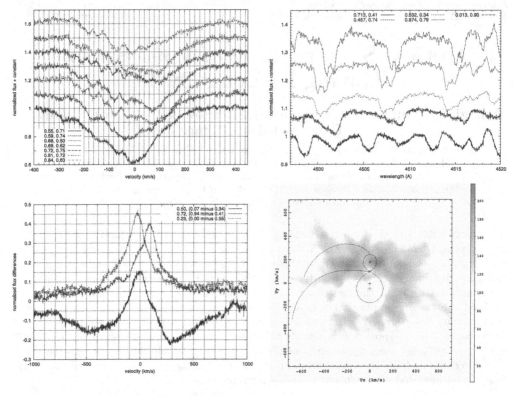

Figure 1. Top left: Example of DACs observed in O I 7773. Hereafter orbital (left) and long (right) phases are given. Top right: Example of emission cores, asymmetries and doubling of metallic lines. Bottom left: Difference spectra for Hβ emission lines at selected epochs. Velocities are with respect to the system center of mass. Bottom right: The back-projection Doppler map for the donor-subtracted Hα line during epochs of long cycle maximum ($0.8 < \Phi_l < 0.3$). Crosses show, from up to down, donor center of mass, L1 point, system center of mass and gainer center of gravity. Roche lobe surfaces for a point mass gainer are shown along with the gas stream path (lower track). The upper track represents the keplerian velocity of a disk if it would exist along the stream.

lines in these optical metallic lines and the detection of an additional metallic spectrum arising from the accretion disk (Fig. 1). Emission cores are visible at main eclipse in the upper spectrum, and disk absorption components are clearly separated from the donor components in the bottom spectrum taken at quadrature.

After removing the donor spectrum by modeling its contribution at each orbital phase, we discovered at some epochs double emission lines in some iron, titanium and carbon lines (not shown here). We measured the radial velocity of these anomalous emissions observed in the metallic lines and found that they roughly follow the radial velocity path of the donor, suggesting an origin somewhere around the donor. The connection of these emission features with the long cycle will be explored in a forthcoming paper.

References

Mennickent, R. E., Pietrzynski, G., Diaz, M., & Gieren, W. 2003, *A&A* 399, L47

Mennickent, R. E., Kołaczkowski, Z., Graczyk, D., & Ojeda, B. 2010, *MNRAS* 405, 1947

From Interacting Binaries to Exoplanets: Essential Modeling Tools
Proceedings IAU Symposium No. 282, 2011   © International Astronomical Union 2012
Mercedes T. Richards & Ivan Hubeny, eds.   doi:10.1017/S1743921311027700

Methods of the Long-term Radial-Velocity Variation Removal and their Application to Detect Duplicity of Several Be Stars

J. Nemravová[1], P. Harmanec[1], P. Koubský[2] and A. Miroshnichenko[3]

[1] Astronomical Institute of the Charles University, Faculty of Mathematics and Physics,
V Holešovičkách 2, CZ–180 00 Praha 8, Czech Republic
email: `janicka.ari@seznam.cz, hec@sirrah.troja.mff.cuni.cz`

[2] Astronomical Institute of the Academy of Sciences, CZ-251 65 Ondřejov, Czech Republic
email: `koubsky@sunstel.asu.cas.cz`

[3] Department of Physics and Astronomy, University of North Carolina at Greensboro,
Greensboro, NC 27402, USA
email: `a_mirosh@uncg.edu`

Abstract. There are several types of binary stars which show non-periodical radial velocity variations with the amplitude larger than those connected with the orbital motion. The non-periodical changes have to be removed in order to study the orbital ones. We propose three removal techniques, two of which are based on the trend modeling with continuous functions and the third one that takes the orbital motion into account.

Keywords. methods: data analysis, techniques: radial velocities, stars: emission-line, Be

1. Introduction

The detection of duplicity of Be stars might be very complicated. Low-amplitude radial-velocity (RV hereafter) orbital changes can be masked by cyclic variations related to changes and asymmetries in the circumstellar matter surrounding such stars. If one wants to study the orbital motion of a Be binary, these cyclic variations have to be removed. We present three techniques that are suitable to the task.

2. Methods

(*a*) The first method is optimal in situations when one has little idea about the value of the orbital period. It is based on a smoothing technique via spline functions developed by Vondrák (1969, 1977). To its practical application, we used the program HEC13, which allows us to smooth either individual observations or normal points averaged over constant time intervals. Cyclic variations modelled with the method are shown in the first panel of Fig. 1.

(*b*) The second method comes in order when one knows the true orbital period. One has to split the RVs into subsets, each being several times shorter than the typical timescale of the long-term changes. Each subset should be as short as possible to obtain the resolution of the long-term changes possible, but it must contain data reasonably defining the orbital RV curve. An individual systemic velocity is then derived for each subset as part of the orbital solution with the SPEL program. This way, the long-term trend is removed in discrete time steps. The resulting 'γ velocities' are shown with empty squares in the first and the third panels of Fig. 1.

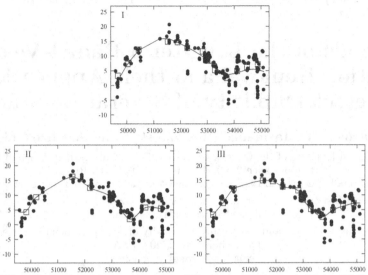

Figure 1. RVs measured on the steep wings of Hα line of BU Tauri. The solid line represents long-term variations estimated with: HEC13 (panel I) and HEC36 (panels II and III). Empty squares represent: γ velocities computed with SPEL (panels I and III) and RVs averaged over constant time interval (panel II).

(*c*) The third method models the long-term changes with Hermite polynomials through suitably chosen normal points. The program HEC36, designed for this task, employs the interpolation subroutine INTEP (see Hill 1982). If one chooses to input normal points computed over constant time intervals, the results are very similar to the ones obtained with the first method. We found that it is better to use the program HEC36 to improve the second method. This can be done by submitting the γ velocities computed with SPEL as the normal points to the program HEC36. The trend estimated with the normal points computed over constant time intervals is shown in panel II and with velocities taken as normal points in panel III of Fig. 1.

3. Comparison

The second method resulted in an orbital RV curve with significantly lower rms for BU Tauri (Nemravová *et al.* 2010) and γ Cassiopeiae (in preparation) than smoothing with the first method. The main reason is that the systemic velocities computed with SPEL properly model the orbital RV curve of the given subset. This does not mean that the second method must always be superior to the first one. We recommend trying all three methods in every particular study and comparing the results. HEC13 and HEC36 written by PH are available for download at *http://astro.troja.mff.cuni.cz/ fpt/hec/*.

Acknowledgements

The research of JN and PH was supported by the grant P209/10/0715 of the Czech Science Foundation and by the grant SVV-263301 of the Charles University of Prague.

References

Hill G. 1982, *Publ. Dominion Astrophys. Obs. Victoria*, 16, 67
Nemravová, J., Harmanec, P.,Kubát, J., Koubský, P., Iliev, L., Yang, S., Ribeiro, J., Šlechta, M., Kotková, L., Wolf, M., & Škoda, P. 2010, *A&AS*, 516, A80
Vondrák J. 1969, *Bull. Astron. Inst. Czechosl.*, 20, 349
Vondrák J. 1977, *Bull. Astron. Inst. Czechosl.*, 28, 84

From Interacting Binaries to Exoplanets: Essential Modeling Tools
Proceedings IAU Symposium No. 282, 2011
Mercedes T. Richards & Ivan Hubeny, eds.
© International Astronomical Union 2012
doi:10.1017/S1743921311027712

Advanced Tools for Exploring Large EB Datasets

E. J. Devinney, A. Prša, E. F. Guinan and M. Degeorge

Villanova University, Dept. of Astronomy & Astrophysics, 800 Lancaster Ave,
Villanova PA 19085, USA; email: edward.devinney@villanova.edu

Abstract. Thanks to OGLE, Kepler, CoRoT and planned new ambitious survey projects, the eclipsing binary (EB) community is beginning to experience a long-predicted data deluge. Beyond the analysis of the many fascinating individual objects yielded by these programs, these complete datasets themselves should yield further insights. Because objects in such datasets are characterized by many parameters, tools that assist in understanding high-dimensional data are acquiring increasing relevance. Chiefly among these are new Advanced Visualization (AV) tools and various methods of clustering data, both approaches complementing each other naturally. We illustrate the use of these tools as applied to OGLE II LMC EB data and respective EBAI light curve solutions.

Keywords. methods: data analysis, methods: numerical, binaries: eclipsing, stars: fundamental parameters, stars: statistics

1. Introduction

GGOBI (www.ggobi.org) and Mondrian (www.rosuda.org/Mondrian) are widely-used AV toolsets, both supporting simultaneous views of multiple, linked two dimensional scatterplots and histograms/barcharts. Objects selected in one plot are highlighted immediately in all the others (brushing), thus helping to identify groups of physically-related objects and offering clues to the nature of outliers. GGOBI also provides for a grand tour of a high-dimensional dataset. This is implemented as a rotating 3D plot into which is projected all the (higher dimensional) data in round-robin fashion. Clusters can then be identified as relatively isolated groupings of points during various phases of the data presentation. While not discounting the hands-on familiarity with a dataset offered by the grand tour process, in recent years the machine-learning community has developed a panoply of automatic clustering algorithms, able to identify clusters of data points in high-dimensional space and thus candidate physical groups. For example, the Waikato University WEKA clustering package (www.cs.waikato/ml/weka) is fast and easy to use, providing a variety of clustering algorithms and the ability to plot the results.

2. Example Application: OGLE II LMC EBs

The two three-part panels, Figs. 1 and 2, illustrate the use respectively of the GGOBI and WEKA packages on OGLE II data from the LMC with light curve solutions using the EBAI neural network system (Prša *et al.* 2008). In each panel, from left to right are: V-I vs I-magnitude; logP[d] vs I-magnitude; and V-I vs logP[d]. In the top GGOBI panel, the middle figure reveals a group of objects showing a period vs I-mag behavior. The left panel indicates these are giants, while the right panel dotted box shows that they are virtually exclusively defined by period and color, as found through interactive exploratory plotting. The WEKA panel below shows the same result, but the cluster of

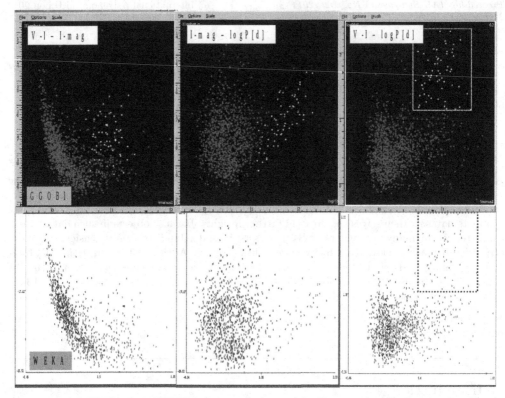

Figure 1. GGOBI (top) and WEKA (bottom) plot panels. Top panel, right, interactive discovery of objects responsible for the period vs I-magnitude trend in the middle plot. Bottom panel, right, the same objects automatically found by the clusterer.

points in the right plot (dotted box) was found by automatic clustering applied to the combined OGLE II LMC and EBAI dataset. This same clustering also automatically delineated main-sequence objects from evolved ones (not shown).

3. Conclusions and New Work

Our experience suggests that AV and clustering tools are destined to become ever more valuable with the increasing data deluge, and that wider effort to exploit them will yield scientific payoffs. We also found the existing AV tools have some deficiencies. Neither GGOBI nor Mondrian are convenient for extended analysis and both have issues in axis scaling. The most serious drawback is that on quitting these applications, all plots and the state of the application are lost. As a result, one of us (MD) is well-advanced on a Python-based AV package PyVU that overcomes these limitations. PyVU allows brushing in multiple plots, has rational axis scaling and labeling, supports display of light curves associated with one or more selected data points and, on quitting, saves the state of the application along with user comments/session summary. With this latter feature, branching explorations of datasets are supported. PyVU will be made available to the user community.

References

Prša, A., *et al.*, 2008, *ApJ*, 687, 542

From Interacting Binaries to Exoplanets: Essential Modeling Tools
Proceedings IAU Symposium No. 282, 2011
Mercedes T. Richards & Ivan Hubeny, eds.
© International Astronomical Union 2012
doi:10.1017/S1743921311027724

Applying the Steepest Descent Method with BINSYN on RY Per Photometry

D. Sudar, H. Božić and D. Ruždjak

Hvar Observatory, Faculty of Geodesy, Kačićeva 26, 10000 Zagreb, Croatia
email: **dsudar@geof.hr**

Abstract. Recent studies of the Algol-type binary RY Per presented strong evidence that there is an accretion disk around the primary in the system. We used new UBV photometry from Hvar Observatory and the BINSYN software package in order to determine the basic parameters of the disk. The search for the best parameter set was performed with a fully automated steepest descent method. The resulting disk is large and visible at all orbital phases. Somewhat surprising is the large mass transfer rate which should be tied with, currently unreported, secular period changes.

Keywords. binaries: eclipsing, stars: emission-line, Be, stars: individual (RY Per)

RY Per (HD 17034) is one of the most massive Algol binaries. Olson & Plavec (1997) determined temperatures, masses and radii of both components. They also presented strong evidence for ongoing mass transfer and the existence of circumstellar matter, although their photometric solution does not contain the disk.

More recently Barai *et al.* (2004) were able to determine radial velocities of the primary in UV spectra. They also presented a time variable but persistent disk model by analyzing $H\alpha$ profiles.

In this work, we will mainly focus on obtaining more information about the accretion disk and related mass transfer from new UBV photometry.

The observations consist of 426 data points in each of the three filters. All measurements were transformed to the standard Johnson UBV system. We adopted the ephemeris derived by Olson & Plavec (1997): $T_0 = \text{HJD}2441655.780 + 6.863569E$.

For light curves analysis of the accretion disk system, we used BINSYN (Linnell & Hubeny 1996). The steepest descent method was used for parameter optimization (see Sudar *et al.* 2011). The geometry of the accretion disk in BINSYN is given by the vertical thickness of the disk, H_V, outer radius of the disk, R_A, and inner radius of the disk, R_B. We allowed for convergence of the first two parameters, while the last one was fixed to the radius of the primary.

The mass transfer rate, $\dot{M}$, determines the temperature profile of the disk's face (cf. e.g. Pringle 1981) and was also allowed to converge. The only other parameter which wasn't fixed was the polar temperature of the primary, T_1; others were taken from previous studies (Olson & Plavec 1997 and Barai *et al.* 2004).

The best fit solution was sought simultaneously in all 3 filters. After several convergence runs, we obtained the best fit paramaters shown in Table 1. The resulting light curves are presented in Fig. 1, and a view of the system at phase $\phi = 0.5$ is shown in Fig. 2. It is readily observable that the disk is visible at all phases, which is in agreement with observed $H\alpha$ and UV emission.

On the other hand, it is rather surprising that the mass transfer rate is so large. In a similar system, UX Mon (Sudar *et al.* 2011), it was found that a smaller transfer rate would cause a secular period change, which should most evidently manifest itself in the $O - C$ diagram of times of primary minima.

Table 1. Best fit values of converged parameters.

Param.	Value
H_V	$1.51 \pm 0.50\ R_\odot$
R_A	$14.12 \pm 1.00\ R_\odot$
$\dot{M}$	$11.31 \pm 0.50\ \mu M_\odot yr^{-1}$
T_1	16702 ± 1000 K

The errors were estimated from several convergence runs.

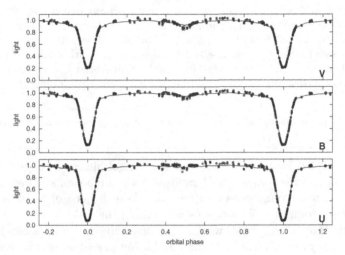

Figure 1. Observed data in UBV filters and their best fit light curves.

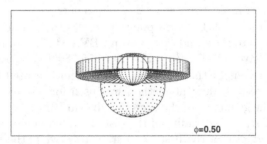

Figure 2. View of the system during secondary minimum.

References

Barai, P., Gies, D. R., Choi, E., Das, V., Deo, R., Huang, W., Marshall, K., McSwain, M. V., Ogden, C., Osterman, M. A., Riddle, R. L., Seymour, Jr., J. L., Wingert, D. W., Kaye, A. B., & Peters, G. J. 2004, *ApJ*, 608, 989

Linnell, A. P. & Hubeny, I. 1996, *ApJ*, 471, 958

Olson, E. C. & Plavec, M. J. 1997. *AJ*, 113, 425

Sudar, D., Harmanec, P., Lehmann, H., Yang, S., Božić, H., & Ruždjak, D. 2011, *A&A*, 528, 146

From Interacting Binaries to Exoplanets: Essential Modeling Tools
Proceedings IAU Symposium No. 282, 2011
Mercedes T. Richards & Ivan Hubeny, eds.
© International Astronomical Union 2012
doi:10.1017/S1743921311027736

The Effects of Eccentric Accretion Structures on the Light Curves of Interacting Algol-type Binary Stars

Phillip A. Reed

Kutztown University of Pennsylvania
Kutztown, PA 19530, USA
email: `preed@kutztown.edu`

Abstract. The light curves of many Algol-type binary stars are complicated with strange variations. Secular variations are due to the transient nature of the accretion structure, while the phase-dependent features, such as outside-of-eclipse dips, are likely geometrical effects of the accretion structure eclipsing the primary star. Presented here is a model of the ultraviolet light curve of R Arae that explains these variations through the combination of an eccentric accretion structure and the system's orbital inclination.

The orbital period of R Ara is 4.4 days, which is too long to allow for direct impact of the mass transfer stream onto the primary star, but not long enough for a stable accretion disk to form. Such intermediate-period Algols are good candidates in which to find transient and eccentric accretion structures. Other examples of interacting Algols that exhibit outside-of-eclipse dips in their light curves include RV Oph ($P_{orb.} = 3.7$ days) and Y Psc ($P_{orb.} = 3.9$ days).

In order to more accurately model eccentric accretion structures with synthetic light curves, especially at visible (and longer) wavelengths, more work must be done to account for emission by the parts of the accretion structure that are not in the line of sight to the primary star. The model presented here accounts only for the eclipsing regions of the accretion structure.

Keywords. accretion, accretion disks, techniques: photometric, (stars:) binaries: eclipsing

---

R Ara's peculiar variations are perhaps a cause for its neglectedness. Badly blended absorption lines were reported by Sahade (1952) and photometric variations were detected by Nield (1991) and Banks (1990). It was not until recently that the first ephemeris curve was constructed and analysed by Reed (2011), which provides strong evidence for a continuous period change due to rapid mass transfer within the system.

The detailed analysis of *IUE* data by Reed *et al.* (2010) relies on mass transfer to explain R Ara's photometric and spectroscopic variations. Since the *IUE* data consist of consecutive images spanning one complete orbital cycle, they eliminate the confusion of secular variability and provide insight into phase-dependent variations.

The *IUE* light curves reveal outside-of-eclipse dips that grow deeper at shorter wavelengths, which indicates they are caused by the primary star being eclipsed by something cooler. Spectroscopic evidence strongly supports the model of an eccentric accretion structure surrounding the primary star. Figure 1 illustrates the accretion structure. The light curve is shown in Figure 2.

The geometry of the eccentric accretion, coupled with the system's orbital inclination of 78°, is quite possibly the cause of the phase-dependent variations in the light curve. The method of Doppler tomography, which requires a large set of high resolution spectra with good phase coverage, could reveal the actual shape of the accretion structure and provide further evidence that we are seeing the effects of an eccentric accretion structure on the light curve of this interacting Algol-type binary star.

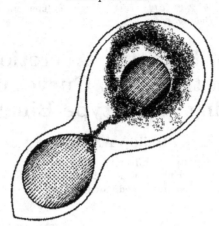

Figure 1: An illustration representing the eccentric accretion structure detected in R Ara. Taken from Reed *et al.* (2010).

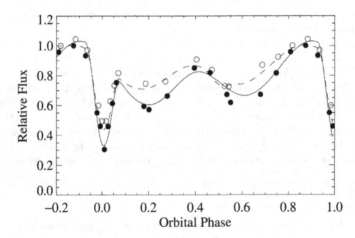

Figure 2: The *IUE* light curve for R Ara. Solid circles are at 1320 Å and open circles are 2915 Å. The solid and dashed lines are the theoretical models. Taken from Reed *et al.* (2010).

References

Banks, T. 1990, *IBVS*, 3455, 1

McCluskey, G. E. & Kondo, Y. 1983, *ApJ*, 111, 1

Nield, K. M. 1991, *Ap&SS*, 180, 233

Reed, P. A., McCluskey, G. E., Kondo, Y., Sahade, J., Guinan, E. F., Giménez, A., Caton, D. B., Reichart, D. E., Ivarsen, K. M., & Nysewander, M. C. 2010, *MNRAS*, 401, 913

Reed, P. A. 2011, *IBVS*, 5975, 1

Richards, M. T. & Ratliff, M. A. 1998, *ApJ*, 493, 326

Sahade, J. 1952, *ApJ*, 116, 27

From Interacting Binaries to Exoplanets: Essential Modeling Tools
Proceedings IAU Symposium No. 282, 2011
Mercedes T. Richards & Ivan Hubeny, eds.

© International Astronomical Union 2012
doi:10.1017/S1743921311027748

Physical Parameters of the Detached Eclipsing Binary KIC3858884 with a δ-Scuti Type Pulsating Component

C.-U. Lee[1], S.-L. Kim[1], J. W. Lee[1], K. Pavlovski[2,3] and J. Southworth[3]

[1]Korea Astronomy and Space Science Institute, Daejeon 305-348, Korea
email: leecu@kasi.re.kr

[2]Department of Physics, University of Zagreb, Bijenička cesta 32, 10 000 Zagreb, Croatia
[3]Astrophysics Group, EPSAM, Keele University, Staffordshire, ST5 5BG, UK

Abstract. We present physical parameters for the detached eclipsing binary KIC3858884 which has a δ-Scuti type pulsating secondary component. To derive orbital elements from the radial-velocity curve, high resolution Echelle spectra were obtained at the Bohyunsan Optical Astronomy Observatory in Korea. The BOES spectra and Kepler photometric data were analyzed with well-known codes: JKTEBOP and Wilson-Devinney model for eclipsing light-curve synthesis, and Period04 for pulsation frequency analysis. After the iterative curve fitting, we determined the physical parameters of KIC3858884 as $M_1 = 2.02 \pm 0.23 M_\odot$, $M_2 = 2.02 \pm 0.16 M_\odot$, $R_1 = 3.61 \pm 0.12 R_\odot$, $R_2 = 2.84 \pm 0.10 R_\odot$, respectively.

Keywords. stars: fundamental parameters, stars: oscillations, binaries: eclipsing

1. Introduction

As the Kepler satellite data archive opened to the public, many interesting light curves with ultra-high precision have been introduced. We compiled interesting targets which show the light variation of a detached eclipsing binary star with a δ-Scuti type pulsation. While Kepler data provide high precision photometry, the data do not contain multi-band photometric information. Therefore, spectroscopic and multi-band photometric follow-up observations are essential to analyze the targets in detail. In this paper, we derive the physical parameters of KIC3858884.

2. Iterative light curve analysis using PERIOD04 and JKTEBOP

We used the public data obtained by Kepler in Quarter 2. The photometric data showed clearly short-period sinusoidal variations at out-of-eclipse phases. From the frequency analysis using Period04 (Lenz & Breger 2005), a total of 198 frequencies which have amplitude signal-to-noise ratio (SNR) of larger than 4 were found at out-of-eclipse phases. Using those frequencies, a synthetic light curve is calculated and then subtracted from the originally observed light curve. The residual light curve was analyzed using JKTEBOP (Southworth *et al.* 2004). Primary and secondary eclipses are shown on the left side of the Figure 1. The black dots represent Kepler observation data with the eclipsing light variation removed. The green dots represent the synthetic pulsation light curve. In the primary eclipses, the amplitude of black dots is larger than green ones, while the amplitude of black dots is smaller than green ones in secondary eclipses. This is strong evidence that the secondary is a pulsating component; minimizing the dilution effect at the primary eclipse and occultation of a pulsator at the secondary eclipse.

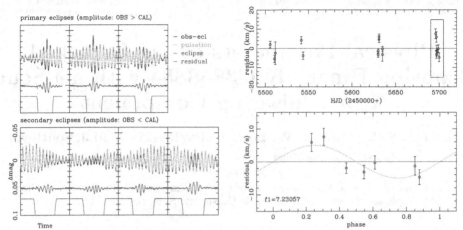

Fig 1. Primary and secondary eclipses are shown with calculated pulsating light curve (left), Residual of secondary RV curve fit and correlation with a dominant pulsation frequency of $f_1 = 7.23057$ which is almost $3^h 20^m$ (right)

3. Radial velocity curve analysis using cross correlation and WD2010

A total of 18 Echelle spectra were obtained from Nov. 3, 2010 to May 16, 2011 at Bohyunsan Observatory, Korea. The highest SNR of observed spectra is $\sim$200 and the lowest is $\sim$50 at near of 6700Å. We used the cross correlation method with a synthetic template spectrum to derive RVs. The optimal parameters (Teff, log g, [Fe/H] and ξ) were determined by minimizing the difference between the observed and synthetic spectra from the iterative process. Balmer lines and strong lines of the observations were compared to the synthetized spectra. To derive RVs, all observed spectra were cross-correlated with the synthetic spectrum for every 100Å width from 4000 to 6400Å. The peak of each cross correlation function was measured from the Gaussian fit. The RVs and errors were determined by median filtering of measurements and standard deviation, respectively. WD2010 (Wilson & Devinney 1971) was used to analyze the RV curve. Scattered residuals were found in the secondary RVs, while no scatter was found in the primary RVs. We suspect this residual is evidence of a pulsating secondary component. We tried to find a relation between the residual of secondary and the dominant pulsation frequency of f_1. The period of the frequency $f_1 = 7.23057$ is almost $3^h 20^m$. The right panel of Figure 1 shows the relation.

4. Physical parameters of KIC3858884

From iterative light curve and radial velocity analysis, physical parameters of KIC3858884 were determined. The parameters are $M_1 = 2.02 \pm 0.23 M_\odot$, $M_2 = 2.02 \pm 0.16 M_\odot$, $R_1 = 3.61 \pm 0.12 R_\odot$, $R_2 = 2.84 \pm 0.10 R_\odot$, respectively. In the empirical mass-radius relation, both components are located above the ZAMS.

References

Lenz P. & Breger M. 2005, *CoAst*, 146, 53
Southworth, J., Maxted, P. F. L., & Smalley, B. 2004, *MNRAS*, 349, 547
Wilson, R. E. & Devinney, E. J. 1971, *ApJ*, 166, 605

From Interacting Binaries to Exoplanets: Essential Modeling Tools
Proceedings IAU Symposium No. 282, 2011
Mercedes T. Richards & Ivan Hubeny, eds.

© International Astronomical Union 2012
doi:10.1017/S174392131102775X

New Light Curve Analysis for Large Numbers of Eclipsing Binaries III. SMC and Galactic Center

Young-Woon Kang, Keong Soo Hong and Pakakaew Rittipruk

Sejong University, Seoul, 143-747, Korea
email: kangye@sejong.ac.kr

Abstract. We improved the method of light curve analysis for large numbers of eclipsing binaries. Current methods require a week to analyze the light curves of an eclipsing binary for its physical and orbital parameters. Therefore, we developed a new method to treat large numbers of light curves of eclipsing binaries. We tested the new method by analyzing more than 14 hundred light curves and 9 hundred light curves discovered by OGLE in the Small Magellanic Cloud and Galactic center, respectively.

Keywords. stars, eclipsing binaries, data analysis, fundamental parameters

1. Introduction

Up to the 1990s, approximately 4000 and 200 eclipsing binaries have been discovered in our galaxy and nearby galaxies, respectively. After several survey observations for microlensing, more than 5000 close binaries were discovered in other galaxies. New surveys are expected in the future with the Large Synoptic Survey Telescope (LSST) and the Panoramic Survey Telescope and Rapid Response system (Pan-STARRS). Currently, Kepler is producing light curves of large numbers of variable stars. We expect those surveys to yield light curves of hundreds of thousands of new variable stars and eclipsing binaries.

The current methods of light curve analyses require technical skill and experience. It takes a week to model a single binary system. Therefore, a new method is needed to analyze large numbers of light curves of eclipsing binary stars. Prša *et al.* (2008) and Guinan (2009) introduced the concept of Eclipsing Binaries with Artificial Intelligence (EBAI) that aims to provide estimates of principal parameters for thousand of eclipsing binaries. Kang (2008, 2010) and Hong & Kang (2009) suggested a new iteration method of light curve analysis using standard eclipsing binaries.

2. Expanded standard eclipsing binaries and initial parameters

We increased the number of standard eclipsing binaries by collecting original data on the light curves and radial velocity curves of approximately 80 eclipsing binary stars. We defined the standard eclipsing binary star as an eclipsing binary star whose multi-color light curves and double-line radial velocity curves were published in good quality. We re-analyzed all light curves and radial velocity curves of these standard eclipsing binaries using the 2005 version of the Wilson and Devinney (WD) differential correction program. The 80 eclipsing binaries will be used as standard eclipsing binaries. The parameters of one of these standard eclipsing binaries will be used as initial parameters for target eclipsing binaries. A morphology of light curves can be characterized by depth and width of the primary eclipse and curvature outside the eclipses. We measure the depth, width, and curvature of the representative observed light curve (usually light curve among multi-color light curves). Then, we select the best similar light curve among the standard eclipsing binaries using the three measurements. We use parameters of the selected standard eclipsing binary star as initial parameters.

3. Improved iteration order

We modified the 2005 version of the WD program to run the differential correction by iterations. First, we adopted theoretical values for initial values of limb darkening coefficients, albedo, gravity darkening exponents. Then we adjusted only L1 to adjust the level of light curves. Then, we adjusted the set of (i,T2,L1), (i,P2,L1), (A2,t2,L1), (A1,X1,L1), (X2,A2,L1), (P1,L1,X2), (i,t2,A1,A2,P1,P2,L1,X1,X2) for each detached system. We basically adjusted 2-3 times for each parameters during whole one cycle. We checked the fitness with deviation between observation and model light curves. If the fitness is not acceptable, then we repeated the iteration of the entire cycle for better adjustment.

4. Application to eclipsing binaries discovered by OGLE

The OGLE project (Wyrzykowski *et al.* 2003) produced I light curves of approximately 1,404 and 934 eclipsing binaries observed in the Small Magellanic Cloud and Galactic center, respectively. They made B and V observations outside of eclipses in order to determine the temperatures of the binary systems. Therefore, we analyzed light curves of a total of 2338 eclipsing binaries in order to test our iteration programs. For SMC binaries, model light curves of 1014 binary systems were successfully conserved to the observations using 7-9 iterations with the mode 2 program, which is for a detached system. Another 243 and 141 light curves were also compared to the observations using 7-9 iterations with mode 4-5 and mode 3, respectively. We discovered that 72 percent of a total 1404 SMC eclipsing binary stars are detached systems. Another 17 and 10 percent are semi-detached and contact binaries, respectively. However, the binary star distribution in the Galactic center direction (Bade window) is different from those of the SMC. The detached, semi-detached and contact binaries are distributed 38 percent, 38 percent and 23 percent, respectively. It is very interesting that 269 of 1014 (27 percent) detached systems in the SMC have eccentric orbits, while only 5 of 357 (2 percent) stars have eccentric orbits.

References

Guinan, *et al.*, 2009, *ASP-CS*, 404, p.361
Hong, K. S. & Kang, Y.W.2009, *JASS*, 26, 295
Kang, Y. W. 2008, *JASS*, 25, 77
Kang, Y. W. 2010, *JASS*, 27, 75
Prša, A., *et al.*, 2008, *ApJ*, 687, 542
Udalski *et al.*, *AcA*, 47, 1
Wilson, R. E. & Devinney, E. J. 1971, *ApJ*, 166, 605
Wyrzykowski, *et al.*, 2003, *AcA*, 53, 1
Wyrzykowski, *et al.*, 2003, *AcA*, 54, 1

From Interacting Binaries to Exoplanets: Essential Modeling Tools
Proceedings IAU Symposium No. 282, 2011   © International Astronomical Union 2012
Mercedes T. Richards & Ivan Hubeny, eds.   doi:10.1017/S1743921311027761

Mass and Orbit Constraints of the Gamma-ray Binary LS 5039

T. Szalai[1], G. E. Sarty[2,3], L. L. Kiss[4], J. M. Matthews[5], J. Vinkó[1], and C. Kiss[4]

[1]Department of Optics and Quantum Electronics, University of Szeged,
Dóm tér 9., Szeged H-6720, Hungary; email: szaszi@titan.physx.u-szeged.hu

[2]Royal Astronomical Society of Canada, Saskatoon Centre,
P.O. Box 317, RPO University, Saskatoon, SK S7N 4J8, Canada

[3]Department of Physics and Engineering Physics, University of Saskatchewan,
Saskatoon, SK S7N 5E2, Canada

[4]Konkoly Observatory of the Hungarian Academy of Sciences,
H-1525 Budapest, P.O. Box 67, Hungary

[5]Department of Physics and Astronomy, University of British Columbia,
6224 Agricultural Road, Vancouver, BC V6T 1Z1, Canada

Abstract. We present the results of space-based photometric and ground-based spectroscopic observing campaigns on the γ-ray binary LS 5039. The new orbital and physical parameters of the system are similar to former results, except we found a lower eccentricity. Our *MOST*-data show that any broad-band optical photometric variability at the orbital period is below the 2 mmag level. Light curve simulations support the lower value of eccentricity and imply that the mass of the compact object is higher than 1.8 $M_\odot$.

Keywords. binaries: spectroscopic, stars: individual (LS 5039)

1. Introduction

LS 5039, the enigmatic high-mass X-ray binary has been intensively observed at various wavelengths in the past years (see Sarty *et al.* 2011, hereafter S11 for a review). Paredes *et al.* (2000) identified relativistic radio jets and also a very high energy (VHE) gamma-ray source at the coordinates of the system; therefore LS 5039 became one of a handful of known gamma-ray binaries. There are several open issues about the system (see S11), but the major question is whether the secondary component orbiting around the O6.5V((f)) star is a black hole or a non-accreting young pulsar.

Hereinafter, we show the results of our spectroscopic and photometric analysis concerning mass and orbit constraints of LS 5039 (see the details in S11).

2. Analysis and parameter determination

Spectroscopic observations were carried out in 2009 with the Echelle spectrograph mounted at ANU 2.3m Telescope (SSO, Australia), and in 2011 using FEROS (Kaufer *et al.* 1999) at MPG/ESO-2.2m telescope at La Silla, Chile. Covering ~40 hours with nearly uniform sampling of the whole orbit between 3900-6750 Å with a resolving power $\lambda/\Delta\lambda \approx 23,000$ at Hα, it is the highest resolution, homogeneous spectral dataset ever obtained for LS 5039.

Radial velocities (RV) of HI and HeI lines show a systematic blueshift with respect to the RVs of HeII lines, therefore, only the latter ones were used to fit eclipsing binary

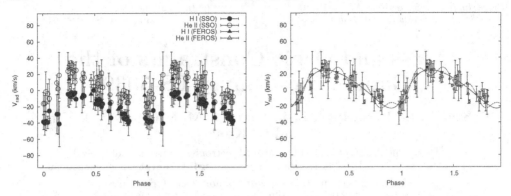

Figure 1. *Left*: Radial velocities based on H Balmer and HeII lines. *Right*: The best-fitting curve to radial velocities of HeII lines (solid line), also with the assumed pulsation of the O star (dotted line).

models with the Wilson–Devinney (WD) code (Wilson & van Hamme 2003). We do not see signs of non-radial pulsations in our data in contrast to the results reported by Casares *et al.* (2010), see Fig. 1.

Our data were analyzed as described in S11, with the addition of the FEROS data. Orbital parameters are close to previous solutions (Casares *et al.* 2005, Aragona *et al.* 2009), but we found the orbital eccentricity ($e=0.24 \pm 0.08$) to be definitely lower than determined previously.

Photometric data, obtained with *MOST* satellite in July of 2009, indicate a variability at the level of 2 mmag. Our light curve (LC) simulations show that the LC amplitude decreases with increasing total mass or decreasing eccentricity. Comparison of data and models suggests that primary mass is at the higher end of estimates based on its spectral type ($\sim$26 M$_\odot$), from which we get a mass for the compact star of at least 1.8 M$_\odot$.

3. Conclusions

We carried out a detailed spectroscopic and photometric analysis to get mass and orbit constraints for the γ-ray binary LS 5039. The new system parameters are close to the previously determined ones, except that we found a lower eccentricity. LC simulations support this result, and imply that the compact object may be a low mass black hole – but do not fully exclude that it may be a neutron star.

Acknowledgements

This work has been supported by the Australian Research Council, the University of Sydney, the Hungarian OTKA Grant K76816, and the "Lendület" Young Researchers' Program of the Hungarian Academy of Sciences.

References

Aragona, C., McSwain, M. V., Grundstrom, E. D., *et al.*, 2009, *ApJ*, 698, 514
Casares, J., Ribó, M., Ribas, I., *et al.*, 2005, *MNRAS*, 364, 899
Casares, J., Corral-Santana, J. M., Herrero, A., *et al.*, 2010, arXiv:1012.4351
Kaufer, A., Stahl, O., Tubbesing, S., *et al.*, 1999, *The Messenger*, 95, 8
Paredes, J. M., Martí, J., & Ribó, M., Massi M. 2000, *Science*, 288, 2340
Sarty, G. E., Szalai, T., Kiss, L. L., *et al.*, 2011, *MNRAS*, 411, 1293 (S11)
Wilson, R. E. & van Hamme, W. 2003, Computing Binary Stars Observables, Ver. 4.

From Interacting Binaries to Exoplanets: Essential Modeling Tools
Proceedings IAU Symposium No. 282, 2011
Mercedes T. Richards & Ivan Hubeny, eds.
© International Astronomical Union 2012
doi:10.1017/S1743921311027773

Spectroscopic Binaries Among λ Bootis-type Stars

Ernst Paunzen[1,2], Luciano Fraga[3], Ulrike Heiter[4], Ilian Kh. Iliev[1], Inga Kamp[5] and Olga Pintado[6]

[1] Rozhen National Astronomical Observatory, Institute of Astronomy of the Bulgarian
Academy of Sciences, P.O. Box 136, BG-4700 Smolyan, Bulgaria
email: ernst.paunzen@univie.ac.at

[2] Institut für Astronomie der Universität Wien, Türkenschanzstrasse 17,
A-1180 Wien, Austria

[3] Southern Observatory for Astrophysical Research, Casilla 603, La Serena, Chile

[4] Department of Physics and Astronomy, Uppsala University, Box 516,
SE-75120 Uppsala, Sweden

[5] Kapteyn Astronomical Institute, University of Groningen, PO Box 9513,
2300 AV Groningen, The Netherlands

[6] Instituto Superior de Correlación Geológica, Av. Perón S/N, Yerba Buena,
4000 Tucumán, Argentina

Abstract. The small group of λ Bootis stars comprises late B to early F-type stars, with moderate to extreme (up to a factor 100) surface under-abundances of most Fe-peak elements and solar abundances of lighter elements (C, N, O, and S). The main mechanisms responsible for this phenomenon are atmospheric diffusion, meridional mixing, and accretion of material from their surroundings. Especially spectroscopic binary (SB) systems with λ Bootis-type components are very important to investigate the evolutionary status and accretion process in more details. Because also δ Scuti type pulsation was found for several members, it gives the opportunity to use the tools of astroseismology for further investigations. We present the results of our long term efforts of detailed abundance analysis, orbital parameter estimation and photometric time series analysis for five well investigated SB systems.

Keywords. stars: chemically peculiar, delta Scuti, circumstellar matter, binaries: spectroscopic

1. Introduction

The small group of λ Bootis stars comprises late B to early F-type stars, with moderate to extreme (up to a factor 100) surface under-abundances of most Fe-peak elements and solar abundances of lighter elements (C, N, O, and S). Since they are found at ages up to 1 Gyr, they are excellent test laboratories for atmospheric processes in the context of stellar evolution. Our international working group has been investigating this star group for about 15 years in a joint effort to shed more light on its nature. Here, we present our results about the SB members.

2. Theories, planets and binary nature

Kamp & Paunzen (2002) and Martinez-Galarza *et al.* (2009) developed a model which is based on the interaction of the star with its local ISM environment. Different levels of under-abundance are produced by different amounts of accreted material relative to the photospheric mass. The small fraction of this star group is explained by the low probability of a star-cloud interaction and by the effects of meridional circulation, which washes out any accretion pattern a few million years after the accretion has stopped. The behavior is suppressed in low temperature stars because of their more massive, difficult to contaminate convection zones. Strong stellar winds prevent the accretion of material for very hot stars. Naturally, both stars of a spectroscopic binary system (SB) pass through the same diffuse cloud. Binary systems with a "normal" and λ Bootis-type companion would rule out this scenario. It is, therefore, essential to study these systems in more detail.

HR 8799 is a λ Bootis, γ Doradus star (the only one known among this group) hosting a planetary system and a debris disk with two rings (Moya *et al.* 2010). Their conclusions is that there are no models with solar metallicity fulfilling the observations, since the stellar luminosity derived from the observations is smaller than any of the possible luminosities of models with solar metallicity. However, the light elements are solar which could lead to the scenario that the star accretes substantial amounts of left-over gas from planet formation at early stages, which would be depleted in metals, but not in C, N, O, and S.

3. The known SB λ Bootis type systems

Up to now, there are several "λ Bootis-type" SB systems known: HD 64491, HD 84948, HD 111786, HD 141851, HD 171948, HD 174005, HD 193256/193281, HD 198160/1, and HD 210111. A detailed abundance analysis was done for HD 84948, HD 171948, and HD 210111 (Heiter 2002, Iliev *et al.* 2002, and Paunzen *et al.* 2002). From our extensive analysis of these SB systems, we conclude

- The ages vary from the zero-age to the terminal-age main sequence
- Both components always show the same λ Bootis-type abundance pattern
- The astroseismic results show, up to now, no differences to normal type δ Scuti stars
- The spacial kinematics are typical for Population I disk members
- For HD 210111, an IR excess was detected which is due to a circumstellar disk

Our current knowledge of the SB λ Bootis type systems is, therefore, consistent with the model of the strong interaction of the objects with their local ISM environment explaining the observed abundance pattern and their widely different evolutionary stages.

Acknowledgements

This work was supported by the financial contributions of the Austrian Agency for International Cooperation in Education and Research (WTZ CZ-10/2010 and HR-14/2010) and the Austrian Research Fund via the project FWF P22691-N16. U.H. acknowledges support from the Swedish National Space Board. I.I. acknowledges partial support from NSF grant DO 02-85.OIP. This paper was also partially supported by PIP0348 by CONICET.

References

Heiter, U. 2002, *A&A*, 381, 959

Iliev, I. K. h., Paunzen, E., Barzova, I. S., Griffin, R. F., Kamp, I., Claret, A., & Koen, C. 2002, *A&A*, 381, 914

Kamp, I. & Paunzen, E. 2002, *MNRAS*, 335, L45

Martinez-Galarza, J. R., Kamp, I., Su, K. Y. L.., Gaspar, A., Rieke, G., & Mamajek, E. E. 2009, *ApJ*, 694, 165

Moya, A., Amado, P. J., Barrado, D., Hernandez, A. G., Aberasturi, M., Montesinos, B., & Aceituno, F. 2010, *MNRAS*, 406, 566

Paunzen, E., Iliev, I. K. h., Kamp, I., & Barzova, I. S. 2002, *MNRAS*, 336, 1030

From Interacting Binaries to Exoplanets: Essential Modeling Tools
Proceedings IAU Symposium No. 282, 2011
Mercedes T. Richards & Ivan Hubeny, eds.
© International Astronomical Union 2012
doi:10.1017/S1743921311027785

Characterizing New Eclipsing Binaries Identified from STEREO Photometry

Harry Markov[1], Zlatan Tsvetanov[2], Ilian Iliev[1], Ivanka Stateva[1] and Nevena Markova[1]

[1]Institut of Astronomy and National Astronomical Observatory Rozhen, Bulgaria,
[2]Johns Hopkins University, Department of Physics and Astronomy, USA
email: hmarkov@astro.bas.bg

Abstract. Since 2010, a program to explore new eclipsing binary systems identified from STEREO photometry has been in progress. Our first results are presented here: light curves and high resolution spectra taken with Coudé spectrograph (National Astronomical Observatory Rozhen) and ARC Échelle spectrometer (ARCES, Apache Point Observatory).

Keywords. techniques: radial velocities, photometric, spectroscopic; binaries: eclipsing

1. Introduction

STEREO is a mission in the NASA Solar Terrestrial Probes program. It uses two nearly identical spacecraft to map coronal mass ejections as they propagate away from the Sun. The continuous series of images obtained by the Heliospheric Imager 1 (HI-1) cameras on the two STEREO S/Cs are well-suited to detect variations in the brightness of the light sources in the explored fields. The STEREO HI-1 photometry provides a complete survey of all bright stars (<10 mag) for 18% of the sky and our project is aimed at detecting transiting exoplanets. As a natural by-product, a substantial number of EBs are detected. In the course of our STEREO project, we obtained and examined aperture photometry of isolated stars extracted from the HI-1A images for four years worth of data (2007-2010). Our input catalogue includes over 70,000 Tycho 2 stars. This coarse examination revealed over 250 EBs, with fully half of them being new. Many of these EBs were recently independently reported by Wright *et al.* (2011), and we note here that our EB list includes about 10% more objects.

Recent available EB models, combined with radial velocity (RV) measurements, can provide a wealth of useful astrophysical information. The goal of our program is to supplement available STEREO photometric light curves with accurate RV curves. For this purpose, high resolution spectroscopic observations are already in progress with the Coude spectrograph (R=15000 and R=30000) of the National Astronomical Observatory Rozhen (Bulgaria) and the ARC Echelle spectrometer (ARCES, R=30000) of the Apache Point Observatory (New Mexico, USA). RV measurements were derived through the cross-correlation method, with a mean standard deviation of about 1.5 km/sec estimated by using different template spectra from the Montes *et al.* (1997) library and RV stable stars taken on the observing night. Particular attention will be paid to systems including small components (late K and M dwarfs) since the observational uncertainties in this part of the stellar mass function are the largest.

2. Results and notes on individual objects

In Fig. 1, we demonstrate the light curves for some EB systems derived from STEREO observations. The data presented here are photometry time series extracted from HI-1A images from four years of observing (2007-2010). We use the latest available calibration

336   H. Markov *et al.*

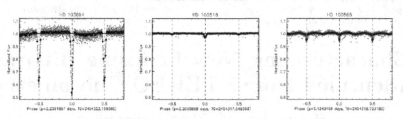

Figure 1. STEREO photometry light curves of some EB systems.

from the STEREO pipeline plus additional calibration developed by us. Low frequency trends were taken out by fitting low order polynomials. The stars shown here were our first targets for the follow-up spectroscopic observations.

HD103694 (see Fig. 2) shows double lined spectra (early to late) which allowed us to derive the separate RV curves of the components and their physical and kinematic characteristics, assuming circular orbits.

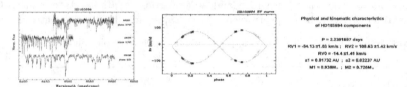

Figure 2. Spectra (left) and RV curves (center) of both HD103694 components. Some physical and kinematic characteristics are shown on the right

HD100565 (Fig. 3, left panel): The spectral lines measured on two spectra are shifted by about 54 km/s in less than 24 hours. This clearly shows the double nature of this system. Although the photometric light curve (Fig.1 above) does not demonstrate strong evidence of a second component, we announce HD100565 as a newly found SB1 type EBS.

HD100518 (HR4454) (Fig. 3, right panel): The spectra show at least two different spectral line sets. Measurements of the stronger one revealed relatively low velocities shifts ($\sim$ 3 km/s) with the phase, but the weaker one (marked by vertical lines) showed significant displacement. Griffin (2006) suspected the system is a multiple. Here, for the first time, we show the spectral features of another component. The STEREO light curve revealed the presence of an EB component with a new period (2.2 d) different from those measured by Griffin (2006) in his extensive RV study.

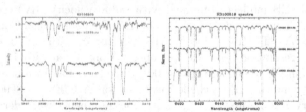

Figure 3. Spectra of HD100565 (left) and HD100518 (=HR4454) (right).

Acknowledgements: This work is supported by the Bulgarian National Science Fund grant DO 02-85, and also NASA grant NNX10AM33G (ZT). This work made use of the SIMBAD and VIZIER data bases, operated at CDS, Strasbourg, France.

References

Griffin R. F. 2006, *Observatory*, 126, 1
Montes D., Martin E. L., Fernandez-Figueroa M. J., *et al.*, 1997, *A&AS*, 123, 473
Wright, K. T. *et al.*, 2011, arXiv:1103.0911

From Interacting Binaries to Exoplanets: Essential Modeling Tools
Proceedings IAU Symposium No. 282, 2011
Mercedes T. Richards & Ivan Hubeny, eds.
© International Astronomical Union 2012
doi:10.1017/S1743921311027797

Modelling Light Curves of Systems with Non-Circular Accretion Disks: KU Cyg

S. Zola[1,2] and T. Szymanski[1]

[1] Astronomical Observatory, Jagiellonian University,
ul. Orla 171, PL-30-244, Krakow, Poland
email: szola@oa.uj.edu.pl

[2] Mt. Suhora Observatory, Pedagogical University,
ul. Podchorazych 2, PL-30-084 Krakow, Poland
email: green@oa.uj.edu.pl

Abstract. We present a code which can be used to simulate the light curve of a binary system harbouring a non-circular accretion disk, and show that such a system geometry can produce light curves with asymmetric minima and different heights of the maxima. We have applied this code to check for the possible existence of a non-circular accretion disk, as suggested by Smak & Plavec (1997), in the Algol-type binary KU Cyg. We compare the best solutions for two sets of multicolour light curves, obtained within circular and non-circular disk models.

Keywords. binaries:eclipsing, accretion disks, stars: individual (KU Cyg)

1. Non-circular disk model and its effects on the light curve

If there is an accretion disk in a binary system, the simplest approach to simulate a theoretical light curve is to assume that the disk is circular in shape. Such a model was described by Zola (1992), and applied to solve the light curves of several long period Algols with accretion disks. In this work, we describe a modification to that code which accounts for non-circular disk effects. The properties of the model are essentially the same: it is based on the Wilson-Devinney code, the system configuration is semi-detached and the disk surrounds the star eclipsed at phase 0. The disk is optically thick, and its geometrical thickness grows linearly with radius. The new modification requires two additional parameters to describe how the disk is "disturbed" from the circular shape: one to define the elliptical shape, while the other to define the disk tilt. The code accounts for geometrical effects arising from the presence of the disk: eclipses of the mass-losing star by the disk, partial or total obscuration of the mass-gaining star, and self-obscuration of the disk at high inclinations. In our model, the disk becomes thicker for parts of the disk with larger radii. This property causes a range of effects in the light curve which are absent when a circular accretion disk is considered.

To display the effects, we have chosen two sets of parameters and created theoretical light curves: one for a high inclination system ($i = 89°$) and the other for a lower ($i = 85°$) inclination system. The angle β, defining the disk thickness, has been chosen to be 5° for the lower inclination and 20° for the higher one. While the system mass ratio, temperatures of the components, and the outer disk temperature were the same for both sets of data, the radius of the primary, mass-gaining star has been chosen such that it was hidden in the disk for the case of the higher inclination system and partly obscured (part of the upper hemisphere visible) for the lower inclination one. Both light curves exhibit a difference in level of the maxima. This is entirely due to the non-circular shape of the disk, which is also responsible for a non-uniform light contribution from the disk outside of the phases when the disk is eclipsed. An additional feature is the asymmetry of the secondary minimum. Both effects are caused by the changing disk surface projection alone, in the case of high inclination system, and the combined effects of the projection (light rise) and the obscuration of the mass gaining star by the disk (decrease of light at the same phase).

2. Application of the model to KU Cyg

KU Cyg is a long period, Algol-type system showing double-peaked Balmer emission lines (Olson 1988, Olson et al. 1995) visible throughout the entire orbital period. The analysis of photometric data with the W-D code (Wilson 1979) have been done by Olson et al. (1995) and by Zola (1992), but using a code accounting for the presence of a disk. There is some controversy about the disk shape: Nguyen and Etzel (1999) found it to be circular while Smak and Plavec (1997), based on their analysis of a single H_α profile, argued for a significant distortion from the circular shape.

This work is aimed at checking the model of Smak and Plavec (1997), applying the code described in previous sections. We performed light curve modelling of two available sets of data: Olson's $vbyI$ and more recent, $BVRI$ data collected at Mt. Suhora Observatory. The Monte Carlo search method (Zola et al. 2004) was applied to find the best solution within the circular and non-circular disk models, separately for both sets of light curves. Two crucial parameters were fixed at the values derived by Smak and Plavec: the system inclination: $i = 86°$ and the mass ratio $q = 0.13$. In addition, we set the secondary temperature to 3750 K, appropriate for its spectral type.

3. Results

The light curve solutions of the two available sets of KU Cyg photometric light curves within a circular disk model resulted in a worse fit and very different parameters for the primary star. The non-circular disk model solutions turned out to fit better and to be consistent with regard to the parameters of the mass-gaining component. We derived 9400–10100 K for its temperature and 0.052–0.053 for its size, in separation units. We found slightly different disk properties (disk size and geometrical thickness) at the epochs when the two sets of data were gathered. The solution of Olson's data resulted in a disk somewhat bigger and thicker but less eccentric than that derived from data taken at Mt. Suhora. However, the position angle values were found to be very different: 336° versus 157° respectively for the Olson and Mt. Suhora data. Our results derived from photometric data seem to support the Smak and Plavec (1997) prediction that tidal effects, similar to those observed in cataclysmic variables (Osaki 1989), would cause the accretion disk in KU Cyg to be unstable and precess with a timescale of a few years.

References

Nguyen Q. T. & Etzel P. B. 1999, *AAS*, 195,7613
Olson E. C. 1988, *AJ*, 96, 1430
Osaki Y. 1989, *PASJ*, 41, 1001
Olson E. C., Etzel P. B., & Dewey M. R. 1995, *AJ*, 110, 2378
Smak J. I. & Plavec M. J. 1997, *AcA*, 47, 345
Wilson R. E. 1979, *ApJ*, 234, 1054
Zola S. 1992, *AcA*, 42, 355
Zola S., Rucinski S. M., Baran A., et al., 2004, *AcA*, 54, 299

From Interacting Binaries to Exoplanets: Essential Modeling Tools
Proceedings IAU Symposium No. 282, 2011
Mercedes T. Richards & Ivan Hubeny, eds.

© International Astronomical Union 2012
doi:10.1017/S1743921311027803

Towards an Increased Accuracy of Fundamental Properties of Stars: Proposing a Set of Nominal Astrophysical Parameters and Constants

A. Prša[1] and P. Harmanec[2]

[1] Villanova University, Department of Astronomy and Astrophysics, 800 E Lancaster Ave,
Villanova, 800 E Lancaster Ave, Villanova, PA 19085
[2] Astronomical Institute of the Charles University, Faculty of Mathematics and Physics, V
Holešovičkách 2, CZ-180 00 Praha 8, Czech Republic

Abstract. With the precision of space-borne photometers better than 100 ppm (i.e. MOST, CoRoT and Kepler), the derived stellar properties often suffer from systematic offsets due to the values used for solar mass, radius and luminosity, and to fundamental astrophysical constants. Stellar parameters are often expressed in terms of $L_\odot$, $M_\odot$ and $R_\odot$, but the actual values used vary from study to study. Here, we propose to adopt a nominal set of fundamental solar parameters that will impose consistency across published works and eliminate systematics that stem from inconsistent values. We further implore the community to rigorously use the official values of fundamental astrophysical constants set forth by the Committee on Data for Science and Technology (CODATA).

1. Motivation

It is customary to express stellar luminosities, masses and radii in terms of solar values $L_\odot$, $M_\odot$, $R_\odot$. Slightly different values of the adopted solar luminosity, mass and radius lead to measurable systematic differences in the determination of basic stellar parameters. For example, the Kepler mission (Borucki *et al.* 2011) routinely finds binaries with periods of the order of several months and longer. If we were to compute a separation between equal 1 solar mass components in a circular 200-day binary from different values of $M_\odot$ and $R_\odot$ adopted in the literature, the relative systematic error would be $\sim 4 \times 10^{-5}$. If we were to compute a separation by timing the eclipses in Kepler short cadence data, the stochastic uncertainty would be $\sim 5 \times 10^{-7}$, smaller by two orders of magnitude.

2. Precedent

The use of outdated and inappropriate physical constants that cause significant systematics in astronomy and geodesy has been seen before with the speed of light. In 1983, at the 17th Conférence Générale des Poids et Mesures (CGPM), the speed of light in vacuum was set to the exact value of 299,792,458 m/s. Consequently, the length of 1 meter was redefined as the distance traveled in vacuum in 1/299,792,458 s. In 1997, the IAU resolution redefined the bolometric magnitude. The zero point is no longer defined w.r.t. the bolometric luminosity of the Sun; now it corresponds to the exact value of $L = 3.055 \times 10^{28}$ W. This introduces an absolute scale of bolometric magnitudes, with continued convenience of $M_{\mathrm{bol},\odot} = 4.75$.

3. Proposal

We propose to:

(a) replace the solar luminosity $L_\odot$ and radius $R_\odot$ by the nominal values $\mathcal{L}_\odot^{\mathrm N}$ and $\mathcal{R}_\odot^{\mathrm N}$ that are by definition exact and expressed in SI units;

(b) compute stellar masses in terms of $M_\odot$ by noting that the measurement error of the product $GM_\odot$ is 5 orders of magnitude smaller than the error in G;

(c) compute stellar masses and temperatures in SI units by using the derived values $M_\odot^{2010}$ and $T_\odot^{2010}$;

(d) clearly state the reference for the values of fundamental physical constants used.

SOLAR PARAMETER VALUES:		FUNDAMENTAL CONSTANTS:	
$1\mathcal{R}_\odot^{\mathrm N}$	$= 6.95508 \times 10^8$ m	c	$= 299{,}792{,}458\,\mathrm{ms}^{-1}$
$1\mathcal{L}_\odot^{\mathrm N}$	$= 3.846 \times 10^{26}$ W	G	$= 6.67384(80) \times 10^{-11}\ \mathrm{m}^3\mathrm{kg}^{-1}\mathrm{s}^{-2}$
$1GM_\odot^{2010}$	$= 1.32712442099(10) \times 10^{20}\ \mathrm{m}^3\mathrm{s}^{-2}$	σ	$= 5.670400(40) \times 10^{-8}\,\mathrm{Wm}^{-2}\mathrm{K}^{-4}$
$1M_\odot^{2010}$	$= 1.988547 \times 10^{30}$ kg		
$1T_\odot^{2010}$	$= 5779.57$ K		

Table 1. Recommended values for solar parameters and fundamental astrophysical constants.

4. G vs. GM

The universal gravitational constant G is one of the least precisely determined fundamental constants in nature. In contrast, the error in the product $GM_\odot$ is 5 orders of magnitude smaller. While the value of the solar mass is, thus, inevitably uncertain due to the value of G, the stellar masses $M_1/M_\odot$ and $M_2/M_\odot$ may be readily written as a function of $GM_\odot$:

$$M_1/M_\odot = PK_2(K_1 + K_2)^2 \left(1 - e^2\right)^{3/2}/(2\pi GM_\odot \sin^3 i)$$

$$M_2/M_\odot = PK_1(K_1 + K_2)^2 \left(1 - e^2\right)^{3/2}/(2\pi GM_\odot \sin^3 i).$$

The same can be done for planet masses in exoplanetary research: GM_{Jup}, $GM_\oplus$, and GM_{Moon} are determined to a much higher accuracy and those values should be used for computations of exoplanet masses.

5. Further information

A full list of references and details has been published in PASP (Harmanec & Prša, 2011). The radii of planets (most notably, Jupiter and Earth) should also be made nominal, and the following values used for extra-solar planet comparison:

$$\text{Jupiter:}\ \ R = 71492\ \text{km},\ \ GM_{\mathrm{J}} = 126686535(2)\ \mathrm{km}^3\mathrm{s}^{-2}$$
$$\text{Earth:}\ \ \ \ R = 6371\ \text{km},\ \ \ GM_\oplus = 398600.4418(8)\ \mathrm{km}^3\mathrm{s}^{-2}$$

References

Borucki *et al.*, 2011, *ApJ*, **728**, 117
Harmanec & Prša 2011, *PASP*, **123**, 976

M. Richards: The conference participants voted in favor of this proposal and recommended that an official IAU resolution be submitted at the 2012 IAU General Assembly to be held in China.

From Interacting Binaries to Exoplanets: Essential Modeling Tools
Proceedings IAU Symposium No. 282, 2011 © International Astronomical Union 2012
Mercedes T. Richards & Ivan Hubeny, eds. doi:10.1017/S1743921311027815

Panel Discussion II

Panel: F. Allard, A. Batten, E. Budding, E. Devinney, P. Eggleton, A. Hatzes, I. Hubeny, W. Kley, H. Lammer, A. Linnell, V. Trimble, and R. E. Wilson

Discussion

I. HUBENY: Today, the discussion will be open to the general audience. In Sessions C, D, and E, we have talked about models and modelling techniques so I expect the discussion will focus on these topics.

V. TRIMBLE: Comparing codes is enormously important - we learned that decades ago with stellar structure and evolution. And it's important that groups comparing codes be sure that they have the same physics before they decide whether they are agreeing or disagreeing. If you have different physics in your models, of course you get different answers. That requires possibly more openness in the early stages of developing codes than is usual. The separate codes need to be compared with each other, with the observations, and with the tables. If you cannot reproduce earlier analytic results, something is wrong.

A. HATZES: For someone who doesn't do any light curve modelling you said, "Compare the codes." Are there standard tests? I mean they should all be compared to the exact same thing.

H. LAMMER: I agree with this completely because this was done in the case of planetary atmosphere interaction studies, at least for the Solar system. Several teams got together to apply hybrid codes for studying the loss of Mars' atmosphere. You also need people who write the codes to discuss their codes with each other, and you need also the exact similar input, because if you don't know what the input is: let's say if you've got 10 codes, you will get 10 different results. I don't know how difficult it will be. You can never find a detailed description of the codes in any paper.

A. BURROWS: At Los Alamos, for example - Livermore, they have standard verification and validation protocols for various types of codes. It's easiest in spherical symmetry, it's easiest in the planar case. There are hydro-calculations you can do in 1D and 2D, in particular, that are almost analytic; some are analytic and some are almost analytic, and have been tested for a long time. You can do resolution studies, but there is a cultural problem as well, that Virginia alluded to, that people don't want to really share their errors or the fact that they have them. So over time, people in the field of numerical simulation, particularly hydro, but also radiative transfer to some degree, have reached a comfort level where this is done much more, and it's also expected much more. So, the number of people doing these calculations has gone up significantly in the last 30 years. In the 1980s for example, people weren't that confident, and there was very little collaboration and communication and verification and validation.

S. MOCHNACKI: When we first did these codes, we used the tables that were developed for spherical stars, and I also worked on doing the reflection effect, the geometrical aspects of it. And it turned out there was a fellow named Napier who worked it out analytically for spherical stars, and got it right! Many people like Kopal had never gotten it right before because they got lost in the analysis and made too many algebraical errors. So, we do stand on the shoulders of giants, and that early work is vital to test the things we write ourselves.

I. HUBENY: Obviously, different codes need different comparisons. I have experience with testing and comparing stellar atmosphere codes. Once we organized a session at a conference in Trieste, we assigned reasonably the same model, meaning the same parameters. People were asked to compute models, which didn't help at all, because the models were very similar, but there were differences impossible to resolve at the conference because of the vast amount of data you need from an atmosphere. So, I did an independent comparison with Klaus Werner and Stefan Dreizler in Tuebingen. I invited them for three weeks to Goddard (when I was working there), and it took three weeks to work out various small things, and after three weeks we finally ended up having a very reasonable agreement. We then did the same thing with John Hiller with CMFGEN that worked for the static approximation. Again, it took us a lot of time, a lot of effort, to tweak all those too. It's an extremely valuable experience, not only an exercise, not only for comparison, but you learn a lot about your mistakes and the other party learns about their mistakes and you get much more confidence in the codes. So, I highly recommend it. Now, for different codes we will need different sorts of comparisons, but it's an extremely important and valuable exercise to do. We also did a very detailed comparison with Detlev Koester and Pierre Bergeron on white dwarf models, because people use them for calibrations. It was again a huge work and we agreed to publish it and never did.

A. HATZES: I remember back in my Doppler imaging days when there were several codes being used to get models of spots, Klaus Strassmeier had a very clever idea. He got all the groups together, he sent them all the same data set, they modelled it not knowing what the other one was getting. He assembled the paper, wrote it and compared the techniques; and it's a highly cited work. So, I would encourage everyone to just do it! Get the groups together, give the people the same data set, and publish a paper making a direct comparison of the result.

R. WILSON: According to my not very specific recollection, there actually have been a number of papers published comparing the light curve models. There typically weren't highly quantitative comments. We ran our program with someone else's program and got basically the same answers. We should do this on a uniform basis including a whole bunch of programs, and not just one or two.

P. EGGLETON: I think an early comparison was made by Galileo between the Copernican system and the Ptolemaic system. I think he rather loaded the argument when he called the supporter of Copernicus by the Latin word for 'clever' while the supporter of the Ptolemaic system was the Latin word for 'stupid.' But, there was also an unbiased observer who, being a servant of the Pope, was prepared to accept the stupid ones. My point is, I think it is very important that we all get on with our own work and we should not always be looking over our shoulders at other people, only occasionally. Otherwise, we will be coming to a stop because we will all be so busy looking at everybody else's work. Ultimately, I think the best work will show up by itself.

A. BATTEN: Since history has been mentioned, I enjoyed Stefan Mochnacki's paper very much because he gave us context and showed us what had to be done without going into too much detail. I did not entirely agree, however, that the introduction of light-curve synthesis was a paradigm shift from the graphical methods of Russell & Merrill. In between them were the iterative methods developed by Kopal & Piotrowski. (Incidentally, considering the part of the world in which we are meeting, it is surprising that Kopal has been mentioned only occasionally and Piotrowski not at all until now. They were leaders in the field in their day.) Those methods are not much used today obviously because light curve synthesis is so much better now. Kopal's methods were more used in Europe but never really caught on in North America, perhaps because of the influence of Russell and his intellectual heirs. Possible nationalist considerations sometimes affect the way in which we do science. As Kopal's graduate student, I had no choice but to use his methods in my first paper. However, I would like to draw a parallel between Kopal and Russell. Stefan said that he's beginning to feel like John Merrill. I have often felt that towards the end of his life, Kopal was beginning to feel like Henry Norris Russell. Russell resisted Kopal's iterative methods and was strongly opposed to them. If you have read Kopal's autobiography, you know that story pretty well. Then, towards the end of his life, Kopal equally resisted the light curve synthesis methods. I think he was wrong in doing so, but until the end of his days he wouldn't accept that light curve synthesis was the way to go. And as you all know, he developed his Fourier transform methods, which one or two people have used, but they don't seem to have caught on. I think it's interesting to draw this parallel between Henry Norris Russell on one hand and Kopal on the other hand.

R. WILSON: Well, Kopal had really three ways of supposedly improving on the analysis problem. The first one would be focusing entirely on the eclipses and ignoring the rest of the light curve. They were purely simple geometrical operations with spheres and so on. Then, there were his corrected spheres model and that would've been useful if they had caught on; useful, I would say for about 5 years, 10 at the most. It didn't catch on, so it missed the technological niche. And then, there is the third one that you just mentioned about the Fourier methods. I realized that it was always flying against its limits from the beginning, because if you had too many Fourier terms, you wound up measuring the noise, and if you didn't have enough terms then you were not fitting the function. So, this method was generally not very effective.

E. BUDDING: Alan, myself, and at least three of us here on this august panel were pupils of the late Zdenek Kopal. One of the things that comes to my mind most strongly about him was his remarkable sense of humor. I recall as a young postgraduate student showing him one of my first light-curve solution efforts, which included things like 'inclination = 80.1234 deg.' After he had a careful look through my sheets, came the response: "Err, Ed, the 4 may well be correct – but the 8 is doubtful."

This morning, we had discussions about the alternative methods of fitting spectra, which I felt was a complimentary and non-adversarial, constructive sort of approach to how different codes can be suited to different contexts and can be complementary. This seems to contrast with the rather gladiatorial or adversarial atmosphere seen in photometric modelling.

Someone cited a quotation at today's meeting, I think from von Neumann, "If you give me four parameters I can model an elephant, and if you give me five parameters I can make him wiggle his tail." It seems that the situation we have is like an elephant seen through a glass darkly. We don't really know if that elephant wants to wiggle its tail. It knows, but we should not say we know what that elephant wants to do. What I am trying to say is that we should not try to get more parameters from a model than the information can really tell you.

P. HARMANEC: While I was listening to the talks here for several days, I was wondering: Are we not moving for some time in a kind of vicious circle? I do not doubt that analysis, let's say we have radial velocity curves or light curves which are very accurate, so formally the accuracy of the parameters (masses or radii) which you get from them are really very accurate. But, we also heard about models which were calculated for 1.5 years and yet there is one parameter which relates to the still not ideal theory for the mixing length parameter. I was surprised that the mixing length used for cool stars was set at 1 or 1.25. But, if you look in recent models, if you try to reproduce our Sun, you need to use a mixing length parameter close to 2. Will you produce also a series of computations for one model again for several months to try also this parameter? Because I'm interested in hot stars, then you have to include rotation, and now the problem of how to compare observations with the data becomes more serious. Also, if you want to compare results with the data, we have to make several steps. For instance, how accurate are the bolometric corrections? Basically, they are sometimes based on emission-lined stars which were measured with an intensity interferometer. So, are you sure that we are not getting the answers which we sometimes put into the data in advance?

I. HUBENY: I don't think we are moving in circles. I think we are really moving forward. For instance, as far as convection is concerned, there was some sort of confusion or misunderstanding because the models like COBOLD or Asplund/Nordlund models reproduce convective motions from first principles. It means that you can translate their results into language using the mixing length parameter. Similarily, some simulations for accretion disks done on the box can reproduce the value of the alpha parameter of Shakura-Sunyaev, which doesn't mean that they use it. The real endeavour is to use physics from first principles to do simulations and to derive previously *ad hoc* parameters, which are no more *ad hoc*. So, I think the program is very sound and in this respect I'm almost sure that progress proceeds in a very right way. Now, of course doing hydro is very difficult, therefore something always has to be sacrificed. There is always something which is unknown. So the question is: What are we sacrificing more? Or are we sacrificing something more important than what we want to determine? Only experience will tell. But, as far as model atmospheres for cool stars is concerned, I think we are generally as a community in the right thread, because there was a lot of good progress toward understanding motions and understanding convection. A very similar situation occurred in accretion disk modeling, because there was a long stagnation after the Shakura-Sunyaev alpha parameter was introduced. For three decades, people were using it, knowing that it was sort of an *ad hoc* parameter, and after three decades finally people did simulations where one can get rid of it. So, I agree with you that part of the simulation is always uncertain because something has to be sacrificed, you cannot model everything, you cannot put in your hydro models everything. But, overall, I think that progress is being made.

S. RUCINSKI: I would like to continue with something that Petr actually brought up. It is this big discrepancy between observational results; a difference of more than 3 sigma or 14 sigma was mentioned. What I'm concerned about is that in many cases we don't know how to do error analysis correctly. Let's admit this. This sigma is a concept that should slowly be put into the annals as useful concepts, like the names of our distinguished predecessors - on whose shoulders we are sitting. But, we have better tools for doing statistical analysis of data. I'm very much for just improving our error analysis including systematic errors and such things that astronomers produce and can do it so well. Let's look at different concepts like uncertainty levels and so on, instead of sigma.

W. KLEY: Coming back to the point which was raised before. You answered it a bit already. I think the situation is bad and good on both sides. It's clear that to produce stellar evolution models or stellar atmosphere models you need to have some quantities like the alpha (mixing length) parameter, which summarizes the uncertainties you have for processes which are 3D, MHD etc. People continue to use that and I think this is of course very good. The bad thing is that you don't really know the value of the full parameter. The other thing is that you don't even know whether the parametrization is correct in the first place. I don't mean the parameter itself, but the whole theory on it. It may not be correct. The same thing applies to the alpha parameter in accretion disk theory since people have been using the alpha parameter for a long time. But, all the uncertainties about the turbulence, whether it's driven by the hydrodynamic turbulence or by full MHD turbulence, has been summarized in one constant number. Presently, it's not even clear if this approach is correct at all. This is the downside of these approaches. The positive side of this is that while we have learned a lot, we will be able to clarify the situation with advanced modelling. Then we can say in hindsight basically: "Well, this was a good idea" or "This was a bad idea." There may be corrections, but one has to keep this in mind. There is progress.

V. TRIMBLE: There may be something to be learned from the medical community, who have been working hard on statistics now for a long time, and whose conclusions about whether something is statistically significant are perhaps more important than ours. But, they define their criteria of what will be regarded as agreement or disagreement before they start doing the calculations or collecting the data. And, when they decide that something is significant or is not, then they stop.

A. LINNELL: Back to Slavek's points. I'd like to bring up again the value of numerical experiments. I think that one of the very valuable things that can be done in testing a particular routine is to generate synthetic data, which you then solve. You can modulate the data with assignable errors and then determine whether or not your solution technique can recover the data to within the error that you expect. You can vary all sorts of parameters, but if you construct a model which you test, where you control the parameters, then you're in a position to say whether you can recover them to within the accuracy that you would expect. I think that is a very valuable test.

P. BONIFACIO: Unfortunately, in many cases, this underestimates your error. We were always very good at taking our models, adding some noise plus something, and recovering the correct parameters. Except when you get the real data, there's not only Poisson noise, it's more complex. But, I agree that it's a very useful experiment.

P. STEE: Another way to do this is to do it in a 'blind mode.' I mean to distribute the same data to other people. We have done this for example with interferometry. We call it the 'beauty contest.' You can identify the best code if you have data that you don't know and you try to reconstruct the images from the data with the different algorithms.

R. WILSON: If you wanted to know how well the models compare, then you have to run the various models for the identical parameters. If you want to know how much the answers depend on the physical constants that go into the code, then you vary the physical constants and see what you get. If you want to know the overall errors after you know those two questions and several other questions, then there are simple little formulas. The simplest method is to just take the rms result and put the results together, and if it's more complicated than that, you may have to take correlations into account but there are simple little statistical formulas for doing that. You have to look at the problems one by one and then combine them at the end.

A. BATTEN: I would like to make some comments on Mercedes Richards' paper of yesterday morning. This is beautiful work and fills me with admiration tinged with envy. That was the sort of thing I wanted to do 40 years ago with U Cephei and could not. I also suspected that magnetic forces were involved somehow, as Mercedes has clearly shown. A few of us here were also at the 1975 conference in Cambridge where the criticism was made that none of the trajectories proposed for gas-streams obeyed Newton's laws of motion. I think that implicit in that remark was the assumption that only gravitational forces were acting on the stream. We did not know then, and are only now beginning to find out, how other forces may act. I hope Mercedes will study U Cephei soon, while I am still alive to see the result!

E. DEVINNEY: A few comments: I was almost blown away by the ambition of people, what kind of problems are being attacked these days. It seems like these problems are growing in complexity with many features: disks, clouds, etc., including new physics. I think that makes the idea of testing and verifying codes much more difficult than it has been in the past. I also really enjoyed Jan Budaj's talk particularily, because of the explicit link between the exoplanets and binary paradigms.

V. TRIMBLE: The IAU met in Prague in 1967, a few people who were there are here. But there was the first set of talks about interacting binaries and the evolutionary scenarios that they predicted. From Kippenhahn, from Mirek Plavec and from Bohdan Paczynski, who in the end took the heat when Anne Underhill stood up and said "There are more models that are not stars than there are stars that are not models." This is probably still true.

K. BJORKMAN: We've heard a lot today about how our models and our observations are both outstripping the fundamental data on which we need to base our interpretation. I'm curious as to your thoughts on whether that message is getting through to the funding agencies. I also worry a lot when I see a lot of our fundamental atomic data specialists retiring and not being replaced. So, I'm wondering where these improved atomic data are going to come from?

E. BUDDING: Someone asked me to mention that it is good to have more communications like this where these points can be made. Perhaps through the medium of something like EVO where people with these needs in mind can form a collective voice to emphasize them and do so repeatedly. Such meetings are held for the Australian SKA Pioneer project at least once a month, if not more often. When the requirements are reiterated, people can't avoid finding out about it.

V. TRIMBLE: IAU Working Groups can be assembled for that precise kind of purpose. It requires two or three people who feel that this is important. They could easily in Beijing put forward a proposal for a Working Group that would probably communicate mostly online but would get together when possible.

I. HUBENY: There are conferences in the US organized every two years about laboratory astrophysics and atomic data. This group has some voice but never a real impact. If you sit on an Astrophysics Theory Panel (ATP) for NASA, there are about a hundred proposals and only five of them will be funded. So, the probability that even a very good proposal for atomic or molecular data will be funded is very low. When I was on the panel, I tried very hard to push them, but I must say I didn't succeed very often, actually only once. But, it is very hard to get funding because it's not astrophysics. Unfortunately, it is falling between the cracks, even if we need it.

A. BURROWS: I was going to say the same thing and add a comment. This comes up all the time. Laboratory astrophysics and doing the fundamental microphysical calculations are central to just about everything we do. But, that effort is not well-credited. There are very few tenured people who know how to do any of this. The chemistry community that might be able to help in some aspects doesn't care about this sort of thing at all. A lot of the calculations that are done, are done in national labs by people on soft money. It's not quite the same in Europe, because Europeans do support this a little bit more, but on both continents it's not supported as much as needed, and I don't think the situation is going to change at all. Voices could be raised, there is an awful lot of data that is necessary, but it's deemed fundamentally as a drudgery by tenure committees and people who are interested in advancing astrophysics. And, it is just going to get worse. I don't want to be too pessimistic, but we've seen this situation evolving over the years. Many people have talked about this, even the decadal surveys have talked about this. They always mention or give lip service to laboratory astrophysics getting microphysical data. Every once in a while there are some advances, for example, opacities for iron. You can add a few things, but just think of how many ionic states, how many molecules, how many line transitions, how many collisional rates for processes, we have to think about. Also, think how difficult it is to do the experiment to just measure one of those. It's going to be difficult.

P. SKODA: There is already a project that is well-funded by the European Union, it's called the Virtual Atomic and Molecular Data Center. It is exactly the attempt to concentrate all the available atomic and molecular databases together and provide a unified interface to access the data. For example, in 2012 there will be a Summer School in Serbia about teaching the students how to operate with such new databases. There are in fact billions of the transitions already available through these systems. So, it's a question of the informatics approach to get all the particle data in the proper way in a machine readable format.

I. HUBENY: My understanding is that they will collect what's already done: the opacity project. However, we need NEW data.

J. HILLIER: One of the things in stellar atmospheres that you can't get from theory are the wavelengths. Theory just cannot predict wavelengths accurately enough. We've talked about the oxygen problem today and it's a nickel line that was one of the contributions to that problem. If we don't know a line is there, you can't correct for one. It's as simple as that. As Ivan mentioned, the High Energy Laboratory Astrophysics Group meets every two years. There are people in that group who are willing to work with you to get the laboratory data if you can demonstrate a need for a particular problem. That is sometimes difficult, so it takes time.

P. BONIFACIO: These things are good, but it will not happen unless we make the laboratory astrophysics and theoretical computations attractive for young people to enter in the field; that means you must provide careers, money, and infrastructure. So, we're losing expertise. For quantum-chemical computations, one thing I want very desperately are computations for cross-sections with collisions with hydrogen atoms. We have lithium, sodium, now magnesium, and it takes years to get the next element.

I. HUBENY: I agree also with John's comments. I heard that too. However, when people ask me for a model, I say "Sure, I can do that." I can in principle, but it takes me sometimes weeks or months, to be honest. Most people also have their own projects and everything. So, in principle, they are willing, but in reality it all goes very slowly.

A. BATTEN: The picture of terrestrial "brown dwarfs" shown this morning reminded me of a story Willem Luyten loved to tell. He was a great authority on white dwarfs. For many years, he submitted proposals to the National Science Foundation and received grants for his research on white dwarfs. One day, a new bureaucrat, who knew neither Luyten nor astronomy, sent him a letter asking, "Does this proposal involve research on human subjects?"

V. TRIMBLE: Alan has just said it, but if you knew Willem Luyten, the answer to that to a certain extent was "Yes."

M. ZEJDA: We live in time when we are hunting for better accuracy in measurements of brightnesses, fluxes, and radial velocities. We use new measurements and also archival values to study long-term variations. The time-base for these long-term studies is usually given in Julian Date with the heliocentric correction (HJD) based on the Coordinated Universal Time (UTC). People use data spread over decades to find differences on a scale smaller than one minute. However, they do not take into account that the accuracy of the timeline based on UTC is not the same! There are leap seconds which caused discontinuous differences of tenths of seconds for data spread over intervals longer then several decades. The solution is quite simple. All authors should publish new studies of any astronomical events at least in the fields of variable stars and exoplanets in the Barycentric Julian Date in the Barycentric Dynamical Time (BJDTBC). Referees should insist on using BJDTBC. The Barycentric Julian Date derived from the Barycentric Dynamical Time should become a new standard. All dates should be recalculated into BJDTBC when old data from an archive are used. As an excellent example of correct work with data, see the paper by Potter *et al.* 2011, MNRAS, 416, 2202. Another hidden problem in using data archives is that the format of time is different in each archive, so users should be very careful when processing archival data. For more details, see the Appendix in Eastman, Siverd & Gaudi, 2010, PASP 122, 935; or Zejda & Domingo, 2011, IBVS 5996.

Part 6

Analysis of Spectra and Light Curves

From Interacting Binaries to Exoplanets: Essential Modeling Tools
Proceedings IAU Symposium No. 282, 2011　　　　　© International Astronomical Union 2012
Mercedes T. Richards & Ivan Hubeny, eds.　　　　　doi:10.1017/S1743921311027827

The Disentangling of Stellar Spectra

P. Hadrava

Astronomical Institute, Academy of Sciences,
Boční II 1401, CZ - 141 31 Praha 4, Czech Republic
email: had@sunstel.asu.cas.cz

Abstract. The techniques of disentangling were originally developed to separate spectra of individual components from time series of spectra of binaries and, simultaneously, to determine either the corresponding radial velocities or directly to solve for orbital parameters.

Generalizations of the disentangling method enable us to include also intrinsic line-profile variability of the component spectra into the underlying model, and thus to solve for additional physical parameters of the stars (either single or components of multiple systems). Depending on the problem in question, it may also be helpful to constrain the space of separated spectra by templates or to bound the solution in the parameter space by photometry, interferometry or other observational data.

Keywords. Methods: data analysis, techniques: spectroscopic, stars: atmospheres, binaries: spectroscopic, line: profiles

1. Introduction

The aim of science is to systematize the knowledge gathered from observations and experiments in order to comprehend the natural laws. The understanding of basic principles of nature yields a tool for its exploitation, however, the theoretical conclusions must be tested by new observations to verify, to modify, or to completely disprove and change the starting assumptions and methods of reasoning. In practice, we usually build a theoretical model predicting the expected results of the observations and, from the comparison with real observations, we find values of some free parameters of the model to fit the observations best. An agreement of the observation with the model, however, does not necessarily mean a proof of the theory; there is a danger of a circle in proof because the interpretation of observations is model-dependent. We never expect a perfect agreement with sufficiently decisive observational tests either due to simplifications in the model or due to observational errors. A relatively good agreement thus does not exclude that another, even a completely different model could fit the given observational data better, not to speak about some other data, which could decide on the validity of alternative models. It is thus safer to test a model by a maximum of the available observational data, and to avoid unnecessary limitations of generality of the model. In any case, it is necessary to keep in mind the underlying assumptions of the model and critically judge their consequences.

A good example of the described general problem are the observations of binary stars and their spectroscopy in particular. The photometry yields relative dimensions of eclipsing binaries and the spectroscopy enables us to scale them in absolute physical units (cf. Kallrath & Milone 1999, 2009 for a recent general review). In the classical approach, radial velocities (RVs) of the component stars are measured from time series of spectral observations and the obtained RV-curves are solved for the orbital parameters – cf. Fig. 1. (It is preferable to solve the RV-curves together with the light-curves, which may better constrain some of the orbital parameters, typically the period or the conjunction

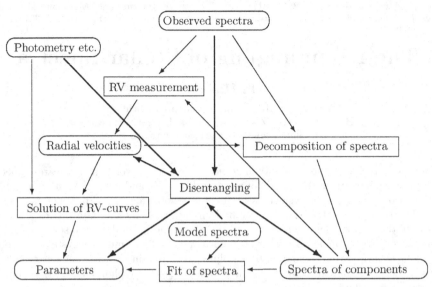

Figure 1. Flow diagram of standard data processing (thin lines) and the disentangling (bold lines) of binaries observations

epoch.) To measure the RVs, we need some model of the component spectra – either the simple assumption that the bottoms of distinguishable line-profiles correspond to the wavelengths of the lines Doppler-shifted for velocity of mass-centers of the component stars, or, in more sophisticated methods like the cross-correlation (cf. Hill 1993) suitable also for blended lines, we need some template spectra of the components. There have been developed methods like the tomographic separation by Bagnuolo & Gies (1991) enabling us to decompose the observed spectra if the RVs of the component stars are known at each exposure. The information about both the RVs and the component spectra is thus entangled in the same set of observed spectra and the original idea of the method of disentangling was to extract it simultaneously without any unsubstantiated *ad hoc* assumptions.

It turns out that, if the RVs can be supposed to obey the laws of Keplerian motion in a multiple stellar system, as we used to assume in the solving the RV-curves, we do not need to include in the fitting of the observed spectra the intermediate step of determining their values, which may be in different phases subjected to different errors (also dependent on S/N), and we can solve directly for the orbital parameters. Regarding the above mentioned restrictions imposed on the orbital parameters from the photometry and/or other data, we can fit these data simultaneously with the observed spectra. Certainly, the RV-curves and their scatter yield an insight into the solution, and for this reason RVs can also be computed, however, it is up to the user which physical model he wants to test with the observational data, eventually, one may try different models and to compare their results. Similarly, if some of the separated component spectra correspond well with computed model spectra, e.g., for plane-parallel stellar atmospheres, it may be preferable to impose such a restriction on the solution already at the stage of the disentangling, to optimize the solution directly with respect to the free physical parameters of the stellar atmosphere and to decrease significantly the number of free parameters to arrive at a safer shape of spectra of the other, possibly peculiar components.

In this generalized view, the disentangling is a versatile tool for testing physical theoretical models with (not-only) spectroscopic observational data, which is open to an additional sophistication in future. Its complete description exceeds the available space and consequently intention of the present contribution and readers may find more details in Hadrava (2004b) and Hadrava (2009b). Here, only the basic principles and some improvements of the method of disentangling will be given.

2. Mathematical principle of the disentangling

The spectrum $I(x,t)$ of a multiple system of n stars observable at time t is a superposition of Doppler shifted spectra $I_j|_{j=1}^{n}$ of all the components. In the simplest case when these spectra do not change with time t, it can be given by expression

$$I(x,t) = \sum_{j=1}^{n} I_j(x) * \Delta_j(x,t,p), \tag{2.1}$$

where $x = c\ln\lambda$ is the logarithmic wavelength scale and

$$\Delta_j(x,t,p) = \delta(x - v_j(t,p)) \tag{2.2}$$

is the Dirac delta-function shifted for v_j, i.e. the instantaneous RV of the component j which depends also on the orbital parameters p. To disentangle a set of N spectra $(N > n)$ exposed at times $t_l|_{l=1}^{N}$ means to fit them by minimizing the residual noise $(O-C)^2$,

$$0 = \delta \sum_{l=1}^{N} \int \left| I(x,t_l) - \sum_{j=1}^{n} I_j(x) * \Delta_j(x,t_l,p) \right|^2 dx \tag{2.3}$$

with respect to the parameters p (which is an equivalent to the measurement of RVs) as well as with respect to the intrinsic component spectra I_j (which is the separation of spectra of the components). While the fitted expression (2.1) is non-linear with respect to p which, however, represents a modest number of degrees of freedom, it is linear with respect to the highly numerous degrees of freedom of $I_j(x)$. For this reason, it is advantageous to use different numerical methods for the optimization with respect to these variables and to apply them iteratively.

The existing techniques of the disentangling can be divided into two basic groups according to the method of solving for I_j. The methods performing the separation of the spectra in the direct wavelength space (x) can be represented by the method introduced by Simon & Sturm (1994), which is based on the SVD-solution (i.e. the singular value decomposition) of the corresponding set of linear equations given by a huge but sparse matrix. Also, the tomographic separation by Bagnuolo & Gies (1991) or the practically equivalent iterative subtraction developed by Marchenko *et al.* (1998) performs the decomposition of the spectra in their wavelength-domain representation.

Fourier disentangling. The alternative method introduced by Hadrava (1995) profits from the fact that the convolution in Eq. (2.1) is transformed into a simple product

$$\tilde{I}(y,t) = \sum_{j=1}^{n} \tilde{I}_j(y)\, \tilde{\Delta}_j(y,t,p) \tag{2.4}$$

in the Fourier domain (y) and consequently the huge set of coupled linear equations in the optimization reduces into independent equations of dimension n only for each Fourier mode y of the separated spectra $\tilde{I}_j$. The solution is thus much faster in this representation not only for the Doppler shifts (2.2) for which

$$\tilde{\Delta}_j(y,t,p) = \exp(iyv_j(t,p)), \tag{2.5}$$

but for any broadening function $\Delta_j(x,t,p)$.

Owing to the Parseval theorem, the residual noise can be expressed equivalently to Eq. (2.3) as an integral in the Fourier domain. The optimization with respect to the parameters p can thus be performed by a standard algorithm like the simplex method with the $(O–C)^2$ evaluated in each step by integration in the y-space of the squared residua with respect to the already optimized Fourier modes of the component spectra. Moreover, it is possible to assign different weights to different Fourier modes. This is of great practical use not only for filtering a high-frequency noise but mainly because it is sometimes advantageous to filter out the lower modes which are poorly conditioned (e.g. the constant mode for $y = 0$ is completely singular as follows from Eq. (2.5)), and their instability caused, e.g., by uneven continua may mislead the convergence of orbital parameters, which are determined mainly by the higher Fourier modes corresponding to the widths of the spectral lines. This advantage fairly prevails over the possibility to weight individual pixels of the input spectra in the wavelength-domain disentangling advocated by Ilijić (2004) which, however, may be substituted by careful data-processing before the disentangling is applied.

It should be emphasized that for all methods of disentangling and separation of spectra, a sharp signal with dumped noise is extracted from all spectra at once similarly as explained by Rucinski (2002) for his broadening-function technique (BFT), in contrast to the various cross-correlation techniques where the noise is blurred together with the signal for each spectrum separately, collecting the information at maximum from different lines in the same exposure.

3. Generalized disentangling

The methods of spectra separation as well as the above described simple disentangling are based on the assumption (2.2) of invariability of the component spectra. This, however, is not always satisfied in nature. For instance, the relative strength of component spectra is changed during the eclipses or even the line-profile varies due to the rotational Schlesinger – Rossiter – McLaughling effect (Schlesinger 1909 etc.). The proximity effects like the ellipticity or reflection effect influence the line-profiles in the interacting binaries or there may take place intrinsic variations of the component spectra, e.g. due to pulsations or activity of the component stars. One possibility is to exclude the most peculiar exposures (e.g. those taken during the eclipses) from the analysis, but then we lose a very valuable source of information which enables us to test e.g. the structure of the stellar atmospheres. Another possibility is to neglect these effects first, to apply the simple disentangling and then to study the deviations of individual exposures from the simplified model and to discuss the possible influence of the simplification on the results (suppressing the weights of the peculiar exposures, e.g. at eclipses, may be helpful for this approach). However, the best way is to generalize the model which we fit to the data and to disentangle also the free parameters characterizing these additional effects (naturally, keeping in mind that the results are conditioned by an appropriate choice and sophistication of the model).

Fortunately, many of the above mentioned effects may be well-approximated in terms of a convolution of the stellar spectrum with some broadening function (e.g. a rotational broadening) or as a superposition of a few spectral functions (corresponding, e.g., to different layers of the atmosphere) convolved with different broadenings. The basic Eq. (2.1) is thus sufficiently general and the whole procedure of the Fourier disentangling can be applied as before with a generalized form of the broadening functions $\Delta_j(x, t, p)$, which will in addition to the Doppler shift given by Eq. (2.2) also include an intrinsic broadening of the line profile (their Fourier transforms are to be multiplied). From the point of view of the underlying physics, the generalized disentangling is related to the BFT designed by Rucinski (2002) for RV measurements. However, the difference is in the question we ask, i.e. the variables we want to solve. In the BFT, a template spectrum is to be chosen (the same for all components) and the broadening function is solved for from the observations, without any *a priori* limitations, to find from it the RVs and possibly also the intrinsic broadening. In the disentangling, we want to solve for the component spectra, hence, we must restrict the space of possible broadening functions by some physical model with a few free parameters only (which can be directly disentangled). It is obvious that if the component spectra are variable, we can solve for some reference spectrum only, which is not uniquely defined because any constant part $b(y)$ of the broadening $\tilde{\Delta}_j(y, t, p) = b(y)\tilde{\Delta}'_j(y, t, p)$ may also be taken as a part of the intrinsic spectrum $\tilde{I}'_j(y) = b(y)\tilde{I}_j(y)$.

Already, the very simple generalization consisting of multiplication of the right-hand side of Eq. (2.2) by a scalar line-strength factor $s_j(t)$ increases significantly possibilities of the method of disentangling (Hadrava 1997). To mention at least some of the examples, it facilitates the fit of spectra for exposures taken during eclipses and thus not only to improve their RVs but also to find limb-darkening in different lines, which show the structure of the stellar atmosphere. It also enables us to disentangle the telluric lines which have variable line-strengths dependent on the atmospheric conditions and air-mass. (The molecular components of the Earth's atmosphere with different variability of abundance can also be mutually separated.) The profit is not only in increased reliability of the disentangled spectra, but also in the possibility to check and to improve the precision of the wavelength scale for the purpose of high-precision RV measurements. (For this purpose also the enhancement of precision to the sub-pixel resolution is important, cf. Hadrava 2009a.) Similarly to the natural telluric lines, the lines from an iodine cell or interstellar and circumstellar lines can also be disentangled. The disentangling of absorption interstellar lines from spectra of binaries or Cepheids yields a unique constraint on the depth structure of the interstellar mass (and simultaneously information about the interstellar extinction needed for a photometric determination of distances of the stars).

A large variety of other models of line-broadening can be included into the generalized disentangling for various purposes. For instance, the already mentioned rotational effect during the eclipses can be disentangled as described by Hadrava (2007). The standard approach to treatment of this effect is to measure RVs by some classical method (e.g., from bisectors or moments of the stellar lines) and to fit the resulting RV-curves with a model which includes a correction of RVs with respect to the Keplerian orbital motion. However, the rotational distortion of the line-profiles depends on the limb-darkening within the profile, and is thus different for different lines. The generalized disentangling offers an alternative possibility to fit the observed spectra directly by appropriately broadened disentangled spectra of the component stars.

Another interesting example is the line-profile variability due to pulsations (Hadrava *et al.* 2010). The RV-variations due to radial pulsations of Cepheids can be measured approximately by disentangling with free RVs (i.e. with switching-off the condition of RVs subjected to an orbital motion). However, the contribution of the parts of atmosphere seen under a non-zero angle results in a distortion of the line-profile, which is moreover dependent again on the limb-darkening in the line. It is thus preferable to model these line-profile variations (LPVs) by an appropriate broadening function and to disentangle the mean intrinsic spectrum of the Cepheid atmosphere in its rest-frame together with the instantaneous pulsational velocities. The variations of line-strengths caused by the changes of the temperature must also be disentangled; eventually the spectrum can be disentangled as a linear superposition of two or more spectral functions corresponding to the different temperatures. This application is important because it enables us to perform the Baade – Wesselink calibration of the period – luminosity relation of Cepheids without a need to establish the projection factor (cf. e.g. Nardetto *et al.* 2004). Similar to the application of disentangling to eclipsing binaries (cf. Wilson 2008), this disentangling of single-star spectra also provides primary distance markers on the extragalactic scale. Challenging is its application to Cepheids in binaries (cf. Pietrzyński *et al.* 2010), which could yield a comparison of both these methods and a better insight into the physics of Cepheids. For the non-radially pulsating stars, which are much more common in binaries, the disentangling can be used to solve for the orbital motion, neglecting first the LPVs which are usually of lower amplitude and can be found as residuals from the mean disentangled spectra. However, a true disentangling of non-radial pulsations should include the application of a proper model of the corresponding LPVs.

Disentangling with templates. The general advantage of disentangling including the separation of the component spectra without any *a priori* assumption on their form may turn into a disadvantage in the cases when we have a good reason for accepting such an assumption. For instance, the shape of the telluric spectrum is basically known (up to the possible variations of relative line-strengths of water vapour and other molecules) but, in systems which have a third or a circumstellar component with small RV- amplitude or some peculiar variability, a part of this stellar spectrum may penetrate into the telluric spectrum in the numerical solution. It is thus better to fix some component spectra I_j to appropriate templates J_j if we know them, and to minimize the expression (2.3) with respect to the unknown components only and with respect to the parameters p of the broadening functions Δ_j (including those belonging to the component spectra constrained by the templates).

This option may also be used temporarily to decrease the number of degrees of freedom of the solution before approximate values of orbital parameters are found, and the final tuning of the solution may be performed without constraining by the template. Alternatively, if an unconstrained disentangling gives a component spectrum closely resembling some standard spectral type, it can be constrained by a corresponding model spectrum to disentangle additional parameters of the component (e.g. its rotational broadening and physical parameters of the atmosphere) and to force the solution to distribute the residua between the other, possibly peculiar components. The interpretation of the separated spectra and determination of the physical parameters of the components is the purpose of the whole process and it can be performed on several levels – either by trial-and-error fitting of the decomposed spectra with models which include the possible broadening (cf. e.g. Zverko *et al.* 1997, Pavlovski & Hensberge 2010), or using BFT (Rucinski 2002), or directly in the process of the disentangling. It should be noted that only after a template spectrum is compared with the separated spectra, the systemic (i.e. the γ-) RV can be determined, and that (due to the above mentioned singularity of the zeroth Fourier mode) the continuum cannot be divided between the components from the spectroscopy alone, and hence the separated spectra must be scaled in strength either from the light-curve solution or just from fitting by model spectra.

Disentangling with constrained parameters. As mentioned in the Introduction, different observational data put different limitations on the free parameters of our models. Although it may be encouraging if we arrive at similar results from independent data of different kinds, there is a danger that their independent fits will yield inconsistent values of parameters for which the sensitivity differs. One way is thus to converge in each solution only those parameters on which the data may put stringent limitations and to fix the others to values found from data which are more sensitive to them. However, quite often the solutions provide some bounding condition on values of several parameters p, and we should search for the solution of other data in a subspace of the parameter space given by an equation $F(p) = 0$. (For example, photometry of eclipsing binaries usually gives a precise epoch of conjunctions, but the epoch of periastron is correlated with the longitude of periastron for eccentric orbits.) To find numerically the minimum of Eq. (2.3) bounded by such a condition or several conditions, we should add to its right-hand side a term $\sum_k \lambda_k F_k^2(p)$, where $\lambda_k > 0$ are Lagrange multiplicators, and to minimize the overall sum for $\lambda_k \to \infty$. Because in practice the bounding conditions result from some other observations, they are not sharp but admit some scatter given by the $(O\text{--}C)^2$ of the data. The disentangling constrained by other observations is thus a simultaneous solution of all available data (e.g. light curves, or published RVs from unavailable spectra, interferometry – either reduced to mutual positions of component stars, or directly the visibility functions, etc.) in addition to the direct fit of the spectra by minimizing a properly weighted sum of squared residua of all the data.

Such a simultaneous solution of all kinds of data enables a direct fit of additional parameters like the distance in the above mentioned solution, together with light-curves or with astrometry (cf. Zwahlen *et al.* 2004). It may also have the advantage of a significant increase in the precision of the obtained parameters values. The errors have to be determined from the Bayesian probability, which should be mapped in the vicinity of the derived best solution in the parameter space. The precision of a common solution for different data may be much higher than that given by an overlap of area restricted by solutions of individual subsets of the data. This is obvious from the example of period searches, where the solutions of data distant in time mutually interfere and their common solution may have an accuracy higher for orders in comparison with the individual solutions.

4. Conclusion

The method of disentangling has been succesfully applied to many studies of individual spectroscopic binaries and multiple stellar systems. Its wavelength-domain versions were independently programmed by several users according to the published descriptions. The Fourier versions of the code: FD-BINARY (Ilijić *et al.* 2004) and KOREL (Hadrava 1995) are available. The latter, which enables us to disentangle up to five components in a two-level hierarchical structure of orbits with their line-strength factors and possible constraining by templates, is now available in the framework of Virtual Observatory (the VO-KOREL at `http://stelweb.asu.cas.cz/vo-korel`, cf. Škoda & Hadrava 2010). There are also versions of KOREL with the pulsational and rotational broadening included, and an implementation of other broadening functions is in progress. The controlling of this version is quite complicated; it is not yet settled to a user-friendly form and hence not yet publically evailable. The same is true for the new code BAŽANT which is a blend of KOREL with the code FOTEL for solution of light-curves and other data (cf. Hadrava 2004a).

The disentangling simplifies the interpretation of spectroscopic observations of binaries, which is quite laborious in the standard way. This could predestinate this method for an automated application to massive data gathered in large space- or ground-based surveys. At the same time, the simplification of the data-processing enables us to sophisticate the method from the point of view of the involved physics, which, however, increases demands on insight of its users. The future development should thus follow both ways. The latter one actually crosses the borders between the often independent fields of theoretical modelling and methods of interpretation of the data. The development of future tools for astrophysics should follow and extrapolate this way to a physical sophistication and versatile applicability.

Acknowledgements. This work has been supported by the Center for Theoretical Astrophysics of the Czech Republic (ref. LC06014) and grant project GAČR 202/09/0772.

References

Bagnuolo, W. R. & Gies, D. R. 1991, *ApJ*, 376, 266
Hadrava, P. 1995, *A&AS*, 114, 393
Hadrava, P. 1997, *A&AS*, 122, 581
Hadrava, P. 2004a, *Publ. Astron. Inst. ASCR*, 92, 1
Hadrava, P. 2004b, *Publ. Astron. Inst. ASCR*, 92, 15
Hadrava, P. 2007, *ASP-CS*, 370, 164
Hadrava, P. 2009a, *A&A*, 494, 399
Hadrava, P. 2009b, *arXiv:* 0909.0172
Hadrava, P., Šlechta, M., & Škoda, P. 2010, *A&A*, 507, 397
Hensberge, H. 2007, *ASP-CS*, 364, 275
Hill, G. 1993, *ASP-CS*, 38, 127
Ilijić, S. 2004, *ASP-CS*, 318, 107
Ilijić, S. Hensberge, H. *et al.*, 2004, *ASP-CS*, 318, 111
Kallrath, J. & Milone E. F. 1999[1], 2009[2], *Eclipsing binary stars: modeling and analysis*, Springer-Verlag, New York
Marchenko, S. V., Moffat, A. F. J., & Eenens, P. J. R. 1998, *PASP*, 110, 1416
Nardetto, N., Fokin, A. *et al.*, 2004, *A&A*, 428, 131
Pavlovski, K. & Hensberge, H. 2010, *ASP-CS*, 435, 207
Pietrzyński, G., Thompson, I. B. *et al.*, 2010, *Nature*, 468, 542
Rucinski, S. 2002, *AJ*, 124, 1746
Schlesinger, F. 1909, *Publ. of Allegheny Obs.*, 1, 123
Simon, K. P. & Sturm, E. 1994, *A&A*, 281, 286
Škoda, P. & Hadrava, P. 2010, *ASP-CS*, 435, 71
Wilson, R. E. 2008, *ApJ*, 672, 575
Zverko, J. & Žižňovský, J., Khokhlova V. L. 1997, *Contrib. Astron. Obs. Skalnaté Pleso*, 27, 41
Zwahlen, N., North, P. *et al.*, 2004, *A&A*, 425, L45

Discussion

S. ZUCKER: How does spectral disentangling deal with gaps in the spectrum? Specifically with multi-order spectra. The question arises because of the Fourier steps that require equidistant sampling in $\log(\lambda)$.

P. HADRAVA: A smooth merging of the orders in Echelle spectra is a subtle problem studied, e.g., by Hensberge (2007). It is important for disentangling of wide spectral regions where some unevenesses may complicate disentangling especially of lower Fourier modes. Fortunately, for an accurate determination of radial velocities, we can use narrow regions with higher sampling within separate orders. Equidistant sampling is more critical for the wavelength-domain disentangling than for the Fourier one, but it can be achieved by interpolation or an appropriate primary data reduction.

From Interacting Binaries to Exoplanets: Essential Modeling Tools
Proceedings IAU Symposium No. 282, 2011   © International Astronomical Union 2012
Mercedes T. Richards & Ivan Hubeny, eds.   doi:10.1017/S1743921311027839

Quantitative Spectroscopy
of Close Binary Stars

K. Pavlovski[1,2] & J. Southworth[2]

[1]Department of Physics, Faculty of Science, University of Zagreb, Croatia

[2]Astrophysics Group, Keele University, Staffordshire, ST5 5BG, UK

Abstract. The method of spectral disentangling has now created the opportunity for studying the chemical composition in previously inaccessible components of binary and multiple stars. This in turn makes it possible to trace their chemical evolution, a vital aspect in understanding the evolution of stellar systems. We review different ways to reconstruct individual spectra from eclipsing and non-eclipsing systems, and then concentrate on some recent applications to detached binaries with high-mass and intermediate-mass stars, and Algol-type mass-transfer systems.

Keywords. binary stars, spectroscopy, stellar atmospheres, chemical composition

---

1. Introduction

The structure and evolution of a star is determined by its mass and chemical composition. Hence, the position of the star in the HR diagram is uniquely defined only when its mass and bulk metallicity are known. Eclipsing and double-lined spectroscopic binaries in a detached configuration remain the primary source of directly measured fundamental stellar quantities: mass and radius. Modern observational techniques are currently able to reach precisions of 1–2% in these parameters, a prerequisite for testing theoretical stellar evolutionary models. In their critical survey and compilation of available results, Torres, Andersen & Gimenez (2010) selected 95 detached eclipsing binaries (dEBs) which satisfied a reasonable criterion of 3% in precision for both quantities. The effective temperatures of the component stars cannot be determined directly from the analysis of light or radial velocity curves, making $T_{\rm eff}$ a less well-determined quantity.

The metallicity has been determined for the components of only about 21 of the 95 binaries in the sample of Torres *et al.* (2010). Usually, metallicity is derived from fitting the overall spectrum of the stellar system. Detailed abundance determinations have been accomplished for only *four* binaries: directly from the spectrum of β Aur (Lyubimkov *et al.* 1996) and CV Vel (Yakut *et al.* 2007), and from disentangled component spectra of V578 Mon (Pavlovski & Hensberge 2005) and V453 Cyg (Pavlovski & Southworth 2009).

Tremendous advances in observational stellar spectroscopy, both in spectral resolution, quantum efficiency and detector linearity, allow new opportunities in binary star research. Reconstructing the individual spectra of the components opens a new window on these stars: detailed spectral analysis, atmospheric diagnostics, and determination of atmospheric chemical compositions (Pavlovski 2004, Pavlovski & Hensberge 2005). This enables a fine probing of stellar evolutionary models and proper calibrations of fundamental stellar quantities (Pavlovski & Southworth 2009, Pavlovski *et al.* 2009, 2011). The advantages of analysing disentangled spectra are manifold.

2. Renormalisation of disentangled component spectra

More than a dozen methods have been invented to obtain separate spectra of the components of multiple systems from a series of composite spectra obtained at a range of orbital phase. Pavlovski & Hensberge (2010) divided them into three basic categories: (i) spectral separation; (ii) spectral disentangling; and (iii) spectroastrometric splitting. In all cases, the isolation of individual spectra exploits the variable Doppler shift of the stars.

In *spectral separation* techniques, radial velocities (RVs) are input data obtained from another source (e.g. cross-correlation). The method most widely used in this category is Doppler tomography, developed by Bagnuolo & Gies (1991). The technique of *spectral disentangling* (SPD; Simon & Sturm 1994, Hadrava 1995) refers to the situation where both the individual spectra and the orbital parameters of the components are measured, simultaneously and self-consistently. The *spectroastrometric splitting* method accesses the spatial information present in long-slit spectra of partially resolved components (see Bailey 1998; Wheelwright, Oudmaijer & Schnerr 2009, and references therein).

The principles of these methods are outlined in the review papers by Hadrava (2004, 2009), Hensberge & Pavlovski (2007), and Pavlovski & Hensberge (2010). Without going into details, we record here some new developments by Konacki *et al.* (2010), Folsom *et al.* (2010) and Kolbas & Pavlovski (this Volume).

Spectral disentangling or separation techniques return individual spectra which are either in the common continuum of the system, or on an arbitrary (generic) continuum. The reason for this is a basic principle of SPD: there is no information on the continuum light ratio of the stars if disentangling is performed on spectra where the light ratio is constant. The absolute spectral line depths are then unknown because the continuum level is unknown. The solution is to obtain spectra during eclipse, as information on the continuum level is then available. Spectra during eclipse can be difficult to secure for reasons including the scheduling of observations and the lower brightness of the system at eclipse times. However, the main difficulty lies in the Rossiter-McLaughlin effect. This distorts the line profiles and thus violates a basic assumption of spectral disentangling. The Rossiter-McLaughlin effect can be only avoided by studying totally-eclipsing systems. In practice, we usually have a situation in which the continuum light ratios do not vary between spectra, and SPD has to be performed with arbitrary or generic light dilution factors. Renormalisation then requires light factors from other sources.

It is expected that the most reliable light dilution factors come from analysis of the light curves of dEBs. This is true, but it depends on particular cases, as discussed below. When eclipses do not occur, or observational data do not cover them, we must rely on information contained in the disentangled spectra themselves. Here, we list several different approaches recorded in the literature:

• For eclipsing binaries, light factors may be available from the time-independent dilution of spectral lines or from light curves (e.g. Hensberge, Pavlovski & Verschueren 2000). The quality of the photometry, the configuration and geometry of the binary system, or the presence of a third star, could make the light factors uncertain.

• The light factors can be determined from the time-dependent dilution of the spectral lines as additional free parameters in SPD calculations (*line photometry*, Hadrava 1997). Any intrinsic line profile changes (pulsations, spots, Rossiter-McLaughlin effect, etc.) violate a basic assumption of spectral disentangling, and can cause erroneous results.

• If the system is not eclipsing, some *physical considerations* can be used to renormalise individual disentangled spectra. In the study of the non-eclipsing triple system DG Leo, Frémat *et al.* (2005) successfully used the very deep Ca II K line profiles. The requirement that the core should not dip below zero flux for any of the component spectra (in this particular case all components are of similar spectral type) imposed very strong constraints on the light factors. Light dilution factors can also be estimated from line depths, or equivalent width ratios, once the atmospheric characteristics are fixed (c.f. Mahy *et al.* 2011). Use of line depth or line intensity ratios require chemical abundances to be specified *a priori*, making this option somewhat uncertain.

• Separated or generic disentangled spectra contain information on the intrinsic spectra. Just as in the way in which information on effective temperature (T_{eff}) and surface gravity ($\log g$) are extracted from renormalised spectra, it is possible to recover also light factors by constrained multi-parameter line-profile fitting of both disentangled spectra. Tamajo, Pavlovski & Southworth (2011) have implemented this idea in the code GEN-FITT. Simulations using synthetic spectra showed that for a reasonable signal-to-noise ratio of S/N $\geqslant$ 100 *constrained genetic fitting* returns reliable T_{eff} and $\log g$ for each star, plus their light factor. It is so called because the optimal fitting is constrained by requiring the sum of the light factors to be unity. $\log g$ can be derived for the stars in close binaries with precisions of 0.01 dex, a crucial advantage in breaking the degeneracy between T_{eff} and $\log g$ for Balmer line profiles. Comparison of the light factors derived by constrained genetic fitting and from light curve analyses is given by Pavlovski *et al.* (2009), Tamajo *et al.* (2011), and Southworth *et al.* (2011b) for some real-world cases.

• The procedure described in the previous item could be generalised in the way that part or whole of the disentangled spectrum is fitted by theoretical spectra. Tkachenko, Lehmann & Mkrtichian (2009, 2010) have used this idea and fitted a large portion of spectra by gridding with precomputed theoretical spectra for a large range of T_{eff}, $\log g$ and metallicity. A similar technique has been applied by Torres *et al.* (2011) but with restriction to solar abundances only.

3. Chemical composition from reconstructed spectra

To determine the chemical composition of a stellar atmosphere, it is first neccessary to specify an appropriate model atmosphere. The model is described by T_{eff}, $\log g$, and metallicity ([M/H]). For most eclipsing binaries, $\log g$ can be determined from the analysis of light and velocity curves, with a precision and accuracy up to an order of magnitude larger than for single stars. This considerably facilitates the determination of T_{eff}, and side-steps the degeneracy between T_{eff} and $\log g$ for hot stars. Moreover, a reliable estimate of T_{eff} is possible directly from *constrained fitting* when the light factors are not well-determined from external sources (Pavlovski *et al.* 2009, Tamajo *et al.* 2011). This also makes it possible to estimate T_{eff} and $\log g$ for non-eclipsing binaries, and stipulate additional constraints in complementary solutions with interferometric observations. The first estimate of T_{eff} can be used in an iterative cycle for fine-tuning light curve and SPD solutions, as was described by Hensberge *et al.* (2000) and further elaborated by Clausen *et al.* (2008).

Detailed abundance work makes possible further improvements in T_{eff} determination through ionisation balance. We have successfully employed the Si II / Si III ionisation balance in high-mass binaries (Pavlovski & Southworth 2009), and the Fe I / Fe II balance in intermediate-mass binaries (work in preparation). Such an improvement in determination of T_{eff} is an important ingredient if one wants to use binaries for distance determination, in particular for galaxies in the Local Group (e.g. Harries *et al.* 2003, North *et al.* 2011).

3.1. *Detached binary stars*

High-mass stars. Despite considerable theoretical and observational efforts, some important pieces of the jigsaw of stellar structure and evolution remain unclear or missing. Meynet & Maeder (2000) and Heger & Langer (2000) found that rotationally-induced mixing and magnetic fields could cause substantial changes in theoretical predictions. Some of these concern evolutionary changes in the chemical composition of stellar atmospheres. In close binaries, tidal effects further complicate this picture (De Mink *et al.* 2009), and pose a big challenge for observational confirmation.

Our observational project on the chemical evolution of high-mass stars in close binaries is directed toward tracing predicted changes in the photospheric abundance pattern due to rotational mixing. In Pavlovski *et al.* (in preparation), we summarise our results for fourteen high-mass stars in eight dEBs, plus some additional high-mass stars in binaries studied from disentangled spectra (Simon *et al.* 1994, Sturm & Simon 1994, Southworth & Clausen 2007). Of these, V380 Cyg (Pavlovski *et al.* 2009), V621 Per (Southworth *et al.* 2004), and V453 Cyg (Pavlovski & Southworth 2009) are the most informative as their primary components are evolved either close to or beyond the terminal-age main sequence. No abundance changes relative to unevolved MS stars of the same mass have been detected for these components, probably due to their relatively long orbital periods (De Mink *et al.* 2009). The study of HD 48099, an O5.5 V((f)) + O9 V binary system, by Mahy *et al.* (2010) reveals a nitrogen enhancement in the primary star, but a solar abundance for the secondary. The estimated masses are 55 and 19 $M_\odot$ for the primary and secondary component, respectively. Determination of chemical composition from disentangled spectra is an important way to constrain theoretical models.

Intermediate-mass and solar type stars. We have recently constructed detailed abundance studies of late-B and A-type stars in the close dEBs AS Cam and YZ Cas (work in preparation). Abundances are also available for the δ Scuti pulsating components in the binaries DG Leo (Frémat *et al.* 2005) and HD 61199 (Harater *et al.* 2008).

Systematic research in FGK stars in binaries, concerning also their chemical composition and an empirial evaluation of their metallicity, was initialized by the late Jens Viggo Clausen and collaborators. A comprehensive study of three F-type binaries has shown the full power of testing and comparing recent stellar evolutionary models using eclipsing binaries, provided their abundances are known (Clausen *et al.* 2008). The same methodology was extended to the solar-type binary systems V636 Cen (Clausen *et al.* 2009) and NGC 6791 V20 (Grundahl *et al.* 2008; Brogaard *et al.*, 2011).

3.2. *Algol systems*

One of the many consequences of the first and rapid phase of mass transfer in close binary systems, and the eventual mass reversal and formation of Algols, is the change in chemical composition of the stars involved. In fact, Algols offer an unique opportunity to probe into stellar interiors since detailed abundance studies of the layers which were once deep inside the star can give important information on the thermonuclear and mixing processes taking place during core hydrogen burning (Sarna & De Greve 1996). Carbon should be depleted in the CNO cycle even during a star's MS lifetime, and observational studies have aimed at testing these predictions (c.f. Tomkin 1981 and references therein).

We have started a new observational programme with the aim of deriving detailed abundances from high-resolution and high-S/N Échelle spectra using the SPD technique. We intend to substantially extend both the number of the elements studied, and the number of lines for each element.

Algol (β Per) is the prototype of the class of binary systems in a semidetached configuration, where the initially more massive and more evolved component fills its inner Roche lobe and transfers material to its now more-massive companion. Algol is one of the most frequently studied objects in the sky, and has been observed at wavelengths ranging from X-rays to radio (c.f. Richards et al. 1988). However, due to difficulties in ground-based observations of such a bright object, and a lack of modern high-resolution spectroscopy, its stellar and orbital parameters were somewhat uncertain.

Since 2007, we have secured 140 high-S/N Échelle spectra of Algol using the FIES spectrograph at the Nordic Optical Telescope and BOES at Bohyunsan Optical Astronomy Observatory in Korea. The available light curves are not on their own sufficient to allow the precise quantification of the contribution of the third component to the total light, which is needed for proper reconstruction of the disentangled spectra of the components. Therefore, we rely only on spectroscopic information. Abundances are derived for 15 elements, and are generally close to solar (Kolbas *et al.*, this volume). We are currently undertaking non-LTE calculations for helium and the CNO elements. We corroborate the weakness of the Ca II lines in the spectrum of the third component, and a slight underabundance of scandium, both classical indicators of a metallic-lined star. We will be performing a detailed abundance of our disentangled spectrum for this candidate Am star.

The importance of an abundance study for understanding stellar evolution in binary systems is nicely shown by Mahy *et al.* (2011) in a study of the semidetached system LZ Cep, an O9 III + ON9.7 V binary. They have found the secondary component, now the less massive star in the system, to be chemically more evolved than the primary, which barely shows any sign of CNO processing. Also, considerable changes in the chemical composition which corroborate predictions have been found for the components of Plaskett's star (54 + 56 $M_\odot$) by Linder *et al.* (2008). Plaskett's star is in a post-case-A Roche lobe overflow stage.

Cool Algols. The chemical composition of the primaries of two oEA stars, TW Dra and RZ Cas, have been derived by Tkachenko, Lehmann & Mkrtichian (2009, 2010). This subclass of Algols is known for cooler A-type primaries which are pulsating with δ Scuti characteristics (Mkrtichian *et al.* 2002). Their analyses of disentangled spectra have shown that these stars are normal A-stars with a chemical composition close to solar. Understanding the chemical composition of the pulsating components in close binaries is an important condition for proper asteroseismologic diagnostics. In combination with precise stellar parameters derived from complementary photometric and spectroscopic observations (c.f. Southworth *et al.* 2011a), this is a very powerful way for probing modern models of stellar structure and evolution.

Acknowledgements. KP acknowledges receipt of a Leverhulme Visiting Professorship which enabled him to work at Keele University, UK, where a part of this work was performed. JS acknowledges support from the STFC in the form of an Advanced Fellowship. This research is supported by a grant to KP from the Croatian Ministry of Science and Education.

References

Bagnuolo, A. & Gies, D. R. 1991, *ApJ*, 376, 266

Bailey, J., 1998, *MNRAS*, 301, 161

Brogaard, K., Bruntt, H., Grundahl, F., Clausen, J. V., Frandsen, S., Vandenberg, D. A., & Bedin, L. R. 2011, *A&A*, 525, A2

Clausen, J. V., Torres, G., Bruntt, H., Andersen, J., Nordström, B., Stefanik, R. P., Latham, D. W., & Southworth, J. 2008, *A&A*, 487, 1095

Clausen, J. V., Bruntt, H., Claret, A., Larsen, A., Andersen, J., Nordström, B., & Giménez, A. 2009, *A&A*, 502, 253

Cugier, H. & Hardrop, I. 1988, *A&A*, 202, 101

De Mink, S., Cantiello, M., Langer, N., Pols, O. R., Brott, I., & Yoon, S.-Ch. 2009, *A&A*, 497, 243

Folsom, C. P., Kochukhov, O., Wade, G. A., Silvester, J., & Bagnulo, S. 2010, *MNRAS*, 407, 2383

Frémat, Y., Lampens, P., & Hensberge, H. 2005, *MNRAS*, 356, 545

Grundahl, F., Clausen, J. V., Hardis, S., & Frandsen, S. 2008, *A&A*, 492, 171

Hadrava, P. 1995, *A&AS*, 114, 393

Hadrava, P. 1997, *A&AS*, 122, 581

Hadrava, P. 2004, *ASP Conf. Ser.*, 318, 86

Hadrava, P. 2009, arXiv:0909.0172

Hareter, M., Kochukhov, O., Lehmann, H., Tsymbal, V., Huber, D., Lenz, P., Weiss, W. W., Matthews, J. M., Rucinski, S., Rowe, J. F., Kusching, R., Guenther, D. B., Moffat, A. F. J., Sasselov, D., Walker, G. A. H., & Scholtz, A. 2008, *A&A*, 492, 185

Harries, T. J., Hilditch, R. W., & Howarth, I. D. 2003, *MNRAS*, 339, 157

Heger, A. & Langer, N. 2000, *ApJ*, 544, 1016

Hensberge, H. & Pavlovski, K. 2007, *in IAU Symp.*, 240, 136

Hensberge, H., Ilijić, S., & Torres, K. B. V. 2008, *A&A*, 482, 1031

Hensberge, H., Pavlovski, K., & Verschueren, W. 2000, *A&A*, 358, 553

Konacki, M., Muterspaugh, M. W., Kulkarni, S. R., & Helminiak, K. G. 2010, *ApJ*, 719, 1293

Linder, N., Rauw, G., Martins, F., Sana, H., De Becker, M., & Gosset, E. 2008, *A&A*, 489, 713

Lyubimkov, L. S., Rachkovskaya, T. T., & Rostopchin, S. I. 1996, *Astronomy Reports*, 40, 802

Mahy, L., Rauw, G., Martins, F., Nazé, Y., Gosset, E., De Becker, M., Sana, H., & Eenens, P. 2010, *ApJ*, 708, 1537

Mahy, L., Martins, F., Machado, C., Donati, J.-F., & Bouret, J.-C. 2011, *A&A*, 533, A9

Meynet, G. & Maeder, A. 2000, *A&A*, 361, 101

Mkrtichian, D. E., Kusakin, A. V., Gamarova, A. Y., & Nazarenko, V. 2002, *ASP Conf. Ser.*, 259, 96

North, P., Gauderon, R., Barblan, F., & Royer, F 2010, *A&A*, 520, A74

Pavlovski, K. 2004, *ASP Conf. Ser.*, 318, 201

Pavlovski, K. & Hensberge, H. 2005, *A&A*, 439, 309

Pavlovski, K. & Hensberge, H. 2010, *ASP Conf. Ser.*, 435, 207

Pavlovski, K. & Southworth, J. 2009, *MNRAS*, 394, 1519

Pavlovski, K., Kolbas, V., & Southworth, J. 2010, *ASP Conf. Ser.*, 435, 247

Pavlovski, K., Southworth, J., & Kolbas, V. 2011, *ApJ*, 734, L19

Pavlovski, K., Tamajo, E., Koubsky, P., Southworth, J., Young, S., & Kolbas, V. 2009, *MNRAS*, 400, 791

Richards, M. T., Mochnacki, S. W., & Bolton, C. T. 1988, *AJ*, 96, 326

Sarna, M. J. & De Greve, J.-P. 1996, *QJRAS*, 37, 11

Simon, K. & Sturm, E. 1994, *A&A*, 281, 286

Southworth, J. & Clausen, J. V. 2007, *A&A*, 461, 1077

Southworth, J., Zucker, S., Maxted, P. F. L., & Smalley, B. 2004, *MNRAS*, 355, 986

Southworth, J., Maxted, P. F. L., & Smalley, B. 2005, *A&A*, 429, 645

Southworth, J., et al., 2011a, *MNRAS*, 414, 2413

Southworth, J., Pavlovski, K., Tamajo, E., Smalley, B., West, R. G., & Andersen, D. R. 2011b, *MNRAS*, 414, 3740

Tamajo, E., Pavlovski, K., & Southworth, J. 2011, *A&A*, 526, A76

Tkachenko, A., Lehmann, H., & Mkrtichian, D. E. 2009, *A&A*, 504, 991

Tkachenko, A., Lehmann, H., & Mkrtichian, D. E. 2010, *AJ*, 139, 1327

Tomkin, J. 1989, *SSRev*, 50, 245

Torres, G., Andersen, J., & Giménez, A. 2010, *A&ARev*, 18, 67

Torres, K. B. V., Lampens, P., Frémat, Y., Hensberge, H., Lebreton, Y., & Škoda, P. 2011, *A&A*, 525, A50

Yakut, K., Aerts, C., & Morel, T. 2007, *A&A*, 467, 647

Wheelwright, H. E., Oudmaijer, R. D., & Schnerr, R. S. 2009, *A&A*, 497, 487

Zavala, R. T., Hummel, C. A., Boboltz, D. A., Ojha, R., Shaffer, D. B., Tycner, C., Richards, M. T., & Hutter, D. J. 2011, *ApJ*, 715, L44

From Interacting Binaries to Exoplanets: Essential Modeling Tools
Proceedings IAU Symposium No. 282, 2011
Mercedes T. Richards & Ivan Hubeny, eds.
© International Astronomical Union 2012
doi:10.1017/S1743921311027840

The Broadening Functions Technique

Slavek M. Rucinski

Department of Astronomy & Astrophysics, University of Toronto
50 St. George Street, Toronto, Ontario, Canada, M5S 3H4
email: rucinski@astro.utoronto.ca

Abstract. Essential assumptions and features of the Broadening Function (BF) technique are presented. A distinction between BF determination and the BF concept and utilization is made. The BF's can be determined in various ways. The approach based on linear deconvolution involving stellar templates, as used during the DDO program (1999 – 2008) is described, but the LSD technique would also give excellent results. The BF concept to prove and/or verify photometric light-curve solutions has so far been very limited to only a few W UMa-type binaries, with AW UMa giving particularly unexpected results.

Keywords. line: profiles, methods: data analysis, techniques: spectroscopic, techniques: radial velocities

1. Introduction

"Solving eclipsing binary light curves" was my occupation for a long time, in fact since 1965.† After many years of experience, I have been generally disappointed by the relatively low information content of light curves, particularly of W UMa-type binaries with partial eclipses. An obvious remedy would be to seek this information in spectroscopic data, by going beyond traditional determination of mass-centre radial velocities (RV). This would mean to utilize the Doppler broadening of spectral lines where information on *shapes* of the components is available, particularly during the (normally avoided in RV work) eclipse phases. Effectively, in place of single brightness measurements taken versus time, one should be able to utilize many little images – Broadening Functions (BF) – of the binary star, as projected into the RV space (Figure 1). Switching from the 1-D (time domain) light curve information to 2-D (time and RV) information should permit a much better binary-star description. The idea was sketched early (Rucinski 1971), but it took a long time to reach a point of maturity in years 1992 – 2008, during the David Dunlap Observatory binary program (see Section 3).

Utilization of the Broadening Functions splits into two aspects: (1) How to determine the BF's? (2) How to use information contained in them? Most of the work has gone into the necessary step of *BF determination* (Section 2). This is not a simple matter by itself, but much less effort has been devoted to utilization of the BF's in geometric element determination of close binaries (Section 4). So far, mostly W UMa-type binaries have been the subject of investigations involving the BF's, which is understandable because there is no need to confirm the shapes of spherical stars. But for W UMa-type binaries, it is hard to prove validity of Lucy's contact model (Lucy 1968a, Lucy 1968b) using

---

† I did an analysis of the photometric data for the close binary DI Pegasi as my MSc project (Rucinski 1967). What may be of relevance for this symposium – in view of Dr. Batten's panel session remarks – is that I used the Russell-Merrill and the Kopal-Piotrowski iterative techniques. Frankly, I did not believe in my results because the binary shows partial eclipses and I had to invoke a third light. It was a great relief when Wen Lu discovered a third star in the system (Lu 1992).

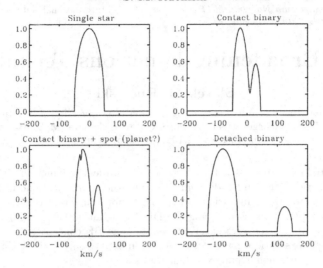

Figure 1. Four schematic examples of Broadening Functions.

photometric information alone: Light curve synthesis codes provide excellent fits to the data; all small details of the light curves can be beautifully explained and one may feel very satisfied. But is this really a true picture? The well-observed AW UMa provides an important, perhaps even sobering case (Section 4) strongly suggesting that light curve synthesis analyses – in spite of excellent light curve fits – may give an inadequate picture.

2. How to determine Broadening Functions

Convolution is an operation that nature does for us. In astronomy, we seldom see "naked" atomic or molecular spectra; instead, we are usually confronted with convolutions. They can be convolved with a spectrograph instrumental profile, with a radial component of the micro-turbulence velocity field in the stellar atmosphere, with rotational broadening; for a distant cluster the combined spectrum is convolved with the velocity dispersion of contributing stars. Thus, instead of a sharp-line spectrum $s(x)$, we observe $p(x)$ which is a convolution with the broadening function $b(x)$:

$$p(x) = \int_{-\infty}^{+\infty} b(u) \, s(x - u) \, du = b(x) * s(x)$$

In binary stars, the BF is simply a projection of the shape of the binary components into the radial velocity (RV) space (Figure 1). If the stars rotate as solid bodies, then their BF's are identical to a simple mapping of their shapes into the respective projected distances from the axis of rotation or of revolution. Purists will note that the observed (flux) spectrum originates by integration of the atmospheric *emergent intensity*, not of the *emergent flux*, so an approximation is involved here. However, with many lines in the spectral window, the differences in line formation mechanisms (absorption and scattering), which reflect in different dependencies on the angle of emergence, become less unimportant (but this matter remains to be studied in the BF context).

Although the above idea is simple, it is not easy to determine the BF's. One of the most obvious ways is to utilize the convolution properties and note that the Fourier transform can be used to separate the two functions: $\mathcal{F}(p) = \mathcal{F}(b) \times \mathcal{F}(s)$. The next step is to apply the inverse transform to the quotient: $b = \mathcal{F}^{-1}(\mathcal{F}(p)/\mathcal{F}(s))$; $s(x)$ is a spectrum of a sharp-line template or a model spectrum. While $b(x)$ can indeed be determined that way, this approach fails in most cases because of the amplification of the high frequency noise in the division step. Frequency filtering becomes necessary and the results tend to depend on how the data processing is done. The only brave attempt of this approach, and the first one to demonstrate utility of the BF's for contact binaries was by Anderson, Stanford & Leininger (1983) who analyzed AW UMa, which was always perceived as a crucial object for our understanding of contact binaries.

The cross-correlation function (CCF) is another popular method to evaluate an approximation of the Broadening Function. The CCF, or $c(x)$ here, can be computed easily in many software packages (note the different symbol):

$$c(x) = \int_{-\infty}^{+\infty} p(u)\, s(u+x)\, du = p(x) \otimes s(x)$$

The problem is that $c(x)$ *is not equal* to $b(x)$ because:

$$c(x) = b(x) * s(x) \otimes s(x) \neq b(x)$$

Broadening of $s(x)$ remains present in the result; the equality would happen only for $s(x) = \sum \delta(x - x_0)$. Still, the CCF is a very useful technique when we are sure that $b(x)$ is symmetric, as then the geometric centre of $c(x)$ coincides with that of $b(x)$.

The most straightforward approach is through representation of the convolution by a system of linear equations:

$$p_i = \sum_{j=0}^{m-1} b_j s_{i+m-j} \qquad i = 0, \ldots, n-1 \qquad n > m$$

(for details, see Rucinski 1992 or Rucinski 2002). For spectra of several thousand pixels (n), this means solving several thousand equations; fortunately, even a rather strongly broadened BF can be typically represented by a few hundred points (m). This leads to very large, over-determined systems ($n > m$, typically $n/m \simeq 10\times$) which are solvable using the least-squares formalism. In 1992, I realized (Rucinski 1992) that such solutions became feasible using a moderate-size computer using the powerful Singular Value Decomposition method. This resulted in the first BF-based analysis of AW UMa just confirming – similarly to the work of Anderson *et al.* (1983) – the previous light curve results. These early attempts have been largely superseded by the later, much improved results of Pribulla & Rucinski (2008), which – surprisingly – showed that spectroscopically, AW UMa is not a contact binary (Section 4).

It should be stressed that the BF's determined through linear equations have the nice and obvious property of *linearity*. This is important because using them one can determine relative luminosities of components directly. This is not a trivial matter because for example the CCF approach requires calibrations which may always leave room for additional uncertainties.

3. The DDO program of close binary star orbits, 1999 – 2008

Broadening Functions have been extensively used during the David Dunlap Observatory (DDO) program of radial velocity orbits of very close ($P < 1$ day) binaries in years 1999 – 2008. The last numbered paper of the 15-paper series was of Pribulla *et al.* (2009) although several additional papers addressed individual interesting systems (such as W Crv, V471 Tau or AW UMa). The DDO program was described at the Brno and Mykonos symposia on binary stars (Rucinski 2010a, 2010b).

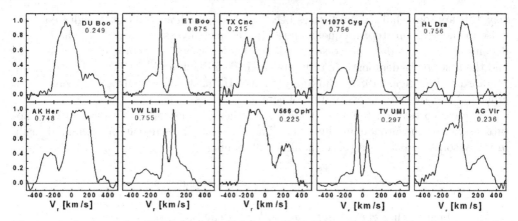

Figure 2. An example of BF's from the DDO-11 (Pribulla *et al.* 2006) paper showing results at orbital quadratures for ten W UMa systems. Among these binaries, one is a triple and four are quadruple systems.

Thanks to the reasonably stable spectrograph on the DDO 1.9m telescope, the moderately fast turn-around time in terms of reductions, but – mainly – to the superior properties of the BF technique, the DDO program produced excellent RV data for 162 binaries with as many as 145 of them being SB2 systems. Of note is that previously totally intractable triple and quadruple systems suddenly became solvable (Figure 2).

The DDO program did not utilize the full potential of the BF's: We used them in a "traditional" way, similarly to measurements of spectral line shifts, by just determining the component light centroids. Some improvements in these measurements have been done half way through the program (we introduced rotational profiles instead of Gaussians), but the great advantage of the BF's of encoding eclipse effects and asymmetries introduced by variable geometric projections was not used (the few exceptions: see the next section). Also, we were constantly under the pressure of time due to the imminent closure of the observatory, so we observed mostly at the predicted RV extremes (orbital quadratures), explicitly avoiding the eclipses. In the end, the published RV data for 162 binaries with $P < 1$ day correspond to the 92% completeness to $V = 10$ mag. We may have erred in determination of the K_i semi-amplitudes, but their ratios giving the important mass ratios $q = M_2/M_1 = K_1/K_2$ are probably the best one can currently achieve for very close binaries, particularly those of the W UMa type.

4. Geometrical element determination: The strange case of AW UMa

So far, the idea of simultaneous fitting of many BF's taken at different phases (including eclipses) has been used to analyze very few binaries. In addition to the first applications to AW UMa by Anderson *et al.* (1983) and Rucinski (1992), the early uses were for AH Vir (Lu & Rucinski 1993) and W UMa (Rucinski *et al.* 1993). Towards the end of the DDO program, we returned again to AW UMa (Pribulla & Rucinski 2008). Because the results for AW UMa were entirely unexpected, we analyzed in parallel the W UMa system V566 Oph.

It is not by accident that AW UMa has been always in the very centre of attention. It is a bright ($V_{max} = 6.8$) W UMa-type binary with a very small mass ratio q. Several photometric analyses based on Lucy's contact model have consistently given $q = 0.080 \pm 0.005$, where the error is not a formal estimate but rather the dispersion of various individual determinations confirming the high stability of the photometric solutions.

The success of Lucy's model in predicting the light curve of AW UMa so well has been generally assumed to signify the confirmation of the model.

Our spectroscopic BF analysis (Pribulla & Rucinski 2008) forced us to revise our original thinking about AW UMa. It appears that it is some sort of a semi-detached or detached binary with the mass ratio around $q = 0.11$. We fitted the BF's in various ways and we could not reduce q below 0.10, which is significantly different from the photometric result. Besides, the primary is smaller than its Roche lobe, but otherwise looks like an ordinary, rapidly rotating star. In contrast, the tiny secondary is strange: It seems to change its shape at various orbital phases and does not look like a normal star; possibly it is a small accretion disk or a fragment of it. The whole system seems to be engulfed in dense stellar material of the temperature of the primary component. Interestingly, V566 Oph with a somewhat larger mass ratio looks like it obeys Lucy's contact model.

AW UMa is telling us something very important: That solutions of photometric light curves may be deceiving. The need for techniques utilizing the BF's is very obvious.

5. BF versus LSD

The method of deriving broadening functions with the use of the Least Squares Deconvolution (LSD) was proposed in 1997 (Donati & Collier Cameron 1997, Donati 1997). The only real difference relative to the BF derivation is in the use of model atmosphere spectra as sharp-line templates (the ones called s above). Since these such spectra are formed into an array which is subsequently inverted, low noise in the template spectra is an important factor. So, why did we use stellar spectra? There are two reasons, both very simple: (1) to permit direct tie-ins to the international system of RV standards, (2) because we did not have access to good model atmospheres and there was always pressure of time during the DDO program.

The template spectra can be made almost arbitrarily smooth, but an additional important advantage of the LSD technique is its easy adaptability for Stokes parameter imaging, mostly to derive magnetic spot geometry on active stars. This is because model spectra can be computed not only in the normal light flux parameter I, but also in the Stokes parameters U, V, Q; such spectra – to be of use – are practically impossible to obtain for standard stars using our BF approach. Although the LSD broadening-function determination technique was so far almost exclusively used for active, spotted stars, there is absolutely no reason why it could not be used to study binaries. The very useful information content of broadening functions can be derived in various ways and the LSD could play an important role here.

References

Anderson, L., Stanford, D., & Leininger, D. 1983, *ApJ*, 270, 200

Donati, J.-F. & Collier Cameron, A., 1997, *MNRAS*, 291, 1

Donati, J.-F., Semel, M., Carter, B. D., Rees, D. E., & Collier Cameron, A., 1997, *MNRAS*, 291, 658

Lu, W., 1992, *AcA*, 42, 73

Lu, W.-X. & Rucinski, S. M. 1993, *AJ*, 106, 361.

Lucy, L. B. 1968a, *ApJ*, 151, 1123

Lucy, L. B. 1968b, *ApJ*, 153, 877

Pribulla, T., Rucinski, S. M., Lu, W., Mochnacki, S. W., Conidis, G., Blake, R. M., DeBond, H., Thomson, J. R., Pych, W., Ogłoza, W., & Siwak, M. 2006, textitAJ, 132, 769

Pribulla, T. & Rucinski, S. M. 2008, *MNRAS*, 386, 377

Pribulla, T., Rucinski, S. M., Blake, R. M., Lu, W., Thomson, J. R., DeBond, H., de Ridder, A., Croll, B., Karmo, T., Ogłoza, W., Stachowski, G., & Siwak, M. 2009, *AJ*, 137, 3655

Rucinski, S. M. 1967, *AcA*, 17, 271

Rucinski, S. M. 1971, *Bull.AAS*, 3, 327

Rucinski, S. M. 1992, *AJ*, 104, 1968

Rucinski, S. M. 2002, *AJ*, 124, 1746

Rucinski, S. M. 2010a, *ASP Conf.*, 435, 195

Rucinski, S. M. 2010b, *AIP Conf.*, 1314, 29

Rucinski, S. M., Lu, W.-X., & Shi, J. 1993, *AJ*, 106, 1174

Discussion

P. NIARCHOS: Is your non-contact solution for AW UMa a result of spectroscopic analysis or a result of a simultaneous LC and RV analysis? What would be the result if you used the spectroscopic mass-ratio in a light curve synthesis program?

S. RUCINSKI: The spectroscopic mass ratio $q = 0.11 \pm 0.01$ with the contact model gives a terrible light curve fit. But, we have enough indications that neither a contact nor a semi-detached model is appropriate for AW UMa, so synthesis light-curve solutions cannot give us a good set of parameters.

R. WILSON: Have you made any dynamical simulations of mini-disks in AW UMa to see if they have long-term stability? I would expect them to fall onto the stars very quickly.

S. RUCINSKI: If there is any real disk-like structure in AW UMa, it must be very small and confined to a small volume of the secondary. More likely is something which mimics the contact envelope, perhaps a dense (must be optically thick) stream of matter originating in the primary, confined to the equatorial region and fully engulfing the secondary.

P. HARMANEC: Let me to congratulate you as well as Petr Hadrava and Krešimir Pavlovski for your excellent talks with fine general background. My two questions are:
1. Which is the maximum magnitude difference between the primary and secondary, for which you are still able to measure the RV's of the secondary?
2. Any idea what criterion of the fit should be used to increase the sensitivity to weak signal from the secondary, comparable to the noise of the spectra used?

S. RUCINSKI: The DDO program achieved a very high percentage of SB2 systems in our unbiased survey of binaries with $V < 10$ and $P < 1$ day partly because most were W UMa-type systems with secondaries much brighter than low-mass stars on the MS; e.g., for AW UMa, $\Delta m \simeq 2.5$ mag. With the short period limit, we did not include any MS binaries so we could not establish a magnitude difference limit. My guess is that for spectra such as those at DDO with $S/N < 100$, the limit would be around $\Delta m \simeq 5$ mag.

From Interacting Binaries to Exoplanets: Essential Modeling Tools
Proceedings IAU Symposium No. 282, 2011
Mercedes T. Richards & Ivan Hubeny, eds.
© International Astronomical Union 2012
doi:10.1017/S1743921311027852

TODCOR – Two-Dimensional Correlation

Shay Zucker

Department of Geophysics and Planetary Sciences
Raymond and Beverly Sackler Faculty of Exact Sciences
Tel Aviv University, 69978 Tel Aviv, Israel
email: shayz@post.tau.ac.il

Abstract. TODCOR is a TwO-Dimensional CORrelation technique to measure radial velocities of the two components of a spectroscopic binary. Assuming the spectra of the two components are known, the technique correlates an observed binary spectrum against a combination of the two spectra with different shifts. TODCOR measures simultaneously the radial velocities of the two stars by finding the maximum correlation. The main use of the technique has been to turn single-lined binaries into double-lined systems. This helps to explore the binary mass-ratio distribution, especially the low-mass regime, where the secondaries are usually very faint and therefore hard to detect. The technique has been generalized to study multi-order spectra, and also triple- and quadruple-lined systems. It has several applications in studying extrasolar planets and in the future may even help to dynamically measure stellar masses of binaries through relativistc effects.

Keywords. Celestial mechanics, methods: data analysis, techniques: radial velocities, techniques: spectroscopic, binaries: spectroscopic, stars: low mass, brown dwarfs, planetary systems

1. Introduction

Since the seminal works of Simkin (1974) and Tonry & Davis (1979) the cross-correlation technique to measure astronomical Doppler shifts has become extremely popular. The advent of digitized spectra and computers made it the preferred method. It has been applied in all the astronomical fields that required the measurement of radial velocities from observed spectra, ranging from binary and multiple stellar systems to cosmology. The most prominent impact of radial velocities measured through cross-correlation was the detection of extrasolar planets (e.g., Mayor & Queloz 1995).

The calculation of the cross-correlation function is quite simple. Let $g(n)$ denote the zero-averaged, continuum-subtracted, observed spectrum, whose Doppler shift is to be found. Let $t(n)$ denote the corresponding 'template' spectrum of zero shift. Both the stellar spectrum and the template are assumed to be described as functions of the bin number, n, where $n = A \ln \lambda + B$. Thus, the Doppler shift results in a uniform linear shift of the spectrum (Tonry & Davis 1979). The cross-correlation function is then given by the expression:

$$c(n) = \frac{1}{N\sigma_g\sigma_t} \sum_m g(m)t(m - n),\tag{1.1}$$

where σ_g and σ_t are the RMS values of the corresponding spectra and N is the length of the spectra (for simplicity, I will ignore here the problem of edge effects). The location of the maximum of the function $c(n)$ is used as an estimate of the Doppler shift of the spectrum. Figure 1 presents a sample cross-correlation function with a well-defined peak.

Single-lined spectroscopic binaries (SB1) are characterized by the periodic variation of the star's radial velocity, corresponding to its projected orbital motion. Double-lined spectroscopic binaries (SB2) are characterized by the presence of two stellar spectra,

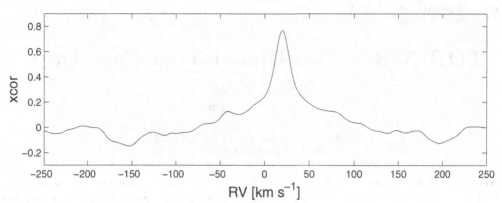

Figure 1. A typical cross-correlation function. The spectrum is a noisy Doppler-shifted synthetic spectrum of a G-star and the template is the original spectrum. The simulated radial velocity is $20\,\mathrm{km\,s^{-1}}$.

which are manifested in the cross-correlation function as two peaks (Figure 2.A). The locations of the two peaks can serve to estimate the velocities of the two components. However, in certain cases, or in certain orbital phases, the peaks might blend, which prevents a simple estimate of the two velocities (Figure 2B). Even when two separate maxima can be detected, the blend can affect their location and introduce systematic bias.

The TwO-Dimensional CORrelation technique – TODCOR, presented by Zucker & Mazeh (1994), offers a way to overcome these difficulties. Some details of the technique are presented in Section 2. Section 3 reviews some current applications of TODCOR and I conclude in Section 4.

2. The Two-Dimensional CORrelaion

The Two-Dimensional Correlaion is a straightforward generalization of the conventional cross-correlation technique. Instead of using a template spectrum – $t(m-n)$, TODCOR uses a combination of two templates, using a different Doppler shift for each template:

$$t_1(m-n_1) + \alpha t_2(m-n_2)$$

where α denotes the relative weight (light ratio) of the two components.

Substituting the above combined template in the expression in Equation 1.1, and after some algebraic manipulation, we obtain:

$$c(n_1, n_2) = \frac{c_1(n_1) + \alpha c_2(n_2)}{\sqrt{1 + 2\alpha c_{12}(n_2 - n_1) + \alpha^2}} \, .$$

In cases where the light ratio – α – is not known in advance, the value of α that would give the best results can be searched for. Alternatively, a value of α that would maximize $C(n_1, n_2)$ can be found analytically:

$$\hat{\alpha}(n_1, n_2) = \frac{c_1(n_1)c_{12}(n_2 - n_1) - c_2(n_2)}{c_2(n_2)c_{12}(n_2 - n_1) - c_1(n_1)} \, .$$

Incorporating this value in the correlation expression, we obtain the symmetric

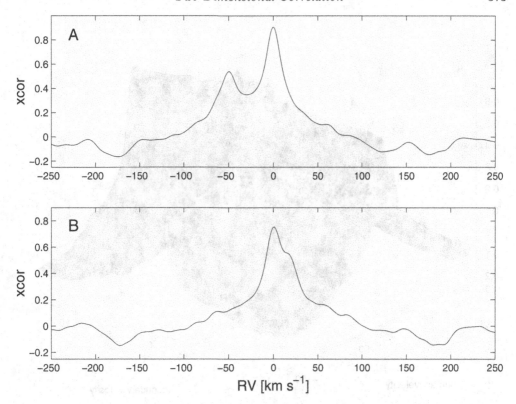

Figure 2. Cross-correlation function for simulated composite spectra. A: the simulated velocities are 0 and $-50\,\mathrm{km\,s^{-1}}$ and two separate peaks can be spotted at these velocities. B: the simulated velocities are 0 and $20\,\mathrm{km\,s^{-1}}$ and the two peaks are severly blended.

expression:

$$R(n_1, n_2) = \sqrt{\frac{c_1^2(n_1) - 2c_1(n_1)c_2(n_2)c_{12}(n_2 - n_1) + c_2^2(n_2)}{1 - c_{12}^2(n_2 - n_1)}}$$

In any case, the correlation is now a function of two variables – the two Doppler shifts n_1 and n_2, or, equivalently, the two radial velocities v_1 and v_2. Figure 3 presents a plot of this two-dimensional function for the same spectrum whose cross-correlation is presented in Figure 2.B. This spectrum was simulated using a primary velocity of 0 and a secondary velocity of $20\,\mathrm{km\,s^{-1}}$. One can see a clear maximum, whose location serves as an estimate of these two velocities.

A closer look at Figure 3 reveals an illuminating topographical structure: the maximum seems to be located at the intersection of two 'ridges'. Each ridge corresponds to 'freezing' one velocity and varying only the other one. Figure 4 shows the function when the secondary velocity is fixed (4.A) and when the primary velocity is fixed (4.B). These two 'cuts' can be regarded as kinds of cross-correlation functions, where we look for the best estimate for one velocity, after already having dealt with the other velocity. Comparing these two plots to Figure 2 highlights the advantage of TODCOR in measuring the two velocities. In these two 'cuts' the corresponding peak (the primary peak in A and the secondary peak in B) is accentuated and the second one is attenuated.

TODCOR is conceptually tailored to analyse composite spectra. The use of two templates allows better chances to detect secondary spectra that are significantly different

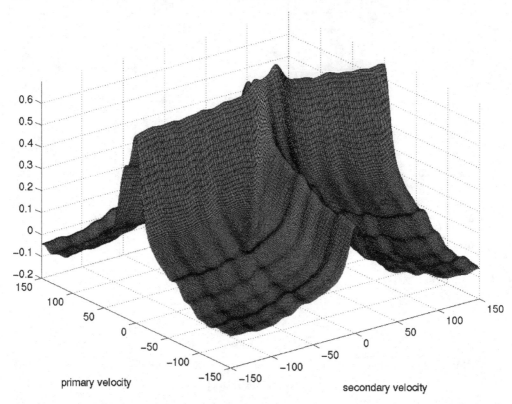

Figure 3. The two-dimensional correlation function for the spectrum whose cross-correlation function is depicted in Figure 2.B. The maximum is very close to the correct velocities – 0 and $20\,\mathrm{km\,s^{-1}}$.

from their primaries, whereas the detection of the secondary peak in cross-correlation relies on some similarity between the two spectra.

Originally, TODCOR was designed to analyse single-order spectra. Nowadays, as the use of Echelle spectrographs is widespread, it became necessary to adapt TODCOR to accept multi-oder spectra as input. Zucker (2003) suggested a statistically optimal formula to combine the correlation function of different orders into a combined function. This practice allows full exploitation of the information in all the orders, thus allowing the detection of very faint secondaries. This led to the complete solution of the system HD 41004, where a brown dwarf orbits the faint companion of a binary system, and a planet orbits the bright companion (Zucker *et al.* 2003).

Another generalization of the original TODCOR approach is its application to analyze 'triple-lined' and 'quadruple-lined' spectra, of triple and quadruple systems. The codes that implement this approach are based on the generalizations of the above formulae. Zucker *et al.* (1995) first developed a TODCOR generalization for spectroscopic triples, and Torres *et al.* (2007) repeated this feat for quadruple stars.

3. Applications

In the years since the introduction of TODCOR, many groups used it in their studies. The most common application is in the study of eclipsing binaries. The common practice is enumeration over a grid of templates. This grid of templates may comprise synthetic

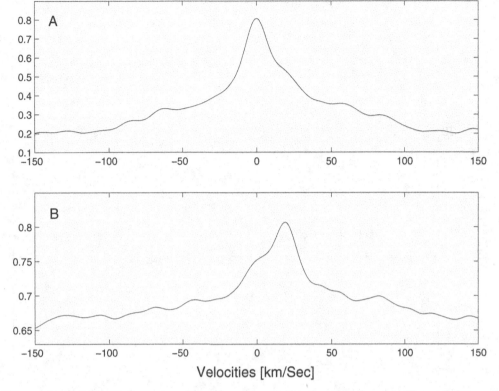

Figure 4. 'Cuts' through the two-dimensional correlation function shown in Figure 3. In panel
A the secondary velocity is fixed while in panel B the primary velocity is fixed.

spectra or alternatively high-S/N observed spectra. The criterion for the optimization is
usually the value of the correlation at the maximum.

Several studies pointed to the fact that residual systematics may still remain in the
velocities TODCOR found. These systematics can be estimated by using the templates
to simulate the observed spectra without noise and applying TODCOR on them (e.g.,
Torres *et al.* 2009).

A common application of TODCOR is in the search of faint companions. Basically,
the goal is to turn an SB1 into an SB2. From the orbital elements of an SB1, one can
calculate only the so-called 'mass function':

$$f(M_2) = M_1 \frac{(q \sin i)^3}{(1+q)^2},$$

where M_1 and M_2 are the binary component masses, i is the orbital inclination, and q is
the mass ratio. Thus, only a lower bound for q can be deduced, by substituting $\sin i = 1$.
If even only a few measurements of the secondary velocities are available, we can measure
the radial velocitiy semi-amplitudes of both components - K_1 and K_2, whose inverse ratio
is the mass ratio:

$$q = \frac{K_1}{K_2}.$$

Figure 5 demonstrates this use of TODCOR. This Figure is taken from Simon & Prato
(2004), who applied TODCOR to study the PMS binary Haro 1-14c. As the figure demonstrates, they needed only a few (eight) successful measurements of the secondary velocity
to constrain K_2. Most of the primary velocities were determined earlier by Reipurth *et al.*

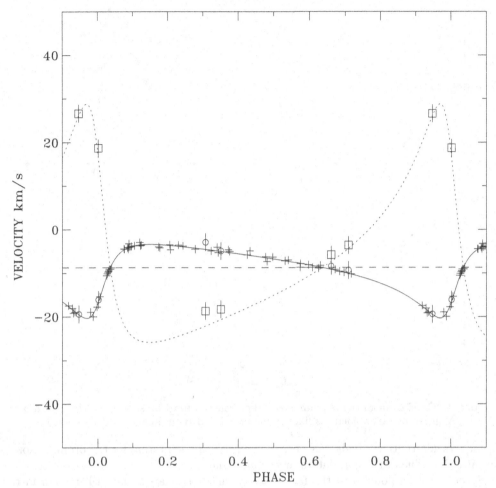

Figure 5. Radial velcity orbital solution of Haro 1-14c. Crosses represent the primary velocities obtained by Reipurth *et al.* (2002). Open symbols represent velocities obtained by Simon & Prato (2004) obtained using TODCOR (image taken from Simon & Prato 2004).

(2002) and served to constrain the other orbital elements. Simon & Prato obtained a mass ratio of $q = 0.310 \pm 0.014$. Schaefer *et al.* (2008) later confirmed this result through interferometry.

The study of extrasolar planets utilizes TODCOR for several purposes. As TODCOR was originally tailored to study binary stars, the most obvious application is the search for planets in spectroscopic binaries. Konacki *et al.* (2009) introduced the TATOOINE survey for that purpose, using TODCOR to get precise radial velocities that may show the signature of a hypothetical planet. TATOOINE uses the Iodine cell technique to calibrate the radial velocities, which required some modification of the original TODCOR approach.

Another use is to exclude the stellar nature of a candidate planet. In principle, the RV motion detected in cross-correlation may be the result of the combined opposite motions of the primary star and a faint companion. Researchers use TODCOR to verify that there's no secondary spectrum detectable in the observed spectra. Konacki *et al.* (2003) used this approach to rule out the stellar nature of the transiting planet OGLE-TR-56.

TODCOR is also being used to identify blend scenarios in candidate extrasolar planetary transits. Thus, for example, Mandushev *et al.* (2005) used TODCOR to prove that what seemed like a transiting brown dwarf around the F star GSC 01944-02289, was actually an eclipsing binary comprising G0V and M3V stars, orbiting the F5V star.

4. Concluding remarks

A few works presented in this meeting use Spectral Disentagling (SD) techniques (Simon & Sturm 1994; Hadrava 1995; Ilijic, Hensberge, & Pavlovski 2002) to analyze SBs. The SD techniqhe was also the topic of the talk by Petr Hadrava. SD aims to solve a different problem than TODCOR: whereas TODCOR assumes the spectra themselves are known or at least constrained, SD tries to estimate them as well as the orbital elements. Thus, the range of applicability is different for the two approaches: when the spectra are not well constrained, SD can be used to provide the individual spectral features. When they are relatively known, trying to estimate them would unnecessarily increase the errors of the orbital elements, and it would be better to use all the available information. Another important difference between the two techniques is the fact that TODCOR works on individual spectra, whereas SD needs a good coverage of orbital phases to obtain satisfactory results in estimating the spectra. Once again, this relates to the fact that SD tackles a much more difficult problem.

I concluded my talk by referring to a future application of SB2s in general, which is mass detemination through relativistic effects. We presented this idea in 2007 (Zucker & Alexander 2007), and it seems to be quite a challenge, as it requires very precise RVs of SB2s. Whether TODCOR is the right way to solve this problem or not, I believe it is a challenge that is worth exploring, as the return of having more dynamically determined stellar masses is significant.

References

Hadrava, P. 1995, *A&AS*, 114, 393
Ilijic, S., Hensberge, H., & Pavlovski, K. 2002, *Fizika B*, 10, 357
Konacki, M., Muterspaugh, M. W., Kulkarni, S. R., & Helminiak, K. G. 2009, *ApJ*, 704, 513
Konacki, M., Torres, G., Jha, S., & Sasselov, D. D. 2003, *Nature*, 421, 507
Mandushev, G., *et al.*, 2005, *ApJ*, 621, 1061
Mayor, M., Queloz D. 1995, *Nature*, 378, 355
Reipurth, B., Lindgren, H., Mayor, M., Mermillod, J.-C., & Cramer, N. 2002, *AJ*, 124, 2813
Schaefer, G. H., Simon, M., Prato, L., & Barman, T. 2008, *AJ*, 135, 1659
Simkin, S. M. 1974, *A&A*, 31, 129
Simon, K. P. & Sturm, E. 1994, *A&A*, 281, 286
Simon, M. & Prato, L. 2004, *ApJ*. 613, L69
Tonry, J. & Davis, M. 1979, *AJ*, 84, 1511
Torres, G., Claret, A., & Young, P. 2009, *ApJ*, 700, 1349
Torres, G., Latham, D. W., & Stefanik, R. P. 2007, *ApJ*, 662, 602
Zucker, S. 2003, *MNRAS*, 342, 1291
Zucker, S. & Alexander, T. 2007, *ApJ*, 670, 1326
Zucker, S. & Mazeh, T. 1994, *ApJ*, 420, 806
Zucker, S., Mazeh, T., Santos, N. C., Udry, S., & Mayor, M. 2003, *A&A*, 404, 775
Zucker, S., Torres, G., & Mazeh, S. 1995, *ApJ*, 452. 863

Discussion

P. HARMANEC: Would you agree that there is also another difference between the correlation technique and disentangling? The correlation technique performs better as the studied wavelength interval increases. The danger hidden in this is that if for some reason (e.g., systematic errors in wavelength calibration, unrecognized star spots affecting different lines differently etc.) the velocity amplitude differs from one line to another, you will not know that. For disentangling, you can investigate each stronger line separately (on the premise that you have enough spectra available) and see whether you are obtaining considerable results.

S. ZUCKER: I agree, in the sense that TODCOR and spectral disentangling are not applicable in the same cases. TODCOR is probably not the best approach for cases when one has to analyze individual lines separately.

A. PRŠA: Regarding the relativistic corrections, wouldn't it be more convenient to apply them to the RV curves that were already extracted, i.e., run TODCOR in a classical way and incorporate the corrections in an EB model?

S. ZUCKER: It was probably not clear in my talk. Indeed, there's no need to introduce relativistic effects at the stage of calculating the velocities. You first calculate the velocities in the regular way, and only introduce relativity at the orbital solution.

From Interacting Binaries to Exoplanets: Essential Modeling Tools
Proceedings IAU Symposium No. 282, 2011
Mercedes T. Richards & Ivan Hubeny, eds.
© International Astronomical Union 2012
doi:10.1017/S1743921311027864

The Long History of the Rossiter-McLaughlin Effect and its Recent Applications

Simon Albrecht

Massachusetts Institute of Technology, Kavli Institute for Astrophysics and Space Research,
Cambridge, MA 02139, USA

Abstract. In this talk I will review the Rossiter-McLaughlin (RM) effect; its history, how it manifests itself during stellar eclipses and planetary transits, and the increasingly important role its measurements play in guiding our understanding of the formation and evolution of close binary stars and exoplanet systems.

Keywords. techniques: spectroscopic, stars: rotation, planets and satellites: formation

1. Introduction

The Sun is the only star for which we can obtain detailed information on spacial scales much smaller than its diameter. For some nearby stars or giant stars, optical/infrared long baseline interferometry does give information on scales comparable to the stellar size (e.g. Baines *et al.* 2010). For most stars, however, we are not able to resolve their surfaces. These stars are essentially point sources, even for the biggest telescopes.

This is a pity as many questions in stellar astrophysics and astronomy would benefit from such knowledge. Astronomers have, therefore, developed a number of techniques to overcome this limitation. For example, Doppler imaging (Strassmeier 2002), polarimetry (see K. Bjorkman, these proceedings), or tomography (see M. Richards, these proceedings) let us gain under certain conditions information on small spatial features. Close binary star systems with orbits of only a few days or stars harbouring extrasolar planets (exoplanets) can provide us with an additional opportunity to obtain high spatial resolution, if the line of sight lies in the orbital plane. In such cases, eclipses or transits may be observed.

During eclipses or transits, telescopes integrate not over the complete stellar disk, as parts are hidden from view. Comparing the amount of light obtained at different phases of eclipses with the light received out of eclipse, system parameters like ratios of the radii of the two objects, orbital inclination, and possible inhomogeneities on the stellar surface of the background star, like star spots, can be determined (e.g., C. Maceroni, these proceedings).

What properties can be studied if we are not only to record the amount of light blocked from view, but also record the dimming as a function of the wavelength? In 1893, Holt realized that observing an eclipse with a spectrograph, which has a high enough spectral resolution to resolve stellar absorption lines, will lead to inside knowledge on stellar rotation (Holt 1893). Since stellar lines are broadened by Doppler shifts due to rotation, light emitted from approaching stellar surface areas is blue-shifted and light emitted from receding stellar surface areas is red-shifted. During the eclipse, parts of the rotating stellar surface is hidden, causing a weakening of the corresponding velocity component of the stellar absorption lines. Modeling of this spectral distortion reveals the projected stellar rotation speed ($v \sin i_\star$) and the angle between the stellar and orbital spins projected on the plane of the sky: the projected obliquity.[†]

A claim of the detection of the rotation anomaly was made by Schlesinger (1910), but more definitive measurements were achieved by Rossiter (1924) and McLaughlin (1924) for the β Lyrae and Algol systems, respectively. These researchers reported the change of the first moment of the absorption lines, sometimes called center of gravity, derived from the shape of the absorption line. Struve & Elvey (1931) reported the shape and its change during eclipse in the Algol system. The phenomenon is now known as the Rossiter-McLaughlin (RM) effect. Various aspects of the theory of the effect have been worked out by Hosokawa (1953), Kopal (1959), Sato (1974), Otha et al. (2005), Gimenez (2006), Hadrava (2009), Hirano et al. (2010) and Hirano et al. (2011a).

2. The RM effect and some quantities which can be measured with it

Holt (1893) realised that the rotation anomaly, occuring during eclipses, is an opportunity to measure $v \sin i_\star$ independently from a measurement of the width of absorption lines. Measuring $v \sin i_\star$ from line widths is challenging as these are influenced not only by rotation but also other processes, most notably by velocity fields on the stellar surface and pressure broadening. The strengths of these mechanisms are often not precisely known, introducing a substantial uncertainty in the $v \sin i_\star$ measurement even if the width of the line can be determined with high accuracy (e.g. Valenti & Fischer 2005). The amplitude of the RM effect is not as strongly influenced by these broadening mechanisms, making it an interesting tool for measuring $v \sin i_\star$ in particular cases (e.g. Twigg 1979, Worek et al. 1988, Rucinski et al. 2009). In addition, if differential rotation is present then it might be detected in fortunate cases via the RM effect (Hosokawa 1953, Hirano et al. 2011a). Currently, however, the RM effect is mainly seen as a tool to obtain the projection of stellar obliquity, an observable that is hard or impossible to measure otherwise.

However, not only stellar rotation can be studied. With the help of the differential RM effect, atmospheres of transiting planets may be studied (Snellen 2004, Dreizler et al. 2009). The RM effect might also aid in the search and confirmation of planet candidates (Gaudi & Winn 2007) or even exomoons (Simon et al. 2010). Also, accretion in an interacting binary might be studied via the RM effect (e.g. Lehmann & Mkrtichian 2004).

3. The RM effect and obliquities in extrasolar planetary systems

The properties of exoplanets discovered over the last few years have been very surprising. Many exoplanets orbit their hosts stars on eccentric orbits and giant planets have been found on orbits with periods of only a few days ('Hot Jupiters'). These findings present challenges for planet formation theories as it is thought that giant planets can only form at distances of several AU from their host stars, where the radiation is less harsh and small particles can survive long enough to build a rocky core which attracts the gaseous envelope from the disk.

Different classes of migration processes have been proposed which might transport giant planets from their presumed birthplaces inward to a fraction of an astronomical unit where we find them. Some of these processes are expected to change the relative orientation between the stellar and orbital spin (e.g. Nagasawa et al. 2008, Fabrycky & Tremaine 2007), while others will conserve the relative orientation (Lin et al. 1996), or

---

† This angle is denoted either β after Hosokawa (1953) or λ after Ohta et al. (2005), $\lambda = -\beta$.

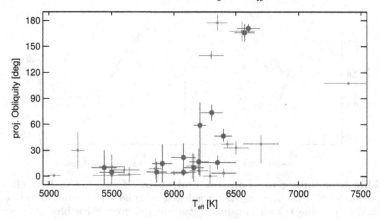

Figure 1. Hotter stars have oblique rotation. The projected obliquity of Hot Jupiter ($M_{planet} > 0.2$ $M_{\rm Jupiter}$; Period < 6 days) systems is plotted as function of the effective temperature of the host star. Using the measurements available 2010 (Winn *et al.* 2010a) noticed that systems with cool stars are aligned, while the obliquities of hot stars tends to be higher (gray small circles). Since then 16 new RM measurements have been reported (red large circles). The systems Kepler-8, CoRoT-1/11, have been omitted (see section 3.2) and the values for WASP-1/2 have been taken from Albrecht *et al.* (2011)

even reduce a possible misalignment (Cresswell *et al.* 2007). Therefore, measuring the obliquity of these systems will lead to inside knowledge of the formation and evolution of these systems.

3.1. *Results of RM measurements*

The first measurement of a projected obliquity in an extrasolar system was made by Queloz *et al.* (2000). They found that HD 209458 has an aligned spin. Over the following years, the angle between the stellar and orbital spins have been measured in about 30 systems. It was found that for some of these systems, the orbits are inclined or even retrograde with respect to the rotational spins of their host stars (see e.g. Hébrard *et al.* 2008, Winn *et al.* 2009, Triaud *et al.* 2010, Simpson *et al.* 2011). Winn *et al.* (2010a) found that close-in giant planets tend to have orbits aligned with the stellar spin if the effective temperature ($T_{\rm eff}$) of their host star is $\lesssim$ 6250 K and misaligned otherwise. Schlaufman (2010) obtained similar results measuring the inclination of spin axes along the line of sight. Winn *et al.* (2010a) further speculated that this might indicate that *all* giant planets are transported inward by processes which randomize the stellar spin. In this picture, tidal waves raised on the star by the close-in planet realign the two angular momentum vectors. The realignment time scale would be short for planets around stars with convective envelopes ($T_{\rm eff} \lesssim$ 6250 K), but long, compared to the lifetime of the system, if the star does not have a convective envelope ($T_{\rm eff} \gtrsim$ 6250 K). Over the last year, the RM effect was measured in another 16 systems, and the predictions made by Winn *et al.* (2010a) were confirmed for these systems (see Fig. 1)†.

3.2. *Challenges*

When analysing RM measurements, there are challenges which need to be overcome before a robust estimation of the stellar spin can be derived. Stellar rotation is not the only mechanism that affects the meaured stellar absorption lines. They are also broadened by stellar rotation fields and the point spread function of the spectrograph.

† Rene Heller maintains a webpage with updated information of obliquity measurements: `http://www.aip.de/People/rheller/content/main_spinorbit.html`

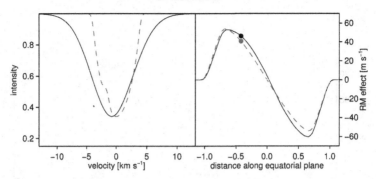

Figure 2. Line broadening mechanisms and their effect on the RM effect. The left panel shows a model of a absorption line broadened by solid body rotation only (red) dashed line and a model of a line taking also macro turbulence, convective blue shift and solar like differential rotation in account. The right panel shows the RM effect for both models. The circles indicate the transit phase when the snapshot of the absorption lines on the left side have been taken. On can see how the lines as well as the expected RM effect differ.

Lines are also not strictly symmetric due to the convective blue shift (Shporer & Brown 2011). See Fig. 2 for an illustration of this effect. In addition, it is not the line center that is measured (the quantity most often used by descriptions of the RM effect), but a cross correlation between a template and the spectrum recorded during transit (Hirano et al. 2011a). For the measurement process additional complications can arise.

- Similar to transit photometry, observations before and after transit are important. The RM effect needs to be isolated from other sources of RV variations (orbital movements, star spots, unknown companions, etc.). We, therefore, expect the uncertainty in the Kepler-8 system to be greater then reported by Jenkins et al. (2010).
- Analyzing low SNR RV data can lead to results which are systematically biased. This was the case for WASP-2 for which a retrograde orbit was reported by Triaud et al. (2010), but it was later found that from the currently available data no information on the obliquity can be derived. See Albrecht et al. (2011b) for details.
- For systems nearly edge on (i.e. low impact systems), there exists a strong degeneracy between $v \sin i_\star$ and the projected obliquity and care has to be taken when applying photometric and spectroscopic priors. This is the case for WASP-1 (Simpson et al. 2011, Albrecht et al. 2011b).

4. Eclipsing binaries

Although it has been more than 80 years since the first RM measurements in binaries, there are relatively few quantitative analyses of the RM effect in these systems. In the past, observing the RM effect was generally either avoided (as a hindrance to measuring accurate spectroscopic orbits) or used to estimate stellar rotation speeds. Almost all authors explicitly or implicitly assumed that the orbital and stellar spins were aligned. This lack of measurements is a pity as the knowledge of obliquity might guide our understanding of binary formation, in particular the formation of close binaries (e.g. Fabrycky & Tremaine 2007, Albrecht et al. 2011a).

There is a complication in the RM measurement relative to the low mass companion or exoplanet case, if one wants to measure the RM effect in double-lined binaries. Also, the foreground object emits light and contributes to the observed spectrum. Measuring the center of gravity of absorption lines would lead to erroneous results. Albrecht et al. (2007), therefore, developed a method to model the stellar absorption lines during occultations. A similar method was also employed in exoplanet systems (Collier Cameron et al. 2010).

The BANANA project (see Albrecht *et al.*, these proceedings) aims to measure the projected obliquities in a number of eclipsing binaries to understand what sets systems with spin-orbit alignment apart from systems where the spins are not aligned. They find that alignment is not a simple function of orbital separation or eccentricity.

Anther project led by Amaury Triaud aims to measure obliquities in binaries with F-star primaries and late-type secondaries (A. Triaud, these proceedings).

5. Outlook

The future for RM-measurements looks bright. Not only will the number of known eclipsing binaries and transiting exoplanets increase thanks to missions like *Kepler*, but these missions will also discover long-period systems and systems with multiple transiting planets. Also, the obliquities in systems with smaller planets, likely to have a different formation history, can be probed (Winn *et al.* 2010b, Hirano *et al.* 2011b). With an improved understanding of the RM effect, we might also be able to measure in a few systems some second order effects, as described above.

Stellar obliquities will also be measured by other techniques, like the method employed by Schlaufman (2010), which is not as accurate as RM measurements, but has the virtue that it does not require transit observations. For slowly rotating stars, the crossing of star-spots can be used as a tracer of stellar obliquity (e.g. Sanchis-Ojeda & Winn 2011). For fast rotating stars, which exhibit gravity darkening, the projected obliquity can be estimated from high-quality photometry (Szabo *et al.* 2011). Having very precise photometry further opens the possibility to measure obliquities via the photometric RM effect (Groot 2012, Shporer *et al.* 2012). Finally, optical interferometry is now able to measure the projected obliquity for some nearby systems (Le Bouquin *et al.* 2009). Therefore, there is the chance that our understanding of stellar obliquity, so far an elusive quantity, will be greatly improved over the coming years.

Acknowledgements

I am grateful to the scientific and local organising committees for organizing a very interesting and stimulating meeting. I am grateful to Josh Winn for valuable comments on the manuscript. I am thankful to Dan Fabrycky, Teru Hirano, John Johnson, Roberto Sanchis-Ojeda, Andreas Quirrenbach, Sabine Reffert, Johny Setiawan, Ignas Snellen, Willie Torres and Josh Winn for interesting discussions on stellar obliquities.

References

Albrecht, S., Reffert, S., Snellen, I., Quirrenbach, A., & Mitchell, D. S. 2007, *A&A*, 474, 565
Albrecht, S., Winn, J. N., Carter, J. A., Snellen, I. A. G., *et al.*, 2011a, *ApJ*, 726, 68
Albrecht, S., Winn, J. N., Johnson, J. A., Butler, *et al.*, 2011b, *ApJ*, 738, 50
Baines, E. K., Döllinger, M. P., Cusano, F., Guenther, E. W. *et al.*, 2010 *ApJ*, 710, 1365
Collier Cameron, A., Bruce, V. A., Miller, G. R. M., Triaud, A. H. M. J., *et al.*, 2010, *MNRAS*, 403, 151
Cresswell, P., Dirksen, G., Kley, W., & Nelson, R. P. 2007, *A&A*, 473, 329
Dreizler, S., Reiners, A., Homeier, D., & Noll, M. 2009, *A&A*, 499, 615
Fabrycky, D. & Tremaine, S. 2007, *ApJ*, 669, 1298
Gaudi, B. S. & Winn, J. N. 2007, *ApJ*, 655, 550
Giménez, A. 2006, *ApJ*, 650, 408
Groot, P. J. 2012, *ApJ*, 745, 55
Hadrava, P. 2009, arXiv:0909.0172

Hébrard, G., *et al.*, 2008, *A&A*, 488, 763

Hirano, T., Suto, Y., Winn, J. N., Taruya, A., *et al.*, 2011, *ApJ*, 742, 69

Hirano, T., Narita, N., Shporer, A., Sato, B., *et al.*, 2011, *PASJ*, 63, 531

Hirano, T., Suto, Y., Taruya, A., Narita, N., *et al.*, 2010, *ApJ*, 709, 458

Holt, J. R. 1893, *A&A*, 12, 646

Hosokawa, Y. 1953, *PASJ*, 5, 88

Jenkins, J. M., Borucki, W. J., Koch, D. G., Marcy, *et al.*, 2010, *ApJ*, 724, 1108

Kopal, Z. 1959, Close binary systems (The International Astrophysics Series, London: Chapman & Hall, 1959)

Le Bouquin, J., Absil, O., Benisty, M., Massi, F., *et al.*, 2009, *A&A*, 498, L41

Lin, D. N. C., Bodenheimer, P., & Richardson, D. C. 1996, *Nature*, 380, 606

Lehmann, H. & Mkrtichian, D. E. 2004, *A&A*, 413, 293

McLaughlin, D. B. 1924, *ApJ*, 60, 22

Nagasawa, M., Ida, S., & Bessho, T. 2008, *ApJ*, 678, 498

Ohta, Y., Taruya, A., & Suto, Y. 2005, *ApJ*, 622, 1118

Queloz, D., Eggenberger, A., Mayor, M., Perrier, C., *et al.*, 2000, *A&A*, 359, L13

Rossiter, R. A. 1924, *ApJ*, 60, 15

Rucinski, S. M. 2009, *MNRAS*, 395, 2299

Sanchis-Ojeda, R. & Winn, J. N. 2011, *ApJ*, 743, 61

Sato, K. 1974, *PASJ*, 26, 65

Schlaufman, K. C. 2010, *ApJ*, 719, 602

Schlesinger, F. 1910, Pub. of the Allegheny Observatory of the University of Pittsburgh, 1, 123

Shporer, A. & Brown, T. 2011, *ApJ*, 733, 30

Shporer, A., Brown, T., Mazeh, T., & Zucker, S. 2012, *New Astron.*, 17, 309

Simon, A. E., Szabó, G. M., Szatmáry, K., & Kiss, L. L. 2010, *MNRAS*, 406, 2038

Simpson, E. K., Pollacco, D., Cameron, A. C., & Hébrard, G., *et al.*, 2011, *MNRAS*, 414, 3023

Snellen, I. A. G. 2004, *MNRAS*, 353, L1

Strassmeier, K. G. 2002, *AN*, 323, 309

Struve, O. & Elvey, C. T. 1931, *MNRAS*, 91, 663

Szabó, G. M., Szabó, R., Benkő, J. M., Lehmann, H., *et al.*, 2011, *ApJ* (Letters), 736, L4

Triaud, A. H. M. J., Collier Cameron, A., Queloz, D., Anderson, D. R., *et al.*, 2010, *A&A*, 524, 25

Twigg, L. W. 1979, PhD thesis, Florida Univ., Gainesville.

Valenti, J. A. & Fischer, D. A. 2005, *ApJS*, 159, 141

Winn, J. N., Fabrycky, D., Albrecht, S., & Johnson, J. A. 2010a, *ApJ* (Letters), 718, L145

Winn, J. N., Johnson, J. A., Howard, A. W., Marcy, G. W., *et al.*, 2010b, *ApJ* (Letters), 723, L223

Winn, J. N., Johnson, J. A., Albrecht, S., Howard, A. W. *et al.*, 2009, *ApJ* (Letters), 703, L99

Worek, T. F., Zizka, E. R., King, M. W., & Kiewiet de Jonge, J. H. 1988, *PASP*, 100, 371

Discussion

P. BONIFACIO: Since you translated effective temperatures to masses in your Teff-Obliquity relation, I assume all your stars are dwarfs. If the physical parameter determining the trend is really mass, you should be able to find some cool massive giants with a high-obliquity planet.

S. ALBRECHT: That is correct. We only have R-M (Rossiter-McLaughlin) measurements for dwarf stars. Unfortunately, it is very difficult to detect transiting planets around giants as the radius ratio is so big. Also the R-M measurements would be very difficult.

From Interacting Binaries to Exoplanets: Essential Modeling Tools
Proceedings IAU Symposium No. 282, 2011
Mercedes T. Richards & Ivan Hubeny, eds.
© International Astronomical Union 2012
doi:10.1017/S1743921311027876

The Rossiter-McLaughlin Effect for Planets and Low-Mass Binaries

Amaury H. M. J. Triaud

Observatoire Astronomique de l'Université de Genève, Chemin des Maillettes, 51, CH-1290
Sauverny, Switzerland
email: `Amaury.Triaud@unige.ch`

Abstract. The Rossiter-McLaughlin effect occurs during the eclipse or transit of an object in front of another one. In our case, it appears as an anomaly on the radial velocity Doppler reflex motion. The modelling of that effect allows one to measure the sky-projected angle between the rotation spin of the primary and the orbital spin of the secondary. In the case of exoplanets, it gave clues about the formation of the *hot Jupiters*. In this paper, I will talk about how the data are acquired, how models are adjusted to them, and which results have been made.

Keywords. planetary systems, binaries: eclipsing, binaries: spectroscopic, methods: data analysis.

1. Introduction

The discovery of the first hot Jupiter, 51 Peg b, by Mayor & Queloz (1995) was a surprise. Gas giants such as this planet were supposed to form beyond the so-called snow-line, where ices and dust can coagulate to form cores large enough to attract a large amount of gas onto them (e.g. Pollack *et al.* 1996 and Alibert *et al.* 2005). Orbital migration had to happen.

Two main pathways were quickly proposed. One required planet-disc interactions which apply a torque on the planet leading to inward migration. The ideas have been put forward in Goldreich & Tremaine (1980), Lin *et al.* (1996) and Ward (1997). Since it was first proposed, the study of disc migration has progressed a lot. One of its main predictions is that planets stay in the plane of the protoplanetary disc, which is aligned with the equatorial plane of its star. The second pathway relied on multi-body gravitational interactions as outlined in Rasio & Ford (1996) for the case of planet-planet interactions. The main point is that a planet is placed via those interactions on an inclined, highly eccentric orbit. At periastron passage, tidal dissipation leads to angular momentum loss which shrinks the semimajor axis and circularises the planet's orbit (see Fabrycky & Tremaine 2007 or Nagasawa *et al.* 2008).

In order to estimate the stellar rotation, Holt (1893) proposed that a colour effect should be visible when two stars eclipse each other. This colour effect was measured as a radial velocity anomaly by Rossiter (1924) and McLaughlin (1924) (see also Simon Albrecht's contribution in these proceedings). This effect was modelled notably by Kopal (1942) and Hosokawa (1953) using the α functions.

In the context of extrasolar planets, the Rossiter-McLaughlin effect was first observed by Queloz *et al.* (2000). Since then, a fast growing number of measurements have been collected which are refining our ideas about the formation of hot Jupiters. In order to interpret the data correctly, one has to build a statistical picture. This requires a large number of subjects. Hot Jupiters are rare planets: they occur around about one star in 200 (eg. Naef *et al.* 2005 and Howard *et al.* 2011). Fortunately, wide angle ground-based

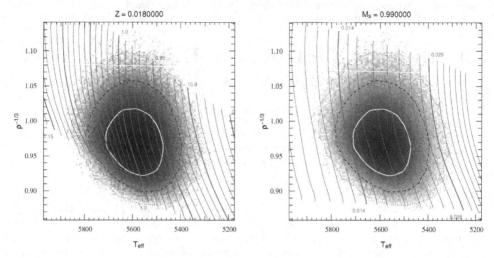

Figure 1. The joint probability distribution, output of an MCMC interpolating within the Geneva stellar evolution tracks in a modified H-R diagram using $\rho_\star^{-1/3}$ against $T_{\rm eff}$. The aim is use the property that we measure $\rho_\star^{-1/3}$ from a planetary transit to estimate the planet's host mass and age. 1, 2 and 3 σ confidence intervals are overplotted. On the left we see the result on mass tracks of constant metallicity, on the right we see tracks for a constant mass at different metallicities. Thus our host star, here WASP-8, is about $0.99 \pm 0.05\,M_\odot$.

transit surveys have now found more than 100 such objects. One of those is the WASP survey (Pollacco *et al.* 2006). The aim of the WASP survey is to find transiting planets on stars bright enough so one could study them more in detail. The survey aimed at stars between V = 9 and V=13. Across the years, we have found close to 60 planets. In addition, in our search for transiting -eclipsing- brown dwarfs, we have detected a large number of low-mass eclipsing binaries consisting mostly of F+M and G+M pairs.

2. Characterising planets

The photometric transit is modelled using Mandel & Agol (2002) or Giménez (2006a). The Rossiter-McLaughlin effect is adjusted using the formalism of Giménez (2006b) based on Kopal's α functions. The adjustment of those models to the data is done using a Markov Chain Monte Carlo algorithm (MCMC). This is a powerful method which allows us to solve for highly non-linear models, on datasets combining multi-band, multi-instrument, multi-epoch photometry and spectroscopy. The interest of such a combined analysis is to produce the best possible compromise between all datasets, which describe the same physical phenomenon, observed differently. For instance, long time-base photometry will very precisely determine the period and phase of the object we will study. When adjusting the radial velocity, there really is only one parameter which is effectively variable: the semi-amplitude. This allows us to refine our error bars on that parameter and, consequently, on the mass of our planet. Similarly, the photometric transit will determine the ratio of radii and impact parameter, so when a model is adjusted for the Rossiter-McLaughlin effect, the fit only varies two quantities: the $V \sin I$ and the spin/orbit angle β. This allows us to detect lower signal-to-noise since we have fewer free parameters. By combining all information, we make sure that all parameters are consistent with each other and with the entire dataset.

Another interest of the MCMC is that it produces probability density distributions marginalised for each parameter instead of χ^2 maps, which makes it very easy to produce

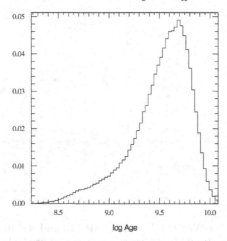

Figure 2. Marginalised probability density function for stellar age, estimated from the distribution over the stellar evolution tracks presented in Figure 1. WASP-8 is 5 ± 2.5 Gyr old.

robust confidence intervals. The concept of the MCMC is well-described in Gregory (2005). They have been in use by the WASP team and are presented in papers such as Cameron *et al.* (2007), Gillon *et al.* (2009), and Triaud *et al.* 2009.

Sozzetti *et al.* (2007) remarked that a planetary photometric transit measures directly the mean density of the host star, $\rho_\star$. This parameter can be used instead of the more traditional log g to interpolate within stellar evolution tracks in order to get an estimate of the stellar mass $M_\star$. To do that, one also needs the T_{eff} and the metallicity Z. This process has now been included in the MCMC in the same spirit as before: that of getting the best compromise out of the data. For each step that the MCMC proposes, the Geneva stellar evolution tracks (Mowlavi *et al.* 2012) are interpolated in $(\rho_\star^{-1/3}, Z, T_{\mathrm{eff}})$ space in order to determine $M_\star$ and the stellar age (Triaud 2011). We thus end up with probability density functions for those two values. Errors in the models can also be accounted for. An example is given in Figure 1 and an age estimate given in Figure 2. We see that the tracks are evolving almost parallel to the value of $\rho_\star^{-1/3}$. This means that for a given T_{eff} and Z, the stellar age is almost solely determined by $\rho_\star^{-1/3}$. In the case presented in the figures, the fit included only one high-precision photometric transit. Let's note here that GAIA, through distance measurements, will produce independent stellar radii, which added as priors in the above analysis should improve our accuracy and precision on stellar masses and ages.

3. The spin/orbit angle survey

Having detected a large number of planets, the WASP consortium endeavoured to learn the most possible about them. Part of that effort was to attempt to measure systematically the Rossiter-McLaughlin effect for each of our planets mostly using the HARPS spectrograph on the 3.6m at La Silla, Chile. This led to a series of papers which showed that hot Jupiters are located on a large variety of angles, including some on retrograde orbits, instead of being coplanar as previously assumed. It also showed that tidal circularisation timescales are shorter than the realignment timescale as planets on circular, retrograde orbits were found. Some of the measurements and an interpretation of those results are presented in Triaud *et al.* 2010. Two effects are also shown in Figure 3.

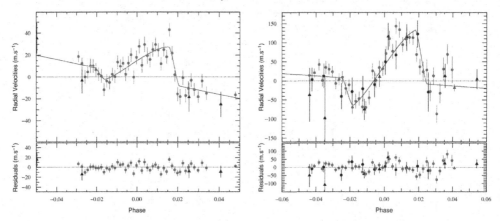

Figure 3. The R-M effect for WASP-15b and WASP-17b and residuals as appearing in Triaud *et al.* 2010. Circles are HARPS observations, triangles are CORALIE observations.

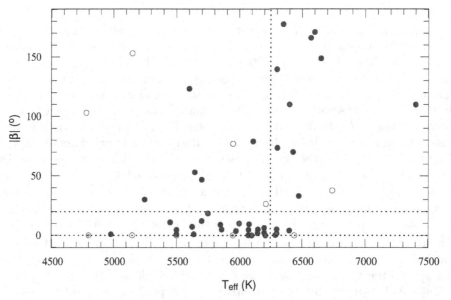

Figure 4. Sky-projected spin/orbit angle β against T_{eff}. This plot is an update from that presented in Winn *et al.* (2010). Filled symbols show secured measurements. Open symbols are numbers likely to change dramatically due to low signal-to-noise detections or other issues. Little dots within symbols show as of yet unpublished measurements.

Winn *et al.* (2010) showed that when the spin/orbit angle β (called λ in their paper) is plotted agains the star's T_{eff}, a structure appears in the data. Aligned planets around hot stars are rare. Since that paper was published, a number of measurements have been made. An update of that plot is presented in Figure 4. With almost twice more data than when originally proposed, the pattern is essentially confirmed.

In addition to the planets that were discovered with WASP, we started characterising the orbits of single line F+M and G+M eclipsing binaries. We have now close to 50 well-sampled orbits. On 15 of those, we have also measured the Rossiter-McLaughlin effect using the CORALIE spectrograph, mounted on the Swiss 1.2 m *Euler* telescope at La

Silla. The same tools used for analysing planets have been used, since those eclipsing M-dwarfs have sizes very similar to planets ($< 2\,R_{\mathrm{Jup}}$) while remaining very faint compared to their primary. So far, all spin/orbit angles have been found aligned in a marked difference with the planet population. According to Fabrycky & Tremaine (2007) (and references there-in), high mass ratio binaries such as our F+M or G+M pairs would form via Kozai-Lidov oscillations caused by a tertiary leading the M component on an inclined, eccentric orbit which would then decay due to tidal friction. If this scenario is justified, our observations show that orbital realignment is shorter than orbital circularisation, *a contrario* with the planet case. All this will soon be published in a series of papers.

Acknowledgements

This is work conducted under supervision by my PhD surpervisor, Didier Queloz, and in collaboration with a large number of people, notably Andrew Collier Cameron, Don Pollacco, Coel Hellier, Michaël Gillon, David Anderson, Barry Smalley, Nami Mowlavi and Leslie Hebb, to name only a few.

References

Alibert, Y., Mordasini, C., Benz, W., & Winisdoerffer, C. 2005, *A&A*, 434, 343
Bayliss, D. R., Winn, J. N., Mardling, R. A., & Sackett, P. D. 2010, *ApJ*, 722, L224
Cameron, A. C., Bouchy, F., Hébrard, G. *et al.*, 2007, *MNRAS*, 375, 951
Fabrycky, D. & Tremaine, S. 2007, *ApJ*, 669, 1298
Gillon, M., Smalley, B., Hebb, L. *et al.*, 2009, *A&A*, 496, 259
Giménez, A. 2006, *A&A*, 450, 1231
Giménez, A. 2006, *ApJ*, 650, 408
Goldreich, P. & Tremaine, S. 1980, *ApJ*, 241, 425
Gregory, P. C. 2005, in *Bayesian Logical Data Analysis for the Physical Sciences*
Holt, J. R. 1893, *A&A*, XII, 646
Hosokawa, Y. 1953, *PASJ*, 5, 88
Howard, A. W., Marcy, G. W., Bryson, S. T. *et al.*, 2011, *ApJ*, submitted
Kopal, Z. 1942, *ApJ*, 96, 399
Lin, D. N. C., Bodenheimer, P., & Richardon D. C. 1996, *Nature*, 380, 606
Mandel, K. & Agol, E. 2002, *ApJ*, 580, L171
Mayor, M. & Queloz, D. 1995, *Nature*, 378, 355
McLaughlin, D. B. 1924, *ApJ*, 60, 22
Moutou, C., Rodrigo, R. F., Udry, S. *et al.*, 2011, *A&A*, submitted
Mowlavi, N., Eggenberger, P., Meynet, G. *et al.*, 2012, *A&A*, submitted (arXiv:1201.3628)
Naef, D., Mayor, M., Beuzit, J.-L. *et al.*, 2005, in *Proceedings of the 13th Cambridge Workshop on Cool Stars*, 833
Nagasawa, M., Ida, S., & Bessho, T. 2008, *ApJ*, 678, 498
Pollack, J. B., Hubickyj, O., Bodenheimer, P. *et al.*, 1996, *Icarus*, 124, 62
Pollacco, D., Skillen, I., Cameron, A. C. *et al.*, 2006, *PASP*, 118, 1407
Queloz D., Eggenberger, A., Mayor, M. *et al.*, 2000, *A&A*, 359, L13
Rasio, F. A. & Ford, E. 1996, *Science*, 274, 954
Rossiter, R. A. 1924, *ApJ*, 60, 15
Sozzetti, A., Torres, G., Charbonneau, D. *et al.*, 2007, *ApJ*, 664, 1190
Triaud, A. H. M.. J., Queloz, D., Bouchy, F. *et al.*, 2009, *A&A*, 506, 377
Triaud, A. H. M.. J., Collier Cameron, A., Queloz, D. *et al.*, 2010, *A&A*, 524, 25
Triaud, A.H.M.J., *PhD Thesis* submitted
Ward, W. R. 1997, *Icarus*, 126, 261
Winn, J., Fabrycky, D., Albrecht, S. *et al.*, 2010, *ApJ*, 178, L145

Discussion

A. VIDOTTO: You mention that M dwarfs transiting/eclipsing in binary systems show $\beta \sim 0$. Are there similar measurements concerning transiting brown dwarfs?

A. TRIAUD: Yes. There are four transiting brown dwarfs: CoRoT-3b, CoRoT-15b, WASP-30b and LHS-6343c. There are two Rossiter-McLaughlin effect measurements, to my knowledge: CoRoT-3b which is found slightly misaligned but is probably aligned - the data are badly sampled due to the faintness of the target (see Triaud et al. 2009); and WASP-30b, which is found aligned (Triaud et al. in prep).

W. KLEY: Is there any correlation between planetary mass and misalignment angle?

A. TRIAUD: Yes. At the moment no planet $> 5\,M_{\mathrm{Jup}}$ is found retrograde (Moutou et al. submitted). Following on the previous comments, the known brown dwarfs and M dwarfs eclipsing F and G primaries are all found aligned at the moment, regardless of the eccentricity, spectral type of the primary or the orbital periods ranging up to 15 days. All those combined point to a correlation with mass.

E. BUDDING: Some of the low mass, rapidly rotating cool dwarfs should be affected by spots. Since the Rossiter-McLaughlin effect will be following the motion of the centre of light for the eclipsed component, won't the question of possible misalignment of the rotation axis of the eclipsed star become complicated by the presence of such spots?

A. TRIAUD: Spots can induce two effects: the first is when the planet crosses a spot, then, since the Rossiter-McLaughlin effect is a change in the light centroid, the planet is being hidden and we lose that information. This, if the planet covers exactly a spot, we lose the radial velocity anomaly and recover the Keplerian velocity that is observed in the out-of-transit data. The second case is when spots cover the star inhomogeneously. Let's imagine we have one spot, located in the blue-shifted hemisphere, out of the transit chord. Then, the planet will hide a comparatively larger area on the blue-shifted hemisphere, compared to the red-shifted hemisphere. This would lead to an asymmetry in the Rossiter-McLaughlin effect, which could be confused for a misalignment. Such an occurrence may be responsible for the different results obtained independently on WASP-17 (see Triaud et al. 2010 and Bayliss et al. 2010).

E. GUINAN: Is there a relation between orbital alignment and time?

A. TRIAUD: Yes. This is work which is still unpublished, while being submitted to A&A. I am reluctant to talk publicly about it, preferring to wait for the referee to give some feedback. To answer the question, I will point out that hot Jupiters found around more massive stars are younger and detectable only for a shorter time (after the star is too big) than those around lower mass stars. Therefore, the observed misaligned systems (found primarily around hot stars) are on average younger systems than the aligned systems which are found around the colder stars.

From Interacting Binaries to Exoplanets: Essential Modeling Tools
Proceedings IAU Symposium No. 282, 2011 © International Astronomical Union 2012
Mercedes T. Richards & Ivan Hubeny, eds. doi:10.1017/S1743921311027888

Period Analyses Without O-C Diagrams

Zdeněk Mikulášek[1,2], Miloslav Zejda[1] and Jan Janík[1]

[1] Department of Theoretical Physics and Astrophysics, Masaryk University,
Kotlářská 2, CZ 611 37 Brno, Czech Republic
email: mikulas@physics.muni.cz

[2] Observatory and Planetarium of J. Palisa, VŠB – Technical University,
Ostrava, Czech Republic

Abstract. We present a versatile method appropriate for the period analyses of observations containing phase information of all kinds of periodic or nearly periodic variable stars on the basis of phenomenological modelling of their phase curves and phase functions. The approach is based on rigorous application of a non-linear weighted least-squares method exploiting all available observational data and does not need an O-C diagram as an intermediate stage for period analyses. However, this approach enables us to determine precise times of extrema of light curves, to calculate ephemerides and construct plausible O-C diagrams. We substantiate the general applicability of the method on eclipsing binaries research.

Keywords. methods: statistical, binaries: eclipsing, variable stars: other

1. Introduction

Most variable stars shows periodic or cyclic changes. Light, spectroscopic or spectropolarimetric changes repeat with one period P or multiple periods. The basic mechanisms of the periodic or nearly periodic stellar variability are the rotation of an anisotropically radiating star, the orbital motions of components in stellar systems, and pulsations or oscillations of various kinds. Basic period analysis consists of finding a reliable ephemeris of the main periodic variation and modelling of the first order effects.

This should be followed by a refined period analysis providing additional pieces of information on the variable star physics, e.g.: 1) Further bodies in the system (stars, planets). 2) Mass exchange between components of interacting binaries. 3) Spin-orbital interactions in the system. 4) Inner stellar structure and general relativity testing by means of the apsidal motion rate. 5) Angular momentum loss through stellar winds. 6) Gravitational waves. 7) Oscillations in the rotation of some chemically peculiar stars. 8) Differential rotation of cool stars.

These period variations are usually delicate; consequently a refined tool is needed for their analysis!

2. The standard O-C diagram method and its afflictions

The majority of period analyses were done using O-C diagrams or their modifications. O is the observed time of a light curve (LC) extremum, C is the time of the same extremum calculated by means of an ephemeris (as a rule linear). Historical O times and their uncertainties can be taken from the literature or specialised databases.

The credibility of obtained astrophysical information strictly depends on the reliability of O times and their uncertainties. Extrema of LCs used to be rather badly defined and even small imperfections in observations in the vicinity of the extremum may cause large deflections in the individual O determinations. The majority of published times of variable stars' extrema were obtained by means of the notorious Kwee-van Woerden method and its modifications. Unfortunately, the method is often used as a black box in an inappropriate way for data where authors themselves ruled out usage of this method. Furthermore, we know from our experience that this method exploits only a minor part of the information hidden in observations that results in the enlargement of the O scatter by two or three times. Quite unusable is the uncertainty estimate which used to be several times underestimated.

The crucial affliction of O values determined by standard methods is based on the fact that they do not take into account that most variable star changes are repeating. Knowledge of the form of light curves derived from past observations can enhance the credibility of O determinations considerably. Then, we can use the method of the phase shift of an observed light curve (LC) with respect to a template LC which provides more reliable results and estimates of the uncertainties. Unfortunately, such methods are utilised relatively rarely.

Standard O-C diagrams are not a very reliable tool for period analysis since they have their specific limitation. What shall we do to improve the reliability of such demanding period analyses? We can improve their input using reliable O values using the periodicity LC. But even better: We can cancel the O-C intermediate step completely! We can analyse period variations directly using original observations of all kinds!

The need for original observational data (magnitudes) is the only limitation of the method. In several instances, we are unable to obtain the original data, so we have to be satisfied with the published O value.

3. Method of direct period analyses and modelling of period variations

The techniques used to analyse the data are based on the rigourous application of the non-linear, weighted, least-squares methods used simultaneously for all relevant data containing phase information. Our technique does not utilise an O-C diagram as an intermediate stage of data processing; O-C diagrams are used only as a visual check on the adequacy of the models. The method can be successfully applied to continuous time series and surveys like ASAS or Hipparcos.

Let us assume that all observed *phase curves* of a star are well-described by the unique general model function $F(\vartheta, \mathbf{a})$, described here by g_a parameters contained in a parameter vector $\mathbf{a}$, $\mathbf{a} = (a_1, ..., a_j, ..., a_{g_a})$. In our computation, we assume that the form of all the phase variations is constant and the time variability of the observed quantities are given by a *phase function* $\vartheta(t, \mathbf{b})$, which is a continuous monotonic function of time t. The fractional part of it corresponds to the common phase, the integer part of it being the so called epoch (E). We can express the phase function by means of a simple model quantified by g_b parameters $\mathbf{b}$, $\mathbf{b} = (b_1, ..., b_k, ..., b_{g_b})$. The instantaneous period $P(t)$ is connected with the phase function by the following simple relations:

$$P(t, \mathbf{b}) = \left(\frac{d\vartheta}{dt}\right)^{-1} , \quad \Rightarrow \quad \vartheta(t, \mathbf{b}) = \int_0^t \frac{d\tau}{P(\tau)}. \tag{3.1}$$

For a realistic modeling of phase variations for all data types, we need g_a free parameters for the description of the model function $F(\vartheta, \mathbf{a})$, and g_b free parameters for the description of the phase function $\vartheta(t, \mathbf{b})$. The computation of the free parameters is iterative under the basic condition that the weighted sum $S(\mathbf{a}, \mathbf{b})$ of the quadrates of the difference Δy_i of the observed value y_i and its model prediction is minimal (w_i being the individual weight of the i-th measurement).

$$\Delta y_i = y_i - F(\vartheta_i); \quad S = \sum_{i=1}^{n} \Delta y_i^2 w_i; \quad \delta S = \mathbf{0}; \quad \Rightarrow \qquad (3.2)$$

$$\sum_{i=1}^{n} \Delta y_i \, \frac{\partial F(\vartheta_i, \mathbf{a})}{\partial a_j} \, w_i = 0; \quad \sum_{i=1}^{n} \Delta y_i \, \frac{\partial F}{\partial \vartheta_i} \, \frac{\partial \vartheta(t_i, \mathbf{b})}{\partial b_k} \, w_i = 0.$$

We obtain here $g = g_a + g_b$ equations of g unknown free parameters. The weights of individual measurements w_i are inversely proportional to their expected uncertainty. The system of equations is always non-linear; we have to determine the parameters iteratively. Nevertheless, with a good initial estimate of the parameter vectors a and b, the iterations converge very quickly. Usually, we need only several tens of iterations to complete the whole iteration procedure.

3.1. *Application to eclipsing binaries*

Most of the studies dealing with orbital period changes of eclipsing variables are based on the O-C diagram techniques, even when several advanced physical codes for EB light curves solutions as PHOEBE, FOTEL or Wilson-Devinney codes offer direct period analysis in the sense of the above outlined method. We recommend the use of these utilities as you can eventually at least use the simulated light curve of an EB as a template phase curve.

Even more simple is to use the phenomenological models of the light curve. We developed a bundle of EB LC models which are able to describe the majority of real cases more than satisfactorily using a minimum of free parameters. For example, the following phenomenological model can be applied to close eclipsing binaries with zero eccentricity (minima at phases $\varphi = 0, 0.5$). The model satisfactorily well describes the primary and secondary eclipse of various forms (even U shape minima), proximity effects, and O'Connell's effect using only seven free parameters.

$$y(t) \simeq y_0 + a_1 \left\{ 1 - \left\{ 1 - \exp\left[-(\varphi_{\mathrm{I}}/d)^2 \right] \right\}^C \right\} + a_2 \left\{ 1 - \left\{ 1 - \exp\left[-(\varphi_{\mathrm{II}}/d)^2 \right] \right\}^C \right\}$$
$$+ a_3 \cos(4\pi\vartheta) + a_4 \left[\sin(2\pi\vartheta) - \tfrac{1}{2}\sin(6\pi\vartheta) + \tfrac{1}{10}\sin(10\pi\vartheta) \right], \qquad (3.3)$$

where

$$\vartheta = (t - M_0)/P; \quad \varphi_{\mathrm{I}} = \vartheta - \mathrm{round}(\vartheta); \quad \varphi_{\mathrm{II}} = \vartheta - \mathrm{floor}(\vartheta) - \tfrac{1}{2}. \qquad (3.4)$$

y_0 is the mean magnitude outside of eclipses, a_1 and a_2 are the depths of the primary and the secondary eclipses, d is the parameter describing the width of eclipses, C determines the sharpness of eclipses, a_3 is the semi-amplitude of changes outside eclipses and a_4 is the amplitude of the O'Connell effect.

4. Instead of conclusions - something for O-C diagram lovers

It would be unreasonable to abandon O-C diagrams completely. The O-C diagram can help us even if we use the method of direct period analyses, namely in the stage of finding adequate models for the the period changes. Fortunately, we can create the classical O-C diagram using so called virtual O-C values for any subsets of observational data. In addition, we can easily construct a diagram illustrating changes of the instant period $P(t)$.

Using the residuals of the observed data Δy_i, we can create individual values of the phase shifts expressed in days $(O-C)_j$ with adapted individual weight W_j for each observed datum, and averages of the phase shifts defined for arbitrarily selected groups of measurements $\overline{\text{(O-C)}}_i$, or deflection of the mean period from the instant model period $\Delta P_k(t_k)$:

$$\text{(O-C)}_j = -P(t_j)\,\Delta y_j \left(\frac{\partial F}{\partial \vartheta}\right)^{-1}; \quad W_j = \left(\frac{\partial F}{\partial \vartheta}\right)^2 w_j; \qquad (4.1)$$

$$\overline{\text{(O-C)}}_k = \frac{\sum_{j=1}^{n_k} \text{(O-C)}_j\, W_j}{\sum_{j=1}^{n_k} W_j}; \quad \Delta P_k = \frac{\sum_{j=1}^{n_k} \text{(O-C)}_j\, \vartheta_j\, W_j}{\sum_{j=1}^{n_k} \vartheta_j^2\, W_j}.$$

Computations of $\overline{\text{(O-C)}}_k$ and ΔP_k follow after finding of the model parameters; consequently, they have no influence on the model solution. They are used only to visualise the solution. Similarly, we can compute virtual 'observed' values of the instant period from a group of observations to generate the model curves in our figures.

Acknowledgements

This work was supported by the grants GAAV IAA 301630901, GAČR 205/08/0003, MEB 0810095, and MUNI/A/0968/2009. The authors thank S. N. de Villiers for his kind improvement of the manuscript language.

Discussion

C. Chambliss: O-C diagrams are sometimes misused. Sinusoidal O-C diagrams have been used in some cases to infer the existence of third light components that later have proven not to exist. [AK Herculis is an example.] Also how good are old photographic or visual estimates?

Z. Mikulášek: In general, visual estimates of eclipsing binary minima timings are not a very reliable source of astrophysical information because of the subjective character of observation and processing. The credibility of visual observations could be enhanced if we treat the complete time series of visual estimates of brightness. Such observations could be processed by the method of direct period analysis. Unfortunately, we generally only have estimates of times of minima at our disposal, not original observation data.

On the contrary, time series of photographic surveys used to be a relatively solid source, namely if we have at our disposal several hundreds of magnitudes individually derived from photographic plates. Such material is optimal for direct period analysis. Times of minima determined from several measurements are relatively reliable. However, we must be suspicious in the case of so called 'weakenings' which are not the time of the photographic minimum but only information that the star is dimmer than usual at a given moment. Such estimates of times of minima used to be even worse sources of information than visual minima time estimates.

R. Wilson: I like this scheme very much. The procedure has been in the WD program for about 12 years, and has been applied in several papers by myself and by others. As formulated in WD, it can supply the usual ephemeris parameters (reference time, period (P), and dP/dt) in combination with apsidal motion ($d\omega/dt$), third body parameters, and all other EB parameters. It can be used with mixed light and velocity curves so as to fill gaps in coverage.

Z. Mikulášek: We thank you for this note; we hope that our paper will contribute so that period analyses will be done more properly.

From Interacting Binaries to Exoplanets: Essential Modeling Tools
Proceedings IAU Symposium No. 282, 2011
Mercedes T. Richards & Ivan Hubeny, eds.
© International Astronomical Union 2012
doi:10.1017/S174392131102789X

Renormalization of KOREL-Decomposed SB2 Spectra

Holger Lehmann[1] and Andrew Tkachenko[2]

[1] Thüringer Landessternwarte Tautenburg, D-07778 Tautenburg, Germany
email: lehm@tls-tautenburg.de

[2] Instituut voor Sterrenkunde, Celestijnenlaan 200B, B-3000 Leuven, Belgium
email: andrew@ster.kuleuven.be

Abstract. We introduce a new computer program that is able to determine the flux ratio between the components of SB2 stars from time series of composite spectra. The spectra of the components are decomposed using the KOREL program and compared to synthetic spectra computed on a grid of fundamental stellar parameters. In the result, we obtain the optimized stellar parameters together with the flux ratio. The program is tested and applied to the oscillating eclipsing binary KIC 10661783 observed by the Kepler satellite mission.

Keywords. methods: data analysis, stars: binaries: spectroscopic, stars: binaries: eclipsing

1. The method and its test

Like other programs for spectral disentangling, the Fourier-transform-based KOREL program (Hadrava 1995) computes the decomposed spectra of multiple stellar systems normalized to the common continuum of all components. One needs information about the continuum flux ratio between the components to be able to renormalize the decomposed spectra accordingly. There exists already one method called GENFITT (Tamajo *et al.* 2011) that computes the flux ratio directly from the spectra itself using a genetic algorithm. We show in this investigation that it is possible to determine the wavelength dependent flux ratio between the two stars of a binary directly from a comparison between observed and synthetic spectra by solving a simple linear system.

We use a library of synthetic spectra computed on a grid in T_{eff}, $\log g$, [M/H], and $v \sin i$ (Lehmann *et al.* 2011) and define a χ^2 that measures the combined accuracy from comparing the decomposed spectra separately with the synthetic ones. Assuming a linear trend of the flux ratio of star k with wavelength, $f_k = f_{k,0} + f_{k,1}\lambda$, and setting the partial derivatives of χ^2 with respect to the $f_{k,j}$ to zero, one gets a system of linear equations that can directly be solved. A detailed derivation will be given in a forthcoming paper.

The accuracy of the derived parameters will depend on the S/N of the composite spectra, the flux ratio and the $v \sin i$ of the components, and on the accuracy of the continuum normalization of the observed spectra. We tested the program on artificial spectra computed for two components of [M/H]=0.1 and T_{eff}=7900/6500 K, $\log g$=3.6/4.3, and $v \sin i$=80/50 km s^{-1} for the primary/secondary, respectively. For the flux ratio of the secondary, we used 0.1 and 0.5. The local continuum of the observed spectra is not *a priori* known. Thus, we compared the results obtained for a fixed (true) continuum of the composite spectra and for a continuum that was corrected by adapting it to those of the synthetic spectra as it is the preferred method in practice. The calculations were done on the wavelength range from 3850 to 5670 Å, the different S/N (100 to 400) have been realized by adding scaled, Poisson-distributed noise to the artificial spectra.

The results show that our program works well for high-S/N spectra. It computes well-constrained flux ratios (error below 2%) in all test cases. The influence of the accuracy of

Table 1. Comparison of fundamental parameters of KIC 10661783. $T_{\text{eff}\,2}$ and $\log g_2$ are given with reflection, the spectroscopic flux ratio f_2/f_1 without/with reflection.

		Kepler light curve		spectroscopic & combined		
P	(d)	1.2313622(2)	1.2313622 fixed	$\log g_{1/2}$		3.6(1)/4.3(5)
T	(K)	2455065.77701(5)	2455065.778(6)	$M_{1/2}$	$(M_\odot)$	2.1(2)/0.18(2)
q		0.0626(8)	0.089(3)	$v_{1/2} \sin i$	(km s^{-1})	81(4)/48(4)
i	(°)	82.8(2)		$R_{1/2}$	$(R_\odot)$	2.0(1)/0.88(5)
f_2/f_1		0.086(9)	0.074(4)/0.03(3)	a	$(R_\odot)$	6.3(1)
$T_{\text{eff}\,1/2}$	(K)	8000/6500 (assumed)	7890(25)/6760(720)			

the continuum normalization is remarkable. For the faint secondary ($f=0.1$), the accuracy in T_{eff} and $\log g$ drops down by a factor of two if we allow for adjusting the continuum as we have to do in practice. This effect will be the stronger, the more the spectra are blended, i.e. the bluer the spectral region and the later the spectral type. In our case, we need at least S/N of 200 to determine its T_{eff} within ± 200 K and its [M/H] within ± 0.15 dex and S/N of 400 for its $\log g$ within ± 0.25 dex. When both components are of about the same brightness, T_{eff} can be determined by ± 70 K for S/N$\geqslant$100, $\log g$ by ± 0.1 for S/N$\geqslant$200, and [M/H] by ± 0.1 for S/N$\geqslant$200.

2. First application to KIC 10661783

The test parameters had been chosen in a way to fit the suspected stellar properties of KIC 10661783, an eclipsing binary that shows more than 60 δ Sct-like oscillations in its light curve observed by the Kepler satellite (Southworth *et al.* 2011). We determine its fundamental stellar parameters from a time series of high-resolution spectra taken with the HERMES spectrograph at the Mercator Telescope on La Palma.

Our analysis results in a flux ratio of 0.074 between secondary and primary and about 8000 K for T_{eff} and 3.6 for $\log g$ of both components. The identical parameters are in agreement with the fact that the decomposed spectra of both components are almost identical if we up-scale the line depths of the secondary by a factor of about 14. Although the flux ratio agrees with that obtained from the light curve analysis, the parameters derived for the secondary are not compatible with the photometric results and in particular not with the derived mass ratio of 0.089. As a possible explanation, we assume that the light in the decomposed spectrum of the secondary contains a remarkable contribution of reflected light from the primary. This would also explain that during both eclipses the RVs of the secondary show the Rossiter effect and its line strengths are strongly weakened. We extended our program accordingly, using a very simple model including synchronized rotation and pure reflection. Table 1 compares our results with those of the light curve analysis. We end up with the very small flux ratio of 0.033 that prevents us from obtaining more accurate values. Independent of reflection, we derived the masses and radii of the components and their separation of only 3.2 R_1. The hypothesis of reflected light has to be checked using the Wilson-Devinney code with the Kepler light curve.

References

Hadrava, P. 1995, *A&AS*, 114, 393

Lehmann, H., Tkachenko, A., Semaan, T., *et al.*, 2011, *A&A*, 526, 124

Southworth, J., Zima, W., Aerts, C., *et al.*, 2011, *MNRAS*, 414, 2413

Tamajo, E., Pavlovski, K., & Southworth, J. 2011, *A&A*, 526, 76

From Interacting Binaries to Exoplanets: Essential Modeling Tools
Proceedings IAU Symposium No. 282, 2011
Mercedes T. Richards & Ivan Hubeny, eds.
© International Astronomical Union 2012
doi:10.1017/S1743921311027906

The BANANA Survey: Spin-Orbit Alignment in Binary Stars

Simon Albrecht[1], J. N. Winn[1], D. C. Fabrycky[2], G. Torres[3] and J. Setiawan[4]

[1] Massachusetts Institute of Technology, Kavli Institute for Astrophysics and Space Research,
Cambridge, MA 02139, USA
[2] University of California, Santa Cruz, CA 95064, USA
[3] Harvard-Smithsonian Center for Astrophysics, Cambridge, MA 02138, USA
[4] Max-Planck-Institut für Astronomie, 69117 Heidelberg, Germany

Abstract. Binaries are not always neatly aligned. Previous observations of the DI Herculis system showed that the spin axes of both stars are highly inclined with respect to one another and the orbital axis. Here, we report on our ongoing survey to measure relative orientations of spin-axes in a number of eclipsing binary systems.

These observations will hopefully lead to new insights into star and planet formation, as different formation scenarios predict different degrees of alignment and different dependencies on the system parameters. Measurements of spin-orbit angles in close binary systems will also create a basis for comparison for similar measurements involving close-in planets.

Keywords. techniques: spectroscopic, binaries: eclipsing, stars: rotation

1. Introduction

While many stars form in binary systems, binary formation is still not well understood. This is particularly true for close binary systems with orbital periods of a few days. If the orbital characteristics of these systems would not have changed since pre-main-sequence then the stars would have formed from one entity, as during the pre-main-sequence phase their sizes were bigger. As fission seems unlikely (Tohline 2002), the orbit probably shrunk and therefore the angular momentum of these systems must have undergone a complex history. One part of the angular momentum distribution which is seldom probed is the stellar spin.

Close binaries and star-planet systems might be expected to have well-aligned orbital and spin angular momenta, since all of the components trace back to the same portion of a molecular cloud. However, good alignment is not guaranteed. Disks around young stars might become warped during the last stage of accretion. This warp could torque the orbit by a large angle while maintaining the orientation of the spins (Tremaine 1991). More generally, star formation may be a chaotic process, with accretion from different directions at different times (Bate 2010).

There are also processes that could alter the stellar and orbital spin directions after their formation. A third body orbiting a close pair on a highly inclined orbit can introduce large oscillations in the orbital inclination and eccentricity of the close pair (Kozai 1962). Tidal dissipation during the high-eccentricity phases can cause the system to free itself of these "Kozai oscillations" and become stuck in a high-obliquity state (Fabrycky & Tremaine 2007). However, if dissipation is sufficiently strong then the system will evolve into the double-synchronous state, characterized by spin-orbit alignment (e.g. Hut 1981). Therefore, whether a close binary or a star-planet system is well-aligned or misaligned depends on its particular history of formation and evolution. Even though this issue is

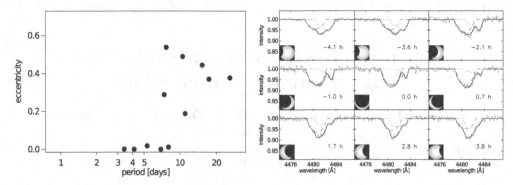

Figure 1. The left panel shows the orbital period and eccentricity of our current sample in the BANANA project. The nine panels on the right show spectra obtained during primary eclipse in the DI Herculis systems. Each panel shows the Mg II lines (4481 Å) of the two stars. The gray line represents the data, the dashed (blue) line the model for the foreground secondary, the (dotted) red line the model for the eclipsed primary and the black line the combined model. Each panel has a inset illustration showing the uncovered part of the primary. The time from mid eclipse is also given. If the primary spin would have been aligned then at mid-eclipse the primary absorption line should be near symmetric, which is not the case.

important for a complete understanding of star formation, there has been very little observational input.

For these reasons, we are conducting measurements of the relative orientations of the rotational and orbital axes in close binary star systems, most of which harbor early-type stars. Our name for this undertaking is the BANANA project, an acronym chosen to remind us that binaries are not always neatly aligned (Albrecht *et al.* 2009, 2011).

Our aims and measurement approach are similar to the efforts undertaken in the exoplanet community (see e.g. Albrecht, these proceedings), but there is one difference in the analysis method. When observing double-lined binaries during eclipses to measure the Rossiter-McLaughlin effect (Rossiter 1924, Mclaughlin 1924), the light from the occulting foreground object cannot be ignored. We, therefore, cannot simply measure the center of lines during eclipses, but have to model the spectra of both stars and compare these to the observations. See Albrecht *et al.* (2007) for details.

The left panel in Fig. 1 shows the orbital periods and eccentricities of the systems in our program. The right panels show observations of an eclipse of the primary star in the DI Herculis system, in which rotation is strongly misaligned with the orbital spin (Albrecht *et al.* 2009).

In our current sample, we find evidence for aligned systems and evidence for spin-orbit misalignment. Furthermore, we find that stellar obliquity seems not to be a simple function of orbital distance or eccentricity.

References

Albrecht, S., Reffert, S., Snellen, I., Quirrenbach, A., & Mitchell, D. S. 2007, *A&A*, 474, 565
Albrecht, S., Reffert, S., Snellen, I. A. G., & Winn, J. N. 2009, *Nature*, 461, 373
Albrecht, S., Winn, J. N., Carter, J. A., Snellen, I. A. G., *et al.*, 2011a, *ApJ*, 726, 68
Bate, M. R., Lodato, G., & Pringle, J. E. 2010, *MNRAS*, 401, 1505
Fabrycky, D. & Tremaine, S. 2007, *ApJ*, 669, 1298
Kozai, Y. 1962, *AJ*, 67, 591
McLaughlin, D. B. 1924, *ApJ*, 60, 22
Rossiter, R. A. 1924, *ApJ*, 60, 15
Tohline, J. E. 2002, *ARAA*, 40, 349
Tremaine, S. 1991, *Icarus*, 89, 85

From Interacting Binaries to Exoplanets: Essential Modeling Tools
Proceedings IAU Symposium No. 282, 2011
Mercedes T. Richards & Ivan Hubeny, eds.
© International Astronomical Union 2012
doi:10.1017/S1743921311027918

A Search for the Secondary Spectrum of ε Aurigae

P. D. Bennett[1,2], P. Harmanec[3], P. Chadima[3] and S. Yang[4]

[1]Eureka Scientific, Inc., 2452 Delmer Street, Suite 100, Oakland, CA 94602-3017, USA
[2]Department of Astronomy & Physics, Saint Mary's University, Halifax, NS, Canada B3H 3C3
email: pbennett@ap.smu.ca
[3]Astronomical Institute of the Charles University, Faculty of Mathematics and Physics,
V Holešovičkách 2, CZ-180 00 Praha 8, Czech Republic
[4]Dept. of Physics and Astronomy, University of Victoria, Victoria, BC, Canada V8W 3P6

1. The observations

The enigmatic long period ($P = 27.1$ yr) eclipsing binary, ε Aurigae, recently emerged from its 2009–2011 eclipse. We have analyzed out-of-eclipse observations (Chadima *et al.* 2010) obtained over the past 17 years: 306 medium-resolution, high S/N, spectroscopic observations from 6300–6700 Å. Of these, 105 spectra were obtained at the Dominion Astrophysical Observatory (DAO) near Victoria, Canada, from 1994–2010, and 201 spectra were obtained at Ondřejov Observatory (OND), from 2006–2010. Analyzing these data, Chadima *et al.* (these proceedings) reported on a positive, but ultimately spurious, detection of a secondary spectrum. Their attempts at disentangling the binary spectra were foiled by line profile variations of the F star primary. The 6300-6700 Å spectral region contains several strong stellar lines but space limitations allow us to present only the results for Si II 6347 Å. We examine the centroids and higher moments of this prominent F star spectral line for any evidence of a secondary spectrum. Even if secondary contributions are blended with the F star lines, contamination by the secondary star should produce a centroid shift that is anti-correlated with the orbit of the F star primary.

2. The analysis

The stellar line profiles observed in ε Aur are quite asymmetric (Figure 1a), and vary on timescales of 50–100 d (Chadima *et al.* 2011), complicating the detection of radial velocity (RV) variability from the companion. Spectra were transformed to the F star reference frame using the Chadima *et al.* (2010) orbital solution. Observations acquired after the start of spectroscopic eclipse, taken to be JD 245 4850, were discarded because of strong absorption from the companion's disk. A small number (∼10%) of low S/N observations were omitted. Also, two DAO observations with highly anomalous shifts of tens of km s^{-1} were rejected. This left 163 spectra: 76 DAO and 87 OND observations. These spectra were then cross-correlated with DAO or OND reference standards, using a spectral region of weak stellar lines (6375–6447 Å), to remove intrinsic stellar radial velocity (RV) variations (Figure 1a). The velocity corrections found were typically small (~ 3 km s^{-1}). Then, the first four central moments $\mu_1, \cdots, \mu_4$, were evaluated, and the line centroid μ_1 (Figure 1b), width $\sigma = \mu_2^{0.5}$, and skewness $\gamma_1 = \mu_3/\mu_2^{1.5} = \mu_3/\sigma^3$ were derived for each observed profile. We also evaluated mean profiles for each quintile of orbital RV, because any profile anomalies associated with the companion should be negatively correlated with the F star's orbital RV, and thus with the RV quintiles.

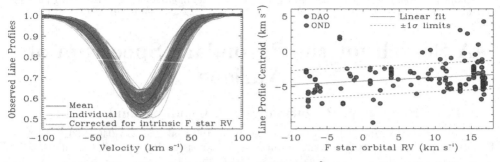

Figure 1. (a) Left: The 163 individual profiles of Si II 6347 Å observed from 1994–2010. Green profiles are in the F star frame, red are corrected for intrinsic RV variations, the black curve is the mean profile. (b) Right: Centroids of DAO (blue) and OND (red) out-of-eclipse observations.

3. The results

The mean line absorption profile of Si II 6347 Å is significantly asymmetric. One possibility is that the intrinsic F star line profiles are symmetric, and these asymmmetries are produced by additional absorption or emission from the companion star. Since any absorption or emission from the companion should be negatively correlated with F star orbital phase, a phase-dependent asymmetry should be present in the observed line profiles. To reduce the observed RV scatter, mean line profiles were computed for observations in each of five equal bins, or quintiles, of the F star's orbital RV (Fig. 2).

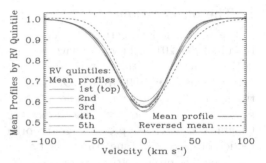

Figure 2. Mean profiles of ε Aur Si II 6347 Å, computed over F star RV quintiles. The overall mean profile is solid black; the mirror-reversed mean profile is the dashed black curve.

4. Conclusions

In the far ultraviolet, ε Aur has an emission line spectrum (Bennett, Ake, & Harper 2005). However, the quintile centroid shifts of Si II 6347 are too small, and do not show the expected proportional relation to the F star RV, to be consistent with infilling emission. Neither is circumstellar absorption a probable cause, because the line cores are shallowest near the onset of eclipse, when any circumstellar absorption should be large.

We conclude that the line profile variations in the optical spectrum of ε Aur do not arise from contamination by the companion. The asymmetric shape of the F star line profiles is probably intrinsic, as is the variability of these profiles.

Acknowledgements

PDB and SY acknowledge the support of the NRC (Canada) Herzberg Institute of Astrophysics in the ongoing operation of the DAO 1.2 m McKellar telescope. PH was supported by grant P209/10/0715 of the Czech Science Foundation.

References

Bennett, P. D., Ake, T. B., & Harper, G. M. 2005, *BAAS*, 37, 495
Chadima, P., Harmanec, P., Yang, S. *et al.*, 2010, *IBVS*, No. 5937
Chadima, P., Harmanec, P., Bennett, P. D. *et al.*, 2011, *A&A*, 530, A146

From Interacting Binaries to Exoplanets: Essential Modeling Tools
Proceedings IAU Symposium No. 282, 2011
Mercedes T. Richards & Ivan Hubeny, eds.
© International Astronomical Union 2012
doi:10.1017/S174392131102792X

An Unexpected Outcome from Disentangling

P. Chadima[1], P. Harmanec[1], P. D. Bennett[2,3], and S. Yang[4]

[1] Astronomical Institute of the Charles University, Faculty of Mathematics and Physics,
V Holešovičkách 2, CZ-180 00 Praha 1, Czech Republic
email: `Pavel.Chadima@gmail.com`; `hec@sirrah.troja.mff.cuni.cz`
[2] Eureka Scientific, Inc., 2452 Delmer Street, Suite 100, Oakland, CA 94602-3017, USA
[3] Department of Astronomy & Physics, Saint Mary's University, Halifax, NS, Canada B3H 3C3
[4] Dept. of Physics and Astronomy, University of Victoria, Victoria, BC, Canada V8W 3P6

1. A surprising result...

During our recent spectroscopic study of ϵ Aur (Chadima *et al.* 2011), we made an attempt to detect weak spectral lines of the secondary, hidden in a dark disk, using the spectral disentangling technique of Simon & Sturm (1994) and Hadrava (1995, 1997, 2004). We used the Dominion Astrophysical Observatory (DAO) and Ondřejov (OND) red electronic spectra, which cover more than one half of the orbital period. To our surprise, two different programs that disentangled the spectrum in Fourier space, KOREL (Hadrava 1995, 2004) and FDBINARY (Ilijić *et al.* 2001), both yielded apparently good, similar reconstructions of *two* well-defined spectra for mass ratios near unity. The results (Solution 1) are shown in Fig. 1 (left panels) and Table 1 (left column). This result is hard to accept as real given the existing knowledge about the system: ϵ Aur is an F-type star with an unseen companion embedded in a cool, dark disk (temperature $\sim$ 500–600 K). A detailed search for any trace of spectral signatures of the secondary in the spectra was carried out (see Bennett *et al.*, these proceedings). Although they found line profile variations that were correlated with orbital phase, these variations were not consistent with the presence of a secondary.

2. ...its verification...

To shed more light on the whole problem, we carried out two further tests. First, we attempted to disentangle the spectra for an arbitrarily-chosen, fictitious orbital period of 700 days. All six orbital elements were allowed to converge freely in KOREL. The disentangling procedure was then repeated using the alternate program FDBINARY, for the same orbital elements found by KOREL. Comparable disentangled profiles were obtained, this time with a strong secondary spectrum (Solution 2), shown in Fig. 1 (middle panels) and Table 1 (middle column). Next, we fixed the period at the value found for the physical variations (66.21 d: Chadima *et al.* 2011) and allowed the remaining five elements in KOREL to converge freely. An even stronger secondary spectrum was found (Solution 3), presented in Fig. 1 (right panels) and Table 1 (right column).

3. ... and a possible explanation

What can be the reason for such unusual behaviour from a technique that usually returns excellent results? Our application violates one of the basic principles of disentangling: the assumption that the line profiles of both binary components vary in intensity but not in shape. The line profiles of ϵ Aur do exhibit changes on a timescale of weeks. However, disentangling has been successfully applied to several binaries with components

401

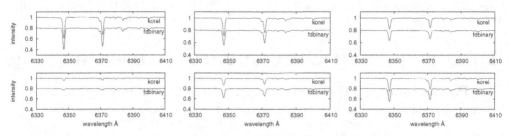

Figure 1. Disentangled spectra of ϵ Aur from KOREL and FDBINARY solutions. Spectra of the primary appear in the top panels; those of the secondary in the bottom panels. Solution 1 is shown on the left, solution 2 is in the middle and solution 3 is on the right.

also exhibiting line profile changes on timescales shorter than their orbital periods, so this is probably not the only reason. We believe that the problem occurs because the amplitude of the ϵ Aur radial velocity changes is small compared to the width of the spectral lines. Therefore, in the model, spectral lines of the primary as well as of the putative secondary remain heavily blended at all orbital phases, and the line width changes are misinterpreted by both programs as arising from the lines of two binary components.

Our result represents a methodological warning that one should not accept even a very satisfactory result from the disentangling procedure without carrying out additional tests to see how realistic the result is. We suggest first mapping the space of the key orbital elements and checking the run of the sum of squares of the residuals.

Table 1. The resulting orbital elements obtained in the three trials discussed in the text. Elements freely converged in a given solution are denoted by asterisks.

element	solution 1	solution 2	solution 3
$T_{\mathrm{per.}}$ (HJD-2454000)	609.0	577.6*	639.5*
P (d)	9890.62	796.7*	66.21
e	0.252	0.295*	0.305*
ω (°)	42.3	114.8*	49.97*
K_1 (km s^{-1})	14.35	14.23*	15.00*
$q = M_2/M_1$	0.994*	0.998*	0.900*

Acknowledgements

We acknowledge the use of the public versions of the programs KOREL of Dr. P. Hadrava and FDBINARY of Dr. S. Ilijić. The research of PH was supported by the grant P209/10/0715 of the Czech Science Foundation and from the Research Program MSM0021620860.

References

Chadima, P., Harmanec, P., Yang, S. *et al.*, 2010, *IBVS*, No. 5937
Chadima, P., Harmanec, P., Bennett, P. D. *et al.*, 2011, *A&A*, 530, A146
Hadrava, P. 1995, *A&AS*, 114, 393
Hadrava, P. 1997, *A&AS*, 122, 581
Hadrava, P. 2004, *Publ. Astron. Inst. Acad. Sci. Czech Rep.*, 89, 15
Ilijić, S., Hensberge, H., & Pavlovski, K. 2001, *Lecture Notes in Physics*, 573, 269
Simon, K. P. & Sturm, E. 1994, *A&A*, 281, 286

From Interacting Binaries to Exoplanets: Essential Modeling Tools
Proceedings IAU Symposium No. 282, 2011
Mercedes T. Richards & Ivan Hubeny, eds.
ⓒ International Astronomical Union 2012
doi:10.1017/S1743921311027931

VO-KOREL: A Fourier Disentangling Service of the Virtual Observatory

Petr Škoda, Petr Hadrava and Jan Fuchs

Astronomical Institute, Academy of Sciences,
Fričova 298, CZ-251 65 Ondřejov, Czech Republic
email: skoda@sunstel.asu.cas.cz

Abstract. VO-KOREL is a web service exploiting the technology of the Virtual Observatory for providing astronomers with the intuitive graphical front-end and distributed computing back-end running the most recent version of the Fourier disentangling code KOREL.

The system integrates the ideas of the e-shop basket, conserving the privacy of every user by transfer encryption and access authentication, with features of laboratory notebook, allowing the easy housekeeping of both input parameters and final results, as well as it explores a newly emerging technology of cloud computing.

While the web-based front-end allows the user to submit data and parameter files, edit parameters, manage a job list, resubmit or cancel running jobs and mainly watching the text and graphical results of a disentangling process, the main part of the back-end is a simple job queue submission system executing in parallel multiple instances of the FORTRAN code KOREL. This may be easily extended for GRID-based deployment on massively parallel computing clusters.

The short introduction into underlying technologies is given, briefly mentioning advantages as well as bottlenecks of the design used.

Keywords. Methods: data analysis, techniques: spectroscopic, line:profiles, Virtual Observatory

1. Virtual Observatory

The Virtual observatory (hereafter VO) is a global infrastructure of distributed astronomical archives and data processing services enabling the standardized discovery and access to the astronomical data worldwide as well as a large set of powerful tools for scientific analysis and visualization (Solano 2006). VO supports mainly the multi-wavelength research or discovery of rare objects by cross-matching huge catalogues.

For interaction with the user as well as other computers, the VO uses the modern technology of Web Services (WS). The WS is typically a complex processing application using the web technology to transfer input data to the main processing back-end and the results (after intensive number crunching) back to user. All this can be done using only an ordinary web browser (and in principle the science may be done on the fast palmtop or advanced mobile phone). An example of WS is the e-shop or ticket reservation system.

2. Universal worker service

The Universal Worker Service — UWS (Harrison & Rixon 2010) is one of many standards of VO describing the exact operation patterns of WS supporting the execution of multiple jobs in an asynchronous way, allowing the simple control of jobs (e.g. changing their execution limits, timeouts and expiration) as well as easy access to the results using the web technology. The current version of UWS is based on a newly re-discovered RESTful technology (Fielding 2000).

The technology of WS allows us to provide the complicated computing code as a service running in the virtualized infrastructure of a large commercial provider or on a dedicated server of the software provider instead of providing only the source code. This idea has been promoted by the large IT companies under the terms of "Software as a service" and it is part of a wider business called *Cloud computing* (Foster *et al.* 2008). The advantages of this approach from the developer's point of view are clear:

- The only SW needed by the user is only a tiny web browser
- There is the only one, current, well tested version of the code (and documentation), maintained and updated often directly by its author
- The computing of various models may be controlled even from the smart mobile phone over a slow connection, as the large data are uploaded only once and most of the investigation requires only changing several numbers in a parameter file and re-submission of a job directly from web browser.
- The web technology provides an easy way of interaction (forms) and graphics output (in-line images)

3. VO-KOREL web service

The principal use of VO-KOREL is similar to e-shop portal, starting with user registration. Every set of input parameters creates a job, which may be run in parallel with others. The user may stop or remove them, can return to the previous versions etc. All user communication is encrypted and the user can see only his/her jobs. The service may be accessed from the KOREL portal at the Astronomical Institute in Ondřejov at address http://stelweb.asu.cas.cz/vo-korel.

The VO-KOREL web service requires the upload of the data file korel.dat and parameter file korel.par (and optionally the template(s) for template-restricted disentangling korel.tmp) in the format described in the KOREL manual (last chapters in Hadrava 2009).

Several sets (they may also be compressed in the form of a tar.gz file) can be uploaded to the server (within the disk quota allowed for the user) and the execution of the jobs may be postponed even until next login of the user. The user can decide on priorities about which jobs to compute, as the system can run only several jobs of every user; the others are queued until the computing resources (memory, CPU) are available.

4. Conclusions

The VO-KOREL service is not only providing a comfortable environment for Fourier disentangling of spectra, but it is a test-bed of the general cloud infrastructure for execution of most scientific computationally intensive codes, like models of stellar atmospheres or special processing of complex data sets.

References

Fielding, R. T. 2000, Architectural Styles and the Design of Network-based Software Architectures, *Ph.D. Dissertation*, University of California, Irvine, retrieved from http://www.ics.uci.edu/~fielding/pubs/dissertation/rest_arch_style.htm
Foster, I., Zhao, Y., Raicu, I., & Lu, S. 2008, *Cloud Computing and Grid Computing 360-Degree Compared*, arXiv:0901.0131
Hadrava, P. 2009, *Disentangling of spectra — theory and practice*, arXiv:0909.0172
Harrison, P. & Rixon, G. 2010, *Universal Worker Service Pattern Version 1.0*, IVOA Recommendation, retrieved from http://ivoa.net/Documents/UWS/index.html
Solano, E. 2006, *Lecture Notes and Essays in Astrophysics*, 2, 71

From Interacting Binaries to Exoplanets: Essential Modeling Tools
Proceedings IAU Symposium No. 282, 2011   © International Astronomical Union 2012
Mercedes T. Richards & Ivan Hubeny, eds.   doi:10.1017/S1743921311027943

Constrained Genetic Disentangling
of Close Binary Spectra

V. Kolbas[1] and K. Pavlovski[1,2]

[1] Department of Physics, University of Zagreb, 10 000 Zagreb, Croatia

[2] Astrophysics Group, EPSAM, Keele University, Staffordshire, ST5 5BG, UK

Abstract. The spectral disentangling technique makes possible separation of individual component spectra in binary or multiple systems, and determination of the orbital elements in a self-consistent way. Since its introduction, a number of variants of their basic idea have been implemented. We present yet another 'direct' approach using optimization by genetic algorithm. Starting with an initial random flux distribution representing individual spectra, genetic optimization returns both individual component spectra and an optimal set of orbital parameters only constrained by time-series of the observed composite spectra of the binary system. Benchmark tests on V453 Cyg, which is an eclipsing binary with total eclipse, as well as tests on the artificial time-series spectra, have proven that 'constrained genetic disentangling' is performing correctly and efficiently, albeit with high demand on CPU time. Since genetic optimization can be easily parallelized, we expect our second release to run on cluster in a less time-consuming way.

Keywords. genetic algorithm, spectral disentangling

---

1. Introduction

The spectral disentangling technique, as invented by Simon & Sturm (1994) and Hadrava (1995), makes possible separation of individual component spectra in binary or multiple systems, and determination of orbital elements in a self-consistent way. One only needs a time series of binary star spectra. There is no need for template spectra as in the technique of cross-correlation, since they are the main source of bias in measuring RVs of the components, and hence determination of orbital elements (see detailed discussion in the review paper by Hensberge & Pavlovski 2007). In spectral disentangling (SPD, here-in-after), the role of templates is overtaken by the spectra of components themselves. The advantage of SPD is obvious, besides the set of orbital elements, individual spectra of the components are calculated. These separated spectra of the components then can be analyzed by all means as single star spectra and a variety of important astrophysical informations can be extracted (c.f. Pavlovski & Hensberge 2010, Pavlovski & Southworth 2011, these proceedings).

A number of different techniques have been implemented to separate individual spectra of components from complex binary (or multiple) star spectra. They go from a simple subtraction technique to very sophisticated numerical methods. The reader is referred to review papers by Hadrava (2004, 2009) and Pavlovski & Hensberge (2010) where details on these techniques can be found, as well as references to original works (c.f. Hadrava, these proceedings).

2. Genetic forward disentangling

SPD in the formulation of Simon & Sturm (1994) is based on solving the matrix equation $Ax = y$, where vector y contains all observed spectra, and vector x contains the spectra of the components. Matrix A has elements (blocks) corresponding to Doppler shifts and light dilution factors. The system is an over-determined system of linear equations (more equations than unknowns) since usually in practice we have more observed spectra than stellar components. Therefore, a least squares solution is required in order to minimize the norm of the residuals $r = ||Ax - y||$. On the computational side, this method is very demanding. Fourier disentangling as invented by Hadrava (1995) overcomes this problem. The method is not limited by the number of input spectra or the length of the spectral interval. The only limitation is that both ends of the spectral range to be disentangled should be exactly on the continuum.

A binary star spectrum is a linear combination of two (or more, for a multiple system) individual spectra of components which are shifted for appropriate RVs in a given phase of the orbital cycle. Therefore, we can perform direct disentangling seeking for two spectra which, for a given set of orbital phases, would be constrained by time series of observed spectra. Like in the SPD method in velocity space of Simon & Sturm (1994), we are seeking a least square solution and minimize squared residuals. In this direct disentangling technique, we have used optimization by genetic algorithm (Charbeonneau 1995).

We have performed a number of different tests on simulated and real data. Special attention has been given to reproduce disentangled component spectra in the benchmark binary star V453 Cyg (Simon & Sturm 1994, Pavlovski & Southworth 2009) which has a total eclipse, and the spectrum of one of the components can be directly observed. An important aspect of *genetic constrained disentangling* is the excellent performance in disentangling difficult cases like broad Balmer lines or, even more significant, disentangling a spectral range that is cut in the line wing(s). This is a severe problem for Fourier disentangling since weights can be assigned only to spectra, and not to pixels. With growing numbers of faint binaries observed spectroscopically in our galaxy, or in galaxies of the Local Group (North *et al.* 2010), which contain nebular emission in Balmer lines, and/or strong interstellar bands, these fine characteristics are of particular importance. Now, we are working on improving the performance of the genetic algorithm code since CPU time is not yet competitive with other methods.

References

Charbonneau, P. 1995, *ApJS*, 101, 309
Hadrava, P. 1995, *A&AS*, 114, 393
Hadrava, P. 2004, *ASP-CS*, 318, 86
Hadrava, P. 2009, *arXiv*:0909.0172
Hensberge, H. & Pavlovski, K. 2007, *in IAU Symp.*, 240, 136
North, P., Gauderon, R., Barblan, F., & Royer, F., 2010 *A&A*, 520, A74
Pavlovski, K. & Hensberge, H. 2010, *ASP-CS*, 435, 207
Pavlovski, K. & Southworth, J. 2009, *MNRAS*, 394, 1519
Simon, K. P. & Sturm, E., 1994, *A&A*, 281, 286

Part 7
Formation and Evolution of Binary Stars, Brown Dwarfs, and Planets

From Interacting Binaries to Exoplanets: Essential Modeling Tools
Proceedings IAU Symposium No. 282, 2011
Mercedes T. Richards & Ivan Hubeny, eds.
© International Astronomical Union 2012
doi:10.1017/S1743921311027955

Binary Star Formation Simulations

C. J. Clarke

Institute of Astronomy, Madingley Rd, Cambridge, U.K., CB3 OHA
email: cclarke@ast.cam.ac.uk

Abstract. Binary stars provide an excellent calibration of the success or otherwise of star formation simulations, since the reproduction of their statistical properties can be challenging. Here, I summarise the direction that the field has taken in recent years, with an emphasis on binary formation in the cluster context, and discuss which observational diagnostics are most ripe for meaningful theoretical comparison. I focus on two issues: the prediction of binary mass ratio distributions and the formation of the widest binaries in dissolving clusters, showing how in the latter case the incidence of ultra-wide pairs constrains the typical membership number of natal clusters to be of order a hundred. I end by drawing attention to recent works that include magnetic fields and which will set the direction of future research in this area.

Keywords. stars: formation, hydrodynamics, stellar dynamics

1. Introduction

It is often stated that the wealth of data describing the properties of binary stars (e.g. binary and higher order multiplicity fractions, separation, eccentricity, and mass ratio distributions as a function of primary mass) provides some of the most stringent tests of star formation simulations. In this contribution, I review the progress of such simulations, from early calculations which studied fragmentation in isolated cores to the cluster simulations that have dominated the field over the last decade, wherein binaries arise as a natural by-product of the turbulent fragmentation process. I take a critical look at those aspects of the simulations that can be most reliably compared with observations and focus on two areas where the gap between theory and observation is closing, i.e. the mass ratio distribution and the formation mechanism for the widest binaries. I conclude by mentioning recent work which highlights how current models are challenged by the inclusion of magnetic fields at a realistic level.

2. Simulation developments

During the 1980s and 1990s the major focus of binary formation simulations was the study of isolated cores (e.g. Boss & Bodenheimer 1979) and the determination of core properties (e.g. density profile, rotation rate, equation of state, amplitude of perturbation) that would give rise to fragmentation (note that since the initial perturbations were generally of the form m=2, the initial fragmentation was predisposed to involve two objects). Such studies were highly valuable for calibrating different numerical codes and for identifying the stage of the collapse process at which binary fragmentation is liable to occur. They were, however, far removed from their ultimate goal, i.e. of creating binaries whose properties could be compared with observed systems, owing to the fact that it was computationally prohibitive to follow the evolution far beyond the point of initial fragmentation. With typically 10% of the core mass involved in the initial binary fragments, it was impossible to trace out the further evolution of mass and orbital elements that would accompany the proto-binaries' subsequent assembly. A further feature of such simulations was that since they typically involved isolated two-body systems, they did not include the possible role of multi-body dynamics in shaping the ultimate binary population.

The simulations of Bate *et al.* (2002a), Bate *et al.* (2002b) and Bate *et al.* (2003), as well as a large body of broadly similar simulations (Delgado *et al.* 2004a, Delgado *et al.* 2004b, Goodwin *et al.* 2004a, Goodwin *et al.* 2004b, Offner *et al.* 2009), changed the emphasis from isolated binary studies to those in which binaries are formed in the context of cluster simulations. The key technical development here is the implementation of 'sink particles' (Bate *et al.* 1995) which allow collapsed regions to be removed from the domain of detailed computation, replacing them instead by point masses that can gravitationally interact and accrete from the surrounding gas. This permits the creation of binaries whose formation can run 'to completion' (i.e. to the point where they have accreted the bulk of their circumstellar environment) and which can in principle be compared with observed systems. These simulations also emphasise the importance of multi-body dynamics in the early stages of binary formation: the basic unit of star formation in the simulations is the small N cluster, within which binary exchange interactions, and the creation of complex high order multiples, is common.

These simulations generate a wealth of synthetic binary data for comparison with observations. Many have been encouraged to believe that the simulations are essentially 'correct' (despite, in most cases, their omission of key physical effects such as magnetic fields or realistic thermal feedback from the forming stars) since they effortlessly reproduce some basic binary statistics (for example, the increase of binary fraction with primary mass and the wider separation distributions of more massive binaries (see Bate 2009a): both these effects relate to the relative fragility of lower mass systems in the small N cluster environment). Nevertheless, it is worthwhile considering how reliable are the various predictions that arise from such simulations before selecting which are the areas where observational comparison is most meaningful.

In the early days of star cluster simulations (also known as 'turbulent fragmentation' simulations owing to the fact that the initial cloud structure is seeded by realistic levels of supersonic turbulence), the field was monopolised by studies using Smoothed Particle Hydrodynamics (SPH) whose Lagrangian nature was well-adapted to the complex geometries and large dynamic range of densities that develop in such simulations. At this stage, there was much discussion of the physical reality of the copious disc fragmentation that is seen in such simulations (and which creates the small N cluster environment that dominates the subsequent dynamics). Fortunately, the recent development of Adaptive Mesh Refinment (AMR) techniques, and the incorporation of sink particles within AMR (Krumholz *et al.* 2004) has opened up the opportunity for detailed code calibration studies. The results have been encouraging: broadly speaking, provided that the respective resolution criteria of the two classes of technique are respected, the Lagrangian (SPH) and Eulerian (AMR) codes give very similar results (Federrath *et al.* 2010). This encouraging state of affairs is not to say that the copious fragmentation mentioned above is necessarily *physically* correct. Indeed, it is now realised that this behaviour is rather sensitive to the thermodynamical behaviour of the gas and that fragmentation is enhanced in the case that one assumes the simplest (isothermal) equation of state. More recent calculations improve on this barotropic treatment of the equation of state by employing flux-limited diffusion to capture the effect of thermal feedback from the protostars (Bate 2009b, Offner *et al.* 2009). These studies find that a more realistic cooling treatment inhibits disc fragmentation and demonstrate that the early (isothermal) calculations over-predicted the harvest of very low mass stars and brown dwarfs. Note, however, that these simulations do still produce binaries, including close binaries. Typically, close binaries in the simulations are systems that form at large separations and are hardened by a combination of dynamical interactions, accretion, and star-disc interactions.

Having discussed the fidelity with which the initial binary fragmentation is captured, we now turn to how well the simulations follow the evolution of the binary orbital elements as the binary accretes from its environment. We will discuss the accuracy with which the mass ratio distribution is modeled in Section 3 below, but note for now that the binary mass ratio is of course fixed during the time that the bulk of the gas is accreted onto the protobinary and cannot change much once the circumstellar gas is a small fraction of the binary mass. This is, however, not the case for the evolution of the binary separation and eccentricity, where it is well known that discs can be highly efficient agents of orbital evolution when they contain only a small fraction of the binary mass (Lubow & Artymowicz 2000). (This is because discs with internal angular momentum transport processes can convey large fluxes of angular momentum to large radii with relatively little mass). This means that the evolution of separation and eccentricity are also sensitive to the late time evolution of the circumstellar environment and, therefore, depend on how well discs are modeled in the late stages of system evolution. This is an area of relative weakness for the simulations: by focusing on large ensembles of stars, the number of SPH particles per disc is typically only of order 10^4 even when the discs are relatively massive. As discs are depleted by accretion, they become correspondingly more poorly resolved: in particular, once the SPH smoothing length becomes greater than the disc's vertical scale height there can be a large erroneous effect associated with the action of artificial viscosity in the disc which can spuriously accelerate disc evolution. It should be stressed that this is not likely to be an important factor in modeling the initial formation and mass acquisition of binaries but does raise a question mark over using such simulations for the detailed study of eccentricity and separation distributions, for example.

3. The mass ratio distribution of binaries

As noted by Clarke (1996), binary pairing statistics for primaries in different mass ranges allow one to draw some rather general conclusions about the processes that dominate binary formation. If, for example, it were the case that the dominant binary formation mechanism was by the fragmentation of isolated cores into two components then, in the absence of any physical effects that make such a process dependent on the core mass, one might expect the splitting process to be *scale free*: in this case the mass ratio distribution would be independent of primary mass. On the other hand, if the dominant process instead involved dynamical capture within small N clusters, then it turns out that it is the shape of the companion mass function (*not mass ratio distribution*) that should be independent of primary mass.

Although the theoretical prediction is here clear cut, it has turned out to be remarkably difficult to test this observationally, due to the lack of high quality data for systems with widely differing primary masses. The situation for G dwarfs is relatively well characterised, following the pioneering work of Duquennoy & Mayor (1991) and its recent update and extension by Raghavan et al. (2010), but it is only recently that the M dwarf samples have been revisited following the study of Fischer & Marcy (1992) (see Reggiani & Meyer 2011, Bergfors et al. 2010). There are, however, two factors that complicate the interpretation in the M dwarf case, i.e. both the wide range of primary masses that enter the M dwarf samples and the fact that, since the lower mass primaries are close to the hydrogen burning mass limit, the mass ratio distribution can only be studied over a limited dynamic range. Moreover, it turns out that the companion mass distributions are rather smooth and featureless in both the G dwarf and M dwarf case - in the absence of a particular feature at given companion mass or given mass ratio, it is hard to distinguish the mode of binary formation involved.

Nevertheless, the ongoing improvements in the characterisation of the M dwarf binary population have demonstrated the continuity of binary properties as a function of mass, following previous claims (Thies & Kroupa 2007) that there is a discontinuous change in binary statistics between low mass (M) stars and brown dwarfs (such that the latter apparently show a more marked predilection for equal mass pairs: however, see below). In fact, it turns out that this apparently abrupt change in binary pairing statistics across the hydrogen burning mass limit can be understood in terms of the broad range of primary masses that enter the M dwarf sample. Now that sample sizes have been increased to the point where they can be meaningfuly sub-divided by mass, it turns out that the trend towards more equal mass companions increases smoothly as the primary mass decreases (Bergfors et al. 2010). This is as expected for any dynamical process and argues against the 'special' brown dwarf formation mechanism proposed by Thies & Kroupa (2007).

Before leaving the issue of binary mass ratio distributions, it is, however, worth taking a critical look at the claim that brown dwarf binaries have a strong tendency towards equal mass companions (e.g. Siegler et al. 2005). This claim is largely based on the results of placing components on colour magnitude diagrams and derived masses using model evolutionary tracks. It is, however, notable that where the component masses are derived dynamically (see the astrometric orbital solutions of Konopacky et al. 2010) this tendency to equal mass ratios disappears: indeed the dynamical masses are quite different (and generally much less evenly paired) from the mass estimates deriving from fitting the components to evolutionary models. However, as stressed by Konopacky et al. (2010), there are still relatively large error bars on the dynamically derived mass ratios. Thus, any conclusions about the mass ratio distributions in brown dwarf binaries are necessarily very preliminary.

4. Is there an extreme mass ratio problem?

We have seen in the preceding section that there is some uncertainty about the dependence of the binary mass ratio distribution on primary mass. In the case of solar type primaries, however, Duquennoy & Mayor (1991) found a mass ratio distribution that rises towards low $q(= M_2/M_1)$. Although this conclusion was somewhat revised by the re-analysis of closer spectroscopic pairs (with periods less than 3000 days; Mazeh et al. 1992), this did not affect the result for the bulk of (wider) pairs: apparently, therefore, binary components typically have rather disparate masses apart from the closest pairs.

This is in stark contrast to the results of numerical simulations which always show a marked preference for the production of more nearly equal mass pairs (e.g. Bate 2009a). This tendency is remarkably insensitive to initial conditions, since it does not derive from the mass ratio of the pair at the time of first fragmentation but instead on the effect of subsequent accretion onto the protobinary pair. This subsequently accreted material usually has higher specific angular momentum than the initial pair and, therefore, first intercepts the secondary's Roche lobe, which is further from the binary centre of mass. The standard result from SPH and ballistic particle simulations dedicated to the evolution of q as a result of accretion (e.g. Artymowicz 1983, Bate & Bonnell 1997, Bate 2000) is that the net effect of accretion is an increase in the binary mass ratio. Thus, it is accretion that drives the mass ratios of binaries in simulations to near unity.

This conclusion has, however, been challenged by more recent grid-based studies (e.g. Ochi et al. 2005, Hanawa et al. 2010; see contribution by Fateeva et al., this volume). Although these investigations agree that gas preferentially enters the secondary's Roche lobe, it is then found to flow via the L1 point and be finally accreted by the primary. In this case, accretion may drive the mass ratio downwards, and low q pairs are expected to be abundant. These authors suggested that the reason that this effect was not found in previous SPH simulations was a consequence of the excessively viscous nature of SPH, whereby artificial dissipation could allow particles entering the secondary's Roche lobe to avoid escape via the L1 point.

This latter interpretation is, however, unlikely, since subsequent convergence tests of accretion onto protobinaries (Delgado *et al.* in prep.) with SPH demonstrated that - away from shocks - the Jacobi constant (i.e. the Bernoulli function in the rotating frame) is well conserved. Thus, the avoidance of the L1 point cannot be a consequence of dramatic levels of artificial viscosity in the flow's first orbit of the secondary. These convergence tests found that the accretion of high angular momentum material always increases the mass ratio, although they demonstrated that the *magnitude* of this effect is resolution dependent: under-resolution means that the flow in the outer parts of the primary's Roche lobe is not well modeled as a coherent fluid but instead consists of discrete particle orbits which can be captured onto the secondary through L1. This flow from primary to secondary at low resolution means that the *magnitude* of increase of q is likely to have been over-estimated (by about a factor two) in low resolution cluster simulations.

The discrepancy with grid-based codes probably instead derives, at least in large part, from the different flow geometries and temperatures employed. The simulations of Ochi *et al.* (2005) and Hanawa *et al.* (2010) are two-dimensional and rather warm (with isothermal sound speed in the gas being $0.25\times$ the binary orbital velocity). When the SPH simulations are also run with the same temperature and in two dimensions, the results essentially replicate the grid-based findings - i.e. a flow from secondary to primary and a net decrease of q; although the grid-based calculations of Ochi *et al.* (2005) and Hanawa *et al.* (2010) cannot, unfortunately, treat the more realistic cold, three dimensional flows modeled with SPH, this result suggests that q may rise due to accretion after all (see also Val-Borro *et al.* (2011) for a recent grid based study showing accretion driving q upwards).

At first sight, this apparently leaves the problem of creating low q pairs unsolved. However, there are three factors that actually make it questionable whether there is an 'extreme mass ratio problem' after all. Firstly, as noted above, it is almost certain that cluster simulations (in which the accretion flow is relatively poorly resolved) will have over-estimated the rate of q increase. Secondly, Moeckel & Bate (2010) found that when they followed the evolution of hydrodynamically-created binaries in a cluster with an N-body code, following gas expulsion, the reconfiguration of complex multiple systems led to some evolution of the q distribution towards lower q pairs. Finally, the most recent re-evaluation of the observed ratio statistics of solar type binaries (Raghavan *et al.* 2010) has demonstrated a flatter distribution than that inferred by Duquennoy & Mayor (1991), owing to an over-generous application of incompleteness corrections at the low q end in the earlier work. The combination of these three factors (i.e. relatively small shifts in both the predicted and observed distributions) means that it is not obvious that there really is an 'extreme mass ratio problem' after all.

5. The formation of wide binaries

The separation distribution of binaries around solar type stars is very broad (Raghavan *et al.* 2010, Duquennoy & Mayor 1991) and extends at the wide end to binaries that are close to the separations where they are likely to be disrupted by dynamical interactions in the Galactic field (Jiang & Tremaine 2009). It is, thus, no mystery why the binary distribution rolls over at around 10^5 AU. What is more puzzling, from the point of view of their creation, is why there is a significant population of slightly closer but still ultra-wide pairs (with separations in the range $10^4 - 10^5$ AU). Stars form from Jeans unstable molecular cloud cores with typical dimensions $\sim 10^4$ AU, and so a binary of that separation can only be created from such a core if it is rotating at break-up velocity (whereas cores in reality rotate at a very small fraction of break-up velocity, Goodman *et al.* 1993). Clearly, the creation of binaries at even wider separations is even more problematical.

There are several possible creation routes. For example, the re-configuration of multiple systems can lead to one member being ejected into a wide but still bound orbit (Delgado *et al.* 2004b). Nevertheless, angular momentum conservation implies that very wide outliers are necessarily of very low mass, in contrast to the observed situation. Alternatively, it is possible in principle for a field star to be captured into wide orbit around a binary, though the incidence is expected to be far lower than the observed incidence of ultra-wide pairs. It is notable that in either of these scenarios, the primary of the ultra-wide pair is expected to be itself a binary; however the recent survey of ultra-wide M dwarf binaries by Law *et al.* (2011) finds a normal multiplicity of the primaries. In 50% of cases, the primary is apparently single so that high-order multiplicity is not a pre-requisite for the creation of ultra-wide pairs.

The only mechanism that creates wide pairs without the involvement of a third bound component is that noted in the N-body simulations of Kouwenhoven *et al.* (2010) and Moeckel & Bate (2010) which tracked the N-body evolution of binaries following gas expulsion. In both cases, snapshots of the simulations revealed the existence of instantaneously bound ultra-wide pairs. This is superficially puzzling, since - even by the standards of the expanding cluster - the pairs are extremely 'soft' . This raises questions both about the survivability of such binaries and about their creation mechanism, since conventional three-body capture within clusters involves three stars whose trajectories are mutually gravitationally focused and produce binary pairs that are 'hard' (i.e. with orbital velocity exceeding the local cluster velocity dispersion). Although intriguing, these studies raised a number of questions about the longevity of the binaries thus produced, as well as the mechanism for their creation.

Moeckel & Clarke (2011) recently undertook a large suite of N-body simulations of clusters which dissolve due to two-body relaxation following stellar dynamical core collapse, taking care to assess the permanence of the binary pairs produced. They found that there are indeed two populations of long-lived binaries: the expected population of hard binaries formed in the cluster core by three-body capture and a comparable population of ultra-wide pairs formed in the outer cluster as it dissolves.

Detailed examination has revealed the mechanism for the creation of the ultra-wide pairs: in any cluster, there is always a population of nearest neighbours that are instantaneously bound. Typically, these are readily broken up by perturbations by other cluster members and their significance is only that a very small fraction of them are eventually hardened by interactions so as to provide a supply of long-lived hard binaries (Goodman & Hut 1993). However, the case of a dissolving cluster is different in that the local stellar density may decline on a timescale that is shorter than the expected timescale for soft binary disruption at that density. Since we have similarity solutions for the decline in cluster density, we can estimate where in the cluster we expect this condition to be fulfilled (in the outer cluster, as observed in the simulations) and can predict how the separation distribution of the permanent soft binaries evolves (at any time, binaries are created at a separation comparable with mean interstellar separation). Furthermore, one can show through simple analytic argument that one expects of order one such binary to form per cluster per decade of separation *independent of N*, a result that is confirmed by analysis of the simulation results.

This last result implies that, in order to explain the fact that a few percent of all solar type stars are in ultra-wide pairs (separation $10^4 - 10^5$ AU), we would need the 'typical' natal cluster to harbour of order ~ 100 stars, this number being compatible with estimates from nearby star forming regions (Lada *et al.* 1991). This may explain why a relatively populous young star cluster, the Orion Nebula Cluster, which contains a few thousand stars, has been observed to be under-abundant in wide binaries compared with the field (Scally *et al.* 1999).

6. An afterword on magnetic fields

We stated at the outset that the last decade has seen a switch from simulations of isolated star forming cores to cluster wide simulations and that this has inevitably involved some compromise in resolution and in the range of physical effects explored. Although all of the results described above relate to unmagnetised simulations, there have been several recent attempts to include magnetic fields (Hennebelle & Teyssier 2008, Price & Bate 2009).

The former study (which was restricted to an isolated magnetised core) posed the provocative question 'Is there a fragmentation crisis?' since it was found that the growth of toroidal fields had important effects in inhibiting fragmentation and binary formation, even in the case of relatively weak fields. For example, it was found that in the absence of large initial fluctuations, there is no fragmentation when the magnetic field exceeds 5% of the 'critical' value (i.e. where magnetic fields can prevent gravitational collapse) and that rapid magnetic braking can even prevent disc formation if the field exceeds $\sim 20\%$ of its 'critical' value. Observed star forming cores, however, have magnetic fields that are close to being 'critical' (Crutcher 1999), leading Hennebelle & Teyssier (2008) to question whether the assumption of ideal MHD is correct or whether there is not a mechanism for accelerated field decoupling during the collapse process. On the other hand, the (cluster scale) simulations of Price & Bate (2009) do produce binaries in the presence of magnetic fields that are globally close to 'critical', suggesting perhaps that differences in the amplitude of initial perturbations and magnetic field morphology may explain the differences with the isolated core simulations. This is clearly an important question, given the undeniable existence of strong magnetic fields in star forming regions, and points to an important new direction for binary formation calculations in the coming years.

References

Artymowicz, P. 1983, *AcA*, 33, 223
Bate, M. 2000, *MNRAS*, 314, 33
Bate, M. 2009a, *MNRAS*, 392, 590
Bate, M. 2009b, *MNRAS*, 393, 1362
Bate, M. & Bonnell, I. 1997, *MNRAS*, 288, 1060
Bate, M., Bonnell, I., & Bromm, V. 2002a, *MNRAS*, 332, L65
Bate, M., Bonnell, I., & Bromm, V. 2002b, *MNRAS*, 336, 705
Bate, M., Bonnell, I., & Bromm, V. 2003, *MNRAS*, 339, 577
Bate, M., Bonnell, I., & Price, N. 1995, *MNRAS*, 277, 362
Bergfors, C. *et al.*, 2010, *A&A*, 520, 54
Boss, A. & Bodenheimer, P. 1979, *ApJ*, 234, 289

Clarke, C. 1996, *MNRAS*, 282, 353

Crutcher, R. 1999, *ApJ*, 520, 706

Delgado-Donate, E., Clarke, C., Bate, M., & Hodgkin, S. 2004a, *MNRAS*, 347, 759

Delgado-Donate, E., Clarke, C., Bate, M., & Hodgkin, S. 2004b, *MNRAS*, 351, 617

Duquennoy, A. & Mayor, M. 1991, *A&A*, 248, 485

Federath, C., Banerjee, R., Clark, P., & Klessen,R. 2010, *ApJ*, 713, 269

Fischer, D. & Marcy, G. 1992, *ApJ*, 396, 178

Goodman, A., Benson, P., Fuller, G., & Myers, P. 1993, *ApJ*, 406, 528

Goodman, J. & Hut, P. 1993, *ApJ*, 403, 271

Goodwin, S., Whitworth, A., & Ward-Thompson, D. 2004a, *A&A*, 414, 633

Goodwin, S., Whitworth, A., & Ward-Thompson, D. 2004a, *A&A*, 423, 169

Goodwin, S., Whitworth, A., & Ward-Thompson, D. 2006, *A&A*, 452, 487

Hanawa, T., Ochi, Y., & Ando, K. 2010, *ApJ*, 708, 485

Hennebelle, P. & Teyssier, R. 2008, *A&A*, 477, 25

Jiang, Y. & Tremaine, S. 2009, *MNRAS*, 401, 977

Konopacky, Q. *et al.*, 2010, *ApJ*, 711, 1087

Kouwenhoven, M., Goodwin, S., Parker, R., Davies, M., Malmberg, D., & Kroupa, P. 2010,
    MNRAS, 404, 1835

Krumholz, M., McKee, C., & Klein, R. 2004, *ApJ*, 611, 399

Lada, E. A., DePoy, D., Evans, N., & Gatley, I. 1991, *ApJ*, 371, 171

Law, N., Dhital, S., Kraus, A., Stassun, K., & West, A. 2011, *arXiv1101.0630*

Lubow, S. & Artymowicz, P. 2000, *Protostars & Planets IV, Univ. Arizona Press*, 731

Mazeh, T., Goldberg, D., Duquennoy, A., & Mayor, M. 1992, *ApJ*, 401, 265

Moeckel, N. & Bate, M. 2010, *MNRAS*, 404, 721

Ochi, Y., Sugimoto, K., & Hanawa, T., 2005, *ApJ*, 623, 922

Offner, S., Klein, R., McKee, C., & Krumholz, M. 2009, *ApJ*, 703, 131

Price, D. & Bate, M. 2009, *MNRAS*, 398, 33

Raghavan, D. *et al.*, 2010, *ApJS*, 190, 1

Reggiani, M. & Meyer, M. 2011, *ApJ*, 738, 60

Scally, A., Clarke, C., & McCaughrean, M. 1999, *MNRAS*, 306, 253

Siegler, N., Close, L., Cruz, K., Martin, E., & Reid, N. 2005, *ApJ*, 621, 1023

Thies, I. & Kroupa, P. 2007, *ApJ*, 671, 767

Val-Borro, M., Gahm, G., Stempels, H., & Pepinski, A. 2011, *MNRAS*, 413, 2679

Discussion

W. KLEY: You mentioned circumbinary disk simulations. I assume that they have been isothermal. How would a change in the equation of state alter the results?

C. CLARKE: Nobody has looked at this in detail. However, I would guess that the most important parameter in determining the flow morphology is the ratio of the local sound speed to the binary orbital speed for material at the inner edge of the circumbinary disc.

A. BURROWS: What is the efficiency of star formation in these simulations?

C. CLARKE: That varies from simulation to simulation, depending on how gravitationally bound is the parent cloud. This latter parameter affects the clustering parameters of the resulting stars but seems to have a rather minimal effect on the binary properties.

From Interacting Binaries to Exoplanets: Essential Modeling Tools
Proceedings IAU Symposium No. 282, 2011
Mercedes T. Richards & Ivan Hubeny, eds.
© International Astronomical Union 2012
doi:10.1017/S1743921311027967

Non-Conservative Evolution of Binary Stars

Christopher A. Tout

Institute of Astronomy, The Observatories, Madingley Road, Cambridge CB3 0HA, England
email: cat@ast.cam.ac.uk

Abstract. Various processes can lead to non-conservative evolution in binary stars. Under conservative mass transfer, both the total mass and the orbital angular momentum of the system are conserved. Thus, the transfer of angular momentum between the orbit and the spins of the stars can represent one such effect. Stars generally lose mass and angular momentum in a stellar wind so, even with no interaction, evolution is non-conservative. Indeed, a strong wind can actually drive mass transfer. During Roche lobe overflow itself, mass transfer becomes non-conservative when the companion cannot accrete all the material transferred by the donor. In some cases, material is simply temporarily stored in an accretion disc. In others, the companion may swell up and initiate common envelope evolution. Often the transferred material carries enough angular momentum to spin the companion up to break-up, at which point it could not accrete more. We investigate how this is alleviated by non-conservative evolution.

Keywords. binaries: close, stars: mass loss, planetary systems

In 1848, Édouard Roche formulated what came to be known as the Roche limit and began the theory of the Roche potentials. In the twentieth century, these became used to study the shapes of interacting binary stars and, with the explanation of the Algol paradox, that in a coeval binary the fainter component could be larger, redder and less massive than its companion (Hoyle 1955; and Crawford 1955), the concept of mass transfer by Roche lobe overflow became established. To understand how a mass-transferring close binary star evolves, we must know whether mass, and indeed angular momentum, are lost from the system in significant quantities during the mass transfer.

1. Conservative Evolution

Let us first define conservative mass transfer to be without mass loss or angular momentum loss from the system. That is, both the total mass and the total angular momentum are conserved. We shall restrict the definition further by assuming the stars behave as point masses, material flows between them uniformly over an orbital period, and no angular momentum can be stored in either the stars or circumstellar matter. For illustration, we shall consider circular orbits. Fig. 1 shows this simple set up in which the two stars with masses M_1 and M_2 orbit their centre of mass in circular orbits of radii a_1 and a_2 at angular velocity Ω. The total mass of the system $M = M_1 + M_2$ and the separation of the stars $a = a_1 + a_2$, so that $a_1 = aM_2/M$. The total orbital angular momentum is then

$$J = \frac{M_1 M_2}{M} a^2 \Omega \tag{1.1}$$

and the orbit obeys the generalised form of Kepler's third law

$$\left(\frac{2\pi}{P}\right)^2 = \Omega^2 = \frac{GM}{a^3}, \tag{1.2}$$

where P is the orbital period and G is Newton's gravitational constant.

Figure 1. A simple binary star consisting of two point mass stars in a circular orbit.

For conservative evolution, we then take the rate of change of total mass $\dot{M} = 0$ but allow mass transfer so that $\dot{M}_2 = -\dot{M}_1$ and $\dot{J} = 0$. Taking a logarithm of equation (1.1) and differentiating, we arrive at

$$\frac{\dot{J}}{J} = \frac{\dot{M}_1}{M_1} + \frac{\dot{M}_2}{M_2} + 2\frac{\dot{a}}{a} + \frac{\dot{\Omega}}{\Omega} = 0. \tag{1.3}$$

We do the same with Kepler's third law and integrate to obtain the familiar results

$$P(M_1 M_2)^3 = \text{const} \quad \text{and} \quad a(M_1 M_2)^2 = \text{const}. \tag{1.4}$$

Thus, for mass transfer from star 1 to star 2 both the period and separation decrease while $M_1 > M_2$, and subsequently increase.

2. Non-Conservative Evolution

When evolution is non-conservative, both $\dot{M} < 0$ and $\dot{J} < 0$. It is hard to envisage substantial mass loss without that mass carrying off angular momentum; and angular momentum cannot be lost without mass to transport it, though this mass may be very small in the case of magnetic braking or gravitational radiation. First, let us consider an isotropic stellar wind from one of our two point masses, star 1, when there is no mass transfer. So, star 1 loses mass at a rate $-\dot{M} = -\dot{M}_1$. This mass carries away its specific orbital angular momentum so that

$$\dot{J} = \dot{M} a_1^2 \Omega. \tag{2.1}$$

We can divide this by J and proceed as before to obtain the, again familiar, results

$$PM^2 = \text{const} \quad \text{and} \quad aM = \text{const}. \tag{2.2}$$

So, the period and separation increase. The solar system will grow as the Sun loses mass.

3. Stable Roche Lobe Overflow

Fig. 2 shows a star that only slightly overfills its Roche lobe. Subsonic flow accelerates to supersonic as it passes through the rocket-like nozzle at the L_1 point. The ballistic stream this creates is then confined to the Roche lobe of its companion. If the stream hits the companion before encircling it, material may accrete directly. If the companion is too small so that the stream has too much angular momentum to fall on directly, it encircles the star, collides with itself and spreads out to form an accretion disc. Viscosity in the disc allows a small amount of the mass to carry angular momentum outwards while most of the mass drifts inwards. The disc is truncated by tides within the Roche lobe of the accreting star and these same tides return angular momentum to the orbit. Thus, we do not expect material flowing through the L_1 point to be lost during the overflow. Any material that is lost must be driven away by some energy source associated with the secondary star or the accretion flow on to it.

However there is, at least circumstantial, evidence that binary systems do not evolve conservatively. All Algol-like systems observed today have mass ratios that are significantly inverted, in that the mass of the star that evolved first, and so was originally

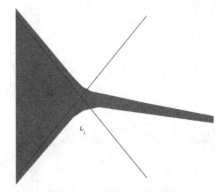

Figure 2. The region of the inner Lagrangian point L_1 for a star that only just overfills its Roche lobe. The equipotential surfaces at the L_1 point form a rocket-like nozzle through which material flows. On the left flow remains subsonic, so that the star can be treated as if in hydrostatic equilibrium, while on the right flow is supersonic and so ballistic.

the more massive, is now somewhat less than that of its accreting companion. Morton (1960) deduced that this meant that all Algols should have gone through a rapid mass transfer phase that was very short-lived compared to the state in which we observe them now. This can be explained if the initially more massive star first filled its Roche lobe in the Hertzsprung gap, as a subgiant, when loss of mass would lead the star to expand and transfer mass on a thermal timescale. Even then, the process of overflow ought to be conservative except that the secondary star cannot itself accrete the material fast enough. Indeed, Paczyński & Ziółkowski (1967) found that the resulting Algol systems are more realistic if half the mass transferred is actually lost from the system.

The need for angular momentum loss from binary stars is confirmed by the cataclysmic variables. Most of these consist of a low-mass red dwarf transferring mass to a somewhat more massive white dwarf. The red dwarf evolves far too slowly to be driving the mass transfer by its own nuclear evolution, and simple mass loss alone would widen rather than shrink the orbit. Indeed, the white dwarf in these systems must have evolved as the core of a red giant whose entire envelope has been lost from the system.

4. The Spins of the Stars

Before considering true non-conservative evolution, let us first consider relaxing the assumption that the stars are point masses. As a star grows to fill its Roche lobe, tides tend to lock its spin with the orbit, and angular momentum flows between the star and the orbit. Typically, a giant star has spun down as it expanded and so requires more angular momentum to spin it up. From Kepler's third law and our expression for orbital angular momentum (equation 1.1), transfer of angular momentum from the orbit to the star shrinks the orbit and so increases its angular velocity. Conversely, a hot main-sequence star may need to spin down to reach corotation with its orbit, and transfer of angular momentum from the star to the orbit expands the orbit and reduces its angular velocity. Thus, either both the star and orbit spin up or both spin down.

To estimate the angular momentum that can be stored in a star, we consider star 1 of radius R_1 to be extended and to have a radius of gyration $k_1 R_1$, and so moment of inertia $I_{star} = M_1 k_1^2 R_1^2$ and angular momentum $J_{spin,1}$ such that

$$\frac{J_{spin,1}}{J_{orb}} = \frac{M_1 (k_1 R_1)^2 M}{(M_1 M_2 a^2)}. \tag{4.1}$$

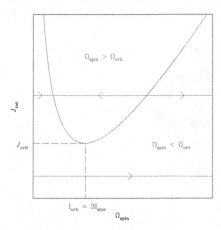

Figure 3. The total angular momentum of a binary system with angular velocity Ω_{orb} one expanded star spinning at Ω_{spin}. The solid curve is the tidal equilibrium. Horizontal lines are the direction of evolution when not in equilibrium. Though most detached and semi-detached binaries lie in regions where they can evolve to stable equilibrium, some contact systems and many close orbiting planetary systems are ultimately unstable.

For two $1\,M_\odot$ main-sequence stars, with $k_1^2 \approx 0.1$, and a separation of $10\,R_\odot$

$$\frac{J_{\text{spin},1}}{J_{\text{orb}}} \approx 0.1 \times \left(\frac{1}{10}\right)^2 \times 2 = 2 \times 10^{-3} \tag{4.2}$$

and this we might neglect. However, for a $5\,M_\odot$ giant with $k_1^2 \approx 0.2$ for its envelope, a core of $M_c = 0.6\,M_\odot$, envelope radius of $R = 300\,R_\odot$, almost filling its Roche lobe when $a = 500\,R_\odot$ and a companion of $0.5\,M_\odot$,

$$\frac{J_{\text{spin},1}}{J_{\text{orb}}} \approx 0.2 \times \left(\frac{300}{800}\right)^2 \times \frac{4.4 \times 5.5}{5.0 \times 0.5} \approx 0.27 \tag{4.3}$$

which becomes significant. Thus, the orbit of a system can be very much reduced by the time that a giant fills its Roche lobe. Indeed for large mass ratios, spinning up the star can take more angular momentum than is available in the orbit. Close binary systems with extreme mass ratios are most prone, and planets that can spin up their host star are very vulnerable. The total angular momentum of a system with one expanded star is

$$J_{\text{tot}} = I_{\text{star}}\Omega_{\text{spin}} + \frac{M_1 M_2}{M}a^2\Omega_{\text{orb}}. \tag{4.4}$$

Fig. 3 shows how this varies with the angular velocity of the star's Ω_{spin}. The left branch of the equilibrium, with $3I_{\text{star}} < I_{\text{orb}} = M_1 M_2 a^2/M$, is stable but the right, with $3I_{\text{star}} > I_{\text{orb}}$, is unstable. Our typical giant spinning more slowly than its orbit lies to the left of the stable branch. As tides spin it up, it moves to the right. However, as it evolves, tidally locked I_{star} increases so that J_{tot} effectively falls, and it descends the stable track. Our example system (equation 4.3) is dangerously close to the minimum J_{crit} beyond which there is no stable equilibrium. Most contact binaries eventually suffer this fate. A hot star spinning faster than its orbit typically lies between the two equilibria and tidally evolves towards the stable branch to the left. After the discovery of 51 Peg B, Rasio *et al.* (1996) soon pointed out that, if the planet were able to raise tides on the star sufficient to spin it up, the planet would fall into the star because $J_{\text{tot}} < J_{\text{crit}}$. The fact that it doesn't, tells us about the strength of the tides it can raise on the star.

5. Consequential Angular Momentum Loss

Most often, non-conservative mass transfer is taken to mean mass and angular momentum loss associated with the process of Roche lobe overflow itself. This needn't be due to part of the stream leaving the system at the L_1 point. Indeed, it is more likely associated with the accretion process on to the companion. When accretion is via a disc there must be some consequential angular momentum loss. Consider star 2 of radius R_2, mass M_2 and spin Ω, accreting from the inner edge of a Keplerian disc. Its angular momentum $J_2 = I_2\Omega \approx k_2^2 M_2 R_2^2 \Omega$ while material accretes at the star's break-up spin of $\Omega_{\mathrm{max}} = \sqrt{GM_2/R_2^3}$ so that $\delta J_2 = \delta M R_2^2 \Omega_{\mathrm{max}}$. Then, the maximum mass that can be accreted conservatively is about $\delta M_2 \approx k_2 M_2$ because stars accreting from a disc are easily spun up. However, white dwarfs in cataclysmic variables spin slowly and primary stars in Algol systems wide enough for an accretion disc spin at most at about four tenths of break-up. Tides are not sufficiently strong in the hot Algol primaries to prevent rapid spin up and so some extra angular momentum loss is necessary. Dervişoğlu, Tout & İbanoğlu (2010) argued that this is most likely by the excitation of a magnetic dynamo generated precisely because the star has been spun up by accretion.

Once strong large scale magnetic fields have been generated, angular momentum loss by magnetic braking (Schatzman 1962) is inevitable. Suppose mass in a stellar wind is forced to effectively corotate to some Alfvén radius R_A (Mestel 1968). Then

$$\dot{J} = \dot{M}a_1^2\Omega + KJ, \tag{5.1}$$

with

$$K = \frac{2}{3}\left(\frac{R_A}{R}\right)^2\left(\frac{R}{a}\right)^2\frac{M}{M_1 M_2}\dot{M}. \tag{5.2}$$

Note that $R_A/R \geqslant 1$ always, and even this minimum is important for extreme mass ratios when $R_1 > a_1$.

Similarly, rapidly spinning magnetised accretors can eject the transferred mass by a propeller effect. When the magnetic field of an accreting star is so high that the magnetic energy density in the stream is similar to its kinetic energy, the transferred material is forced to flow along the field lines. This is the case in the polars, and intermediate polars, cataclysmic variables with a highly magnetic white dwarf that accretes at the magnetic poles. In the polars, the spin of the white dwarf is locked to the orbit, but in cases where the white dwarf has been spun up beyond this, perhaps by an episode of disc accretion, material can be accelerated by the magnetic field and flung out of the system. A probable example is the cataclysmic variable AE Aquarii (Eracleous & Horne 1996) and similar effects are expected for accreting magnetic neutron stars.

6. Common Envelope Evolution

Perhaps the most extreme case of non-conservative evolution is that of a common envelope phase. Paczyński (1976) described how such a process is necessary to account for the very short orbital periods of cataclysmic variables and their immediate progenitors. The white dwarf must originally have been the core of a giant of some many tens to hundreds of solar radii. Thus, a substantial amount of both angular momentum and energy must have been lost from the orbit along with the giant's massive envelope. Fig. 4 illustrates the process which remains one of the most important, but least understood, in binary star evolution. The onset of common envelope evolution follows a giant filling its Roche lobe and transferring mass to a less massive companion. The giant responds to mass loss by expanding its convective envelope, while the orbit responds by shrinking.

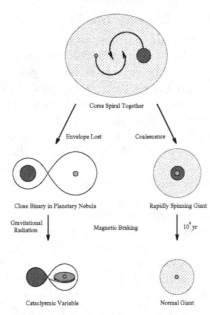

Figure 4. Common-envelope evolution. After dynamical mass transfer from a giant, a common envelope enshrouds the relatively dense companion and the core of the original giant. These two spiral together as their orbital energy is transferred to the envelope until either the entire envelope is lost or they coalesce. In the former case a close white-dwarf and main-sequence binary is left, initially as the core of a planetary nebula. Magnetic braking or gravitational radiation may shrink the orbit and create a cataclysmic variable. Coalescence results in a rapidly rotating giant which will very quickly spin down by magnetic braking.

This positive feedback means that the giant overfills its Roche lobe yet more, and the rate of mass transfer rises towards a dynamical limit. The companion certainly cannot accrete at such a rate and the material forms a common envelope.

Some wide Algol-like systems ought not to have avoided this catastrophe. Tout & Eggleton (1988) argued that mass loss before Roche lobe overflow can be sufficiently enhanced by a companion that the mass ratio can invert well before the onset of mass transfer so that the Roche lobe grows faster than the star and mass transfer can proceed on a stable nuclear timescale. They proposed that

$$\dot{M} \to \dot{M}\left\{1 + B\min\left[\frac{R}{R_L}, \frac{1}{2}\right]\right\}^6 \tag{6.1}$$

and from RS CVn systems with inverted mass ratios $B \approx 10^4$. Tout & Hall (1991) went on to say that, because red giants expand on losing mass, a strong stellar wind can actually drive mass transfer. To see this, we may approximate the radius of a giant by $R_1 = f(L_1)M_1^{-n}$, where $f(L_1)$ is some function of the star's luminosity L_1, which varies only as its core grows on a nuclear timescale. Typically, $n \approx 0.27$. Thus, when the timescale for mass transfer is fast compared to nuclear evolution

$$\frac{\dot{R}_1}{R_1} = -n\frac{\dot{M}_1}{M_1}. \tag{6.2}$$

Combining this with Paczyński's formula for the Roche lobe radius,

$$\frac{R_L}{a} = \frac{2}{3^{4/3}}\left(\frac{M_1}{M}\right)^{\frac{1}{3}}, \qquad 0 < M_1/M_2 < 0.8, \tag{6.3}$$

we find that when $R_1 = R_L$ then $\dot{R}_1 > \dot{R}_L$ when

$$q = \frac{M_1}{M_2} < \frac{1 + 3n}{3(1 - n)}, \tag{6.4}$$

so that mass loss actually drives the giant to overflow its Roche lobe and mass transfer begins. When such mass transfer is stable, $R_1 \approx R_L$ and then

$$\dot{M}_2 = \frac{1 + 3n - 3(1 - n)q}{(1 + q)(5 - 3n - 6q)} \dot{M}. \tag{6.5}$$

If $6q > 5 - 3n$, mass transfer is dynamically unstable. At lower q, $\dot{M}_2 \approx -0.5\dot{M}$ and about half as much mass as is lost in the wind is transferred to the companion as well.

7. Stellar Evolution Codes

At least two versions of the Cambridge STARS code (Eggleton 1971) can be used to evolve both components of a binary star simultaneously, taking care of mass and angular momentum transport by various physical mechanisms. These are the TWIN code (Nelson & Eggleton 2001) and the BS code (Stancliffe & Eldridge 2009). Versions of both are available from http://www.ast.cam.ac.uk/ stars/index.html for those who wish to experiment with them. Users are, however, warned that neither version has been written with the casual user in mind, and it would be beneficial to seek out an existing user before engaging in a project. Those listed on the web pages are often very happy to assist.

8. Conclusions

All binary star evolution is non-conservative to some extent. There is both direct and indirect evidence for both mass and angular momentum loss. However, there is very little agreement on the physics and its implementation. Common envelope evolution is the most extreme, but perhaps least understood, non-conservative process.

We have only picked out a few of the drivers of non-conservative evolution. There are many others. Nova eruptions on the surface of white dwarfs in cataclysmic variables occur because hydrogen ignites degenerately when a shell of $10^{-4} M_\odot$ or so has accumulated. The violence of the thermonuclear runaway throws off most of this accreted material, presumably along with the specific angular momentum of the white dwarf's orbit. Supernovae explosions can expel more than half the mass in a binary system, perhaps with a violent kick. This surely unbinds most systems, but the existence of X-ray binaries and binary millisecond pulsars shows that neutron stars can often be retained. Most accretion discs are associated with mass loss in jets or outflows, with typically one tenth of the accreted mass being lost. Gravitational radiation is probably the best understood mechanism for angular momentum loss, with negligible mass loss. General relativity describes precisely how fast it should operate and this is backed up by observations of many close binary systems.

We have only briefly touched on the uncertainties of stellar mass loss rates. Even for single stars it is mass loss that determines the end points of stellar evolution. But when does a star lose most of its mass? Is it a relatively gradual process that slowly increases over its evolution or is it all lost at the end in an asymptotic giant's superwind (Vassiliadis & Wood 1993). In a binary, it matters how fast the material leaves the star because, if it is slow enough, a significant fraction of the stellar wind can be accreted by a companion as it orbits through the expanding gas. This is just yet another uncertainty in the evolution of binary stars still to be fully understood.

CAT is very grateful to Churchill College Cambridge for his Fellowship and to the STFC for funding his attendance at this symposium.

References

Crawford, J. A. 1955, *ApJ*, 121, 71

Dervişoğlu, A., Tout, C. A., & İbanoğlu, C. 2010, *MNRAS*, 406, 1071

Eracleous, M. & Horne, K. 1996, *ApJ*, 471, 427

Hoyle, F. 1955, *Frontiers of Astronomy*, (London: Heinemann)

Montez, R., Jr., De Marco, O., Kastner, J. H., & Chu, Y.-H. 2010, *ApJ*, 721, 1820

Mestel, L. 1968, *MNRAS*, 138, 359

Morton, D. C. 1960, *ApJ*, 132, 146

Paczyński, B. 1976, in: P. P. Eggleton, S. Mitton S. & J. Whelan (eds), *Structure and Evolution of Close Binary Systems*, Proc. IAU Symposium No. 73, (Dordrecht: Reidel), p. 75

Paczyński, B. & Ziółkowski, J. 1967, *AcA* 17, 7

Rasio, F. A., Tout, C. A., Lubow, S. H., & Livio, M. 1996, *ApJ*, 470, 1187

Schatzman, E. 1962, *Ann. Astrophys.*, 25, 18

Tout, C. A. & Eggleton, P. P. 1988, *MNRAS*, 231, 823

Tout, C. A. & Hall, D. S. 1991, *MNRAS*, 253, 9

Vassiliadis, E. & Wood, P. R. 1993, *ApJ*, 413, 641

Discussion

O. DE MARCO: Common envelope simulations/theory indicate that envelope mass does not accrete on to the secondary star. However, X-ray observations of post common envelope binaries (Montez *et al.* 2010) are consistent with a spun up secondary (which has an X-ray bright corona generated by the spin). What is your opinion about accretion on to the secondary?

C. TOUT: The length of the dynamical phase is so short that there is no time to accrete. So, it would follow that there is no accretion. However, I think that if the CE interaction evolved over a longer time, there could be time for accretion. The observations you mention may argue for such a longer phase.

P. HARMANEC: Is there evidence for much mass loss from Algol systems?

C. TOUT: There is no quantitative measure of mass loss from Algol systems, all of which are observed now in a slow phase of mass transfer. I expect mass loss from the system at a rate similar to the mass transfer rate could easily escape unnoticed. Some evidence might be forthcoming from orbital period variations but the only star to show a measurable long term change is U Cep, the mass ratio of which places it so close to dynamical instability that only a very small systemic mass loss can drive substantial mass transfer and so it is hard to distinguish between nuclear evolution and wind-driven mass transfer.

G. PETERS: The amount of mass lost in an Algol system is unknown but we can identify a number of sites of mass loss that include, (1) a splash region caused by the gas stream impact in direct-impact systems, (2) the commonplace mass loss near phase 0.5 and (3) bipolar flows. Furthermore, Walter van Rensbergen has shown that hot accretion spots can drive mass out of the system. Given the uncertainties in the area over which the mass loss occurs and the outflow velocities (we usually derive a projected velocity), the extent of the total mass loss remains an open question.

From Interacting Binaries to Exoplanets: Essential Modeling Tools
Proceedings IAU Symposium No. 282, 2011
Mercedes T. Richards & Ivan Hubeny, eds.
© International Astronomical Union 2012
doi:10.1017/S1743921311027979

Direct Imaging of Bridged Twin Protoplanetary Disks in a Young Multiple Star

Satoshi Mayama[1], Motohide Tamura[1,2], Tomoyuki Hanawa[4],
Tomoaki Matsumoto[5], Miki Ishii[3], Tae-Soo Pyo[3], Hiroshi Suto[2],
Takahiro Naoi[2], Tomoyuki Kudo[3], Jun Hashimoto[2],
Shogo Nishiyama[2], Masayuki Kuzuhara[6] and Masahiko Hayashi[6]

[1] The Graduate University for Advanced Studies (SOKENDAI),
Shonan International Village, Hayama-cho, Miura-gun, Kanagawa, 240-0193, Japan
email: mayama_satoshi@soken.ac.jp
[2] National Astronomical Observatory of Japan, 2-21-1, Osawa, Mitaka, Tokyo 181-8588 Japan
[3] Subaru Telescope, National Astronomical Observatory of Japan,
650 North A'ohoku Place, Hilo, Hawaii, 96720, USA
[4] Center for Frontier Science, Chiba University, Inage-ku, Chiba, 263-8522, Japan
[5] Faculty of Humanity and Environment, Hosei University,
Fujimi, Chiyoda-ku, Tokyo 102-8160, Japan
[6] University of Tokyo, Hongo, Tokyo 113-0033, Japan

Abstract. Protoplanetary disks are ubiquitously observed around young solar-mass stars and are considered to be not only natural by-products of stellar evolution but also precursors of planet formation. If a forming star has close companions, the protoplanetary disk may be seriously influenced. It is important to consider this effect because most stars form as multiples. Thus, studies of protoplanetary disks in multiple systems are essential to describe the general processes of star and planet formation.

We present the direct image of an interacting binary protoplanetary system. We obtained an infrared image of a young multiple circumstellar disk system, SR24, with the Subaru 8.2-m Telescope. Both circumprimary and circumsecondary disks are clearly resolved with a 0.1 arcsecond resolution. The binary system exhibits a bridge of infrared emission connecting the two disks and a long spiral arm extending from the circumprimary disk. A spiral arm would suggest that the SR24 system rotates counter-clockwise. The orbital period of the binary is 15,000 yr. Numerical simulations reveal that the bridge corresponds to gas flow and a shock wave caused by the collision of gas rotating around the primary and secondary stars. The simulations also show that fresh material streams along the spiral arm, confirming the theoretical proposal that gas is replenished from a circum-multiple reservoir. These results reveal the mechanism of interacting protoplanetary disks in young multiple systems. Furthermore, our observations provide the first direct image that enables a comparison with theoretical models of mass accretion in binary systems. The observations of this binary system provide a great opportunity to test and refine theoretical models of star and planet formation in binary systems.

Keywords. stars:binaries, stars:formation, stars:planetary systems: protoplanetary disks

---

In a binary system, both the primary and secondary stars in orbit around each other have circumprimary and circumsecondary disks, respectively, and the entire system is surrounded by a circumbinary disk. Numerical simulations demonstrate that the stability of a protoplanetary disk in a multiple system is seriously jeopardized (Artymowicz & Lubow 1994). In simulations, despite the dynamical interactions between disks and stars, individual circumstellar disks can survive and large gaps are produced in the circumbinary disk. A circumbinary disk can supply mass to the circumstellar disks through a gas stream that penetrates the disk gap without closing it. Therefore, this infalling material through the spiral arm plays an important role in the formation of circumstellar disks.

However, such circum-multiple disks and spiral arms in multiple systems have never been directly imaged or resolved to date. Here, we report (Mayama *et al.* 2010) and analyse an infrared image of a young multiple system with bridged twin disks obtained with the adaptive optics (AO) (Takami *et al.* 2004) coronagraph CIAO (Tamura *et al.* 2000) mounted on the Subaru 8.2-m Telescope. In our study, we investigate the geometry of a multiple circumstellar disk system to understand its nature based on observation and numerical simulation. We study the hierarchical multiple SR24, located in the Ophiuchus star forming region at a distance of 160 pc (Chini 1981), composed of the low-mass T Tauri type stars SR24S (primary) and SR24N (secondary). SR24S is a class II source (stellar age of 4 Myr; Andrews & Williams 2007) of spectral type K2, with a mass of $>1.4 \, M_\odot$ (Correia *et al.* 2006), where $M_\odot$ is the mass of the Sun. SR24N is located 810 AU north of SR24S (Correia *et al.* 2006) and is itself a binary system composed of SR24Nb and SR24Nc, with a projected separation of 30 AU (Correia *et al.* 2006). The spectral type and mass of SR24Nb are K4-M4 and $0.61 \, M_\odot$, respectively (Correia *et al.* 2006). Those of SR24Nc are K7-M5 and $0.34 \, M_\odot$, respectively (Correia *et al.* 2006). Because the separation between SR24Nb and SR24Nc is comparable to the angular resolution and is much smaller than that between SR24N and SR24S, we handle SR24Nb and SR24Nc together in this paper as SR24N, with a mass of $0.95 \, M_\odot$. Accordingly, we regard the SR24 system as a binary with a primary to secondary mass ratio of 0.68, assuming the mass of SR24S to be $1.4 \, M_\odot$.

The left panel of Figure 1 shows the H-band (1.6 μm) image of SR24 overlaid with the Roche lobe contours, showing the effective gravitational potential of the binary. SR24S was occulted by a coronagraphic mask during the observations. The image shows the residual emission after the point spread function (PSF) was subtracted from SR24S and SR24N, revealing faint near-infrared nebulosity at a resolution of 0.1 arcsecond. The emission arises from dust particles mixed with gas in the circumstellar structures scattering the stellar light. Both circumprimary and circumsecondary disks are clearly resolved, the first time these have been imaged for a young stellar binary. The primary disk has a radius of 420 AU and is elongated in the northeast-southwest direction. The secondary disk has a radius of 320 AU and is elongated in the east-west direction. Both disks overflow the inner Roche lobes (dotted contours in the figure), which show the regions gravitationally bound to each star, suggesting that the material outside the lobes can fall into either of the inner lobes. A curved bridge of emission is seen, connecting the primary and secondary disks. This emission begins at the southeast of the secondary disk, extending to the south while curving to the west, and reaches the north edge of the primary disk. This suggests a physical link, such as a gas flow between the two disks. Another salient feature is a broad arc starting from the southwestern edge of the primary disk, extending to the southeast through the Lagrangian point L3. Its tail is at least 1600 AU from SR24S. This emission is most likely a spiral arm and demonstrates that the SR24 system rotates counter-clockwise. The orbital period of the binary is 15,000 yr. The arm also implies replenishment of the twin disk gas from the circumbinary disk.

We performed 2D numerical simulations of accretion from a circumbinary disk to identify the features seen in the coronagraphic image. The right panel of Figure 1 shows a snapshot of the 2D simulations. The mass of SR24S is assumed to be $1.4 \, M_\odot$ and the orbit is assumed to be circular, for simplicity. Although the gas flow is not stationary, especially inside the Roche lobes, the stage of the 2D simulations shown in Figure 1 shares common features with the observed image in the following two points. (1) A bridge is seen connecting the primary and secondary disks. It runs through the Lagrange point L1. (2)

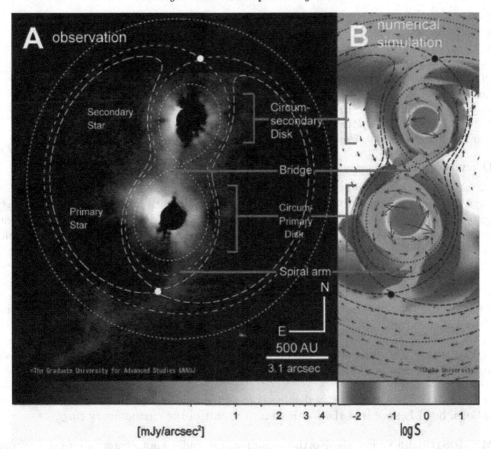

Figure 1. Observed and simulated images of the young multiple star, SR24. (A) H-band (1.6 μm) coronagraphic image of SR24. The PSFs of the final images have sizes of 0.1 arcsecond (FWHM) for the H-band. L1, L2, and L3 represent the inner Lagrangian point, outer Lagrangian point on the secondary side, and outer Lagrangian point on the primary side, respectively. (B) Snapshot of accretion onto the binary system SR24 based on 2D numerical simulations. The color and arrows denote the surface density distribution and velocity distribution, respectively.

A long spiral arm runs through the Lagrange point L3, with a pitch angle consistent with that of the observed spiral arm. These agreements suggest the following interpretations of the bridge and spiral arm associated with the SR24 system: The bridge corresponds to gas flow and a shock wave caused by the collision of gas rotating around the primary and secondary stars. The arm corresponds to a spiral wave excited in the circumbinary disk. Note that the bridge and spiral arm seen in the simulations are wave patterns and their shapes fluctuate with time.

Our observations provide the first verification of the theory that gas is replenished from the circumbinary disk to circumstellar disks, which was originally proposed by Artymowicz & Lubow (1996) but has not been confirmed by direct observations. Our direct imaging observations visualized the subsistent structures associated with a young multiple system which cannot be reproduced by spectroscopic observations or SED-model studies. Our research is the first observation of the mechanism of interacting protoplanetary disks in young multiple systems, and thus will contribute to a better understanding of star and planet formation in binary systems on the basis of subsistent morphology.

References

Andrews, S. M. & Williams, J. P. 2007, *ApJ*, 659, 705
Artymowicz, P. & Lubow, S. H. 1994, *ApJ*, 421, 651
Artymowicz, P. & Lubow, S. H. 1996, *ApJ*, 467, L77
Chini, R. 1981, *A&A*, 99, 346
Correia, S., Zinnecker, H., Ratzka, T., & Sterzik, M. F. 2006, *A&A*, 459, 909
Mayama, S., *et al.*, 2010, *Science*, 327, 306
Takami, H., *et al.*, 2004, *PASJ*, 56, 225
Tamura, M., *et al.*, 2000, *Proc. SPIE*, 4008, 1153

Discussion

C. CLARKE: This is beautiful work. Note that your simulations (which have rather a large ratio of sound speed to binary orbital velocity) are well matched to your observed system, which is wide (has a low orbital velocity). One would not expect to see the same bridge phenomenon if one was able to image close binary systems. So, this result should not be interpreted as evidence in favour of flow from secondary to primary in general.

C. LEE: How did you derive the masses of components?

S. MAYAMA: The masses of both components used in this study were from a previous observational study.

S. DAEMGEN: The simulations used a circumbinary disk. Are you able to see this disk in any of your observations?

S. MAYAMA: The circumbinary disk was out of the field of view of the SUBARU observations, but I believe that the southern arm is part of the circumbinary ring.

M. MONTGOMERY: For the Northern Source, the radio knot center is off-center from the IR. Have they followed up with X-ray observations to see if jet is losing mass from observations?

S. MAYAMA: The literature to date says no.

S. HINKLEY: How do you explain the brightness asymmetries in the post-subtraction image of your stars while the numerical model is very symmetric?

S. MAYAMA: The eastern part is brighter in both the circumprimary and circumsecondary disks. We interpret this as due to disk inclinations. We suggest that eastern part is the near side of the disk if we assume that forward scattering dominates, as is the case for Mie scattering of dust grains in the disks.

From Interacting Binaries to Exoplanets: Essential Modeling Tools
Proceedings IAU Symposium No. 282, 2011   © International Astronomical Union 2012
Mercedes T. Richards & Ivan Hubeny, eds.   doi:10.1017/S1743921311027980

Formation and Orbital Evolution of Planets

Wilhelm Kley

Institut für Astronomie & Astrophysik
Universität Tübingen, Morgenstelle 10, 72076 Tübingen, Germany
email: `wilhelm.kley@uni-tuebingen.de`

Abstract. The formation of planetary systems is a natural byproduct of the star formation process. Planets can form inside the protoplanetary disk by two alternative processes. Either through a sequence of sticking collisions, the so-called sequential accretion scenario, or via gravitational instability from an over-dense clump inside the protoplanetary disk. The first process is believed to have occurred in the solar system. The most important steps in this process will be outlined. The observed orbital properties of exoplanetary systems are distinctly different from our own Solar System. In particular, their small distance from the star, their high eccentricity and large mass point to the existence of a phase with strong mutual excitations. These are believed to be a result of early evolution of planets due to planet-disk interaction. The importance of this process in shaping the dynamical structure of planetary systems will be presented.

Keywords. planetary systems: formation, accretion disks, hydrodynamics

1. Introduction

Already in ancient times, it was recognized that planets are special objects in the heavens that are distinguished from the stars through their motion across the sky. Only much later, it was appreciated that planets are in orbit around our Sun, and the dynamical structure of the whole planetary system has been analyzed subsequently in much more detail. The most important facts are, that the Solar System planets are orbiting the Sun in nearly circular orbits, in the same direction and basically all in one plane, the ecliptic. Based on this information, Kant and Laplace developed in the 1750s a formation theory for the Solar System including the Sun and the planets. This model implied the collapse/contraction of a previously larger cloud, its subsequent flattening, and the condensation of the planets in the forming disk. Interestingly, these early ideas still form the basics of today's models of planet formation.

In these disks, planets can form along two different pathways. In the sequential accretion scenario, the growth proceeds bottom-up from small to large particles; this will be explained in more detail below. An alternative scenario of planet formation is the direct fragmentation of an unstable disk. This occurs if a local density perturbation is strong enough such that it begins to contract under its own self-gravity (Toomre 1964). Important for the occurrence of this gravitational instability is the mass of the disk and its cooling ability. For typical conditions around young stars, it is expected that only distant planets (e.g. around HR 8799) can form via this way.

2. The formation of the solar system

The protosolar nebula consisted of 99% gas and about 1% dusty material. In today's terms, this would be described as an accretion disk which transfers material inward to be accumulated by the protostar and transports angular momentum outward. The angular momentum transport in these disks is believed to be driven by magneto-rotational turbulence (Balbus & Hawley 1993). The details and efficiency is calculated typically via 3D magneto-hydrodynamical simulations employing finite volume methods. For an up-to-date simulation involving full radiative transport, see Flaig *et al.* (2010) and Fig. 1.

The standard scenario for planet formation in the solar system in such disks, follows the *sequential accretion scenario*, i.e. the small dust particles in the gaseous disk collide with each other driven by Brownian and turbulent motions in the gas disk. For small relative velocities, the particles stick to each other through Van der Waals forces and form larger, fluffy and fractal aggregates. For a review on this early phase of planet formation see Blum & Wurm (2008). In the subsequent growth phase, the particles decouple from the gas because they move with a different velocity, since they do not feel the gas pressure. This leads to larger relative collision velocities such that mutual collisions might result in fragmentation rather than growth of objects. Additionally, these particles show a relatively fast inward drift towards the star. Together, these two obstacles to planetesimal formation (fragmentation and drift) are often referred to as the meter size barrier (Weidenschilling 1977, Nakagawa *et al.* 1986). Possible solutions to circumvent this problem are accumulation of particles in long-lived vortices (Klahr & Bodenheimer 2006), or turbulent eddies (Johansen *et al.* 2007), or different material properties. Concerning the last option, it was noticed some time ago that, for typical relative velocities, the collision of the rocks (basalt spheres) will lead to full fragmentation and destruction of the two bodies (Benz & Asphaug 1999). But, since it is known that pristine objects in the solar system (asteroids, comets) can have average densities far below the corresponding bulk density of the underlying material, it has been suggested that the initial growth might be facilitated through a high porosity of the material (Dominik *et al.* 2007). Indeed, recent simulations of aggregate collisions show that under suitable conditions this might lead to sticking and growth. A very suitable numerical method is a modified version of *SPH* which includes solid state physics (Libersky *et al.* 1993, Schäfer *et al.* 2007). Using this technique, Geretshauser *et al.* (2010) have first calibrated the code with experimental laboratory data and have then analyzed the outcomes of mutual collisions; see Geretshauser *et al.* (2011).

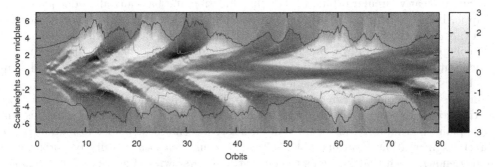

Figure 1. Space-time evolution of the horizontally averaged azimuthal magnetic field (B_y) for local fully radiative 3D magneto-hydrodynamical simulations of accretion disk turbulence. The green line denotes the location of the photosphere ($\tau = 1$ surface) and the blue line the location of the magnetosphere ($p_{\mathrm{gas}} = p_{\mathrm{mag}}$) of the disk. Figure adapted from Flaig *et al.* (2010).

After having overcome these initial obstacles, planetary cores may form via gravity-assisted growth where the geometrical cross section for the colliding partner can be greatly enhanced through gravitational focussing. Following the so called core-instability model, the gas giants can then form through rapid, runaway accumulation of gas onto the planetary cores (Perri & Cameron 1974), forming, for example, Jupiter and Saturn in the Solar System which are believed to have solid cores in their centers. During the whole growth phase of planets, they remain embedded in the gaseous disk from which they form. This leads to gravitational interactions between the planet and the disk, which will be described in more detail further below. After reaching a critical mass of about $1/2\, M_{\rm Jup}$, an annular gap forms at the location of the planet, due to angular momentum transfer from the planet to the gas, which limits its further mass growth. This effect has been known for some time in the context of disk-satellite interaction (Goldreich & Tremaine 1980). Altogether, the sequential accretion scenario describes the formation and present physical and dynamical structure of the Solar System (inner rocky, outer gaseous planets, flatness, angular momentum, maximum mass) very well.

3. Extrasolar planetary systems

The orbital elements of the observed extrasolar planets are distinctly different from the Solar System. While major Solar System planets have a nearly coplanar configuration and orbits with small eccentricity, the exoplanet population displays large eccentricities, and many planets orbit their host star on very tight orbits. Recently, it has been discovered that inclined and retrograde orbits are quite frequent as well, at least for close-in planets. Historically, it was exactly the relatively 'calm' dynamical structure of the solar system that led to the hypothesis that planets form in protoplanetary disks. The discovery of the hot planets, which could not have been formed in-situ described by the above scenario due to the hot temperatures and limited mass reservoir, gave rise to the exploration of dynamical processes that are able to change the location of planets in the disk, accompanying the regular formation process.

The density disturbances in the disk induced by the growing protoplanet back-react onto the planet and exert gravitational torques which can change the orbital elements of the planet. It has been suggested that it is possible to bring a planet that has formed at large distances from the star to its close proximity (Ward 1986, Lin *et al.* 1996). This migration process has provided a natural explanation for the population of hot planets, and their existence has been considered as evidence for the migration process.

Another indication for a planetary migration process comes from the high fraction (nearly 20%) of configurations in a low order mean-motion resonance, within the whole sample of multi-planet systems. As the direct formation of such systems seems unlikely, only a dissipative process that changes the energy (semi-major axis) is able to bring planets from their initial non-resonant configuration into resonance. Since resonant capture excites the planetary eccentricities typically to large values in contradiction to the archetypical system GJ 876, it has been inferred that planet-disk interaction should lead to eccentricity damping (Lee & Peale 2002, Crida *et al.* 2008).

4. Planetary migration

An embedded object disturbs the ambient disk dynamically in two important ways: First, it divides it into an inner and outer disk separated by a co-orbital (horseshoe) region. Secondly, the propagating sound waves that are sheared out by the Keplerian differential rotation generate density waves in the form of spiral arms in the disk. The induced density structures in the co-orbital region and the spiral arms back-react on the planet and cause a change in its orbital elements. The main important parameter to be studied is the semi-major axis, i.e. the migration of the embedded planet.

Spiral arms: To put it simply, the spiral arms can be considered as density enhancements in the disk that 'pull' gravitationally on the planet. This gives rise to so-called *Lindblad torques* that change the planet's angular momentum. For circular orbits, the disk torque exerted on the planet is directly a measure of the speed and direction of migration. The inner spiral forms a leading wave that causes a positive torque, while the outer wave generates a negative contribution. The combined effects of both spirals determine then the sign and magnitude of the total torque. A positive total torque will add angular momentum to the planet and cause outward migration. On the other hand, a negative torque will induce inward migration. It turns out that under typical physical disk conditions the contributions of the inner and outer spiral arm are comparable in magnitude. However, the effect of the outer spiral quite generally wins over the inner one causing the planet to migrate inward.

Corotation region: As viewed in the corotating frame, material within the corotation region performs so-called horseshoe orbits. Here, the gas particles come close to the planet, at the two ends of the horseshoe orbit and perform U-turns. They are periodically shifted from an orbit with a semi-major axis slightly larger than the planetary one to an orbit with slightly smaller value, and vice versa. Hence, at each close approach with the planet, there is an exchange of angular momentum between (co-orbital) disk material and the planet. The total corotation torque is then obtained by adding the contributions from both ends of the horseshoe. To obtain a net, non-zero torque, an asymmetry between the two U-turns has to exist. This requires non-vanishing radial gradients of vortensity and entropy across the corotation region (Baruteau & Masset 2008). For an ideal gas without friction or heat diffusion, mixing effects within the horseshoe tend to flatten out these gradients yielding a vanishing corotation torque, or so-called torque saturation.

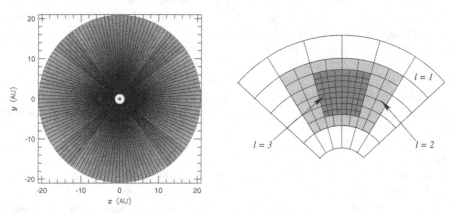

Figure 2. The numerical grid structure employed in numerical planet-disk simulations. **Left**: The global grid structure for a 128×128 grid. Indicated is the Roche-lobe of a Jupiter-type planet, located at $(-5.2, 0.0)$. **Right**: The structure of a nested grid, shown here with two sub-grid levels. The planet is located in the middle of the finest grid (G. D'Angelo).

4.1. *Migration in isothermal disks*

Small mass planets do not alter the global disk structure significantly, in particular they do not open a gap within the disk. Hence, the combined effect of Lindblad and corotation torques can be calculated for small planetary masses using a *linear analysis*. The outcome of such linear, no-gap studies has been termed *Type-I migration*. Due to the complexity of considering heat generation and transport in disks, these linear studies have relied nearly exclusively on simplified, locally isothermal disk models. Here, the temperature is assumed to be independent of height and is given by a pre-described function of radius, $T = T(r)$. Typically, it is assumed that the relative scale height H of the disk is a constant, $H/r = const.$, yielding $T \propto r^{-1}$. The total torques Γ_{tot} is given as the sum of Lindblad and corotation torques $\Gamma_{\text{tot}} = \Gamma_{\text{L}} + \Gamma_{\text{CR}}$. The speed of the induced linear, type-I migration scales inversely with the disk temperature (i.e. disk thickness) as $\propto (H/r)^{-2}$, linear with the planet mass $\propto m_{\rm p}$, and with the disk mass $\propto m_{\rm d}$. Linear models have been calculated for flat 2D disks as well as full 3D configurations. The problem of 2D simulations lies in taking into account approximately the neglected vertical stratification of the disk, which is typically done through a smoothing of the gravitational potential near the planet. Additional problems arise when considering radial gradients.

The linear studies have been supplemented recently by many numerical studies in 2 and 3 dimensions. Here, the disk is typically modelled using as a viscous gas, and the Navier-Stokes equations are solved using grid-based methods. To verify the correctness of the results, a detailed *code comparison* project was conducted a few years ago; see De Val Borro *et al.* (2006). A very important ingredient to speed up such simulations is the FARGO-method introduced by Masset (2000). To increase resolution around the planet, a nested-grid structure has been employed successfully in 2D and 3D; see D'Angelo *et al.* (2002, 2003) and Fig. 2. New full 3D, nested grid locally isothermal hydrodynamic simulations of planet-disk interaction give very good agreement with previous 3D linear results (Tanaka & Ward 2004) and yield the following form for the total torque for small mass planets below about 10 $M_{\rm Earth}$ (D'Angelo & Lubow 2010)

$$\Gamma_{\rm tot} = -(1.36 + 0.62\alpha_\Sigma + 0.43\alpha_T) \left(\frac{m_{\rm p}}{M_*}\right)^2 \left(\frac{H}{r_{\rm p}}\right)^{-2} \Sigma_{\rm p} \, r_p^4 \, \Omega_{\rm p}^2. \qquad (4.1)$$

In eq. (4.1), the index p refers to the planet, α_Σ and α_T refer to the radial variation of density and temperature, such that $\Sigma(r) \propto r^{-\alpha_\Sigma}$ and $T(r) \propto r^{-\alpha_T}$. One should keep in mind that these results depend on the magnitude of the viscosity.

For larger planet masses, a gap is opened in the disk because the planet transfers angular momentum to the disk, positive exterior and negative interior to the planet. The depth of the gap that the planet carves out depends for given disk physics (temperature and viscosity) only on the mass of the planet. Because the density in the co-orbital region is reduced, the corotation torques are strongly affected and are no longer of any importance for larger planet masses. For very large masses, even the Lindblad torques are reduced yielding a slowing down of the planet. This non-linear regime has been coined the *Type-II* regime of planetary migration; here the drift of the planet is dominated by the disk's viscous evolution.

4.2. *Migration in radiative disks*

The migration rates obtained through the analyses mentioned above have resulted in approximate formulae for the migration speed $\dot{a}$ of a planet (e.g. from eq. 4.1) that are frequently used in *population synthesis models*. These are growth models of planets that include disk evolution and planet migration in parameterized form, which can be used to calculate the final location of many planets in the mass/semi-major axis diagram and compare the results statistically with the observed distribution. The results indicate in particular, that the migration for small mass planets in the type-I regime is by far too fast to account for the observed distribution; the majority of planets would have been lost to the star, see Ida & Lin (2008) and Mordasini *et al.* (2009). Only a significant reduction in the type-I migration speed gives satisfactory results.

Here, we shall concentrate on a *physical* improvement to the models, that represent a possible solution to the type-I migration problem: the inclusion radiative effects. Extending the previous isothermal models requires the incorporation of a heating and cooling mechanism. The importance of radiative diffusion has first been pointed out by Paardekooper & Mellema (2006), and recent papers quote approximate formulas for viscous and diffusive disks (Masset & Casoli 2010, Paardekooper *et al.* 2011). To demonstrate the effect for two-dimensional flat disks, we show results for a planet-disk simulation where we include viscous heating, local radiative cooling, as well as diffusive radiative transport in the disk's plane. The energy equation then reads

$$\frac{\partial \Sigma c_v T}{\partial t} + \nabla \cdot (\Sigma c_v T \mathbf{u}) = -p \nabla \cdot \mathbf{u} + D - Q - 2H \nabla \cdot \vec{F} \qquad (4.2)$$

where Σ is the surface density, T the midplane temperature, p the pressure, D the viscous dissipation, Q the radiative cooling, and $\vec{F}$ the radiative flux in the midplane. Models where the various contributions on the right hand side of eq. (4.2) were selectively switched off and on, have been constructed by Kley & Crida (2008).

 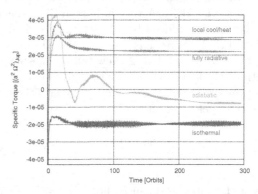

Figure 3. The evolution of a 20 M_{Earth} planet embedded in a disk with 0.01 solar masses. **Left**: Spiral arm structure (Courtesy, F. Masset). **Right**: Time evolution of the torque acting on the planet. From bottom to top the curves indicate simulations where *i*) no energy equation (isothermal), *ii*) only the first term on the rhs. of eq. (4.2) (adiabatic), *iii*) all terms on the rhs. (fully radiative), and *iv*) all but the last term on the rhs. (heating/cooling), have been used (after Kley & Crida, 2008).

The effect of this procedure on the resulting torque is shown in the right panel of Fig. 3. The basis for all the models is the same equilibrium disk model constructed using all terms on the rhs. of eq. (4.2) and no planet. Embedding a planet of $20 M_{\text{Earth}}$ yields in the long run *positive torques* only for the radiative disks, where the maximum effect is given when only viscous heating and local radiative cooling are considered. The left hand panel of Fig. 3 shows the global spiral arms structure of the embedded planet. The strength of this positive corotation effect also scales with the square of the planet mass up to about 20 to 25 M_{Earth}. Beyond this mass, gap opening begins and only the Lindblad torques remain, and the planets begin to migrate inwards again. In full 3D radiative simulations, the results are qualitatively the same (Kley *et al.* 2009), but interestingly the full 3D results show an even stronger effect. New population synthesis models based on the modified migration rates seem to show better agreement with the observational data set (Mordasini *et al.* 2011).

5. Eccentricity and inclination

In addition to a change in semi-major axis, planet-disk interaction will modify the planetary eccentricity (*e*) and inclination (*i*) as well. For small *e* and *i*, linear analysis has indicated exponential damping (Ward 1988, Ward & Hahn 1994). Extending previous isothermal numerical studies by Cresswell *et al.* (2007), fully 3D radiative disk simulations have been performed recently to study the evolution *e* and *i*. The results indicate that both are damped for all planetary masses (Bitsch & Kley 2010, 2011). For small values of e and i that are below about $2H/r$, the damping occurs exponentially on timescales comparable to the linear estimates by Tanaka & Ward 2004. For larger values, the damping is slowed down and follows approximately $\dot{e} \propto e^{-2}$, and for the damping of i, an identical relation holds. Interestingly, the presence of outward migration is coupled to the magnitude of e and i. Outward migration only occurs for eccentricities smaller than about 0.02, and inclinations below about 4°. This reduction is due to the fact that for non-circular orbits the flow structure in the corotation region becomes strongly time dependent and no stationary corotation torque can develop.

6. Summary

The formation of planets takes place in protoplanetary disks via sequential accretion or the gravitational instability. In the first process, the initial growth is not yet fully understood. Close in planets have migrated inwards from their birthplace either through disk migration or tidal shrinking. The eccentric and inclined population of planets has been created by dynamical scattering processes.

Acknowledgement: This work has been supported by DFG grant FOR 759: *The formation of planets: The critical first growth phase.*

References

Balbus, S. A. & Hawley, J. F. 1993, *ApJ*, 367, 214

Baruteau, C. & Masset, F. 2008, *ApJ*, 672, 1054

Benz, W. & Asphaug, E. 1999, *Icarus*, 142, 5

Bitsch, B. & Kley, W. 2010, *A&A*, 523, A30 ; 2011, *A&A*, 530, A41

Blum, J. & Wurm, G. 2008, *ARA&A*, 46, 21

Cresswell, P., Dirksen, G., Kley, W., & Nelson, R. P. 2007, *A&A*, 473, 329

Crida, A., Sándor, Z., & Kley, W. 2008, *A&A*, 483, 325

D'Angelo, G. & Lubow, S. H. 2010, *ApJ*, 724, 730

D'Angelo, G., Henning, Th., & Kley, W. 2002, *A&A*, 385, 647 ; 2003, *ApJ*, 599, 548

De Val Borro *et al.*, 2006, *MNRAS*, 370, 529

Dominik, C., Blum, J., Cuzzi, J. N., & Wurm, G. 2007, *Protostars and Planets V*, Eds. B. Reipurth, D. Jewitt, and K. Keil, p783-800

Flaig, M., Kley, W., & Kissmann, R. 2010, *MNRAS*, 409, 1297

Geretshauser, R. J., Meru, F., Speith, R., & Kley, W. 2011, *A&A*, 531, 166

Geretshauser, R. J., Speith, R. Güttler, C., Krause, M., & Blum, J. 2010, *A&A*, 513, 58

Goldreich, P. & Tremaine, S. 1980, *ApJ*, 241, 425

Ida, S. & Lin, D. N. C. 2008, *ApJ*, 673, 487

Johansen, A., Oishi, J. S., Mac Low, M.-M., Klahr, H., Henning, T., & Youdin, A. 2007, *Nature*, 448, 1022

Klahr, H. & Bodenheimer, P. 2006, *ApJ*, 639, 432

Kley, W., Bitsch, B., & Klahr, H. 2009, *A&A*, 506, 971

Kley, W. & Crida, A. 2008, *A&A*, 487, L9

Lee, M. H. & Peale, S. J. 2002, *ApJ*, 567, 596

Libersky, L. D., Petschek, A. G., Carney, T. C., Hipp, J. R., & Allahdadi, F. A. 1993, *Journal Computational Physics*, 109, 67

Lin, D. N. C., Bodenheimer, P., & Richardson, D. C. 2006, *Nature*, 380, 606

Masset, F. S. 2000, *A&AS*, 141, 165

Masset, F. S. & Casoli, J. 2010, *ApJ*, 723, 1393

Mordasini, C., Alibert, Y., Benz, W., & Naef, D. 2009, *A&A*, 501, 1161

Mordasini, C., Dittkrist, K., Alibert, Y., Klahr, H., Benz, W., & Henning, T. 2011, *ArXiv*, 1101.3238

Nakagawa, Y., Sekiya, M., & Hayashi, C. 1986, *Icarus*, 67, 375
Paardekooper, S.-J. & Mellema, G. 2006, *A&A*, 459, L17
Paardekooper, S.-J., Baruteau, C., & Kley, W. 2011, *MNRAS*, 410, 293
Perri, F. & Cameron, A. G. W.. 1974, *Icarus*, 22, 416
Schäfer, C., Speith, R., & Kley, W. 2002, *A&A*, 470, 733
Tanaka, H. & Ward, W. R. 2004, *ApJ*, 602, 388
Toomre, A. 1964, *ApJ*, 139, 1217
Ward, W. R. 1986, *Icarus*, 67, 164
Ward, W. R. 1988, *Icarus*, 73, 330
Ward, W. R. & Hahn, J. M. 1994, *Icarus*, 110, 95
Weidenschilling, S. J. 1977, *MNRAS*, 180, 57

Discussion

J. GRYGAR: Are there estimates how much time elapses to migrating Jupiters from their birthplaces to close orbits of hot Jupiters?

W. KLEY: Yes, the migration formulae and the numerical simulations yield timescales of typically 10^4 orbital periods.

M. MONTGOMERY: Have you considered adding asteroids in the gap which will eventually meet at L4 and L5 and thus may help in taking away energy for planet migration?

W. KLEY: Asteroids have not been taken into account in migration simulations, as their mass is too small to prevent or slow down migration significantly. Note, that the whole coorbital horseshoe region with all its mass is dragged along with the inward motion of the planet.

J. SOUTHWORTH: I think it is premature to suggest that many planetary systems are coplanar. Kepler-II tells us that they can be coplanar, but planetary systems which are not coplanar will not be picked up by Kepler as the planets in questions will not transit. On the plus side, as Kepler data continue to accrue we will be able to detect such planets by their dynamical effects on those planets in the system which do transit.

W. KLEY: Yes, it is difficult to decide presently what fraction of planetary systems will be nearly coplanar. But the existing data indicate that at least some systems will have undergone migration.

S. HINKLEY: Is it a coincidence that HR8799, Fomalhaut and Pictoris (the systems with directly imaged planets and presumably formed through gravitational instability) are all found orbiting A-stars which presumably possess more massive disks?

W. KLEY: Yes, the mass of a disk is important as the instability scales with the disk mass. But the cooling and heating (irradiation) of the disk have to be taken into account.

K. BJORKMAN: The requirement for forming a planet via gravitational instability is that the surface density be large enough, so if it is possible to have an unusual surface density distribution (say by infall into the outer disk) then a large total disk mass may not be required.

W. KLEY: That's correct, but typically it is assumed that the disk structure (and infall) vary relatively smoothly with radius, and then the total mass is again a suitable criterion.

From Interacting Binaries to Exoplanets: Essential Modeling Tools
Proceedings IAU Symposium No. 282, 2011
Mercedes T. Richards & Ivan Hubeny, eds.
© International Astronomical Union 2012
doi:10.1017/S1743921311027992

Towards a Theory for the Atmospheres, Structure, and Evolution of Giant Exoplanets

Adam Burrows

Department of Astrophysical Sciences, Princeton University, Princeton, NJ USA 08544
email: `burrows@astro.princeton.edu`

Abstract. In this short review, I summarize some of the salient features of the emerging theory of exoplanets in general, and of giant exoplanets in particular. A focus is on the characterization of transiting planets at primary and secondary eclipse, but various other related topics are covered, if only briefly. A theme that clearly emerges is that a vibrant new science of comparative exoplanetology is being born.

Keywords. planets and satellites: general; (stars:) planetary systems; stars: low-mass, brown dwarfs; stars: atmospheres

1. Introduction

It can be said, without fear of hyperbole, that we are living in the heroic age of exoplanetary studies. Since the discovery of 51 Peg b by Mayor & Queloz (1995), astronomers have discovered more than 1500 planets beyond the solar system. The majority of the latter are gas giants discovered from the ground and with the *Kepler* (Borucki *et al.* 2010) and CoRoT (Baglin *et al.* 2006) space telescopes, but more than a hundred "Neptunes" are now known, and many so-called "super-Earths" have been detected. In the next few years, we are likely to determine the statistics of the orbits, radii, and masses of exoplanets, constraining not only their origins and dynamics, but their atmospheres. In the process, we will determine, if only as a byproduct, the galactic context of our own planetary cohort.

Importantly, more than 120 transiting planets have been discovered from the ground and most of these are close enough to be followed-up to obtain radial-velocity masses. Excitingly, many of these can be studied photometrically and spectroscopically at secondary and primary eclipse to derive temperatures and compositions, both from the ground and using *Spitzer* and HST/NICMOS space assets. What is more, the phase orbital light curves of some of the closest giant exoplanets have now been measured at various wavelengths. Collectively, with such data we are learning a great deal about the structures, chemistry, and atmospheres of exogiants. Indeed, the pace with which this field is moving is outstripping all predictions.

This extraordinary pace is such that theory and theoretical studies will have a full smorgasbord of issues, problems, and puzzles to address for the foreseeable future. Moreover, new theory will be crucial to interpret the flood of incoming data. Clearly, summarizing this effort would be a daunting task. Rather, for this short review, I cherry-pick from some of my own efforts at interpretation and understanding to communicate just a fraction of the findings and thoughts that have emerged in the last few years in response to the need to explain these new objects. The reader can be assured that there is much more.

2. The Radii of Brown Dwarfs and Giant Exoplanets

Observers have used theoretical evolutionary models of brown dwarfs for many years to determine the physical properties, in particular the effective temperatures, surface gravities, masses, ages, and compositions, of the objects they find and study. Theory is the essential tool with which to convert data into meaning (Burrows *et al.* 2001). One of the best techniques to constrain physical theory is with eclipsing/transiting systems (Johnson *et al.* 2011; Bouchy *et al.* 2011; Deleuil *et al.* 2008) in the brown dwarf realm, rare as they are, since in this way radii can be determined and compared with published models. However, model radii depend upon more than the equation of state (EOS) (Saumon, Chabrier, & Van Horn 1995). In fact, a brown dwarf radius at a given age and mass is a function of atmospheric metallicity, bulk helium fraction, and the cloud models employed by the theorist. The latter is the most problematic aspect of brown dwarf theory, for though silicate and iron clouds can dominate the atmospheres of L dwarfs, the specific particle size and spatial distributions and particle optical properties are barely understood (Helling *et al.* 2001). Given this, there is a range of radii expected for brown dwarfs of a given age and mass, and this ambiguity had not been properly appreciated.

Recently, however, Burrows, Heng, & Nampaisarn (2011) have found that the spread in radius at a given mass and age can be as large as $\sim10\%$ to $\sim25\%$, with higher-metallicity, higher-cloud-thickness atmospheres resulting quite naturally in larger radii. For each 0.1 dex increase in [Fe/H], radii increase by $\sim1\%$ to $\sim2.5\%$, depending upon age and mass. They also find that, while for smaller masses and older ages brown dwarf radii decrease with increasing helium fraction (Y) (as expected), for more massive brown dwarfs and a wide range of ages they increase with helium fraction. The increase in radius in going from $Y = 0.25$ to $Y = 0.28$ can be as large as ~0.025 R_J ($\sim2.5\%$). Furthermore, they find that for very-low-mass (VLM) stars, an increase in atmospheric metallicity from 0.0 to 0.5 dex increases radii by $\sim4\%$, and from -0.5 to 0.5 dex by $\sim10\%$. They suggest that opacity due to higher metallicity might naturally account for the apparent radius anomalies in some eclipsing VLM systems. Ten to twenty-five percent variations in radius exceed errors stemming from uncertainties in the equation of state alone. This serves to emphasize that transit and eclipse measurements of brown dwarf radii recently published using *Kepler* and CoRoT constrain numerous effects collectively, importantly including the atmosphere and condensate cloud models, and not just the equation of state. At all times, one is testing a multi-parameter theory, and not a universal radius–mass relation.

This lesson is all the more important when studying transiting giant exoplanets. Though much lower in mass than the average brown dwarfs, they have similar radii. However, the differences are revealing. A close-in transiting exoplanet is irradiated by its primary and this radically changes its atmosphere on both its day and night sides. The profiles on these atmospheres match onto their convective cores and it is the loss of energy and the concomitant decrease in core entropy that determines a gas giant's radius. For each irradiated giant planet, one must perform custom models that take into account stellar irradiation, planet mass, system age, and, if possible, bulk and atmospheric composition. The latter is by and large unknown, and the stellar age and metallicity are not reliably obtained for the vast majority of stars. In addition, proximity to its primary and even slight orbital eccentricity can lead to tidal heating of unknown magnitude and convection and rotation can generate magnetic fields of significance that could play a role in day/night heat redistribution and in both core and atmospheric heating. Finally, giant planets can have central cores of denser material (ice and rock) of unknown mass, as well as envelopes enriched in heavy elements.

The result is an imperfect theory with which to interpret planet transit data. Nevertheless, there are two trends that bear mentioning. The first is that when one generates the requisite custom models for a large number of irradiated, transiting planets and introduces a dense core to improve the fits, a trend with stellar metallicity emerges. For no giant planet orbiting a lower-metallicity star do Burrows *et al.* (2007a) or Guillot *et al.* (2006) infer a large inner core. Conversely, for no giant planet orbiting the highest-metallicity stars do these authors infer a small inner core. Intriguingly, the core masses Burrows *et al.* (2007a) find for exogiants transiting near-solar metallicity stars are close to those estimated for Jupiter and Saturn. The upshot is that a roughly montonically-increasing relationship between stellar metallicity and estimated core mass emerges from the study of transiting giant planets collectively. Note that stellar metallicity was not used in the planet modeling. Hence, these twin correlations may speak to the mechanism of giant planet formation and are in keeping with the core-accretion model of their origin.

The second trend is the more well-known. There are a number of close-in transiting giants with radii that are too large to be explained by the default theory. Examples are WASP-12b ($\sim$1.736 R_J), TreS-4 ($\sim$1.706 R_J), WASP-19b ($\sim$1.991 R_J), and HAT-P-32b ($\sim$2.037 R_J). While measurement errors are certainly possible, anomalously large radii seem indicated in an interesting subset of transiting giant exoplanets. Culprits could be core or deep-atmosphere heating (tidal or magnetic), extreme age errors (a much younger planet is bigger), large planet atmospheric opacities (similar to the effect for the brown dwarfs alluded to earlier), or, again, radius measurement errors. Whatever the reason, resolving this apparent anomaly is one of the most important goals of those studying the newly-discovered close-in giant planets.

3. On Using Deuterium Burning to Distinguish Giant Exoplanets from Brown Dwarfs

Gas giant planets and brown dwarfs share many characteristics. They both have molecular atmospheres, whose mass is dominated by hydrogen and helium. Their atmospheric opacities are dominated by a small set of compounds, notably water, methane, carbon monoxide, and often various cloud species. The most important constituents of their cores are hydrogen and helium and their EOS is in a realm in which Coulomb and degeneracy effects compete. The upshot is that, over two orders of magnitude in mass from $\sim$0.3 to $\sim$100 M_J, the cold radius of such objects differs by no more than $\sim$30-40% from 1 R_J.

Nevertheless, different origins and astronomical sociology seem to require that one be able to distinguish one family from another. I think it sensible to distinguish giant planets from brown dwarfs by their origins, but an origin is not an observable and we don't yet know how either class of objects forms, nor what their expected mass functions might be. The latter are likely to be different, but to overlap. In fact, despite the possibility that the mass functions of these families could overlap, what has emerged to distinguish one family from the other is a simple mass cut. Burrows *et al.* (1997) published evolutionary curves for H_2/He-rich objects with masses spanning a broad range from Saturn's to above 0.2 $M_\odot$. In their Figure 7, they (seemingly) arbitrarily distinguished "planets," and "brown dwarfs" by whether they burned deuterium. This border mass was near 13 M_J and, as a result, many began to use 13 M_J as the boundary between the giant planets and brown dwarfs. This was not the original intent of the authors, but it is a simple, one-dimensional condition that has stuck.

However, those who use such a simple criterion should be aware that this one number (13 M_J) ignores the fact that the burned fraction (e.g., whether 10%, 50%, or 90%), metallicity, and helium fraction all come into play when defining a deuterium-burning mass. Indeed, Spiegel, Burrows, & Milsom (2011) have shown that when these considerations are accounted for "the" deuterium-burning mass can vary from $\sim$12 M$_J$ to $\sim$14 M$_J$. Hence, even this simple discriminant is not absolute. When one can at last separate these two families on the basis of origin via orbit, rotation, composition, presence or absence of a dense core, or whatever characteristics emerge in a statistical sense to distinguish them, a truly astronomically relevant naming convention will finally be available.

4. The Wavelength Dependence of the Transit Radii of Exoplanets

The transit method for exoplanet discovery and characterization hinges upon measurments of the periodic photometric variations in the stellar light caused by passage of the orbiting planet in front of the star. The magnitude of the fractional diminution in the stellar light is $(R_p/R_*)^2$, where R_p and R_* are the planet and star radius, respectively. Given R_*, R_p can be determined and, since the planet is in transit and the orbital inclination (i) is therefore measured, radial velocity measurements, which yield $m_p \sin(i)$, provide the planet's mass directly. With both mass and radius, one can do physics and constrain structural models.

However, the radius measured by this method is the "transit radius," which is the impact parameter from the planetary center of the stellar rays intercepted at the Earth that traverse a chord near the terminator for which the optical depth in the planet's atmosphere is of order unity. The optical depth in the radial direction, so important for planet "emission" spectra, is irrelevant here. Importantly, since giant planets have extended atmospheres and the opacities that go into determining the impact parameter are composition- and wavelength-dependent, the transit radius itself is a function of wavelength. The upshot is that the spectra of transit radii reflect atmospheric composition and scale heights and can be used, with profit, to identify atmospheric atoms and molecules. This is how Charbonneau et al. (2002) detected sodium in the atmosphere of HD 209458b, and how water, carbon monoxide, and, perhaps, carbon dioxide have been claimed in the atmospheres of other transiting planets. This technique is complementary to the traditional direct planetary spectral measurement technique for probing atmospheres and is more composition-dependent than the latter. Fortney et al. (2003) estimated that the potential radius variation in and out of water features in the near infrared could be as much as a few percent, a result echoed by Burrows et al. (2011b), who derived values near $\sim$5%. Hence, measurements of the spectral variation of transit radii provide direct and useful diagnostics of planet composition.

5. Interpretation of Planet Flux Measurements at Secondary Eclipse

Just before the secondary eclipse of a transiting planet by a star, astronomers can measure their summed light. During secondary eclipse, however, only the stellar light contributes to the signal. Therefore, the difference between these two measurements can yield the planet's flux. In the mid-infrared, this difference can be a few parts in 10^3 or 10^4 of the stellar flux and is measurable by $Spitzer$. Without needing to image separately planet from star, the planet's emissions can be obtained! True, this is a severely irradiated planet, and not one all-but-isolated from its parent star. Nevertheless, such data provide a wealth of information about the planet's atmosphere (its temperature, composition, and temperature profile), as well as its wind dynamics. Super-rotational flows and jet streams can advect heat deposited by the star downstream of the substellar point before it is re-radiated. The angular (and, hence, temporal) shift this causes with respect to the orbital ephemeris reflects, among other things, the wind speeds. Hence, measurement of secondary eclipse spectra and photometry has inaugurated the era of remote sensing and detailed characterization of exoplanets. The *Spitzer* space telescope has been the workhorse of these studies, and if JWST flies, it will provide an order of magnitude improvement over the still-precious *Spitzer* data.

5.1. *Inversions and Hot Upper Atmospheres*

In the course of such secondary eclipse studies, it was found that the spectra of some highly-irradiated giant planets show signs of superheated upper atmospheres and/or thermal inversions. The signature of the latter is the flipping from the classical absorption spectra of normal atmospheres into emission spectra − spectral troughs became peaks, and vice versa (Hubeny *et al.* 2003; Burrows *et al.* 2007b; Knutson *et al.* 2008). Inverted spectra (positive temperature gradients going outward) were indicated by the switch of the ratio of the IRAC1/IRAC2 flux ratio seen using *Spitzer* from greater than one (normal) to less than one (inverted). The upper atmosphere heating inferred could amount to an increase in atmospheric temperatures near the $\sim 10^{-2}$ bar pressure level of ~ 1000-1500 K and an increase in the IRAC3 band flux of as much as a factor of three! The origin and cause of this inferred severe heating and anomalous thermal profile is unknown, but an as-yet-unidentified absorbing molecule is being sought. The effect is not small, for as much as tens of percent of the total intercepted stellar flux is implicated. This constitutes one of the great mysteries to emerge from the recent study of transiting exoplanets.

6. Albedos

A venerable tradition in Solar System science is the study of the reflected optical light from its cold planets, moons, and asteroids. When the optical and mid-infrared spectral "bumps" from such objects are well-separated and distinct, the interpretation of such bumps as optical "reflection" and "thermal emission" is well-justified. The darkness or lightness of such bodies in the optical, and the optical colors of their reflected light can indicate their compositions, and help determine their radii. The geometric albedo is a measure of the strength of this reflection, with high values below (but near) one indicating highly-reflective surfaces and low values below ~ 5-10% indicating highly-absorbing surfaces. Cloudy atmospheres frequently have high albedos.

This tradition of reflection photometry and spectroscopy has been carried over into exoplanet research. However, the objects best measured in this manner, by *Kepler*, CoRoT, and MOST for instance, are the close-in, transiting, hot giants. Such objects can be self-luminous, and their optical reflection and thermal emission components can be quite close and overlap. Under these circumstances, the concept of a reflection albedo is ambiguous. Nevertheless, at times one can infer optical albedos and compare with theory. When such comparisons are performed, we find that the albedos of giant transiting exoplanets are quite low, below ~ 10%, and likely often below ~ 5%. Such low albedos were predicted (Sudarky, Burrows, & Pinto 2000; Burrows, Ibgui, & Hubeny 2008, and references therein) and are likely due to the presence and dominance in the optical of the broad sodium doublet. The same sodium feature is seen to dominate in the optical spectra of T dwarf brown dwarfs; another physical and chemical correspondence between brown dwarfs and hot giant planets.

7. Light Curves as a Function of Wavelength

Before the recent explosion in exoplanet research, the method most discussed with which to discover planets was by high-contrast imaging. The planet would be separated spatially on a plate or CCD from its bright parent star and probed individually. This classic approach, requiring high-contrast capabilities better than $\sim 10^{-4}$ to $\sim 10^{-6}$ for giant planets and $\sim 10^{-9}$ to $\sim 10^{-10}$ for exoEarths, is very technologically challenging, but there has been some recent success with the discovery of HR 8799bcd (Marois *et al.* 2008) and e, Fomalhaut b (Kalas *et al.* 2008), and βPic b. Specifically, the HR 8799 planets have contrast ratios of $\sim 10^{-4}$ and are at wide separations (far beyond a Jupiter orbit) from their parent, an A star. These discoveries are exciting and promise much more in the future to complement what is being learned from the plethora of transiting planets now known.

8. Conclusions

The pace of exoplanet research is truly astonishing and shows no signs of abating soon. This data-rich subject is creating a generation of theorists poised to challenge past orthodoxies and write new textbooks. Though interpretations of the extant data may be fraught with ambiguity, and many mistakes have no doubt been made in characterizing exoplanetary atmospheres and structure, the current efforts might be considered a training exercise with which a new generation of theorists is cutting its teeth, in preparation for an astonishing future.

References

Baglin, A., Auvergne, M., Boisnard, L., Lam-Trong, T., Barge, P., Catala, C., Deleuil, M., Michel, E., & Weiss, W. 2006, *36th COSPAR Scientific Assembly*, 36, 3749.
Borucki, W. *et al.*, 2010, *Science*, 327, 977
Bouchy, F. *et al.*, 2011, *A&A*, 525, 68 (arXiv:1010.0179)
Burrows, A., Marley, M., Hubbard, W. B., Lunine, J. I., Guillot, T., Saumon, D., Freedman, R., Sudarsky, D., & Sharp, C. 1997, *ApJ*, 491, 856
Burrows, A., Hubbard, W. B., Lunine, J. I., & Liebert, J. 2001, *Rev. Mod. Phys.*, 73, 719
Burrows, A., Sudarsky, D., & Hubeny, I. 2006, *ApJ*, 640, 1063
Burrows, A., Hubeny, I., Budaj, J., & Hubbard, W. B. 2007a, *ApJ*, 661, 502
Burrows, A., Hubeny, I., Budaj, J. Knutson, H. & Charbonneau, D. 2007b *ApJ*, 668, L171
Burrows, A., Ibgui, L., & Hubeny, I. 2008, *ApJ*, 682, 1277
Burrows, A., Heng, K., & Nampaisarn, T. 2011, *ApJ*, 736, 47
Burrows, A., Rauscher, E., Spiegel, D. & Menou, K. 2011 *ApJ*, 719, 341-350, 2010
Deleuil, M. *et al.*, 2008, *A&A*, 491, 889
Fortney, J. J., Sudarsky, D., Hubeny, I., Cooper, C. S., Hubbard, W. B., Burrows, A., & Lunine, J. I. 2003, *ApJ*, 589, 615
Guillot, T., Santos, N.C., Pont, F., Iro, N., Melo, C., & Ribas, I. *A&A*, 453, L21, 2006
Helling, Ch., Oevermann, M., Lüttke, M. J. H., Klein, R., & Sedlmayr, E. 2001, *A&A*, 376, 194
Hubeny, I., Burrows, A., & Sudarsky, D. 2003, *ApJ*, 594, 1011
Johnson, J. A., Apps, K., Gazak, J. Z., Crepp, J., Crossfield, I. J., Howard, A. W., Marcy, G. W., Morton, T. D., Chubak, C., & and Isaacson, H. 2011, *ApJ*, 730, 79
Kalas, P., *et al.*, 2008, *Science*, 322, 1345
Knutson, H. A., Charbonneau, D., Allen, L. E., Torres, G., Burrows, A., & Megeath, S. T. 2008, *ApJ*, 673, 526
Macintosh, B. *et al.*, 2006, *Proc. of the SPIE*, Vol. 6272, 62720L
Marois, C., *et al.*, 2008, *Science*, 322, 1348
Mayor, M. & Queloz, D. 1995, *Nature*, 378, 355
Saumon, D., Chabrier, G., & Van Horn, H. 1995, *ApJS*, 99, 713
Spiegel, D., Burrows, A., & Milsom, J. A. 2011, *ApJ*, 727, 57
Sudarsky, D., Burrows, A., & Pinto, P. 2000, *ApJ*, 538, 885

Discussion

S. HINKLEY: Are the lack of points at low metallicity and high core mass another slam dunk for core accretion?

A. BURROWS: Rather, I would say that this correlation is "quite suggestive" of the core-accretion scenario. It is still possible that heavy elements in the envelope of the planet, and not the core, can explain the correlation (though this too could be a signature of core accretion) and there are those who claim that the direct instability model can account for such envelope enrichment. Personally, I think those claims are a bit contrived, but one can't yet be definitive on this point.

From Interacting Binaries to Exoplanets: Essential Modeling Tools
Proceedings IAU Symposium No. 282, 2011   © International Astronomical Union 2012
Mercedes T. Richards & Ivan Hubeny, eds.   doi:10.1017/S1743921311028006

Eclipse Timing Variations of Planets in P-Type Binary Star Systems

Richard Schwarz[1], Nader Haghighipour[2], Barbara Funk[1], Siegfried Eggl[1], and Elke Pilat-Lohinger[1]

[1] Institute for Astronomy, University of Vienna,
Türkenschanzstrasse 17, 1180 Austria
email: schwarz@astro.univie.ac.at

[2] Institute for Astronomy and NASA Astrobiology Institute, University of Hawaii
2680 Woodlawn Dr., Honolulu, HI, 96822 USA

Abstract. In close eclipsing binaries, measurements of the variations in the binary's eclipse timing may be used to infer information about the existence of planets in P-Type motion. To study the possibility of detecting such planets with CoRoT and Kepler, we calculated eclipse timing variations (ETV) for different values of the mass and orbital elements of the perturbing planet. These investigations are a continuation of the work of Schwarz *et al.* (2011).

Keywords. methods: numerical, binaries: eclipsing, planetary systems, planets: detection

1. Introduction

Today (June 2011), we know 53 extrasolar planets in 39 binary systems and 7 extrasolar planets in 7 triple star systems. In general, one may distinguish different types of stable orbits for planets in binary systems (Dvorak 1988):
- **S-Type:** The planet moves around one of the two stars.
- **P-Type:** The planet moves around the entire binary.
- **L-Type:** The planet moves close to the equilibrium points (L_4 and L_5) of a binary like Jupiter's Trojans. Possible candidates were presented in the work of Schwarz *et al.* (2009).

In the study by Schwarz *et al.* (2011), the dynamics of a binary star system with a close P-Type planet, has been analysed and the eclipse timing variations (ETV) were calculated for different values of the mass ratio and orbital elements of the binary system. In this study, we investigated P-Type planets farther away from the binary (up to 5 AU).

2. Model and methods

Our numerical simulations were carried out using the Lie-Series-method. As dynamical model, we used the full three-body problem and considered the binary system in circular motion with an initial separation of 0.05 AU and a mass of $m_1 = m_2 = 1M_{sun}$ (m_1 is the mass of the primary star and m_2 the mass of the secondary star). The planet moves in a far away circular orbit around both stars and has masses of 1 and 10 Jupiter-masses (M_J). Gravitational interactions between the binary and the planet will perturb the star's motion, resulting in a deviation from their Keplerian orbits. In an eclipsing binary, these deviations result in variations in the time and duration of the eclipse. We compared the expected amplitude of the ETV ($\sigma = (d_{tmax} - d_{tmin})/2$) with the detection limit of current space observatories. With the amplitude $\sigma = 16sec$ for CoRoT (at L=15.5

444

mag) and for Kepler $\sigma = 4sec$ (at L=14.5 mag), we may show if the ETV signals of our calculations are detectable or not.

3. Results

If the ETV signal is higher than the observational threshold, we can assume that a planet will be indirectly detectable by the eclipse timing measurements of the secondary star (Schwarz *et al.* 2011). There are 2 dynamical effects which change the ETV signal, caused by the perturbation of an additional planet:

(*a*) Short term variations in the binary's period. Former investigations showed (Schwarz *et al.* 2011) that the planets have to be in circular orbits very close to the secondary star $a = 0.1$ AU or, in the case of planets with a larger distance $a > 0.1$ AU, they have to be more massive ($m > 5M_J$) to produce detectable ETV signals.

(*b*) The binary stars also perform an orbit around the common barycenter, leading to different light travel times.

The results of our investigations are presented in Figure 1 for a mass of the planet of $1M_J$ (left graph) and $10M_J$ (right graph). We compared the amplitude (σ) of the ETV signal for the two different effects as follows: the solid line marks the signal with the first effect and the dashed line represents the second effect. Thus, we can conclude that planets which are very close ($a \leqslant 0.3$ AU) to the binary system are dominated by the first effect (Schwarz *et al.* 2011), whereas planets which are far ($a > 0.3$ AU) from the binary are dominated by the second effect, this is true for both masses of the planet.

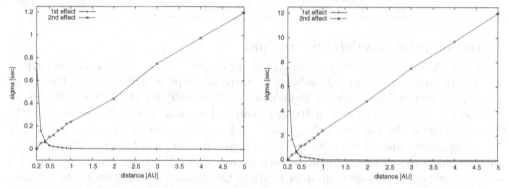

Figure 1. Amplitude (σ) of the ETV signals for planets with 1 M_J (left graph) and 10 M_J (right graph) for different distances to the binary.

Acknowledgement

Support was received from the ÖFG projects 06/12290 for R.S. and 06/12277 for B.F., the NASA Astrobiology Institute for N.H., the Austrian FWF project no. P20216 for S.E. and the Austrian FWF projects no. P20216 and P22603 for E. P-L.

References

Dvorak, R., 1984, *CeMDA*, 34, 369

Schwarz, R., Süli, Á., & Dvorak, R. 2009, *MNRAS*, 398, 2085

Schwarz, R., Haghighipour, N., Eggl, S., Pilat-Lohinger, E., & Funk, B., 2011, *MNRAS*, 414, 2763.

From Interacting Binaries to Exoplanets: Essential Modeling Tools
Proceedings IAU Symposium No. 282, 2011
Mercedes T. Richards & Ivan Hubeny, eds.
© International Astronomical Union 2012
doi:10.1017/S1743921311028018

New Insights into the Dynamics of Planets in P-Type Motion Around Binaries

Barbara Funk, Siegfried Eggl, Markus Gyergyovits, Richard Schwarz and Elke Pilat-Lohinger

Institute for Astronomy, University of Vienna
Türkenschanzstrasse 17, A-1180 Vienna, Austria
email: funk@astro.univie.ac.at

Abstract. Up to now, more than 500 extra-solar planets have been discovered. Many of these extrasolar systems consist of one star and only one giant planet. However, recently more and more different types of systems have become known, including also extrasolar planets in binaries. In our study, we will concentrate on such systems, since a large percentage of all G-M stars are expected to be part of binary or multiple stellar systems. Therefore, these kinds of systems are worthy of investigation in detail. In particular, we will concentrate on planets in P-Type motion, where the planet orbits around both stars. During the last few years, four such systems (NN Ser, HW Vir, HU Aqr and DP Leo) have been discovered. In our study, we performed dynamical studies for three multi-planetary systems in binaries (NN Ser, HW Vir, HU Aqr), and compared simulated eclipse timing variations (ETV) to current observational data.

Keywords. celestial mechanics, methods: n-body simulations, eclipses, binaries: eclipsing

1. Introduction, model, and methods

This work is dedicated to planets and planetary systems in P-Type motion, where we investigated the dynamical stability of three multi-planetary systems. The initial conditions used can be found in Beuermann *et al.* (2010) for the system NN Ser, in Qian *et al.* (2011) for the system HU Aqr, and in Lee *et al.* (2009) for the system HW Vir. In all cases, the motion of the planets were considered in the framework of purely Newtonian forces and all the celestial bodies involved were regarded as point masses. As the dynamical model for all investigations, we used the full n-body problem, consisting of two stars and two planets. In order to study the dynamical evolution of the different systems, we applied the Lie-Series Integration Method (see e.g. Eggl & Dvorak 2010) and checked these results with a Gauss-Radau integration method (see e.g. Everhart 1974) - both were in good agreement. All calculations were done for an integration time of 500,000 years.

Since all four planetary systems were discovered via eclipse timing variation measurements (ETV, perturbations of the stellar eclipse timing, caused by an unseen planet), we investigated also ETV curves.

2. Dynamical study

The two systems HW Vir and HU Aqr turned out to be unstable after a short time for the given initial conditions. Hence, we integrated best case scenarios, with the most favorable initial conditions, that were compatible with the given error bars. These best cases also lead to unstable motion. In a second step, we investigated the influence of relative orbital inclination, where the two stars define the plane (i = 0) and the planets are inclined to this plane. We changed the inclination of both planets (i = 0° - 50°,

$\Delta i = 10°$) and calculated the corresponding mass ($m' = m \cdot sin(i)$). Again, for the systems HU Aqr and HW Vir, none of these configurations lead to stable motion of the planets. Nevertheless, up to now, not the whole possible parameter space has been tested, hence stable motion may be found for some specific initial conditions, which we will investigate in the future.

For the binary system NN Ser, the authors give two possible planetary configurations. In the first case (marked with (2a) in Beuermann *et al.* 2010), the system turned out to be stable just for some specific initial conditions (the eccentric orbit needs to have a mean anomaly (M) between $\approx 120°$ and $310°$). The second configuration (marked with (2b) in Beuermann *et al.* 2010), turned out to be less sensitive to the initial conditions, but again here the mean anomaly plays an important role for the long-term stability.

In a second step, we investigated the influence of relative orbital inclination for M = 0, which corresponds to a worst case scenario. Even for this worst case, we could find stable configurations (up to an integration time of 500,000 years) up to relatively high inclinations (i = 40°) of the two planets. So, we can conclude that at least the system NN Ser could be long-term stable. A detailed description of the whole work including all results can be found in Eggl *et al.* (2011).

3. Eclipse timing variations (ETV)

Since some of the proposed planetary systems seemed to be unstable, we investigated whether a detection with the ETV method is possible at all. Thus, we integrated the given systems and measured the ETV signal caused by circumbinary planets. The first results showed that short term variations in the binary's period are minimal. Yet, the long term effects caused by the binary's motion around the system's center of mass combined with a finite light travel time indeed produce measurable signals.

Acknowledgements

B. Funk acknowledges support by the ÖFG project no. 06/12277. S. Eggl and M. Gyergyovits acknowledge support by the Austrian FWF project no. P20216-N16. R. Schwarz acknowledges support by the ÖFG project no. 06/12290. E. Pilat-Lohinger acknowledges support by the Austrian FWF project no. P22603-N16.

References

Beuermann, K., Hessman, F. V., Dreizler, S., *et al.*, 2010, *A&A*, 521, L60
Eggl, S. & Dvorak, R. 2011, *LNP*, 790, 431
Eggl, S., Funk, B., Schwarz, R., *et al.*, 2011, *MNRAS*, submitted
Everhart, E. 1974, *CMDA*, 10, 35
Lee, J. W., Kim. S.-L., Kim, C.-H., *et al.*, 2009, *AJ*, 137, 3181
Qian, S.-B., Liu,L., Liao, W.-P., *et al.*, 2011, *MNRAS*, 414, L16

From Interacting Binaries to Exoplanets: Essential Modeling Tools
Proceedings IAU Symposium No. 282, 2011   © International Astronomical Union 2012
Mercedes T. Richards & Ivan Hubeny, eds.   doi:10.1017/S174392131102802X

Dependence of Circumsubstellar Disk SEDs on System Inclination

Olga Zakhozhay

Main Astronomical Observatory, National Academy of Sciences of Ukraine,
27, Akad. Zabolotny St., Kyiv, 03680, Ukraine
email: `zkholga@mail.ru`

Abstract. The spectral energy distributions (SEDs) have being simulated for 1120 systems that contain brown dwarfs with different physical parameters and protoplanetary disks that are inclined at different angles. The SED's shape dependence on disk inclination toward the observer is discussed.

Keywords. circumstellar matter, stars: low-mass, brown dwarfs, planetary systems: protoplanetary disks.

1. Introduction

The spectral energy distribution (SED) of a protoplanetary disk and hosted star (or substar) depends on the astrophysical properties of the central object, age of the system, and disk inclination. The results presented take into account all of these parameters and based on the: (1) physical model for substellar evolution (Pisarenko *et al.* 2007); (2) temperature distribution calculation model for passive flat disks (Chiang & Goldreich 1997); (3) geometrical model for SED's calculation (Zakhozhay 2011, Zakhozhay 2011, Zakhozhay *et al.* 2011); (4) disk size dependence from central objects mass (Zakhozhay 2005).

2. Results

An atlas of 1120 SEDs was created for systems with different parameters:
- substellar masses within the range 0.01-0.08 M_{sun};
- protoplanetary disks with different inclination angles (0°-80°);
- system ages are 1-30 Myr;
- substars and protoplanetary disks irradiate like a black body;
- distance from Sun to substar is 10 pc;
- disk's inner radius is the central object radius and sublimation radius at 1Myr.

Figure 1 shows how the shape of the SED depends on inclination of the systems, without and with an inner hole. Figure 2 shows how the system's SED is affected by the presence of an inner hole for systems located face on and inclined on 60° toward the observer. On small inclination angles, flux from the gapless systems will always be bigger because the emitting area of such a disk is bigger. But high inclination angles correspond to higher flux from the systems with inner holes. The explanation for this is the following: when the disk has no inner hole, its inner edge starts to cover a part of the central object at the moment when the inclination becomes $j > 0°$. When the disk inner hole is present, the inner edge of the disk starts to cover the part of the central object over large angles (the exact value for which depends on the system's geometrical parameters; it is $\sim 75°$ for the systems shown here). As flux from the central object always gives a dominant

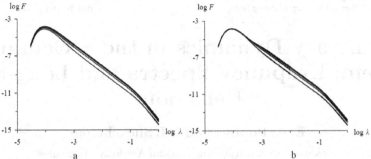

Figure 1. SEDs for substars with disks without inner holes (a) and with it (b), with substellar mass equals to 0.08 M_{sun} and age 1 Myr. Different lines correspond to different system inclinations (from top to bottom): 0°- 80° with step 20°. F - radiant flux, $erg/(cm^2 \cdot s \cdot cm)$, λ - wavelength, μm.

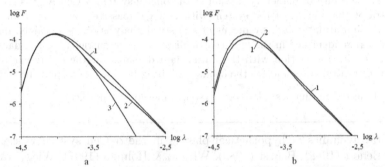

Figure 2. SEDs for the systems that contain a substar with mass 0.01 M_{sun} and age 1 Myr. (a) - The SEDs for systems that are located face on (0°), and (b) - systems that are inclined on 60°. On each panel are shown (1) - SED for gapless disk, (2) - SED for disk with inner hole and (3) - substellar back body irradiation. F - radiant flux, $erg/(cm^2 \cdot s \cdot cm)$, λ - wavelength, μm.

contribution to the total flux, the SED for a system with an inner hole will have a bigger maximum.

3. Conclusions

The shape of the SED for a brown dwarf with a protoplanetary disk strongly depends on the system inclination. Systems that are located face on have a maximum flux and when the inclination angle toward the observer is growing, the flux intensity decreases. The shape of the SED strongly depends on the presence of an inner hole, thus, different geometry calculation models should be used for gapless systems and systems with and without inner holes. The correct determination of the disk inclination relating to the observer permits us to identify the geometrical parameters for the true disk and central object and, thus, to obtain the real physical parameters of the central object such as mass, age, and size.

References

Pisarenko, I. A., Yatsenko, A. A., & Zakhozhay V. A. 2007, *Astron. Rep.*, 51, 605
Chiang, E. I. & Goldreich, P. 1997, *ApJ*, 490, 368
Zakhozhay, O. V. 2011, *Radio Physics and Radio Astronomy*, 2, 125
Zakhozhay, O. V. 2011, *Radio Physics and Radio Astronomy*, 2, 211
Zakhozhay, V. A., Zakhozhay, O. V., & Vidmachenko, A. P. 2011, *Kin. Phys. Cel. Bodies*, 27, 140
Zakhozhay, O. V. 2005, *Vysnyk astronomichnoi shkoly (in Russian)*, 4, 55

From Interacting Binaries to Exoplanets: Essential Modeling Tools
Proceedings IAU Symposium No. 282, 2011   © International Astronomical Union 2012
Mercedes T. Richards & Ivan Hubeny, eds.   doi:10.1017/S1743921311028031

Planetary Dynamics in the α Centauri System: Lyapunov Spectra and Long-term Behaviour

E. A. Popova and I. I. Shevchenko

Pulkovo Observatory of the Russian Academy of Sciences
Pulkovskoye ave. 65, Saint Petersburg 196140, Russia
email: m02pea@hotmail.com

Abstract. The stability of planetary motion in the binary system α Cen A–B is studied. Lyapunov spectra of the motion of the system with a single massive planet are computed on a fine grid of the initial data, and, by means of statistical analysis of the obtained data arrays, chaotic domains are identified in the "pericentric distance — eccentricity" initial data space for the planetary orbit. Association with the initial data domains for the orbits exhibiting close encounters with central stars and for the orbits exhibiting long-term escape is investigated.

Keywords. Celestial mechanics, planetary systems, methods: numerical

---

The orbital dynamics of hypothetical planets in the α Cen system was extensively studied by Benest (1988), Benest (1989), Wiegert & Holman (1997), Wiegert & Holman (1999), Benest & Gonczi (1998), Benest & Gonczi (2003), and other authors. We consider the double star α Cen A–B, setting the component masses $M_1 = 1.1 M_\odot$, $M_2 = 0.91 M_\odot$, semi-major axis $a = 23.4$ AU and eccentricity $e = 0.52$ (Pourbaix *et al.* 1999). We perform computations of the planetary orbits in the two cases: the planar elliptic restricted three-body problem (the planet mass is zero) and the planar general three-body problem (the planet mass is nonzero; it is set equal to one Jupiter mass). At the initial moment of time, the three bodies are located on the "B–A–planet" straight line at the pericenters of their unperturbed orbits. We integrate on a fine grid of the initial data for the planet orbit, varying the pericentric distance q and the eccentricity e.

To explore the stability problem for the planetary motion, we use two stability criteria. The first criterion is the "escape-collision" one: the orbit is stable if the distance between the planet and one of the stars does not become less than 10^{-3} AU and does not exceed 10^3 AU. The second criterion is the value of the maximum Lyapunov characteristic exponent (maximum LCE). We perform the computations of the full LCE spectra using the algorithms and codes by von Bremen *et al.* (1997), Shevchenko & Kouprianov (2002), Kouprianov & Shevchenko (2003), Kouprianov & Shevchenko (2005). The LCE spectra are obtained for each point on the grid of the initial data.

The statistical method, used here for the separation of regular and chaotic orbits, was proposed by Melnikov & Shevchenko (1998) (see also Shevchenko 2002, Shevchenko & Melnikov 2003). It consists of 4 steps. (*i*) Two differential distributions of the maximum LCE are constructed for the initial data grid using two different integration time intervals. (*ii*) Each of these distributions has at least two peaks. The peak that shifts (moves in the direction of the LCE smaller values), when the integration time interval is increased, corresponds to the regular trajectories. The fixed peak (peaks) corresponds to the chaotic ones. (*iii*) The maximum LCE value in the middle between the peaks (in the distribution built taking the maximum computation time) gives the numerical

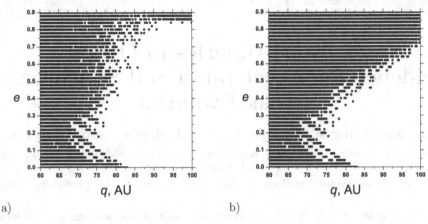

q, AU

a)

q, AU

b)

Figure 1. Stability diagrams constructed by the "escape-collision" (a) and LCE (b) criteria.

criterion for the separation of the regular and chaotic trajectories. (*iv*) In the further computations on finer data grids, the obtained criterion is used to separate the regular and chaotic trajectories using much smaller time intervals of integration.

Examples of the constructed stability diagrams are shown in Fig. 1. These are two diagrams for the outer orbits of the system in the general three-body problem. Fig. 1a is constructed by the "escape-collision" criterion, and Fig. 1b by the maximum LCE criterion. The chaotic domains are shown in black.

Our basic conclusions are as following. (*i*) For the initially circular planetary orbits, the outer borders of the chaotic domains in the stability diagrams correspond to the semi-major axis ∼ 80 AU, whereas the inner borders correspond to ∼ 5 AU. (*ii*) The chaotic domains in the diagrams expand approximately linearly with the eccentricity. (*iii*) The representative values of the Lyapunov times in the chaotic domains are: ∼ 500 yr for the outer orbits, and ∼ 60 yr for the inner orbits. (*iv*) The LCE criterion, applied for the construction of the stability diagrams, provides a better resolution of the chaos-order borders in the stability diagrams in comparison with the "escape-collision" criterion.

References

Benest, D. 1988, *A&A*, 206, 143

Benest, D. 1989, *A&A*, 223, 361

Benest, D. & Gonczi, R. 1998, *Earth, Moon, and Planets*, 81, 7

Benest, D. & Gonczi, R. 2003, *Earth, Moon and Planets*, 93, 175

von Bremen, H. F., Udwadia, F. E., & Proskurowski, W. 1997, *Physica D*, 101, 1

Kouprianov, V. V. & Shevchenko, I. I. 2003, *A&A*, 410, 749

Kouprianov, V. V. & Shevchenko, I. I. 2005, *Icarus*, 176, 224

Melnikov, A. V. & Shevchenko, I. I. 1998, *Sol. Sys. Res.*, 32, 480 [*Astron. Vestnik*, 32, 548]

Pourbaix, D., Neuforge-Verheecke, C., & Noels, A. 1999, *A&A*, 344, 172

Shevchenko, I. I. 2002, in: *Asteroids, Comets, Meteors 2002*. Ed. by Warmbein B., Berlin: ESA, 2002. P. 367–370.

Shevchenko, I. I. & Kouprianov, V. V. 2002, *A&A*, 394, 663

Shevchenko, I. I. & Melnikov, A. V. 2003, *JETP Lett.*, 77, 642 [*Pis'ma ZhETF*, 77, 772]

Wiegert, P. A. & Holman, M. J. 1997, *AJ*, 113, 1445

Wiegert, P. A. & Holman, M. J. 1999, *AJ*, 117, 621

From Interacting Binaries to Exoplanets: Essential Modeling Tools
Proceedings IAU Symposium No. 282, 2011
Mercedes T. Richards & Ivan Hubeny, eds.
© International Astronomical Union 2012
doi:10.1017/S1743921311028043

T Tauri Binaries in Orion:
Evidence for Accelerated and Synchronized Disk Evolution

Sebastian Daemgen[1], Monika G. Petr-Gotzens[1], and Serge Correia[2]

[1] European Southern Observatory, Karl-Schwarzschildstr. 2, 85748 Garching, Germany
email: sdaemgen@eso.org

[2] Astrophysikalisches Institut Potsdam, An der Sternwarte 16, 14482 Potsdam, Germany

Abstract. In order to trace the role of binarity for disk evolution and hence planet formation, we started the currently largest spatially resolved near-infrared photometric and spectroscopic study of the inner dust and accretion disks of the individual components of 27 visual, 100–400 AU binaries in the Orion Nebula Cluster (ONC). We study the frequency of Brackett-γ (2.165μm) emitters to assess the frequency of accretion disk-bearing stars among the binaries of the ONC: only 34±9% of the binary components show signs of accretion and, hence, the presence of gaseous inner disks—less than the fraction of gas accretion disks among single stars of the ONC of ~50%. Additionally, we find a significant difference between binaries above and below 200 AU separation: no close systems with only one accreting component are found. The results suggest shortened disk lifetimes as well as synchronized disk evolution.

Keywords. stars: late-type, stars: formation, circumstellar matter, binaries: visual

1. The Project & Observations

Twenty-seven binaries in the ONC were selected from the binary census in Petr(1998) and Köhler *et al.* (2006). The sample consists primarily of low-mass stars with spectral types K3–M6; separations are in a range of 0.25″–1″, which is equivalent to ~100–400 AU at the distance of the ONC. All binaries were spatially resolved with Adaptive Optics assisted VLT/NACO+Gemini/NIRI Near-Infrared (JHK) photometry and (K-band, 2.2μm) spectroscopy. Stellar parameters (spectral types, ages, masses, luminosities, and others) and dust disk and accretion signatures (Brackett-γ equivalent widths, NIR color excess, accretion luminosities, mass accretion rates) were derived for each binary component. Active accretion from the inner disk around each stellar component was inferred from the presence of emission in the Brackett-γ (2.166μm) Hydrogen line, the presence of dust in the inner disk is measured from color excess in $H-K$.

2. Results

We find an accretion disk fraction of 34±9% among the ONC binary components. This is lower than the 50% accretion fraction found for singles in the ONC (Hillenbrand *et al.* 1998). We also see a reduced—compared to single stars in the ONC—number of binary components with $H-K$ color excess. A reason for this deficiency might be accelerated disk dispersal: since the disks of binary components are truncated to radii of ~0.3–0.5 times the binary separation (Armitage *et al.* 1999), viscous evolution at the outer radius of the disk can reduce the disk lifetime to scales ≪ 1 Myr.

At separations smaller than 200 AU, we see a relatively high abundance of pairs of two classical T Tauri stars (with both components accreting) and two weak-line T Tauri stars

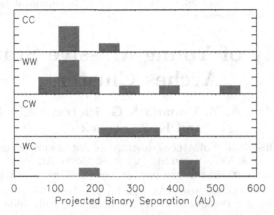

Figure 1. Histograms of binary separation as a function of component accretion activity: Pairs of classical T Tauri Stars (CC) and pairs of weak-line T Tauri Stars (WW) appear in tighter binary systems than mixed pairs with classical primary and weak-lined secondary (CW) or the other way around (WC). This is statistically significant (K-S test) with 99.5%.

(no accreting component; see Fig. 1). In contrast, mixed pairs of accreting/non-accreting components have significantly larger separations with 99.5% confidence (K-S test). Hints for an under-density of close mixed pairs have been observed in other star forming regions (Monin *et al.* 2007), however, never with such high significance. A possible explanation: independent disk evolution in wide pairs while disks in close binaries are synchronized.

The mass accretion rates of binary components of the ONC are comparable to those of other star forming regions, in the same mass range. Despite the considerably denser stellar environment of the ONC and reduced disk masses compared to star forming regions, like e.g. Taurus, the magnitude of accretion from the inner disk is not significantly reduced.

3. Discussion

Our results suggest shortened disk lifetimes and provide statistically significant evidence for synchronized evolution of the individual disks around medium-separated binary components—with important implications for planet formation scenarios. For instance, among the more than 40 planets in orbit around a stellar binary component (Eggenberger & Udry 2010), it is mainly the more massive component hosting the planet. While this could be explained by shorter-lived disks around the less massive binary components, in our data, mixed systems with either component hosting a disk are equally abundant. Likewise, only few binaries with each component orbited by its own planet are known. If this is not entirely due to selection effects, it will most likely be a consequence of differential disk evolution, which we, however, observe to be attenuated in <200 AU binaries.

References

Petr 1998, *PhDT*, U Heidelberg, 1998

Köhler, R., Petr-Gotzens, M. G., McCaughrean, M. J., Bouvier, J., Duchêne, G., Quirrenbach, A., & Zinnecker, H. 2006, *A&A*, 458, 461

Hillenbrand, L. A., Strom, S. E., Calvet, N., Merrill, K. M., Gatley, I., Makidon, R. B., Meyer, M. R., & Skrutskie, M. F. 1998, *AJ*, 116, 1816

Armitage, P. J., Clarke, C. J., & Tout, C. A. 1999, *MNRAS*, 304, 425

Monin, J.-L., Clarke, C. J., Prato, L., & McCabe, C. 2007, *PPV*, 295

Eggenberger & Udry 2010, in *Planets in Binary Systems*, ed: Haghighipour, Springer 2010

From Interacting Binaries to Exoplanets: Essential Modeling Tools
Proceedings IAU Symposium No. 282, 2011   © International Astronomical Union 2012
Mercedes T. Richards & Ivan Hubeny, eds.   doi:10.1017/S1743921311028055

Variability of Young Massive Stars in the Arches Cluster

K. Markakis[1], A. Z. Bonanos[1], G. Pietrzynski[2], L. Macri[3], and K. Z. Stanek[4]

[1] National Observatory of Athens, Institute of Astronomy & Astrophysics,
I. Metaxa & Vas. Pavlou St., P. Penteli 15236, Athens, Greece
email: markakis@astro.noa.gr, bonanos@astro.noa.gr

[2] Warsaw University Observatory, Al. Ujazdowskie 4, 00-478 Warszawa, Poland
Universidad de Concepción, Departamento de Astronomia, Casilla 160-C, Concepción, Chile

[3] Department of Physics & Astronomy, Texas A&M University, College Station, TX 77842-4242, USA

[4] The Ohio State University, 140 West 18th Avenue, Columbus, OH 43210, USA

Abstract. We present preliminary results of the first near-infrared variability study of the Arches cluster, using adaptive optics data from NIRI/Gemini and NACO/VLT. The goal is to discover eclipsing binaries in this young (2.5 ± 0.5 Myr), dense, massive cluster for which we will determine accurate fundamental parameters with subsequent spectroscopy. Given that the Arches cluster contains more than 200 Wolf-Rayet and O-type stars, it provides a rare opportunity to determine parameters for some of the most massive stars in the Galaxy.

Keywords. Galaxy: center, infrared: Stars, open clusters and associations: individual (Arches cluster), binaries: eclipsing, stars: variables, stars: Wolf-Rayet

1. Introduction

One of the most important questions is how massive can the most massive stars in the Universe be today. In other words, what is the upper limit of the Initial Mass Function in the Universe. The Arches Cluster provides us with a unique opportunity to address this question. Being a young massive cluster which lies near the Galactic center, it is bound to contain massive eclipsing binary systems, which provide the means to accurately measure parameters of massive stars (Bonanos 2009).

2. Datasets & Reduction

We used two datasets in the K_s band. The first dataset was obtained with Gemini's NIRI infrared camera covering 8 nights from April to July of 2006. The NIRI data have undergone a linearity correction. The second dataset was obtained with the VLT's NACO infrared camera on 29 nights between June of 2008 and March of 2009. The reduction of the NIRI images was performed with the IRAF† Gemini v1.9 package while the reduction of the NACO images was performed via the NACO reduction pipeline, based on ESO's Common Pipeline Library.

3. Image Subtraction & Photometry

We tested four different methods in order to achieve accurate photometry. Initially, we began with the image subtraction package ISIS (Alard & Lupton 1998, Alard 2000), which

---

† IRAF is distributed by the NOAO, which are operated by the Association of Universities for Research in Astronomy, Inc., under cooperative agreement with the NSF.

is optimal for detecting variables in crowded fields and IRAF's DAOPHOT (Stetson 1987) package. Both of these software packages use a mathematical PSF model in order to model the stellar line profile, and produce large photometric errors (the order of 2-5 magnitude differences for the same object from frame to frame which is not physically acceptable) primarily because of the speckles that are being introduced by the use of adaptive optics. We concluded that it is impossible to fit a mathematical function on these speckles. In order to solve this problem, we tried a different approach with the use of an empirical PSF. For this reason, we used the StarFinder code (Diolaiti *et al.* 1999, 2000). With the original version of the code, we saw a big improvement as far as the photometric errors are concerned (the magnitude differences have dropped below 1 magnitude which is physically acceptable). However, we were not able to identify any non-variable stars. The reason for this is that the StarFinder code does not allow for a spatially variable PSF option, which is crucial in our case since the already imperfect correction by the adaptive optics degrades rapidly with increasing distance from the AO guide star. In order to improve our results further, we used the version of the StarFinder code developed by Schoedel (2010). The main difference in this version is that it uses a local PSF by dividing the frame into several subframes with large overlap with each other. The PSF is considered to be stable across these subframes. Moreover, the code performs photometry on each object with more than one PSF model (on most occasions the same object appears on more than one frame due to the large overlap), which helps the statistics of the actual counts value. Another interesting and helpful feature of Dr. Schoedel's approach is that the code performs photometry on a Wiener deconvolved version of the original image, which favors the deblending of nearby sources in dense fields. After applying this method on our frames, we saw further improvement on our results. The magnitude differences have now dropped even further (in the range of 0.2 to 0.6 magnitudes), but the problem of not finding stable stars remains. This behavior of our data may be explained by the underestimated errors produced by StarFinder (an issue that has been discussed by several researchers), by a possible contamination of our data (aborted nights, bad performance of the AO system etc.) or by a combination of the above.

4. Future Work & Conclusions

Currently, we are trying to check our sample for possible contamination with bad quality frames and to better estimate the errors that the code produces. We conclude that the use of an empirical PSF is mandatory for accurate photometry on AO data. Moreover, the use of Wiener deconvolution is very helpful when one works on crowded regions as it favors the deblending of nearby sources.

Acknowledgement

K.M. and A.Z.B. acknowledge support from the IAU and the European Commission for an FP7 Marie Curie International Reintegration Grant.

References

Alard C. & Lupton R. H. 1998, *ApJ*, 503, 325
Alard C. 2000, *A&AS*, 144, 363
Bonanos A. 2009, *ApJ*, 691, 407
Diolaiti E. *et al.*, 1999, *ESOC*, 56, 175
Diolaiti E. *et al.*, 2000, *A&AS*, 147, 335
Schoedel 2010, *A&A*, 509, 58
Stetson P. B. 1987, *PASP*, 99, 191

From Interacting Binaries to Exoplanets: Essential Modeling Tools
Proceedings IAU Symposium No. 282, 2011
Mercedes T. Richards & Ivan Hubeny, eds.

© International Astronomical Union 2012
doi:10.1017/S1743921311028067

The Evolution of Low Mass Contact Binaries

Kazimierz Stępień[1] and Kosmas Gazeas[2,3]

[1] Warsaw University Observatory, Al. Ujazdowskie 4, 00-478 Warszawa, Poland
email: kst@astrouw.edu.pl

[2] Department of Astrophysics, Astronomy and Mechanics, Faculty of Physics, University of Athens, GR-157 84, Zografos, Athens, Greece
[3] European Space Agency, ESTEC, Mechatronics and Optics Division, Keplerlaan 1, 2200AG, Noordwijk, The Netherlands
email: kgaze@physics.auth.gr, Kosmas.Gazeas@esa.int

Abstract. We discuss the origin and evolution of low mass contact binaries with P_{orb} shorter than 0.3 d that have properties somewhat different from the rest of the contact binaries. A comparison of an evolutionary model set with observations shows that both components are on the main sequence, the age of the binaries is at least several Gyr, while the contact phase lasts only less than 1 Gyr.

Keywords. contact binaries, W UMa stars, stellar evolution

1. Evolution of contact binaries, and the suggested new model

The problem of the origin and evolution of cool contact binaries is still far from full understanding. The TRO theory explains the situation quite well. However, it encounters several problems and some of its predictions are at odds with observations (Webbink 2003, Stępień 2011). A different theory has been suggested recently (Stępień 2006, 2009, Gazeas & Stępień 2008), which assumes that mass transfer following Roche lobe overflow (RLOF) proceeds in a similar way as in classical Algols, i.e. until mass ratio reversal and angular momentum (AM) loss, which lead to a contact configuration.

The new model is based on the assumption that the initial P_{orb} of progenitors of cool contact binaries is close to a couple of days. It takes several Gyr until enough AM is lost and RLOF occurs. This time is sufficient to terminate the main sequence (MS) evolution, having components with mass higher than 1 solar mass. After mass exchange, the low mass components or LMCs (former high mass components or HMCs) of such binaries have hydrogen depleted cores. A new set of models of LMCBs was calculated by Stępień (2011), The initial parameters of these models are: 0.9+0.3(2.5), 0.9+0.4(2.5), 0.9+0.5(2.0), 0.9+0.5(1.5), 0.9+0.7(2.0), 1.1+0.5(2.0) and 1.1+0.5(1.5), where the first two numbers give the initial component masses in solar units and the number in parentheses gives the initial P_{orb} in days. It takes, on average, about 7 Gyr for each binary to reach the contact phase, but they live in a contact configuration for about 0.8 Gyr. After this phase, the orbit is so tight, that both stars overflow their outer critical surfaces. The binary loses mass and AM through the L_2 point, which results in its coalescence. Both components are still on the MS, although the LMC may be close to (or slightly beyond) the terminal age MS (TAMS) region.

Observations show that there are several low mass contact binaries (LMCB) with the total mass close to 1-1.4 $M_\odot$ (Gazeas & Niarchos 2006, Gazeas & Stępień 2008). The LMCB compared with our models include the first 9 stars from Table 1 of Gazeas & Stępień (2008). These systems have P_{orb} shorter than 0.3 d and their orbital AM is low (significantly less than 3×10^{51} in cgs units), while the component masses and radii indicate that they are both on the MS (Fig. 1). The HMC in all LMCB appear to have lower average surface brightness than their lower mass companions, which classifies them as W-type systems, according to the classification made by Binnendijk (1970). The heavily covered surface of the HMC by cool spots is the most likely explanation of the so-called $W - phenomenon$ (Stępień 1980, Eaton, Wu & Rucinski 1980, Zola et al. 2010).

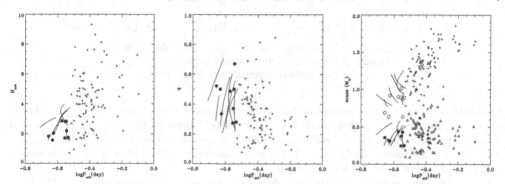

Figure 1. The plots present the comparison of the computed models with the observations for the H_{orb} vs P_{orb} (left) and the mass ratio vs P_{orb} (center) of massive contact binaries (crosses) and LMCB (filled circles). Each model (solid line) corresponds to the time evolution of one of the seven model binaries, from the time when the components reach contact till they both overflow the outer critical surface, which results in their quick merging (Webbink 1977). Direction of evolution is from higher to lower values of H_{orb} and the same for q. Separate component masses of LMCB are shown in right panel. Here, crosses and triangles correspond, respectively, to HMC and LMC of massive CBs, whereas open and filled circles correspond to the same components of LMCB. The solid lines show evolution of the component masses from right to left.

2. Conclusions

Low-mass contact binaries have P_{orb} shorter than 0.3 d, total mass lower than about 1.4 $M_\odot$, orbital AM less than about 3×10^{51} (in cgs units), radii corresponding to MS stars, relatively high mass ratios, and none of them is of A-type. According to our models, they originate from detached binaries with total initial mass lower than 1.6 solar mass and initial P_{orb} of 1.5-2 d. Evolution is driven mostly by mass transfer and AM loss via the magnetized wind, which shrinks the orbit and makes both components overflow their outer Roche lobes. All LMCB are old, with a typical age of 7-8 Gyr, although their contact phase lasts less than 1 Gyr, leading into coalescence.

References

Binnendijk, L., 1970, *Vistas in Astron.*, 12, 217
Eaton, J., Wu, C. C., & Rucinski, S. M., 1980, *ApJ*, 239, 919
Gazeas, K. D. & Niarchos, P. G., 2006, *MNRAS*, 370, L29
Gazeas, K. & Stępień, K., 2008, *MNRAS*, 390, 1577
Stępień, K., 1980, *AcA*, 30, 315
Stępień, K., 2006, *AcA*, 56, 199
Stępień, K., 2009, *MNRAS*, 397, 857,
Stępień, K., 2011, *AcA*, 61, in press
Webbink, R. F., 1977, *ApJ*, 211, 881
Webbink, R. F., 2003, in: 3D Stellar Evolution, *ASP-CS*, Vol. 293, S. Turcotte et al. (eds), p. 76
Zola, S., Gazeas, K., Kreiner, J. M., et al. 2010, *MNRAS*, 408, 474

From Interacting Binaries to Exoplanets: Essential Modeling Tools
Proceedings IAU Symposium No. 282, 2011
Mercedes T. Richards & Ivan Hubeny, eds.
© International Astronomical Union 2012
doi:10.1017/S1743921311028079

Kinematic Properties of Chromospheric Active Binary Stars

M. Tüysüz[1], F. Soydugan[1], S. Bilir[2], and O. Demircan[1]

[1]Çanakkale University Observatory, Çanakkale, Turkey
[2]Istanbul University Observatory, Istanbul, Turkey
email: mehmettuysuz@comu.edu.tr

Abstract. The kinematic behaviour of 362 chromospherically active binary stars (CABs) in the solar neighbourhood were investigated. The Third CABs Catalog by Eker *et al.* (2008) was used as the main source. The spatial distribution and the components of the Galactic space velocities of the programme stars were determined. The effects of differential rotation and Local Standard of Rest (LSR) were corrected for all systems.

Forty probable moving group (MG) members were determined by Eggen's criteria. The kinematic age of the young systems, which are probable members of MGs, was calculated as 0.79 (0.21) Gyr and the rest of 322 field stars were found to have a kinematic age of 4.38 (1.1) Gyr. Field CABs were separated into two sub-groups: dwarf systems, which were formed by main sequence (dwarf) stars, and evolved systems included at least one evolved (giant or sub-giant) component. The kinematic age of 134 dwarf systems was calculated as 4.69 (0.75) Gyr and 4.15 (1.29) Gyr for 188 evolved CABs.

Keywords. active stars, kinematic, kinematic age

1. Introduction

Chromospheric Active Binary Stars (CABs), whose one or two components with late spectral types (F-G-K-M) are giant, sub-giant or Main Sequence stars, are detached binary systems. Emission in the centre of the Ca II H and K, and sometimes, H lines is the most basic indicator of chromospheric activity. Another evidence for magnetic activity is photometric variations, which are caused by large stellar spots. The activity is usually explained by the "dynamo model" (Hall 1989)

2. Kinematic Properties

Galactic space velocity components (U, V, W) of CABs and their errors are calculated using the algorithm given by Johnson and Soderblom (1987). In order to determine the space velocities, the equatorial coordinates, proper motion components, parallaxes of the stars, radial velocities of the mass centre of binary systems, and also the errors of all these, the data must be available. Taking into account the Third CAB Catalogue and the literature, radial velocity data of 362 CABs stars have been collected. Astrometric data of these systems (components of proper motion and trigonometric parallax) were obtained from the new Hipparcos satellite data. Correction for galactic differential rotation was made on the space velocities. Besides this, for kinematic age determination, the solar velocity correction was also made. Space velocity values given for effects of differential rotation and Local Standard of Rest (LSR) were used. The galactic velocity components of CABs were calculated and velocity diagrams of U-V and W-V were formed (Fig. 1). Applying Eggen's criteria to the targets, moving group (MG) members were identified (N=40 CABs) and population analysis was performed.

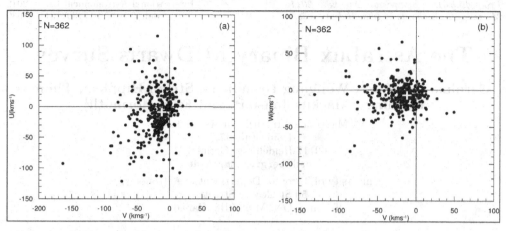

Figure 1. Velocity dispersions of the sample of the CABs (a) on the U, V plane and (b) on the W, V plane. The velocities are heliocentric. The position of the LSR is marked by +.

3. Kinematic Ages

All CAB's were divided into three sub-groups: (i) MG members, (ii) dwarf systems, (iii) evolved systems. For each sub-group, kinematic age determination was made by using Wielen's (1977) tables of age-space velocity dispersion formula. Ages of 4.69, 4.15 and 0.79 Gyr were calculated for 134 dwarf systems, 188 evolved systems, and the MG members, respectively.

Acknowledgements

This study is a part of the Ph.D. thesis of M. Tüysüz. This study was partly supported by the Turkish TUBITAK under the Grant No. 111T224.

References

Eker Z., Ak N. Filiz, Bilir S., Doğru D., Tüysü M., Soydugan E., Bakş H., Uğraş B., Soydugan F., Erdem A., & Demircan O. 2008, *MNRAS*, 389, 1722
Eggen O. J. 1958, *MNRAS*, 118, 65
Eggen O. J. 1989, *PASP*, 101, 366
Hall Douglas S. 1989, *SSRv*, 50, 219
Pols Onno R, Schroder, Klause-Peter, Hurley Jarrod R, Tout Christpoher A., & Eggleton Peter P. 1998, *MNRAS*, 298, 525
Tüysüz M. 2011, *Çanakkale Onsekiz Mart University Graduate School Thesis of PhD of Science. Investigation of Kinematic, Dynamic and Rotation Properties of the Chromospheric Active Binary Stars*, p. 46.
van Leeuwen F. 2007, *A&A*, 474, 653
Wielen R. 1977, *A&A*, 60, 263

From Interacting Binaries to Exoplanets: Essential Modeling Tools
Proceedings IAU Symposium No. 282, 2011
Mercedes T. Richards & Ivan Hubeny, eds.
© International Astronomical Union 2012
doi:10.1017/S1743921311028080

The AstraLux Binary M Dwarfs Survey

Carolina Bergfors[1], Wolfgang Brandner[1], Stefan Hippler[1], Thomas Henning[1], Markus Janson[2], and Felix Hormuth[1]

[1] Max-Planck-Institut für Astronomie,
Königstuhl 17,
69117 Heidelberg, Germany
email: bergfors@mpia.de

[2] University of Toronto, Department of Astronomy,
50 St George Street,
Toronto, ON, M 5S 3H8, Canada

Abstract. Binary/multiple properties provide clues to the formation of stars. In the AstraLux binary survey, we use the Lucky Imaging technique to search for companions to a large sample of young, nearby M dwarfs. We present results from observations of the first sub-sample, consisting of 124 M dwarfs in the southern sky.

Keywords. techniques: high angular resolution – binaries: visual – stars: low-mass, brown dwarfs

1. Introduction

M dwarfs form a link between solar-type stars on one side and very-low-mass stars and brown dwarfs on the other, and may represent a transition between different formation modes. Multiplicity characteristics of M dwarfs, such as the binary fraction and distribution of mass-ratio and separations can, therefore, provide clues to the formation of very-low-mass stars and brown dwarfs.

While multiplicity characteristics are well known for solar-type stars, they are less well constrained for M dwarfs. The binary fraction, $f_{bin} = N_{bin}/N_{total}$, decreases with decreasing mass from $\approx$57% for Sun-like stars (Duquennoy & Mayor 1991) to only 10-30% for very-low-mass stars and brown dwarfs (e.g., Burgasser *et al.* 2007). The distributions of mass-ratio, $q = M_2/M_1$, and binary separation also appear different for very-low-mass stars and brown dwarfs, compared to those of solar-type stars. These differences may indicate different formation scenarios (Thies & Kroupa 2007).

2. The survey

The AstraLux M dwarfs survey is the largest survey for binary/multiple M dwarfs to date. We observed ~ 800 young, nearby, early- to mid-M type stars using the two Lucky Imaging instruments *AstraLux Norte* at the 2.2 m telescope at Calar Alto, Spain, and *AstraLux Sur* at the 3.5 m New Technology Telescope (NTT) at La Silla, Chile (Hormuth *et al.* 2008; Hippler *et al.* 2009).

The aim of the survey is to investigate multiplicity properties of M dwarfs from a statistically large sample, and to find and characterize young, close binaries containing very-low-mass stars and brown dwarfs. Follow-up orbital monitoring with AstraLux combined with near-infrared spectra of selected systems discovered in the survey will provide observational calibration of the mass-luminosity relation and evolutionary models for very-low-mass stars and brown dwarfs.

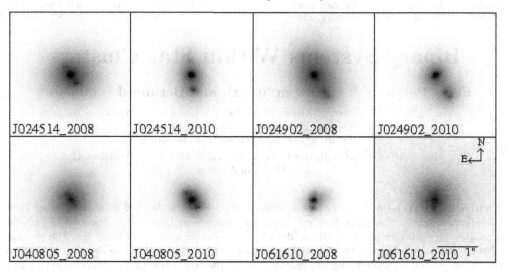

Figure 1. Three of the close binaries and one triple system discovered in the AstraLux M dwarfs survey. The figure shows SDSS z'-band observations obtained with *AstraLux Sur* in November 2008 and January 2010. The separation between the two close companions of (2MASS) J02490228-1029220 is $\rho \approx 0.15$ arcsec. These systems are being continuously monitored with *AstraLux Sur*, and near-infrared spectra have been obtained with SINFONI at the VLT.

3. The first results

In the first part of the southern sky survey, 124 young, nearby M dwarfs were observed with *AstraLux Sur* at NTT in November 2008 (Bergfors *et al.* 2010). This subsample is in itself the largest M dwarf multiplicity survey to date.

- 34 new and 17 previously known companions were identified in this sample. Most of these were separated by less than 1 arcsec, and would thus have been missed in a seeing-limited survey.

- We found a multiplicity fraction of $32 \pm 6\%$ for a set of 108 M0–M6 dwarfs within 52 pc from the Sun and with angular separations 0.1–6 arcsec, which corresponds to projected separations of 3–180 AU at the median distance of 30 pc.

- Late-type M dwarfs ($\geqslant$ M3.5) seem to prefer more equal-mass binaries than early-type M dwarfs. They are also mainly found in closer binaries, with the projected separation being closer than 20 AU for more than half of the sample.

References

Bergfors, C., Brandner, W., Janson, M. *et al.*, 2010, *A&A*, 520, A54

Burgasser, A. J., Reid, I. N., Siegler, N. *et al.*, 2007, in Protostars and Planets V, ed. B. Reipurth, D. Jewitt, & K. Keil, 427

Duquennoy, A. & Mayor, M. 1991, *A&A*, 248, 485

Hippler, S., Bergfors, C., Brandner, W. *et al.*, 2009, *The Messenger*, 137, 14

Hormuth, F., Hippler, S., Brandner, W. *et al.*, 2008, *SPIE*, 7014

Thies, I. & Kroupa, P. 2007, *ApJ*, 671, 767

From Interacting Binaries to Exoplanets: Essential Modeling Tools
Proceedings IAU Symposium No. 282, 2011   © International Astronomical Union 2012
Mercedes T. Richards & Ivan Hubeny, eds.   doi:10.1017/S1743921311028092

Binary Systems Within Star Clusters

Ernst Paunzen[1,2], Christian Stütz[2], and Bernhard Baumann[2]

[1]Rozhen National Astronomical Observatory, Institute of Astronomy of the Bulgarian
Academy of Sciences, P.O. Box 136, BG-4700 Smolyan, Bulgaria
email: ernst.paunzen@univie.ac.at

[2]Institut für Astronomie der Universität Wien, Türkenschanzstrasse 17,
A-1180 Wien, Austria

Abstract. WEBDA (http://www.univie.ac.at/webda) is a site devoted to observational data of stellar clusters in the Milky Way and the Small Magellanic Cloud. It is intended to provide a reliable presentation of the available data and knowledge about these objects. The success of WEBDA is documented by its worldwide usage and the related acknowledgements in the literature: more than 650 refereed publications within the last twelve years acknowledged its use. It collects all published data for stars in open clusters that may be useful either to determine membership, or to study the stellar content and properties of the clusters.

The database content includes astrometric data in the form of coordinates, rectangular positions, and proper motions, photometric data in the major systems in which star clusters have been observed, but also spectroscopic data like spectral classification, radial velocities, and rotational velocities. It also contains miscellaneous types of supplementary data like membership probabilities, orbital elements of spectroscopic binaries, and periods for different kinds of variable stars as well as an extensive bibliography. Several powerful tools help to plot, query and extract the data, which can be directly retrieved via http. At the time of writing, about four million individual measurements have been included in the database. The Star Clusters Young & Old Newsletter (SCYON), a bi-monthly newsletter devoted to star cluster research with about 600 subscribers, is hosted in parallel with the database.

We present the current and upcoming new interface and tools, which are needed to visualize and analyze the increasing amount of data from all-sky surveys, and deeper investigations of binary systems, low mass dwarfs, as well as planet-hosting stars.

Keywords. astronomical data bases: miscellaneous, catalogs, open clusters and associations: general, binaries: spectroscopic

1. New developments within WEBDA

The current version of the WEBDA open cluster database is comprised basically from files in RDB standard, a clever folder hierarchy and various perl scripts and gnuplot interfaces. At present, a consistent set of tools is developed, to treat all the data in WEBDA homogeneously.

This new suite fuses operations requiring direct access on site with the extraction and charting capabilities provided for the community. Already in coding stage are the following routines:

• Coordinate transformations, automatic scanning for, and refining of, center coordinates.

• The possibility to link and cross check all different datasets for individual clusters.

• Automatic quality and completeness checks to support the previous manually maintained database.

• Charting features we are testing right now include:

• Displaying coordinates and measurements in charts that allow zooming, filtering objects based on their position, brightness, on the fly.

• Selection of objects within the charts and extracting all available data for these objects.

• Interactive fitting of any displayed, filtered, zoomed sets of data is currently developed.

• Interactive semi-automatic isochrone fitting.

On the technical side, we restrict the suite to freeware or open source tools only. Another requirement we decided on is that there is not anything to be installed on the client side except a web browser capable of interpreting HTML.

We started a statistical analysis of the currently available positional and kinematic data. The analysis is naturally connected with a quality check of the database content and the identification of erroneous entries.

We are confronted with the situation that the availability of B1950 coordinates does not imply that there are also J2000 coordinates for the same object and vice versa. So, we developed several Perl programs which perform the transformation and the calculation of the mean values of stars with available multiple coordinates.

The outcome of this statistical analysis provides us with about 150 000 new J2000 coordinates. For about 100 000 data points of stars in 446 open clusters, we were able to check the differences of the calculated versus the published J2000 coordinates. From that comparison, we can conclude the proper motions of cluster members and, therefore, the open cluster's proper motion. So, we compiled a new catalogue of 722 cluster proper motions to verify and test our results.

These new coordinates, together with the additional data points in Webda, will help us to create new star charts and e.g. to determine the star cluster's mass.

2. Binary systems included in WEBDA

Binary stars are perfect astrophysical laboratories for a number of most important fundamental studies. They represent a significant portion of all stellar systems, and are, therefore, found throughout our Milky Way. Star clusters represent samples of stars of constant age and homogeneous chemical composition, suited for the study of processes linked to stellar structure and evolution, and to fix lines or loci in several most important astrophysical diagrams such as the colour-magnitude diagram, or the Hertzsprung-Russell diagram. Studying binary systems in star clusters is the ideal combination to

• Estimate the binary fraction in correlation with the age and metallicity

• Test the current models of stellar formation and evolution

• Test the current sets of isochrones

• Establish and test the current distances of star clusters with eclipsing binary systems.

Within WEBDA, all published known binary systems within the areas of star clusters are listed in a convenient way. The extraction of all available data for these objects can be easily performed. Currently, there are 1346 spectroscopic and eclipsing binary systems within 134 star clusters available. The whole range of evolutionary stages and metallicities are covered with this sample.

Any feedback to further improve the database would be greatly appreciated.

Acknowledgements

This work was supported by the financial contributions of the Austrian Agency for International Cooperation in Education and Research (WTZ CZ-10/2010 and HR-14/2010) and the Austrian Research Fund via the project FWF P22691-N16.

From Interacting Binaries to Exoplanets: Essential Modeling Tools
Proceedings IAU Symposium No. 282, 2011
Mercedes T. Richards & Ivan Hubeny, eds.

© International Astronomical Union 2012
doi:10.1017/S1743921311028109

Age Dependent Angular Momentum, Orbital Period and Total Mass of Detached Binaries

O. Demircan[1], M. Tüysüz[1] F. Soydugan[1], and S. Bilir[2]

[1] Çanakkale University Observatory, Çanakkale, Turkey
[2] Istanbul University Observatory, Istanbul, Turkey
email: demircan@comu.edu.tr

Abstract. The orbital angular momenta OAM (J) of detached binaries (including both cool and hot binaries) were estimated and nine subgroups were formed according to their OAM (J) distribution. The mean kinematical ages of all subgroups have been estimated by using their space velocity distributions and, thus, the age dependent variations of the mean OAM (J), orbital period (P), and total mass (M) of all subgroups were investigated. It was discovered that: i) The orbital period of detached binaries with radiative components decrease very slowly during the main sequence (MS) evolution. It is interesting that the large amount of mass loss is almost balanced by the OAM loss, and not much change in the orbital periods is observed. ii) The nuclear evolution of radiative components beyond the MS initiates the increase of the periods until the components have convective upper layers, i.e. until they become later than F5 IV, and the system becomes a cool binary with sub-giant or giant components. iii) The large co-rotating distance of the magnetically-driven wind in cool binaries (CAB) carries out a large amount of OAM and then the periods of such binaries decrease significantly, and the orbits shrink until another effect such as mass transfer dominates the period changes.

Keywords. detached binaries, kinematic age, angular momentum, orbital period, mass

1. Introduction

The absolute dimensions, proper motions, and radial velocities of known detached binaries were collected from the literature. The subgroups of detached binaries and the number of systems with collected data are as follows: 1. Detached binaries with radiative atmospheres (dbrs; 108 systems) 2. Detached binaries with convective atmospheres (dbcs; 179 systems) 2a. Chromospherically active binaries (CABs 149 systems) 2aa. Dwarf CABs (DCABs; with main sequence components; 69 systems) 2ab. Evolved CABs (ECABs; with at least one component subgiant or giant; 80 systems) 2b. Chromospherically inactive binaries (CIABs; 30 systems).

2. Data Analysis

Stellar population and moving group (MG) analysis were carried out for all systems; the MG members were extracted and only the systems with thin disk population were considered. The orbital angular momentum (OAM) of each system was estimated and each group dbrs, DCABs, ECABs and CIABs were listed in order of increasing OAMs. Each group was divided into five subgroups with increasing OAM, except CIABs which were considered to be a single subgroup. The mean values of OAM, the mean orbital period P, the mean total mass M of the component stars, and the mean kinematical age τ for each subgroup were estimated together with standard deviations.

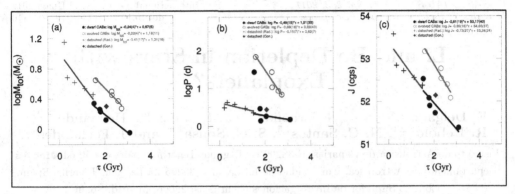

Figure 1. Age dependent J, P, and M of detached binaries.

3. Results

The results are presented in the following three diagrams of Fig 1. In Fig. 1, different decreasing rates of J, P, and M of all subgroups with increasing age are visible. Such strong dependences may be related with the origin of the systems, or may well be caused by some selection effects. For example (i) a certain kinematical age group contains also some younger systems. (ii) The short period systems have more chance to be observed. (iii) the chromospheric activity is an additional chance for observability. Different subgroups of detached binaries in Fig. 1 are definitely evolved towards increasing age but at different rates of J, P, and M changes, which is not clear in the diagram. However, having comparable masses of dbrs and ECABs in Fig. 1a implies evolution from dbrs to ECABs which requires an increase in P, as seen in Fig. 1b. Such an observational discovery is in line with the result of the theory of isotropic mass loss from detached binaries (see e.g. Pringle 1985, Demircan *et al.* 2006). The rate of P increase and related J, and M changes can be estimated by a detailed study. Since there are no evolved counterparts of DCABs in Fig. 1, the rates of the J, P, and M changes of this subgroup cannot be estimated. However, there is other evidence that the periods of DCABs decrease due to a large amount of J loss until another effect such as mass transfer dominates the period changes (see e.g. Van't Veer 1993, Demircan 1999, Guinan and Bradstreet 1988, Eker *et al.* 2006).

Acknowledgements

This study is a part of the Ph.D. thesis of M. Tüysüz. This study was partly supported by the Turkish TUBITAK under the Grant No. 111T224.

References

Demircan O. 1999, *Tr. J. of Physics*, 23, 425
Demircan O., Eker Z., Karataş Y., & Bilir S. 2006, *MNRAS*, 366, 1511
Eker Z., Demircan O., Bilir S., & Karataş Y. 2006, *MNRAS*, 373, 1483
Guinan E. F. & Bradstreet D. H. 1988, *felm. conf.*, 345
Karataş Y., Bilir S., Eker Z., & Demircan O. 2004, *MNRAS*, 349, 1069
Pringle J. E. 1985, *Interacting Binary Stars*, eds. J. Pringle & R. Wade (Cambridge: Cambridge University Press), p. 1

From Interacting Binaries to Exoplanets: Essential Modeling Tools
Proceedings IAU Symposium No. 282, 2011
Mercedes T. Richards & Ivan Hubeny, eds.

© International Astronomical Union 2012
doi:10.1017/S1743921311028110

Li and Be Depletion in Stars with Exoplanets?

E. Delgado Mena[1,2], G. Israelian[1,2], J. I. González Hernández[1,2], R. Rebolo[1,2,3], N. C. Santos[4,5], S. G. Sousa[4,5], and J. Fernandes[6]

[1] Instituto de Astrofísica de Canarias, E-38200 La Laguna, Tenerife, Spain. email: edm@iac.es

[2] Departamento de Astrofísica, Universidad de La Laguna, 38205 La Laguna, Tenerife, Spain.

[3] Consejo Superior de Investigaciones Científicas, 28006, Madrid, Spain

[4] Centro de Astrofísica, Universidade do Porto, Rua das Estrelas, 4150-762 Porto, Portugal.

[5] Departamento de Física e Astronomia, Faculdade de Ciências, Universidade do Porto, Portugal.

[6] Centro de Física Computacional, Universidade de Coimbra, Portugal.

[7] Observatório Astronómico e Departamento de Matemática, Universidade de Coimbra, Portugal.

Abstract. It is well known that stars with orbiting giant planets have a higher metallic content than stars without detected planets. In addition, we have found that solar-type stars with planets present an extra Li depletion when compared with field stars. On the other hand, Be needs a greater temperature to be destroyed, so we may find such a relation in cooler stars, whose convective envelopes are deep enough to carry material to layers where Be can be burned. We present Li and Be abundances for an extensive sample of stars with and without detected planets, covering a wide range of effective temperatures (4700-6500 K) with the aim of studying possible differences between the abundances of both groups. The processes that take place in the formation of planetary systems may affect the mixing of material inside their host stars and hence the abundances of light elements.

Keywords. stars: abundances — stars: atmospheres — stars: fundamental parameters — stars: planetary systems — stars: planetary systems: formation

1. Lithium

In a previous work, we reported an extra Li depletion in planet-host solar-type stars when compared with stars without planets (Israelian *et al.* 2009). Here, we present new Li abundances for the HARPS metal-poor sample of stars with and without planets (Santos *et al.* 2011, Sousa *et al.* 2011). In the left panel of Figure 1, these new Li abundances are plotted with those of the stars in the HARPS GTO sample and some planet hosts from other surveys, as a function of age. Lithium is expected to decrease with age, therefore, if planet host stars were on average older than comparison sample stars they would have depleted more lithium. However, we can see that comparison sample stars are still equally distributed at both sides of $\log N(Li) = 1.5$. Moreover, age does not seem to play a role in this behaviour (Sousa *et al.* 2010), except for the younger objects (age < 2 Gyr). Indeed, 50% of comparison stars present high Li abundances for a wide range of ages, although most of the planet-hosts show a severe Li destruction regardless of their age.

We propose that the low Li abundance of planet-host solar-analogue stars is directly associated with the presence of planets due to the effect of these in the angular momentum evolution of the star and the surface convective mixing (Bouvier 2008, Israelian *et al.*

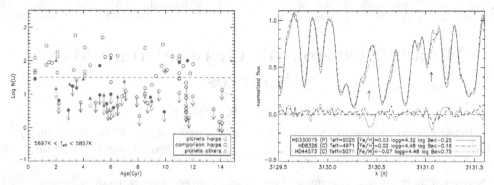

Figure 1. *Left panel:* Li abundances as a function of age for solar analogue stars with (red filled circles) and without (blue open circles) planets from HARPS GTO and HARPS metal-poor samples. Green triangles are stars with planets from other surveys. *Right panel:* Real spectra around Be II lines (located at the arrows) for the planet host HD330075 (red line) and the comparison stars HD8326 (blue dashed line). The black and green lines represent the differences in flux between spectra.

2004). To understand this Li extra depletion in stars with planets, observations of younger stars and proper modelling are required.

2. Beryllium

The determination of Be abundances for the coolest stars ($T_{eff} < 5200$ K) is very difficult since the Be II line at 3131.06 Å is blended with a Mn I line which becomes stronger as the temperature decreases, therefore, in this T_{eff} regime the uncertainties in Be abundances are large. To determine possible differences between stars with and without planets, we propose to compare directly the spectra of stars with very similar parameters, T_{eff}, [Fe/H], log g, v sin i, and Mn abundance. Therefore, if there is a difference between the spectra of two stars, it should be due to a difference in Be abundance. In a previous study (Delgado Mena *et al.* 2011), we found two planet hosts which showed a clear depletion of Be when compared to several stars without planets. One of them, HD330075, is shown in the right panel of Figure 1 together with two new comparison stars. HD44573 shows stronger Be lines than the planet host although HD8326 shows a similar level of depletion. In this work, we analyze more pairs of analogue cool stars but all of them present weak Be lines and no differences are observed between their fluxes. Therefore, the effect of extra Li depletion in solar-type stars with planets when compared with stars without detected planets does not seem to be present for Be, although the number of cool stars is still too small to reach a final conclusion.

References

Bouvier, J. 2008, *A&A*, 489, 53

Israelian, G., Santos, N. C., Mayor, M., & Rebolo, R. 2004, *A&A*, 414, 601

Delgado Mena, E., Israelian, G., González Hernández J. I., Santos, N. C., & Rebolo. R. 2011, *ApJ*, 728, 148

Israelian, G., Delgado Mena, E. Santos. N. C., Sousa, S. G., Mayor, M., Udry, S., Domínguez Cerdeña, C., Rebolo, R., & Randich, S. 2009, *Nature*, 462, 189

Santos, N. C., *et al.*, 2011, *A&A*, 526, A112

Sousa, S. G., Fernandes, J., Israelian, G., & Santos. N. C. 2010, *A&A*, 512, L5

Sousa, S. G., Santos, N. C., Israelian, G., Lovis, C., Mayor, M., Silva, P. B., & Udry, S. 2011, *A&A*, 526, A99

From Interacting Binaries to Exoplanets: Essential Modeling Tools
Proceedings IAU Symposium No. 282, 2011
Mercedes T. Richards & Ivan Hubeny, eds.
© International Astronomical Union 2012
doi:10.1017/S1743921311028122

Binary Systems As Gravitational Wave Sources

O. Köse[1] and K. Yakut[1,2]

[1]Department of Astronomy and Space Sciences, University of Ege, 35100, İzmir, Turkey

[2]Institute of Astronomy, University of Cambridge, Madingley Road, Cambridge CB3 0HA, UK

Abstract. Binary systems with compact components and some newly discovered systems with relatively short orbital periods are studied as gravitational-wave sources. The gravitational wave amplitudes of these systems have been compared with the limit of gravitational wave interferometers (e.g. LISA, LIGO, and VIRGO).

Keywords. gravitation, gravitational waves, stars: binaries : close

1. Introduction

Gravitational waves (GWs), unlike electromagnetic waves, are propagating ripples in space-time, according to Einstein's General Relativity (GR) (Einstein 1916). GR predicts the existence of GWs that travel at the speed of light with two polarization states, h(+) and h(x). Relativistic binary systems (e.g. neutron binary systems) can propagate GW in the observable frequency range (Phinney 1991, Nelemans *et al.* 2001). GWs, therefore, can be detected with ground-based GW detectors such as LIGO, VIRGO, GEO 600, TAMA 300, and future planned space-based laser interferometer, LISA.

2. Gravitational wave amplitudes and frequencies of binaries

Starting from the linearized Einstein field equation, one can obtain the gravitational field equation. Following straightforward calculations, we obtain the GW amplitude (h)

$$h \simeq 2.5 \times 10^{-22} M_1 M_2 M^{-1/3} D^{-1} f^{2/3} \tag{2.1}$$

where f is the frequency, D is the distance of source in Mpc, and M is the total mass in solar units.

Close binary systems with compact components lose energy by GWs (Paczynski & Sienkiewicz 1981, Verbunt & Zwaan 1981, Yakut *et al.* 2008, Kalomeni 2010). Studying GWs, it is possible to detect binary systems and double black hole binaries, if they exist. A few detectors have been improved to detect binaries consisting of ultra compact objects. The amplitude of a binary can be measured with Eq. (1). Gravitational frequencies and amplitudes of some binaries are listed in Table 1. In Fig. 1, we show frequencies and amplitudes of these binary systems and detectors. Apart from AM CVn systems and binaries with neutron star components, some hypothetical binary systems are also added. These systems are supposed to be NS+NS, NS+BS, and BH+BH binaries. It is apparent from the figure that these objects lies within the observation limit of LISA satellite. Massive black hole binary systems are also added.

Fig. 1 shows that high mass binary systems with small separation (short period) emit more waves that make them easy to detect. The distance of the target is also crucial for the detection limit of detectors. We still continue to analyze close binaries that are

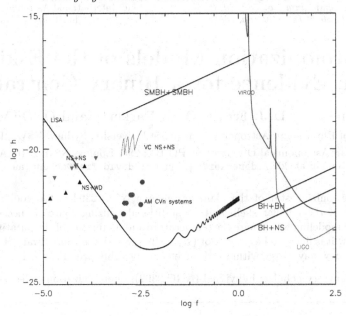

Figure 1. Evolution of some binaries and sensitivities of ground based (LIGO, VIRGO) and space based (LISA) detectors in frequency (f) vs. GW amplitude (h) diagram.

Table 1. GW amplitude and frequency of selected binary systems.

	Type	P(min)	M_1, M_2	$\log(f)$	$\log(h)$
V523 Cas	MS+MS	337	0.75, 0.38	1.2	-20
J18073024+4551325	MS+WD	279	0.60, 0.40	-3.9	-21
AM CVn	WD+WD	17.1	0.68, 0.12	-3.0	-20.5
XTE J2123-058	NS+MS	357	1.46, 0.53	-4.0	-20.9
PSR J0737-3039	NS+NS	147	1.25, 1.34	-3.6	-20.3
SS433	BH+MS	18835	16, 22	-5.8	-19.8

observed with the Kepler satellite in order to catalogue possible sources to study with this technique (Köse and Yakut 2012).

Acknowledgements

This study was supported by the Turkish Scientific and Research Council (TÜBİTAK 109T047 and 111T270) and Ege University Research Fund. KY acknowledges support by the Turkish Academy of Sciences (TÜBA).

References

Einstein, A. 1916, *Annalender Physik*, 354, 769
Kalomeni, B.: 2010, *IAU Symposium*, 262, 362.
Köse, O. & Yakut, K.: 2012, in preparation.
Nelemans, G., Yungelson, L. R., & Portegies Zwart, S. F.: 2001, *A&A*, 375, 890.
Paczynski, B. & Sienkiewicz, R.: 1981, *ApJ*, 248, L27.
Phinney, E. S.: 1991, *ApJ*, 380, L17.
Verbunt, F. & Zwaan, C.: 1981, *A&A*, 100, L7.
Yakut, K., Kalomeni, B., & Tout, C. A.: 2008, *ArXiv e-prints*, arXiv:0811.0455.

From Interacting Binaries to Exoplanets: Essential Modeling Tools
Proceedings IAU Symposium No. 282, 2011
Mercedes T. Richards & Ivan Hubeny, eds.

© International Astronomical Union 2012
doi:10.1017/S1743921311028134

Photoionization Models of the Eskimo Nebula: Evidence for a Binary Central Star?

A. Danehkar[1], D. J. Frew[1], Q. A. Parker[1,2] and O. De Marco[1]

[1]Department of Physics and Astronomy, Macquarie University, Sydney, NSW 2109, Australia

[2]Australian Astronomical Observatory, PO Box 296, Epping, NSW 1710, Australia
email: ashkbiz.danehkar@mq.edu.au; david.frew@mq.edu.au

Abstract. The ionizing star of the planetary nebula NGC 2392 is too cool to explain the high excitation of the nebular shell, and an additional ionizing source is necessary. We use photoionization modeling to estimate the temperature and luminosity of the putative companion. Our results show it is likely to be a very hot ($T_{\rm eff} \simeq 250$ kK), dense white dwarf. If the stars form a close binary, they may merge within a Hubble time, possibly producing a Type Ia supernova.

Keywords. Planetary nebulae: individual (NGC 2392); photoionization codes; shock models

1. Introduction

NGC 2392 is a bright, double-envelope planetary nebula (PN), nicknamed the Eskimo nebula, with a bright hydrogen-rich central star (CSPN). The effective temperature, derived from spectral-line fitting, is 43 kK (Méndez *et al.* 2011). However, the surrounding PN exhibits high-excitation emission lines, such as He II λ4686 and [Ne V] λ3426, which cannot be produced by the visible star. In particular, the presence of [Ne V] implies $T_{\rm eff} > 100$ kK for the ionizing source. It seems that a hot companion is needed to supply the hard-UV photons, likely to be a white dwarf. If this is the case, the Eskimo will be a valuable addition to the small sample of PNe with (pre-)double-degenerate nuclei.

In this work, we aim to estimate the luminosity, temperature and mass of the optically invisible secondary star in NGC 2392 through photoionization modeling. We use the 3-D code MOCASSIN (Ercolano *et al.* 2003) to model the PN emission lines. We also investigate an alternative hypothesis, i.e. that the high-excitation lines are due to shocks produced by a fast bipolar outflow from the CSPN. We use the 1-D shock ionization code Mappings-III (Sutherland & Dopita 1993) to test the feasibility of this hypothesis.

2. Modeling

We initially adopted the observed nebular line intensities and abundances (except log S/H = −5.16) from Pottasch *et al.* (2008), and used plasma diagnostics to derive the electron temperature and density in the usual way. We adopted a distance of 1.8 kpc following Pottasch *et al.* (2011). As expected, the photoionization modeling showed it was necessary to use a model with a heterogeneous density distribution. This was constructed from narrow-band images and kinematic data following O'Dell *et al.* (1990), and we adopted densities of 3000 and 1300 cm^{-3} for the inner prolate spheroid and outer zone, respectively. We used NLTE model atmosphere fluxes from the grid of Rauch (2003). Our first attempt to determine the characteristics of the putative companion shows that a very hot, high-gravity white dwarf with $T_{\rm eff} = 250$ kK and $L/L_\odot = 650$ is a plausible source of the additional ionizing photons. We compare our results with the observed spectrum in Table 1, and compare the model output to the [O III] image in Figure 1.

Table 1. Photoionization model output versus observations.

Ion	λ(Å)	Obs.	Mod.	Ion	λ(Å)	Obs.	Mod.
[Ne v]	3426	4.0	2.3	[N ii]	5755	1.6	2.6
[O ii]	3727	110	107	He i	5876	7.4	7.5
[Ne iii]	3869	105	130	Hα	6563	285	282
Hγ	4340	47	47	[N ii]	6584	92	129
[O iii]	4363	19	13	[S ii]	6717	6.7	3.2
He ii	4686	37	35	[S ii]	6731	8.6	4.6
Hβ	4861	100	100	[Ar iii]	7135	14	12
[O iii]	5007	1150	1143	[S iii]	9532	91	94

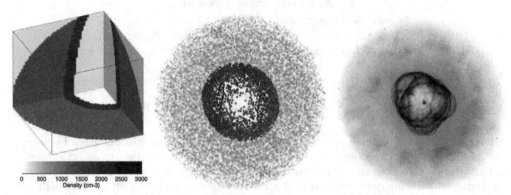

Figure 1. (Left) Cross-section of the density distribution used for NGC 2392. (Right) Computed surface brightness of NGC 2392 in [O iii] λ5007 compared with the *HST* image.

While various shock models can roughly reproduce the [Ne v] flux, they fail to reproduce the ionization structure of the other lines. Our photoionization model predictions generally agree with the observations, adopting a volume filling factor of 0.07 (see Boffi & Stanghellini 1994). We note that Guerrero *et al.* (2011) have discovered a very hard X-ray source coincident with the CSPN. If the stars form a close binary (see Méndez *et al.* 2011), it is possible that the X-rays are produced from accretion of material on to the companion. The high effective temperature could also be explained by re-heating of this star. We also estimate the stellar masses from evolutionary tracks. The WD mass is $\approx 1\,M_\odot$, and the total mass is $\approx 1.6\,M_\odot$, which exceeds the Chandrasekhar limit. Hence, the system is a potential Type Ia SN progenitor. Further observations are needed to better understand the nature of this very interesting system.

Acknowledgements

AD acknowledges receipt of an MQRES PhD Scholarship and an IAU Travel Grant.

References

Boffi, F. R. & Stanghellini, L. 1994, *A&A*, 284, 248
Ercolano, B., Barlow, M. J., Storey, P. J., & Liu, X.-W. 2003, *MNRAS*, 340, 1153
Guerrero, M. A., Chu, Y.-H., & Gruendl, R. A. 2011, *IAUS* 283, submitted.
Méndez, R. H., Urbaneja, M. A., Kudritzki, R. P., & Prinja, R. K. 2011, *IAUS* 283, submitted.
O'Dell, C. R., Weiner, L. D., & Chu, Y.-H. 1990, *ApJ*, 362, 226
Pottasch, S. R., Bernard-Salas, J., & Roellig, T. L. 2008, *A&A*, 481, 393
Pottasch, S. R., Surendiranath, R., & Bernard-Salas, J. 2011, *A&A*, 531, A23
Rauch, T. 2003, *A&A*, 403, 709
Sutherland, R. S. & Dopita, M. A. 1993, *ApJS*, 88, 253

From Interacting Binaries to Exoplanets: Essential Modeling Tools
Proceedings IAU Symposium No. 282, 2011
Mercedes T. Richards & Ivan Hubeny, eds.
© International Astronomical Union 2012
doi:10.1017/S1743921311028146

The Keck I/HIRES and TNG/SARG Radial Velocity Survey of Speckle Binaries

Milena Ratajczak[1], Maciej Konacki[1,2], Shrinivas R. Kulkarni[3] and Matthew W. Muterspaugh[4]

[1] Nicolaus Copernicus Astronomical Center, Polish Academy of Sciences,
ul. Rabiańska 8, 87-100 Toruń, Poland,
email: milena@ncac.torun.pl

[2] Astronomical Observatory of Adam Mickiewicz University,
ul. Sloneczna 36, 60-286 Poznań, Poland

[3] California Institute of Technology, Division of Physics, Mathematics and Astronomy,
Pasadena, CA 91125, USA

[4] Tennessee State University, Department of Mathematics and Physics,
College of Arts and Sciences,
Boswell Science Hall, Nashville, TN 37209, USA

Abstract. A sample of about 160 speckle binary stars was observed with the Keck I telescope and its Échelle HIRES spectrograph over the years 2003-2007 in an effort to detect substellar and planetary companions to components of binary and multiple star systems. This data set was supplemented with the data obtained at the TNG telescope equipped with the SARG Échelle spectrograph over the years 2006-2007. The high-resolution (R = 65000 for HIRES and R = 86000 for SARG) and high signal-to-noise (typically 75-150) spectra were used to derive radial velocities of the components of the observed speckle binaries. Here, we present a summary of this effort, which includes the discovery of new triple star systems and improved orbital solutions of a few known binaries.

Keywords. stars: binaries: visual, techniques: radial velocities

1. Method

Precise radial velocities (RV) of stars are commonly obtained using an iodine (I2) absorption cell (Marcy & Butler 1992). This classic approach is well-suited to measure RVs of single stars only. We developed a novel variant employing an I2 absorption cell which allows us to measure RVs of both components of double-lined binaries (SB2; Konacki 2005) and used it on a sample of ~160 speckle binaries to detect new stellar and substellar companions to members of known binary systems. The classical approach with the iodine cell cannot be used in the case of binary stars because it is not possible to observationally obtain two separate template spectra of binary components. The procedure is as follows (Konacki 2005):

(*a*) two exposures, with and without the I2 cell, are always taken to obtain an instantaneous template used to model the immediate exposure taken with the cell;

(*b*) a least squares fit is carried out to determine the parameters: $\Delta\lambda_s$ (shift of the star spectrum), $\Delta\lambda_{I2}$ (shift of the iodine transmission function), and PSF using the observed (with and without the I2 cell) spectra. An extracted stellar spectrum has an accurate wavelength solution and it is free of the I2 lines and the influence of a varying PSF;

(*c*) two-dimensional cross correlation TODCOR (Zucker & Mazeh 1994) is used on the extracted spectrum to obtain RVs of the components. To this end, synthethic spectra computed with ATLAS9 and ATLAS12 (Kurucz 1995) are used. The formal errors of RVs are derived using the TODCOR formalism.

2. Example results

Below, we present the analysis of 2 speckle binary systems. HD88417 (Fig. 1) is a newly discovered triple system, HD185082 (Fig. 2) is a triple system for which we derived the tertiary's orbit for the first time. For many other observed systems, we were able to obtain improved orbits of the components (the rms of our orbital fits in a few cases is even 100 times lower compared with the literature data).

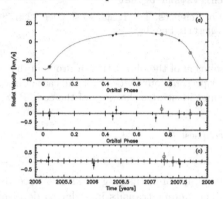

HD88417

WDS J10121+2118AB, HIP 49970 (V=8.8 mag), is a speckle binary with an orbital period of 97 years, semi-major axis of 0.2 arcsec and an eccentricity of 0.26 (Heintz 1976; Hartkopf *et al.* 2001). Our measurements enable us to establish the binarity of the secondary and clasify it as a triple system.

Figure 1. RV (*a*) and rms (*b*): as a function of orbital phase, (*c*): as a function of time, of our orbital fit of the newly discovered component of HD88417.

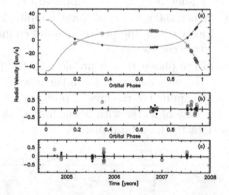

HD185082

WDS J19351+5038, HIP 96321 (V=8.3 mag), is a speckle binary with an orbital period of 103 years, semi-major axis of 0.3 arcsec and an eccentricity of 0.58 (Hartkopf *et al.* 2001). The binarity of the primary was established by Smekhov 1999 who derived its spectroscopic orbit. Our RV measurements allow us to improve the orbital parameters significantly. In particular, the rms of the orbital fit for the primary is 0.1 km/s compared to 1.3 km/s of Smekhov 1999. Additionally, we were able to detect the lines of the tertiary (the spectroscopic companion of the primary) and derive its orbit as well.

Figure 2. RV (*a*) and rms (*b*): as a function of orbital phase, (*c*): as a function of time, of our orbital fit of both components of the primary of HD185082.

References

Hartkopf, W. I., Mason, B. D., & Worley, C. E. 2001a, *6th Catalog of Orbits of Visual Binary Stars*

Heinz, W. D. 1976, *ApJ*, 208, 474

Konacki, M. 1995, *ApJ*, 626, 431

Kurucz R. L. 1995, *ASP-CP*, 78

Marcy, G. W. & Butler R. P. 1992, *PASP*, 104, 270

Smekhov, M. G. 1999, *Astron. Lett.*, 25, 536

Zucker, S. & Mazeh, T. 1994, *ApJ*, 420, 806

From Interacting Binaries to Exoplanets: Essential Modeling Tools
Proceedings IAU Symposium No. 282, 2011
Mercedes T. Richards & Ivan Hubeny, eds.
© International Astronomical Union 2012
doi:10.1017/S1743921311028158

New Approach for Solution of the Planet Transit Problem

Diana P. Kjurkchieva[1] and Dinko P. Dimitrov[2]

[1] Dept. of Astronomy, Shumen University, 9700 Shumen, Bulgaria;
email: **d.kyurkchieva@shu-bg.net**

[2] Institute of Astronomy, Bulgarian Academy of Sciences, Tsarigradsko shossee 72, 1784 Sofia
email: **dinko@astr0.bas.bg**

Abstract. We propose a new approach for a solution of the planet transit problem that is based on numerical calculation of integrals. The paper presents our method for the case of linear limb-darkening law and orbital inclination $i = 90°$, and illustrates the work of the code PTS written on the basis of our approach.

Keywords. methods: analytical, numerical, eclipses

There has been a sharp rise in the detections of transiting planets in recent years. This requires simple and easy modeling of the planet transits in order to determine the parameters of the planetary systems. The known codes for stellar eclipses are not applicable for this aim due to different reasons (non-effective convergence of the differential corrections for observational precisions poorer than 1/10 the depth of planet transit, etc.). New solutions of the direct problem for planet transits were proposed (Mandel & Agol 2002, Seager & Mallen-Ornelas 2003, Gimenez 2006, Pal *et al.* 2009, Kipping 2008). Their formulae contain special functions and as a result the derived results cannot be used directly to solve the inverse problem.

We propose a new approach for a solution of the planet transit problem that is based on numerical solution of integrals. Some of the derived analytical formulae as well as the code written on the numerical calculations can be applied to the inverse problem solution. This paper presents briefly our method for the case of a star with linear limb-darkening law and an inclination $i = 90°$ of the line-of-sight to the orbital plane.

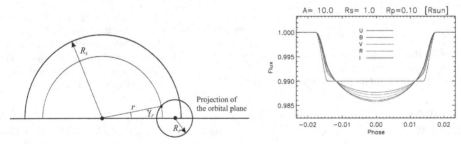

Figure 1. Left: Geometry of the transit; Right: UBVRI synthetic transits for a star with T=5000 K and $\log g$=4.5 obtained by the code PTS (the black curve corresponds to u=0)

We describe the relative decrease of the stellar normalized flux during the transit of a planet with radius R_p orbiting a star with radius R_s on a circle orbit with radius A by the expression

$$J(\varphi) = 1 - \frac{\int\limits_{r_{min}(\varphi)}^{r_{max}(\varphi)} \left[1 - u + u\sqrt{1 - (r/R_s)^2}\right] 2\gamma_r(\varphi)rdr}{\pi R_s^2(1 - u/3)}. \tag{1.1}$$

The stellar area covered by the planet at phase φ is a sum of arc-like rings with areas $2\gamma_r(\varphi)rdr$ (Fig. 1, Left). We obtained analytical expressions for $\gamma_r(\varphi)$ and the extreme radii $r_{min}(\varphi)$ and $r_{max}(\varphi)$ of the stellar brightness' isolines covered by the planet.

Particularly, the phases of the outer and inner contact planet-star are

$$\varphi_1 = \frac{1}{2\pi}\left(\frac{\pi}{2} - \arccos\frac{R_s + R_p}{A}\right); \varphi_2 = \frac{1}{2\pi}\left(\frac{\pi}{2} - \arccos\frac{R_s - R_p}{A}\right) \tag{1.2}$$

The transit is partial in the phase ranges $[-\varphi_1, -\varphi_2]$ and $[\varphi_2, \varphi_1]$ while in the phase range $[-\varphi_2, \varphi_2]$ it is total.

The integral in eq. (1.1) has analytical solution only at the center of the transit ($\varphi = 0$)

$$J(0) = 1 - \frac{3(1 - u)}{(3 - u)}\frac{R_p^2}{R_s^2} + \frac{2u}{3 - u}\left[1 - \left(1 - \frac{R_p^2}{R_s^2}\right)^{3/2}\right]. \tag{1.3}$$

The expressions (1.2-1.3) can be applied directly for the inverse problem' solution.

We made numerical solutions of the integral in eq. (1.1) at the phases of the planet transit. For this aim, we wrote a code PTS (Planet Transit Simulator) with input parameters: A, R_s, R_p and limb-darkening coefficient u. Figure 1 (Right) illustrates the result of our numerical solution of the direct problem of a planet transit.

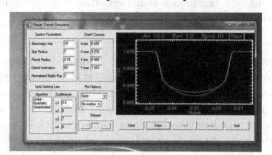

Figure 2. Graphical User Interface of the code PTS

The code PTS enables rapid and interactive calculation of planet transit light curves. The graphical possibilities (Fig. 2) make the code PTS user-friendly. Generalizations of our approach and code (for arbitrary limb-darkening law, arbitrary orbital inclination, flattened planets, etc.) as well as its application for a solution of the inverse problem by trials and errors and quantitative estimation of the fit quality are forthcoming.

Acknowledgements
The research was supported partly by funds of project DO 02-362 of the Bulgarian Ministry of Education and Science.

References
Gimenez A. 2006, *A&A*, 450, 1231
Kipping, D. 2008, *MNRAS*, 389, 1383
Mandel, K. & Agol, E. 2002, *ApJ*, 580, L171
Pal, A., Bakos, G., Noyes, R., & Torres, G. 2009, *in IAUS 253: Transiting Planets*, 428
Seager, S. & Mallen-Ornelas, G. 2003, *ApJ*, 585, 1038

From Interacting Binaries to Exoplanets: Essential Modeling Tools
Proceedings IAU Symposium No. 282, 2011   © International Astronomical Union 2012
Mercedes T. Richards & Ivan Hubeny, eds.   doi:10.1017/S174392131102816X

The Origin and Evolution of the Black Hole Binary XTE J1118+480

Jonay I. González Hernández[1,2], Rafael Rebolo[1,2] and Jorge Casares[1,2]

[1]Instituto de Astrofísica de Canarias, Vía Láctea, s/n, E-38205 La Laguna, Tenerife, Spain
email: jonay@iac.es

[2]Departamento de Astrofísica, Universidad de La Laguna, E-38206 La Laguna, Tenerife, Spain

Abstract. Black hole X-ray binaries with large mass ratios and short orbital periods are expected to change their orbital period due to magnetic breaking, mass loss, gravitational radiation, or mass evaporation of the black hole in alternative descriptions of gravity, like in braneworld gravity scenarios.

The black hole X-ray binary XTE J1118+480, consisting of a late-type secondary star orbiting a $\sim 8\ M_\odot$ black hole in a 4.1-hr period, offers a unique opportunity to test these models. New spectroscopic data allow us to determine the time of the inferior conjunction of the secondary star at different epochs. Observations over a 10 year span will provide constraints on the rate of any orbital period change.

We present here a preliminary radial velocity curve obtained with the 10.4m GTC telescope equipped with OSIRIS medium-resolution spectrograph, as part of an ongoing long-term program to study the orbital period evolution in this binary.

Keywords. black hole physics, techniques: radial velocities, stars: individual (XTE J1118+480), X-rays: binaries, supernovae: general

---

The black hole X-ray binary XTE J1118+480 is located in Galactic halo regions and its galacto-centric motion is similar to those of halo stars (Mirabel *et al.* 2001). However, the high metal content of the secondary star in this system suggests that the system probably originated in the Galactic plane (González Hernández *et al.* 2006, 2008b), and was launched into its current location via an asymmetric kick in a supernova/hypernova explosion (Gualandris *et al.* 2005). The galacto-centric orbit of the system crosses the Galactic plane backwards in time in roughly 11 Myr. This particularity has been used to derive an upper-limit to the asymptotic radius curvature radius, $L \leqslant 80\ \mu m$, in the Anti-de-Sitter (AdS) braneworld Randall-Sundrum gravity model (Psaltis 2007).

During the last decade, we have been collecting high-quality spectroscopic data of black-hole X-ray binaries (González Hernández *et al.* 2004, 2008a, 2008b, 2010, 2011) and neutron-star X-ray binaries (Casares *et al.* 2007, 2010; González Hernández *et al.* 2005). In particular, medium-resolution spectroscopy has been obtained with Keck II/ESI which allowed us to derive radial velocities (RVs) of the secondary star in XTE J1118+480 (González Hernández *et al.* 2008b). These RV measurements give an accurate determination of the time of the inferior conjunction of the secondary star, T_0, in this system. By comparing the T_0 values obtained at different epochs, one can determine the orbital period derivative. This provides not only important information on the variation of the angular momentum of the system but also places strong constraints on the size of the extra dimensions L in the context of AdS braneworld gravity theory (Johannsen 2009).

We have performed spectroscopic observations with the OSIRIS medium-resolution spectrograph, attached to the 10.4m GTC telescope at the Observatorio del Roque de los Muchachos in La Palma (Spain). In Fig. 1, we display the RV measurements obtained

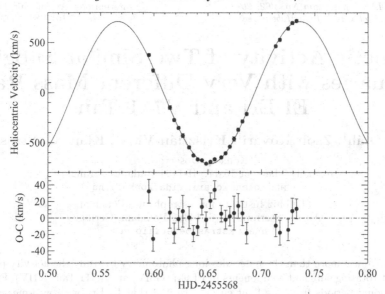

Figure 1. *Top panel*: radial velocities of the secondary star in the black hole X-ray binary XTE J1118+480 obtained from GTC/OSIRIS spectroscopic data taken over one night in January 2011, folded on the best-fitting orbital solution. *Bottom panel*: residuals of the fit, with a rms of $\sim 15 \mathrm{km\ s^{-1}}$.

in one night which allow us to derive the time of the inferior conjunction of the companion star. This value of T_0 will be used, in combination with previous T_0 measurements, to derive the orbital period variation with time and the implications on the curvature radius L, which gives the rate at which the black hole evaporates in the AdS braneworld gravity.

References

Casares, J., Bonifacio, P., González Hernández, J. I., *et al.*, 2007, *A&A*, 470, 1033

Casares, J., González Hernández, J. I., Israelian, G., & Rebolo, R. 2010, *MNRAS*, 401, 2517

González Hernández, J. I., Rebolo, R., Israelian, G., *et al.*, 2004, *ApJ*, 609, 988

González Hernández, J. I., Rebolo, R., Israelian, G., *et al.*, 2005, *ApJ*, 630, 495

González Hernández, J. I., Rebolo, R., Israelian, G. *et al.*, 2006, *ApJ*, 644, L49

González Hernández, J. I., Rebolo, R., & Israelian, G. 2008a, *A&A*, 478, 203

González Hernández, J. I., Rebolo, R., Israelian, G., *et al.*, 2008b, *ApJ*, 679, 732

González Hernández, J. I. & Casares, J. 2010, *A&A*, 516, A58

González Hernández, J. I., Casares, J., Rebolo, R., *et al.*, 2011, arXiv:1106.4278

Gualandris, A., Colpi, M., Portegies Zwart, S., & Possenti, A. 2005, *ApJ*, 618, 845

Johannsen, T. 2009, *A&A*, 507, 617

Psaltis, D. 2007, *Phys. Rev. Lett.*, 98, 181101

Mirabel, I. F., Dhawan, V., Mignani, R. P., *et al.*, 2001, *Nature*, 413, 139

From Interacting Binaries to Exoplanets: Essential Modeling Tools
Proceedings IAU Symposium No. 282, 2011
Mercedes T. Richards & Ivan Hubeny, eds.
© International Astronomical Union 2012
doi:10.1017/S1743921311028171

Magnetic Activity of Two Similar Subgiants in Binaries with Very Different Mass Ratios: EI Eri and V711 Tau

Katalin Oláh[1], Zsolt Kővári[1], Krisztián Vida[1], Klaus G. Strassmeier[2]

[1]Konkoly Observatory,
Konkoly Thege út 15-17., H-1121 Budapest, Hungary
email: olah, kovari, vida@konkoly.hu

[2]Leibniz Institute for Astrophysics Potsdam,
An der Sternwarte 16, 14482 Potsdam, Germany
email: kstrassmeier@aip.de

Abstract. We use more than three decades-long photometry to study the activity patterns on the two fast-rotating subgiant components in EI Eri (G5IV) and V711 Tau (K1IV). From yearly mean rotational periods from the light curves, we find that EI Eri, with well-measured solar-type differential rotation, always has spots from the equator to high latitudes. The measured differential rotation of V711 Tau is controversial, and in any case is very small. The spots on the K1IV star in V711 Tau seem to be tidally locked. The physical parameters of the two systems are similar, with one remarkable difference: EI Eri has a low mass M4-5 dwarf companion, whereas V711 Tau has a G5V star in the system, thus their mass centers are in very different positions. This may modify the whole internal structure of the active stars, causing marked differences in their surface features.

Keywords. stars: activity, stars: imaging, stars: individual (EI Eri, V711 Tau), stars: spots, stars: late-type

1. Introduction

Magnetic activity on stars under the influence of a close binary companion is an important but not well-studied topic. For decades, it has been known that the measure of activity is closely related to the parameters of the companion star and the orbit of the binary. Yet, apart from some statistical analysis of the strength of activity in different binaries (e.g., Schrijver & Zwaan 1991) and a first attempt of modelling the behaviour of magnetic flux tubes in the gravitational field of a companion star in a binary for a simplified case (Holzwarth & Schüssler 2003a, Holzwarth & Schüssler 2003b), not much have been done in this interesting field. It is thought as well, that the strength and even the orientation of the differential rotation is modified by the binary companions. Surface patterns of active stars could best be depicted through Doppler Imaging. Unfortunately, the available Doppler maps are too few, both for a time-sequence of a star and for the number of stars, due to the known restrictions such as brightness, inclination and rotational velocity of the objects. On the other hand, much less restricted long-term photometric monitoring of active stars is carried out for decades (cf. Strassmeier *et al.* 1997), and those datasets, apart from the long-term, cyclic changes, contain information on the rotational behaviour of the spotted stars. In this work, we analyse two similar, post-main sequence subgiants, which have companions of very different masses, consequently, the mass centers are in different positions. The results are verified through existing Doppler images.

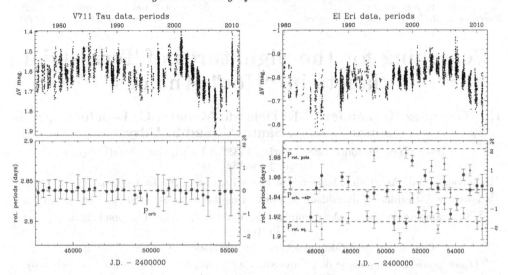

Figure 1. Observed data and yearly period values for EI Eri (left, blue dots: periods with the highest amplitude, red triangles: additional periods) and V711 Tau (right). The y-scale on the right sides show the difference between the rotational and orbital periods, in percents. See text.

2. Basic stellar and orbital parameters, and results

EI Eri: sp. types G5IV+dM4-5, T_{eff} ≈5500K, masses: 1.09/0.25 $M_\odot$, radii: 2.37/0.3 $R_\odot$, $v\sin i$=51 km/s, $i \approx 56^o$, P_{orb}=1.947232 days, $a = 5.0 \times 10^6$km, (Washüttl *et al.* 2009), and V711 Tau: sp. types K1IV+G5V, T_{eff} ≈4750/5500K, masses: 1.45/1.14 $M_\odot$, radii: 4.12/1.32 $R_\odot$, $v\sin i$=41 km/s, $i \approx 33^o$, P_{orb}=2.83774 days, $a = 8.06 \times 10^6$km (Garcia-Alvarez *et al.* 2003). Both systems are circularized and synchronized.

Observations of about three decades, plotted in Fig. 1, upper panels, show that the two active subgiants have very similar cyclic behaviour and light curve amplitudes. Yet, a marked difference appear when we plot the yearly mean rotational period(s) in Fig. 1, lower panels. The multiple yearly periods of EI Eri, reaching ±2% deviation from the orbital period. is a clear signature of a differential rotation, verified by Doppler Imaging (Kővári *et al.* 2009). V711 Tau has just one rotational period close to the orbital one each year, and its first harmonic, indicating two spotted regions opposite to each other (cf. Lanza *et al.* 2006). This difference originates very possibly from the difference between the secondaries.

Acknowledgements

The authors acknowledge support from the Hungarian Research Grant OTKA K-81421 and the "Lendület" Program of the Hungarian Academy of Sciences.

References

Garcia-Alvarez, D., Foing, B. H., Montes, D. *et al.*, 2003, *A&A*, 397, 285
Holzwarth, V. & Schüssler, M. 2003a, *A&A*, 405, 291
Holzwarth, V. & Schüssler, M. 2003b, *A&A*, 405, 303
Schrijver, C. J. & Zwaan, C. 1991, *A&A*, 251, 183
Kővári, Zs., Washüttl, A., Foing, B. H. *et al.*, 2009, AIP Conference Proceedings, 1094, 676
Lanza, A. F., Piluso, N., ò, M. *et al.*, 2006, *A&A*, 455, 595
Strassmeier, K. G., Boyd, L. J., Epand, D. H., & Granzer, Th. 1997, *PASP* 109, 697
Washüttl, A., Strassmeier, K. G., Granzer, T., Weber, M., & Oláh, K. 2009, *AN*, 330, 27

From Interacting Binaries to Exoplanets: Essential Modeling Tools
Proceedings IAU Symposium No. 282, 2011
Mercedes T. Richards & Ivan Hubeny, eds.
© International Astronomical Union 2012
doi:10.1017/S1743921311028183

Searching for the Signatures of Terrestrial Planets in "Hot" Analogs

J. I. González Hernández[1,2], **E. Delgado Mena**[1,2], **G. Israelian**[1,2], **S. G. Sousa**[3,4], **N. C. Santos**[3,4], **and S. Udry**[5]

[1] Instituto de Astrofísica de Canarias, E-38200 La Laguna, Tenerife, Spain.
email: jonay@iac.es

[2] Departamento de Astrofísica, Universidad de La Laguna, 38205 La Laguna, Tenerife, Spain.

[3] Centro de Astrofísica, Universidade do Porto, Rua das Estrelas, 4150-762 Porto, Portugal.

[4] Departamento de Física e Astronomia, Faculdade de Ciências, Universidade do Porto, Portugal.

[5] Centro de Física Computacional, Universidade de Coimbra, Portugal.

[6] Observatoire Astronomique de l'Université de Genève, 51 Ch. des Maillettes, Sauverny, Ch1290 Versoix, Switzerland

Abstract. The Sun has been suggested to have a slightly low refractory-to-volatile abundance ratio when compared with field solar twins. This result may be interpreted as due to the fact that the refractory elements were trapped in rocky planets at the formation of the Solar System.

A detailed and differential chemical abundance study was already performed in order to investigate this hypothesis in solar analogs with and without detected planets using high-resolution and high-S/N HARPS and UVES spectra of a relatively large sample of solar analogs with and without planets. We obtained very similar behaviours for both samples of stars with and without planets, even for two stars with super-Earth-like planets, which may indicate that this solar trend may not be related to the presence of terrestrial planets.

The depletion signature should be imprinted once the convection zone reaches the current size. This suggests that stars hotter than the Sun should show this effect enhanced, due to their narrower convective zone. However, to avoid non-LTE, 3D, and other effects, we need to identify "hot" analogs with a $T_{\rm eff} \sim 6100$ K, to perform a differential analysis.

Here, we present the preliminary results of our analysis using HARPS and UVES high-resolution and high-S/N spectra of a sample of ~ 60 "hot" analogs with and without planets, trying to search for some "hot" reference analogs.

Keywords. stars: abundances — stars: atmospheres — stars: fundamental parameters — stars: planetary systems — stars: planetary systems: formation

1. Introduction: [X/Fe] vs. T_C

In the last few years, the chemical abundances of heavy elements in solar analogs have been extensively investigated. Meléndez *et al.* (2009) found in a sample of 11 solar twins a well-defined trend [X/Fe] with respect to the condensation temperature, T_C, which was interpreted as the result of the formation of terrestrial planets in the solar system (see also Ramírez *et al.* 2009, 2010).

A detailed chemical analysis of very high-quality HARPS and UVES spectroscopic data did not show any significant difference between the abundance ratios of a sample of 95 solar analogs with and without planets (see González Hernández *et al.* 2010). In addition, two stars of the HARPS sample hosting Super-Earth like planets do not seem to provide clear evidence of terrestrial planets in their abundance patterns relative to iron (see González Hernández *et al.* 2011). In particular, after correcting for the Galactic chemical evolution effects, both stars do not seem to show any trend. This may indicate that there is no signature of terrestrial planets in these two stars even if there has been already detected one Super-Earth-like planet orbiting each of these stars.

2. Identifying the "hot" reference analog

According to Meléndez *et al.* (2009), the formation of terrestrial planets in a stellar planetary system would cause a depletion of refractory (with high T_C values) elements with respect to volatiles (with low T_C values) in the convective envelope of the star. However, the convective zone would have to reach the present state early enough (i.e. after $\sim$ 20 Myr for a solar-type star) to be able to show the depletion signature.

This possible signature would, in principle, be stronger in stars hotter than the Sun, due to their shallower convective zones. Thus, we have inspected the HARPS GTO sample of stars in the temperature range 5950–6350 K and found about 60 stars with and without planets with very high-quality UVES and HARPS spectroscopic data.

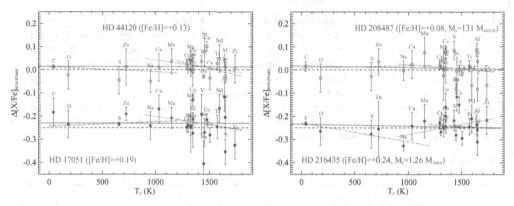

Figure 1. *Left panel*: Abundance differences, Δ[X/Fe]$_{\text{SUN}-\text{STARS}}$, between the Sun, and 2 "hot" analogs without planets from the HARPS GTO sample. Each element abundance ratio has been corrected using a linear fit to the Galactic chemical trend of the corresponding element at the metallicity of each star. Linear fits for different T_C ranges to the data points weighted with the error bars are also displayed. We note the different slopes derived when choosing the range $T_C > 1200$ K (dashed-dotted line) as in Meléndez *et al.* (2009) and González Hernández *et al.* (2010), and $T_C > 900$ K (dashed-three-dotted line) as in Ramírez *et al.* (2009, 2010). *Right panel*: Same as left panel of this figure but for two "hot" planet-host stars. An arbitrary shift of -0.25 dex has been applied, in both panels, to the abundances of the element abundances in the stars HD 17051 and HD 216435.

In the left panel of Fig. 1, we display the abundances of two "hot" stars without planets at $T_{\text{eff}} \sim 6100$ K and some linear fits for different T_C ranges. We have already removed the Galactic chemical evolution effects with the aim to find stars with similar abundance trends than the Sun to use them as references. In the right panel of Fig. 1, we display two possible "hot" reference stars hosting planets.

References

González Hernández, J. I., Israelian, G., Santos, N. C., *et al.*, 2010, *ApJ*, 720, 1592
González Hernández, J. I., *et al.*, 2011, *IAU Symposium 276* (arXiv:1011.6125)
Meléndez, J., Asplund, M., Gustafsson, B., & Yong, D. 2009, *ApJ Letters*, 704, L66
Ramírez, I., Meléndez, J., & Asplund, M. 2009, *A&A Letters*, 508, L17
Ramírez, I., Asplund, M., Baumann, P., Meléndez, J., & Bensby, T. 2010, *A&A*, 521, A33

From Interacting Binaries to Exoplanets: Essential Modeling Tools
Proceedings IAU Symposium No. 282, 2011
Mercedes T. Richards & Ivan Hubeny, eds.
© International Astronomical Union 2012
doi:10.1017/S1743921311028195

Calibrating Ultracool Atmospheres with Benchmark Companions from WISE+2MASS

Joana I. Gomes[1], David Pinfield[1], Avril Day-Jones[1,2], Hugh Jones[1], Ben Burningham[1], Federico Marocco[1], ZengHuan Zhang[1], and Lieke van Spaandonk[1]

[1] Centre for Astrophysics Research, University of Hertfordshire,
College Lane AL10 9AB, Hatfield, United Kingdom
email: j.gomes@herts.ac.uk

[2] Departamento de Astronomia, Universidad de Chile, Santiago, Chile

Abstract. The complexities of ultracool atmospheres are best confronted by observations of ultracool dwarfs (UCDs) with well known physical properties (luminosity, mass, T_{eff}, log(g), [M/H]), so-called "benchmark objects." We present two discoveries from a new WISE+2MASS search for benchmark wide companions to Hipparcos and Gliese stars. This survey combination provides a powerful tool to confirm new companions using color-magnitude and common proper motion selections, and also yield full NIR-MIR measurements of the ultracool emission. These primary companions are providing important constraints on the age and composition of the benchmark brown dwarf, and the new discoveries add to our growing population of benchmarks that is providing crucial tests of ultracool physics.

Keywords. ultracool dwarfs, binaries

1. Introduction

For field stars and brown dwarfs in particular, ages are one of the most difficult parameters to estimate. This is true for brown dwarfs especially, because they cool over time and it is extremely difficult to break the degeneracy between some parameters like age, mass, T_{eff}, and luminosity. Currently, the radii of brown dwarfs has to be calculated using evolutionary models, and their T_{eff} strongly depends on how accurately this can be done. Also, metallicity effects on the spectral energy distribution (SED) of UCDs is still not well understood. All this combines to make it extremely important to study ultracool dwarfs with well constrained ages and composition, and these objects we refer to as benchmark objects. Wide binaries that contain a UCD and a main sequence (MS) star are ideal to study how UCD SEDs depend on these properties, as we assume both components are coeval and we can infer the properties of the secondary by studying the primary.

2. New binaries

Our first binary is a newly confirmed L dwarf with a K8 star as a companion. The proper motion for this binary was calculated using 2MASS and WISE epochs, with a 10 yr baseline. The primary star has a measured parallax distance of 35.87 ± 1.47 pc. Using the color-magnitude M_k vs V-K and the relations in Johnson & Apps (2009), we have estimated a metallicity of 0.03 dex for the K8 star.

An optical low resolution spectrum was obtained for the L dwarf with DOLORES at the TNG, La Palma. We have compared our spectrum with two templates and using a reduced χ^2 fit, we conclude that this is an L1± 0.5 dwarf. To determine the bolometric luminosity, we have combined our optical spectrum, with wavelengths ranging from 5.0 μm to 1.1μm, with the 2MASS and WISE W1 and W2 photometry. We also considered that the L dwarf was at the same distance as its companion. For a surface gravity log(g)=4.5 and metallicity of -0.3 dex, we have estimated a bolometric luminosity $L_{bol} = 1.67 \pm 0.17 \times 10^4$ $L_{\odot}$.

The second new binary is part of a triple system as the companion is actually a double system with an M4 and K5 star. The UCD companion is a known L4 $\pm$ 2 dwarf. The proper motion for this object has been measured by Jameson *et al.* (2008). We used these values to calculate that it has a common proper motion with the binary system. Taking the parallax distance of the binary system, we estimated the absolute magnitude M_J for the L dwarf, and used the Marocco *et al.* (2010) relation to estimate a spectral type of L6 for this UCD. This result is still in agreement with previous classifications if we take into account the error bars.

We also obtained an optical spectrum for the L dwarf and are now in the process of reducing the data and estimating a more accurate spectral type. The bolometric luminosity will also be calculated for this UCD, in order to estimate an accurate T_{eff}.

3. Conclusions and future work

Ongoing efforts are taking place to find more of these benchmark binaries. Exploring the WISE database will allow us to find more, and colder UCD candidates. These will be combined with main sequence stars in the PPMXL catalog, where proper motion measurements for the primaries are available. We will also include the two binaries presented here along with future discoveries in the current sample of benchmark UCD systems. Constraining the UCD's properties (e.g., age and metallicity) from their companion stars will help improve the current models and extend our knowledge of ultracool atmospheric physics.

References

Jameson, R. F., Casewell, S. L., Bannister, N. P., Lodieu, N., Keresztes, K., Dobbie, P. D. & Hodgkin, S., T. 2008, *MNRAS*, 384, 1399

Johnson, J. A. & Apps, K. 2009, *ApJ*, 699, 933

Marocco, F., Smart, R. L., Jones, H. R. A., Burningham, B., Lattanzi, M. G., Legget, S. K., Lucas, P. W., Tinney, C. G., Adamson, A., Evans, D. W., Lodieu, N., Murray, D. N., Pinfield, D. J. & Tamura, M. 2010, *A&A*, 524, 38

Pinfield D. J., Jones, H. R. A., Lucas, P. W., Kendall, T. R., Folkes, S. L., Day-Jones, A. C., Chappelle, R. J., & Steele I. A. 2006, *MNRAS*, 368, 1281

From Interacting Binaries to Exoplanets: Essential Modeling Tools
Proceedings IAU Symposium No. 282, 2011   © International Astronomical Union 2012
Mercedes T. Richards & Ivan Hubeny, eds.   doi:10.1017/S1743921311028201

First Catalogue of Optically Variable Sources Observed by OMC Onboard INTEGRAL

Julia Alfonso-Garzón, Albert Domingo, and José Miguel Mas-Hesse

Departamento de Astrofísica, Centro de Astrobiología (INTA-CSIC)
POB 78, 28691 Villanueva de la Cañada, Spain
email: julia@cab.inta-csic.es

Abstract. In this work, we present the first catalogue of optically variable sources observed by the Optical Monitoring Camera (OMC), with information about the variability of more than 5000 objects and periodicity of ~ 1000 sources.

Keywords. catalogs, methods: data analysis, (stars:) binaries: general, (stars: variables:).

1. Introduction

The INTEGRAL Optical Monitoring Camera, OMC Mas-Hesse *et al.* 2003), observes the optical emission from the prime targets of the gamma ray instruments on-board the ESA mission INTEGRAL: SPI (gamma ray spectrometer) and IBIS (gamma ray imager), with the support of the JEM-X monitor in the X-ray domain. OMC provides photometry in the Johnson V band (centred at 5500 Å) and it is able to monitor sources from V $\simeq$ 7 mag (for brighter sources saturation effects appear) to V $\simeq$ 16-17 mag (magnitude limit for 3σ source detection). Typical observations are done performing a sequence of different integration times, allowing for photometric uncertainties below 0.1 magnitude for objects with V $\leqslant$ 16. At this moment, the OMC database Gutierrez *et al.* 2004) contains light curves for more than 60 000 sources (with more than 50 photometric points each).

2. Data Analysis

Selection of the sources Sources with more than 300 photometric points have been selected from the OMC database. In order to include only high-quality data, some selection criteria have been applied to individual photometric points, rejecting those ones that present saturation, low signal-to-noise, or are affected by cosmic rays.

Detection of variability We have fitted a constant to the data in the light curve (supposing the source is not variable). Then, we calculate the χ^2 and the significance. This value gives the probability of being wrong when rejecting the null hypothesis (the source is constant). We have considered as variable those sources with $\alpha < 0.05$ (probability of being variable of 95%).

Study of the periodicity To determine which sources are periodic and to derive their periods, an algorithm based on the PDM technique (Phase Dispersion Minimization; Stellingwerf 1978) has been developed. This method divides the time-folded data into a series of bins and computes the variance of the amplitude within each bin with respect to a mean curve. This mean curve is obtained doing linear interpolations between the means of the bins. The ratio between the sum of the bin variances and the overall variance of the data set is called Θ, and the period that minimizes this value will be the best estimate (see Fig. 1). Once this have been done, a visual inspection is needed to determine if the period is good enough. To calculate the error of the period, we have fitted a parabola to the peak of the periodogram corresponding to the minima. We have estimated the error as the distance from the x-value at minima to the x-value corresponding to a height in Θ equal to the deviation of the fit (see dashed orange lines in Fig. 1).

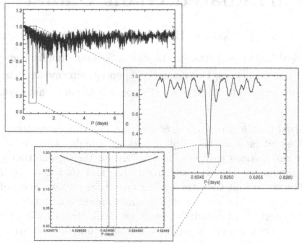

Figure 1. Visual description of the process of period determination for IOMC 0460000022. Top: Periodogram from the full process. Middle: Zoom in the peak of the periodogram. Bottom: Detail of the parabolic fit in the peak of the minima used to determine the error of the period.

3. Some results

This first catalogue of optically variable sources observed by the OMC provides information about the variability of 5518 sources. When possible, we have studied the periodicity too, so we have determined good periods for approximately 1000 sources. The distribution of typical periods and object types can be found in Fig. 2. We have computed periods for several objects whose periodicity was unknown and, in many other cases, we have improved the results with respect to those found in the literature.

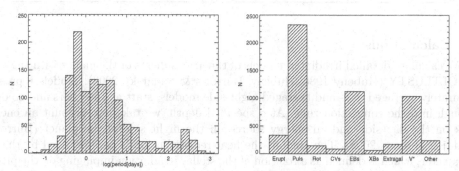

Figure 2. Left: Histogram of the periods derived. Typical values vary between a few hours and 10 days, with a peak of frequency in 15 hours. Right: Histogram of the type of objects identified in the catalogue. There are a big number of unclassified objects.

References

R. Gutiérrez, E. Solano, A. Domingo, & J. García 2004, *Astronomical Society of the Pacific Conference*, Vol. 314, Astronomical Data Analysis Software and Systems XIII, 153

J. M. Mas-Hesse, A. Giménez, J. L. Culhane *et al.*, 2003, *A&A*, 411, L261

R. F. Stellingwerf 1978, *ApJ*, 224, 953

From Interacting Binaries to Exoplanets: Essential Modeling Tools
Proceedings IAU Symposium No. 282, 2011
Mercedes T. Richards & Ivan Hubeny, eds.
© International Astronomical Union 2012
doi:10.1017/S1743921311028213

Day-Night Side Cooling of a Strongly Irradiated Giant Planet

Ján Budaj[1], Adam Burrows[2] and Ivan Hubeny[3]

[1] Astronomical Institute, Tatranská Lomnica, Slovakia, email: `budaj@ta3.sk`

[2] Dept. of Astrophysical Sciences, Princeton, USA, email: `burrows@astro.princeton.edu`

[3] Dept. of Astronomy, University of Arizona, Tucson, USA, email: `hubeny@as.arizona.edu`

Abstract. The internal heat loss or cooling of a planet determines its structure and evolution. We address in a consistent fashion the coupling between the day and the night sides by means of model atmosphere calculations with heat redistribution. We assume that a strong convection leads to the same entropy on the day and night side and that the gravity is the same on both hemispheres. We argue that the core cooling rate from the two hemispheres of a strongly irradiated planet may not be the same and that the difference depends on several important parameters. If the day-night heat redistribution is very effective, or if it takes place at a large optical depth, then the day-side and the night-side cooling may be comparable. However, if the day-night heat transport is not effective, or if it takes place at a shallow optical depth, then there can be a big difference between the day-side and the night-side cooling and the night side may cool more effectively. If the stellar irradiation gets stronger e.g. due to the stellar evolution or migration, this will reduce both the day and the night side cooling. Enhanced metallicity in the atmosphere acts as a "blanket" and reduces both the day- and the night-side cooling. However, the stratosphere on the day side of the planet can enhance the day-side cooling since its opacity acts as a "shield" which screens the stellar irradiation. These results might affect the well known gravity darkening and bolometric albedo effects in interacting binaries, especially for strongly irradiated cold objects.

Keywords. binaries: eclipsing, stars: atmospheres, planets and satellites: general, convection

1. Calculations

We calculate detailed irradiated models of the atmospheres of the planet with the code COOLTLUSTY (Hubeny 1988; Hubeny, Burrows & Sudarsky 2003). Models represent separately averaged day- and averaged night-side models, start at 10^{-5} bars and go deep enough into the convection zone. At a specified depth, we take into account an energy sink on the day side and an energy source on the night side of the planet (Burrows, Budaj & Hubeny 2008). The amount the heat redistribution is parametrized by the P_n parameter, which is defined as a fraction of the stellar irradiation impinging on the planet that is transferred from the day to the night side and radiated out from there (Burrows, Sudarsky & Hubeny 2006).

We consider here a well known planet HD209458b. If not stated otherwise, we assume the solar chemical composition of the planetary atmosphere, energy sink/source at 0.03-0.3 bar, and $P_n = 0.3$. TiO and VO opacities were not considered. The opacities we use are those of Sharp & Burrows (2007), assuming chemical equilibrium composition with a rain-out but no cloud opacity. The Kurucz (1993) spectrum of the parent star HD209458 was used as a source of irradiation. The parameters of the star and planet were taken from Henry *et al.* (2000), Charbonneau *et al.* (2000), and Knutson *et al.* (2007).

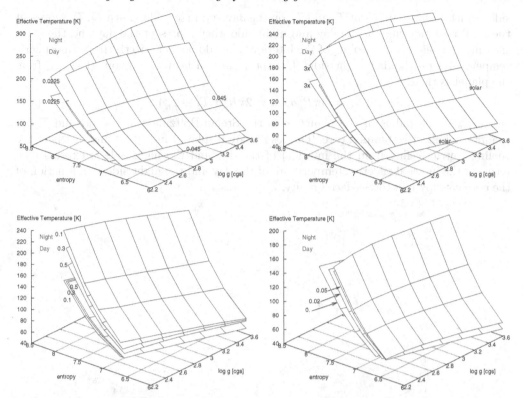

Figure 1. Top-Left: The effect of the planet-star distance or irradiation on the day-night side cooling of the planet. The cooling is expressed as the internal effective temperature in K as a function of the core entropy and the surface gravity. The day-side is red and night-side is blue. Calculated for two planet-star distances $0.045, 0.0225 \; AU$. The cooling from the day side is decreasing with the stellar irradiation (shorter distance). The cooling from the night side behaves in a similar way. Thus for the higher irradiation the total heat loss is lower. **Top-Right: The effect of varying the metallicity in the planetary atmosphere on the day-night side cooling.** Calculated for two metallicities: solar and 3x solar metallicity. The cooling from the day side is decreasing with the metallicity. The cooling from the night side behaves in a similar way. Thus for the higher metallicity the total heat loss is lower. **Bottom-Left: The effect of the P_n parameter (the effectiveness of day-night heat transfer), on the day - night side cooling.** Calculated for three different values of $P_n = 0.1, 0.3, 0.5$. The cooling from the night side is decreasing with the amount of the heat redistribution (P_n). The cooling from the day side behaves in the opposite way and is increasing with increasing P_n. However, the nights side is more sensitive to the P_n parameter and governs the total heat loss which is decreasing with increasing P_n. The difference between the night and the day side cooling is largest for smaller values of P_n. **Bottom-Right: The effect of the extra opacity in the stratosphere of the day side of the planet on the day - night side cooling.** Calculated for three different values of extra opacity $\kappa = 0., 0.02, 0.05 \; cm^2 g^{-1}$. (Extrasolar planets may have stratospheres due to high opacity at very low optical depth (Hubeny, Burrows & Sudarsky 2003; Fortney *et al.* 2008). The cooling from the day side is increasing with the value of the extra opacity. The cooling from the night side is the same since only the day side is changing. The difference between the night and the day side cooling is thus largest for smaller values of extra opacity. The total heat loss is increasing with increasing the extra opacity.

We calculate a grid of models with/without the irradiation corresponding to the day/night side of the planet for a range of day and night side internal effective temperatures (T_d, T_n) and surface gravities ($\log g$). Each model has a certain entropy in the convection zone. In the next step, we match the entropy and gravity of the day and night sides which result in different T_d and T_n on the day and night side. Since T_d, T_n represent the total radiation flux on the day and night side, they represent the day and the night side internal heat loss or cooling of the interior. They do not refer to the real atmospheric temperatures on the day/night side. The total internal heat loss (cooling), L_{cool}, from the planet is then

$$L_{\text{cool}} = 4\pi R_p^2 \sigma T_{\text{eff}}^4 = 2\pi R_p^2 \sigma (T_d^4 + T_n^4), \qquad (1.1)$$

where R_p is the radius of the planet, σ is the Stefan-Boltzmann constant, and T_{eff} is the internal effective temperature. The results are displayed in Figs. 1 and 2, where the cooling (intrinsic flux from the interior) of the day and the night side is expressed in terms of the internal effective temperature of the day and the night side as a function of the core entropy and the surface gravity.

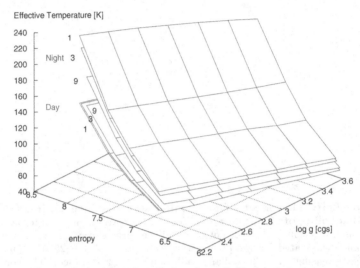

Figure 2. The effect of the depth of the day-night heat transport on the day - night side cooling. The day-side (red) and the night-side (blue) cooling of the planet for three different values of the day side Rosseland optical depth of the heat redistribution region $\tau_{\text{ross}} = 1, 3, 9$. On the day side, the heat loss is increasing with increasing optical depth. On the night side, the heat loss is decreasing with increasing optical depth and this trend seems to overwhelm the opposite trend from the day side. As a result, the night-to-day difference is smaller for larger optical depth and the total heat loss decreases with increasing optical depth.

2. Conclusions

We studied the effects of several processes on the day- and night-side cooling of a strongly irradiated planet. The main conclusions are as follows:
- We have demonstrated that the cooling of a strongly irradiated giant planet is different on the day and night sides.
- An increased amount of planet irradiation leads to both reduced day- and night-side cooling and thus to the reduced total heat loss.
- An increased metallicity throughout the atmosphere leads to both reduced day- and night-side cooling and thus to the reduced total heat loss.
- An increased efficiency of the day-night heat redistribution reduces the heat loss from the night side while increasing the heat loss from the day side. The night side is more sensitive to the effect and the total heat loss is reduced.

- A possible extra opacity in the stratosphere of the day-side of the planet acts as a shade which screens the irradiation from the star. Consequently, it increases the day-side cooling as well as the total heat loss.

- If the day-night heat redistribution takes place at larger optical depths it increases the day-side cooling while the night-side cooling is suppressed. The total heat loss is also reduced.

- The night-side cooling is generally more efficient than the day-side cooling. However, this does not rule out a situation that there is a combination of parameters when the opposite case would take place.

These results may affect the evolution of the cold object and the well known gravity darkening and bolometric albedo in interacting binaries (von Zeipel 1924; Lucy 1967, Rucinski 1969, Vaz & Norlund 1985, Claret 1998) especially for strongly irradiated cold objects. It would be interesting to study these effects using data from the Kepler mission.

Acknowledgements

JB thanks the VEGA grants Nos. 2/0074/9, 2/007810, 2/0094/11. This work was partially supported also by SAIA.

References

Burrows, A., Budaj, J., & Hubeny, I. 2008, *ApJ*, 678, 1436
Burrows, A., Sudarsky, D., & Hubeny I. 2006, *ApJ*, 650, 1140
Charbonneau, D., Brown, T. M., Latham, D. W., & Mayor, M. 2000, *ApJ*, 529, L45
Claret, A. 1998, *A&AS*, 131, 395
Fortney, J. J., Lodders, K., Marley, M. S., & Freedman, R. S. 2008, *ApJ*, 678, 1419
Henry, G. W., Marcy, G. W., Butler, R. P., & Vogt, S. 2000, *ApJ*, 529, L41
Hubeny, I. 1988, *Computer Physics Comm.*, 52, 103
Hubeny, I., Burrows, A., & Sudarsky, D. 2003, *ApJ*, 594, 1011
Knutson, H. A., Charbonneau, D., Noyes, R. W., Brown, T. M., & Gilliland, R. L. 2007, *ApJ*, 655, 564
Kuricz R. L. 1993, CD-ROM 13, ATLAS 9 Stellar Atmosphere programs and 2km/s Grid (Cambridge: SAO)
Lucy, L. B. 1967, *ZfA*, 65, 89
Rucinski, S. M. 1969, *AcA*, 19, 245
Sharp, C. & Burrows, A. 2007, *ApJS*, 168, 140
Vaz, L. P. R. & Norlund, A. 1985, *A&A*, 147, 28
von Zeipel, H. 1924, *MNRAS*, 84, 665

From Interacting Binaries to Exoplanets: Essential Modeling Tools
Proceedings IAU Symposium No. 282, 2011
Mercedes T. Richards & Ivan Hubeny, eds.
© International Astronomical Union 2012
doi:10.1017/S1743921311028225

NSVS 01031772 Cam: A New Low-Mass Triple?

M. Wolf[1], P. Zasche[1], K. Hornoch[2], M. Chrastina[3], J. Janík[3], and M. Zejda[3]

[1] Astronomical Institute, Faculty of Mathematics and Physics, Charles University Prague, CZ-180 00 Praha 8, V Holešovičkách 2, Czech Republic, email: wolf@cesnet.cz
[2] Astronomical Institute, Academy of Sciences, CZ-251 65 Ondřejov, Czech Republic
[3] Institute of Theoretical Physics and Astrophysics, Masaryk University, Brno, Czech Republic

Abstract. We present a photometric study of the newly discovered low-mass eclipsing binary NSVS 01031772 Cam based on observations obtained at Ondřejov observatory from 2007 – 2011.

Most determinations of the fundamental parameters of low-mass stars using eclipsing binaries indicate a strong discrepancy between theory and observations. The measurements clearly indicate that the observed radii are generally larger than the predictions by stellar models. The eclipsing binary NSVS 01031772 Cam (GSC 4561.647, V=12.6 mag) was discovered by McIntyre & Shaw (2005) as a low-mass, double-lined and detached eclipsing binary in the Northern Sky Variability Survey (NSVS). The comprehensive study was later presented by López-Morales *et al.* (2006, hereafter LM06), who found the precise masses of both components and discovered that the radius of each component exceeds the evolutionary model by about 8.5% on average. In this paper, we present improved system parameters based on our new photometric observations.

Our new CCD photometry was obtained at Ondřejov Observatory, Czech Republic, during 2007-2011. We used mostly the CCD camera G2-3200 of Moravian Instruments attached to the 0.65-m telescope, and VR photometric filters. We used the same comparison star GSC 4561.0787 as done by previous investigators. The processing of CCD frame series was done routinely by the APHOT software package. The supplementary CCD photometry was obtained at Masaryk University Observatory, Brno, during 2009-2011. The 0.62-m reflecting telescope, the CCD camera SBIG ST-8, and VR filters were used. The additional measurements were secured by JJ during one night at Mt. Suhora Observatory, Poland, using the 0.65-m telescope and BVR filters in November 2009.

The new times of mid-eclipse were determined using the profile fitting method; the complete list will be presented elsewhere. Using published minima and our new timings, we improved the linear light elements: Pri.Min. = HJD 24 53456.68023(8) + 0.36814052(6)E.

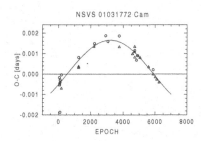

Figure 1. The $O-C$ diagram of NSVS 01031772. The primary and secondary minima are denoted by circles and triangles. The curve denotes the possible LITE with the 11-yr period.

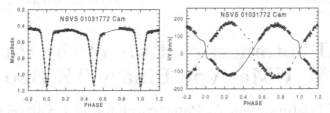

Figure 2. Left panel: The PHOEBE solution of the V light curve obtained at Ondřejov with one spot on the primary component ($T_f = 0.95$, latitude 90°, longitude 180° and radius 25°). **Right panel**: The RV curve solution. Circles for primary, triangles for secondary RVs.

Table 1. Physical parameters of NSVS 01031772 Cam.

Element	this paper	McIntyre & Shaw (2005)	LM06
P [day]	0.36814052(6)	0.368154	0.368141(fixed)
i [deg]	85.7 ± 0.1	88 ± 0.5	85.91 ± 0.03
M_1 [$M_\odot$]	0.555 ± 0.005	–	0.543 ± 0.003
M_2 [$M_\odot$]	0.495 ± 0.005	–	0.498 ± 0.003
T_1 [K]	3500 (fixed)	2885 ± 5	3615 ± 72
T_2 [K]	3455 ± 25	2895 ± 5	3513 ± 31
R_1 [$R_\odot$]	0.592 ± 0.005	–	0.526 ± 0.003
R_2 [$R_\odot$]	0.492 ± 0.005	–	0.509 ± 0.003

A significant sinusoidal period change is clearly visible on the current $O-C$ diagram (see Fig. 1). We tried to fit the $O-C$ values by a light-time effect (hereafter LITE) caused by the orbiting third body. The preliminary solution of the third body's circular orbit gives a minimal period of about 3900 days (11 years) and a semi-amplitude of about 0.00165 days (2.4 min). The minimal mass of the third body in a coplanar orbit follows as 0.063 $M_\odot$ or 65 Jupiter mass only.

Our own VR light curves and radial velocity curves of LM06 were reduced using the current PHOEBE computer code, version 31a, developed by Prša & Zwitter (2005). The gravity darkening exponents were set 0.045 according to LM06. The bolometric albedo $A = 0.5$ corresponding to stars with a convective envelope was taken for each star. Since the light curve of the system has a wave-like distortion, we included spots in our solution, which are parameterised in the same way as in the Wilson and Devinney code: the temperature factor T_f, the latitude and longitude of the spot centre, and the angular radius of the spot. The resulting elements given in Table 2 are compared with the previous solutions of McIntyre & Shaw (2005) and LM06.

Adopting masses of both components given in Table 2, the radii according to theoretical models of Baraffe *et al.* (1998; Z=0.02, 0.35 Gyr isochrone) should be about 0.50 and 0.46 $R_\odot$, which are lower than our results. The current spot activity could be well reproduced by one spot on the primary component. New high-accuracy timings of this low-mass eclipsing binary are necessary in the near future to improve the LITE parameters derived in this paper and to confirm the 11-year orbital period of the possible third body.

This research was supported by grants GAČR 205/08/H005 and P205/10/0715; Research Program MSM0021620860 of the Ministry of Education (to MW, PZ); and GAAV IAA301630901, MEB051018 and MUNI/A/0968/2009 (to MCh, JJ, MZ). We thank Dr. M. López-Morales for the use of their original data.

References

Baraffe, I., Chabrier, G., Allard, F., & Hauschildt, P. H. 1998, *A&A* 337, 403
López-Morales, M., Orosz, J. A., Staw, J. S., Havelka, L. *et al.*, 2006, astro-ph/0610225v1, LM06
McIntyre, T. & Shaw, J. S. 2005, *IAPPP Comm.* 101, 38
Prša, A. & Zwitter T. 2005, *ApJ* 628, 426

From Interacting Binaries to Exoplanets: Essential Modeling Tools
Proceedings IAU Symposium No. 282, 2011   © International Astronomical Union 2012
Mercedes T. Richards & Ivan Hubeny, eds.   doi:10.1017/S1743921311028237

CCD Photometric Study of the Puzzling W UMa-type Binary TZ Boo

P.-E. Christopoulou, A. Papageorgiou and I. Chrysopoulos

Astrophysical Laboratory, Department of Physics, University of Patras, 26500, Patra, Greece

Abstract. The puzzling W-UMa type binary TZ Boo has been monitored at the University of Patras, Observatory from March to July 2010. Photometric solutions were determined through an analysis of the complete BVRI light curves with PHOEBE (Wilson-Devinney code) and published spectroscopic data. This low mass ratio binary turns out to be a deep overcontact system with $f = 52.5\%$ of A-subtype. A conservative spot model has been applied to fit the particular features of light curves. Based on our 7 new light minimum times and all others compiled from the literature over 70 years, we studied the orbital period from the O-C curve.

1. Introduction

The system TZ Bootis (BD $+40^o 2857$=HIP 074061=GSC 3045.00893, $\alpha_{J2000} = 15^h 08^m$ 09.1^s and $\delta_{J2000} = +39^o 58' 13''$, $V_{max} = 10.45$ mag) is one of the most interesting and unusual W-UMa binaries because it has exhibited several changes in the period and in the depth of both minima (Awadalla *et al.* 2006) since its discovery by Guthnick & Prager (1927). Nevertheless there exists no modern photometric solution.

2. New CCD photometric observations and analysis

The system was observed on 2010 April 12, May 27 and June 7, 10, 11, 16 with the 35.5 *cm* $f/6.3$ Schmidt-Cassegrain telescope at the University of Patras Observatory and its SBIG ST-10 XME CCD camera and standard Johnson-Cousin-Bessel set of BVRcIc filters. GSC 3045 959 and GSC 3045-01495 were selected as comparison and check star, respectively. Seven light minima times were calculated with the software Minima25c (Nelson2005) using the K-W (Kwee & van Woerden 1956) fitting method. In our data, the appearance of the eclipses shows that TZ Boo is likely an A-Type rather than W-Type system. For the LC analysis, we used the Wilson & van Hamme (2007) code as implemented in the software PHOEBE (Prša & Zwitter 2005). The mean temperature value for star 1 (star eclipsed at primary light minimum) was taken to be 5890 *K* and the spectroscopic mass ratio $q = 0.207$ according to the last spectroscopic analysis (Pribulla *et al.* 2009), but during the final calculations it was an adjustable parameter. Considering the hypothesis that the orbital period variation on the system is due to the hypothetical third body, the third light was also included. In order to effectively model the flat secondary minimum where there is an obvious slope, a cool star spot was introduced on the primary component. The synthetic and observed light curves are shown in Fig. (a). We compiled all available light minimum timings up to date, from 1926 to 2010, which together with our seven new CCD observations, make a total of 412 (60 visual, 176 photoelectric and 176 CCD). We note that many of the published minimum times were misidentified for their types, so the O-C residuals computed by the ephemeris MinI=2452500.188+0d.2971604xE of Kreiner (2004) are incorrect. It might be caused by

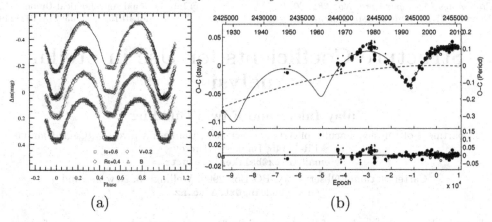

(a)   (b)

the varying depth of both primary and secondary eclipse and the continuously changing orbital period. After corrections for the type of eclipse, the computing code of Zasche *et al.* (2009) was used for the analysis whereas weight 1 was assigned to the reliable visual and photographic observations and weight 10 to the precise photoelectric and CCD. The general trend of (O-C) residuals can be described by a parabolic curve superposed with the sinusoidal oscillations (LITE, Irwin 1952). The final fit and the residuals of the analysis are presented in Fig. (b).

3. Conclusions

Our photometric results suggested that TZ Boo is a low mass A-subtype W-UMa type overcontact binary system with an over-contact degree $f = 52\%$. Combining our photometric solutions with the total projected mass from spectroscopy (Pribulla *et al.* 2009) and the orbital period $P = 0^d.2971599$ from the new ephemeris, the primary and the secondary components were found to have masses 0.99 $M_\odot$ and 0.21 $M_\odot$, respectively, and radii 1.08 $R_\odot$ and 0.56 $R_\odot$, respectively. The steady decrease of its orbital period $dP/dE = -0.17 \times 10^{-10}$ days/cycle is probably due to mass transfer from the more to the less massive component and/or angular momentum loss by the magnetic breaking which would cause the overcontact degree to increase, and finally the binary will evolve into a single rapidly-rotating star. The sinusoidal variation can be fit with an eccentric orbit ($e = 0.63$) and period 31.18 years. According to our photometric solution, its contribution is about 2.1%, in agreement with the upper limit of the spectroscopic results.

References

Awadalla N. S., Hanna M. A., Saad A. S. & Morcos A. B. 2006, *Contrib. Astron. Obs. Skalnaté Pleso*, 36, 47

Berry R. & Bumell J. 2000, *The Handbook of Astromonical Image Processing*, Willmann-Bell, Richmond (Virginia)

Guthnick P. & Prager R. 1927, *K1. Veröff. Berlin-Babelsberg*, 4, 8

Irwin J. B. 1952, *AJ*, 116, 211

Kreiner J. M. 2004, *AcA*, 54, 207

Kwee K. K. & van Woerden H. 1956, *Bull. Astron. Inst. Netherlands XII*, No 464, 327

Pribulla T. & Rucinski S.M. 2006, *ApJ*, 131, 2986

Pribulla T. *et al.*, 2009, *AJ*, 137, 3646

Prša A. & Zwitter T. 2005, *ApJ*, 628, 426

Wilson R. E. & van Hamme W. 2007, *Computing Binary Stars Observables*, U. Florida, Gainesville, FL

Zasche P. *et al.* 2009, http ://sirrah.troja.mff

From Interacting Binaries to Exoplanets: Essential Modeling Tools
Proceedings IAU Symposium No. 282, 2011
Mercedes T. Richards & Ivan Hubeny, eds.
© International Astronomical Union 2012
doi:10.1017/S1743921311028249

Structure Coefficients for Use in Stellar Analysis

Gulay Inlek[1] and Edwin Budding[2]

[1] Department of Physics, Faculty of Arts and Sciences, Balikesir University, Çağış Campus,
TR-10145, Balikesir, Turkey
email: `inlek@balikesir.edu.tr`

[2] Carter National Observatory, PO Box 2909, Wellington, New Zealand
email: `budding@xtra.co.nz`

Abstract. We give new values of the structural coefficients η_j, and related quantities, for realistic models of distorted stars in close binary systems. Our procedure involves numerical integration of Radau's equation for detailed structural data for stellar models taken from the EZWeb compilation of the Department of Astronomy, University of Wisconsin-Madison.

Keywords. Stellar structure, structural coefficients, close binary systems

1. Introduction

The classical approach to finding the shape of a body distorted by rotation and tides utilized equipotential surfaces (Kopal 1959). This approach permits inroads into the solution of the relevant Poisson's equation, if contributory effects can be regarded as additive perturbations upon simpler, more basic forms having spherical symmetry. The perturbations are expressed in terms of suitable harmonic expansions. The equipotentials satisfy Clairaut's equation, and this becomes tractable, due to the orthogonality conditions for products of harmonics in an integral (MacRobert 1927).

2. Equations

We set out the main underlying equations for this work. More background can be found in Kopal (1959). First, we have a series expansion for the potential, thus:

$$V = \Sigma_0^\infty r^{-(n+1)} G \int r'^n P_n \left(\cos \gamma \right) dm',$$ (2.1)

with mass element dm'

$$dm' = \int \int \int \rho r'^2 dr' \sin \theta' d\theta' d\phi'.$$ (2.2)

Clairaut's equation for surface perturbation can be set out in first-order form:

$$\frac{G}{(2j+1)a_1^{j+1}} \int_0^{a_1} \left(ja^j Y_j^i + a^{j+1} \frac{\partial Y_j^i}{\partial a} \right) dm' =$$ (2.3)

$$= c_{i,j} a_1^j P_j^i (\theta, \phi).$$

We write now, for the perturbation coefficients,

$$\eta_j(a) = \frac{a}{Y_j^i} \frac{\partial Y_j^i}{\partial a}.$$ (2.4)

494

which also satisfy 'Radau's equation':

$$a\frac{d\eta_j}{da} + \frac{6\rho}{\bar{\rho}}(\eta_j + 1) + \eta_j(\eta_j - 1) = j(j+1), \tag{2.5}$$

3. Procedure and Results

We wrote a small FORTRAN program to carry out numerical integration of Radau's equation. That was first combined with a separate program used to integrate polytropic models of stars. This was comparable to the method of Brooker & Olle (1995; hereafter BO), except that, with modern computers, steps can easily be made suitably small to avoid the numerical problems mentioned by BO, and still return reliable results in a short time. The Lane-Emden equation is rearranged as two simultaneous first-order difference equations, while Radau's equation becomes a first-order difference equation for η_j applying to each layer. We confirmed numerical agreement with BO to 8 digits with this program (RADAU).

Replacing the Emden equation integrator with the numerical tables of internal structure downloaded from the EZWeb compilation

http://www.astro.wisc.edu/~townsend/static.php?ref=ez-web,

we could apply RADAU to derive corresponding structural parameters for these more realistic stellar models (of mass M). We computed also representative polytropic indices (n_1, n_2) for such models for comparison with historic treatments. A few examples follow.

Table 1. Zero Age Solar Composition Models
$M = 0.5$; $n_1 = 2.52, n_2 = 2.19$

j	2	3	4	5	6	7
η_j	2.83417	3.39029	4.77155	5.03932	6.23161	7.37262
Δ_j	1.76418	1.29864	1.15808	1.09570	1.06282	1.04366
k_j	0.38209	0.14932	0.07904	0.04785	0.03141	0.02183

$M = 3.0$; $n_1 = 2.75, n_2 = 2.80$

j	2	3	4	5	6	7
η_j	2.97626	3.99355	4.99743	5.99875	6.99931	7.99959
Δ_j	1.00478	1.00092	1.00028	1.00012	1.00006	1.00002
k_j	0.00239	0.00046	0.00014	0.00006	0.00003	0.00001

$M = 10.0$; $n_1 = 2.67, n_2 = 2.47$

j	2	3	4	5	6	7
η_j	2.89750	3.97376	4.98982	5.99509	6.99730	7.99837
Δ_j	1.02092	1.00376	1.00114	1.00044	1.00020	1.00010
k_j	0.01046	0.00188	0.00057	0.00022	0.00010	0.00005

4. Conclusion

This kind of result should have increasing importance with the improved photometric accuracies of the post-Kepler Mission era, i.e. light curves of mmag accuracy or better. Proximity effects associated with ellipticity are typically of order 0.1 mag in the majority of normal close binary light curves. Our results show that stellar type dependent structural variations affecting the principal terms of the ellipticity are significant at the 1% level, i.e. ~0.001 mag, and therefore should receive attention.

References

Kopal, Z. 1959, *Close Binary Systems*, Chapman & Hall
Brooker, R. A. & Olle, T. W. 1995, *MNRAS 115, 101*
MacRobert, T. M. 1927, *Spherical Harmonics*, E.P. Dutton, London

From Interacting Binaries to Exoplanets: Essential Modeling Tools
Proceedings IAU Symposium No. 282, 2011
Mercedes T. Richards & Ivan Hubeny, eds.
© International Astronomical Union 2012
doi:10.1017/S1743921311028250

Surface Brightness Variation of the Contact Binary SW Lac: Clues From Doppler Imaging

Hakan Volkan Şenavcı

University of Ankara, Faculty of Science, Department of Astronomy and Space Sciences,
TR-06100 Tandoğan-Ankara, TURKEY
email: hvsenavci@ankara.edu.tr

Abstract. In this study, we present the preliminary light curve analysis of the contact binary SW Lac, using B, V light curves of the system spanning 2 years (2009 - 2010). During the spot modeling process, we used the information coming from the Doppler maps of the system, which was performed using the high resolution and phase dependent spectra obtained at the 2.1m Otto Struve Telescope of the McDonald Observatory, in 2009. The results showed that the spot modeling from the light curve analysis are in accordance with the Doppler maps, while the non-circular spot modeling technique is needed in order to obtain much better and reliable spot models.

Keywords. techniques: photometric, (stars:) binaries: eclipsing, stars: spots

1. Introduction

The light variability of the short-period contact binary SW Lac (P $0.^d32$, $V_{max}=8.^m91$) is very well known and studied by several investigators since its discovery by Ashall (Leavitt 1918). The first photoelectric UBV light curves of the system were obtained by Brownlee (1956), who also pointed out the light curve asymmetries from cycle to cycle. These asymmetries were confirmed and attributed to the existence of cool spot regions by several authors (see Albayrak *et al.* 2004, Alton & Terrell 2006, and references therein). The spectral studies of the system including spectral classification, mass ratio determination and UV/X-ray region spectral analysis were carried out by several investigators, who revealed that the system is a W-type contact binary showing chromospheric and coronal activity (see Şenavcı *et al.* 2011, and references therein, for details).

The aim of this study is to perform the light curve analysis with the spot modeling, using the 2009 and 2010 light curves of the system with the help of the information from the Doppler maps obtained by Şenavcı *et al.* (2011).

2. Observations and Data Reduction

The 2009 and 2010 BV band light curves of the contact binary SW Lac were obtained at the Ankara University Observatory, using an Apogee Alta U47 CCD camera attached to a 40 cm Schmidt-Cassegrain telescope. BD+37° 4715 and BD+37° 4711 were chosen as comparison and check stars, respectively. The nightly extinction coefficients for each passband were determined by using the observations of the comparison star. A total of 700 and 995 data points were obtained in each passband, while the probable error of a single observation point was estimated to be ±0.003/0.004 and ±0.004/0.004 for the 2009 and 2010 BV bands, respectively.

3. The Light Curve Analysis

The 2009 and 2010 BV light curves were analysed simultaneously with the radial velocity curves of the system obtained by Rucinski *et al.* (2005) using the interface version of the Wilson-Devinney code (Wilson & Devinney 1971), PHOEBE (Prsa & Zwitter 2005). Since the surface reconstructions of the system were performed using the time series spectra obtained in 2009, we first adopted the spot modeling, as three main circular spot regions, to the 2009 light curves and carried out the LC modeling (see Fig.1). The results from the LC and spot modeling were represented in Fig.2.

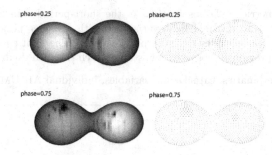

Figure 1. The Doppler maps and the adopted spots for LC modeling of the system.

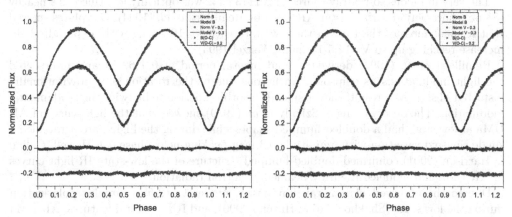

Figure 2. Observational and theoretical light curves with O-C residuals for 2009 and 2010.

4. Conclusion

The analysis showed that the theoretical light curves are compatible with the observed ones, even though the circular spot modeling was performed. However, in order to perform more reliable spot modeling, a code with a non-circular shaped spot approximation is needed as the Doppler maps clearly show us the spots are not circular.

References

Albayrak, B., Djurasevic, G., Erkapic, S., & Tanrıverdi, T. 2004, *A&A*, 420, 1039

Alton, K. B. & Terrell, D. 2006, *JAVSO*, 34, 188

Leavitt, H. 1918, *Harvard Obs. Circ.*, No. 207

Brownlee, R. R. 1956, *AJ*, 61, 2

Prsa, A. & Zwitter, T. 2005, *AJ*, 628, 426

Rucinski, S. M., Pych, W., Ogloza, W., DeBond, H., Thomson, J. R., Mochnacki, S. W., Capobianco, C. C., Conidis, G., & Rogoziecki, P. 2005, *AJ*, 130, 767

Şenavcı, H. V., Hussain, G. A. J., O'Neal, D., & Barnes, J. R. 2011, *A&A*, 529, 11

Wilson, R. E. & Devinney, E. J. 1971, *AJ*, 166, 605

From Interacting Binaries to Exoplanets: Essential Modeling Tools
Proceedings IAU Symposium No. 282, 2011
Mercedes T. Richards & Ivan Hubeny, eds.
© International Astronomical Union 2012
doi:10.1017/S1743921311028262

An Unusual Low State of the Polar AR UMa

Diana P. Kjurkchieva and Dragomir V. Marchev

Dept. of Astronomy, Shumen University, 9700 Shumen, Bulgaria
email: d.kyurkchieva@shu-bg.net; d.marchev@shu-bg.net

Abstract. Our photometric CCD observations of the short-period cataclysmic star AR UMa in 2008 during its low state revealed light variability with a bigger period (by around 10%) and considerably smaller amplitudes than the previous ones. The light curve had a single-wave shape, opposite to the previously observed low states which revealed doubled-humped shape.

Keywords. stars: close binaries, cataclysmic variables, individual AR UMa

1. Introduction

The very luminous soft X-ray source 1ES 1113+432 was optically identified as a nearby (88 pc) short-period variable star AR UMa by Remillard *et al.* (1994). Old plates showed that it spent most of the time in the low state with V=16.5 mag with a sporadic light increase reaching up to V=14.5-15 mag (Wenzel 1993).

Remillard *et al.* (1994) derived a fundamental period of 0.9662 hr and interpreted the light modulations as ellipsoidal variations due to the rotation of a gravitationally distorted M star. As a result, they assumed the orbital period to be twice the fundamental modulation. The observations of Szkody *et al.* (1999) showed that the light curve of AR UMa at low state had a doubled-humped shape, while during the high state it revealed a single sinusoid variation with amplitude 0.3 mag peaking near phase 0.4. Howell, Gelino & Harrison (2001) confirmed doubled-humped structure of the low-state IR light curves of AR UMa and modeled them by beamed cyclotron radiation.

Due to the observed high X-ray luminosity, moderate (2-5%) circular polarization during the low state (Shakhowskoj & Havelin 2000), and IUE spectral features, AR UMa was classified as a polar but the confirmation of its extremely strong magnetic field of around 230 MGs was provided by the detection of Zeeman components of Lα and presence of forbidden lines (Schmidt *et al.* 1996, Gansicke *et al.* 2001).

The model of AR UMa includes a red secondary filling its Roche lobe and transferring mass to the more massive WD. Its magnetic field diverts the matter from its ballistic trajectory and funnels it along magnetic flux tubes. This confinement of the accreted material results in the formation of accretion columns at the WD's magnetic poles. Mass transfer appears to cease when the binary passes into a low state.

2. Observations and analysis of the data

We observed AR UMa during 3 almost consecutive nights in 2008 March (7-10) with the 60 cm telescope of the Mt. Suhora observatory and CCD camera Apogee ALTA U47UV. The observations were taken in V filter with exposures 60 sec. The standard procedure for reduction of the photometric data was used. For transition from the instrumental system to standard photometric system, we used 3 standard stars AR UMa-1, AR UMa-6, and AR UMa-12 that were constant within 0.005 mag during the all observations.

The periodogram analysis of all our photometric data performed by the Period04 software (Lenz & Breger 2005) led to the ephemeris

$$HJD(Min) = 2454530.618503 + 0.08850625 * E. \qquad (2.1)$$

The obtained period of 2.12415 hr is around 10% bigger than the known orbital period of 1.932 hr.

We phased our photometric data with the ephemeris (2.1). Figure 1 (left) presents the corresponding folded curve. The mean V magnitude of AR UMa during our observations is 16.55 mag, which means the polar has been at its low state. The folded curve has a single-wave shape, while the previous light curves of AR UMa in low state have a double-humped shape and bigger amplitude (Fig. 1).

3. Conclusions

Our observations of the polar AR UMa at low state revealed several peculiarities:

(a) a period of variability $\sim 10\%$ bigger than the previously determined value;

(b) considerably smaller amplitude of light variability than those of the previously observed low states;

(c) a single-wave shape of the light curve, in opposition to the doubled-humped light curves observed at the previous low states.

These results show the complex type of the light variability of the polar AR UMa as well as the necessity for additional photometric observations (especially in the optical range) in order to improve this model.

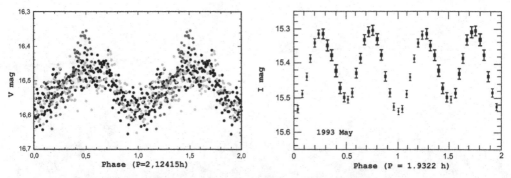

Figure 1. Left: The new V light curve of AR UMa at low state; Right: The old I light curve of AR UMa at low state (Remillard *et al.* 1994)

Acknowledgements

The research was supported partly by funds of project DO 02-362 of the Bulgarian Ministry of Education and Science.

References

Gansicke, B., Schmidt, G., Jordan, S., & Szkody, P. 2001, *ApJ*, 555, 380

Howell S., Gelino D., & Harrison T. 2001, *AJ*, 121, 482

Lenz P. & Breger M. 2005, *Comm. in Asteroseismology*, 146, 53

Remillard, R., Schachter, J., Silber, A., & Slane, P. 1994, *ApJ*, 426, 288

Shakhowskoj N. & Havelin A. 2000, *IBVS*, 4858

Schmidt, G., Szkody, P., Smith, P., Silber, A., Tovmassian, G., Hoard, D. W., Gansicke, B. T., & de Martino, D. 1996, *ApJ*, 473, 483

Szkody, P., Vennes, S., Schmidt, G., Wagner, R., Fried, R., Shafter, A., & Fierce, E., 1999, *ApJ*, 520, 841

Wenzel, W. 1993, *IBVS*, 3890

From Interacting Binaries to Exoplanets: Essential Modeling Tools
Proceedings IAU Symposium No. 282, 2011   © International Astronomical Union 2012
Mercedes T. Richards & Ivan Hubeny, eds.   doi:10.1017/S1743921311028274

Panel Discussion III

Panel: F. Allard, A. Batten, E. Budding, E. Devinney, P. Eggleton, A. Hatzes, I. Hubeny, W. Kley, H. Lammer, A. Linnell, V. Trimble, and R. E. Wilson

Discussion

I. HUBENY: Does anyone from the panel have a theme question to start with today?

V. TRIMBLE: It's another one-liner: From an active galaxy meeting many years ago when people talked about spiral structure. I was reminded by Dr. Rucinski's talk of Lodewijk Woltjer's remark: "The larger our ignorance, the stronger the magnetic field."

A. BATTEN: The last two talks this afternoon (by Wilhelm Kley and Adam Burrows) left my mind reeling! Some years ago, I read (in translation) Kant's *Allgemeine Naturgeschicle mit Theorie des Himmels*, the book in which he presented both the idea of "island universes" and his theory of the origin of the Solar System. The latter is purely qualitative and the tone of Kant's presentation is disturbingly similar to that of the many crank letters all of us receive. I thought to myself that the only thing lacking was a statement that the author did not have the mathematical ability to work out the details himself but offered those ideas to those who could do so. Sure enough, just a page or two later Kant made such a remark! Then, of course, Laplace came along and took up the challenge, but even his treatment of the problem proved inadequate. Wilhelm Kley has shown us that the mathematics required goes far beyond the techniques available to Laplace. If someone had told me when I was a graduate student, more than half a century ago, that I would live to see not only the detection of planets around other stars but also the probing of the constitution of those planets' atmospheres, I would have supposed that he or she was joking. If he or she convinced me that the prediction was serious, I would have dismissed him or her as a crank, but Adam Burrows has shown us the evidence. The only conclusion I can draw is that we should never say to ourselves that any kind of observation will be "impossible."

V. TRIMBLE: Possibly everybody else has already heard this, but NASA announced a fourth moon of Pluto today. If you have four moons, you're a planet; I'm sorry.

A. LINNELL: This conference shows the result of deliberate planning by the organizers to bridge two different but related areas. A number of us from the U.S. can remember the wrenching discussion within the AAS over the issue of initiating parallel sessions at meetings. There is the hazard of increasing compartmentalization. We need to fight against the sort of thing that happened in physics, where people in one sub-field simply don't understand things that are going on in other sub-fields. So, I believe the meeting organizers deserve thanks from all of us for the way they have planned this meeting.

I. HUBENY: That was exactly the point we realized, and in fact was one of main reasons for organizing this conference. For instance, only a few astronomers knew about the Rossiter-McLaughlin effect beyond the group of binary star astronomers, and now it is a common term.

I. HUBENY: Any more comments?

D. J. HILLIER: We have the planets in the Solar System. How well do we do modeling them, especially if we limit our knowledge base to what we might infer for them (now and in the next decade) if they existed around an external star?

A. BURROWS: We assumed equilibrium methane abundances at super-solar levels for the giant planets. We didn't really put much into that study, and that paper was published in 2000. For the other planets, the compositions are clearly non-equilibrium (e.g., the Earth's atmosphere), we need to know what that is in advance. But, if you're given the composition, then it's not too hard to simulate the atmosphere. That's been done for Mars very simply, and for the Earth, in particular. People have been focusing a lot on the Earth, for a variety of obvious reasons. So, there have been some benchmarks like that. There are some subtleties for Jupiter at 8 microns. We see methane in emission, so you do see some weak inversions in the upper atmosphere and we need to be able to reproduce that. There is some heating in the upper atmosphere of Jupiter that's not completely explained.

There's a nice summary on the atmospheres of Jupiter and Saturn that was done by Tristan Guillot and Didier Saumon. There are also many other workers who have focused on Jupiter and Saturn. And what they do is to use the best equations of state and a variety of equations of state, just to explore the range, to try to include the rotation that's observed on the surface and assume rotation on cylinders. And they try to fit the gravitational moments that are measured by fly-bys, etc. So, they see the J_2, J_4, J_6 moments of inertia. There are some ambiguities with J_6, but they want to be able to figure out the atmospheric structure and be able to infer for Saturn that there's a definite presence of a core of about 15 Earth masses. For Jupiter it's a little ambiguous. For both planets, the structures are consistent with super-solar abundances, consistent with the measurements of the atmospheres. There's a depletion in helium in Saturn, which is quite significant in the upper atmosphere, and that's consistent with theories that were developed for the miscibility of helium and hydrogen and the settling of helium. The evolution of Jupiter was examined to see whether it has the radius and the temperature that we measure now after 4.6 billion years. And that works. For Saturn, it doesn't work unless you include the helium drop rates. If you include the amount of helium rain-out that is inferred from the atmosphere (the depletion in the atmosphere of Saturn, which is only about 18% by mass instead of 25% by mass of helium), and if you include the gravitational energy contribution that would heat up Saturn and keep it hotter longer, then instead of being the current temperature it is now of 95 K at 2.5 Gyr, it's the current temperature of 95 K at 4.5 Gyr with the emissibility of that heat source. So, there has been some attempt to use those objects as benchmarks or as launching pads to adventure beyond the Solar System. It's not perfect. We know far too much about the Solar System not to be humble. Exoplanets are easier in that sense. There have been these campaigns, and they're pretty good.

E. BUDDING: I would like to ask Dr. Burrows to explain something about the term 'disequilibrium chemistry' that was mentioned once or twice during his talk. The physics alone seemed complex.

A. BURROWS: Welcome to my world. For brown dwarfs, there are some good spectral indications that the carbon and nitrogen chemistry is out of equilibrium. The equilibrium is modeled by comparing in a simple way the timescale for the chemical equilibration of methane and CO, and the timescale for the upwelling by convection of those same materials. In brown dwarfs, it turns out in many circumstances that the upwelling timescale is shorter than the chemical equilibration timescale, which is a very stiff function of temperature; you get the low temperatures too quickly and it punches out the CO abundance. You see a superfluidity of CO and a deficit of methane. You see the same thing with ammonia in the nitrogen chemistry. It can be reproduced quite nicely. You see this for brown dwarfs and more importantly you see this in Jupiter; this an old story for Jupiter. So, that's the way it's handled. People also try to include photolysis, particularly when they're talking about winds from HD209458b. We measured winds coming off these planets. We have scandium, magnesium, atomic hydrogen, all sorts of indications of species that are coming off at reasonable speeds with interesting mass losses. You have to do that sort of thing on equilibrium. Those models are starting to be developed for photolysis. I would be one of the last people to say that we really have a handle on all of this. It's going to be at least as complicated as any of the non-LTE stuff we've heard about. We're just starting; these are the early-days; we're trying to do the simple things first.

V. TRIMBLE: This is probably a question to France. Do we need magnetic fields for anything with exoplanets?

F. ALLARD: Nice question. Well, like Adam has been saying and as I have been saying, we need more precise cloud models before we need magnetic fields. But the answer is "Yes."

A. BURROWS: I would like to follow-up on that comment. There is an issue of what determines the speeds of super-rotational flows in the planets of the Solar System, or jet streams. If you look at Jupiter, you can watch them moving around in the belts, but we don't know what determines those speeds. You heat from one side and you cool from the other side, and so you have an engine, but that would accelerate the flow. You need to have some dissipation. For Jupiter, a number of years ago, it was suggested that magnetic fields in Jupiter would give you magnetic torques, and there could be ohmic dissipation as well. This was work done by Liu and Schneider, in particular, and Dave Stevenson. Recently, people have used those ideas, not for Jupiter but for exoplanets, to try to determine what is limiting the speeds of these winds, which we don't understand that well, but we know some process has to happen. Also, perhaps this type of magnetic dissipation, joule heating, might contribute to the puffing up of some of these planets. That was suggested by Batygin and Stevenson recently, with an interesting set of ideas that were very poorly developed; it needs a lot more work. But in both Jupiter and in the large exoplanets, magnetic fields are starting to be invoked in some recent work by Kristen Menou and collaborators. We're also doing some work with it. Magnetic fields may be much more important much earlier than we hoped.

R. WILSON: When I first heard about people discovering sodium and potassium in these planets, I thought maybe it was a misprint because they're so reactive. Maybe they can exist because it's very hot there, but potassium is not really light, so I don't know how they get up that high where you can see them. So, are there conditions in which we could observe sodium and potassium in Solar System planets?

A. BURROWS: We do see sodium and potassium prominent in the Earth's atmosphere that's coming from dust coming in, and they are used in AO systems to produce artificial stars. But that's not something you can calculate *ab initio*. It's just there, it's just one of those things you have to measure.

R. WILSON: These elements have some very nice strong spectral lines, but I imagine that concentrations are actually very very low. It's just because the conditions are good for finding them that we are able to see them at all.

A. BURROWS: I'll just call your attention, and I know France would as well, to the fact that brown dwarf spectra are dominated by sodium and potassium; the two lines dominate from 0.4 microns to 1 micron. The wings are so broad, whatever their shapes, they cover the entire region. There are other features there, but the two lines determine the entire slope of the spectrum. We measure this for many, many objects. It's just the chemistry. They're hot enough that they haven't condensed out into sulfides or chlorides into which they would otherwise condense. The Solar System objects are just too cold.

I. HUBENY: I remember you made the point that brown dwarfs cannot be brown, because if you do the color synthesizer it absorbs all the red part of the spectrum, so they will be at best magenta or magenta-brown instead.

C. CHAMBLISS: If you toss even a single salt crystal into a fire you will get the characteristic Na D lines. So their resonance lines are exceedingly easy to excite, and it takes only moderate temperatures of approximately 800 K or 1500 F. But, 800 K will do a very nice job of producing the D lines. The potassium lines aren't quite as obvious, because their resonances are over in the near-infrared; or at least deep red. But the sodium is bright yellow. Well, I use streetlights for that too, and those are easy to turn on, although astronomers don't always like them.

A. BURROWS: To answer Bob's question: We use solar luminosities with the potassium in the atmosphere. These large strengths can easily explain what we see, within a factor of two – the saturated lines.

N. BOCHKAREV: When we talk about astrobiology, mainly original life in the universe, it is important to know how the evolution of the planetary system depends on the chemical composition of matter. Are there minimal abundances of heavy elements for the origin of planetary systems?

P. EGGLETON: I just wonder sometimes how comfortable we are with the fact that the general public is probably led to expect us to find life on other planets. Are there many people here who expect to find life on other planets, and are they looking forward to it? My personal answers would be 'No' and 'No.'

V. TRIMBLE: There is of course a considerable literature on this: from science fiction, from theology, from ordinary people who worry a lot. The chance of finding anybody we could talk to without needing 10^6 years for the messages to go back and forth is quite small. If you happen to like stromatolites, that is pond scum, there may well be pond scum or something equivalent fairly close. Would I like other intelligent life? Yes. Do I know they would be friendly? No, but I think it's worth finding out. Consider the cultures that were destroyed when Europe reached Australia, sub-Saharan Africa, Native America; if it hadn't happened just then with those people it would have happened not much later with other people. So, if it's out there, we will find it, or it will find us. Sir Martin Ryle was terribly worried that we had sent a message to a globular cluster from which the round trip travel time is 10^5 years or something. There is no use in being afraid of it. One has to think about the chances and what kind of communication could take place. For some people, silver rockets land in their yards and take them for rides, but these people generally have other problems as well.

P. EGGLETON: I can't help thinking that the public is going to be rather disappointed with our progress if we don't find intelligent life. I wonder whether the public has been led to believe that it is rather likely we will find such things. I am personally very worried about it. I do wonder whether you get into the taxi and say "take me to the astronomy tower" and they'll say "oh yes you must be looking for life on other planets."

V. TRIMBLE: If we disappoint long enough they'll stop sending money. But at the moment, at least in the US, there is still a considerable, well not enthusiasm but, willingness at least in Congress to continue funding unmanned (unpersoned) missions to Mars. There have been several near misses with Mars rocks that had interesting structures and some of those interesting structures may have been alive. That's one bit of astrophysics and astronomy research that Congress and the public still seem to be willing to support even though all they've got is at best rocks with old stromatolites.

A. BATTEN: I find myself halfway between the skeptics and the enthusiasts. I recently wrote a book in which I quoted W. R. Inge, an Anglican clergyman who was Dean of St. Paul's Cathedral in London during the early part of the twentieth century. In the year that I was born, he wrote: "There is, I think, something derogatory to the Deity in supposing that he made this vast universe for so paltry an end as the production of ourselves and our friends."

V. TRIMBLE: There was also a very famous cartoon in the United States which maybe half of you would be old enough to remember, if you are Americans. It showed a couple of animals in a swamp looking up the sky and saying: "Either we're the most intelligent creatures in the Universe or we are not. And it's pretty sobering either way."

P. NIARCHOS: I would like to ask how many of the panel believe in extraterrestrial intelligence?

A. BATTEN: I believe that once per galaxy is a reasonable guess for the frequency of the emergence of intelligent life. The Drake equation does not help us very much, since we do not know very much more about the quantities on the right-hand side than that none of them is zero, but we do know that the left-hand side is at least one. Even if intelligent life occurs only once in a hundred galaxies, the cosmos could be teeming with life, but it would be difficult in that case to envisage the various communities making contact with each other. However, I will not say that contact would be impossible!

R. WILSON: My feeling would be that, at a given moment, the chance of finding a technical civilization that we could communicate with is pretty close to zero for our Galaxy. If you would integrate over a billion years, there could be several other civilizations.

E. BUDDING: I think the issue of extraterrestrial intelligence depends on the definition of life, about which it is difficult to be categorical. Regarding 'intelligence,' this can perhaps be related basically to the operation of a feedback mechanism dealing with information about the environment surrounding an organism. This could operate at a very low level in a wide variety of feasible situations, but that probably does not concern the type of thing Panos has in mind. From the empirical point of view, however, I would think that the most practical steps that could be taken at the present time relate to those studies of 'disequilibrium chemistry' discussed by Dr. Burrows.

I. Hubeny: Actually, when the Terrestrial Planet Finder (TPF) mission was still alive there were a number of conferences asking that question, and it relates to the previous point about the expectations of the public. For example, the detection of life would be considered as the simultaneous detection of ozone and methane in the atmosphere in the spectrum of an exoplanet. Those two gases cannot really exist in large concentration to be able to produce spectral features. Ozone is a proxy for oxygen. That would be a detection and a big discovery, but it would be a sort of non-equilibrium chemistry, because this additional oxygen would have been created by life of any sort. Of course, that life could mean bacteria on the level which inhabited the Earth during the first 2.5 billion years. That would've been a big discovery, but the public would certainly be disappointed.

F. Allard: Long before we have spectra of extraterrestrial Earths at 1 AU, perhaps polarization would be the way to see something on another planet. One life characteristic is monochirality.

K. Bjorkman: A couple of years ago, there was actually a very interesting and speculative poster at the AAS in which Wolstencroft actually had done some calculations of the polarization that would be produced by various types of plant life on Earth-like exoplanets. It was quite entertaining, and I chatted with him a little bit and he said: "Well, I figured I might as well go ahead and do some calculations and we'll see what we might be able to see."

A. Hatzes: I think life is very easy to form, but intelligent life? As an infamous Secretary of Defense once said: "There are so many unknown unknowns." You have the wonderful Drake equation, and there are a lot of factors that probably should go in there that we don't know about. We need plate tectonics. We need something colliding with the Earth and produce the Moon to stabilize its inclination axis. You need a Jupiter outside to clean out the inner debris, so you don't have as many impacts. There are a lot of things we don't know that we don't know. The question is what happens when you put in all these probabilities? I think the probability is one in a hundred billion! That's why I say it's one per galaxy.

V. Trimble: It can be considerably less than 1 per galaxy, if you agree with panspermia.

N. Bochkarev: This week was launched a radioastronomy mission, which can measure other condensations with very high angular resolution. There are some predictions of measurements of protoplanetary disks and protoplanets. It's an actual problem now.

V. Trimble: I think that's more important news than we perhaps felt as Dr. Bochkarev said this the first time. At least three groups: US, Japan, Russia have been talking about doing radio interferometry with the baseline larger than the diameter of the Earth. This is a step toward that. I think it's very important and I'm ashamed that we haven't heard about it as quickly as we've heard about the fourth moon of Pluto.

I. Hubeny: That's all for today. Thanks for your participation in this discussion.

Part 8
Hydrodynamic Simulations of Exoplanets and Mass Transfer in Interacting Binaries

From Interacting Binaries to Exoplanets: Essential Modeling Tools
Proceedings IAU Symposium No. 282, 2011
Mercedes T. Richards & Ivan Hubeny, eds.
© International Astronomical Union 2012
doi:10.1017/S1743921311028286

Flow Structure in Magnetic CVs

Dmitry V. Bisikalo* and Andrey G. Zhilkin

Institute of Astronomy of the Russian Acad. of Sci., 48 Pyatnitskaya str., Moscow, Russia
*email: bisikalo@inasan.ru

Abstract. We present a review of the modern concept of physical processes which go on in magnetic CVs with the mass transfer between the components. Using results of 3D MHD simulations, we investigated variations of the main characteristics of accretion disks depending on the value of the magnetic induction on the surface of the accreting star. In the frame of a self-consistent description of the MHD flow structure in close binaries, we formulate conditions of the disk formation and find a criterion that separates two types of flows corresponding to intermediate polars (intermediate magnetic field) and polars (strong field).

The influence of asynchronous rotation of the accretor on the flow structure in magnetic close binaries is also discussed. Simulations show that the accretion instability arising in binaries with rapid rotation of accretor ("propeller" regime) can explain the mechanism of quasi-periodic dwarf nova outbursts observed in DQ Her systems.

Keywords. binaries: close, MHD, accretion disks

---

1. Introduction

Magnetic cataclysmic variables (CVs) are close binary stars where the matter of the donor-star (low-mass late-type star) flows through the inner Lagrangian point L_1 onto a white dwarf (see e.g. Warner 2003). There are two main types of magnetic CVs, intermediate polars and polars. In polars (AM Her type), the white dwarf possesses a strong proper magnetic field ($\sim 10^7 - 10^8$ G on the surface). There are no accretion disks observed in such systems. The rotation of components in polars is synchronized. It is accepted that in polars the matter flowing from the donor-star forms a collimated stream that moves along the field lines onto one of the magnetic poles of the accretor (Campbell 1997; Warner 2003). Intermediate polars are binary systems with a relatively weak magnetic field ($\sim 10^4 - 10^6$ G on the accretor's surface). They occupy an intermediate stage between polars and non-magnetic cataclysmic variables. Intermediate polars demonstrate a great variety of orbital periods ranging from several hours to tens of hours. Proper rotations of accretors in such systems are, as a rule, much faster (ten to thousands times) than orbital rotations of the systems (Norton *et al.* 2004). The asynchronous rotation in intermediate polars is explained by the interaction of the dwarf's magnetic field with the matter of the accretion disk that takes place near the boundary of the magnetosphere (Campbell 1997; Warner 2003).

The first numerical studies of the influences of a magnetic field on the flow structure were made in the early 1990s. However, since the problem is very complicated, the investigations have been performed either in the frame of simplified models (King 1993; Wynn & King 1995; Wynn *et al.* 1997; King & Wynn 1999; Norton *et al.* 2004; Ikhsanov *et al.* 2004; Norton *et al.* 2008), or in a limited region of the stellar magnetosphere (Koldoba *et al.* 2002; Romanova *et al.* 2003, 2004a, 2004b). Only over the last few years, the authors have managed to develop a comprehensive 3D numerical model to calculate the flow structure in close binaries (Zhilkin & Bisikalo 2009, 2010a, 2010b, 2010c). In our approach, we use a complete system of MHD equations that allows one to describe all the main dynamic effects concerned with the magnetic field. In the numerical model, we take into account the following phenomena: radiative heating and cooling and diffusion of the magnetic field due to dissipation of currents in turbulent vortexes, magnetic buoyancy, and wave MHD turbulence. It is important to note that in the developed model the disk formation and evolution happen in a natural way due to the mass transfer through the inner Lagrangian point. Thus, we have pioneered a self-consistent description of the MHD flow structure in close binary systems that may be used to interpret observations.

2. Morphology of the flow pattern

Let us investigate how magnetic field influences the flow structure using as an example a close binary star with parameters of SS Cyg (see, e.g., Giovannelli *et al.* 1983). The donor-star (red dwarf) in this system has a mass $M_d = 0.56 M_\odot$ and effective temperature of $4000\ K$. The mass of the accretor (white dwarf) is $M_a = 0.97 M_\odot$ and the temperature is of $37000\ K$. The orbital period of the system is $P_{orb} = 6.6$; the separation is $A = 2.05 R_\odot$. According to morphological properties, some scientists refer this star to the U Gem subtype. However, there are a number of features allowing one to consider this system as an intermediate polar with the value of magnetic field $B_a = 10^4 - 10^6$ G (Fabbiano *et al.* 1981; Kjurkchieva *et al.* 1999). For the first approach, we assume that the accretor's rotation is synchronous, i.e. $P_{spin} = P_{orb}$.

Fig. 1 (left panel) demonstrates the morphology of the flow pattern in the considered system. In this figure, one can see the distribution of the surface density (in units of $\rho(L_1)*$ A) and velocity vectors in the equatorial plane of the system. Shock waves forming in the disk are seen in the figure as condensations of the density isolines. The condensations of the isolines near the edge of the disk correspond to the sharp drop of the gas density from the values common for the disk to the density of the circumbinary envelope gas. The tidal influence of the donor-star leads to occurrence of a spiral shock wave. This wave has two arms located in outer regions of the disk. One can see in the presented results that the interaction of the circumdisk halo and the stream from the L_1 point has all the features of an oblique collision of two flows. A structure consisting of two shocks and tangential discontinuity between them that occurs due to this collision has a complex shape. Outer regions of the circumdisk halo have low density and the shock caused by the interaction between them and stream is located along the edge of the stream. As the gas density increases, the shock curves and, finally, locates itself along the disk edge. The formed shock is rather extended and may be called a "hot line".

In inner regions of the disk which do not undergo gas dynamical perturbations, particle orbits demonstrate retrograde precession due to the gravitational influence of the donor star. Since flow lines cannot intersect each other in a gas dynamical disk, the equilibrium forms with time and all the flow lines start to move with the same angular velocity, i.e. the precession becomes of quasi solid-body type. Since the velocity of the precession depends on the specific size of the orbit, gas more distant from the accretor flow lines must be turned by a larger angle in the direction opposite to the motion of matter, since the precession is retrograde. The precession velocity is in a range between the velocities of the outer (fast) and inner (slow) orbits. Formation of spiral structures in accretion disks was considered in Lyubarskij *et al.* (1994), Ogilvie (2001), and Bisikalo *et al.* (2004). Analysis of the results of the 3D numerical simulations shown in Fig.1 (left panel) completely proves the hypothesis that a spiral density wave can form in inner regions of a cold accretion disk. The results of the calculations show that the wave has a low velocity in the observer's frame. In the non-inertial frame rotating with the binary, the period of its rotation is a little larger than the orbital one. We should note that the precession

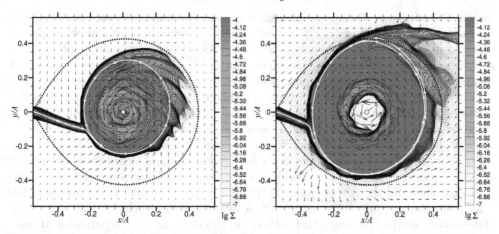

Figure 1. The surface density distribution and velocity vectors in the equatorial plane of the system with synchronous (Left panel) and asynchronous (Right panel) rotation. The flow lines edging the accretion disk are shown by white solid lines. The Roche lobes are depicted by the dashed line.

wave found in the simulations is a density wave. Nonetheless, its presence in the disk leads to significant redistribution of the angular momentum. The accretion rate caused by the increase of the radial flux behind the precessional wave grows approximately by an order of magnitude, compared to a solution with no wave. Despite the significant role of the magnetic field, all the main features of the flow structure found in the purely gas dynamic solution (see e.g. Bisikalo & Matsuda 2007) exist in the "magnetic" case. It may be explained in the way that formation of these structures is mainly due to gravitational effects.

Let us note the most essential differences of MHD solutions in comparison with HD case. The magnetosphere is formed near the accretor and the accretion goes on through the funnel flows. Besides, due to magnetic braking of rotation in the disk, the accretion rate approximately doubles in comparison with the purely HD case where it was about 20-30% from the mass transfer rate.

3. Changes in the flow structure with increasing of the magnetic field

To investigate how the value of the magnetic field influences the flow structure, we have performed calculations with various values of B_a in a range from 10^5 to 10^8 G (Zhilkin & Bisikalo 2010c). Let us consider seven models, with values of the magnetic induction equal to: 10^5 G (model 1), 5×10^5 G (model 2), 10^6 G (model 3), 5×10^6 G (model 4), 10^7 G (model 5), 5×10^7 G (model 6), and 10^8 G (model 7). These models can be divided into two groups. The first group contains models 1, 2 and 3 with relatively low magnetic fields and accretion disks formed. These models correspond to the case of intermediate polars. The second group consists of models 4, 5, 6 and 7 with strong fields. There are no disks formed in these models. They correspond to polars.

The 3D flow structure of intermediate polars (model 3) is shown in Fig. 2 (left panel). This figure demonstrates isosurfaces of the decimal logarithm of the density $log\rho = -4.5$ (in units of $\rho(L_1)$) and field lines which start from the surface of the accretor. In the case of $B_a = 10^5$ G (model 1), the flow pattern corresponds to that described in the previous section. If $B_a = 5 \times 10^5$ G (model 2), the outer radius of the accretion disk becomes much smaller (about 0.15A). The efficiency of the magnetic braking and angular momentum transfer increases. The magnetosphere becomes larger. The accretion columns become more distinguishable near the magnetic poles of the accretor. Finally, in the case of $B_a = 10^6$ G (model 3, Fig. 2 left panel), the accretion disk almost disappears. The matter makes only 1-2 rounds before falling onto the accretor. To describe such a structure, the term "spiralo-disk" is appropriate, since the velocity in this structure is significantly non-Keplerian. The outer radius of this "spiralo-disk" is about $0.1A$. It is almost entirely in the region of the magnetosphere. A significant portion of this disk includes the accretion columns. We can conclude that this variant is the ultimate case of intermediate polars. It is necessary to note that the numerically calculated size of the magnetosphere is well correspondent with that estimated analytically.

Fig. 2 (right panel) demonstrates isosurfaces of the decimal logarithm of the density $log\rho = -5$ (in units of $\rho(L_1)$) and field lines for the model 5. In models 4-7, we observe no disk formed and the flow structure is the funnel flow. In model 7, the magnetic field is so strong that it controls the flow structure over the entire Roche lobe of the accretor. Matter is captured by the magnetic field very close to the inner Lagrangian point. It starts to move along the field line toward the magnetic poles of the star forming a powerful southern stream and a weaker northern stream. In this case, the magnetosphere is larger than the Roche lobe of the accretor and partially penetrates into the envelope of the donor.

Thus, in our simulations of the flow structure in a close binary system whose parameters correspond to those of SS Cyg, in models where the value of magnetic field $B_a \leqslant 10^6$G, we observe accretion disks formed. If the field is stronger, no disks are formed. To explain this result, we propose the following simple ideas. The matter motion in the stream issuing from the inner Lagrangian point is supersonic, therefore, the stream can be investigated in a ballistic approach with no pressure and magnetic field taken into account (Boyarchuk et al. 2002; Fridman & Bisikalo 2008). The analysis of the trajectories (Lubow & Shu 1975) shows that the stream approaches close to the surface of the accretor at a minimal distance $R_{min} = 0.0488\, q^{-0.464} A$. If R_{min} is larger than the radius of the magnetosphere $r_m = [(B_a^4 \times R_a^{12})/(8 \times G \times M_a \times \dot{M}_a^2)]^{(1/7)}$, the magnetic field will have no strong influence on the matter motion. If R_{min} is smaller than the radius of the magnetosphere, then in a certain region of its trajectory the stream will be strongly influenced by the magnetic field. Electromagnetic forces acting in this zone brake the stream and make it lose its angular momentum. As a result, the stream will not be able to round the star and form an accretion disk. So a "boundary" between intermediate polars (with accretion disks) and polars (without disks) is determined by the expression $r_m = R_{min}$. If we use parameters of SS Cyg to calculate R_{min}, we obtain a corresponding value of the magnetic field $B_a \approx 10^6$ G that separates these two regimes. However, this estimate of the "boundary" induction of the magnetic field separating cases of intermediate polar and polars is rather common, since in cataclysmic variables q varies only a little and is equal approximately to 0.5.

Let us consider the accretion rate $\dot{M}_a$ as a function of the magnetic field induction B_a on the surface of the accretor. The main feature of this function is that it is non-monotonic. For $B_a < 10^6$ G, the accretion rate grows when the magnetic field grows; while the amplitude of the variations decreases. At the point $B_a = 10^6$ G, the maximal value of the accretion rate is achieved. If the magnetic field induction continues to grow, the accretion rate decreases. This dependence is completely correspondent to the ideas we described above. If $B_a < 10^6$ G, an accretion disk forms and the accretion rate is determined by processes of the angular momentum transfer in the disk. An increase in the magnetic field leads to increasing efficiency of the magnetic braking in the disk. Thus, while $B_a < 10^6$ G, the function $\dot{M}_a(B_a)$ grows. When the value of the magnetic field is above $B_a = 10^6$ G, no disk is formed and a funnel flow takes place. In this case, the

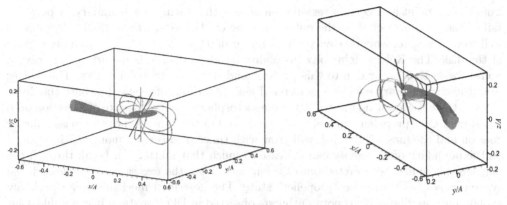

Figure 2. The isosurfaces of the decimal logarithm of the density and field lines which start from the surface of the accretor for the model 3 (Left panel, $lg\rho = -4.5$ in units of $\rho(L_1)$) and for the model 5 (Right panel, $lg\rho = -5.0$ in units of $\rho(L_1)$). The rotation (straight thin line) and magnetic (bold tilted line) axes of the accretor are also shown.

accretion rate is determined by the throughput capacity of the stream. If the magnetic field induction grows, the cross-section of the stream decreases and the accretion rate decreases as well.

4. Asynchronous rotation of the accretor

The influence of the proper rotation velocity of the accretor on the MHD flow structure may be specified by the ratio between the magnetosphere radius r_m and corotation radius r_c. The corotation radius is the distance where the rotation velocity of the magnetic lines coincides with the velocity of matter motion in the accretion disk. If we assume the rotation of the magnetic lines to be of solid-body type with the angular velocity Ω_* and the angular velocity of matter to be ω_K, then (according to Lipunov 1987): $r_c = (G \times M_a/\Omega_*^2)^{1/3}$.

If the accretor rotates slowly ($r_c > r_m$), the velocity of the magnetic field rotation on the boundary of the magnetosphere is lower than the Keplerian one. Thus, matter can freely fall onto the surface being captured by the field lines. This regime may be called a regime of "accretor". If the rotation of the accretor is fast ($r_c < r_m$), a centrifugal barrier occurs on the boundary of the magnetosphere that prevents matter from falling onto the surface. This regime is known as the "propeller regime". In this regime, the accretion process is very non-stationary (Romanova *et al.* 2004b, 2005; Ustyugova *et al.* 2006). In the equilibrium rotation, the condition $r_c = r_m$ is satisfied. Analysis shows (Lipunov 1987) that the interaction of the disk and magnetosphere makes the system evolve toward the equilibrium rotation.

To investigate how the asynchronous rotation of the accretor influences the flow structure in the magnetic close binary, we have performed 3D numerical simulations with different ratios P_{spin}/P_{orb}. In Fig. 1 (right panel), we show distributions of the density and velocity in the equatorial plane for models "propeller" by the moment when the stationary flow regime has been achieved (in about 17 orbital periods). In the "accretor" model, the flow structure is similar to the structure observed in the case of synchronous rotation (see Fig. 1, left panel). The rotation of the accretor leads to a more powerful tail of the matter outflowing through the L_3 point. However, in the "propeller" model, the flow structure has significant differences. The accretion disk is larger. A flow line issuing from the L_1 point at a certain region even crosses the Roche lobe boundary. A powerful tail of the matter ejected to the outer envelope of the system through the L_3 point is well seen. A magnetospheric cavity with a radius of $0.05 - 0.1A$ is formed in inner regions of the disk. The Kelvin-Helmholtz instability develops on the boundary of this cavity. Spiral waves that occur due to this process tend to turn into shock waves. The matter from the circumbinary envelope is accreted mainly not onto the star but onto the disk.

Since there is almost no matter in the magnetospheric cavity, the rate of accretion onto the star in the "propeller" model is almost zero. On the other hand, the mass transfer goes on and the mass of the disk will grow with time. At a certain moment, the density on the boundary of the cavity can increase so much that matter can break through the magnetosphere and be accreted onto the surface. When the excess mass is released, the system comes back into the "propeller" state. The described mechanism can probably explain quasi-periodic dwarf nova outbursts observed in DQ Her stars. It is notable, that the SS Cygni system demonstrates outbursts approximately every 200 orbital periods. The amplitude of the outbursts reaches 4.5^m. If one assumes that most of the flux from the system is due to accretion, then the increase of the system brightness observed during outbursts corresponds to the increase of the accretion rate by about 60 times. The flow structure in SS Cyg during an outburst was investigated in Kononov et al. (2008) using observed spectra and Doppler tomograms.

In the "superpropeller" model (AE Aqr systems), very rapid rotation of the accretor prevents the disk from formation. The matter issuing from the L_1 point is captured by the magnetosphere, acquires additional angular momentum, and is ejected from the Roche lobe. It leads to formation of a long tail twisting around the system and forming its common envelope.

5. Conclusions

The main results of the work may be summarized as follows.

1. It is found that in models with a relatively weak magnetic field, $B_a \leqslant 10^6$ G, an accretion disk is formed in the system. The disk has all the specific features: "hot line" shock, tidal spiral shocks, precessional wave etc. When B_a grows, the disk radius becomes smaller and the size of the magnetosphere increases. These types of flow correspond to intermediate polars.

In models where $B_a > 10^6$ G, no accretion disk is formed. The flow is a funnel flow starting from the L_1 point and ending on the magnetic poles of the accretor. These types of flow, in accordance with morphology, correspond to polars.

A value $B_a = 10^6$ G that separates two types of flows is determined by the ratio of R_{min}, a minimal distance at which the stream approaches the accretor, and r_m a radius of the magnetosphere. An accretion disk is formed when $R_{min} > r_m$. Otherwise, the stream undergoes strong influence of the magnetic field and no disk is formed. This estimate of the magnetic field induction is rather common, since it is not strongly dependent on the parameters of a system.

2. The flow structure in magnetic close binaries is strongly influenced by asynchronous rotation of the accretor. If the rotation is slow ($P_{spin} > 0.033P_{orb}$, "accretor" regime), the structure is similar to that formed when the rotation is synchronous.

If the rotation is rapid ($P_{spin} < 0.033P_{orb}$, "propeller" regime), a magnetosphere cavity is formed near the accretor and the accretion rate falls down to almost zero. Subsequent growth of the mass of the disk leads to a break of matter through the magnetosphere and sharp jump in the accretion rate. In accordance with the results of the simulations, this mechanism can be used to explain quasi-periodic dwarf nova outbursts observed in DQ Her systems.

If the accretor rotates very rapidly ($P_{spin} \approx 0.001 P_{orb}$), no accretion disk in the system is formed. The matter is pushed by the rapidly rotating magnetosphere ejected from the accretor's Roche lobe and forms a tail spiraling around the system. This type of flow ("superpropeller") is observed in AE Aqr systems.

Acknowledgements

This work has been supported by the Basic Research Program of the Presidium of the Russian Academy of Sciences, Russian Foundation for Basic Research (projects 09-02-00064, 09-02-00993, 11-02-00076, 11-02-01248), Federal Targeted Program "Science and Science Education for Innovation in Russia 2009-2013".

References

Bisikalo, D. V., Boyarchuk, A. A., Kaigorodov P. V., Kuznetsov, O. A., & Matsuda T. 2004, *Astron. Rep.*, 48, 449

Bisikalo, D. V. & Matsuda, T. 2007, W. I. Hartkopf, E. F. Guinan & P. Harmanec (eds.), *Proceedings of IAU Symposium 240 Binary Stars as Critical Tools & Tests in Contemporary Astrophysics* (Cambridge: Cambridge University Press), p. 356

Boyarchuk, A. A., Bisikalo, D. V., Kuznetsov, O. A., & Chechetkin, V. M. 2002, *Mass transfer in close binary stars*, Taylor and Francis, London

Campbell, C. G. 1997, *Magnetohydrodynamics in binary stars* Dordrecht: Kluwer Acad. Publishers

Fridman, A. M. & Bisikalo, D. V. 2008, *Phys. Usp.*, 51, 551

Fabbiano, G., Hartmann, L., Raymond, J., Branduardi-Raymont, G., Matilsky, T., & Steiner, J. 1981, *ApJ*, 243, 911

Giovannelli, F., Gaudenzi, S., Rossi, C., & Piccioni, A. 1983, *AcA*, 33, 319

Ikhsanov, N. R., Neustroev, V. V., & Beskrovnaya, N. G. 2004, *A&A*, 421, 1131

King, A. R. 1993, *MNRAS*, 261, 144

King, A. R. & Wynn, G. A. 1999, *MNRAS*, 310, 203

Kjurkchieva, D., Marchev, D., & Ogloza, W. 1999, *Ap&SS*, 262, 53

Koldoba, A. V., Romanova, M. M., Ustyugova, G. V., & Lovelace, R. V. E.. 2002, *ApJ*, 576, L53

Kononov, D. A., Kaigorodov, P. V., Bisikalo, D. V., Boyarchuk, A. A., Agafonov, M. I., Sharova, O. I., Sytov, A. Y.u., & Boneva, D. 2008, *Astron. Rep.*, 52, 835

Lipunov, V. M. 1992, *Astrophysics of Neutron Stars* Heidelberg: Springer

Lubow, S. H. & Shu, F. H. 1975, *ApJ*, 198, 383

Lyubarskij, Yu.E., Postnov, K. A., & Prokhorov, M. E. 1994, *MNRAS*, 266, 583

Norton, A. J., Wynn, J. A., & Somerscales, R. V. 2004, *ApJ*, 614, 349

Norton, A. J., Butters, O. W., Parker, T. L., & Wynn, G. A. 2008, *ApJ*, 672, 524

Ogilvie, G. I. 2001, *MNRAS*, 325, 231

Romanova, M. M., Ustyugova, G. V., Koldoba, A. V., Wick, J. V., & Lovelace, R. V. E.. 2003, *ApJ*, 595, 1009

Romanova, M. M., Ustyugova, G. V., Koldoba, A. V., & Lovelace, R. V. E. 2004a, *ApJ*, 610, 920

Romanova, M. M., Ustyugova, G. V., Koldoba, A. V., & Lovelace, R. V. E. 2004b, *ApJ*, 616, L151

Romanova, M. M., Ustyugova, G. V., Koldoba, A. V., & Lovelace, R. V. E. 2005, *ApJ*, 635, L165

Ustyugova, G. V., Koldoba, A. V., Romanova, M. M., & Lovelace, R. V. E. 2006, *ApJ*, 646, 304

Warner, B. 2003, *Cataclysmic Variable Stars* Cambridge: Cambridge Univ. Press

Wynn, G. A. & King, A. R. 1995, *MNRAS*, 275, 9

Wynn, G. A., King, A. R., & Horne, K. 1997, *MNRAS*, 286, 436

Zhilkin A. G. & Bisikalo, D. V. 2009, *Astron. Rep.*, 53, 436
Zhilkin A. G. & Bisikalo, D. V. 2010a, *Advances in Space Research*, 45, 437
Zhilkin A. G. & Bisikalo, D. V. 2010b, *Astron. Rep.*, 54, 840
Zhilkin A. G. & Bisikalo, D. V. 2010c, *Astron. Rep.*, 54, 1063

Discussion

W. KLEY: To calculate the light curve (e.g., OY Car), you need the surface temperature. How is this determined in your model?

D. BISIKALO: In our gas dynamic model, the radiative heating and cooling processes are taken into account in an approximate manner. We just add the heating and cooling terms in the energy equation. Parameters of these terms are chosen for the equilibrium temperature to correspond to the observed temperature ($T \sim 10^4$ K). However, relative temperature variations over the disk (e.g. due to shock waves) are calculated in an accurate manner. So, the shape of the synthetic light curve takes into account all the features of the solution and may be used to interpret observations.

V. TRIMBLE: What would count as an exotic mechanism in this context?

D. BISIKALO: I have talked about models which are used to analyze light curves but they do not take into account results of numerical simulations of gas dynamics of the matter in the close binary. I call them exotic since, as a rule, they even do not take into account well known facts of classical physics.

R. WILSON: I have a simple question about terminology. All mentions of "precessional" spiral waves on your slides have "precessional" in quotes, so apparently you do not really mean precession. Of course, precession involves two planes that slide around in a conical motion (of their normals), as in the precession of a gyroscope, but here there is only one plane. Would it be fair to call the phenomenon rotational or rotating waves? Or perhaps another name as a replacement for "precession"? Often, we see a similar usage of "precession" for the periastron motion of planet Mercury and other orbiting bodies, which of course is not precession.

D. BISIKALO: You are absolutely right. Indeed, the motion of the wave, I talked about, is the counter-clockwise rotation of the apsidal lines of the elliptic flow lines. Their apastrons form the wave. I agree that the term "rotating density wave" is more appropriate.

P. HARMANEC: It is interesting to note that what you called "a pre-eclipse" dip in the light curves of cataclysmic variables, i.e. another light minimum seem near phase 0.75 (counted from the primary minimum), is also observed in the light curves of some Be binaries. For the examples, see Harmanec and Kříž (1976, in IAU Symp. 70 on Be and Shell Stars, ed. A. Slettebak), and maybe R Arae shown in the poster here might be another example. Our qualitative explanation at that time was basically the same as what your hydrodynamic solution shows.

D. BISIKALO: Great! This is one more example of how one can formulate an adequate conceptual model using the main physical facts in a proper way. It is evident that if one uses more sophisticated gas dynamical calculations, he will only prove (or improve) this model.

From Interacting Binaries to Exoplanets: Essential Modeling Tools
Proceedings IAU Symposium No. 282, 2011
Mercedes T. Richards & Ivan Hubeny, eds.
© International Astronomical Union 2012
doi:10.1017/S1743921311028298

How Common Envelope Interactions Change the Lives of Stars and Planets

O. De Marco[1,2], J.-C. Passy[3,4], F. Herwig[3], C. L. Fryer[5], M.-M. Mac Low[4] and J. S. Oishi[6]

[1]Macquarie University Research Centre in Astronomy, Astrophysics & Astrophotonics
[2]Dept. of Physics and Astronomy, Macquarie University, Sydney, Australia
email: orsola.demarco@mq.edu.au

[3]Dept. of Physics and Astronomy, University of Victoria, Victoria, BC, Canada
[4]Astrophysics Department, American Museum of Natural History, New York, NY, USA
[5]Los Alamos National Laboratory, Los Alamos, NM, USA
[6]Kavli Institute, Stanford University, Palo Alto, CA, USA

Abstract. The common envelope interaction between a giant star and a stellar or substellar companion is at the origin of several compact binary classes, including the progenitors of Type Ia SN. A common envelope is also what will happen when the Sun expands and swallows its planets as far out as Jupiter. The basic idea and physics of the common envelope interaction has been known since the 1970s. However, the outcome of a common envelope interaction - what systems survive and what their parameters are - depends sensitively on the details of the engagement. To advance our knowledge of the common envelope interaction between stars and their stellar and substellar companions, we have carried out a series of simulations with Eulerian, grid-based and Lagrangian, smoothed particle hydrodynamics codes between a 0.88-$M_\odot$, 85-$R_\odot$, red giant branch star and companions in the mass range 0.1-0.9 $M_\odot$. In this contribution, we will discuss the reliability of the techniques, the physics that is not included in the codes but is likely important, the state of the ejected common envelope, and the final binary separation. We also carry out a comparison with the observations. Finally, we discuss the common envelope efficiency parameter, α and the survival of planets.

1. Introduction

Stars with initial mass from 1 to 8 $M_\odot$ go through two major phases of expansion during which interaction with stellar and planetary companions can occur.

Approximately 20% of F and G type stars have companions orbiting within 10 AU, so this type of interaction is not uncommon (Duquennoy & Mayor 1991). Common envelope interactions (Paczynski 1976) between stars and their stellar-mass companions are intensely studied because they give rise to about two dozen binary and single star (merged) classes, both at the low and high ends of the mass spectrum (e.g., cataclysmic variables [Warner 1995], close binary central stars of planetary nebula [De Marco 2009], low and high mass X-ray binaries [e.g., Verbunt 1993]). They are also supposedly at the origin of many stellar phenomena such as type Ia supernovae (e.g., Ruiter *et al.* 2011) and, possibly, gamma ray bursts (Fryer *et al.* 1999).

In addition, preliminary studies report that as many as ~30% of common stars may have Jupiter-type planets within 3 AU (Lineweaver & Grether 2003, Bowler *et al.* 2010). This fraction may be even higher for more massive and more metal-rich stars (Fisher & Valenti 2005). Therefore, star-planet interactions may be common in the universe and, from studies of planets around subdwarf B stars (e.g., Setiawan *et al.* 2011), can affect the life of the star. For example, Soker (1998) suggested that giant-planet interactions are at the origin of subdwarf B stars.

CE interactions are complex physical phenomena to model because of the vast range of time and size scales that need to be resolved. The most modern simulations of star-star interactions are those of Sandquist *et al.* (1998), De Marco *et al.* (2003), Ricker & Taam (2008), Passy *et al.* (2012), and Ricker & Taam (2012). In this contribution, we discuss the simulations of Passy *et al.* (2012), which were carried out with two different modelling techniques.

2. Results

We here summarise the results of our common envelope binary interaction simulations and refer the reader to the paper by Passy *et al.* (2012) for more details. We have simulated the interaction between a 0.88-$M_\odot$, 85-$R_\odot$, non-rotating, RGB star and companions with masses 0.9, 0.6, 0.3, 0.15, and 0.1 $M_\odot$. We also present here, for the first time, a simulation with a 10-M_J-mass companion. The core of the giant had a mass of 0.39 $M_\odot$ and was represented by a point mass as was the companion. The giant star envelope physical parameters of density, temperature, pressure, internal energy, etc., were obtained by a 1-dimensional stellar structure calculation with the code EVOL (Herwig 2000) and relaxed into the computational domain. The companion was placed on the surface of the star and imparted a Keplerian orbital velocity (we also carried out simulations where the companion was given a larger velocity or was deposited 5% farther away, with a resulting small eccentricity of the initial orbit).

We have used two techniques in parallel. The grid technique implemented by the code *Enzo* (O'Shea *et al.* 2004), which we have modified to include an analytically-calculated potential for the point masses (this has resulted in higher degree of precision in the orbital calculation). *Enzo* is an adaptive mesh refinement technique, but we have for now used it in uni-grid mode with a resolution that was at best 256 cell on a side. Since the computational domain was 20 AU, this resulted in a resolution of 17 $R_\odot$. The second technique was the smooth particle hydrodynamic technique developed by Fryer, Rockefeller & Warren (2006) and known as SNSPH, using with 500,000 particles (for similar equivalent resolutions). The results obtained with these two techniques were very comparable.

The companion spirals rapidly inward in all cases and, after ∼200 days, its orbit becomes stable with a much reduced orbital separation. All simulations last ∼600 days (Figure 1). The halting of the in-spiral is due to the removal of mass from the space within the orbit. However, interestingly, while most of the mass is "lifted" to a substantial distance from the giant core (∼100 $R_\odot$), most of it remains lightly bound. One energy source missing from our simulation and which may help unbind the envelope could be recombination energy (Han *et al.* 1995). In addition, a non-zero initial primary stellar spin may also help (Ricker & Taam 2008), although this idea was tested by Sandquist *et al.* (1998) and shown not to have much of an effect.

A second result was that the separation at the end of the dynamical in-spiral was relatively large and a strong function of $q = M_2/M_1$. This is expected from an energy conservation point of view, as more massive companions have more orbital energy to deliver and do not need to spiral in as much. However, a comparison with known post-common envelope systems (see for instance the compilations of De Marco *et al.* 2011, or Davis *et al.* 2012), reveals that most observed systems have homogeneously small final separations. There appears to be a mechanism by which most-to-all systems spiral in farther than predicted by our models. It is possible that if material is not unbound during the rapid in-fall phase, it may fall back towards the binary forming a circumbinary disk that can further reduce the binary separation (Kashi & Soker 2011).

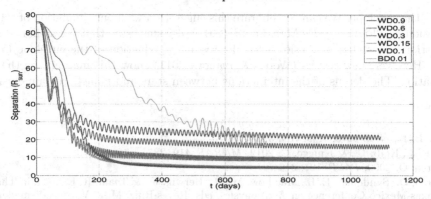

Figure 1. The orbital separation between the giant core and the companion for *Enzo* simulations. This figure is a modified version of the one presented by Passy *et al.* (2012), where we have added the orbital evolution of a 10-M$_J$ companion. This simulation corroborates that smaller companions take longer to in-spiral, a thing that may contribute to their survival. For a colour version of this figure see Passy *et al.* (2012).

3. The common envelope efficiency and implications for low mass companions

The common envelope efficiency parameter, also known as α, allows one to determine the final separation of a post-common envelope system in population synthesis models (e.g., Politano & Weiler 2007). These models use, e.g., SN delay times to identify their progenitors. The common envelope efficiency is the ratio of the energy available to unbind the envelope (i.e., the orbital energy of the companion-primary system) to the work that needs to be done to unbind the envelope, (i.e., the binding energy of the primary's envelope). Energy sinks, which would make α small, are heating the gas and radiating this energy away, or affecting the kinetic energy of the ejected envelope. In our simulations, which do not lose heat, it is only the ejection speed that can make $\alpha < 1$. Since the envelopes of our simulated interactions are not fully unbound, it makes no sense to calculate the values of α.

In a semi-analytical study, De Marco *et al.* (2011) analysed a set of post common envelope systems for which primary and secondary masses, as well as orbital separations are known. Assuming that the post-common envelope primary mass is the same as the core mass of the giant primary at the time of the common envelope, and assuming the evolutionary stage of the primary at the time of the common envelope is known, one can reconstruct the stellar parameters at the time of the common envelope interaction. From this, the value of α can be calculated. By doing so, an inverse trend of increasing α with decreasing q emerged (albeit at low statistical significance, see also Davis *et al.* 2012), implying that higher mass companions sink similarly into the potential well as lower mass ones. This is in qualitative agreement with the observations but not with the simulations. A "cleaner" dataset, where fewer assumptions are made in the reconstruction would help answer the question of the relationship between q and α; such a dataset could be one where all systems are central stars of planetary nebula, guaranteeing an AGB common envelope (De Marco 2009).

Post-common envelope systems with low mass companions (including brown dwarfs [Qian *et al.* 2009] and planets) appear to have values of α larger than unity, indicating that their orbital energy is insufficient to eject the envelope of the primary. De Marco *et al.* (2011) suggested that low mass companions may take a longer time to spiral in towards the core of the primary and that this may favour a stellar reaction that aids in ejecting the envelope. We can corroborate this suggestion with our simulation of a 10-M$_J$ companion (see Figure 1). De Marco *et al.* (2011) also suggested that the stellar response in question is a stellar expansion with the resulting reduction of the envelope binding energy. However, new results (Woods & Ivanova 2011) show this may not be the right explanation. The details of the interactions between stars and planets remain elusive.

References

Bear, E. & Soker, N. 2010, *New Astronomy*, 15, 483
Bowler, B. P., *et al.* 2010, *ApJ*, 709, 396
Davis, P. J., Kolb, U., & Knigge, C. 2012, *MNRAS*, 419, 287
De Marco, O. 2009, *PASP*, 121, 316
De Marco, O., Sandquist, E. L., Mac Low, M.-M., Herwig, F., & Taam, R. E. 2003, in The VIII Texas-Mexico Conference on Astrophysics, eds. Reyes-Ruiz, M. & Vázquez-Semadeni, E., *Revista Mexicana de Astronomia y Astrofísica Conf. Ser.*, Vol. 18, pp. 24
De Marco, O., Passy, J.-C., Moe, M., Herwig, F., Mac Low, M.-M., & Paxton, B. 2011, *MNRAS*, 28
Duquennoy, A. & Mayor, M. 1991, *A&A*, 248, 485
Fischer, D. A. & Valenti, J. 2005, *ApJ*, 622, 1102
Fryer, C. L., Woosley, S. E., & Hartmann, D. H. 1999, *ApJ*, 526, 152
Fryer, C. L., Rockefeller, G., & Warren, M. S. 2006, *ApJ*, 643, 292
Han, Z., Podsiadlowski, P., & Eggleton, P. P. 1995, *MNRAS*, 272, 800
Herwig, F. 2000, *A&A*, 360, 952
Lineweaver, C. H. & Grether, D. 2003, *ApJ*, 598, 1350
O'Shea, B. W., Bryan, G., Bordner, J., Norman, M. L., Abel, T., Harkness, R., & Kritsuk, A. 2004, astro-ph/0403044
Paczynski, B. 1976, in *IAU Symposium 73, Structure and Evolution of Close Binary Systems*, eds. Eggleton, P., Mitton, S., & Whelan, J., p. 75
Passy, J.-C., De Marco, O., Fryer, C. L., Herwig, F., Diehl, S., Oishi, J., Mac Low, M.-M., Bryan G. L., & Rockefeller, G. 2012, *ApJ*, 744, 52
Politano, M. & Weiler, K. P., 2007, *ApJ*, 665, 663
Qian, S., *et al.* 2009, *ApJ* (Letters), 695, L163
Ricker, P. M. & Taam, R. E., 2008, *ApJ* (Letters), 672, L41
—. 2012, *ApJ*, 746, 74
Ruiter, A. J., Belczynski, K., Sim, S. A., Hillebrandt, W., Fryer, C. L., Fink, M., & Kromer, M. 2011, *MNRAS*, 417, 408
Sandquist, E. L., Taam, R. E., Chen, X., Bodenheimer, P., & Burkert A. 1998, *ApJ*, 500, 909
Setiawan, J., Rainer, K., Henning, T., Rix, H.-W., Boyke, R., Rodmann J., & Schultze-Hartung, T. 2010, *Science*, 330, 1642
Soker, N. 1998, *ApJ*, 496, 833
Verbunt, F. 1993, *ApJ*, 31, 93
Warner, B. 1995, *Cataclysmic Variable Stars*
Woods, T. E. & Ivanova, N. 2011, *ApJ*, 739, L48

Discussion

V. TRIMBLE: I am glad that you read my paper (Trimble & Ceja 2007; Astronomische Nachrichten, Vol. 328, p. 983) and even gladder that you chose to ignore it!

W. KLEY: Very nice talk. You showed that all companions stop after a very short time (few hundred years). So, what made them stop? The loss of the envelope?

O. DE MARCO: Companions stop after a much shorter time than that, one to a few years! The reason why the companion stops its in-spiral in the simulations is that very little envelope mass remains within the orbit (approximately 10^{-2} M$_\odot$). As a result, the orbit becomes stable once again.

From Interacting Binaries to Exoplanets: Essential Modeling Tools
Proceedings IAU Symposium No. 282, 2011
Mercedes T. Richards & Ivan Hubeny, eds.

© International Astronomical Union 2012
doi:10.1017/S1743921311028304

Hydrodynamics of Young Binaries with Low-Mass Secondaries

Tatiana Demidova[1], Vladimir Grinin[1,2] and Nataliya Sotnikova[2]

[1]Pulkovo Astronomical Observatory of the Russian Academy of Sciences,
196140, Pulkovskoye chaussee 65/1, Saint-Petersburg, Russia
email: proxima1@list.ru

[2]Saint-Petersburg State University, V.V. Sobolev Astronomical Institute,
198504, Universitetskij pr. 28, Petrodvorets, Saint-Petersburg, Russia

Abstract. The model of a young star with a low-mass secondary component ($q = M_2/M_1 \leqslant 0.1$) accreting matter from a circumbinary (CB) disc is considered. It is assumed that the orbit and the CB disc can be coplanar and non-coplanar. The model parameters were varied within the following ranges: the component mass ratio q ranged from 0.1 to 0.003, the eccentricity e varied from 0 to 0.7, the inclination of the orbit plane to the CB disc ranged from 0 to 10 degrees, and the parameter that defines the viscosity of the system was also varied. A number of hydrodynamics models of such a system have been calculated by the SPH method and then the variations of the circumstellar extinction and phase brightness curves were determined. The calculated brightness curves differ in shape and amplitude and it depends on the model parameters and the orientation of the system relative to the observer. The results were used to analyze the cyclic activity of UX Ori type stars.

Keywords. accretion, accretion discs, hydrodynamics, numerical method, close binaries

1. Introduction

The variable circumstellar extinction is one of the reasons for the photometric activity of young stars. This variability can be easily recognized in the UX Ori type stars because of the "optimal" orientation of their circumstellar discs with respect to the line of sight (Grinin *et al.* 1991). The photometric observations of these stars show that there are cyclic large-scale brightness variations along with chaotic minima of brightness (Schevchenko *et al.* 1993, Grinin *et al.* 1998, Rostopchina 1999). The cyclic variability reflects the existence of stable dust structures around the stars. Grinin *et al.* (1998) and Bertout (2000) suggested that such structures could be density waves and streams of matter caused by the motion of a small companion or a giant planet. This was first predicted by Artymowicz & Lubow (1996). We considered models of a young binary system accreting matter from a circumbinary disc. It is shown that tidal forces create a gap in the innermost part of a disc. Under certain circumstances, two unequal streams of matter penetrate the gap due to viscosity forces and gravitational perturbations. They feed the accretion discs around the binary components. Artymowicz & Lubow (1996) showed that the accretion rate in an eccentric binary varies strongly in time and depends strongly on the phase of the orbital period.

What would be the behavior of such a system if it was observed edge-on or at a slight angle to the line of sight? For the first time, this question was discussed in the paper Sotnikova & Grinin (2007). These authors showed that the column density of matter along the line of sight can be a complicated function of the phase of the orbital period. Similar calculations have been continued in our subsequent papers (Demidova *et al.*

2010, Grinin *et al.* 2010a). We considered two types of binary orbits. In the first case, an orbit was chosen coplanar with a circumbinary disc plane. In the second one, it was inclined at a small angle to the disc plane. The latter case was investigated earlier by Mouillet *et al.* (1997) and Larwood & Papaloizou (1997). The authors have modelled the circumstellar disc of β Pic, which has an inner part inclined by a few degrees with respect to the outer one (Burrows *et al.* 1995). It was shown that the inclination was caused by the motion of a planet on an orbit inclined at a small angle to the circumstellar disc. Recently, the same situation was observed in the circumstellar disc of CQ Tau star (Eisner 2004, Chapillon 2008). Both these events allow us to suggest that such situations are not so rare and more detailed calculations are needed. In this paper, we model the brightness curves of binary systems, orbiting at a slight angle to the circumbinary disc plane.

2. Overview

We used the SPH method to calculate hydrodynamic flows in a series of models of young binaries embedded into a gas-dust disc. The column density of matter in the direction to the line of sight was calculated in the course of simulations. It turned out that it depends on the inclination of the CB disc to the line of sight (ϕ) and the position of the apsidal line relative to an observer (θ). Input parameters were: the component mass ratio - $q = 0.1 - 0.003$, eccentricity - $e = 0 - 0.7$, the period of the binary - $P = 1 - 5$ years, parameter c (the speed of sound), which characterizes the turbulent viscosity parameter, the inclination of the binary orbit to the circumbinary disc plane - ψ, the accretion rate - $\dot{M}_a$, and opacity per gram of matter - κ. The dust-to-gas ratio was taken to be 1:100, close to the value in the interstellar medium. The computations were performed for a few hundred orbital periods. The number of SPH particles involved in the simulations ranged from 60,000 to 200,000. We used the input parameters to transform the column density of SPH particles to optical depth. Then, we calculated the theoretical brightness curves (see Demidova *et al.* (2010) for more details).

The calculations show that the brightness variations with the orbital period can be observed even if the mass of the companion is one hundredth of the mass of the primary star. The amplitudes of variations ranged from 0.8 to 2 magnitudes in the models with the mass ratio $q = 0.01$ and $q = 0.003$, provided moderate values of accretion rate ($\dot{M}_a = 10^{-9} M_\odot/yr$).

For the eccentric binaries, the form and amplitude of brightness variations strongly depend on the orientation of the system with respect to an observer. The same effect arises in the case of circular orbits if the binary's orbital plane is misaligned with the plane of the CB disc. Relatively small changes in the system orientation lead to considerable changes of the brightness curve form. Depending on the position of the line of sight, two types of cyclic activity are possible. The first one is described by two-component brightness curves (see Fig. 1). A prolonged eclipse is caused by the extinction in the stream of matter, moving to the primary component. The second, more compact, eclipse is due to extinction in the denser stream of matter directed to the secondary component. Cyclic activity of the second type demonstrates the simple almost sinusoidal brightness curve (Fig. 2). In this case, there is only one stream of matter, which moves to the primary component and crosses the line of sight, provided an appropriate orientation of the system. Our analysis shows that the first type of activity is more common for the misaligned models. The latter one is observed if the binary's orbital plane is coplanar with the CB disc plane.

3. Implications

We compared the observed brightness curves with theoretical ones and found many common features. For example, some of UX Ori stars show two-component cycles. Fig. 1 demonstrates such a case for BF Ori (Grinin *et al.* 2010b). We obtained a similar brightness curve for the simple model of the binary with the CB disc inclined at small angle to the line of sight.

However, most of UX Ori type stars demonstrate the simple cycles, for instance CO Ori (Rostopchina 2007). It should be noted that the simple brightness curves can be obtained in models with noticeably different parameters. Therefore, we need some additional information on the system parameters (the inclination of the disc plane to the line of sight, radial velocity variations, etc.) to chose the "best fit model".

Unfortunately, the photometric cycles of UX Ori type stars are not studied well enough because the cycle periods are rather long, but the series of observations, in contrast, are short. Nevertheless, our calculations show that at least photometric cycles of some UX Ori stars can be caused by the binarity.

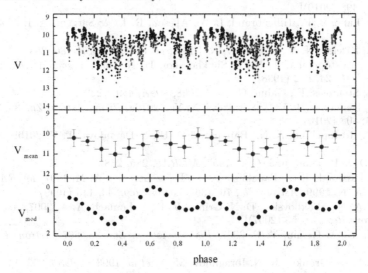

Figure 1. Top and center are the brightness curve of BF Ori, folded with period 12 yr. Bottom is the model brightness curve: $c = 0.05$, $q = 0.01$, $e = 0.5$, $\psi = 5°$, $\theta = 180°$, $\phi = 7°.5$, $\dot{M}_a = 10^{-9} M_\odot/\text{yr}$.

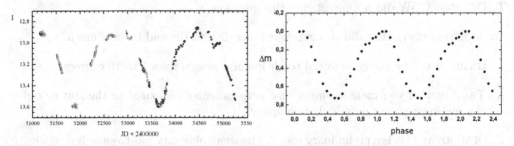

Figure 2. Left is the brightness curve of V718 Per. Right is the brightness curve of the model: $c = 0.05$, $M_2 = 6 M_J$, $e = 0.0$, $\psi = 10°$, $\theta = 45°$, $\phi = 7°$, $\dot{M}_a = 2 \cdot 10^{-10} M_\odot/\text{yr}$.

The results of our simulations can also be applied to young stars with long-lasting eclipses. One of such stars is the weak-lined T Tauri star V718 Per. The duration of its eclipses is about 3.5 yrs and the period is about 4.7 yrs. The star does not show the presence of the secondary component within the errors of the observations: according to (Grinin *et al.* 2008), the mass of an expected planet has to be less than 6 Jupiter masses. We obtained a similar brightness curve in a model with the mass of secondary about 6 Jupiter masses (Fig. 2). Thus, the discussed models qualitatively describe the cyclic activity of UX Ori type stars and can be used to investigate young stars with abnormal prolonged eclipses.

References

Artymowicz, P. & Lubow, S. H. 1996, *ApJ*, 467, L77

Bertout, C. 2000, *A&A*, 363, 984

Burrows, C. J., Krist, J. E., Stapelfeldt, K. R. & WFPC2 Investigation Definition Team 1995, *BAAS*, 27, 1329

Chapillon, E., Guilloteau, S., Dutrey, A., & Pie'tu, V. 2008, *A&A*, 488, 565

Demidova, T. V., Grinin, V. P., & Sotnikova, N. Ya. 2010, *Pisma Astron. Zh.*, 36, 526 [*Astron. Lett.*, 36, 498 (2010)]

Eisner, J. A., Lane, B. F., Hillenbrand, L. A., Akeson, R. L., & Sargent, A. I. 2004, *ApJ*, 613, 1049

Grinin, V. P., Kiselev, N. N., Minikulov, N. Kh. *et al.* 1991, *Ap&SS*, 186, 283

Grinin, V. P., Rostopchina, A. N., & Shakhovskoi, D. N. 1998, *Pisma Astron. Zh.*, 24, 925 [*Astron. Lett.*, 24, 802 (1998)]

Grinin, V. P., Stempels, E., Gahm, G. *et al.* 2008, *A&A*, 489, 1233

Grinin, V. P., Demidova, T. V., & Sotnikova, N. Ya., 2010a, *Pisma Astron. Zh.*, 36, 584 [*Astron. Lett.*, 36, 808 (2010)]

Grinin, V. P., Rostopchina, A. N., Barsunova, O. U., & Demidova, T. V. 2010b, *Astrophysics* 53, 367

Larwood, J. D. & Papaloizou, J. C. P. 1997, *MNRAS*, 285, 288

Mouillet, D., Larwood, J. D., Papaloizou, J. C. B., & Larange, A. M. 1997, *MNRAS*, 292, 896

Rostopchina, A. N. 1999, *Astron. Zh.*, 76, 136 [*Astron. Rep.*, 43, 113 (1999)]

Rostopchina, A. N., Shakhovskoi, D. N, Grinin, V. P., & Lomach, A. A. 2007, *Astron. Zh.*, 84, 60 [*Astron. Rep.*, 51, 55 (2007)]

Sotnikova, N. Ya. & Grinin, V. P. 2007, *Pis'ma Astron. Zh.*, 33, 667 [*Astron. Lett.*, 33, 594 (2007)]

Schevchenko, V. S., Grankin, K. N, Ibragimov, M. A. *et al.* 1993, *Ap&SS*, 202, 137

Discussion

R. BALUEV: Is it more likely to form a planet around a binary or a single star?

T. DEMIDOVA: We did not investigate this question.

M. MONTGOMERY: How did you get your inner disk to tilt and by how much?

T. DEMIDOVA: The tilt is an initial condition and ranges from 1 to 10 degrees.

A. TRIAUD: Can you make an inner disc perpendicular compared to the star or outer disc by placing a planet on a higher obliquity?

T. DEMIDOVA: This is a preliminary result. The small obliquity was to show how viscosity affects the disc. Higher obliquities have not been tried.

From Interacting Binaries to Exoplanets: Essential Modeling Tools
Proceedings IAU Symposium No. 282, 2011   © International Astronomical Union 2012
Mercedes T. Richards & Ivan Hubeny, eds.   doi:10.1017/S1743921311028316

Exoplanet Upper Atmosphere Environment Characterization

Helmut Lammer[1], Kristina G. Kislyakova[2,3], Petra Odert[3], Martin Leitzinger[3], Maxim L. Khodachenko[1], Mats Holmström[4], Arnold Hanslmeier[3]

[1] Austrian Academy of Sciences, Space Research Institute
Schmiedlstr. 6, A-8042, Graz, Austria
email: helmut.lammer@oeaw.ac.at, maxim.khodachenko@oeaw.ac.at

[2] N.I. Lobachevsky State University, University of Nizhnij Novgorod,
23 Prospekt Gagarina, 603950 Nizhnij Novgorod, Russian Federation
email: kislyakova.kristina@gmail.com

[3] Institute for Physics/IGAM, University of Graz, Universitätsplatz 5, 8010 Graz, Austria
email: petra.odert@uni-graz.at, martin.leitzinger@uni-graz.at,
arnold.hanslmeier@uni-graz.at

[4] Swedish Institute of Space Physics, Box 812, SE-98128 Kiruna, Sweden
email: matsh@irf.se

Abstract. The intense stellar SXR and EUV radiation exposure at "Hot Jupiters" causes profound responses to their upper atmosphere structures. Thermospheric temperatures can reach several thousands of Kelvins, which result in dissociation of H_2 to H and ionization of H to H^+. Depending on the density and orbit location of the exoplanet, as a result of these high temperatures the thermosphere expands dynamically up to the Roche lobe, so that geometric blow-off with large mass loss rates and intense interaction with the stellar wind plasma can occur. UV transit observations together with advanced numerical models can be used to gain knowledge on stellar plasma and the planet's magnetic properties, as well as the upper atmosphere.

Keywords. Exoplanets, Roche lobe, mass loss, characterization, ENAs, magnetospheres

---

1. Introduction

Exoplanetology is one of the fastest growing fields in present-day space science. Sixteen years after the discovery of 51 Peg b, more than 550 exoplanets (August 2011) have been detected. The discovery of Jupiter-type planets at orbital distances $d < 0.05$ AU soon opened questions regarding atmospheric mass loss. In early studies of close-in exoplanets, the radiative effective temperature T_{eff} was used to estimate evaporation rates (Guillot *et al.* 1996; Konaki *et al.* 2003). In reality, the exobase temperature T_{exo}, which results from the absorbtion of the stellar SXR and EUV (XUV) radiation in the upper atmosphere controls thermal escape and is $\gg$ than T_{eff}. To estimate T_{exo} of "Hot Jupiters", Lammer *et al.* (2003) applied a scaling law which is based on an approximate solution of the heat balance equation in the thermosphere and found that the upper atmosphere of "Hot Jupiters" will be heated to several thousands of Kelvins so that hydrostatic conditions are not anymore valid.

In this work, we discuss relevant physical processes and modeling techniques, which can be used together with present and future high resolution UV observations for characterizing the upper atmosphere structure, the magnetic field, and the stellar plasma environment around close-in exoplanets. Because HD 209458b is a well-studied and well-observed gas giant, we focus our discussions on that particular planet. In Sect. 2, we discuss the upper atmosphere structure of HD 209458b obtained from hydrodynamic and empirical models. In Sect. 3, we investigate the mass loss of H-rich hot gas giants. In Sect. 4, we discuss how one can infer knowledge related to the stellar wind plasma and magnetic properties around exoplanets from UV observations and advanced modeling techniques. In Sect. 5, we give a brief outlook to the future of such observations and studies.

2. Upper atmosphere structure of H-rich "Hot Jupiters"

2.1. *Hydrodynamic models*

HD 209458b is a hot gas giant with a visual radius $R_{pl} = 1.38 R_{Jup}$ (Southworth 2010) and a mass $M_{pl} = 0.64 \pm 0.09 M_{Jup}$ (Snellen *et al.* 2010) which orbits a $\sim$4 Gyr old solar like G-type star at 0.047 AU. After the discovery of HD 209458b, several hydrodynamic models were applied to investigate the response of the stellar XUV radiation to the planet's thermosphere (Yelle 2004, 2006; Tian *et al.* 2005; García Muñoz 2007; Penz *et al.* 2008; Murray-Clay *et al.* 2009; Guo 2011). These models solved the set of the 1D hydrodynamic fluid equations for mass, momentum and energy conservation

$$\frac{\partial n}{\partial t} + \frac{1}{r^2} \frac{\partial n v r^2}{\partial r} = 0, \tag{2.1}$$

$$n \frac{\partial v}{\partial t} + n v \frac{\partial v}{\partial r} + \frac{1}{m} \frac{\partial p}{\partial r} = n \left[-\frac{GM_{pl}}{r^2} + \frac{GM_*}{(d-r)^2} + \frac{G(M_* - M_{pl})}{d^3}(s-r) \right], \tag{2.2}$$

$$nm \left(\frac{\partial E}{\partial t} + v \frac{\partial E}{\partial r} \right) = q - p \frac{1}{r^2} \frac{\partial r^2 v}{\partial r} + \frac{1}{r^2} \frac{\partial}{\partial r} \left(r^2 \chi \frac{\partial T}{\partial r} \right), \tag{2.3}$$

with particle number density n, gas velocity v, mass m of atomic H, temperature T, Boltzmann constant k, XUV volume heating rate q, thermal pressure $p = nkT$, total energy density $E = p/[nm(\gamma - 1)]$, heat conductance χ, and adiabatic index γ. Roche lobe effects due to tidal interaction as shown in Eq. (2.2) were included in García Muñoz (2007) and Penz *et al.* (2008), where G is the gravitational constant, M_{pl} the planet's mass, M_* the host star's mass, d is the orbital distance, and s the distance of the center of mass to the center of the planet. By solving these equations, one obtains the velocity, density and temperature profiles of the dynamically expanding and outward flowing bulk atmosphere. Yelle (2004) was the first to apply a photochemical model to HD 209458b within his 1D hydrodynamic code up to 3.3 R_{pl}. The second hydrodynamic model, which included photochemical and ionization processes and for the first time also Roche lobe effects, was applied by García Muñoz (2007). Both model simulations indicate that the majority of the XUV radiation is absorbed or deposited between 1.05–1.5R_{pl}. The hydrodynamic models of Tian *et al.* (2005) and Penz *et al.* (2008) do not include photochemistry, but use an energy deposition function as modeled by Yelle (2004) or García Muñoz (2007).

The modeled neutral H atom density in the simulation of Yelle (2004) between 2–3.3R_{pl} is $\sim$7$\times$10^{12}–10^{12} m^{-3}. The temperature at the altitude range is $\sim$1.2 $\times$ 10^4 K. At larger distances the temperature becomes lower due to expansion and adiabatical cooling. The outflow velocity of the bulk H atoms at 3R_{pl} is $\leqslant$2.5 km s^{-1} and the mass loss rate for atomic H in the model by Yelle (2004) is $\sim$4.7 $\times$ 10^{10} g s^{-1}.

Tian *et al.* (2005) applied a 1D time-dependent hydrodynamic model together with a 2D energy deposition model to HD 209458b where the inner boundary is the visual planetary radius 1R_{pl} and the upper boundary is 10R_{pl}. This choice makes the Yelle model more realistic compared to Tian *et al.* (2005) because extending the hydrodynamic

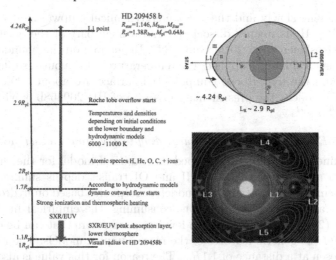

Figure 1. Left: Illustration of the upper atmosphere structure of HD 209458b or similar "Hot Jupiters" according to hydrodynamic and empirical models (Yelle 2004; Garzia Muñoz 2007; Penz *et al.* 2008; Koskinen *et al.* 2010). Right: The Roche lobe plays an important role in the mass loss of close-in exoplanets. If the XUV-heated and expanded thermosphere reaches the first Roche equipotential surface geometric blow-off occurs.

model to distances where collisions become negligible is questionable, even for the purpose of calculating the escape rate. The neutral H atom density in the model simulation of Tian *et al.* (2005) between 2–3.3$R_{\rm pl}$, depending on the assumed heating efficiency η, is about 10^{13}–10^{12} m^{-3}. The temperature at this altitude range is $\sim 10^4$–3×10^4 K. The outflow velocity of the bulk H atoms at 3$R_{\rm pl}$ is ~ 10 km s^{-1}. The maximum mass loss rate for H atoms in their model runs is $\sim 6 \times 10^{10}$ g s^{-1}.

García Muñoz (2007) assumed slightly different input parameters in the lower atmosphere close to 1$R_{\rm pl}$ compared to Yelle (2004) and Tian *et al.* (2005), the temperature was 1200 K instead of 750 K. The upper boundary was as in Tian *et al.* (2005) at large distances $\sim 15 R_{\rm pl}$. As pointed out in Yelle (2004), such large distances may most likely not give accurate results. The H gas bulk flow velocities and number densities at $\sim 3R_{\rm pl}$ depending on the uncertainties in the assumed initial conditions are between 5–10 km s^{-1} and 10^{13}–10^{14} m^{-3}. This model yields temperatures $< 10^4$ K at about 3$R_{\rm pl}$ comparable to Yelle (2004) and Tian *et al.* (2005). Depending on the heating efficiency and XUV flux chosen, this author obtains a loss rate of 6×10^{10} g s^{-1}.

Penz *et al.* (2008) applied a hydrodynamic model to HD 209458b by using the calculated XUV energy deposition rates of Yelle (2004) which are also in agreement with García Muñoz (2007), and obtained temperatures and mass loss rates for heating efficiencies between 20–30 % at 2–3$R_{\rm pl}$ between 6000–8000 K and 1.5–2.5 $\times 10^{10}$ g s^{-1}, respectively. A temperature of $\sim 10^4$ K is obtained for a heating efficiency of 60% at an altitude range of ~ 1.5–1.8 $R_{\rm pl}$. After this distance, the temperature decreases due to adiabatic expansion and the non-availability of heating sources. The velocities of the outward flowing H atoms in this model are < 10 km s^{-1} at 3$R_{\rm pl}$ and the density at 3$R_{\rm pl}$ is $\sim 4 \times 10^{13}$ m^{-3}.

From these hydrodynamical and photochemical model simulations, we conclude that the upper atmosphere of HD 209458b and similar "Hot Jupiters" is structured as illustrated in Fig. 1. The stellar XUV radiation is absorbed mainly in the lower thermosphere $\leqslant 1.5 R_{\rm pl}$, where strong ionization, dissociation and heating take place. According to all hydrodynamic models discussed above, the atomic H gas at altitude levels between ~ 1.5–

$1.7R_{\rm pl}$ starts moving slowly and than expands dynamically upwards at a higher speed. Above ~1.7–$2R_{\rm pl}$, H_2 is mainly dissociated and a huge part of the outflowing gas becomes ionized (Koskinen *et al.* 2010; Guo 2011). Depending on the boundary conditions, η and hydrodynamical models applied, the temperatures of the outward flowing H atoms between $2R_{\rm pl}$ and the first Roche equipotential surface are about 6000–11000 K. According to these models, the present mass loss rate of HD 209458b is within a range of $\sim 10^{10}$–5×10^{10} g s^{-1}.

2.2. *Empirical model results according to UV transit observations*

Recently, Koskinen *et al.* (2010) developed an empirical model for the thermosphere of HD 209458b to analyze the observed HI and OI transit depths summarized by Ben-Jaffel and Hosseini (2010). Because the model atmospheres based on hydrodynamics and photochemistry are complicated and time-consuming, the empirical model applied by Koskinen *et al.* (2010) is only based on a few free parameters that can be constrained by observations and the generic features of the more complex models. The lower boundary condition is chosen at a distance of $1.1R_{\rm pl}$. The reason for this value is also in agreement with the previously discussed hydrodynamic models that most of the XUV radiation is absorbed above 0.1 μbar and the peak absorption of the radiation occurs at a distance between ~1.1–$1.5R_{\rm pl}$.

At higher altitudes, in agreement with Yelle (2004) and García Muñoz (2007), mainly H atoms populate the upper thermosphere. The atmosphere below is opaque to FUV radiation which results in a temperature of ~1300 K. The ionization of the H atoms and other species occurs between $2R_{\rm pl}$–$5R_{\rm pl}$, this is also in agreement with the photochemical models which include ion-chemistry. According to the pressure level in Koskinen *et al.* (2010), the bulk flow velocities of the H atoms at $3R_{\rm pl}$ are between 1–10 km s^{-1}, which is also in agreement with the model results of Yelle (2004), García Muñoz (2007) and Penz *et al.* (2008). According to Koskinen *et al.* (2010), two of their case studies can reproduce the HI and OI Hubble Space Telescope (HST) observations of HD 209458b best if they adopt a temperature at $2.9R_{\rm pl}$ of 11000 K or 8000 K at $2.72R_{\rm pl}$. The density of the H atoms at the first Roche lobe equipotential distance for 11000 K is $\sim2.6 \times 10^{13}$ m^{-3} and, for the cooler temperature case of 8000 K, the number density is $\sim3.1 \times 10^{13}$ m^{-3}. According to Koskinen *et al.* (2010), the recent non-detection of auroral and dayglow emissions of H_2 from HD209458b (France *et al.* 2010) can be seen as an additional constraint and that the H_2/H dissociation front is deeper than the 0.1 μbar level. In that case, an upper atmosphere $T \sim8000$ K yields the best-fitting model.

From the hydrodynamical and empirical models discussed in the previous sections, we can summarize that the upper atmosphere H atom number density $n_{\rm H}$, and temperature $T_{\rm H}$ at $\sim2.9R_{\rm pl}$ of HD 209458b are most likely $\sim3 \times 10^{13}$ m^{-3} and ~8000–10000 K, respectively.

3. Mass loss and the relevance of the Roche lobe

The importance of tidal forces for close-in exoplanets was first discussed by Lecavelier des Etangs *et al.* (2004). Erkaev *et al.* (2007) found that the critical temperature calculated for the modified potential barrier approaches zero when the exobase expands to the Roche lobe boundary from below, which indicates that the effect of the Roche lobe can enhance the possibility that "Hot Jupiters" may reach blow-off conditions more easily compared to similar planets in orbit locations where Roche lobe effects are negligible. It was also shown by Erkaev *et al.* (2007) that hydrodynamically modeled mass loss rates can be well approximated by a modified "energy-limited" mass loss formula

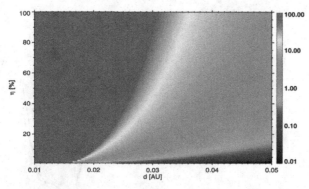

Figure 2. Mass loss of HD 209458b in planetary mass modeled over the host stars lifetime of 4 Gyr in percentage of the planets present mass as a function of orbital distance d and heating efficiency η. The XUV evolution with time of the solar-like G-type star HD 209458 is taken from the power law given in Ribas *et al.* (2005).

$dM/dt = [3\eta F_{XUV}(t)]/[4G\rho_{pl}K(\xi)]$, which includes a mass loss enhancement factor $K(\xi)$ which depends non-linearly on a dimensionless Roche lobe boundary distance $\xi = R_L/R_{pl}$ (Erkaev *et al.* 2007) and realistic $\eta \ll 100\%$ (Lammer, *et al.* 2009) for the stellar XUV radiation. $F_{XUV}(t)$ is the XUV flux corresponding to the stellar age at the orbital distance of the planet, ρ_{pl} is the planetary density and G Newton's gravitational constant.

Fig. 2 shows the modeled mass loss of HD 209458b over the host star's lifetime of 4 Gyr in percentage of the planetary mass at various orbital locations and values of η. One can see that the mass loss is highly overestimated if one applies the energy-limited approach with $\eta = 100\%$. In reality, the ratio of the net heating rate to the rate of stellar energy absorption is $\sim$15–20% (e.g., Lammer *et al.* 2009; Murray-Clay *et al.* 2009). At HD 209458b's orbit location of 0.047 AU, the planet lost a negligible fraction of its mass. If a planet with the same initial density would be at d <0.025 AU, such a planet may shrink to an Uranus-type body or even evaporate to its core mass due to the Roche lobe effect. Depending on a stellar spectral type of the host star, planetary density, η, d, and the related Roche lobe effect, we expect that at distances between 0.01–0.025 AU Jupiter-class and sub-Jupiter-class exoplanets can lose several percent of their initial masses and planets which originated with low densities may even evaporate down to their cores.

4. Characterization of the upper atmosphere-magnetosphere environment by UV observations and advanced numerical models

HST observations during transits of HD 209458b in the UV show absorption in the stellar Lyman-α line at 1215.67Å. Vidal-Madjar *et al.* (2003) used the G140M grating during the observation of 3 transits with a spectral resolution of $\sim$ 0.08Å, which allowed a detailed analysis of the line profile of the H Lyman-α emission line, where they obtained a transit depth of 15 $\pm$ 4% from the ratio of the fluxes in two wavelength (λ) regions around the core of the H Lyman-α line to the flux in the wings of the line during transit. After the observation and data interpretation of Vidal-Madjar *et al.* (2003) and several debates regarding the analysis of these observations, two subsequent observations with lower spectral resolution with the HST STIS/ACS instruments (Ehrenreich *et al.* 2008; Vidal-Madjar *et al.* 2008) were carried out and convinced now also sceptics that the transit depth in the stellar Lyman-α line is significantly greater than the transit depths expected from the planetary disk alone (Vidal-Madjar *et al.* 2008; Ben-Jaffel and Sona Hosseini 2010). Besides these Lyman-α observations, additional transits of HD 209458b

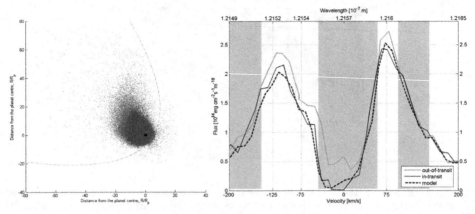

Figure 3. Left: Modeled stellar wind interaction with HD 209458b with an assumed magnetopause (dashed green line) sub-stellar obstacle at $\sim 4.7 R_{\mathrm{pl}}$. Shown from above, perpendicular to HD 20948b's orbital plane as seen from Earth, along the direction of the x-axis. The stellar wind protons are not plotted but flow around the magnetopause obstacle and interact with the planetary hydrogen exosphere (blue dots) and produce energetic neutral atoms (ENAs: red dots). Right: Modeled attenuation spectra and comparison with the HST/STIS observations of Vidal-Madjar *et al.* (2003). Observed profile before transit (bold line); observed profile during transit (thinner line). The dashed line is the modeled profile which is computed at the instant of mid-transit for stellar wind density and velocities $n_{\mathrm{sw}} \approx 2.5\text{--}3.5 \times 10^9$ m^{-3} and 100 km s^{-1} $< v_{\mathrm{sw}} < 250$ km s^{-1} and a sub-stellar stand-off distance $R_{\mathrm{s}} \approx 4.5 R_{\mathrm{pl}}$.

were observed with the HST/STIS and COS spectrographs in the λ range 1180–1710 Å and indicated absorptions of O, C and Si multiply charged ions (Vidal-Madjar *et al.* 2004; Linsky *et al.* 2010).

There are three important points which should also be mentioned related to UV observations and the empirical model described in Sect. 2 of Koskinen *et al.* (2010). First, the upper atmosphere temperatures between 8000–10000 K are based on the assumption that the observed O, C and Si particles are of planetary origin, so that they are dragged upward by the planetary H wind. One should note that the upper atmosphere temperature could be $\leqslant 8000$ K if these particles originate within the stellar wind plasma, as now expected for Mg ions around WASP-12b (Llama *et al.* 2011). Second, Koskinen *et al.* (2010) fitted their model to the transit depth measurements covering the full width of the stellar H Lyman-α profile in the λ range 1212–1220 Å with absorption depths of 6.6 $\pm$ 2.3% (Ben-Jaffel and Hosseini 2010) and not to the early Vidal-Madjar *et al.* (2003) observations. One should note that the empirical model atmospheres can more easily fit the lower absorption depths without introducing an external H source. Third, large column densities along the lines-of-sight through the atmosphere are necessary, which depend on the pressure level of the model-dependent H_2/H dissociation levels. For explaining the high velocity H atoms in the flanks of the HST attenuation spectra, Koskinen *et al.* (2010) argue that, due to natural and thermal broadening and large column densities along the lines-of-sight through the atmosphere, the optical depth in the wings of the line profile can be significant even in the absence of actual bulk flows towards or away from the observer.

However, this interpretation gets into trouble due to another observation of an extended upper atmosphere by Lyman-α absorption during transits of HD 189733b (Lecavelier des Etangs *et al.* 2010). Guo (2011) studied the upper atmosphere structure of HD 189733b, which is exposed to a higher XUV flux, and found that in contrast to HD 209458b, $\sim 80\%$ of its dynamically expanding upper atmosphere is ionized so that its planetary wind is almost composed of H^+ ions. Under this circumstance, the remaining neutral bulk atmosphere cannot produce the detected adequate absorption in Lyman$-\alpha$. Because of the steep decline of the number density of neutral H atoms, the optical depth in the wing of the line is very low. If the amount of atomic H is not adequate to fit the observations, the fact that the transits of HD 189733b in H Lyman$-\alpha$ have been observed imply that external processes, such as the production of energetic neutral atoms (ENAs) as suggested by Holmström *et al.* (2008) and Ekenbäck *et al.* (2010), play an important role. One should also note that even in case the assumption of Koskinen *et al.* (2010) is right, ENAs will be produced if the upper atmosphere interacts with the stellar wind plasma. This agrees with another finding of Koskinen *et al.* (2010) that the upper boundary of their model is close to both the boundary of the Roche lobe and an ionopause of a weakly magnetized planet and that the density of the outward flowing neutral atoms decreases sharply above this boundary. If this is the case, the production of ENAs in the Lyman-α attenuation spectra cannot be neglected. Shematovich (2010) studied the production and escape of dissociated hot H atoms from HD 209458b, which have velocities < 45 km s^{-1} with the majority of the atoms at ~ 20 km s^{-1}. Thus, it is difficult that thermal and non-thermal planetary atoms reach velocities $\geqslant 100$ km s^{-1} which were observed in the high energy part of the spectrum.

ENAs will be produced by charge exchange with stellar wind protons and neutral H atoms from the planetary upper atmosphere. In Fig. 3, we show two modeled hydrogen corona and ENAs as well as the attenuation spectra in comparison with the HST/STIS in and out off transit observations of Vidal-Madjar *et al.* (2003). We apply the same plasma flow, exosphere, ENA production and Lyman-α attenuation model described in detail in Holmström *et al.* (2008) and Ekenbäck *et al.* (2010) to the planetary parameters favored by Koskinen *et al.* (2010). The results shown in Fig. 3 indicate that the sub-stellar magnetopause obstacle should be located between 4.5–6$R_{\rm pl}$ corresponding to a magnetic moment of $\approx 40\%$ that of Jupiter's.

5. Future outlook

Our studies show that future high resolution UV observations of exoplanets by space observatories like the Russia-led World Space Observatory-UV (WSO-UV) in combination with discovered transiting exoplanets around "bright" stars as planned with ESA's PLATO mission would open a promising field for the characterization of the stellar plasma environment and its interaction with exoplanet upper atmospheres and magnetospheres. From these observations, we would obtain knowledge on the stellar wind and magnetic properties of the exoplanet as well as its upper atmosphere structure and mass loss.

Acknowledgements

H. L., K. G. K., M. H., and M. L. K. acknowledge the ISSI team "Characterizing stellar- and exoplanetary environments." H. L., K. G. K & M. L. K. thank the RFBR-FWF project 09-02-91002-ANF_a / I199-N16, the FWF project P21197-N16. A.H, P. O. and M. L. acknowledge the FWF project P22950-N16. K. G. K. also acknowledge the RFBR project 08-02-00119_a, the NK-21P and the Russian Education Ministry. The authors also thank the EU FP7 project IMPEx (No.262863) and the EUROPLANET-RI projects, JRA3/EMDAF and the Na2 science WG4 and WG5 for support.

References

Ben-Jaffel, L. & Hosseini, Sona, S. 2010, *ApJ*, 709, 1284

Ekenbäck, A., Holmström, M., Wurz, P., Grießmeier, J.-M., Lammer, H., Selsis, F., & Penz, T. 2010, *ApJ*, 709, 670

Ehrenreich, D., Lecavelier des Etangs, A., Hébrard, G., Désert, J.-M., Vidal-Madjar, A., McConnell, J. C., Parkinson, C. D., Ballester, G. E., & Ferlet, R. 2008, *A&A*, 483, 933

Erkaev, N. V., Kulikov, Yu. N., Lammer, H., Selsis, F., Langmayr, D., Jaritz, G. F., & Biernat, H. K. 2005, *A&A*, 472, 329

France, K., Stocke, J. T., Yang, H., Linsky, J. L., Wolven, B. C., Froning, C. S., Green, J. C., & Osterman, S. N. 2010, *ApJ*, 712, 1277

García Muñoz, A. 2007, *Planet. Space Sci.*, 55, 1426

Guillot, T., Burrows, A., Hubbard, W. B., Lunine, J. I., & Saumon, D. 1996, *ApJ*, 459, L35

Guo, J. H. 2011, *ApJ*, 733, 98

Holmström, M., Ekenbäck, A., Selsis, F., Penz, T., Lammer, H., & Wurz, P. 2008, *Nature*, 451, 970

Konacki, M., Torres, G., Jha, S., & Sasselov, D. 2003, *Nature*, 421, 507

Koskinen, T. T., Yelle, R., Lavvas, P., & Lewis, N. K. 2010, *ApJ*, 723, 116

Lammer, H., Selsis, F., Ribas, I., Guinan, E. F., & Bauer, S. J. 2003, *ApJL*, 598, L121

Lammer, H., Odert, P., Leitzinger, M., Khodachenko, M. L., Panchenko, M., Kulikov, Yu. N., Zhang, T. L., Lichtenegger, H. I. M., Erkaev, N. V., Wuchterl, G., Micela, G., Penz, T., Biernat, H. K., Weingrill, J., Steller, M., Ottacher, H., Hasiba, J., & Hanslmeier, A. 2009, *A&A*, 506, 399

Lecavelier Des Etangs, A., Vidal-Madjar, A., McConnell, J. C., & Hébrard, G. 2004, *A&A*, 418, L1

Lecavelier Des Etangs, A., Ehrenreich, D., Vidal-Madjar, A., Ballester, G. E., Désert, J.-M., Ferlet, R., Hébrard, G., Sing, D. K., Tchakoumegni, K.-O., & Udry, S. 2010, *A&A*, 514, A72

Linsky, J. L., Yang, H. France, K., Froning, C. S., Green, J. C., Stocke, J. T., & Osterman, S. N. 2010, *ApJ*, 717, 1291

Llama, J., Wood, K., Jardine, M., Vidotto, A. A., Helling, Ch., Fossati, L., & Haswell, C. A. 2011, *MNRAS*, in press

Murray-Clay, R. A., Chiang, E. I., & Murray, N. 2009, *ApJ*, 693, 23

Penz, T., Erkaev, N. V., Kulikov, Yu. N., Langmayr, D., Lammer, H., Micela, G., Cecchi-Pestellini, C., Biernat, H. K., Selsis, F., Barge, P., Deleuil, M., & Léger, A. 2008, *Planet. Space Sci.*, 56, 1260

Shematovich, V. I. 2010, *Solar Sys. Res.*, 44, 96

Snellen, I. A. G., de Kok, R. J., de Mooij, E. J. W., & Albrecht, S. 2010, *Nature*, 465, 1049

Southworth, J. 2010, *MNRAS*, 408, 1689

Tian, F., Toon, O. B., Pavlov, A. A., & Sterck, H. D.e. 2005, *ApJ*, 621, 1049

Vidal-Madjar, A., Lecavelier Des Etangs, A., Désert, J. M., Ballester, G. E., Ferlet, R., Hébrard, G., & Mayor, M. 2003, *Nature*, 422, 143

Vidal-Madjar, A., Désert, J.-M., Lecavelier Des Etangs, A., Hébrard, G., Ballester, G. E., Ehrenreich, D., Ferlet, R., McConnell, J. C., Mayor, M., & Parkinson, C. D. 2004, *ApJ*, 604, L69

Vidal-Madjar, A., Lecavelier Des Etangs, A., Désert, J.-M., Ballester, G. E., Ferlet, R., Hébrard, G., & Mayor M. 2008, *ApJ*, 676, 57

Yelle, R. V. 2004, *Icarus*, 170, 167

Discussion

A. TRIAUD: Do you include in the calculations for mass loss the fact that orbits delay? This is the idea is that "Hot Jupiters" might have circularized from longer period eccentric orbits.

H. LAMMER: This was not taken into account. The difference in the results are smaller relative to the mass loss if one compares it to the uncertainties in the XUV flux or heating efficiencies.

From Interacting Binaries to Exoplanets: Essential Modeling Tools
Proceedings IAU Symposium No. 282, 2011
Mercedes T. Richards & Ivan Hubeny, eds.
© International Astronomical Union 2012
doi:10.1017/S1743921311028328

3D Models of Exoplanet Atmospheres

Ian Dobbs-Dixon

Department of Astronomy, University of Washington
Box 351580, Seattle, WA 98195
Sagan Postdoctoral Fellow
email: `ianmdd@gmail.com`

Abstract. Dynamical models of strongly irradiated gas-giant atmospheres exhibit a range of behavior, the nature of which depends on both the adopted parameters and the adopted numerical model. Discerning the correct choice of physical parameters and modeling philosophy can be difficult. Here, I present a series of wavelength-dependent transmission spectra for the giant planet HD209458b based on 3D radiative hydrodynamical models for a range of kinematic viscosities. While flow patterns and temperature distributions can vary significantly, disk-averaged phase curves mask much of this information. Transmission spectra, on the other hand, probe the day-night transition where advective contributions dominate and differences are often most pronounced. Transmission spectra illustrate noticeable changes, especially when comparing the differences between transmission spectra of eastern and western hemispheres, as might be seen during ingress and egress.

Keywords. hydrodynamics, radiative transfer, techniques: spectroscopic, eclipses

1. Introduction

Within the last decade, there has been an explosion in both transiting extrasolar planets and new observational techniques aimed at determining interior and atmospheric composition, dayside temperature, and efficiency of energy redistribution. In addition to primary transit measurements (Charbonneau *et al.* 2000), these tools include secondary eclipse (Deming *et al.* 2005), differential spectroscopy (Charbonneau *et al.* 2002), phase-curve monitoring (Harrington *et al.* 2006, Cowan *et al.* 2007), Doppler absorption spectroscopy (Snellen *et al.* 2010), and transmission spectra (Brown *et al.* 2002). These observations have motivated many groups to study the atmospheric dynamics of these planets and its role in both redistributing incident stellar energy throughout the atmosphere and (potentially) in influencing the overall evolution of the interior. Multiple groups are working on coupled radiation-hydrodynamical solutions with multiple assumptions and approaches. See Dobbs-Dixon *et al.* (2010) for a list of approaches to both the dynamical and radiative portions of the model.

One common feature of many simulations of the short period planets HD209458b and HD189733b are strong circumplanetary, eastward equatorial jets. Observations of HD189733b (Knutson *et al.* 2007, 2009; Agol *et al.* 2010) appear to have convincingly demonstrated the existence of equatorial super-rotation, showing that the hottest point in the phase curve appears slightly before secondary eclipse, as first predicted by Showman & Guillot (2002). However, as we are able to characterize more planets, we are finding a wide variety of behaviors. For example, observations of ν-Andromeda B (Crossfield *et al.* 2010) indicate a much larger phase shift of ≈ 89°, completely out of the range of current model predictions. In addition, models with varying viscosity (Dobbs-Dixon *et al.* 2010), perhaps due to magnetic drag, sub-grid turbulence, or shocks, indicate that the circumplanetary jet may not be as universal as once thought.

Unfortunately, much of the detailed spatial and temporal structure that is seen in numerical simulations is hidden in the necessarily hemispherically averaged phase curves (Cowan *et al.* 2008). Temperature differences across jets, latitudinal dependence, vortices, and other interesting sub-hemisphere scale phenomena largely disappear. However, one technique that may prove quite useful in this respect is transmission spectroscopy. Taken as the planet transits its host star, transit spectroscopy measures the absorption of stellar light by the upper limbs of the planetary atmosphere yielding a wavelength-dependent radius for the planet (Seager *et al.* 2000). Given the viewing geometry of the star, planet, and observer, such a measurement probes the meridians delineating the day and night hemispheres (the terminators). The variation in opacity with wavelength can cause the planet to vary in absorption radius by $\approx 5h$, where h is the atmospheric scale height, leading to depth variations on the order of $10 R_p h / R_*^2 \approx 0.1\%$ for $5h \approx 3500$ km, $R_p \approx R_J$, and $R_* \approx R_\odot$.

Dynamics play a crucial role in shaping the temperatures across the terminators both at the surface and at depth. It is here that one expects the largest deviations from radiative equilibrium. As high velocity jets advect energy across the terminator to the night-side, the flow will cool radiatively, thus one would expect the largest night-side temperatures to be closest to the terminator. Indeed, simulations show similar behavior, but the exact temperature distribution depends sensitively on the details of the flow structure and radiative and advective efficiencies. Transmission spectra taken both at mid-transit and during ingress and egress have great potential to help distinguish between models and adopted parameters. Several models presented in Dobbs-Dixon *et al.* (2010) show pronounced variability with the largest amplitudes at the terminator region. Targeting the terminator regions with multiple transmission spectral measurements may reveal spectral changes due to such dynamically driven weather.

Here, I utilize the 3D pressure-temperature profiles calculated using the 3D Navier-Stokes equations and a multi-channel flux-limited diffusion. Models differ from Dobbs-Dixon *et al.* (2010) in several important respects; I now allow for advective flow over the polar regions and the stellar energy deposition is now wavelength-dependent. Section 2 briefly describes the methods for calculating transmission spectra from multi-dimensional models. In Section 3, I present several diagnostics utilizing these transmission spectra that may be detectable with the next generation of instruments. I conclude in Section 4 with a discussion of the relevance of this work for compact, mass-transferring binaries.

2. Transit Spectra Calculations

To calculate the pressure and temperature throughout the atmosphere of HD209458b, we utilize a fully non-linear, coupled radiative hydrodynamical code. We solve the fully compressible Navier—Stokes equations throughout the 3D atmosphere together with coupled thermal and radiation energy equations. Direct stellar heating of the planet is taken into account through a spatial and wavelength-dependent source term. Local 3D radiative transfer and cooling are included through flux-limited diffusion. Wavelength-dependent opacities are taken from the atomic and molecular opacity calculations of Sharp *et al.* (2007). For mode detailed information on the dynamical model, see Dobbs-Dixon *et al.* (2010, 2011).

To determine the transmission spectra of HD209458b models, I calculate the wavelength-dependent absorption of stellar light traversing through the limb of the planet. This allows the determination of the effective radius of the planet and a fractional reduction of stellar flux F_*. Neglecting limb-darkening of the star, this can be expressed as $\left(\frac{F_{intransit}}{F_\star}\right)_\lambda = \frac{\int \left(1 - e^{\tau(b,\phi,\lambda)}\right) b \, db \, d\phi}{\pi R_\star^2}$, where $\tau(b, \alpha, \lambda)$ is the total optical depth along a given chord with impact parameter b and polar angle α, defined on the observed planetary disk during transit. The density and temperature needed to calculate τ at each location are interpolated from the values in the 3D models.

Several other groups have also explored the differences between transmission spectra calculated with 1D or 3D models. Fortney *et al.* (2010) and Burrows *et al.* (2010) perform similar calculations utilizing the 3D GCM models of Showman *et al.* (2009) and Rauscher *et al.* (2010), respectively. As mentioned above, the dynamical calculations differ amongst these models and ours, but the method for calculating the resulting spectra from the results is largely equivalent.

3. Calculated Transmission Spectra

In this Section, I present transmission spectra from the 3D models exploring the different flow structures among models with varying viscosity. While several features may influence the transmission spectra at precisions already observed, others may await the next generation of instruments. I highlight two methods for extracting this signal from actual data, exploiting the differences between transmission spectra taken during transit ingress (probing the eastern terminator) and transit egress (probing the western hemisphere).

Transmission spectra calculated from 3D models with varying viscosities are shown in Figure 1. The variations due to varying flow structures (see Figures 2 and 3 of Dobbs-Dixon *et al.* (2010) for an illustration of the differences) can be quite dramatic, including a transition from sub-sonic to supersonic wind speeds as the viscosity is lowered. Simulations S1, S2, S3, and S4 have viscosities of 10^{12}, 10^{10}, 10^8, and 10^4, respectively.

Short period, synchronously rotating, irradiated planets are not spherical due to a number of reasons including rotation, day-night temperature gradients, equator-pole temperature gradients, and the tidal potential of the host star. For a slow rotating planet such as HD209458b, the temperature gradients across the planet (yielding different scale-heights) are the largest factor in changing the planet's shape.

The first method we discuss is wavelength-dependent transit timing. As the western terminator has a smaller scale height and this part of the planet transits first, the ingress will be slightly delayed, while the eastern terminator is more extended causing the end of ingress to be delayed. Likewise, the egress will be delayed as well, so the overall shift

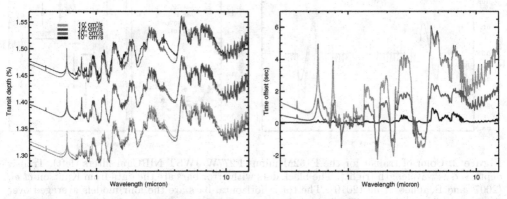

Figure 1. Difference in mid-transit transmission spectra among models with varying viscosity (left). Effective time offset versus wavelength (right). Black, red, green, and blue lines are from S1, S2, S3, and S4 simulations respectively.

will be a delay of the transit relative to the center of mass of the planet. At wavelengths with larger opacity, this asymmetry is stronger so the transit time delay is larger, while at wavelengths with smaller opacity, it is weaker; consequently, the central time of transit will appear to vary with wavelength if one fits the transit with a symmetric planet model. Figure 1 shows the effective transit time offset versus wavelength computed for a model with no limb-darkening for the star and by fitting the transit at each wavelength with a circular planet model (Mandel *et al.* 2002). At some wavelengths, the transit time is offset earlier, which is caused by a change in the opacity at certain wavelengths due to chemistry differences between the terminators. As the viscosity grows, the two hemispheres have smaller temperature differences, and hence the transit shape is more symmetric, causing a smaller time offset.

Another model-independent diagnostic for the planet asymmetry is the color-dependent transit shape. Figure 2 shows the difference in the shape of transit for two wave bands: 1.55-2.42 and 1.70-3.1 micron, corresponding to the JWST NIRCam F162M and F277W filters, respectively. To compute the shape difference, the depths of the transit in each band were divided by the maximum depth, and then subtracted from one another: $C = (D_1/D_{1,max} - D_2/D_{2,max})$ where $D_{1,2}$ is the depth of transit (in dimensionless units: i.e. the out-of-transit flux minus the in-transit flux divided by the out-of-transit flux), and the *max* subscript indicates the maximum depth of transit. The asymmetry in these light curves demonstrates that at different wavelengths the planet absorbs with a different asymmetric cross section. These filters have the advantage that they can be observed simultaneously with JWST using the dichroic and the first band has a low water opacity, while the second has a higher water opacity for this particular model.

3.1. *Comparison to observations*

As a check on the total depth of transit, we have compared the transit depth to observations by Knutson *et al.* (2008) and Beaulieu *et al.* (2010). These observations cover a wavelength from 0.3 to 10 microns, and thus provide an important validation test of these numerical models. We exclude the bands with strong sodium and potassium absorption as there is evidence that potassium is depleted (which is not accounted for in our opacity tables), while the sodium absorption occurs quite high in the atmosphere, and

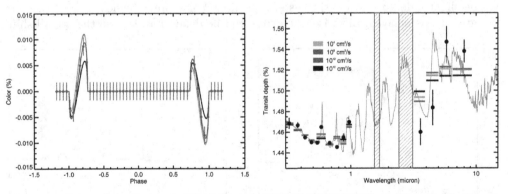

Figure 2. Color of transit for the F162M minus F277W JWST NIRCam filters (left). Transit depth versus wavelength (right). The black dots with error bars are the data from Knutson *et al.* (2007) and Beaulieu *et al.* (2010). The thick horizontal bars are the four models averaged over the measured wave bands. The light red curve is the 10^{10} cm^2 s^{-1} model binned by a factor of 10 in wavelength; this is over-plotted as it had the best fit of all four models (although not statistically significant)

thus cannot be handled in our models. We have averaged the models over the observed bands, and have varied the value of the inner radius of the simulation zone to obtain the best agreement with the observed transit depths. This is not as accurate as carrying out additional simulations in which the model region is varied; however, this would be much more computationally expensive. Furthermore, the assumption of constant **g** in the simulations of Dobbs-Dixon *et al.* (2010) and other groups allows for such a shift as only the curvature terms are affected.

There are twelve wave bands that we use, and with one free parameter, gives eleven degrees of freedom for each model. Figure 2 shows the comparison of the models to the data. We find best-fit chi-squares of 23.9, 24.1, 23.6, and 28.3 for the models with viscosities of $10^4, 10^8, 10^{10}$ and 10^{12} cm^2 s^{-1}, respectively. Qualitatively, the overall agreement of the models with the data is quite good: (1) the transit depth is weakly dependent on the viscosity; (2) there are no discrepancies between the data and model greater than $2 - \sigma$; (3) the observed transit spectrum with wavelength shows the expected features due to water and Rayleigh scattering. In general, the success of this model is comparable to transmission spectra calculated from other models which utilize GCM simulations (Fortney *et al.* 2010, Burrows *et al.* 2010).

However, in detail there are significant discrepancies; in particular, the observed IRAC transit depths appear to vary more strongly with wavelength than the model predicts. This is reflected in the larger chi-square of the fits, of which two-thirds is due to the infrared discrepancies. The fit to the infrared data obtained by Beaulieu *et al.* (2010) has an extremely good chi-square; however, their model was one-dimensional, and allowed the abundances and temperature-pressure profile to float, so it is not surprising that they obtain a good fit with so many degrees of freedom. Another possibility is that systematic errors still exist in the IRAC data reduction. For example, for the transiting planet HD 189733, different groups have obtained markedly different transit depths at infrared wavelengths using the same IRAC data sets; consequently, there may be some remaining systematic error present in the data (Desert *et al.* 2009). The final possibility is that there is still physics that are not included in our models which are causing the discrepancies; for example, varied chemical abundances, non-equilibrium chemistry, and magnetic drag (yielding non-isotropic viscosities) have not been included in these models.

4. Discussion

In these proceedings, I have presented theoretical transmission spectra of HD209458b calculated from full 3D radiative-hydrodynamical models. In general, there is good agreement with observational spectra though some discrepancies remain. I have highlighted 2 new observational techniques for deducing the strength of a possible circumplanetary equatorial jet. These include wavelength-dependent transit timing and color-dependent transit shapes. Both methods may be applicable to the next generation of instruments.

Simulations of highly irradiated atmospheres of exoplanets are relevant for the binary community at large. In particular, the processes occurring on the surfaces of lower-mass, shorter-period binaries will likely resemble the results presented here. Tidal locking in these systems is expected to occur rapidly, resulting in permanent day-night temperature differences. The pressure gradient associated with this will drive strong winds altering both observables (as discussed here) and potentially mass-transfer processes. Although the subject has been studied to some extent, much work remains before a coherent picture of the atmosphere dynamics in binaries emerges.

References

Agol, E., Cowan, N. B., *et al.*, 2010, ApJ, 721, 1861
Beaulieu, J. P., Kipping, D. M., Batista, V., *et al.*, 2010, *MNRAS*, 409, 963
Brown, T. M., Libbrecht, K. G., & Charbonneau, D. 2002, *PASP*, 114, 826
Burkert, A., Lin, D. N. C., *et al.*, 2005 *ApJ*, 618, 512
Burrows, A., Rauscher, E., Spiegel, D. S., & Menou, K. 2010, *ApJ*, 719, 341
Charbonneau, D., Brown, T. M., Latham, D. W., & Mayor, M. 2000, *ApJ*, 529, L45
Charbonneau, D., Brown, T. M., Noyes, R. W., & Gilliland, R. L. 2002, *ApJ*, 568, 377
Cho, J. Y. K., Menou, K., Hansen, B. M. S., & Seager, S. 2003, *ApJ*, 587, L117
Cho, J. Y. K., Menou, K., Hansen, B. M. S., & Seager, S. 2008, *ApJ*, 675, 817
Cooper, C. S. & Showman, A. P. 2005, *ApJ*, 629, L45
Cooper, C. S. & Showman, A. P. 2006, *ApJ*, 649, 1048
Cowan, N. B. & Agol, E. 2008, *ArXiv e-prints*
Cowan, N. B., Agol, E., & Charbonneau, D. 2007, *MNRAS*, 379, 641
Crossfield, I. J. M., Hansen, B. M. S., Harrington, J., *et al.*, 2010, *ApJ*, 723, 1436
Deming, D., Seager, S., Richardson, L. J., & Harrington, J. 2005, *Nature*, 434, 740
Desert, J.-M., Lecavelier des Etangs, A., H ebrard, G., *et al.*, 2009, *ApJ*, 699, 478
Dobbs-Dixon, I., Agol, E., & Burrows, A. 2011, *In Prep.*
Dobbs-Dixon, I., Cumming, A., & Lin, D. N. C. 2010, *ApJ*, 710, 1395
Dobbs-Dixon, I. & Lin, D. N. C. 2008, *ApJ*, 673, 513
Fortney, J. J., Shabram, M., Showman, A. P., *et al.*, 2010, *ApJ*, 709, 1396
Harrington, J., Hansen, B. M., Luszcz, S. H., *et al.*, 2006, *Science*, 314, 623
Knutson, H. A., *et al.*, 2008, *ApJ*, 673, 526
Knutson, H. A., Charbonneau, D., Allen, L. E., *et al.*, 2007, *Nature*, 447, 183
Knutson, H. A., Charbonneau, D., Cowan, N. B., *et al.*, 2009, *ApJ*, 690, 822
Langton, J. & Laughlin, G. 2007, *ApJ*, 657, L113
Langton, J. & Laughlin, G. 2008, *ApJ*, 674, 1106
Mandel, K. & Agol, E. 2002, *ApJ*, 580, L171
Menou, K. & Rauscher, E. 2009, *ApJ*, 700, 887
Rauscher, E. & Menou, K. 2010, *ApJ*, 714, 1334
Rauscher, E., Menou, K., Cho, J. Y.-K., Seager, S., & Hansen, B. M. S. 2008, *ApJ*, 681, 1646
Seager, S. & Sasselov, D. D. 2000, *ApJ*, 537, 916
Sharp, C. M. & Burrows, A. 2007, *ApJS*, 168, 140
Showman, A. P., Cooper, C. S., Fortney, J. J., & Marley, M. S. 2008, *ApJ*, 682, 559
Showman, A. P., Fortney, J. J., Lian, Y., *et al.*, 2009, *ApJ*, 699, 564
Showman, A. P. & Guillot, T. 2002, *AA*, 385, 166
Snellen, I. A. G., de Kok, R. J., de Mooij, E. J. W., & Albrecht, S. 2010, *Nature*, 465, 1049

Discussion

E. DEVINNEY: About the non-axis symmetric flows, are there for tidally-locked planets?

I. DOBBS-DIXON: Yes, all the models I have been discussing have rotating rates that are tidally-locked to their orbital periods.

E. GUINAN: GJ 581 has two super-Earth-type planets in or near the dM star's HZ. Have you carried out modeling of them? The circulation of the atmosphere from the hot (substellar) side to the cooler dark side would be important to habitability.

I. DOBBS-DIXON: No, I have not pushed my models down to Earth-sized planets. There are a number of groups that are doing this, and indeed dynamics will play an important role in determining habitability.

From *Interacting Binaries to Exoplanets: Essential Modeling Tools*
Proceedings IAU Symposium No. 282, 2011
Mercedes T. Richards & Ivan Hubeny, eds.

© International Astronomical Union 2012
doi:10.1017/S174392131102833X

Dynamical Stability and Habitability of Extra-Solar Planets

Elke Pilat-Lohinger

Institute of Astronomy, University of Vienna,
Türkenschanzstrasse 17, A-1180 Vienna, Austria
email: elke.pilat-lohinger@univie.ac.at

Abstract. Observations of about 60 binary star systems hosting exoplanets indicate the necessity of stability studies of planetary motion in such multi-stellar systems. For wide binary systems with separations between hundreds and thousands of AU, the results from single-star systems may be applicable but, in tight double stars systems, we have to take the stellar interactions into account which influences the planetary motion significantly.

This review discusses the different types of planetary motion in double stars and the stability of the planets for different binary configurations. An application to the most famous tight binary system (γ Cephei) is also shown. Finally, we analyze the habitability from the dynamical point of view in such systems, where we discuss the motion of terrestrial-like planets in the so-called habitable zone.

Keywords. Binaries: general, planetary systems, n-body simulations, numerical

1. Introduction

Stability studies of planetary motion in binary star systems are very important, since we expect an increasing number of detected planets in multi-stellar systems in the future – due to the fact that most of the stars in the solar neighborhood form double or multiple star systems (see e.g. Duquenor & Mayor 1991). From the dynamical point of view, we distinguish 3 types of motion in such systems (Dvorak 1984):

(*i*) *S-type motion*, where the planet moves around one stellar component;

(*ii*) *P-type motion*, where the planet surrounds both stars in a very distant orbit; and

(*iii*) *L-type motion*, where the planet moves in the same orbit as the secondary. They are locked in 1:1 mean motion resonance, which makes the system stable. But, the stability is limited to certain mass-ratios of the two stars: $\mu = m_2/(m_1 + m_2) < 1/26$. Therefore, this motion is not so interesting for double stars.

The literature shows that long before the first planet in a binary system was discovered, astronomers, working in Dynamical Astronomy, carried out theoretical and numerical stability studies for the different types of motion (see e.g. Dvorak 1984, 1986; Rabl & Dvorak 1988; Dvorak *et al.* 1989). The discovery of planets in such systems convinced many research groups to examine the planetary formation and evolution in binary star systems, either in general or for selected ones (see e.g. Holman *et al.* 1997; Holman & Wiegert 1999; Ford *et al.* 2000; Pilat-Lohinger & Dvorak 2002; Pilat-Lohinger *et al.* 2003; Dvorak *et al.* 2003a,b; Thébault *et al.* 2010; Musielak *et al.* 2005; Haghighipour 2006; Raghavan *et al.* 2006; Cuntz *et al.* 2007; Kley & Nelson 2008; Paardekooper *et al.* 2008; Takeda *et al.* 2008; Saleh & Rasio 2009; Marzari *et al.* 2010; Haghighipour *et al.* 2010; and many others).

Most of the planets detected so far are in S-type (or circumprimary) motion, so that we will mainly discuss this type of motion in this paper. Although, planets moving in

P-type (or circumbinary) orbits have been detected during the last few months, we learned during this symposium that the existence of these planets is still quite questionable. Dynamical studies of P-type motion by Dvorak *et al.* (1989), Holman & Wiegert (1999), Pilat-Lohinger *et al.* (2003) have shown that a planetary orbit surrounding both stars is only stable for distances (from the mass-center) larger than twice the separation of the two stars. In the case of eccentric motion of the binary, the planet's distance to the mass-center has to be increased in order to be stable.

In the following sections, we will discuss the stability of circumprimary motion in general, taking into account the eccentricity of the binary and of the planet. Moreover, we show the influence of the planet's mass on the stable region and apply the study to the most famous tight binary star system that hosts a planet (i.e. γ Cephei).

2. Dynamical model and computations

Most of the studies mentioned above used for the numerical simulations the elliptic restricted three body problem (ER3BP) that describes the motion of a mass-less body in the gravitational field of two massive bodies, which move in elliptic orbits (Keplerian motion) around their center of mass, without being influenced by the mass-less body. Numerical studies by Rabl & Dvorak (1988; hereafter, RD) and Holman & Wiegert (1999; hereafter, HW) determined the stable region as a function of the binary's mass-ratio and eccentricity, and the motion of the planet was circular The numerical investigation by Pilat-Lohinger & Dvorak (2002; hereafter, PLD) also analyzed the influence of the planet's eccentricity. In these three investigations, the stable regions of planetary motion have been determined in a similar way. The planet (which is considered as mass-less body) moves around star m_1. The distance between m_1 and the second star (m_2) is set to 1, and the eccentricity of the binary varies between 0 and 0.9 with a step of 0.1. There are two starting positions for m_2: the peri-center and the apo-center.

The planet's semi-major axis is taken between 0.1 and 0.9. For each orbit, at least four starting positions were used (i.e. mean anomaly $= 0^o$, 90°, 180°, 270°) and the planet's eccentricity was set to zero in RD and HW, and was varied between 0 and 0.5 (or 0.9 in certain cases) with a step of 0.1 for all mass-ratios in PLD.

Moreover, in PLD, the orbital behavior was determined by means of the Fast Lyapunov Indicator (FLI; see Froeschle *et al.* 1997), which is quite a fast tool to distinguish between regular and chaotic motion. Chaotic orbits can be found very quickly because of the exponential growth of the tangent vector in the chaotic region. For most chaotic orbits, only a small number of primary revolutions is needed to determine the orbital behavior. In order to distinguish between stable and chaotic motion, we defined a critical value for the FLIs depending on the computation time. In the general stability study of S-type motion, the FLIs were computed for 1000 periods of the binary.† A comparison of the results of RD, HW and PLD show them in good agreement. Minor variations are caused by the different methods used to determine the stable region‡. Table 1 shows the border of stable motion for circular S-type orbits in binary star systems for different mass-ratios and different eccentricities of the two stars. This border is defined by the largest distance of the planet to its host-star m_1, where stable motion has been found for all initial positions of the planet.

† Even if the computation time seems to be quite short, one has to take into account that the results are valid for a much longer time due to the application of the FLI, and test-computations of three selected mass-ratios over a longer time (of $10^4, 10^5$ and 10^6 primary periods) did not change the result significantly.

‡ In some cases the FLI results indicates a slightly larger stable region due to the fact that only 4 starting positions were used whereas HW used 8.

Table 1. Stable zone (in dimensionless units) of S-type motion for all computed mass-ratios and eccentricities of the binary. The given size for each μ, e_{binary} pair is the lower value of the studies by HW and PLD.

e_{binary}	\multicolumn{9}{c}{mass-ratio (μ)}								
	0.1	0.2	0.3	0.4	0.5	0.6	0.7	0.8	0.9
0.0	0.45	0.38	0.37	0.30	0.26	0.23	0.20	0.16	0.13
0.1	0.37	0.32	0.29	0.27	0.24	0.20	0.18	0.15	0.11
0.2	0.32	0.27	0.25	0.22	0.19	0.18	0.16	0.13	0.10
0.3	0.28	0.24	0.21	0.18	0.16	0.15	0.13	0.11	0.09
0.4	0.21	0.20	0.18	0.16	0.15	0.12	0.11	0.10	0.07
0.5	0.17	0.16	0.13	0.12	0.12	0.09	0.09	0.07	0.06
0.6	0.13	0.12	0.11	0.10	0.08	0.08	0.07	0.06	0.045
0.7	0.09	0.08	0.07	0.07	0.05	0.05	0.05	0.045	0.035
0.8	0.05	0.05	0.04	0.04	0.03	0.035	0.03	0.025	0.02

As already mentioned, the study by PLD examined also eccentric planetary motion in the different binary configurations and showed that the reduction of the stable zone due to an increase of the binary's eccentricity is certainly stronger than the influence of the planet's eccentricity. Fig. 1 shows a summary of this study for $\mu = 0.2$, where we see for each (e_{Binary}, e_{Planet}) pair on the (x,y) plane the respective extension of the stable zone (z-axis), which is defined by the semi-major axis of the last stable orbit (corresponding to the largest distance of the planet to its host-star). The Gray plane represents the limiting plane for stable motion. Similar 3-D plots for all mass-ratios (from 0.1 to 0.9) as well as a detailed discussion of the results are given in PLD.

Even if the size of the stable region does not show a strong dependence on the eccentricity of the planet, it is not negligible, especially if a planet is close to the border of chaotic motion and moves in a highly eccentric orbit. In a recent study, where we examined the influence of the planetary mass on the size of the stable region, we found out that eccentric motion of massive planets would shrink the stable zone significantly (a paper thereto is in preparation).

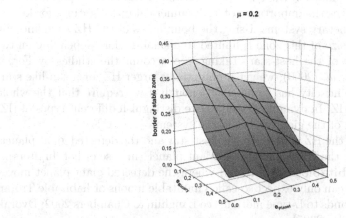

Figure 1. The size of the stable zone of S-type motion in a binary with mass-ratio $\mu = 0.2$ depending on the eccentricity of the binary (x-axis) and of the planet (y-axis) It is clearly seen that the variation of e_{Binary} influences the extension of the stable zone (z-axis) stronger than the variation of e_{Planet}.

In the next section, we show the application of the results for $\mu = 0.2$ given in Fig. 1 to the binary γ Cephei, where a giant planet has been discovered by Hatzes et al. (2003).

3. Planetary motion in the binary γ Cephei

γ Cephei is one of the most interesting double star systems that hosts a planet. It is nearly 14 pc away from our Solar System and consists of a K1 IV star (of 1.4 – 1.6 solar masses) and a M4 V star (of 0.4 solar masses). Thus, the mass-ratio of this system is about 0.2. The detected planet of 1.76 – 1.85 Jupiter-masses orbits the K1 IV star at a distance of about 2 AU.

Applying the general stability study, the border of stable planetary motion is between 3.6 and 3.8 AU, which shows that the detected planet is obviously in the stable region in case of circular planetary motion. If we increase the planet's eccentricity up to 0.5, the stable region will shrink to less than 3 AU. Taking into account different masses for the planet, the reduction of the stable region strongly depends on the mass and the eccentricity of the planet: for a 1 Jupiter-mass (=JM) planet, the stable region is reduced by 1.6 AU; for a 3 JM planet, the reduction is 1.8 AU and for a 5 JM planet, 2.2 AU if we increase the planet's eccentricity from 0 to 0.5. This shows clearly, that a massive planet moving in a high eccentric orbit can easily be dropped out of the stable region determined for circular planetary motion by HW or RD.

4. Binary star systems and habitable planets?

The habitable zone (HZ) is the region around a star, where liquid water is stable on the surface of an Earth-like planet (see Kasting et al. 1993). Another assumption for such a planet is the existence of an appropriate planetary atmosphere. The study of habitability is certainly an interdisciplinary venture including astrophysical, biological, geophysical, and chemical studies. From the astrophysical point of view, studies of the stellar luminosity and its influence on the distance of the HZ as well as the planet's mass (to maintain an atmosphere) and planetary composition (assuming a terrestrial planet) are important contributions to the science of habitability. It is well known, that the evolution of a biosphere is a process over a long time, therefore, it is obvious that long-term orbital stability of a planet in the HZ represents one of the basic requirements for habitability. This emphasizes the importance of such numerical investigations for known and future extra-solar planetary systems. Using the boundaries of the HZ as defined by Kasting et al. (1993), the size of this zone is limited to a small region, depending on the spectral type and the age of the host-star. Taking into account the studies by Forget et al. (1997) or Mischna et al. (2000), we get a potentially larger HZ for a sun-like star. However, the planet's eccentricity has to be small enough if we require that the whole orbit of a planet is in the HZ. In dynamical studies, we distinguish different types of HZ, depending on the location of the giant planet in the system:

(1) The **inner HZ**, where the HZ is between the host-star and the detected giant planet.

(2) The **outer HZ**, where the HZ is outside the giant planet (in case of hot-Jupiters).

(3) The **giant-planet habitable zone (GHZ)** where the detected giant planet moves in the HZ. In this case, we can only expect so-called habitable moons or habitable Trojan planets (the latter corresponds to L-type motion; see Laughlin & Chambers 2002; Dvorak et al. 2004; or Erdi & Sandor 2005).

In the case of a binary star system, the situation is more difficult since we have to study as well the influence of the second star on the planetary motion. Considering a second planet in the binary γ Cephei, which moves between the host-star and the detected

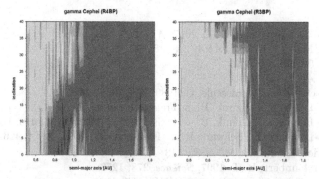

Figure 2. FLI stability maps for a fictitious planet in the vicinity of γ Cephei: left panel shows the result in the restricted 4 body problem (R4BP) (i.e. $m_1 + m_2 +$ detected planet + fictitious planet) and right panel shows the result in the restricted 3 body problem (R3BP) (i.e. $m_1 +$ detected planet + fictitious planet). The dark region shows the chaotic zone and the white area the stable one.

giant planet, the influence of the M4 V star at about 20 AU can be visualized by doing calculations with (left panel of Fig. 2) and without (right panel of Fig. 2) this star. A comparison of the two results shows significant differences. The presence of the perturbing star (see Fig. 2.a) decreases the stable region (i.e. the faint region in the panels) and shows an arc-like chaotic structure with a stable island around 1 AU (which corresponds to the 3:1 mean motion resonance). A first study about this significant difference is given in Pilat-Lohinger (2005), where a variation of the semi-major axis of the detected giant planet shows the following:

When the giant planet is quite close to the host-star (e.g. around 1.3 AU) the faint region in Fig. 2 is mainly perturbed by mean motion resonances with respect to the giant planet. If the giant planet is shifted towards the M4 V star, the curved chaotic structure appears due to secular perturbations, because this star causes a precession of the perihelion of the giant planet. (As all massive bodies were placed in the same plane a precession of the ascending node cannot be modeled.)

Since the host-star γ Cephei is already a sub-giant and the detected giant planet moves in the HZ, we used for the our numerical study of habitability from the dynamical point of view another real binary star system HD41004 AB, where a giant planet has been discovered by Zucker *et al.* (2003 and 2004). Using various configurations (real and fictitious) of this binary and the detected giant planet, and adding a fictitious terrestrial planet between HD41004 A and the giant planet, we have seen that certain configuration would allow the existence of a habitable planet, at least from the dynamical point of view – i.e. long-term stability of the whole system and low-eccentricity motion of the planet in the HZ. A first result is given in Pilat-Lohinger & Funk (2010) and a more detailed study is in preparation.

Of course this result is just "a small part of a big puzzle," but it is well known that the long-term stability of a planetary system is one of the basic requirements for habitability. Since our dynamical studies have shown that potential habitable planets could also exist in binary star systems, the study of habitability in multi-star systems is certainly an interesting topic for future research. Especially, as we know that most of the stars in the solar neighborhood are in double or multiple star systems.

Acknowledgements

EP-L acknowledges the support by the Austrian Science Fund (FWF) – project no. P22603-N16.

References

Cuntz, M., Eberle J., & Musielak, Z. E. 2007, *ApJ*, 669, 105

Duquennoy, A. & Mayor, M. 1991, *A&A*, 248, 485

Dvorak, R. 1984, *CMDA*, 34, 369

Dvorak, R. 1986, *A&A*, 167, 379

Dvorak, R., Froeschlé, Ch., & Froeschlé, C. 1989, *A&A*, 226, 335

Dvorak, R., Pilat-Lohinger, E., Funk, B., & Freistetter, F. 2003, *A&A*, 398, L1

Dvorak, R., Pilat-Lohinger, E., Funk, B., & Freistetter, F. 2003, *A&A*, 410, L13

Dvorak, R., Pilat-Lohinger, E., Schwarz, R., & Freistetter, F. 2004, *A&A*, 426, L37

Érdi, B. and Sándor, Zs. 2005, *CMDA*, 92, 113

Forget, F. & Pierrehumbert, R. T. 1997, *Science*, 278, 1273

Froeschlé, C., Lega, E., & Gonczi, R. 1997, *CMDA*, 67, 41

Haghighipour, N. 2006, *ApJ*, 644, 543

Haghighipour, N., Dvorak, R., & Pilat-Lohinger, E. 2010, *Planets in Binary Star Systems*, ed. Haghighipour, N., ASS Library, 366, 285

Hatzes, A. P., Cochran, W. D., Endl, M., & McArthur, B., et al. 2003, *ApJ*, 599, 1383

Holman, M. J., Touma, J., & Tremaine, S. 1997, *Nature*, 386, 254

Holman, M. J. & Wiegert P. A. 1999, *AJ*, 117, 621

Kasting, J. F., Whitmire D. P., & Reynolds, R. T. 1993, *Icarus*, 101, 108

Kley, W. & Nelson, R. P. 2008, *A&A*, 487, 671

Laughlin, G. & Chambers, J. E. 2002, *AJ*, 124, 592

Marzari, F., Thbauld, P., Kortenkanmp, S., & Scholl, H. 2010, *Planets in Binary Star Systems* ed. Haghighipour, ASS Library 366, 165

Mischna, M. A., Kasting, J. F., Pavlov, A., & Freedman, R. 2000, *Icarus*, 145, 546

Musielak, Z. E., Cuntz, M., Marshall, E. A., & Stuit, T. D. 2005, *A&A*, 434, 355

Paardekooper, S. J., Thbault, P., & Mellema, G. 2008, *MNRAS*, 386, 973

Pilat-Lohinger, E. & Dvorak, R. 2002, *CMDA*, 82, 143

Pilat-Lohinger, E., Funk, B., & Dvorak, R. 2003, *A&A*, 400, 1085

Pilat-Lohinger, E. 2005, *IAU Coll.* 197, eds. Knezevic & Milani, Cambridge Univ. Press, 71

Pilat-Lohinger, E. & Dvorak, R. 2007, *Extrasolar Planets*, ed. Dvorak, Wiley-VCH, 179

Pilat-Lohinger, E. & Funk, B. 2010, *LNP*, 790, p 481

Rabl, G. & Dvorak, R. 1988, *A&A*, 191, 385

Raghavan, D., Henry, T. J., Mason, B. D., Subasavage, J. P., Jao, W.-C., Beaulieu, T. D., & Hambly, N. C. 2006, *ApJ*, 646, 523

Saleh, L. A. & Rasio, F. A. 2009, *ApJ*, 694, 1566

Takeda, G., Kita, R., & Rasio, F. A. 2008, *ApJ*, 683, 549

Thbault, P., Marzari, F., & Augereau, J. C. 2010, *A&A*, 524, 13

Zucker, S., Mazeh, T., Santos, N. C., Udry, S., & Mayor, M. 2003, *A&A*, 404, 775

Zucker, S., Mazeh, T., Santos, N. C., Udry, S., & Mayor, M. 2004, *A&A*, 426, 695

Discussion

S. ZUCKER: In the case of HD41004 AB, we don't know the details of the wide orbit, which may be eccentric.

E. PILAT-LOHINGER: Yes, the eccentricity of HD41004 AB is not known, but with our stability analysis, it was possible to make some restrictions for the eccentricity of the binary. Moreover, we studied for the binary and the detected giant planet different fictitious configurations to examine if there are configurations that could host a habitable planet.

From Interacting Binaries to Exoplanets: Essential Modeling Tools
Proceedings IAU Symposium No. 282, 2011
Mercedes T. Richards & Ivan Hubeny, eds.
© International Astronomical Union 2012
doi:10.1017/S1743921311028341

Gas Dynamic Simulation of the Star-Planet Interaction using a Binary Star Model

D. E. Ionov, D. V. Bisikalo, P. V. Kaygorodov, V. I. Shematovich

Institute of Astronomy of the Russian Acad. of Sci., Pyatnitskaya 48, Moscow, Russia

Abstract. We have performed numerical simulations of the interaction between a "hot Jupiter" planet and gas of the stellar wind using a numerical code developed for investigations of binary stars. With this code, we have modeled the structure of the gaseous flow in the system HD 209458. The results have been used to explain observations of this system performed with the COS instrument on-board the HST.

Keywords. planetary systems, binaries, hydrodynamics

The observations of a "hot Jupiter" planet HD 209458b were carried out using the COS spectrograph mounted aboard HST (Linsky 2010). The results showed that the investigated spectral absorption lines (of CII, Si III) obtained as the difference of the stellar spectra in the transition and out of it have a non-trivial double-peaked shape. The distance between the peaks is about 20 km/s and for the carbon line it is clearly seen that the peaks are asymmetric.

To reveal physical processes that can lead to formation of such spectral lines, we performed gas dynamic simulations of the interaction between a planet and stellar wind. Since the planet is pretty close to the host star ($10.1\ R_{sun}$), the orbital velocity of the planet is so high (V = 143 km/s) that even in the case of a hot stellar wind with temperatures around 10^5 K the motion of the planet is supersonic. It is known that if a gravitating body or a body with an atmosphere moves with a supersonic velocity, a bow-shock must occur. The matter of the stellar wind mixes with the matter of the atmosphere and forms two streams moving in different directions from the head-on collision point. This motion can lead to occurrence of two peaks in the observed spectral lines.

The model we have used for our numerical experiments is similar to that described in (Bisikalo *et al.* 2003a). The flow structure in this model is described by a system of 3D equations of gravitational gas dynamics including non-adiabatic processes of the radiative heating and cooling. To obtain the numerical solution of this system, we have used the Roe-Osher method (Boyarchuk *et al.* 2002; Roe, 1986; Chakravarthy & Osher, 1985; Bisikalo *et al.* 2003b) adapted to perform simulations using multiprocessor computers. We carried out our calculations in a cylindrical coordinate system with the origin at the center of mass of the planet. The size of the computational domain has been limited to 5 planet radii. The stellar wind was simulated through setting a constant inflow on the outer boundary of the computational domain. The matter of the stellar wind is considered as a neutral monatomic gas whose parameters are typical for the Solar wind, i.e.: $\rho = 1,4 \cdot 10^4$ cm^{-3}, $T = 7,3 \cdot 10^5\ K$, $V = 100$ km/s (Withbroe 1988). The planet atmosphere is set to isothermal, with $T_{pl} = 5000\ K$. Its number density has been determined using the barometric formula. The number density at the radius $R_{pl} = 1.4 R_{Jup}$ has been set as $2 \cdot 10^{10}$ cm^{-3}.

The shape of the line is determined by projections of the velocities of matter behind the bow shock wave onto the line of sight. In Fig. 1 (right panel), we show the synthetic absorption line profile calculated using our gas dynamic results. It is seen that the line has two clearly distinguishable asymmetric peaks. The synthetic line profile has the same features as the observed one. Parameters of the line (peak position, their relative intensity and even their number) strongly depend on the accepted parameters of the modeled stellar wind and atmosphere, and can vary in a wide range.

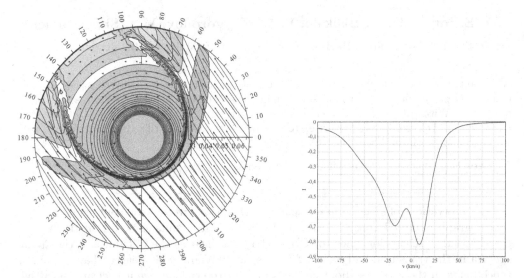

Figure 1. Density distribution and velocity vectors in the equatorial plane (left panel) and synthetic spectral line profile(right panel).

The presented model allows us to conclude that when analyzing observational properties of the atmosphere of a planet, one must take into account gas dynamical processes caused by the interaction of the atmosphere and stellar wind. It is important to note that the considered model allows one to explain the existence of absorption lines of ions having high ionization potentials.

Acknowledgements

This work was supported by the Basic Research Program of the Presidium of the Russian Academy of Sciences, Russian Foundation for Basic Research (projects 09-02-00064, 09-02-00993, 11-02-00076, 11-02-01248), and the Federal Targeted Program "Science and Science Education for Innovation in Russia 2009-2013."

References

Bisikalo, D. V., Boyarchuk, A. A., Kaigorodov, P. V., & Kuznetsov, O. A. 2003, *Astronomy Reports*, 47, 10, 809

Bisikalo, D. V., Boyarchuk, A. A., Kaigorodov, P. V., & Kuznetsov, O. A. 2003, *Matematicheskoye modelirovanie: Problemi i rezultati*, 71

Boyarchuk, A. A., Bisikalo, D. V., Kuznetsov, O. A., & Chechetkin, V. M. 2002, *Mass Transfer in Close Binary Stars* (London: Taylor & Francis)

Chakravarth, S. R. & Osher, S. 1985, *Proceedings of the 23rd Aerospace Sci. Meeting*, p. 363

Linsky, J. L., Hao, Y., France, K., Froning, C. S., Green, J. C., Stocke, J. T., & Osterman, S. N. 2010, *ApJ*, 717, 1291

Roe, P. L. 1986, *Ann. Rev. Fluid Mech.*, 18, 337

Withbroe, G. L. 1988, *ApJ*, 325, 442

From Interacting Binaries to Exoplanets: Essential Modeling Tools
Proceedings IAU Symposium No. 282, 2011
Mercedes T. Richards & Ivan Hubeny, eds.
© International Astronomical Union 2012
doi:10.1017/S1743921311028353

Gas Dynamic Simulations of Inner Regions of Protoplanetary Disks in Young Binary Stars

A. M. Fateeva*, D. V. Bisikalo, P. V. Kaygorodov, A. Y. Sytov

Institute of Astronomy of the Russian Acad. of Sci., 48 Pyatnitskaya, Moscow, Russia
*email: fateeva@inasan.ru

Keywords. protoplanetary disk, accretion, gap, binary stars, T Tauri stars

We have carried out 2D and 3D numerical simulations (Kaigorodov *et al.* 2010, Fateeva *et al.* 2011, Sytov *et al.* 2011) of accretion processes in binary T Tauri stars (TTSs) DQ Tau, UZ Tau E, V4046 Sgr, GW Ori, RoXs 42C using a finite-difference Roe-Osher-Einfeld TVD scheme. The morphology of the flow pattern for UZ Tau E is shown in Fig. 1 (left panel). The flow structure includes accretion disks surrounding the components, bow-shocks in front of both the components, a shock wave ("bridge") between the circumstellar accretion disks and a gap containing rarefied gas in the inner part of the protoplanetary disk.

The performed simulations show that the radii of the gaps which formed due to the bow-shocks fit the observations better than the radii calculated using positions of the Linblad resonances according to Artymowicz & Lubow (1994). For systems with circular orbits, the calculated gap radius is ∼ 3A (A – semi-major axis of the system) and for those with elliptic orbits it is ∼ 3.2 – 3.3A. Thus, the bow-shocks govern the size and shape of the gap in young binary systems.

Analysis of the fluxes demonstrates that the re-distribution of the angular momentum in the envelope due to the bow-shocks leads to occurrence of two flows propagating from the inner edge of the protoplanetary disk to the components (see Fig. 1, right panel). Let us consider streams near the less massive star (secondary). The matter in the gap splits into two streams at the head-on collision point when passing through the bow-shock. The first portion of matter (stream A in Fig. 1, right panel) loses its angular momentum at the shock and starts to move toward the circumstellar accretion disk forming a spiral flow. The second portion of matter

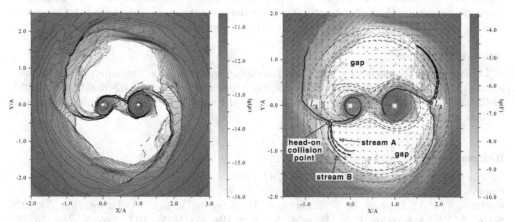

Figure 1. Results of 3D simulations for UZ Tau E. Left panel: the density distribution in the equatorial plane of the system is presented. Density isolines are also depicted. Right panel: the distribution of the matter flux and the velocity field in the equatorial plane of the system are presented. The flow lines corresponding to the main matter streams in the system are also shown by bold solid lines. The Roche equipotentials are depicted by the dashed line.

(stream B) moves from the head-on collision point along the bow-shock to the protoplanetary disk carrying out the excess of the angular momentum. The gap size depends on the power of this flow. The same flow structure exists in the vicinity of the primary component.

The presented analysis of the streams allows us to make conclusions on the rates of accretion onto each component of the system. The secondary star moves faster than the primary one. It means that gas loses more angular momentum at the secondary's stronger shock and finally it leads to the higher flux toward the secondary. Furthermore, the secondary is located closer to the edge of the protoplanetary disk where gas is denser; hence, the flux onto this component must increase even more. Indeed, analysis of the fluxes demonstrates that, starting from the head-on collision point, the matter flux along the bow-shock of the secondary notably exceeds the same flux along the shock of the primary.

However, an accretion disk may not accept more matter than it is allowed by viscosity. The rest of the matter rounds the disk and collides with the same stream of another component. As a result, a bridge-like stationary shock between the circumstellar accretion disks of the components is formed (see Fig. 1, right panel). Due to this collision, the streams lose more angular momentum, since they are partially annihilated. The matter, having lost its angular momentum, is mainly accreted onto the massive star because its gravitational capture radius is larger. Also, the "bridge" is significantly tilted, because the flux of the matter that rounds the secondary is higher. As a consequence, a part of the spiral stream moving along the "bridge" toward the primary directly collides with its accretion disk. As a result, the shock wave is formed on the edge of the accretion disk. It leads to an increase in the loss of angular momentum in the primary accretion disk. Thus, despite the higher matter flux from the protoplanetary disk toward the secondary, the rate of accretion onto the primary is higher.

Conclusions

• Due to the supersonic motion of the components of TTSs, two bow-shocks are formed in the circumbinary disks of these systems. In the inner region of the binary, the velocity distribution is far from the Keplerian one and the flow structure is mainly governed by the bow-shocks.

• Existence of the shocks leads to redistribution of the angular momentum in the protoplanetary disk. As a result, two flows propagating from the inner edge of the protoplanetary disk to the components are formed.

• The flux of matter from the inner edge of the protoplanetary disk toward the less massive component is larger.

• The circumstellar accretion disk cannot accept all the falling matter. This leads to the complex redistribution of matter in the region between the circumstellar accretion disks. As a consequence, despite the higher matter flux from the protoplanetary disk toward the secondary, most of the matter is accreted onto the primary component.

Acknowledgements

This work was supported by the Basic Research Program of the Presidium of the Russian Academy of Sciences, Russian Foundation for Basic Research (projects 09-02-00064, 09-02-00993, 11-02-00076, 11-02-01248), and the Federal Targeted Program "Science and Science Education for Innovation in Russia 2009-2013."

References

Artymowicz, P. & Lubow, S. H. 1994, *ApJ*, 421, 651
Fateeva, A. M., Bisikalo, D. V., Kaygorodov, P. V., & Sytov, A. Y. 2011, *Ap&SS*, 335, 125
Kaigorodov, P. V., Bisikalo, D. V., Fateeva, A. M., & Sytov, A. Y. 2010, *Astron. Rep.*, 54, 1078
Sytov, A. Y., Kaygorodov, P. V., Fateeva, A. M., & Bisikalo, D. V. 2011, *Astron. Rep.*, 55, 793

From Interacting Binaries to Exoplanets: Essential Modeling Tools
Proceedings IAU Symposium No. 282, 2011   © International Astronomical Union 2012
Mercedes T. Richards & Ivan Hubeny, eds.   doi:10.1017/S1743921311028365

Modeling Fluid Flow Effects in Close Binary and Protoplanetary Systems

M. M. Montgomery

Physics Department, University of Central Florida, PY 132, Orlando, Florida, 32816, USA
email: montgomery@physics.ucf.edu

Abstract. Accretion disks around some white dwarfs in Cataclysmic Variables are thought to tilt around the line of nodes by the lift force acting at the disk's center of pressure. We investigate whether protoplanetary disks can also experience disk tilt. We find that lift may be possible by an asymmetric, net uni-directional, in-falling gas/dust stream overflowing a bluff body (e.g., Class I sources) or inner annuli of young Class II sources if gas/dust is still in-falling and the aspect ratio and disk surface area are large enough. However, inner disks of Class II sources LkCa 15, UX Tau A, and Rox 44 are not large enough, and therefore disk tilt is not likely.

Keywords. accretion, accretion disks; hydrodynamics; planetary systems: protoplanetary disks; novae, cataclysmic variables

1. Introduction

Montgomery & Martin (2010) find that an asymmetric gas stream flowing over a disk rim in a Cataclysmic Variable system can result in a lift force acting at the disk's center of pressure, resulting in a disk tilt around the line of nodes. Gravitational forces cause the tilted disk to precess retrogradely, and the source of precession is tidal torques like those by the Moon and Sun on the oblate, spinning, tilted Earth (Montgomery 2009). In this work, we apply this model to Class I and Class II sources to find if these disks can also tilt by the lift force. We use 04108+2803B as our model Class I source and LkCa 15, UX Tau A, and Rox 44 as our model pre-transitional disk Class II sources. We consider in-falling gas/dust particles that turn and flow over/under the disk and mass transfer effects through the disk, but we do not consider effects due to jets and winds.

2. Lift in Protostellar Disks

Observational Data. In Tables 1 & 2, we list the stellar and disk property values assumed in Chiang & Goldreich (1999) and (Espaillat *et al.* 2010) for the Class I source 04108+2803B and for Class II pre-transitional disk sources LkCa 15, UX Tau A, Rox 44 that may have young planets forming, respectively. In these tables, the target star mass M, stellar radii R, mass transfer rate $\dot{M}$ through the disk, disk inner (i) and outer (o) radius r, and disk height z are given. The inner wall radius (r^i_{wall}) and the inner disk radius (r^i_{disk}) differ in values, since the inner wall location is based upon the dust sublimation radius whereas the inner disk location is a constrained value (see Espaillat *et al.* 2010). As Class I sources have no gap, the inner disk rim and disk mass are assigned r^i_{wall} and M^o_{disk}, respectively. The rim disk height z^i_{wall} is taken to be four times the gas scale height, a conservative value, as determined by Equation (3f) of Chiang & Goldreich (1999). The dashed lines indicate no known data and N/A means Not Applicable. Note that the inner disk is geometrically thin relative to the gap width and the outer disk height.

Table 1. Class I and Class II Stellar Property Values

Class Source	Target	Spectral Type	M_1 ($M_\odot$)	R_1 ($R_\odot$)	$\dot{M}$ ($M_\odot \text{yr}^{-1}$)
I	04108+2803B	–	0.5	2.5	–
II	LkCa 15	K3	1.3	1.6	0.33×10^{-8}
II	UX Tau A	G8	1.5	1.8	1.1×10^{-8}
II	Rox 44	K3	1.3	1.6	0.93×10^{-8}

Table 2. Pre-transitional disc Candidates & Assumed disc Property Values

Class Source	Target	r^i_{wall} (AU)	z^i_{wall} (AU)	r^i_{disk} (AU)	M^i_{disk} ($M_\odot$)	r^o_{wall} (AU)	z^o_{wall} (AU)	M^o_{disk} ($M_\odot$)	r^o_{disk} (AU)
I	04108+2803B	0.07	–	N/A	N/A	N/A	268	1.5×10^{-2}	270
II	LkCa 15	0.15	0.017	< 0.19	$< 2 \times 10^{-4}$	58	12.9	10×10^{-2}	300
II	UX Tau A	0.15	0.009	< 0.21	$< 0.6 \times 10^{-4}$	71	13.8	4×10^{-2}	300
II	Rox 44	0.25	0.034	< 0.4	$< 0.8 \times 10^{-4}$	36	9.9	3×10^{-2}	300

Analytical Model. Unequal in-falling gas/dust parcels on different flow paths above and below the disk can result in different pressures on the disk faces and thus a disk tilt. Equating Equations (4) and (7) in Montgomery & Martin (2010), disk tilt results if

$$\frac{A_s}{2} \rho (v^o_{disk})^2 (1 - \beta^2) \; > \; \frac{8 G \Sigma m M_1 \sin \theta}{9 r_d^2} + \frac{3 r_d G m M_2 \sin \theta}{8 (d^2 + \frac{9}{16} r_d^2 - \frac{3}{2} r_d d \cos \theta)^{3/2}}. \quad (2.1)$$

In this relation, A_s is disk face surface area, $\rho = 10^{-5}$ kg m^{-3} is in-falling gas/dust density (Larson, 1969), $\beta = 0.9$ is fraction of the gas velocity flowing under to over the disk, G is the universal gravitational constant, Σm is disk mass, $m \approx M^o_{disk}/100,000$ is mass of gas/dust parcel, r_d is radius of a circular and geometrically thin disk, M_2 is mass of forming planet in disk, d is separation distance, and $\theta = 4^o$ is minimum disk tilt angle.

Results - Class I Sources. From conservation of energy, $|v^o_{disc}|$ is found by assuming a gas/dust parcel falls from rest at infinity to the disk rim. Upon substitution of $A_s = \pi (r^o_{disc})^2$ and $M_2 \ll M_1$, the left hand side of Equation (2.1) is greater than the right hand side and thus, disk tilt seems likely.

Results - Young Class II Sources. In young Class II sources, $M_2 \ll M_1$ since no gap has formed. For hypothetical younger versions of LkCa 15, UX Tau A, and Rox 44, we find in-fall speeds to the disk rim are subsonic and disk surface areas are large and, therefore, a tilt on outer disk annuli is not likely. However, as the gas/dust parcels drop further into the potential well, speeds increase and a net asymmetric unidirectional gas/dust stream flowing over and under the inner annuli of the disk may result in disk tilt.

Results - Class II Sources. In Equation (2.1), we replace r^o_{disc} with $(r^o_{disc} - r^o_{wall})$ and allow $M_2 \ll M_1$. For LkCa 15, UX Tau A, and Rox 44, inner disks do not have enough surface area and, therefore, disk tilt will not occur.

Acknowledgement

The author is grateful for financial support from the IAU to attend this symposium.

References

Chiang, E. L. & Goldreich, P. 1999, *ApJ*, 519, 279
Espaillat, C., D'Alessio, P., Hernández, J., Nagel, E., Luhman, K. L., Watson, D. M., Calvet, N., Muzerolle, J., & McClure, M. 2010, *ApJ*, 717, 441
Larson, R. B. 1969, *MNRAS*, 145, 271
Montgomery, M. M. 2009, *ApJ*, 705, 603
Montgomery, M. M. & Martin, E. L. 2010, *ApJ*, 722, 989

From Interacting Binaries to Exoplanets: Essential Modeling Tools
Proceedings IAU Symposium No. 282, 2011   © International Astronomical Union 2012
Mercedes T. Richards & Ivan Hubeny, eds.   doi:10.1017/S1743921311028377

Panel Discussion IV

Panel: F. Allard, A. Batten, E. Budding, E. Devinney, P. Eggleton, A. Hatzes, I. Hubeny, W. Kley, H. Lammer, A. Linnell, V. Trimble, and R. E. Wilson

Discussion

I. HUBENY: Welcome to the last panel meeting. We invite general comments either from the audience or from the panelists.

V. TRIMBLE: Well, Mercedes started us with a vocabulary item and I think I would like to end with a vocabulary item. When they were first discovered, we called them 'extra solar system planets' which was descriptive and fine, but it's just rather cumbersome. At some point they became 'extra solar planets.' Now I have never seen a planet inside the Sun. And therefore 'extrasolar' is not a good descriptor. 'Exoplanets' is OK, but now that there are so many of them that perhaps they are simply 'the planets.' When you want to specialize to ours, you could say 'solar system planets.' Think how much ink it would save.

W. KLEY: We heard today again in the talk by Ian Dobbs-Dixon that there is a strong similarity between exoplanets and secondary stars in close binaries. The only difference he said is in the chemistry, but I think the disk is also there. Nevertheless, these suggestions are very similar and I think this is a good connection with the theme of the conference: planets and binaries. They are the same thing, and it is clear that the planet is also a secondary object, like the secondary star in a binary. But, one thing which came to my mind yesterday is that when you're looking at the light curves that you analyze for the exoplanets, there are lots of different light curve modeling procedures around. So the question came up, these models that you developed to analyze binary eclipse light curves, are they suitable for planets or are we using different models, and what is the relation between them? Are they completely independent? You find on planets these displaced hot spots and so on, I assume all these results must have been familiar to the binary synthesis light curve modelers. So, I think there is a good connection between binaries and planets in several aspects.

O. DE MARCO: Dr. Kley took the words from my mouth. I'll put an example to this in the other direction. I'm interested in irradiated binaries that come out of common envelopes and then become central stars or planetary nebula. There are very strong irradiation effects, and we model them with some Wilson-Devinney type codes. Now, one thing that worried me a lot is the transfer radiation from one side to the other, and some of these systems are tidally locked, and some probably are not yet. There should be in theory very different light curve properties and, now that Kepler is taking some data of these systems, I don't think we can avoid thinking of these effects of radiation transfer. Now, the 'Hot Jupiter' people, like Ian, did a great job of that; but capacitance will be very different so radiation transfer will be different. So, I would love to have a way, or for somebody to do this for me, to give us a tool in the Wilson-Devinney code with some approximation for these effects that can help with this particular problem. I think the pieces are there.

R. WILSON: I had not even realized the program was being used all that much for planets. For most purposes, it appears to me that there are adequate programs around that the exoplanet community have written that handle the job, since the stars are basically spherical in most cases. In many aspects the problems are simpler, but there are some aspects in which it's more complicated that we just mentioned. The heating problem, reflection, and so on, and Jan Budaj's paper at this meeting on the reflection effect and the other things you've mentioned about radiation coming from heat transfer around the back, those are not in the binary star programs now. So, there's some work to do for the binary star community.

I. HUBENY: I think convergence of modeling approaches is also being seen on the level of synthetic codes, like BINSYN that Al Linnell mentioned earlier, namely spectrum synthesis. It's already being done for close binaries using the same basic codes. TLUSTY was originally a stellar code and now it is being used for planets as well. So from that point of view, there is certainly convergence, and there are those synthetic models not specifically tailored for light curves but for predicting spectra as a function of phase, they certainly exist. As far as spectra are concerned, there is more progress now in planets than in stars.

A. BATTEN: I would like to make a very general comment that the powerpoint presentations worked very well at this conference. There used to be many problems fitting visitor's laptop computers to resident projection systems, and these deterred me from using powerpoint myself until less than a year ago. The problems were not in evidence here and I particularly enjoyed the animations that illustrated several talks, including two this morning. So, powerpoint presentations seem to have come of age.

A story: During coffee, Ed Guinan and I were discussing the question of whether any star was truly constant in velocity, luminosity, etc.. This reminded me that R. M. Petrie, my boss in my early days at the Dominion Astrophysical Observatory, used to quote H. D. Curtis (of the Shapley-Curtis debate). Curtis was very fond of saying: "There is more joy in Heaven over one star that is found to be constant than over ninety-and-nine that are known to be variable!"

A. HATZES: I find that as I get older, I like to get my pearls of wisdom, whether they're wisdom or not. I noticed that there are a lot of young people here and the topic of this conference was about essential tools. I also saw these beautiful codes that people use for modelling the stellar atmosphere have come a long way. But, as a warning to the young people here: there's a tendency to use things that are black boxes. There can be a real danger in that. So, I encourage you to understand the physics behind these black boxes. Don't just use them blindly. Understand how the code works and even better yet, look into the code. And, I give you two examples from my own experience: When we were looking for planets and finding them with Bill Cochran before it was fashionable and we needed to compute orbits, our colleagues in Texas said, "Oh there's a spectroscopic binary orbit, it works, use these!" So we used them and we started to get eccentric orbits and it was producing garbage. We thought, "Well, let's look into the code." Well it's good we looked into the code (it was in FORTRAN, so only dinosaurs like me could read the code), because there was a comment: "Doesn't work for eccentricities above 0.2" So, it was completely invalid to use it!

Another more serious one. My colleague Heike Rauer does work on planetary atmospheres. She got a standard code, was using it and seeing some irregularities, so she decided to look into the code and there were some parameters that went negative which shouldn't. These were physical variables like temperature that should not go negative. What does the code do? It sets the values to zero and merrily goes on and produces an answer. So be careful. You don't want to base your career on someone else's mistakes. They're very clever people but I'm still finding bugs in my own code, years after the fact. So user beware.

E. BUDDING: Continuing from the comment of Dr. Hatzes, I would like to refer to the subject of the Roche lobe, which has been mentioned a number of times at this meeting. Incidentally, Edouard Albert Roche lived in the nineteenth century. Alan here will recall that the person who introduced this term and provided a lot of work for its early development was very sensitive to its use. He would frequently recall that the Roche contours were a very artificial device, relating to equipotential surfaces, i.e. where matter was at zero velocity, and in a situation with two mass points moving in circular orbits, with rigidly rotating massless envelopes. All of these restrictions have been broken in the contexts in which these formulae for the potential were being applied, so it is well to remember that such theoretical modellings, detailed as they may be, are approximations.

On the other hand, as we have seen, there are also accumulating masses of detailed data, but they also still contain errors and have a finite information content. The appropriate matching of these data by suitable models has been the challenge before us, but this matching process requires us to keep in mind the key issues of model adequacy and determinacy.

E. DEVINNEY: Maybe it's because of my age and generation, but I was thinking about how rich nature is as we see the phenomena we've discussed at this meeting. Musing about this to myself in these last few moments and thinking probably there are a lot of us older persons here who, when you talk about exoplanets or when exoplanets were first found, said, "This is incredible." But, maybe younger people are thinking "Yeah, it's interesting." And probably in another ten years, people will just think "Oh, yeah, exoplanets." So, it is great to be at this stage where we can think about these things with some awe and amazement.

I. HUBENY: Thank you. With those wise words we will end our panel discussion.

Summary

From Interacting Binaries to Exoplanets: Essential Modeling Tools
Proceedings IAU Symposium No. 282, 2011
Mercedes T. Richards & Ivan Hubeny, eds.
© International Astronomical Union 2012
doi:10.1017/S1743921311028389

Summary of Observational Techniques

Pavel Koubský

Astronomical Institute, Academy of Sciences of CR, Fričova 298, CZ 251 65 Ondřejov,
Czech Republic
email: koubsky@sunstel.asu.cas.cz

Abstract. The observational techniques applied to the study of binary stars, brown dwarfs, and exoplanets are summarized in this paper.

Keywords. (stars:) binaries: general

1. Introduction

Nearly two years ago, I heard about this meeting for the first time. And now the remarkable conference is over and I have to attempt to summarize the first portion of it. In the past years, a bunch of meetings devoted to binaries took place. I enjoyed the Dubrovnik meeting in 2003, followed by the Kopal Memorial at Litomyšl in 2004. Two binary meetings were organized in 2006: At the spring meeting on "Solar and Stellar Physics through Eclipses," the participants were lucky observers of the total solar eclipse in the Earth-Moon-Sun system on the coast of southern Turkey; and later the IAU Symposium 240 in was held Prague. In 2009, a very useful conference on binaries was held in Brno. Its scope accented the growing interests in exoplanet research. In this Symposium, the exoplanets have already become a rightful part of a binary meeting. So much information has been presented at this conference that I am very grateful not to have to cover it all, but to leave part to Adam Burrows (especially after I heard and saw his fantastic summary performance!). The synopsis I will try to present does not attempt to be objective. It is much biased and reduced to a personal view with some background comments that I wanted to express. I will limit my summary to some oral papers and a selection of posters.

2. Oral Papers

An extensive of the impact of current and planned telescope systems and new technologies on binary and exoplanetary research was presented by Ed Guinan. It was a prefect overture to the Symposium performed in a lucid, Guinan-like style. At present and in the future, progress in the study of binary stars and exoplanets will rely on instrumentation not originally dedicated to these fields. Despite these limitations, the data available for binaries and exoplanet research are enormous. Astronomers in binary and exoplanetary research are harvesting the data from large surveys like OGLE, ASAS, EROS or MA-CHO. And the "widest, fastest, deepest" synoptic sky survey: the LSST with the 3.2-billion-pixel camera is under preparation. Ultrahigh precision light curves of binaries observed by the Kepler satellite show effects predicted but never before observed - beaming binaries (Zucker *et al.* 2007). The phase shifts detected from transit-timing variations and transit-duration variations can lead to discoveries of exomoons. Ultrahigh precision spectroscopy (2 ms^{-1}) is now available on HARPS and HIRES spectrographs mounted on 3.6-m and Keck telescopes, respectively. And the possibility to measure Doppler shifts as small as 10 cms^{-1} should appear soon.

We heard about the evolution of big ground telescopes - E-ELT, TMT, space missions like Gaia, but also endangered projects like TESS or (hopefully not) JWST.

Niarchos reviewed the ground-based and space observations of interacting binaries. He discussed the use of data from microlensing and exoplanets surveys for interacting binary research - SuperMacho - extended use of the Blanco 4-m Telescope, which monitored fields in SMC and LMC, and MOA - Microlensing Observations in Astrophysics. Other techniques discussed were the heliospheric imager instrument on board the two solar STEREO satellites and robotic telescopes such as ROTSE. He also mentioned the WSO satellite which in the near future should resume the regular UV spectroscopic observations of nearby stars, including the interacting binaries as was done by the IUE many years ago.

Bonanos presented the use of OGLE/MACHO surveys for detecting eclipsing binaries in the SMC, LMC, and galaxies in the Local Group. The extracted EB are then targets for follow-up observations with 6-10 m class telescopes equipped with multi-object spectrographs which can reach beyond 1 Mpc, capable of going up to the Sculptor or M81 Groups.

The prospects of the Gaia project for binary research were discussed in several talks during the Dubrovnik meeting in 2003. The meetings which followed, Prague 2006 and Brno 2009, mentioned the project only marginally. It was a good decision of the organizers of this Symposium to include the Gaia project again and invite Eyer, one of the key organizers of the project. The ESA Gaia mission will provide a multi-epoch database for a billion objects, including binaries. Gaia is a successor to the Hipparchos satellite which should, besides its higher sensitivity, provide spectroscopic information too; medium resolution spectra in the near infrared region and low dispersion data in two blue and red channels. Each object will be observed by Gaia an average of 70 times during the 5-year mission. It seems that Gaia will be one of several survey projects. One of these is LSST. The 8.4-meter LSST will survey the entire visible sky deeply in multiple colors every week probing the mysteries of Dark Matter and Dark Energy, and opening a movie-like window on objects that change or move rapidly: exploding supernovae, potentially hazardous near-Earth asteroids, and distant Kuiper Belt Objects. Eyer has shown how Gaia and LSST will complement each other in the research on binaries. While the sampling of Gaia is more suitable for periodic objects, LSST will play an important role in monitoring the irregular phenomena.

Maceroni gave a talk on the impact of CoRoT and Kepler satellites. Since its launch, at the end of 2006, the CoRoT space mission has acquired continuous and high accuracy photometry of a hundred thousand stars in several long (150d) and short (30d) runs. Hundreds of new eclipsing binaries have been discovered and photometry of unprecedented accuracy is becoming available. The CoRoT Public N2 data archive provides an open access to the CoRoT data one year after their delivery to the CoRoT Co-Is, following the CoRoT data policy. Kepler uses a photometer developed by NASA to continuously monitor the brightness of over 145,000 main sequence stars in a fixed field of view. The data collected from these observations will be analyzed to detect periodic fluctuations that indicate the presence of exoplanets that are in the process of crossing the face of other stars. Kepler Q0, Q1, Q2 data are now public data. Thanks to CoRoT and Kepler, excellent photometry research topics such as asteroseismology of EB components are quickly developing, and very fine phenomena like Doppler boosting (van Kerkwijk *et al.* 2010) have been unambiguously detected.

Peters' talk was devoted to the use of various databases in binary star research. She stressed the power of existing databases for data ranging from the gamma ray region to the infrared. She showed some examples of discoveries of secondary components to some

Be binaries, discoveries enabled by the extended use of existing databases of spectro-scopic data. One example: the detection of a hot subdwarf companion to the Be star FY CMa (Peters *et al.* 2008). With some amount of nostalgia, she recalled the International Ultraviolet Explorer as a "discovery machine" in the 1980s. There is a good chance that a new, advanced IUE would appear. The hope is called the World Space Observatory for the Ultra-Violet (WSO-UV), a satellite, which will carry a 1.7-m telescope equipped with spectrographs and camera for the ultraviolet region of the spectrum (Kappelmann *et al.* 2009). The satellite, which was developed in the former Soviet Union in late 1980s under the name Astron, is to be launched in 2015.

Queloz described the state of exoplanet research sixteen years after the discovery of 51 Peg b with the 1.93-m telescope/ELODIE instrumentation (Mayor & Queloz 1995). Since the time of the Elodie spectrometer, much better techniques are available: high precision spectrometers securing spectra with high signal-to-noise ratio and resolution up to 100,000. Now a large population of multiple planetary systems is known. A new paradigm on the formation, structure, and composition of planets is emerging, wider that we had anticipated from the knowledge of the Solar System. Among 500 known exoplanetary systems, Queloz pointed out some very interesting observations: HD 69830 around a K0 star in the constellation Puppis, located at a distance of 12.6 parsecs, has three detected planets of approximately Neptune mass, traveling in orbits of low eccentricity, with a narrow debris ring reminiscent of the Asteroid Belt. This system is remarkable for at least two reasons: it is part of the small group of known exoplanetary systems without any known gas giants, and belongs to a still smaller group with a field of colliding debris that produces warm dust. A multi-planet system orbiting the star HR 8799 now contains four bodies and a debris disk. The four planets were discovered the old-fashioned way: they were directly imaged.

Pasternacki presented a homogeneous reanalysis of the currently known CoRoT-exoplanets. This way, the refined planetary and stellar parameters are to be obtained. A preliminary result suggests that hot stars may be more active than solar-type stars.

Allers talked about brown dwarfs in binaries. Brown dwarfs were predicted in the 1970s, and called then black dwarfs. The first brown dwarfs were discovered in 1988. Now, about 1000 BDs are known. Binary brown dwarfs provide a unique opportunity to empirically determine fundamental parameters, which can be used to test model predictions.

Konacki gave an interesting review on detecting and characterizing exoplanets in bi-nary systems. In principle, there are two methods to detect circumbinary planets: ET eclipse timing and precision RV techniques. At the present time, circumbinary planets represent 10% of the exoplanet population. Two projects were presented in the talk: SO-LARIS – to commemorate the outstanding Polish writer Stanislaw Lem and his novel, and TATOOINE – a wild acronym for The Attempt To Observe Outer-planets In Non-single stellar Environments (Konacki *et al.* 2009). SOLARIS should use a network of robotic telescopes to carry out precision photometry of a sample of eclipsing binaries, while TATOOINE uses the best high resolution spectrographs and a novel iodine cell-based approach to determine radial velocities.

Stee gave a very nice summary of results from interferometric observations of bina-ries and multiple systems. He focused on three interferometers working in the visible and infrared regions - VLTI, CHARA and NPOI. Ten binary orbits and five images of binaries have now been obtained from interferometric studies. In several cases, the dis-tance, brightness ratio of the components, masses of the primary and secondary, effective temperatures, and limb darkening were derived. The results he has shown confirm his statement that interferometry is a mature technique now; but the number of objects ob-served clearly show that interferometry is not a commonly used method. The explanation is obvious: while there are hundreds of telescopes to be used for binary research, there are less than ten interferometers capable of studying binaries.

We saw fine pictures of systems like β Lyr or υ Sgr based primarily on data from the VEGA instrument on CHARA; but in the case of β Lyr, a nostalgic reminiscence of the old GI2T was presented (Harmanec *et al.* 1996). As a masterpiece of the interferometric technique, a detection of the close faint exoplanet companion to HD 59717 was presented. Duvert *et al.* (2010) using the AMBER/VLTI instrument in the K band, detected the 5-mag fainter companion of this star at a distance of 4 stellar radii from the primary. This is one of the highest contrasts detected by interferometry. We hope that exoplanets may be detected this way in the future. A giant planet in the well known system β Pic has now been detected in two IR photometric bands, giving the possibility to better estimate its spectral type, atmospheric parameters, and mass (Bonnefoy *et al.* 2011).

Serabyn's talk was focused on observations close to binary stars using vortex coronography, visible-wavelength adaptive optics, and nulling interferometry. He showed images of a debris disk in HD 32297 obtained using phase-mask coronagraphy on the 1.6m well-corrected subaperture on the 5m Palomar Hale telescope. The image clearly demonstrated the benefits of operating in the extreme adaptive optics regime with a coronagraph able to reach a very small inner working angle. Another example of the progressive new technique was the image of exoplanets in the system of HR 8799 (evidently a very popular object among the high angular resolution people) taken in infrared light with Palomar's Hale Telescope. The image was captured using a 1.5-meter-diameter portion of the Hale telescope's mirror.

Hinkley described the use of several new dedicated observing platforms geared toward high contrast imaging of binaries, brown dwarfs, and exoplanets: The Gemini Planet Imager adaptive optics instrument for the Gemini South telescope, VLT- SPHERE project to gain at least one order of magnitude with respect to the present VLT AO facility NACO, project 1640, and PALM 3000. He showed the image of the ζ Vir system (M dwarf orbiting an A3V star), secured with the coronographic integral field spectrograph (Project 1640). He described the PALM 3000, a high order adaptive optics system at Palomar, which uses 3388 actuators. First light was in summer 2011.

Schmidt talked about the methods of estimating masses of sub-stellar companions around young stars. The observations were secured with the VLT Adaptive Optics (AO) instrument NACO and AO integral field spectroscopy with SINFONI obtained to deduce the physical parameters of the companions.

3. Posters

I have selected five posters from the numerous and diverse collection of posters.

An interesting application of ultra-precise photometric data was presented in the poster by Kiss and Derekas, a discovery of a unique triply-eclipsing triple system HD 181068. The original discovery was published in Science in April 2011.

Whittaker showed interesting results from the twin STEREO satellites, primarily devoted to solar research. The stellar photometric data from onboard imaging cameras were suitable for planetary transit searches and a number of new eclipsing binaries were found.

Hadrava was invited to present a talk on his disentangling code KOREL; such a presentation was deeply missed in some previous binary meetings. It was an excellent and very ripe presentation (the first version of the code has been more than 15 years old now). Hadrava's talk was complemented by a number of posters. Here are the two most interesting.

Škoda presented VO-KOREL, a web service exploiting the technology of the Virtual Observatory to run the most recent version of KOREL. It was one of the most interesting mini-talks at the conference.

The poster by Chadima *et al.* which showed the results of the spectroscopic study of ε Aur, brought a clear warning that disentangling codes both in lambda and the Fourier domain cannot be treated as a black box machine.

And finally: a technique which could mean an important jump in the efficiency of ground-based astronomy and astrophysics. Mkrtichian *et al.* presented a poster on daytime Doppler spectroscopy. They showed nice results of daytime spectroscopy used for astroseismology of bright stars.

4. Last Thoughts

During the first Panel Discussion, Virginia Trimble warned that the HST successor James Web Space Telescope is endangered and mentioned the actions of astronomers to save this centerpiece of U.S. space astronomy for the next two decades. She used this example to call on (European) astronomers to struggle for any threatened project.

The history of efforts to save telescope projects is probably as old as modern astronomy. One example is closely connected with the site of this Symposium and with its dedication. The largest telescope in former Czechoslovakia is a 60-cm Zeiss reflector that was installed in a place called Stará Ďala in southern Slovakia in 1927. This was an important step for the development of astrophysics in Czechoslovakia realized by Dr. Bohumil Šternberk (1897 – 1983). Before the outbreak of WW II when the borders of European states started to change again, it was necessary to move the telescope to the inland of Slovakia. It was Dr. Šternberk who succeeded in saving the telescope for the astronomical community. The 60-cm telescope was later erected at a new observatory at Skalnaté Pleso. As we heard during the Opening Ceremony, the building of the Observatory was initiated by Antonin Bečvář (1901 – 1965), whose anniversary is commemorated by this Symposium. The telescope was very extensively used, partly devoted to binary research, until 1977. Since 1994, the refurbished 60-cm telescope has been used at Modra Observatory of Comenius University in Bratislava. One of the projects run by the telescope is the study of exoplanet transits.

At the end, warm thanks and applauses are addressed to the Scientific Organizing Committee, led by Mercedes Richards and Ivan Hubeny, and the Local Organizing Committee, led by Theo Pribulla and Ladislav Hric, who prepared this memorable meeting.

References

Kappelmann, N., Barnstedt, J., Werner, K., Becker-Ross, H., & Florek, S. 2009, *Ap&SS*, 320, 191

Konacki, M., Muterspaugh, M. Kulkarni, S. R., & Helminiak, K. G. 2009, *ApJ*, 704, 513

Mayor, M. & Queloz, D. 1995, *Nature*, 358, 355

van Kerkwijk, M. H., Rappaport, S. H., Breton, R. P., Justham, S., Podsiadlowski, P., & Han, Z. 2010, *ApJ*, 715, 51

Zucker, S., Mazeh, T., & Alexander, T. 2007, *ApJ*, 670, 1326

Peters, G. J., Gies, D. R., Grundstrom, E. D. & McSwain, M. V. 2008, *ApJ*, 686, 1280

From Interacting Binaries to Exoplanets: Essential Modeling Tools
Proceedings IAU Symposium No. 282, 2011
Mercedes T. Richards & Ivan Hubeny, eds.
© International Astronomical Union 2012
doi:10.1017/S1743921311028390

Summary of Theoretical Techniques

Adam Burrows

Department of Astrophysical Sciences, Princeton University, Princeton, NJ USA 08544
email: burrows@astro.princeton.edu

Abstract. In this summary, I address the next generation of theoretical tools with which it may be necessary to interpret the data anticipated as both stellar and planetary astronomy enter their next decades.

Keywords. planets and satellites: general; stars: atmospheres; hydrodynamics; techniques: miscellaneous; (stars:) binaries: eclipsing; (stars:) binaries: spectroscopic; methods: numerical; methods: n-body simulations; methods: laboratory; methods: n-body simulations

---

One implicit theme of this meeting, bringing together as it does researchers in binary stars, stellar atmospheres, and planets, is that the field of exoplanets is reviving stellar astrophysics. Another is that the young field of exoplanets owes a great debt to the more mature and developed field of binary studies. Without an understanding of stars, we can't characterize the planets that orbit them. Without the tools developed by earlier generations of stellar astronomers, the field of exoplanets would be all but bereft of its modern methodologies. The manifest synergy between the two allied fields is something to exploit, and to celebrate.

The numerical tools crafted to understand binary stars and close-in planets orbiting their parent stars, their spectra, and their hydrodynamics require fast computers with large memories. It is curious to note that the pioneering early developments in efficient methodologies to study stellar atmospheres and interacting binary stars (e.g., Auer and Mihalas 1969 and Wilson and Devinney 1971, respectively) occurred ~40 years ago, just at the start of the ramp up in computational capability epitomized by Moore's Law. There has been a ~10^5-fold increase in the number of transistors per chip since that time and we as a field (or fields) are riding on the coattails of that dramatic exponentiation. Whether Moore's Law can continue through the next few decades is not clear, but what is clear is that progress at the cutting edge of planet and star theory will depend on continuing developments at the frontiers of computational science and applied mathematics. Moreover, there is little doubt that widespread access to increasingly-capable computers at competitive prices has been a key to progress across science, no less so in stellar and planetary astronomy.

I emphasize the importance of being at the cutting-edge of computation because the problems we have been pondering at this meeting, interacting stellar binaries and planet-star interactions, are fundamentally 3D radiation(spectral)-hydrodynamic problems (witness contact binaries and 3D general climate models!). Techniques for full non-LTE spectral synthesis in 1D stellar atmospheres are now at a high level of development. Techniques for 1D, 2D, and 3D hydrodynamics are as well. However, the coupling of the two, particularly when hundreds of thousands of transitions and rates are concerned, and in full 3D, has yet to be accomplished. The next-generation codes must include full time-dependence, most likely should include magnetic interactions, may have to include plasma processes, and most certainly need reliable spectroscopic and rate parameters and coefficients. The latter require complete databases validated by Laboratory Astrophysics, and the necessary compilation, archiving, and validation are by no means assured. Few scientists go into laboratory studies of processes of relevance to astronomers. Few chemists in the know about spectroscopy and chemical rates value the generation of databases that would be useful to astrophysics.

What is more, though atomic processes and spectroscopy have been the focus of astronmers for approximately one hundred years, and this focus has been necessary to elevate stellar atmospheres to its present sophistication, a corresponding effort has not been expended for molecular and exoplanetary atmospheres, their spectroscopy, or chemical rates. While there may be hundreds of thousands of transitions important for atoms, there are billions that are important for molecules. Thermochemical and spectroscopic data for molecules lag in quality far behind those available for atoms, except in temperature regimes of relevance to Earth's atmosphere. Even then, there are many holes in our chemical knowledge. What is worse, at the lower temperatures at which molecules form and come to dominate, clouds can form as well. Extant cloud models hardly deserve the name. Even for the Earth and its water clouds, we do not have a rigorous, predictive theory. Add to this, ammonia, silicate, and iron clouds, and hazes of currently indeterminate composition, and one glimpses the problems before us as we attempt to model, in credible fashion, cool planetary atmospheres.

A similar challenge confronts us as we seek to understand stellar evolution, a bullwark of binary and planetary studies. Almost all stars are convective and convection is a non-linear, 3D problem. Mixing and doubly-diffusive instabilities, convective overshoot into radiative zones, rotational dynamics, angular momentum transport, and magnetic field generation and influences (some pivotal) are all not only 3D processes, but occur on timescales much, much shorter than those of stellar evolution. This makes stellar evolution a numerically stiff 3D magneto-radiation-hydrodynamic problem, with 3D atmospheres (!).

However, the stellar evolution codes used in support of the science we have been discussing this week are one-dimensional, with ad hoc prescriptions to handle 3D effects. Mixing-length "theory" and 1D diffusion approximations to mixing processes are poor substitutes for the real multi-D dynamics executed by real stars. The upshot is that stellar evolution theory is not yet as robust as it needs to be to support star and planet studies that are achieving percent precisions. How stellar evolution theory will make its next leap is not at all clear, but make this leap it must.

With daunting radiation, hydrodynamic, plasma, chaotic dynamical, spectroscopic, atmospheric, and chemical challenges before us as we seek to improve the interpretative tools of exoplanet and stellar research, we must do more than passively ride Moore's Law to advance our science. We must inaugurate best practices in the computational arts, team with chemists and spectroscopists to create the most comprehensive databases of input physics, and foster a generation of theorists comfortable with and competent in computational astrophysics on its frontiers. The goal is a more comprehensive and credible theory of stars and planets as they execute their interactive dance, as well as insight into the distinctive characteristics that make planets planets.

References

Auer, L. H. & Mihalas, D. 1969, *ApJ*, 158, 641
Wilson, R. E. & Devinney, E. J. 1971, *ApJ*, 171, 413

From Interacting Binaries to Exoplanets: Essential Modeling Tools
Proceedings IAU Symposium No. 282, 2011   © International Astronomical Union 2012
Mercedes T. Richards & Ivan Hubeny, eds.   doi:10.1017/S1743921311028407

Closing Remarks

Mercedes T. Richards

The conference organizers are grateful to everyone who came, from all parts of the world, to share this experience with us. There were one hundred and seventy-seven participants from thirty-one countries. We thank the speakers, session chairs, distinguished panelists, the SOC members, the LOC members led by Theo Pribulla and Laco Hric, and Aleš Kucera, Director of the Astronomical Institute of the Slovak Academy of Sciences. Special thanks go to Richard Komžík for ensuring the smooth operation of the computer and audio-visual support for the conference.

We began the meeting by "shaking the pot," and several pertinent issues surfaced as a result of this contemplation. During our panel discussions, a concern was raised that there is a continuing need for fundamental atomic data, even as many specialists retire. Virginia Trimble suggested that we should assemble an IAU Working Group to address this concern, perhaps at the 2012 IAU General Assembly in Beijing. In addition, Petr Harmanec and Andrej Prša will lead the charge to prepare an IAU resolution to adopt a nominal set of astrophysical parameters and constants to improve the accuracy of the fundamental parameters that we assume in our calculations.

In closing, I am reminded of a statement by George Box, the statistician, who stated that *"All models are wrong, but some are useful."* We came to this meeting to discuss the status of our codes and techniques; we recognize that these modeling tools are not perfect and yet they have provided us with many useful results. Even so, we should be mindful that these tools need to be re-examined on a regular basis.

Farewell – Dovidenia.

Author Index

Printed in the United States
by Baker & Taylor Publisher Services